Methods in Enzymology

Volume 309

AMYLOID, PRIONS, AND OTHER PROTEIN AGGREGATES

METHODS IN ENZYMOLOGY

Methods in Enzymology

Volume 309

Amyloid, Prions, and Other Protein Aggregates

EDITED BY

Ronald Wetzel

UNIVERSITY OF TENNESSEE MEDICAL CENTER
KNOXVILLE, TENNESSEE

ACADEMIC PRESS
San Diego London Boston New York Sydney Tokyo Toronto

This book is printed on acid-free paper. ♾

Academic Press
A Harcourt Science and Technology Company
525 B Street, Suite 1900, San Diego, California 92101-4495, USA
http://www.academicpress.com

Academic Press Limited
24-28 Oval Road, London NW1 7DX, UK
http://www.hbuk.co.uk/ap/

International Standard Book Number: 0-12-182210-9

PRINTED IN THE UNITED STATES OF AMERICA
99 00 01 02 03 04 MM 9 8 7 6 5 4 3 2 1

Table of Contents

Section I. Characterization of *in Vivo* Protein Deposition

A. Identification and Isolation of Aggregates

B. Isolation and Characterization of Protein Deposit Components

C. Characterization of Aggregates *in Situ* and *in Vitro*

Section II. Characterization of *in Vitro* Protein Deposition

A. Managing the Aggregation Process

B. Aggregation Theory

C. Monitoring Aggregate Growth by Dye Binding

D. Measurement and Characterization of Assembly Intermediates

E. Monitoring Aggregate Growth by Measuring Solid-Phase Accumulation

F. Monitoring Aggregate Growth and Structure Using Light Scattering

G. Aggregation Inhibitors

Section III. Aggregate and Precursor Protein Structure

A. Aggregate Morphology

B. Molecular Level Aggregate Structure

C. Characterization of Precursor Protein Structure

Section IV. Cellular and Organismic Consequences of Protein Deposition

A. Microbial Model Systems

B. Animal Models of Protein Deposition Diseases

C. Cell Studies on Protein Aggregate Cytotoxicity

Contributors to Volume 309

Article numbers are in parentheses following the names of contributors.
Affiliations listed are current.

YOUSEF AL-ABED (11), *The Picower Institute for Medical Research, Manhasset, New York 11030*

KARIN ANDERSSON (38), *Department of Cell Molecular Biology, Umeå University, S-901 87 Umeå, Sweden*

DENAH M. APPELT (12), *Department of Biomedical Sciences, Philadelphia College of Osteopathic Medicine, Philadelphia, Pennsylvania 19131*

MICHAEL A. BALDWIN (37), *Mass Spectrometry Facility, Department of Pharmaceutical Chemistry, University of California, San Francisco, California 94143-0446*

BRIAN J. BALIN (12), *Department of Pathology, Microbiology, and Immunology, Philadelphia College of Osteopathic Medicine, Philadelphia, Pennsylvania 19131*

GILLIAN P. BATES (43), *Medical and Molecular Genetics, UMDS, Guy's Hospital, London SE1 9RT, United Kingdom*

GEORGE B. BENEDEK (27), *Department of Physics and Center for Materials Science and Engineering, Massachusetts Institute of Technology, Cambridge, Massachusetts 02139*

SCOTT D. BETTS (22), *Department of Biology and Biotechnology Process Engineering Center, Massachusetts Institute of Technology, Cambridge, Massachusetts 02139*

ANJA LEONA BIERE (25), *Amgen, Inc., Thousand Oaks, California 91320-1799*

KURT R. BRUNDEN (46), *Gliatech Inc., Cleveland, Ohio 44122*

RICHARD BUCALA (11), *The Picower Institute for Medical Research, Manhasset, New York 11030*

JOS BUIJS (26), *Department of Chemical Engineering, Mälardalen University, 63105 Eskilstuna, Sweden*

D. ALLAN BUTTERFIELD (48), *Department of Chemistry, Center of Membrane Sciences, and Sanders-Brown Center on Aging, University of Kentucky, Lexington, Kentucky 40506-0055*

JOHN F. CARPENTER (16), *Center for Pharmaceutical Biotechnology, Department of Pharmaceutical Sciences, University of Colorado Health Sciences Center, Denver, Colorado 80262*

ANIL G. CASHIKAR (41), *Howard Hughes Medical Institute, University of Chicago, Chicago, Illinois 60637*

BYRON CAUGHEY (9), *Rocky Mountain Laboratories, National Institute of Allergy and Infectious Diseases, National Institutes of Health, Hamilton, Montana 59840*

BYEONG S. CHANG (16), *Amgen, Inc., Thousand Oaks, California 91320*

STEVEN CLARKE (7), *Department of Chemistry and Biochemistry and Molecular Biology Institute, University of California, Los Angeles, California 90095-1570*

WILFREDO COLÓN (39), *Department of Chemistry, Rensselaer Polytechnic Institute, Troy, New York 12180*

INGRID DACKLIN (38), *Department of Cell Molecular Biology, Umeå University, S-901 87 Umeå, Sweden*

STEPHEN W. DAVIES (43), *Department of Anatomy and Developmental Biology, University College London, London WC1E 6BT, United Kingdom*

FREDERICK C. DE BEER (44), *Department of Internal Medicine, Division of Endocrinology and Molecular Medicine, MN522 Chandler Medical Center, University of Kentucky, Lexington, Kentucky 40536*

ELIANA DE BERNARDEZ CLARK (15), *Department of Chemical Engineering, Tufts University, Medford, Massachusetts 02155*

RÉMI DEMAIMAY (9), *Rocky Mountain Laboratories, National Institute of Allergy and Infectious Diseases, National Institutes of Health, Hamilton, Montana 59840*

TOMAS T. DING (33), *Center for Neurologic Diseases, Brigham and Women's Hospital and Harvard Medical School, Boston, Massachusetts 02115*

PATRICK DOHERTY (43), *Experimental Pathology, UMDS, Guy's Hospital, London SE1 9RT, United Kingdom*

HOLGER EICKHOF (24), *Max Planck Institute for Molecular Genetics, D-14195 Berlin, Germany*

PAULA EMBURY (46), *Gliatech Inc., Cleveland, Ohio 44122*

WILLIAM P. ESLER (23), *Department of Pharmacology and Cell Biophysics, University of Cincinnati College of Medicine, Cincinnati, Ohio 45267*

FRANK FERRONE (17), *Department of Physics, Drexel University, Philadelphia, Pennsylvania 19104*

MARK A. FINDEIS (30), *PRAECIS Pharmaceuticals Inc., Cambridge, Massachusetts 02139-1572*

ANTHONY L. FINK (36), *Department of Chemistry and Biochemistry, University of California, Santa Cruz, California 95064*

PAUL E. FRASER (34), *Centre for Research into Neurodegenerative Diseases, University of Toronto, Toronto, Ontario, Canada M5S 3H2*

GLORIA GALLO (5), *Department of Pathology, New York University Medical Center, New York, New York 10016*

FUMITAKE GEJYO (20), *Department of Medicine* (II), *Niigata University School of Medicine, Niigata 951-8510, Japan*

GEORGE GEORGIOU (3), *Department of Chemical Engineering, University of Texas, Austin, Texas 78712*

GUNDARS GOLDSTEINS (38), *A. I. Virtanen Institute, University of Kuopio, Kuopio, Finland*

JAMES D. HARPER (33), *Center for Neurologic Diseases, Brigham and Women's Hospital and Harvard Medical School, Boston, Massachusetts 02115*

VOLKER HEISER (24), *Max Planck Institute for Molecular Genetics, D-14195 Berlin, Germany*

KEIICHI HIGUCHI (42), *Department of Aging Angiology, Research Center on Aging and Adaptation, Shinshu University School of Medicine, Matsumoto 390-8621, Japan*

VLADIMIR HLADY (26), *Center for Biopolymers at Interfaces, Department of Bioengineering, University of Utah, Salt Lake City, Utah 84112*

CARL HOBBS (43), *Experimental Pathology, UMDS, Guy's Hospital, London SE1 9RT, United Kingdom*

ELLEN HOLM NIELSEN (31), *Anatomy and Neurobiology, IMB, SDU, Odense University, DK-5000 Odense C, Denmark*

ANITA HONG (13), *Anaspec, Inc., San Jose, California 95131*

MOTOHIRO HORIUCHI (9), *Rocky Mountain Laboratories, National Institute of Allergy and Infectious Diseases, National Institutes of Health, Hamilton, Montana 59840*

MASANORI HOSOKAWA (42), *Department of Medical Embryology and Neurobiology, Institute for Frontier Medical Sciences, Kyoto University, Sakyo-ku, Kyoto 606-8397, Japan*

RUDI HRNCIC (5), *Department of Medicine, Human Immunology and Cancer Program, University of Tennessee Medical Center/Graduate School of Medicine, Knoxville, Tennessee 37920-6999*

SADAYUKI INOUE (32), *Department of Anatomy and Cell Biology, McGill University, Montreal, Quebec, Canada H3A 2B2*

ROBERT F. JACOB (19), *Membrane Biophysics Laboratory, Cardiovascular and Pulmonary Research Institute, Allegheny University of the Health Sciences, Pittsburgh, Pennsylvania 15212*

HERBERT P. JENNISSEN (26), *Institut für Physiologische Chemie, Universitätklinikum Essen, D-45122 Essen, Germany*

KENNETH H. JOHNSON (1), *Department of PathoBiology, College of Veterinary Medicine, University of Minnesota, St. Paul, Minnesota 55108*

BATIA KAPLAN (5), *Heller Institute of Medical Research, Chaim Sheba Medical Center, Tel Hashomer, Israel*

APHRODITE KAPURNIOTU (11), *The Picower Institute for Medical Research, Manhasset, New York 11030*

BRENT S. KENDRICK (16), *Amgen, Inc., Thousand Oaks, California 91320*

RITU KHURANA (36), *Department of Chemistry and Biochemistry, University of California, Santa Cruz, California 95064*

MARK S. KINDY (44), *Department of Biochemistry, Stroke Program of the Sanders-Brown Center on Aging, L017 Kentucky Clinic, University of Kentucky, Lexington, Kentucky 40536*

JONATHAN KING (22), *Department of Biology and Biotechnology Process Engineering Center, Massachusetts Institute of Technology, Cambridge, Massachusetts 02139*

ROBERT KISILEVSKY (32), *Department of Pathology, Richardson Laboratory, Queen's University, and The Syl and Molly Apps Research Center, Kingston General Hospital, Kingston, Ontario, Canada K7L 3N6*

WILLIAM E. KLUNK (19), *Department of Psychiatry, Western Psychiatric Institute and Clinic, University of Pittsburgh School of Medicine, Pittsburgh, Pennsylvania 15213*

JUNE KOCSIS-ANGLE (46), *Gliatech Inc., Cleveland, Ohio 44122*

TANUJA KOPPAL (48), *Department of Chemistry and Center of Membrane Sciences, University of Kentucky, Lexington, Kentucky 40506-0055*

CARSTEN KORTH (8), *Department of Neurology, University of California, San Francisco, California 94143-0518*

ANTHONY S. KOWAL (41), *Howard Hughes Medical Institute, University of Chicago, Chicago, Illinois 60637*

YU-MIN KUO (4), *Haldeman Laboratory for Alzheimer Disease Research, Sun Health Research Institute, Sun City, Arizona 85351*

VIRGINIA M.-Y. LEE (6), *Center for Neurodegenerative Disease Research, Division of Anatomic Pathology, Department of Pathology and Laboratory Medicine, University of Pennsylvania School of Medicine, Philadelphia, Pennsylvania 19104-4283*

HANS LEHRACH (24), *Max Planck Institute for Molecular Genetics, D-14195 Berlin, Germany*

HARRY LEVINE III (18, 29), *Department of Neuroscience Therapeutics, Parke-Davis Pharmaceutical Research Division, Warner-Lambert Company, Ann Arbor, Michigan 48105-1047*

SUSAN L. LINDQUIST (41), *Department of Molecular Genetics and Cell Biology and Howard Hughes Medical Institute, University of Chicago, Chicago, Illinois 60637*

ARIEL G. LOEWY (12), *Department of Biology, Haverford College, Haverford, Pennsylvania 19041*

ALEKSEY LOMAKIN (27), *Department of Physics and Center for Materials Science and Engineering, Massachusetts Institute of Technology, Cambridge, Massachusetts 02139, and Center for Neurologic Diseases, Brigham and Women's Hospital, Boston, Massachusetts 02115-5716*

JONATHAN DAVID LOWENSON (7), *Department of Chemistry and Biochemistry and Molecular Biology Institute, University of California, Los Angeles, California 90095-1570*

ERIK LUNDGREN (38), *Department of Cell Molecular Biology, Umeå University, S-901 87 Umeå, Sweden*

KAN MA (13), *Department of Chemistry, Case Western Reserve University, Cleveland, Ohio 44106-7078*

JOHN E. MAGGIO (23), *Department of Pharmacology and Cell Biophysics, University of Cincinnati, Cincinnati, Ohio 45267*

LAURA MANGIARINI (43), *Medical and Molecular Genetics, UMDS, Guy's Hospital, London SE1 9RT, United Kingdom*

MARK C. MANNING (16), *Center for Pharmaceutical Biotechnology, Department of Pharmaceutical Sciences, University of Colorado Health Sciences Center, Denver, Colorado 80262*

PATRICK W. MANTYH (23), *Molecular Neurobiology Laboratory, Department of Preventative Sciences, University of Minnesota Dental School, Minneapolis, Minnesota 55417*

R. PRESTON MASON (19), *Membrane Biophysics Laboratory, Cardiovascular and Pulmonary Research Institute, Allegheny University of the Health Sciences, Pittsburgh, Pennsylvania 15212*

MARK P. MATTSON (47), *Sanders-Brown Research Center on Aging, University of Kentucky, Lexington, Kentucky 40536*

SUSAN M. MOLINEAUX (30), *PRAECIS Pharmaceuticals Inc., Cambridge, Massachusetts 02139-1572, and Praelux Inc., Lawrenceville, New Jersey 08648*

JAHAN J. MOSLEHI (41), *Howard Hughes Medical Institute, University of Chicago, Chicago, Illinois 60637*

CHARLES L. MURPHY (5, 14), *Department of Medicine, Human Immunology and Cancer Program, University of Tennessee Medical Center/Graduate School of Medicine, Knoxville, Tennessee 37920-6999*

DAVID G. MYSZKA (25), *Department of Oncological Sciences, Huntsman Cancer Institute, University of Utah, Salt Lake City, Utah 84132*

HIRONOBU NAIKI (20), *2nd Department of Pathology, Fukui Medical University, Matsuoka, Fukui 910-1193, Japan*

EWAN J. NETTLETON (40), *Oxford Centre for Molecular Sciences, Oxford University, Oxford OX1 3QT, United Kingdom*

MADS NYBO (31), *Immunology and Microbiology, IMB, SDU, Odense University, DK-5000 Odense C, Denmark*

BRUNO OESCH (8), *Prionics AG, 8057 Zürich, Switzerland*

ANDERS OLOFSSON (38), *Department of Cell Molecular Biology, Umeå University, S-901 87 Umeå, Sweden*

GEORGE PERRY (10), *Institute of Pathology, Case Western Reserve University, Cleveland, Ohio 44106*

HAKAN PERSSON (38), *Department of Cell Molecular Biology, Umeå University, S-901 87 Umeå, Sweden*

ROSEMARIE RAFFEN (21), *Biosciences Division, Argonne National Laboratory, Argonne, Illinois 60439*

THEODORE W. RANDOLPH (16), *Center for Pharmaceutical Biotechnology, Department of Chemical Engineering, University of Colorado, Boulder, Colorado 80309*

GREGORY RAYMOND (9), *Rocky Mountain Laboratories, National Institute of Allergy and Infectious Diseases, National Institutes of Health, Hamilton, Montana 59840*

CAROL V. ROBINSON (40), *Oxford Centre for Molecular Sciences, Oxford University, Oxford OX1 3QT, United Kingdom*

ALEX EUGENE ROHER (4, 7), *Haldeman Laboratory for Alzheimer Disease Research, Sun Health Research Institute, Sun City, Arizona 85351*

RAINER RUDOLPH (15), *Institut für Biotechnologie, Martin-Luther-Universität Halle-Wittenberg, 06120 Halle, Germany*

KIRUPA SATHASIVAM (43), *Medical and Molecular Genetics, UMDS, Guy's Hospital, London SE1 9RT, United Kingdom*

LAWRENCE M. SAYRE (10), *Department of Chemistry, Case Western Reserve University, Cleveland, Ohio 44106*

EBERHARD SCHERZINGER (24, 43), *Max Planck Institute for Molecular Biology, Dahlem, D-14195 Berlin, Germany*

JEFFREY D. SCHOLTEN (29), *Department of Biochemistry, Parke-Davis Pharmaceutical Research Division, Warner-Lambert Company, Ann Arbor, Michigan 48105-1047*

ELISABETH SCHWARZ (15), *Institut für Biotechnologie, Martin-Luther-Universität Halle-Wittenberg, 06120 Halle, Germany*

TRICIA R. SERIO (41), *Department of Molecular Genetics and Cell Biology, University of Chicago, Chicago, Illinois 60637*

LOUISE C. SERPELL (34), *Laboratory of Molecular Biology, Medical Research Council Centre, Cambridge CB2 2QH, United Kingdom*

SANGITA SESHADRI (36), *Inhale Therapeutics, San Carlos, California 94070*

HAIYAN SHAO (13), *Department of Chemistry, Case Western Reserve University, Cleveland, Ohio 44106-7078*

MARK S. SHEARMAN (45), *Neuroscience Research Centre, Merck Sharp and Dohme Research Laboratories, Harlow, Essex CM20 2QR, United Kingdom*

ANNIE SITTLER (24), *Max Planck Institute for Molecular Genetics, D-14195 Berlin, Germany*

MARK A. SMITH (10), *Institute of Pathology, Case Western Reserve University, Cleveland, Ohio 44106*

ALAN SOLOMON (5, 14, 28), *Department of Medicine, Human Immunology and Cancer Program, University of Tennessee Medical Center/Graduate School of Medicine, Knoxville, Tennessee 37920-6999*

MARGARET SPEED (22), *Amgen, Inc., Thousand Oaks, California 91320*

FRED J. STEVENS (21), *Biosciences Division, Argonne National Laboratory, Argonne, Illinois 60439*

EVELYN R. STIMSON (23), *Department of Pharmacology and Cell Biophysics, University of Cincinnati, Cincinnati, Ohio 45267*

PETER STREIT (8), *Brain Research Institute, University of Zürich, 8057 Zürich, Switzerland*

MARGARET SUNDE (34), *Department of Biochemistry, Cambridge University, Cambridge CB2 1AG, United Kingdom*

SVEN-ERIK SVEHAG (31), *Immunology and Microbiology, IMB, SDU, Odense University, DK-5000 Odense C, Denmark*

TOSHIO TAKEDA (42), *Council for SAM Research, Institute for Frontier Medical Sciences, Kyoto University, Sakyo-ku, Kyoto 606-8397, Japan*

GLENYS A. TENNENT (2), *Immunological Medicine Unit, Division of Medicine, Imperial College School of Medicine, Hammersmith Hospital, London W12 ONN, United Kingdom*

DAVID B. TEPLOW (27), *Center for Neurologic Diseases, Brigham and Women's Hospital, Boston, Massachusetts 02115-5716*

JOHN TROJANOWSKI (6), *Center for Neurodegenerative Disease Research, Division of Anatomic Pathology, Department of Pathology and Laboratory Medicine, University of Pennsylvania School of Medicine, Philadelphia, Pennsylvania 19104-4283*

PASCAL VALAX (3), *Covance Biotechnology Services, Research Triangle Park, North Carolina 27709-3865*

SRIDHAR VARADARAJAN (48), *Department of Chemistry and Center of Membrane Sciences, University of Kentucky, Lexington, Kentucky 40506-0055*

JONATHAN WALL (14, 28), *Human Immunology and Cancer Program, University of Tennessee Medical Center, Knoxville, Tennessee 37920*

JUN WANG (6), *Center for Neurodegenerative Disease Research, Division of Anatomic Pathology, Department of Pathology and Laboratory Medicine, University of Pennsylvania School of Medicine, Philadelphia, Pennsylvania 19104-4283*

ERICH E. WANKER (24, 43), *Max Planck Institute for Molecular Biology, Dahlem, D-14195 Berlin, Germany*

Deborah T. Weiss (5), *Department of Medicine, Human Immunology and Cancer Program, University of Tennessee Medical Center/Graduate School of Medicine, Knoxville, Tennessee 37920-6999*

David E. Wemmer (35), *Department of Chemistry, University of California, Berkeley, California 94720*

Gunilla T. Westermark (1), *Department of Biomedicine and Surgery, Division of Cell Biology, University of Linköping, Linköping S-581 85, Sweden*

Per Westermark (1), *Department of Genetics and Pathology, Uppsala University, Uppsala S-751 85, Sweden*

Stephen J. Wood (25), *Amgen, Inc., Thousand Oaks, California 91320-1799*

Jing Yang (13), *Department of Chemistry, Case Western Reserve University, Cleveland, Ohio 44106-7078*

Stephen L. Yates (46), *Gliatech Inc., Cleveland, Ohio 44122*

Servet M. Yatin (48), *Department of Chemistry and Center of Membrane Sciences, University of Kentucky, Lexington, Kentucky 40506-0055*

Michael G. Zagorski (13), *Department of Chemistry, Case Western Reserve University, Cleveland, Ohio 44106-7078*

Hong Zeng (13), *Department of Chemistry, Case Western Reserve University, Cleveland, Ohio 44106-7078*

Preface

This volume of *Methods in Enzymology* focuses on a new discipline in the area of protein biochemistry: the study of alternative folded states of polypeptides and, in particular, insoluble protein aggregates formed as a consequence of aberrant folding. While an effort has been made to include many aspects of the study of biological protein deposition, the concentration of the volume is on the biochemical and biophysical characterization of protein aggregation mechanisms and products. Many of these methods have been recently developed and have only been validated in the context of one particular type of protein aggregation. It can be reasonably predicted, however, that a good number of these methods will extend, with appropriate modifications, to other protein aggregation conditions and reactions.

Although folding-related protein aggregation has only recently been recognized as an important field of research, our awareness of the phenomenon itself is hardly new. The ability of many protein molecules to respond to heating *in vitro* by irreversibly aggregating and precipitating was one of the first physical methods used to characterize this class of macromolecules. This is perhaps best illustrated by the realization that the myeloma-associated circulating immunoglobulin light chains encountered by Dr. Henry Bence-Jones 150 years ago were recognized by him as being a different class of proteins precisely because their thermal precipitation response is atypical of globular proteins. *In vitro* thermal aggregation experiments in the first decade of the twentieth century provided some of the first clues to the biophysical individuality of different proteins and to the highly cooperative nature of their folding stability. Alois Alzheimer published his famous note on the unusual staining features in the brain of a dementia patient in 1907. In fact, amyloid deposits in tissue were first reported in the 1850s, although their proteinaceous character was not deduced until much later.

In the early decades of the twentieth century, the sole interest in the process of protein aggregation and precipitation by most protein biochemists was as the ultimate marker for protein "denaturation." In later decades, as more quantitative techniques developed that allowed protein unfolding to be studied in solution, and, ultimately, under reversible conditions, protein aggregation became viewed as an unfortunate but—outside of its nuisance value—unimportant side reaction of the folding process. Aggregation reactions were modeled as the nonspecific, hydrophobically driven association of fully unfolded, random coil states.

The situation began to change in the last decades of this century. The seeds of change were planted in the 1970s with (1) the proposal of Michel Goldberg and colleagues that specific misfolding events could explain at least some protein aggregation phenomena, (2) the recognition that protein folding pathways can include metastable folding intermediates with solution properties different from those of the native state, and (3) the experience of recombinant DNA-based biotechnology companies that overexpression of some normal proteins in bacteria can lead to the formation of dense, highly insoluble, protein-rich inclusion bodies. It was also in the early 1970s that Glenner's group published the important observations that amyloid is composed of particular major proteins and that it can be generated from these proteins *in vitro*. In the 1980s, the work of Jonathan King and coworkers showed that folding-associated protein aggregation could be influenced by subtle amino acid replacements, independent of any effect on overall protein folding stability, and that aggregation therefore can be a much more specific process than previous models indicated. Also in the 1980s, the discovery that many heat shock proteins are molecular chaperones which function in the cell to nudge wayward proteins from the path of aggregation to the path of solubility demonstrated that protein misfolding and aggregation are both biologically important and ubiquitous.

In the decade now ending, we have seen the demonstration of the fundamental role for protein destabilization and misfolding in a number of amyloid conditions, ushering in the concept of human diseases which are fundamentally caused by protein misfolding events. We have also seen solid validation of the overall prion hypothesis by the characterization of a number of microbial prion activities. In the past three years, a number of well-known neurodegenerative disorders, including Huntington's Disease, amyotrophic lateral sclerosis, and various Lewy body-associated dementias including Parkinson's disease, have been added to Alzheimer's disease and prion disorders as neurological conditions with an associated protein aggregation component.

Protein misfolding and aggregation are now viewed as intrinsic features of the energy landscape of the protein folding reaction, as a challenge faced in both the molecular evolution of proteins and in the life of the contemporary cell, and as an emerging area in the study of molecular mechanisms of disease.

Our enhanced appreciation of the importance of protein aggregation would be frustrating and bittersweet, however, were it not for the availability of new techniques for the study of the formation and properties of these protein aggregates. The imaginative studies of the many groups represented in this book make it clear that, while molecular studies on protein aggregation and aggregates represent unique challenges, progress is being made which will lead us to a solid understanding of the mechanistic and structural aspects of protein misassembly processes and products.

The coverage of this book reflects the fact that much of the work on protein misassembly has been driven by the sudy of amyloid formation in the Alzheimer's disease system. At the same time, this is not meant to be a book on the molecular biology of Alzheimer's disease. In addition, because the physical process of protein aggregation is the emphasis of this book, other important aspects of protein aggregation have been treated relatively lightly by inclusion only of representative techniques. In particular, it has not been possible to adequately represent the wealth of viewpoints and techniques currently available for the study of protein aggregate cytotoxicity.

This volume is the result of the selfless donation of time and energy on the part of many of my scientific colleagues. At the top of the list are the authors who have put aside precious time from the myriad pressing demands of contemporary scientific life and worked hard to clearly state the rationales and protocols of their laboratory methods. Their task was aided by the anonymous work of many reviewers who generously gave time to provide feedback on initial drafts of the chapters. Finally, many people provided me with encouragement and valuable suggestions on coverage in the early stages of the organization of this volume; in particular, I gratefully acknowledge the helpful comments of Gloria Gallo, Mark Shearman, Jean Sipe, Fred Stevens, Margaret Sunde, Toshio Takeda, Glenys Tennant, Jonathan Wall, and Michael Zagorski.

RONALD WETZEL

METHODS IN ENZYMOLOGY

VOLUME XVII. Metabolism of Amino Acids and Amines (Parts A and B)
Edited by HERBERT TABOR AND CELIA WHITE TABOR

VOLUME XVIII. Vitamins and Coenzymes (Parts A, B, and C)
Edited by DONALD B. MCCORMICK AND LEMUEL D. WRIGHT

VOLUME XIX. Proteolytic Enzymes
Edited by GERTRUDE E. PERLMANN AND LASZLO LORAND

VOLUME XX. Nucleic Acids and Protein Synthesis (Part C)
Edited by KIVIE MOLDAVE AND LAWRENCE GROSSMAN

VOLUME XXI. Nucleic Acids (Part D)
Edited by LAWRENCE GROSSMAN AND KIVIE MOLDAVE

VOLUME XXII. Enzyme Purification and Related Techniques
Edited by WILLIAM B. JAKOBY

VOLUME XXIII. Photosynthesis (Part A)
Edited by ANTHONY SAN PIETRO

VOLUME XXIV. Photosynthesis and Nitrogen Fixation (Part B)
Edited by ANTHONY SAN PIETRO

VOLUME XXV. Enzyme Structure (Part B)
Edited by C. H. W. HIRS AND SERGE N. TIMASHEFF

VOLUME XXVI. Enzyme Structure (Part C)
Edited by C. H. W. HIRS AND SERGE N. TIMASHEFF

VOLUME XXVII. Enzyme Structure (Part D)
Edited by C. H. W. HIRS AND SERGE N. TIMASHEFF

VOLUME XXVIII. Complex Carbohydrates (Part B)
Edited by VICTOR GINSBURG

VOLUME XXIX. Nucleic Acids and Protein Synthesis (Part E)
Edited by LAWRENCE GROSSMAN AND KIVIE MOLDAVE

VOLUME XXX. Nucleic Acids and Protein Synthesis (Part F)
Edited by KIVIE MOLDAVE AND LAWRENCE GROSSMAN

VOLUME XXXI. Biomembranes (Part A)
Edited by SIDNEY FLEISCHER AND LESTER PACKER

VOLUME XXXII. Biomembranes (Part B)
Edited by SIDNEY FLEISCHER AND LESTER PACKER

VOLUME XXXIII. Cumulative Subject Index Volumes I–XXX
Edited by MARTHA G. DENNIS AND EDWARD A. DENNIS

VOLUME XXXIV. Affinity Techniques (Enzyme Purification: Part B)
Edited by WILLIAM B. JAKOBY AND MEIR WILCHEK

VOLUME XXXV. Lipids (Part B)
Edited by JOHN M. LOWENSTEIN

VOLUME XXXVI. Hormone Action (Part A: Steroid Hormones)
Edited by BERT W. O'MALLEY AND JOEL G. HARDMAN

VOLUME XXXVII. Hormone Action (Part B: Peptide Hormones)
Edited by BERT W. O'MALLEY AND JOEL G. HARDMAN

VOLUME XXXVIII. Hormone Action (Part C: Cyclic Nucleotides)
Edited by JOEL G. HARDMAN AND BERT W. O'MALLEY

VOLUME XXXIX. Hormone Action (Part D: Isolated Cells, Tissues, and Organ Systems)
Edited by JOEL G. HARDMAN AND BERT W. O'MALLEY

VOLUME XL. Hormone Action (Part E: Nuclear Structure and Function)
Edited by BERT W. O'MALLEY AND JOEL G. HARDMAN

VOLUME XLI. Carbohydrate Metabolism (Part B)
Edited by W. A. WOOD

VOLUME XLII. Carbohydrate Metabolism (Part C)
Edited by W. A. WOOD

VOLUME XLIII. Antibiotics
Edited by JOHN H. HASH

VOLUME XLIV. Immobilized Enzymes
Edited by KLAUS MOSBACH

VOLUME XLV. Proteolytic Enzymes (Part B)
Edited by LASZLO LORAND

VOLUME XLVI. Affinity Labeling
Edited by WILLIAM B. JAKOBY AND MEIR WILCHEK

VOLUME XLVII. Enzyme Structure (Part E)
Edited by C. H. W. HIRS AND SERGE N. TIMASHEFF

VOLUME XLVIII. Enzyme Structure (Part F)
Edited by C. H. W. HIRS AND SERGE N. TIMASHEFF

VOLUME XLIX. Enzyme Structure (Part G)
Edited by C. H. W. HIRS AND SERGE N. TIMASHEFF

VOLUME L. Complex Carbohydrates (Part C)
Edited by VICTOR GINSBURG

VOLUME LI. Purine and Pyrimidine Nucleotide Metabolism
Edited by PATRICIA A. HOFFEE AND MARY ELLEN JONES

VOLUME LII. Biomembranes (Part C: Biological Oxidations)
Edited by SIDNEY FLEISCHER AND LESTER PACKER

VOLUME LIII. Biomembranes (Part D: Biological Oxidations)
Edited by SIDNEY FLEISCHER AND LESTER PACKER

VOLUME LIV. Biomembranes (Part E: Biological Oxidations)
Edited by SIDNEY FLEISCHER AND LESTER PACKER

VOLUME 91. Enzyme Structure (Part I)
Edited by C. H. W. HIRS AND SERGE N. TIMASHEFF

VOLUME 92. Immunochemical Techniques (Part E: Monoclonal Antibodies and General Immunoassay Methods)
Edited by JOHN J. LANGONE AND HELEN VAN VUNAKIS

VOLUME 93. Immunochemical Techniques (Part F: Conventional Antibodies, Fc Receptors, and Cytotoxicity)
Edited by JOHN J. LANGONE AND HELEN VAN VUNAKIS

VOLUME 94. Polyamines
Edited by HERBERT TABOR AND CELIA WHITE TABOR

VOLUME 95. Cumulative Subject Index Volumes 61–74, 76–80
Edited by EDWARD A. DENNIS AND MARTHA G. DENNIS

VOLUME 96. Biomembranes [Part J: Membrane Biogenesis: Assembly and Targeting (General Methods; Eukaryotes)]
Edited by SIDNEY FLEISCHER AND BECCA FLEISCHER

VOLUME 97. Biomembranes [Part K: Membrane Biogenesis: Assembly and Targeting (Prokaryotes, Mitochondria, and Chloroplasts)]
Edited by SIDNEY FLEISCHER AND BECCA FLEISCHER

VOLUME 98. Biomembranes (Part L: Membrane Biogenesis: Processing and Recycling)
Edited by SIDNEY FLEISCHER AND BECCA FLEISCHER

VOLUME 99. Hormone Action (Part F: Protein Kinases)
Edited by JACKIE D. CORBIN AND JOEL G. HARDMAN

VOLUME 100. Recombinant DNA (Part B)
Edited by RAY WU, LAWRENCE GROSSMAN, AND KIVIE MOLDAVE

VOLUME 101. Recombinant DNA (Part C)
Edited by RAY WU, LAWRENCE GROSSMAN, AND KIVIE MOLDAVE

VOLUME 102. Hormone Action (Part G: Calmodulin and Calcium-Binding Proteins)
Edited by ANTHONY R. MEANS AND BERT W. O'MALLEY

VOLUME 103. Hormone Action (Part H: Neuroendocrine Peptides)
Edited by P. MICHAEL CONN

VOLUME 104. Enzyme Purification and Related Techniques (Part C)
Edited by WILLIAM B. JAKOBY

VOLUME 105. Oxygen Radicals in Biological Systems
Edited by LESTER PACKER

VOLUME 106. Posttranslational Modifications (Part A)
Edited by FINN WOLD AND KIVIE MOLDAVE

VOLUME 107. Posttranslational Modifications (Part B)
Edited by FINN WOLD AND KIVIE MOLDAVE

VOLUME 108. Immunochemical Techniques (Part G: Separation and Characterization of Lymphoid Cells)
Edited by GIOVANNI DI SABATO, JOHN J. LANGONE, AND HELEN VAN VUNAKIS

VOLUME 109. Hormone Action (Part I: Peptide Hormones)
Edited by LUTZ BIRNBAUMER AND BERT W. O'MALLEY

VOLUME 110. Steroids and Isoprenoids (Part A)
Edited by JOHN H. LAW AND HANS C. RILLING

VOLUME 111. Steroids and Isoprenoids (Part B)
Edited by JOHN H. LAW AND HANS C. RILLING

VOLUME 112. Drug and Enzyme Targeting (Part A)
Edited by KENNETH J. WIDDER AND RALPH GREEN

VOLUME 113. Glutamate, Glutamine, Glutathione, and Related Compounds
Edited by ALTON MEISTER

VOLUME 114. Diffraction Methods for Biological Macromolecules (Part A)
Edited by HAROLD W. WYCKOFF, C. H. W. HIRS, AND SERGE N. TIMASHEFF

VOLUME 115. Diffraction Methods for Biological Macromolecules (Part B)
Edited by HAROLD W. WYCKOFF, C. H. W. HIRS, AND SERGE N. TIMASHEFF

VOLUME 116. Immunochemical Techniques (Part H: Effectors and Mediators of Lymphoid Cell Functions)
Edited by GIOVANNI DI SABATO, JOHN J. LANGONE, AND HELEN VAN VUNAKIS

VOLUME 117. Enzyme Structure (Part J)
Edited by C. H. W. HIRS AND SERGE N. TIMASHEFF

VOLUME 118. Plant Molecular Biology
Edited by ARTHUR WEISSBACH AND HERBERT WEISSBACH

VOLUME 119. Interferons (Part C)
Edited by SIDNEY PESTKA

VOLUME 120. Cumulative Subject Index Volumes 81–94, 96–101

VOLUME 121. Immunochemical Techniques (Part I: Hybridoma Technology and Monoclonal Antibodies)
Edited by JOHN J. LANGONE AND HELEN VAN VUNAKIS

VOLUME 122. Vitamins and Coenzymes (Part G)
Edited by FRANK CHYTIL AND DONALD B. MCCORMICK

VOLUME 123. Vitamins and Coenzymes (Part H)
Edited by FRANK CHYTIL AND DONALD B. MCCORMICK

VOLUME 124. Hormone Action (Part J: Neuroendocrine Peptides)
Edited by P. MICHAEL CONN

Volume 125. Biomembranes (Part M: Transport in Bacteria, Mitochondria, and Chloroplasts: General Approaches and Transport Systems)
Edited by Sidney Fleischer and Becca Fleischer

Volume 126. Biomembranes (Part N: Transport in Bacteria, Mitochondria, and Chloroplasts: Protonmotive Force)
Edited by Sidney Fleischer and Becca Fleischer

Volume 127. Biomembranes (Part O: Protons and Water: Structure and Translocation)
Edited by Lester Packer

Volume 128. Plasma Lipoproteins (Part A: Preparation, Structure, and Molecular Biology)
Edited by Jere P. Segrest and John J. Albers

Volume 129. Plasma Lipoproteins (Part B: Characterization, Cell Biology, and Metabolism)
Edited by John J. Albers and Jere P. Segrest

Volume 130. Enzyme Structure (Part K)
Edited by C. H. W. Hirs and Serge N. Timasheff

Volume 131. Enzyme Structure (Part L)
Edited by C. H. W. Hirs and Serge N. Timasheff

Volume 132. Immunochemical Techniques (Part J: Phagocytosis and Cell-Mediated Cytotoxicity)
Edited by Giovanni Di Sabato and Johannes Everse

Volume 133. Bioluminescence and Chemiluminescence (Part B)
Edited by Marlene DeLuca and William D. McElroy

Volume 134. Structural and Contractile Proteins (Part C: The Contractile Apparatus and the Cytoskeleton)
Edited by Richard B. Vallee

Volume 135. Immobilized Enzymes and Cells (Part B)
Edited by Klaus Mosbach

Volume 136. Immobilized Enzymes and Cells (Part C)
Edited by Klaus Mosbach

Volume 137. Immobilized Enzymes and Cells (Part D)
Edited by Klaus Mosbach

Volume 138. Complex Carbohydrates (Part E)
Edited by Victor Ginsburg

Volume 139. Cellular Regulators (Part A: Calcium- and Calmodulin-Binding Proteins)
Edited by Anthony R. Means and P. Michael Conn

Volume 140. Cumulative Subject Index Volumes 102–119, 121–134

VOLUME 141. Cellular Regulators (Part B: Calcium and Lipids)
Edited by P. MICHAEL CONN AND ANTHONY R. MEANS

VOLUME 142. Metabolism of Aromatic Amino Acids and Amines
Edited by SEYMOUR KAUFMAN

VOLUME 143. Sulfur and Sulfur Amino Acids
Edited by WILLIAM B. JAKOBY AND OWEN GRIFFITH

VOLUME 144. Structural and Contractile Proteins (Part D: Extracellular Matrix)
Edited by LEON W. CUNNINGHAM

VOLUME 145. Structural and Contractile Proteins (Part E: Extracellular Matrix)
Edited by LEON W. CUNNINGHAM

VOLUME 146. Peptide Growth Factors (Part A)
Edited by DAVID BARNES AND DAVID A. SIRBASKU

VOLUME 147. Peptide Growth Factors (Part B)
Edited by DAVID BARNES AND DAVID A. SIRBASKU

VOLUME 148. Plant Cell Membranes
Edited by LESTER PACKER AND ROLAND DOUCE

VOLUME 149. Drug and Enzyme Targeting (Part B)
Edited by RALPH GREEN AND KENNETH J. WIDDER

VOLUME 150. Immunochemical Techniques (Part K: *In Vitro* Models of B and T Cell Functions and Lymphoid Cell Receptors)
Edited by GIOVANNI DI SABATO

VOLUME 151. Molecular Genetics of Mammalian Cells
Edited by MICHAEL M. GOTTESMAN

VOLUME 152. Guide to Molecular Cloning Techniques
Edited by SHELBY L. BERGER AND ALAN R. KIMMEL

VOLUME 153. Recombinant DNA (Part D)
Edited by RAY WU AND LAWRENCE GROSSMAN

VOLUME 154. Recombinant DNA (Part E)
Edited by RAY WU AND LAWRENCE GROSSMAN

VOLUME 155. Recombinant DNA (Part F)
Edited by RAY WU

VOLUME 156. Biomembranes (Part P: ATP-Driven Pumps and Related Transport: The Na,K-Pump)
Edited by SIDNEY FLEISCHER AND BECCA FLEISCHER

VOLUME 157. Biomembranes (Part Q: ATP-Driven Pumps and Related Transport: Calcium, Proton, and Potassium Pumps)
Edited by SIDNEY FLEISCHER AND BECCA FLEISCHER

VOLUME 158. Metalloproteins (Part A)
Edited by JAMES F. RIORDAN AND BERT L. VALLEE

VOLUME 159. Initiation and Termination of Cyclic Nucleotide Action
Edited by JACKIE D. CORBIN AND ROGER A. JOHNSON

VOLUME 160. Biomass (Part A: Cellulose and Hemicellulose)
Edited by WILLIS A. WOOD AND SCOTT T. KELLOGG

VOLUME 161. Biomass (Part B: Lignin, Pectin, and Chitin)
Edited by WILLIS A. WOOD AND SCOTT T. KELLOGG

VOLUME 162. Immunochemical Techniques (Part L: Chemotaxis and Inflammation)
Edited by GIOVANNI DI SABATO

VOLUME 163. Immunochemical Techniques (Part M: Chemotaxis and Inflammation)
Edited by GIOVANNI DI SABATO

VOLUME 164. Ribosomes
Edited by HARRY F. NOLLER, JR., AND KIVIE MOLDAVE

VOLUME 165. Microbial Toxins: Tools for Enzymology
Edited by SIDNEY HARSHMAN

VOLUME 166. Branched-Chain Amino Acids
Edited by ROBERT HARRIS AND JOHN R. SOKATCH

VOLUME 167. Cyanobacteria
Edited by LESTER PACKER AND ALEXANDER N. GLAZER

VOLUME 168. Hormone Action (Part K: Neuroendocrine Peptides)
Edited by P. MICHAEL CONN

VOLUME 169. Platelets: Receptors, Adhesion, Secretion (Part A)
Edited by JACEK HAWIGER

VOLUME 170. Nucleosomes
Edited by PAUL M. WASSARMAN AND ROGER D. KORNBERG

VOLUME 171. Biomembranes (Part R: Transport Theory: Cells and Model Membranes)
Edited by SIDNEY FLEISCHER AND BECCA FLEISCHER

VOLUME 172. Biomembranes (Part S: Transport: Membrane Isolation and Characterization)
Edited by SIDNEY FLEISCHER AND BECCA FLEISCHER

VOLUME 173. Biomembranes [Part T: Cellular and Subcellular Transport: Eukaryotic (Nonepithelial) Cells]
Edited by SIDNEY FLEISCHER AND BECCA FLEISCHER

VOLUME 174. Biomembranes [Part U: Cellular and Subcellular Transport: Eukaryotic (Nonepithelial) Cells]
Edited by SIDNEY FLEISCHER AND BECCA FLEISCHER

VOLUME 175. Cumulative Subject Index Volumes 135–139, 141–167

VOLUME 176. Nuclear Magnetic Resonance (Part A: Spectral Techniques and Dynamics)
Edited by NORMAN J. OPPENHEIMER AND THOMAS L. JAMES

VOLUME 177. Nuclear Magnetic Resonance (Part B: Structure and Mechanism)
Edited by NORMAN J. OPPENHEIMER AND THOMAS L. JAMES

VOLUME 178. Antibodies, Antigens, and Molecular Mimicry
Edited by JOHN J. LANGONE

VOLUME 179. Complex Carbohydrates (Part F)
Edited by VICTOR GINSBURG

VOLUME 180. RNA Processing (Part A: General Methods)
Edited by JAMES E. DAHLBERG AND JOHN N. ABELSON

VOLUME 181. RNA Processing (Part B: Specific Methods)
Edited by JAMES E. DAHLBERG AND JOHN N. ABELSON

VOLUME 182. Guide to Protein Purification
Edited by MURRAY P. DEUTSCHER

VOLUME 183. Molecular Evolution: Computer Analysis of Protein and Nucleic Acid Sequences
Edited by RUSSELL F. DOOLITTLE

VOLUME 184. Avidin–Biotin Technology
Edited by MEIR WILCHEK AND EDWARD A. BAYER

VOLUME 185. Gene Expression Technology
Edited by DAVID V. GOEDDEL

VOLUME 186. Oxygen Radicals in Biological Systems (Part B: Oxygen Radicals and Antioxidants)
Edited by LESTER PACKER AND ALEXANDER N. GLAZER

VOLUME 187. Arachidonate Related Lipid Mediators
Edited by ROBERT C. MURPHY AND FRANK A. FITZPATRICK

VOLUME 188. Hydrocarbons and Methylotrophy
Edited by MARY E. LIDSTROM

VOLUME 189. Retinoids (Part A: Molecular and Metabolic Aspects)
Edited by LESTER PACKER

VOLUME 190. Retinoids (Part B: Cell Differentiation and Clinical Applications)
Edited by LESTER PACKER

VOLUME 191. Biomembranes (Part V: Cellular and Subcellular Transport: Epithelial Cells)
Edited by SIDNEY FLEISCHER AND BECCA FLEISCHER

VOLUME 192. Biomembranes (Part W: Cellular and Subcellular Transport: Epithelial Cells)
Edited by SIDNEY FLEISCHER AND BECCA FLEISCHER

VOLUME 193. Mass Spectrometry
Edited by JAMES A. MCCLOSKEY

VOLUME 194. Guide to Yeast Genetics and Molecular Biology
Edited by CHRISTINE GUTHRIE AND GERALD R. FINK

VOLUME 195. Adenylyl Cyclase, G Proteins, and Guanylyl Cyclase
Edited by ROGER A. JOHNSON AND JACKIE D. CORBIN

VOLUME 196. Molecular Motors and the Cytoskeleton
Edited by RICHARD B. VALLEE

VOLUME 197. Phospholipases
Edited by EDWARD A. DENNIS

VOLUME 198. Peptide Growth Factors (Part C)
Edited by DAVID BARNES, J. P. MATHER, AND GORDON H. SATO

VOLUME 199. Cumulative Subject Index Volumes 168–174, 176–194

VOLUME 200. Protein Phosphorylation (Part A: Protein Kinases: Assays, Purification, Antibodies, Functional Analysis, Cloning, and Expression)
Edited by TONY HUNTER AND BARTHOLOMEW M. SEFTON

VOLUME 201. Protein Phosphorylation (Part B: Analysis of Protein Phosphorylation, Protein Kinase Inhibitors, and Protein Phosphatases)
Edited by TONY HUNTER AND BARTHOLOMEW M. SEFTON

VOLUME 202. Molecular Design and Modeling: Concepts and Applications (Part A: Proteins, Peptides, and Enzymes)
Edited by JOHN J. LANGONE

VOLUME 203. Molecular Design and Modeling: Concepts and Applications (Part B: Antibodies and Antigens, Nucleic Acids, Polysaccharides, and Drugs)
Edited by JOHN J. LANGONE

VOLUME 204. Bacterial Genetic Systems
Edited by JEFFREY H. MILLER

VOLUME 205. Metallobiochemistry (Part B: Metallothionein and Related Molecules)
Edited by JAMES F. RIORDAN AND BERT L. VALLEE

VOLUME 206. Cytochrome P450
Edited by MICHAEL R. WATERMAN AND ERIC F. JOHNSON

VOLUME 207. Ion Channels
Edited by BERNARDO RUDY AND LINDA E. IVERSON

VOLUME 208. Protein–DNA Interactions
Edited by ROBERT T. SAUER

VOLUME 209. Phospholipid Biosynthesis
Edited by EDWARD A. DENNIS AND DENNIS E. VANCE

VOLUME 210. Numerical Computer Methods
Edited by LUDWIG BRAND AND MICHAEL L. JOHNSON

Volume 211. DNA Structures (Part A: Synthesis and Physical Analysis of DNA)
Edited by David M. J. Lilley and James E. Dahlberg

Volume 212. DNA Structures (Part B: Chemical and Electrophoretic Analysis of DNA)
Edited by David M. J. Lilley and James E. Dahlberg

Volume 213. Carotenoids (Part A: Chemistry, Separation, Quantitation, and Antioxidation)
Edited by Lester Packer

Volume 214. Carotenoids (Part B: Metabolism, Genetics, and Biosynthesis)
Edited by Lester Packer

Volume 215. Platelets: Receptors, Adhesion, Secretion (Part B)
Edited by Jacek J. Hawiger

Volume 216. Recombinant DNA (Part G)
Edited by Ray Wu

Volume 217. Recombinant DNA (Part H)
Edited by Ray Wu

Volume 218. Recombinant DNA (Part I)
Edited by Ray Wu

Volume 219. Reconstitution of Intracellular Transport
Edited by James E. Rothman

Volume 220. Membrane Fusion Techniques (Part A)
Edited by Nejat Düzgüneş

Volume 221. Membrane Fusion Techniques (Part B)
Edited by Nejat Düzgüneş

Volume 222. Proteolytic Enzymes in Coagulation, Fibrinolysis, and Complement Activation (Part A: Mammalian Blood Coagulation Factors and Inhibitors)
Edited by Laszlo Lorand and Kenneth G. Mann

Volume 223. Proteolytic Enzymes in Coagulation, Fibrinolysis, and Complement Activation (Part B: Complement Activation, Fibrinolysis, and Nonmammalian Blood Coagulation Factors)
Edited by Laszlo Lorand and Kenneth G. Mann

Volume 224. Molecular Evolution: Producing the Biochemical Data
Edited by Elizabeth Anne Zimmer, Thomas J. White, Rebecca L. Cann, and Allan C. Wilson

Volume 225. Guide to Techniques in Mouse Development
Edited by Paul M. Wassarman and Melvin L. DePamphilis

Volume 226. Metallobiochemistry (Part C: Spectroscopic and Physical Methods for Probing Metal Ion Environments in Metalloenzymes and Metalloproteins)
Edited by James F. Riordan and Bert L. Vallee

VOLUME 227. Metallobiochemistry (Part D: Physical and Spectroscopic Methods for Probing Metal Ion Environments in Metalloproteins)
Edited by JAMES F. RIORDAN AND BERT L. VALLEE

VOLUME 228. Aqueous Two-Phase Systems
Edited by HARRY WALTER AND GÖTE JOHANSSON

VOLUME 229. Cumulative Subject Index Volumes 195–198, 200–227

VOLUME 230. Guide to Techniques in Glycobiology
Edited by WILLIAM J. LENNARZ AND GERALD W. HART

VOLUME 231. Hemoglobins (Part B: Biochemical and Analytical Methods)
Edited by JOHANNES EVERSE, KIM D. VANDEGRIFF, AND ROBERT M. WINSLOW

VOLUME 232. Hemoglobins (Part C: Biophysical Methods)
Edited by JOHANNES EVERSE, KIM D. VANDEGRIFF, AND ROBERT M. WINSLOW

VOLUME 233. Oxygen Radicals in Biological Systems (Part C)
Edited by LESTER PACKER

VOLUME 234. Oxygen Radicals in Biological Systems (Part D)
Edited by LESTER PACKER

VOLUME 235. Bacterial Pathogenesis (Part A: Identification and Regulation of Virulence Factors)
Edited by VIRGINIA L. CLARK AND PATRIK M. BAVOIL

VOLUME 236. Bacterial Pathogenesis (Part B: Integration of Pathogenic Bacteria with Host Cells)
Edited by VIRGINIA L. CLARK AND PATRIK M. BAVOIL

VOLUME 237. Heterotrimeric G Proteins
Edited by RAVI IYENGAR

VOLUME 238. Heterotrimeric G-Protein Effectors
Edited by RAVI IYENGAR

VOLUME 239. Nuclear Magnetic Resonance (Part C)
Edited by THOMAS L. JAMES AND NORMAN J. OPPENHEIMER

VOLUME 240. Numerical Computer Methods (Part B)
Edited by MICHAEL L. JOHNSON AND LUDWIG BRAND

VOLUME 241. Retroviral Proteases
Edited by LAWRENCE C. KUO AND JULES A. SHAFER

VOLUME 242. Neoglycoconjugates (Part A)
Edited by Y. C. LEE AND REIKO T. LEE

VOLUME 243. Inorganic Microbial Sulfur Metabolism
Edited by HARRY D. PECK, JR., AND JEAN LEGALL

VOLUME 244. Proteolytic Enzymes: Serine and Cysteine Peptidases
Edited by ALAN J. BARRETT

VOLUME 245. Extracellular Matrix Components
Edited by E. RUOSLAHTI AND E. ENGVALL

VOLUME 246. Biochemical Spectroscopy
Edited by KENNETH SAUER

VOLUME 247. Neoglycoconjugates (Part B: Biomedical Applications)
Edited by Y. C. LEE AND REIKO T. LEE

VOLUME 248. Proteolytic Enzymes: Aspartic and Metallo Peptidases
Edited by ALAN J. BARRETT

VOLUME 249. Enzyme Kinetics and Mechanism (Part D: Developments in Enzyme Dynamics)
Edited by DANIEL L. PURICH

VOLUME 250. Lipid Modifications of Proteins
Edited by PATRICK J. CASEY AND JANICE E. BUSS

VOLUME 251. Biothiols (Part A: Monothiols and Dithiols, Protein Thiols, and Thiyl Radicals)
Edited by LESTER PACKER

VOLUME 252. Biothiols (Part B: Glutathione and Thioredoxin; Thiols in Signal Transduction and Gene Regulation)
Edited by LESTER PACKER

VOLUME 253. Adhesion of Microbial Pathogens
Edited by RON J. DOYLE AND ITZHAK OFEK

VOLUME 254. Oncogene Techniques
Edited by PETER K. VOGT AND INDER M. VERMA

VOLUME 255. Small GTPases and Their Regulators (Part A: Ras Family)
Edited by W. E. BALCH, CHANNING J. DER, AND ALAN HALL

VOLUME 256. Small GTPases and Their Regulators (Part B: Rho Family)
Edited by W. E. BALCH, CHANNING J. DER, AND ALAN HALL

VOLUME 257. Small GTPases and Their Regulators (Part C: Proteins Involved in Transport)
Edited by W. E. BALCH, CHANNING J. DER, AND ALAN HALL

VOLUME 258. Redox-Active Amino Acids in Biology
Edited by JUDITH P. KLINMAN

VOLUME 259. Energetics of Biological Macromolecules
Edited by MICHAEL L. JOHNSON AND GARY K. ACKERS

VOLUME 260. Mitochondrial Biogenesis and Genetics (Part A)
Edited by GIUSEPPE M. ATTARDI AND ANNE CHOMYN

VOLUME 261. Nuclear Magnetic Resonance and Nucleic Acids
Edited by THOMAS L. JAMES

VOLUME 262. DNA Replication
Edited by JUDITH L. CAMPBELL

VOLUME 263. Plasma Lipoproteins (Part C: Quantitation)
Edited by WILLIAM A. BRADLEY, SANDRA H. GIANTURCO, AND JERE P. SEGREST

VOLUME 264. Mitochondrial Biogenesis and Genetics (Part B)
Edited by GIUSEPPE M. ATTARDI AND ANNE CHOMYN

VOLUME 265. Cumulative Subject Index Volumes 228, 230–262

VOLUME 266. Computer Methods for Macromolecular Sequence Analysis
Edited by RUSSELL F. DOOLITTLE

VOLUME 267. Combinatorial Chemistry
Edited by JOHN N. ABELSON

VOLUME 268. Nitric Oxide (Part A: Sources and Detection of NO; NO Synthase)
Edited by LESTER PACKER

VOLUME 269. Nitric Oxide (Part B: Physiological and Pathological Processes)
Edited by LESTER PACKER

VOLUME 270. High Resolution Separation and Analysis of Biological Macromolecules (Part A: Fundamentals)
Edited by BARRY L. KARGER AND WILLIAM S. HANCOCK

VOLUME 271. High Resolution Separation and Analysis of Biological Macromolecules (Part B: Applications)
Edited by BARRY L. KARGER AND WILLIAM S. HANCOCK

VOLUME 272. Cytochrome P450 (Part B)
Edited by ERIC F. JOHNSON AND MICHAEL R. WATERMAN

VOLUME 273. RNA Polymerase and Associated Factors (Part A)
Edited by SANKAR ADHYA

VOLUME 274. RNA Polymerase and Associated Factors (Part B)
Edited by SANKAR ADHYA

VOLUME 275. Viral Polymerases and Related Proteins
Edited by LAWRENCE C. KUO, DAVID B. OLSEN, AND STEVEN S. CARROLL

VOLUME 276. Macromolecular Crystallography (Part A)
Edited by CHARLES W. CARTER, JR., AND ROBERT M. SWEET

VOLUME 277. Macromolecular Crystallography (Part B)
Edited by CHARLES W. CARTER, JR., AND ROBERT M. SWEET

VOLUME 278. Fluorescence Spectroscopy
Edited by LUDWIG BRAND AND MICHAEL L. JOHNSON

VOLUME 279. Vitamins and Coenzymes (Part I)
Edited by DONALD B. MCCORMICK, JOHN W. SUTTIE, AND CONRAD WAGNER

VOLUME 280. Vitamins and Coenzymes (Part J)
Edited by DONALD B. MCCORMICK, JOHN W. SUTTIE, AND CONRAD WAGNER

VOLUME 281. Vitamins and Coenzymes (Part K)
Edited by DONALD B. MCCORMICK, JOHN W. SUTTIE, AND CONRAD WAGNER

VOLUME 302. Green Fluorescent Protein
Edited by P. MICHAEL CONN

VOLUME 303. cDNA Preparation and Display
Edited by SHERMAN M. WEISSMAN

VOLUME 304. Chromatin
Edited by PAUL M. WASSARMAN AND ALAN P. WOLFFE

VOLUME 305. Bioluminescence and Chemiluminescence (Part C)
Edited by MIRIAM M. ZIEGLER AND THOMAS O. BALDWIN

VOLUME 306. Expression of Recombinant Genes in Eukaryotic Systems
Edited by JOSEPH C. GLORIOSO AND MARTIN C. SCHMIDT

VOLUME 307. Confocal Microscopy
Edited by P. MICHAEL CONN

VOLUME 308. Enzyme Kinetics and Mechanism (Part E: Energetics of Enzyme Catalysis)
Edited by VERN L. SCHRAMM AND DANIEL L. PURICH

VOLUME 309. Amyloid, Prions, and Other Protein Aggregates
Edited by RONALD WETZEL

VOLUME 310. Biofilms
Edited by RON J. DOYLE

VOLUME 311. Sphingolipid Metabolism and Cell Signaling (Part A) (in preparation)
Edited by ALFRED H. MERRILL, JR., AND Y. A. HANNUN

VOLUME 312. Sphingolipid Metabolism and Cell Signaling (Part B) (in preparation)
Edited by ALFRED H. MERRILL, JR., AND Y. A. HANNUN

VOLUME 313. Antisense Technology (Part A: General Methods, Methods of Delivery and RNA Studies) (in preparation)
Edited by M. IAN PHILLIPS

VOLUME 314. Antisense Technology (Part B: Applications) (in preparation)
Edited by M. IAN PHILLIPS

VOLUME 315. Vertebrate Phototransduction and the Visual Cycle (Part A) (in preparation)
Edited by KRZYSZTOF PALCZEWSKI

VOLUME 316. Vertebrate Phototransduction and the Visual Cycle (Part B) (in preparation)
Edited by KRZYSZTOF PALCZEWSKI

VOLUME 317. RNA-Ligand Interactions (Part A: Structural Biology Methods) (in preparation)
Edited by DANIEL W. CELANDER AND JOHN N. ABELSON

Section I

Characterization of *in Vivo* Protein Deposition

A. Identification and Isolation of Aggregates
Articles 1 through 4

B. Isolation and Characterization of Protein Deposit Components
Articles 5 through 7

C. Characterization of Aggregates *in Situ* and *in Vitro*
Articles 8 through 12

[1] Staining Methods for Identification of Amyloid in Tissue

By GUNILLA T. WESTERMARK, KENNETH H. JOHNSON, and PER WESTERMARK

The staining reaction given by amyloid after treatment with iodine was often used in the earlier studies of amyloidosis, and amyloid is still to this date identified by its characteristic histological staining reactions. Despite the enormous amount of knowledge now known regarding the molecular nature of amyloid, histological staining methods are crucial for the diagnosis of amyloidosis and are also used commonly in amyloid research. Also, the introduction of modern immunohistochemical techniques has made it possible to identify normal and abnormal components in tissue. Immunohistochemistry (often used interchangeably with immunocytochemistry) thus has become an important tool in amyloid research.

History

Amyloid was first recognized by tinctorial properties elicited when amyloid-laden tissues were treated with iodine at the autopsy table (Virchow). This reaction is now known to depend on the presence of minor carbohydrate components in the amyloid deposits. Iodine reacts with the amyloid, giving it a mahogany-like color that changes to blue when sulfuric acid is subsequently added. The staining properties of amyloid with rosaniline dyes (e.g., methyl violet and cresyl violet), which were the main staining methods for amyloid before Congo red staining was introduced in the 1920s,[1] are also based on the presence of these same carbohydrate components. Because of their low sensitivity and lack of specificity, these methods are not commonly used any longer.

Most, if not all, dyes used for the identification of amyloid are compounds developed for use by the textile industry. This includes the dye Congo red, which was introduced as the first direct cotton dye in 1884. Much of the background knowledge regarding the properties of these amyloid-associated dyes comes from textile staining.

[1] H. Bennhold, *Münch. Med. Wochenschr.* **44,** 1537 (1922).

0076-6879/99 $30.00

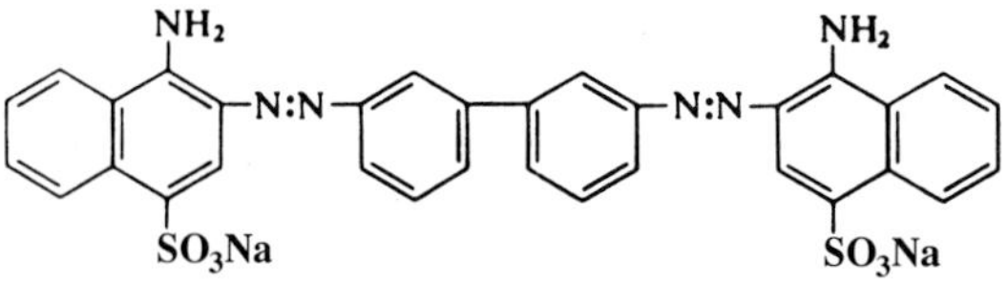

FIG. 1. The chemical structure of Congo red.

Staining Methods

Congo Red

Staining with Congo red is the most universally used method for the demonstration of amyloid. When used in association with polarization microscopy it is considered the most specific of the available amyloid staining methods. This staining procedure is also sensitive and is simple to perform. Congo red staining is, therefore, dealt with in some detail in this article. It is important to emphasize that Congo red-stained histological sections are not always easy to interpret and there are a few pitfalls to be avoided. Additionally, in order to be reliable, the staining procedure must be performed under well-controlled conditions and tissue examination should be made by an experienced individual using appropriate microscopic equipment and light sources.

Theoretical Background

Congo red is a symmetrical sulfonated azo dye with a hydrophobic center (Fig. 1) consisting of a biphenyl group spaced between the negatively charged ends of the dye molecule. When the Congo red molecules are aligned in parallel arrays on fibrils, they induce a green birefringence in polarized light, and this property is extremely important in microscopic studies of amyloid, including practical diagnostic work.[2] The exact binding mode of the Congo red molecule to amyloid is not completely elucidated, but it is clear the Congo red bind to the determining component of amyloid, the amyloid fibril itself. Synthetically made amyloid-like fibrils have the same Congo red binding properties as the native amyloid substance.[3] It therefore seems likely that the binding properties of Congo red to the amyloid fibrils, as well as the polarization microscope characteristics of

[2] P. Ladewig, *Nature* **156,** 81 (1945).

[3] G. G. Glenner, E. D. Eanes, H. A. Bladen, R. P. Linke, and J. D. Termine, *J. Histochem. Cytochem.* **22,** 1141 (1974).

amyloid fibrils, depend on the cross β-pleated sheet organization of the proteins in the amyloid fibrils.[4]

Three main theories have been proposed concerning the binding of Congo red dye to the amyloid fibril. One theory proposes involvement between the electrostatic forces of the charged ends of the Congo red molecule and the β-pleated sheets within the fibril. Another suggests hydrogen bonding between the fibril and the dye molecule. The third and most likely theory focuses on possible hydrophobic interactions between the Congo red dye and the β-pleated sheet fibril.[3–5] The characteristic and very useful green birefringence elicited by amyloid after staining with Congo red is explained by the fact that Congo red dye molecules are highly oriented in a linear and parallel manner on the amyloid fibrils.[6] This organization also provides an explanation as to why green birefringence is so unique for amyloid deposits in tissues.

Any tissue component that binds Congo red in an ordered way, as described earlier, will not only appear red but will also exhibit green birefringence with polarized light. In human tissues, the most common problem in staining occurs with dense collagen fibers,[7–9] including cartilage. In our experience, the alkaline Congo red method[10] applied to sections of tissue fixed in formalin rarely causes this problem. Romhány,[9] however, suggested mounting tissue sections in gum arabic to overcome problems with collagen staining.

In aqueous solution Congo red stains many different structures, although amyloid has a stronger affinity for the dye. The specificity of the staining method can be increased greatly by performing the staining in a high alcohol concentration combined with high ion strength and a high pH. Many Congo red staining methods may be used, but tissues must be stained under strictly controlled conditions because structures or substances other than amyloid can give false-positive results. Probably the most commonly used and well-controlled method is that of Puchtler *et al.*[10] Several other variations have been introduced, but to our knowledge, none has been shown to be superior to the method of Puchtler *et al.*

[4] G. G. Glenner, *N. Engl. J. Med.* **302,** 1283 and 1333 (1980).

[5] J. H. Cooper, *in* "Amyloid and Amyloidosis 1990" (J. B. Natvig, Ø. Førre, G. Husby, A. Husebekk, B. Skogen, K. Sletten, and P. Westermark, eds.), p. 515. Kluwer Academic Publishers, Dordrecht, 1991.

[6] G. G. Glenner, E. D. Eanes, and D. L. Page, *J. Histochem. Cytochem.* **20,** 821 (1972).

[7] R. A. DeLellis, G. G. Glenner, and S. G. Ram, *J. Histochem. Cytochem.* **16,** 663 (1968).

[8] G. Klatskin, *Am. J. Pathol.* **56,** 1 (1969).

[9] G. Romhányi, *Virch. Arch. A* **354,** 209 (1971).

[10] H. Puchtler, F. Sweat, and M. Levine, *J. Histochem. Cytochem.* **10,** 355 (1962).

Method[10]

The type of fixative used is not critical: 10% neutral buffered formalin, absolute ethanol, Carnoy's fixative, Bouin's solution, and 2% paraformaldehyde in phosphate buffer (pH 7.4) are all suitable. Fixation time is also not critical but the prolonged storage of tissues in formalin should be avoided, if possible, because it reduces the stainability with alkaline Congo red.[11] Frozen sections or tissue smears may be treated with the same method.

Solutions

1. Mayer's Hematoxylin. Dissolve 1 g hematoxylin in 1000 ml preheated (55°) distilled water. Add 0.2 g sodium iodate and 50 g aluminum ammonium or aluminum potassium sulfate (alum). Stir until the alum is dissolved and then add 1 g citric acid and 50 g chloral hydrate. This results in a deep red color (like red wine). This solution can be reused until the color changes to purple.

2. Congo Red

a. Stock solution A: Add 10 g NaCl to 1000 ml 80% (v/v) ethanol.
b. Stock solution B: Add 2 g Congo red and 10 g NaCl to 1000 ml 80% ethanol. Replace this stock solution every second month. Let the newly prepared stock solutions stand for 24 hr before use.
c. 1% NaOH in distilled water.

Working Solutions

To 100 ml of stock solution A add 1 ml 1% (w/v) NaOH and filter.
To 100 ml of stock solution B add 1 ml 1% (w/v) NaOH and filter.
The working solutions have limited durability and should be used within 15 min.

Procedure

1. Use about 10-μm-thick sections. Deparaffinize sections to distilled water.
2. Stain nuclei with Mayer's solution for 1 min.
3. Rinse in tap water for 2 min and then in distilled water.
4. Place sections in solution A for 20 min.
5. Transfer the sections directly over to solution B and stain for 20 min.
6. Rinse briefly in two changes of absolute ethanol. Each rinse should not exceed 10 sec.

[11] S. N. Meloan and H. Puchtler, *Histochemistry* **58,** 163 (1978).

7. Clear in three changes of xylene and mount under cover glass in a synthetic mounting medium.

Results

With this method, amyloid appears as pink to red-orange (rarely deep red) material whereas nuclei stain blue. It should be noted that the strength of the staining reaction varies significantly between different amyloids and also sometimes within an amyloid deposit. Typically, in human tissues, no other substances or structures other than amyloid are stained with the alkaline Congo red method. The staining of collagen indicates that the method is not optimal. It has sometimes been claimed that successful staining includes labeling of the lamina elastica interna of arteries. However, when this has occurred, false-positive green birefringence is likely to be seen in some collagen structures. Also, granules in eosinophilic granulocytes may be stained with Congo red but show no green birefringence.

When observed with crossed polarizers in polarization microscopy, amyloid stained with Congo red shows a bright green birefringence, often referred to as "apple green birefringence." The strength of this birefringence, as well as its brightness, depends on several factors, including how intensely the amyloid has been stained with Congo red. Also, amyloid in some structures (e.g., vessel walls) often shows a very bright and intense birefringence, whereas other deposits (e.g., in glomeruli) tend to show a weaker intensity of birefringence. Such differences are likely related to how compactly the amyloid fibrils are packed and how they are oriented. This is probably the reason why newly deposited amyloid [e.g., in the spleen in experimental AA amyloidosis (derived from the apolipoprotein serum amyloid A)] is stained weakly with Congo red and elicits fairly light birefringence.

The strength of amyloid staining with Congo red also varies somewhat with the type of amyloid. Thus, some biochemical forms of amyloid are generally less intensely stained with Congo red than others. Examples of comparably less intensely stained amyloids are senile systemic amyloidosis (derived from wild-type transthyretin) and localized aortic media amyloid (derived from an as yet undetermined precursor). The staining intensity of AL amyloid (derived from monoclonal immunoglobulin light chains) varies significantly between individuals, but the reason for this variability is unknown.

Some intracellular filamentous structures stain with Congo red and subsequently elicit green birefringence with polarization microscopy. These examples include the paired helical filaments present in neurons and intracellular inclusions in the choroid plexus, ependymal cells,[12] and adrenal cortical cells.[13] These several examples of filamentous structures are usually not regarded as amyloid but they otherwise most likely fulfill the criteria for amyloid.[14]

Pitfalls

False-negative results with Congo red staining may occur unless certain precautions are taken. For example, very thin tissue sections are difficult to interpret, and green birefringence will be absent or weak unless the histological sections are of required thickness. It is advisable to cut the tissue sections slightly thicker (5–10 μm) than what is generally used for ordinary histology. Prolonged nuclear staining with hematoxylin may interfere with the identification of weakly stained amyloid. For that reason we usually shorten the routinely used hematoxylin staining time. A critical step in Congo red staining is dehydration before mounting. Because it is important to avoid aqueous solutions, washing and dehydration should be performed only with concentrated (99–100%) ethanol prior to xylene and subsequent mounting. A strong light source is very important in the polarization microscopy

The possible occurrence of false-positive staining of collagen has been described earlier. In human tissues (and in other mammalian tissues), some foreign material may stain like amyloid. These materials include chitin of arthropods, fungal cell walls, and other plant components.[15]

Congo red-stained sections are fairly stable. However, with time there is a tendency for fading to occur, especially in strong light.[16] The potential effect of long tissue storage in formalin has been mentioned previously.

It is our repeated experience that examination of amyloid after staining with Congo red is not a simple task. Considerable experience is needed to avoid false-negative or false-positive results. For the untrained it may be wise to include a positive control section.

Congo Red Fluorescence

Congo red is a fluorescent dye.[17] The alkaline Congo red method for studies in ordinary light is also useful for fluorescence studies if mounting is performed in a nonfluorescent medium. However, this fluorescence is

[12] L. Eriksson and P. Westermark, *Am. J. Pathol.* **125,** 124 (1986).
[13] L. Eriksson and P. Westermark, *Am. J. Pathol.* **136,** 461 (1990).
[14] D. A. Kirschner, C. Abraham, and D. J. Selkoe, *Proc. Natl. Acad. Sci. U.S.A.* **83,** 503 (1986).
[15] R. L. DeLellis and M. C. Bowling, *Hum. Pathol.* **1,** 655 (1970).
[16] F. Sweat Waldrop, H. Puchtler, and L. S. Valentine, *Arch. Pathol.* **93,** 37 (1972).
[17] H. Puchtler and F. Sweat, *J. Histochem. Cytochem.* **13,** 693 (1965).

not as specific for amyloid as is the green birefringence observed with polarized light. Nevertheless, the fluorescence of Congo red in tissue sections may be potentially useful for the quantitation of amyloid in tissue and the staining is more easy to recognize by less experienced individuals.

Use of Congo Red Staining for Differentiation between Different Chemical Forms of Amyloid

Pretreatment of sections prior to the staining with Congo red can be used for differentiation between some chemical types of amyloid. Originally, a method using trypsin pretreatment was described, but the most commonly used method is to pretreat sections with a solution of potassium permanganate prior to staining with Congo red.[18] With this method, some types of amyloid lose their stainability with Congo red after treatment with potassium permanganate, whereas others are resistant and still stain with Congo red. The method was initially created to differentiate between AA amyloidosis (potassium permanganate sensitive) and other forms of amyloid that are potassium permanganate insensitive, particularly AL amyloidosis. Subsequently, several other types of amyloid have been found to be sensitive to the pretreatment with potassium permanganate. These $KMnO_4$-sensitive forms include $A\beta_2$-microglobulin amyloidosis, AApoAI amyloidosis, and the localized amyloid of the seminal vesicles. The theoretical explanation for the variation in sensitivity to permanganate is not known, and the method is now used less commonly due to the increasing availability of antibodies for more specific immunohistochemical methods.

Sirius Red

Sirius red (direct red 80) is related to Congo red, and the former dye has been proposed to have an advantage as an amyloid stain because it gives a more intense staining reaction, which increases the contrast with adjacent tissue. This attribute of Sirius red may be especially valuable when preparing microscopic slides for photography. A combination of Sirius red and Congo red can also be used (K. H. Johnson, unpublished data, 1998). Like Congo red, Sirius red elicits a green birefringence from stained amyloid deposits. However, Sirius red does not elicit fluorescence.[19] For human tissues, Sirius red is comparable to Congo red with regard to specificity for amyloid deposits, but nonspecific staining of connective tissue may occur in animal tissues.[19] It is our experience, as well as the opinion of others,[19]

[18] J. R. Wright, E. Calkins, and R. L. Humphrey, *Lab. Invest.* **36,** 274 (1977).
[19] D. Brigger and T. J. Muckle, *J. Histochem. Cytochem.* **23,** 84 (1975).

that the alkaline Congo red method of Puchtler *et al.*[10] is the best stain to specifically demonstrate amyloid.

Method[20]

The type of fixation is not critical; most commonly used fixatives can be used.

Solutions

1. Sirius red: Dissolve 0.5 g Sirius red in 45 ml distilled water. Add 50 ml absolute ethanol and 1 ml of 1% NaOH. Slowly add 20% NaCl under vigorous shaking until a fine precipitate forms (up to 4 ml). Leave overnight and filter.
2. Mayer's hematoxylin (see Congo red staining method described earlier).

Procedure

1. Deparaffinize sections.
2. Stain nuclei with Mayer's solution.
3. Rinse in tap water and then in 70% ethanol.
4. Place in Sirius red for 1 hr.
5. Wash in tap water for 10 min.
6. Dehydrate, clear, and mount.

Results

Amyloid usually stains intensely red against a weakly stained background and exhibits green birefringence with polarized light. Elastin may also stain. Granules of eosinophilic leukocytes stain strongly, and Paneth cells stain more weakly, but do not show birefringence.

Other Cotton Dyes

Both Congo red and Sirius red are direct cotton dyes. Extensive studies have been conducted to evaluate the usefulness of other cotton dyes with respect to the histological identification of amyloid. Several possible alternatives exist,[16,21] but these dyes have not been used very extensively. There are also some newly developed probes based on the Congo red molecule

[20] B. D. Llewellyn, *J. Med. Lab. Technol.* **27,** 308 (1970).
[21] H. Puchtler, F. Sweat Waldrop, and S. N. Meloan, *Histochemistry* **77,** 431 (1983).

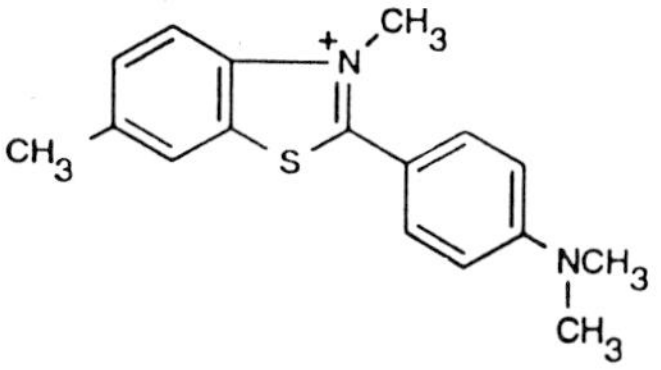

FIG. 2. The chemical structure of thioflavin T.

and with promising staining properties of amyloid.[22,23] There is limited experience with them, however.

Thioflavin Stains

Thioflavin S and thioflavin T are sulfur-containing compounds that bind to amyloid fibrils. Only thioflavin T seems to be fully characterized chemically. While thioflavin S is a mixture of several components with different properties,[24] thioflavin T is a small, positively charged benzothiazole compound with a formula weight of 319 (Fig. 2). The binding mode of thioflavin T to amyloid is not known, but studies with fibrils made *in vitro* from purified amyloid proteins or synthetic peptides indicate that the dye interacts with the specific quaternary structure of the β-pleated sheet fibril and not the monomeric peptides. Thus the binding is not dependent on any specific amino acid sequence.[25] However, the binding motif is not present in all synthetic amyloid-like fibrils.[25] Binding of thioflavin T to amyloid fibrils creates a characteristic 120-nm red shift of its excitation spectrum.[25,26]

Thioflavin staining is currently popular for amyloid studies, not only for the detection of amyloid in tissue sections, but also for the study of fibril formation *in vitro.* Staining with thioflavin is easy to perform, but the requirement of fluorescence microscopy limits the usefulness of these staining methods. For microscopic detection of amyloid in tissue sections, the thioflavin T method by Vassar and Culling[27] and the thioflavin S method by Schwartz[28] are used most commonly. The advantage of thioflavin staining compared to that with Congo red is a higher sensitivity (at least for those

[22] W. E. Klunk, M. L. Debnath, and J. W. Pettegrew, *Neurobiol. Aging* **16,** 541 (1995).

[23] T. T. Ashburn, H. Han, B. F. McGuinness, and P. T. J. Lansbury, *Chem. Biol.* **3,** 351 (1996).

[24] D. Stiller, D. Katenkamp, and K. Thoss, *Acta Histochem.* **38,** 18 (1970).

[25] H. I. LeVine, *Amyloid* **2,** 1 (1995).

[26] H. I. LeVine, *Arch. Biochem. Biophys.* **342,** 306 (1997).

[27] P. S. Vassar and C. F. A. Culling, *Arch. Pathol.* **68,** 487 (1959).

[28] P. Schwartz, *in* "Amyloidosis" (E. Mandema, L. Ruinen, J. H. Scholten, and A. S. Cohen, eds.), p. 400. Excerpta Medica, Amsterdam, 1968.

who are not sufficiently experienced to interpret Congo red results). Thioflavin staining also gives a result that is easier to quantitate as compared with Congo red. The disadvantage of the thioflavin dyes is associated with their greater level of nonspecificity.[21]

Methods: Thioflavin S[28]

The type of fixation is not critical.

Procedure

1. Deparaffinize tissue sections.
2. Incubate sections in 1% (w/v) thioflavin S in distilled water for 5–10 min. We recommend new solution to be made up daily.
3. Differentiate in 80% (v/v) ethanol to remove excess fluorochrome.
4. Mount in gum arabic.

Methods: Thioflavin T[27]

The type of fixation is not critical.

Solutions

Thioflavin: 1% thioflavin T in distilled water. We recommend new solution to be made up daily.
Mayer's hematoxylin (see Congo red staining method).

Procedure

1. Deparaffinize tissue sections.
2. Place sections for 2 min in Mayer's hematoxylin.
3. Wash in water.
4. Incubate sections in thioflavin T for 3 min.
5. Rinse in water and differentiate in 1% acetic acid for 20 min.
6. Wash thoroughly in water, dehydrate, clear in xylene, and mount in nonfluorescent medium.

Results

Amyloid fluoresces with an intense green, yellow, or blue color depending on the filter.

Pitfalls

Thioflavin S has been suggested as a very specific amyloid stain, being as specific as Congo red[28] and more specific than thioflavin T. However, most studies indicate that both thioflavin staining methods are not fully specific; other structures, such as fibrin or collagen, may bind the dye. Thioflavin T is a basic dye and therefore binds to acidic components such as nucleic acids. This staining of nonamyloid components must be quenched by prestaining with hematoxylin. The specificity of staining with thioflavin T may also increase by the use of thioflavin T in 0.1 *N* hydrochloric acid instead of water.[29] In general, we do not recommend thioflavin stains for diagnostic purposes.

Silver Staining

Silver stains are generally not used for the demonstration of amyloid. Silver staining techniques are, however, commonly used for the interpretation of amyloid-associated changes in human kidney biopsies and in neuropathological specimens. These silver-staining procedures thus can be utilized to help visualize certain pathological features in some amyloid related diseases, including Alzheimer's disease. Silver-staining methods are particularly useful for the facilitation of detailed microscopic studies of small amyloid deposits and their relation to other tissue structures. Examples include amyloid plaques and neurofibrillary tangles present in the brain and early amyloid deposits in renal glomeruli.[30] Although useful for some amyloid-associated changes in tissues, it should be emphasized that these silver-staining methods primarily stain nonamyloid structures (e.g., dystrophic neurites and reticular fibers) and thus are not amyloid specific. A silver-staining method used for the study of lesions in Alzheimer's disease is the Palmgren method.[31] Methenamine silver staining is commonly used in renal pathology.

Reaction of Amyloid with Cationic Dyes

Amyloid binds cationic dyes, such as Alcian blue,[32] most likely mainly due to the presence of glycosaminoglycans in amyloid deposits. These staining methods can be used for amyloid staining, and they have even been suggested to be better than the alkaline Congo red method for the demonstration of certain amyloids.[33] However, it should be observed that these staining reactions are not specific for amyloid.[32] The binding of Alcian

[29] J. Burns, C. A. Pennock, and P. J. Stoward, *J. Pathol. Bact.* **94,** 337 (1967).
[30] S. F. Nolting and W. G. J. Campbell, *Hum. Pathol.* **12,** 724 (1981).
[31] R. B. Cross, *Med. Lab. Sci.* **39,** 67 (1982).
[32] R. W. Mowry and J. E. Scott, *Histochemie* **10,** 8 (1967).
[33] A. Pomerance, G. Slavin, and J. McWatt, *J. Clin. Pathol.* **29,** 22 (1976).

blue 8GX at different concentrations of magnesium chloride has been used for the histochemical detection of, and the differentiation between, different glycosaminoglycans in histological sections.[34] This method has also been used for studies of amyloid.[32] However, experiments with amyloid-like fibrils made from synthetic peptides in the absence of glycosaminoglycans show that such fibrils can also bind Alcian blue in the presence of a high concentration of magnesium chloride.

Immunohistochemistry

Immunohistochemistry (for light microscopy) and immune electron microscopy are very widely used methods in histopathological studies and are of increasing importance as our knowledge of compositional variation in normal and pathological tissue structures increases. Immunohistochemistry and immunocytochemistry are also of increasing importance in studies of amyloid. Because the different amyloid forms are of distinct protein nature, antibodies can recognize specific epitopes on amyloid fibrils and on associated components, such as amyloid P-component and proteoglycans. The same antibodies or antisera used for Western blot analyses can often be used in immunohistochemistry and immunocytochemistry.

The principle of immunohistochemistry is basically quite simple. For most methods, tissue sections are incubated with antibodies to a specific tissue component. Bound antibodies are subsequently visualized using one of a variety of methods.

Both polyclonal and monoclonal antibodies are used in immunohistochemistry and immunocytochemistry. Commercial antibodies are presently available for several amyloid fibril proteins, e.g., β-protein and protein AA, but it is often necessary to prepare specific antisera for amyloid studies. Working polyclonal antisera may be made against purified amyloid fibril proteins. However, when the amyloid fibril protein is known, antisera made against synthetic peptides corresponding to an amyloid fibril protein, or to a part thereof, may be better alternatives. Monoclonal antibodies may also be prepared, but making them is a more laborious procedure.

There are several good alternative immunohistochemical and immunocytochemical procedures available for use. These procedures are detailed in several contemporary references and manuals.[35–38] Examples of methods

[34] J. E. Scott and J. Dorling, *Histochemie* **5,** 222 (1965).

[35] L.-I. Larsson, "Immunocytochemistry: Theory and Practice." CRC Press, Boca Raton, FL, 1988.

[36] J. D. Bancroft, A. Stevens, and D. R. Turner, *in* "Theory and Practice of Histological Techniques." Churchill Livingstone, Edinburgh, 1990.

[37] G. C. Howard, *in* "Methods in Nonradioactive Detection." Appelton & Lange, Norwalk, CT, 1993.

[38] C. Röcken, E. B. Schwotzer, R. P. Linke, and W Saeger, *Histopathology* **29,** 325 (1996).

provided here are those that we have found to be especially useful in the study of amyloid.

Methods in Immunohistochemistry

The remainder of the article discusses each step in the immunohistochemistry procedure followed by a schematic drawing that describes different enzyme systems (Fig. 3). Antibodies and detection systems used in immunohistochemistry are available from multiple commercial suppliers. We use the antibodies and detection systems supplied by DAKO (Copenhagen, Denmark). However, we make most primary polyclonal and monoclonal antibodies in our own laboratories.

Fixation

The use of appropriate methods for the preservation of antigenic epitopes is generally vital for immunohistochemical studies. Factors such as the methods used for handling tissue prior to fixation, choice of fixative, and fixation time may influence the preservation of antigenic epitopes significantly. However, amyloid fibrils are very resistant to degradation and

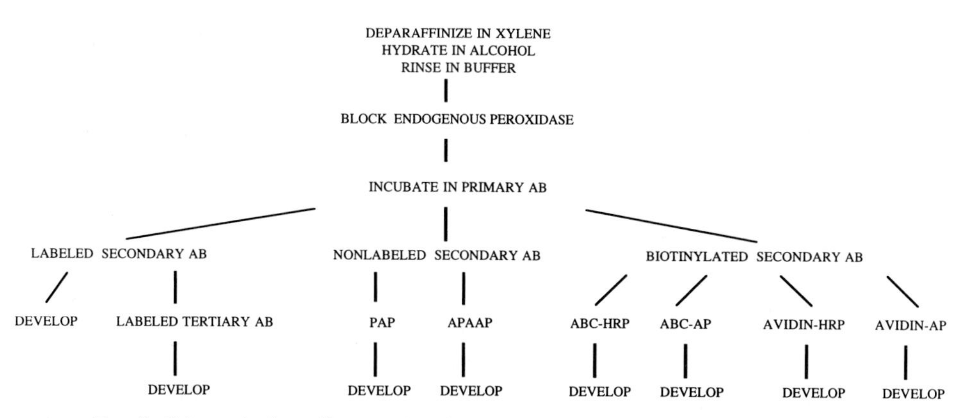

FIG. 3. Schematic flow diagram showing, in simplified form, the different methods in immunohistochemistry. In the examples to the left, secondary antibodies (AB) are labeled, for example, with horseradish peroxidase (HRP). This reaction may be enhanced by the addition of a HRP-labeled tertiary antibody. In the techniques with nonlabeled secondary antibodies (center), these antibodies are directed against immunoglobulins and thereby link the primary antibodies with the labeled immune complexes. PAP, peroxidase–antiperoxidase immune complex; APAAP, alkaline phosphatase–antialkaline phosphatase immune complex. In the techniques to the right, the secondary antibodies are labeled with biotin, a molecule that theoretically can bind four avidin–enzyme complexes, which creates an enhanced reaction. ABC, avidin–biotin complex; AP, alkaline phosphatase. Use antibody at dilution recommended by the manufacturer.

there usually are few or no problems with the use of autopsy material collected for the identification of the various amyloid fibril proteins. However, more detailed precautions may be of great importance if other amyloid related components are to be studied.

Fixatives

Fixatives containing formaldehyde are commonly used.[39] One of the most commonly used for routine histological specimens is neutral buffered formalin. It is a noncoagulating and fast penetrating solution that penetrates a tissue at a rate of approximately 1 cm per hour. Bouin's solution is a useful fixation for fat-containing tissues. It is a fat solvent and the lipid components in the tissues are altered. Therefore, Bouin's solution may not be used for the study of lipids. The preservation of endocrine cells with Bouin's solution is excellent. After fixation in Bouin's solution, the tissue has to be washed in 50% ethanol followed by 70% ethanol to remove excess picric acid. Because antigenic epitopes vary in sensitivity to different fixatives, one must determine (for each tissue and antibody) which fixative will give optimum results. It is also often possible to recover lost immunoreactivity in a tissue by different modes for "antigenic retrieval."

Neutral Buffered Formalin (pH Approximately 7.0): 100 ml concentrated formalin (37–40% formaldehyde), 900 ml distilled water, 6.5 g disodium hydrogen phosphate (Na_2HPO_4), and 4.0 g sodium dihydrogen phosphate (NaH_2PO_4).

Bouin's Solution: 15 parts of picric acid saturated in water final concentration, 5 parts concentrated formaldehyde solution (37–40%), and 1 part acetic acid. Mix the three components immediately before use.

Sectioning and Mounting on Slides

Most commonly, tissue is embedded in paraffin and sectioned. Sections (4 μm thick) are placed on pretreated slides. Pretreatment of the slide is often necessary to increase the adhesion of the tissue and prevent loss of material. "Plus Slides" (Menzel-Gläser, Germany) are commercially available slides treated to increase the adhesiveness.

Three different types of pretreatment for slides are described. There is no major difference in efficiency among these procedures. However, poly (L-lysine)-coated slides are not suitable if a comparison between immunohistochemical reaction and localization of Congo red staining is performed.

[39] D. C. Sheehan and B. B. Hrapchak, "Theory and Practice of Histotechnology." Mosby, St. Louis, 1980.

Poly(L-lysine) (PLL) may take the Congo red stain and thus give rise to a false-positive reaction, causing these specimens to be difficult to interpret.

Commercially available slides are usually clean and can be used directly, but if this is not the case they can be cleaned by soaking in 0.2% Triton X-100 in phosphate-buffered saline (PBS, see later) overnight followed by rinsing in distilled water.

Mounting Agents

Chrome-Gelatin. One gram of gelatin is dissolved and swollen in 190 ml distilled water. This can be speeded up by heating to 80°. When the gelatin is dissolved, add 0.1 g chromalum in 10 ml water. Dip the slides into the solution and dry overnight at 37° and store in a dust-free area.

Aminosilane. Mix 2 ml of 3-aminopropyltriethoxysilane with 100 ml acetone. Place clean slides in the solution for 20 sec. Rinse the slides in two baths of acetone and finally in a bath of distilled water. Leave the slides to dry at 50°. Store the slides in closed boxes.

Poly(L-lysine). Use 0.1% PLL solution. Water is added to PLL to a final concentration of 0.1% (w/v). Store PLL in small aliquots at −20°. Apply 5 μl of PLL on the lower part of one slide and smear with a second slide. In this way two slides are prepared at the same time. It is recommended to mark the treated side with a small pencil mark. The section is dried on to the slide. This can be performed by placing the mounted section in an oven at 60° for 1 to 24 hr.

Deparaffination is performed in xylene (two baths, 30 min each). Hydration should be performed in decreasing (100, 95, 70%) concentrations of ethanol with two 10-min baths at each concentration. This is followed by a rinse in buffer solution. The buffer should be inexpensive and easy to make. Most commonly used are Tris–HCl buffer or phosphate buffer. We use Tris-buffered saline in all immunohistochemical procedures except for immunohistochemistry performed on skin amyloid.

Recipes for Wash Buffers

Tris-Buffered Saline. Stock solution (TBS ×10): Dissolve 60.5 g Tris(hydroxymethyl)aminomethane and 87 g NaCl in 500 ml of water and then adjust pH to 7.2–7.6 with 5 *M* HCl. Add distilled water to 1000 ml. Working solution: one part stock solution + nine parts distilled water. The final concentration of the working solution is 0.05 *M* Tris–HCl in 0.15 *M* NaCl, pH 7.2–7.6.

Phosphate-Buffered Saline. Phosphate-buffered saline stock solution is prepared by mixing the following components to an adjusted final volume of 1000 ml: 83 g NaCl, 11.5 g disodium hydrogen phosphate (anhydrous) (Na_2HPO_4), 2.0 g potassium dihydrogen phosphate (KH_2PO_4), and 2.0 g potassium hydrochloride (KCl). The pH on the stock solution is below pH 7, but when the buffer is diluted it is raised to pH 7.4.

Retrieval of Antigenic Epitopes

Antigenic epitopes are often masked or destroyed after fixation and subsequent embedding in paraffin. Certain epitopes demand frozen sections for immunohistochemical study, but antigenic epitopes can often be retrieved in paraffin-embedded materials. Several methods exist for antigenic retrieval. Antigenic retrieval can be attained through enzymatic digestion, microwave treatment, and, specifically for amyloid, pretreatment with guanidine hydrochloride, urea, or formic acid.

Pepsin. Dissolve 0.25 g of pepsin in 100 ml 0.01 *M* HCl, prewarmed to 37°. The concentration of pepsin and incubation time needed depend on the fixation and type of tissue, but a 10-min digestion usually gives an improved immunostaining without any loss of tissue. To inhibit further digestion, rinse the slides in several changes of wash buffer.

Trypsin. 0.1% (w/v) trypsin solution in 0.05 *M* Tris–HCl buffer, pH 7.6, containing 0.1% (w/v) calcium chloride. Incubate slides for 5–10 min at room temperature. The digestion is stopped after rinsing the slides in cold water.

Proteinase K. Dissolve 0.1 mg proteinase K in 100 ml 0.1 *M* Tris–HCl buffer, pH 8.0, containing 50 m*M* EDTA and prewarmed to 37°. A commonly used digestion time is 15–20 min. Stop the digestion by immersing the sections in 0.1 *M* glycine/PBS buffer for 5 min.

Microwave Treatment. Retrieval of antigen epitopes can also be attained by microwave treatment, and this mode is now one of the most commonly used methods. An often recommended exposure is 750 W in 0.02 *M* sodium citrate buffer, pH 6.0. This procedure often results in loss of tissue from the slides. We have found that incubation of the sections in preheated (90°) 0.01 *M* sodium citrate buffer, pH 6.0, improves immunolabeling without loss of tissue. The deparaffinized sections are placed in the hot solution and allowed to cool to room temperature.[40] Others have found that heating in Tris–HCl buffer at high (pH 10) or low (pH 1) pH is more optimal for many antibodies.[41] The duration of heating is also important.

[40] Z. Ma, G. T. Westermark, Z.-C. Li, U. Engström, and P. Westermark, *Diabetologia* **40,** 793 (1997).

[41] S.-R. Shi, R. J. Cote, B. Chaiwun, L. L. Young, Y. Shi, D. Hawes, T. Chen, and C. R. Taylor, *Appl. Immunohistochem.* **6,** 89 (1998).

Other Treatments. Incubation of sections in concentrated formic acid, 6 *M* guanidine hydrochloride or 8 *M* urea[42] have all been used for the retrieval of antigenic epitopes in amyloid. The microwave technique has, to a great extent, replaced the use of these solvents in immunohistochemistry.[43]

Blocking

Background staining is a common problem in immunohistochemistry, which may have several different causes. Generally, antibody solutions must be diluted as much as possible, and nonspecific binding of antibodies can be reduced by blocking tissue sections with nonreactive substances such as bovine serum albumin or nonfat milk. Depending on the selected enzyme system, endogenous enzyme activity in tissue sections can give rise to undesirable reaction and must also be blocked.

Endogenous Peroxidase. When peroxidase is used as the enzyme in the immunohistochemical procedure it is recommended to block endogenous peroxidase reactivity present in red blood cells and granulocytes. This can be performed in 0.3% hydrogen peroxide (v/v) in TBS or in methanol for 30 min.

Endogenous Biotin. Problems with endogenous biotin are not common but can occur. This nonspecific reaction can be inhibited by incubation in 0.1% (w/v) avidin followed by an incubation in 0.01% biotin.[44]

Endogenous Alkaline Phosphatase. To block endogenous alkaline phosphatase, 1 to 5 m*M* levamisole is included in the developing solution during the development of the antibody reaction. Endogenous intestinal alkaline phosphatase, which is not inhibited by levamisole, may be inhibited by heating the sections to 80° prior to incubation in primary antibody solution if the antigen is resistant.

Dilution and Purification of Antibodies

An important relationship exists among antibody titer, incubation time, and the appearance of nonspecific background staining. Each antiserum is unique and its specific characteristics vary between batches and sources. The dilution factor depends on antibody concentration, affinity, avidity, and choice of detection system. Therefore, an appropriate dilution test has to be performed for each individual antibody. Polyclonal antibodies can

[42] P. P. Costa, B. Jacobsson, V. P. Collins, and P. Biberfeld, *J. Histochem. Cytochem.* **34,** 1683 (1986).

[43] P. P. Liberski, R. Yanagihara, P. Brown, R. Kordek, I. Kloszewska, J. Bratosiewicz, and D. C. Gajdusek, *Neurodegeneration* **5,** 95 (1996).

[44] L. A. Sternberger and N. H. Sternberger, *J. Histochem. Cytochem.* **34,** 599 (1986).

often be used at dilutions of 1:50–1:10,000. Monoclonal antibodies may sometimes have to be used undiluted with hybridoma culture medium applied directly or diluted up to 1:500. The lower dilutions are used in less sensitive systems such as direct methods or when fluorescence-labeled antibodies are used. The higher dilutions are used in more sensitive systems.

No further purification of monoclonal antibodies is needed when these antibodies are used in immunohistochemistry. The same is often true for polyclonal antibodies raised against pure antigen or synthetic peptides. When antisera are raised against more crude antigen preparations, it may be necessary to perform additional purification.

Purification of Antibodies. Commercially available prepacked protein A or protein G columns can easily be used for purification of the immunoglobulin (Ig) fraction. More specific antibodies can be obtained by affinity purification. For this, the antigen can be chemically linked to a support such as cyanogen bromide (CNBr)-activated Sepharose (Pharmacia, Piscataway, NJ). The antiserum or tissue culture medium is passed through the antigen-coupled support. Antibodies that recognize the antigen will bind whereas all other antibodies will be removed. By changing the pH or salt concentration, the bound antibodies can be eluted and concentrated.

Binding may occur between the Ig Fc part and Fc receptors present in the tissue. This reaction can be abolished if the antibodies are cleaved enzymatically by pepsin to remove the Fc part. The remaining antigen-binding part [the $F(ab)_2$ fragment] can then be used for immunolabeling.

Incubation with Primary Antibody

Antibodies are diluted in the buffer of choice. If there is a tendency for nonspecific labeling of the tissue, this can be reduced by the addition of 1–5% bovine serum albumin to the incubation buffer. The slides are placed in a humid chamber and the sections are covered with the antibodies. Incubation may take place at 4° or at room temperature overnight or, for shorter time, at room temperature or at 37°.

Detection of Bound Antibodies

There is a large number of possible immunodefection systems from which one can select. The choice of method depends on the degree of sensitivity that is required, time limitation of the procedure, and the cost of the reagents. The sensitivity of the system increases with the number of antibodies included, and direct detection systems are less sensitive compared to indirect systems.

Direct Methods

In direct systems, the enzyme is linked to the primary antibody. There are relatively easy methods to purify the primary antibodies and to conjugate enzyme to them. Horseradish peroxidase can be covalently conjugated by periodate (or be cross-linked with glutaraldehyde to the antibody), the former being nearly 100-fold more efficient.[45] Alkaline phosphatase can be bound specifically to the Fc region of the immunoglobulin by conjugation to carbohydrate moieties; this procedure prevents interference between the antigen-binding site and the alkaline phosphatase.[46] Fluorescent conjugation of primary antibodies is performed easily with fluorochromes, such as fluorescein isothiocyanate, that bind to free amino groups in proteins.

Indirect Methods

Peroxidase–Antiperoxidase and Alkaline Phosphatase–Antialkaline Phosphatase Methods. Indirect methods include one or more secondary antibodies or enzyme–protein complexes. The simplest indirect immunostaining method is the use of an enzyme-labeled secondary antibody. This secondary antibody is raised against immunoglobulins of the species used for primary antibody production and recognizes and binds to them. The sensitivity of this two-step indirect method can be enhanced by the addition of a second enzyme-labeled antibody that recognizes and reacts with the secondary antibody. In this way the amount of enzyme is increased and substrate precipitation will be intensified.

Soluble enzyme immune complex methods constitute a subgroup among indirect methods. The two most commonly used immune complexes in this group are peroxidase–antiperoxidase and alkaline phosphatase–antialkaline phosphatase. The soluble immune complex is built up by the enzyme and an antibody is produced against the enzyme. The antibodies are mixed with the enzyme in excess to produce soluble enzyme immune complexes. The primary antibody is detected by a secondary unlabeled antibody added in excess so that the binding to the primary antibody only occurs with one of its antigen-binding sites, thus leaving the other free. This free binding site is used for binding to the antibody in the immune complex. Therefore, the primary and immune complex antibodies are derived from the same species. The secondary antibody functions as a bridge and is called the bridge or link antibody. This antibody must be used at a high concentration to prevent binding of both Fab sites to the primary

[45] J. P. Tresca, R. Ricoux, M. Pontet, and R. Engler, *Ann. Biol. Clin. (Paris)* **53,** 227 (1995).
[46] M. Husain and C. Bieniarz, *Bioconjug. Chem.* **5,** 482 (1994).

antibody. If this occurs, the binding of the immune complex antibody and the oxidation of the chromogen will be abolished.

Avidin–Biotin Methods. This system utilizes the high affinity that occurs between the protein avidin (or streptavidin) and the biotin. Each avidin molecule can theoretically bind four biotin molecules; however, due to steric hindrance, fewer biotin molecules are bound. There are two different methods available, the avidin–biotin complex (ABC) method and the labeled avidin–biotin (LAB) method. Both systems use a secondary antibody directed against the primary antibody and to which multiple biotin molecules are conjugated at the Fc part. In the ABC method, avidin and biotinylated peroxidase molecules are mixed in a test tube and large complexes consisting of avidin and biotin–peroxidase are formed. Sections treated with primary and biotinylated secondary antibodies are then incubated with these preformed complexes and will bind to biotinylated secondary antibodies.

In the LAB method, sections after incubation with primary and biotinylated secondary antibodies are incubated with peroxidase-labeled avidin. Biotin–peroxidase and avidin–peroxidase may, in both methods, be substituted with biotin–alkaline phosphatase and avidin–alkaline phosphatase, respectively.

The avidin–biotin systems are more sensitive than the other indirect methods described.

Enzymes and Substrates

Peroxidase and Its Substrates. Peroxidase oxidizes electron donors such as 3,3′-diaminobenzidine tetrahydrochloride (DAB) or 3-amino-9-ethylcarbazole (AEC). These substances become colored and precipitate during oxidation and are therefore called chromogens.

3,3′-Diaminobenzidine Tetrahydrochloride. Fifteen milligrams of DAB is dissolved in 100 ml of 0.05 *M* Tris–HCl buffer, pH 7.4. This substance is classified as carcinogenic. We recommend that a larger amount of substance be prepared as a stock solution (15 mg/ml) and stored frozen in 1-ml aliquots. The reaction is initiated by the addition of 4 μl H_2O_2, which should be performed just before staining. Incubation time varies between 5 and 20 min, and the slides should be checked during the developing time. The reaction is stopped in tap water. It is suitable to counterstain with Mayer's hematoxylin.

Result. The color reaction is brown and is insoluble in alcohol. DAB sections may therefore be stained subsequently with alkaline Congo red for the simultaneous study of amyloid.

It is possible to further enhance the DAB color reaction. If a metal salt such as $NiCl_2$ is present during the oxidation of DAB, the color product will change from brown to black. This increases the contrast between the deposited chromogen and the surrounding tissue and also increases the sensitivity of the detection system. The developing procedure is the same as described earlier but with the addition of 7 mg of $NiCl_2$ to the developing solution.

3-Amino-9-ethylcarbazole. Dissolve 25 mg AEC in 1 ml dimethyl sulfoxide. This can be stored as a stock solution frozen at −20°. Add the dissolved chromogen to 100 ml 0.1 *M* sodium acetate buffer, pH 5.2. Finally, 20 μl H_2O_2 is added to initiate the reaction. The sections should be incubated between 5 and 20 min, and the slides should be checked during the developing time.

Result. AEC will form a red precipitate when it is oxidized. This precipitate fades when exposed to light, is soluble in alcohol, and cannot be stained subsequently with Congo red.

Alkaline Phosphatase and Its Substrates. The enzyme hydrolyzes and transfers a phosphate group from an organic compound such as new fuchsin and a combination of nitro blue tetrazolium chloride (NBT) and 5-bromo-4-chloro-3-indolyl phosphate (BCIP). This enzyme system is especially useful when tissues containing blood and bone marrow cells are studied, as these cells contain endogenous peroxidase and often show increased nonspecific staining with peroxidase.

NBT/BCIP. Prepare a 100-ml developing buffer consisting of 0.1 *M* Tris–HCl buffer, pH 9.5, containing 0.15 *M* sodium chloride and 50 m*M* magnesium chloride. Solubilize 35 mg NTB in 1 ml 70% *N,N*-dimethylformamide and 17 mg BCIP in 1 ml 100% *N,N*-dimethylformamide, respectively. Add 25 mg levamisole and the solubilized NTB and BCIP to the developing buffer. Develop the slides for 20–120 min. The slides can be checked during the developing time. The reaction is stopped by immersing the slides in a stop buffer containing 20 m*M* Tris-HCl and 5 m*M* EDTA.

Results. This reaction appears as purple-blackish precipitation. The reaction product is partially soluble in acidic solution, and nuclear staining in Mayer's hematoxylin is not recommended. Nuclear fast red (Kernechtrot) is a better choice.

New Fuchsin–Naphthol Phosphate. Mix 60 μl 4% new fuchsin in 2 *M* HCl and 60 μl of 4% sodium nitrite in distilled water and shake for 1 min. In a separate tube, dissolve 10 mg naphthol AS-TR phosphoric acid in 100 μl dimethylformamide and add 10 ml 0.2 *M* Tris–HCl buffer, pH 9.0, to this solution. The two solutions are mixed and filtered directly onto the slides. The color development has to be followed under the microscope, and the reaction is stopped with tap water.

Results. This reaction is exhibited as a bright red product. This product is not water soluble.

Mounting Media

Two groups of mounting media, water-based and organic mounting media, are used for mounting histology slides. An example of the former is glycerin–gelatin and it is used together with the substrates AEC and BCIP/NBT. Examples of organic mounting media include Eukitt (Merck, Darmstadt, Germany) and Mountex (HistoLab, Gothenburg, Sweden). These are used for DAB and new fuchsin substrate solution.

Glycerin–Gelatin. Combine 5 g gelatin and 50 ml distilled water in a 250-ml evaporation flask, heat to 80°, and let swell. Add 50 ml glycerin and evaporate with water suction. This mounting medium is stored in smaller aliquots at 4° to prevent mold and bacterial growth.

Controls in Immunohistochemistry

Proper controls are especially important in immunohistochemistry as false negative and positive results are extremely common. It is important to differentiate between nonspecific labeling and unwanted specific labeling. Binding of the antibody to structures other than the antigen causes nonspecific labeling. Unwanted specific labeling may occur due to additional antibodies present in the antiserum.

Omission of the primary antibody or substitution with a nonimmune serum from the same species is often used as a control. When the primary antibody has been omitted, no labeling should occur. Thus this control will show whether any of the components in the detection system cause any reaction.

Substitution of the primary antiserum with serum from an animal that has not been immunized is a very commonly used control. Ideally, this serum should come from the same animal (preimmune serum). However, this can only be achieved in laboratories where the antibody has been raised. If monoclonal antibodies are used, the primary antibody has to be substituted by another nonrelevant monoclonal antibody with the same subclass.

Absorption of Antibodies

The correct way to evaluate the specificity of the immunoreaction is to absorb the antiserum with the antigen. The procedure for absorption of the primary antibody depends on the access to and the purity of the antigen. Antibodies raised against synthetic peptides can be absorbed by the addition of peptide directly to the antibody solution. We absorb our antisera after they have been diluted to their appropriate dilution used in immunohistochemistry. The concentration of antigen added is 10 μg/ml diluted antiserum. If the synthetic peptide is coupled to a carrier protein prior to immunization, the antiserum has to be absorbed with the peptide alone. In this way it is possible to rule out if the reactions occur with antibodies against the peptide or against the carrier protein. Some synthetic peptides are difficult to dissolve. They can then be dissolved in a small volume of dimethyl sulfoxide and the solution diluted with an appropriate buffer. It is necessary to add the same amount of solvent (without the addition of antigen) to the antibody solution and in this way ensure that the antibodies are not destroyed by the solvent. The antiserum should be incubated at room temperature overnight with end-over-end rotation. The antigen–antibody precipitate is pelleted by centrifugation (14,000 rpm in an Eppendorf centrifuge, 15 min at room temperature).

Sometimes it is difficult to absorb an antiserum raised against small synthetic peptides. The reason for this can be the conformation of the antigen or that the antigen–antibody aggregate is too small and will not be pelleted by centrifugation. This problem can be overcome by the following procedure. The antigen is coupled to CNBr-activated Sepharose. The antigen-coated Sepharose particles are subsequently added to the diluted antiserum. The incubation conditions are the same as described earlier.

Antibodies raised against purified proteins can be absorbed in the same way. Problems may occur when the antigen used for immunization is not 100% pure. We have experienced some problems with unwanted specific labeling when using antisera raised against purified amyloid fibril proteins. This problem may be overcome by absorption of the antiserum with a normal tissue extract or with a purified amyloid protein of another biochemical nature.

Acknowledgment

Supported by the Swedish Medical Research Council.

[2] Isolation and Characterization of Amyloid Fibrils from Tissue

By GLENYS A. TENNENT

Amyloid fibrils are abnormal, insoluble, and relatively proteinase-resistant structures, produced by protein misfolding and aggregation of normally soluble autologous proteins. The *in vivo* extracellular deposition of such fibrils accompanies a variety of different acquired or hereditary diseases in humans, other mammals, and birds and contributes directly to their pathology.[1] Approximately 20 precursor proteins are currently known to form amyloid fibrils, the peptide subunits of which are derived from wild-type, variant, or truncated proteins, and which differ in different forms of the disease. Although identification of the fibril protein is mandatory for the classification and management of all patients with amyloidosis, elucidation of the structural basis for protein misassembly into amyloid fibrils may also permit rational approaches to therapy. Indeed, high-resolution X-ray diffraction of many different amyloid fibrils isolated from tissues containing amyloid deposits formed *in vivo* has provided powerful new insights into the structure of amyloid fibrils and to the molecular pathways of amyloid fibrillogenesis.[2,3]

Isolation of Amyloid Fibrils

Amyloid fibrils are isolated from amyloidotic tissues in three principal ways. They were originally extracted by differential centrifugation after homogenization in saline of tissue from patients with systemic AA and AL amyloidosis. The uppermost layer of the tissue sediments, when physically separated from the lower layers, was enriched in amyloid fibrils as monitored by apple-green birefringence after Congo red staining. Moreover, the fibrils maintained their characteristic fibrillar appearance by electron microscopy following prolonged digestion with collagenase.[4] Further purification of these "top layer" fibrils was achieved by sucrose gradient centrif-

[1] M. B. Pepys, *in* "Samter's Immunologic Diseases" (M. M. Frank, K. Austen, H. N. Claman, and E. R. Unanue, eds.), p. 637. Little Brown, Boston, 1994.

[2] M. Sunde, L. C. Serpell, M. Bartlam, P. E. Fraser, M. B. Pepys, and C. C. F. Blake, *J. Mol. Biol.* **273,** 729 (1997).

[3] M. Sunde and C. C. F. Blake, *Quart. Rev. Biophys.* **31,** 1 (1998).

[4] A. S. Cohen and E. Calkins, *J. Cell Biol.* **21,** 481 (1964).

METHODS IN ENZYMOLOGY, VOL. 309 0076-6879/99 $30.00

ugation.[5] In 1984, Glenner and Wong[6] used a similar physical extraction procedure, followed by collagenase digestion, to isolate the fibrils from the cerebrovascular amyloid lesions of patients with Alzheimer's disease.

The second, and most widely used, method for isolation of *ex vivo* amyloid fibrils involves homogenization of tissue in water.[7] After the initial removal of soluble proteins and other components by repeated homogenization in saline, the tissue sediment is homogenized repeatedly in pure water. When the ionic strength is sufficiently low, most types of amyloid fibrils form a solution–suspension and are recovered in the clear supernatant following centrifugation for 1 hr at 20,000*g*. However, their solubility in water is relative because the fibrils are sedimented completely after centrifugation for ≥1 hr at 100,000*g*. The addition of salts, especially divalent cations such as $CaCl_2$, also precipitates the fibrils rapidly. The only significant modifications to this classical water extraction technique followed the discovery that serum amyloid P component (SAP), a universal constituent of amyloid deposits *in vivo,* is calcium dependently bound to amyloid fibrils[8,9] and is therefore removed more efficiently by initial homogenization of tissue in buffers containing citrate[10] or EDTA.[11] Evidence suggests that SAP may contribute to stability and persistence of amyloid deposits *in vivo* and thus to pathogenesis of amyloid-related disease.[12] Sulfated glycosaminoglycans (GAGs) are also ubiquitous constituents of amyloid deposits *in vivo,* being tightly though noncovalently bound to the fibrils. A small proportion of GAGs is removed by calcium chelation, but most are dissociated from the fibrils completely only after exhaustive digestion with proteinases, solubilization with guanidine hydrochloride, or very high concentrations of salt.[13] Serine proteinase inhibitors, complement components, apolipoproteins, and various other extracellular matrix proteins and constituents of basement structure have been detected immunohistochemically in amyloid deposits. Despite their much greater abundance compared with

[5] T. Shirahama and A. S. Cohen, *J. Cell Biol.* **33,** 679 (1967).

[6] G. G. Glenner and C. W. Wong, *Biochem. Biophys. Res. Commun.* **120,** 885 (1984).

[7] M. Pras, M. Schubert, D. Zucker-Franklin, A. Rimon, and E. C. Franklin, *J. Clin. Invest.* **47,** 924 (1968).

[8] M. B. Pepys, A. C. Dash, E. A. Munn, A. Feinstein, M. Skinner, A. S. Cohen, H. Gewurz, A. P. Osmand, and R. H. Painter, *Lancet* **i,** 1029 (1977).

[9] M. B. Pepys, R. F. Dyck, F. C. de Beer, M. Skinner, and A. S. Cohen, *Clin. Exp. Immunol.* **38,** 284 (1979).

[10] M. Skinner, T. Shiraham, A. S. Cohen, and C. L. Deal, *Prep. Biochem.* **12,** 461 (1983).

[11] S. R. Nelson, M. Lyon, J. T. Gallagher, E. A. Johnson, and M. B. Pepys, *Biochem. J.* **275,** 67 (1991).

[12] M. B. Pepys, D. R. Booth, W. L. Hutchinson, J. R. Gallimore, P. M. Collins, and E. Hohenester, *Amyloid Int. J. Exp. Clin. Invest.* **4,** 274 (1997).

[13] J. H. Magnus and T. Stenstad, *Amyloid Int. J. Exp. Clin. Invest.* **4,** 121 (1997).

SAP, many of these proteins are not universally present in amyloid deposits *in vivo* and are barely detectable in preparations of isolated amyloid fibrils *in vitro.*

A consistent feature of fibrils derived from the amyloid precursor proteins serum amyloid A protein (in AA amyloidosis) and immunoglobulin light chains (AL),[14] immunoglobulin heavy chains,[15] apolipoprotein AI (ApoAI),[16] β_2-microglobulin (β_2M),[17] cystatin C[18] (formerly γ trace), gelsolin,[19] fibrinogen Aα-chain,[20] lysozyme (Lys),[21] and transthyretin[22] (TTR) (formerly prealbumin) is that they are recovered predominantly in the supernatant after extraction in water, from most anatomical sites in which the deposits occur. In contrast, fibrils derived from the polypeptide hormones amylin/islet amyloid polypeptide,[23] atrial natriuretic factor,[24] (pro)calcitonin,[25] insulin,[26] and prolactin[27] appear to be sparingly soluble after homogenization in water and remain in the tissue sediment. These "bottom layer" suspensions are usually extracted with chaotropic agents, organic solvents, detergents, acids, or alkali to yield amyloid fibril proteins that are subsequently purified and chemically characterized. Clearly, many factors modulate the *in vitro* solubility of isolated *ex vivo* amyloid fibrils. Amyloid deposits are principally located extracellularly, in spaces normally occupied by structural proteins that are themselves insoluble under physio-

[14] M. Pras, D. Zucker-Franklin, A. Rimon, and E. C. Franklin, *J. Exp. Med.* **130,** 777 (1969).

[15] M. Eulitz, D. T. Weiss, and A. Solomon, *Proc. Natl. Acad. Sci. U.S.A.* **87,** 6542 (1990).

[16] A. K. Soutar, P. N. Hawkins, D. M. Vigushin, G. A. Tennent, S. E. Booth, T. Hutton, O. Nguyen, N. F. Totty, T. G. Feest, J. J. Hsuan, and M. B. Pepys, *Proc. Natl. Acad. Sci. U.S.A.* **89,** 7389 (1992).

[17] F. Gejyo, T. Yamada, S. Odani, Y. Nakagawa, M. Arakawa, T. Kunitomo, H. Kataoka, M. Suzuki, Y. Hirasawa, T. Shirahama, A. S. Cohen, and K. Schmid, *Biochem. Biophys. Res. Commun.* **129,** 701 (1985).

[18] D. H. Cohen, H. Feiner, O. Jensson, and B. Frangione, *J. Exp. Med.* **158,** 623 (1983).

[19] M. Haltia, F. Prelli, J. Ghiso, S. Kiuru, H. Somer, J. Palo, and B. Frangione, *Biochem. Biophys. Res. Commun.* **167,** 927 (1990).

[20] L. H. Asl, J. J. Liepnieks, T. Uemichi, J-M. Rebibou, E. Justrabo, D. Droz, C. Mousson, J-M. Chalopin, M. D. Benson, M. Delpech, and G. Grateau, *Blood* **90,** 4799 (1997).

[21] M. B. Pepys, P. N. Hawkins, D. R. Booth, D. M. Vigushin, G. A. Tennent, A. K. Soutar, N. Totty, O. Nguyen, C. C. F. Blake, C. J. Terry, T. G. Feest, A. M. Zalin, and J. J. Hsuan, *Nature* **362,** 553 (1993).

[22] P. P. Costa, A. S. Figueira, and F. R. Bravo, *Proc. Natl. Acad. Sci. U.S.A.* **75,** 4499 (1978).

[23] P. Westermark, C. Wernstedt, E. Wilander, D. W. Hayden, T. D. O'Brien, and K. H. Johnson, *Proc. Natl. Acad. Sci. U.S.A.* **84,** 3881 (1987).

[24] B. Johansson, C. Wernstedt, and P. Westermark, *Biochem. Biophys. Res. Commun.* **148,** 1087 (1987).

[25] K. Sletten, P. Westermark, and J. B. Natvig, *J. Exp. Med.* **143,** 993 (1976).

[26] F. E. Dische, C. Wernstedt, G. T. Westermark, P. Westermark, M. B. Pepys, J. A. Rennie, S. G. Gilbey, and P. J. Watkins, *Diabetologia* **31,** 158 (1988).

[27] P. Westermark, L. Eriksson, U. Engström, S. Eneström, and K. Sletten, *Am. J. Pathol.* **150,** 67 (1997).

logical conditions. The composition of the microenvironment, in which amyloid fibrils of different chemical types are deposited *in vivo,* may thus be one factor that influences their variable solubility *in vitro.*

The isolation of amyloid fibrils by physical separation and the recovery of amyloid fibril proteins by extraction with denaturing solvents are described elsewhere in this volume with reference to fibrils derived from amyloid-β protein in cerebrovascular amyloid deposits,[28] the amyloid-like paired helical filaments composed of τ protein in neurofibrillary tangles,[29] and to amyloid fibrils embedded in tissues that are chemically fixed with aldehydes.[30] Fibrils are isolated from affected tissues of animals, with naturally occurring or experimentally induced amyloidosis, using similar methods described earlier according to the fibril type. Irrespective of the isolation method used, it is critically important to ensure, if possible, that most, if not all, of the amyloid fibrils are extracted from the tissue before their characterization and to be aware, as in any isolation procedure, that systematic recovery or loss of particular subsets may yield results that do not correctly reflect the situation in the tissues *in vivo.*

The following section includes notes on safety and preparation of tissue and buffers for fibril extraction. In the subsequent section, two water extraction protocols used successfully in our laboratory to isolate amyloid fibrils from homogenates of unfixed, fresh, or frozen tissue are described. These techniques are simple, require no specialized equipment or expertise, and need little further explanation. They preserve the native quaternary structure of amyloid fibrils and allow their recovery with reasonable yield in relatively pure form. Information on histochemical and immunohistochemical staining procedures that we use to identify and to type amyloid deposits *in situ* follows. Although described elsewhere in this volume, these techniques are included here because they are associated closely with the isolation and characterization of fibrils from tissue. The final section of this article describes aspects of identification, characterization, and storage of the isolated fibrils and mentions their use in functional assays involving SAP.

Safety Notes and Preparation of Tissue

Safety

All fresh, unfixed samples of tissue are viewed as potentially hazardous (the handling of neural tissue is not covered in this article). In the absence

[28] A. E. Roher and Y.-M. Kuo, *Methods Enzymol.* **309** [4] (1999) (this volume).
[29] V. M-Y. Lee, J. Wong, and J. Q. Trojanowski, *Methods Enzymol.* **309** [6] (1999) (this volume).
[30] B. Kaplan, R. Hrncic, C. L. Murphy, G. Gallo, D. Weiss, and A. S. Solomon, *Methods Enzymol.* **309** [5] (1999) (this volume).

of evidence to the contrary, the handling of tissue obtained from patients with most types of acquired systemic, localized, or hereditary amyloidosis does not present a serious risk of infection from pathogens. Neither does the handling of amyloid fibrils isolated from this tissue. However, by analogy with prions and amyloid-enhancing factor (AEF), appropriate caution is advised at all times. Amyloid-enhancing factor is a potent activity present at low levels in extracts of most normal nucleated cells and at much higher levels in amyloidotic tissues.[31] Although AEF has eluded biochemical characterization to date, it clearly promotes the formation of AA amyloid. Standard precautions required for the safe working and prevention of infection in clinical laboratories are therefore followed. This includes the wearing of gloves at all times when handling fresh, unfixed tissue. Cutting and homogenization of tissue are performed in a microbiological safety cabinet that provides protection to the user and environment (class I or II containment levels). The cabinet must exhaust through a HEPA filter (single or double) to the outside air. The protocols for isolation of amyloid fibrils are designed so that homogenization and centrifugation of tissue are performed in one container throughout the procedure. This minimizes contact with the tissue homogenate and the dispersion of it, as well as reducing tissue loss. Tissue debris, gloves, scalpels, and other disposable items used when cutting should be sterilized by autoclaving before incineration. Centrifuge tubes, glassware, homogenization tools, plastic containers, and other nondisposable items used during the isolation procedure should ideally be kept aside for the purpose and must be disinfected appropriately after use. The handling and disposal of hazardous chemicals, such as formalin fixatives, enzyme substrates used in immunohistochemistry [3,3′-diaminobenzidine tetrahydrochloride (DAB)], dewaxing, and delipidation solvents, are performed according to the appropriate health and safety regulations.

Buffers

All buffers are made in ultrapure water (low organic content, conductivity $\leq 0.5\ \mu S/cm$, pretreated by reverse osmosis) with analytical grade reagents. The following buffers used in the fibril extraction procedures are of physiological ionic strength, but are pH 8.0 to minimize nonspecific interactions (the pH is adjusted to 7.5 with HCl where necessary). These are made to the final basic Tris-buffered saline (TN buffer) recipe: 10 m*M* Tris–HCl, 140 m*M* NaCl, 0.1% (w/v) NaN_3, pH 8.0; Tris–calcium buffer (TC buffer) contains 138 m*M* NaCl and an additional 2 m*M* $CaCl_2$; Tris–EDTA

[31] R. Kisilevsky, E. Gruys, and T. Shirahama, *Amyloid Int. J. Exp. Clin. Invest.* **2,** 128 (1995).

buffer (TE buffer) contains an additional 10 m*M* EDTA. For immunohistochemistry, 50 m*M* Tris–HCl, 0.85% NaCl, pH 7.6 (TBS buffer) is used as diluent and wash buffer.

Human Tissue

Amyloidotic tissue is obtained, with informed consent, as soon as possible after biopsy, surgery, or autopsy to minimize proteolysis. Parenchymal organs affected most commonly (especially in systemic forms of amyloidosis) include the spleen, liver, kidneys, and heart; the gastrointestinal tract and skin are also frequent sites of amyloid deposition. Excess blood is washed away quickly but thoroughly in large volumes and several changes of ice-cold saline or buffered saline with or without calcium (see earlier). Hemoglobin, immunoglobulins, and albumin are frequent and significant contaminants of fibril isolates, especially if the tissue from which they were extracted was washed inadequately beforehand. The washed tissue is cut into portions of approximately 10–12 g (wet weight), from which smaller specimens (2–10 mm^2) are taken for cryostat sectioning and Congo red staining. Tissue pieces are blotted dry on fiber-free filter paper. Specimens are taken also for immunohistochemistry, which are fixed immediately in 10% (v/v) neutral-buffered formalin, dehydrated through a series of ascending ethanol concentrations to xylene, and embedded in low melting point (54–58°) paraffin wax, according to standard histological techniques. However, it should be noted that shorter fixation and processing times than those used conventionally may provide better antigen preservation and thus more reproducible results in amyloid immunohistochemistry.[32] Tissue removed for cryostat sectioning is embedded in a matrix compound, such as OCT, immediately snap frozen in isopentane supercooled over liquid nitrogen according to standard techniques, and stored at −70° until use. The large portions of tissue are placed in zip-seal plastic bags, labeled clearly and permanently with the patient's name, tissue collection and storage dates, tissue type, and amyloid fibril type if known. These are stored inside sturdy freezer boxes at −70° until use.

Once thawed, certain tissues, including those rich in collagen, may be difficult to homogenize directly in buffer. These can be frozen solid in liquid nitrogen and pulverized to a powder with a pestle and mortar prior to use. Collagen fibrils are removed if necessary by the digestion of tissue homogenates (in TC buffer pH 7.5) with crude collagenase (Type I, Sigma-Aldrich, Dorset, UK) [enzyme : substrate ratio, 1 : 10–100 (w/w) depending on the tissue] at 37° with shaking for 18–24 hr. After separation by centrifu-

[32] E. Arbustini, P. Morbini, L. Verga, M. Concardi, E. Porcu, A. Pilotto, I. Zorzoli, P. Garini, E. Anesi, and G. Merlini, *Amyloid Int. J. Exp. Clin. Invest.* **4,** 157 (1997).

gation, the amyloid fibrils in the pellet are extracted as usual. Adipose tissue, such as abdominal subcutaneous fat, is usually delipidated by solvent extraction in chloroform : methanol [2 : 1 (v/v); to remove neutral lipids and phospholipids] or acetone (neutral lipids only). Following homogenization in saline or buffer and centrifugation, the tissue pellet is homogenized (3–5 min) in the solvent in a suitably resistant centrifuge tube to a final dilution of approximately 20 times the volume of tissue sample. The fibril protein is subsequently extracted from the tissue pellet after centrifugation, with 6 *M* guanidine hydrochloride in TN buffer with or without EDTA and dithiothreitol.[21,33]

Fibril Isolation Protocols

The quantity of fibrils recovered depends principally on the extent of amyloid deposition in the starting tissue used for extraction; the quality of fibril isolates is enhanced by adequate prior washing of the starting tissue. The following protocols should be viewed as guidelines only, with conditions for optimum isolation (especially the volumes of pure water added to extract fibrils) modified according to the extent of amyloid deposition in each tissue.

A maximum of 30 g tissue (wet weight) can be processed in one container using the standard protocol if only small amounts of amyloid are present or if a large supply of fibrils in solution–suspension is required. However, it is preferable to use only 10–20 g and to not exceed 30 g at any one time. During the extensive preliminary homogenizations in saline or buffered saline, significant losses of fibrils can occur. Thus in order to maximize the recovery of fibrils, especially from milligram quantities of tissue obtained at biopsy, we developed a rapid water extraction procedure,[34] a modified version of what is described here. After a single homogenization in buffered saline, containing EDTA to release SAP, the tissue is homogenized directly in water to extract the fibrils immediately. They are then precipitated in pure form with hypertonic saline and EDTA in which all other components of the extracted tissue remain soluble. Although this modified protocol is recommended when there is less than 1 g of tissue available for extraction, the method may also be used with larger quantities of tissue for the recovery of fibrils in suspension. In contrast to other rapid methods for the characterization of amyloid proteins from biopsy specimens, by extraction with aqueous acidic acetonitrile and purification by high-performance liquid chromatography (HPLC)[35] or collagenase digestion and formic acid solubi-

[33] K. Olsen, K. Sletten, and P. Westermark, *Biochem. Biophys. Res. Commun.* **245,** 713 (1998).

[34] S. Y. Tan, I. E. Murdoch, T. J. Sullivan, J. E. Wright, O. Truong, J. J. Hsuan, P. N. Hawkins, and M. B. Pepys, *Clin. Sci.* **87,** 487 (1994).

[35] B. Kaplan, G. German, and M. Pras, *J. Liq. Chromatogr.* **16,** 2249 (1993).

lization,[36] the fibrils obtained by the modified water extraction protocol here are in their native form.

Standard Protocol for Isolation of ex Vivo Amyloid Fibrils (from 1.0- to 30-g Unfixed, Frozen Tissue)

Preparation of Tissue Homogenate

1. Remove a portion(s) of washed tissue from −70° and weigh: 10–30 g (wet weight) will yield abundant fibrils if ≥25% of the tissue section stained with Congo red is positive. Thaw tissue at 37°, dice into smaller pieces, and rinse well by swirling in a glass beaker with large volumes and several changes of ice-cold TC buffer until the supernatant of the final wash is clear and free of erythrocytes (lysed or not). Aspirate excess fluid, gently blot tissue dry on fiber-free filter paper, and reweigh. Transfer tissue to the appropriate size centrifuge container. (Typically a 16-ml round-bottom polycarbonate tube is used for 1.0–1.9 g tissue; a 50-ml round-bottom tube for 2.0–4.9 g tissue; a 150-ml round-bottom tube for 5.0–10.9 g tissue; and a 250-ml centrifuge bottle for 11–30 g tissue.)
2. Add ice-cold TC buffer at 2 ml buffer/g starting tissue. Homogenize tissue directly in the centrifuge container for 1 min at medium speed using a mechanical homogenizer (Ultra-Turrax T25, 8–24,000 rpm; Merck Ltd, Dorset, UK) mounted on a retort stand and the appropriate size dispersing tool. Add a further 5 ml ice-cold TC buffer/g starting tissue to homogenate and repeat homogenization. Remove a 50-μl aliquot of the starting tissue homogenate and keep aside to test by Congo red staining. Centrifuge at 15,000*g* for 30 min at 10° (or 20,000–27,000*g* for 30 min if the tissue pellet does not sediment firmly). Decant the supernatant carefully into a clean container and store at 4°. (The homogenate is very proteinaceous and may overflow from the container if too much buffer is added initially or if homogenization is performed too vigorously.)

Removal of SAP and Other Constituents Calcium Dependently Bound to Tissue/Fibrils

3. Add ice-cold TE buffer to tissue pellet at 7 ml buffer/g starting tissue, homogenize, and centrifuge as described earlier. If the homogenate overflows, add the TE buffer in stages as mentioned in step 2. Aspirate supernatant into a clean container, using a syringe and cannula.

[36] E. M. Castaño, F. Prelli, L. Morelli, A. Avagnina, A. Kahn, and B. Frangione, *Amyloid Int. J. Exp. Clin. Invest.* **4,** 253 (1997).

Monitor the protein concentration in TE washes by measuring A_{280}. Repeat homogenization/centrifugation in TE buffer until the A_{280} of the supernatant is ≤0.05; this often requires 15–30 cycles and is performed as continuously as possible. Store TE washes at 4°. (Homogenate may be left at 4° overnight when necessary. Homogenization in EDTA removes proteins soluble in EDTA and the bulk of SAP from the fibrils. More soluble material is removed from the tissue with EDTA compared with homogenization in citrate or saline.)

Removal of Salt to Recover Amyloid Fibrils in Solution–Suspension

4. Add ice-cold pure water, in stages, to the tissue pellet, at 8 ml water/g starting tissue. Homogenize as described earlier but centrifuge at 20,000*g* for 1 h at 10°; increase to 27,000*g* if necessary. (Distinct layers are often seen in the pellet; amyloid fibrils are usually in the uppermost layer, which is soft and glutinous.)
5. Aspirate the supernatant carefully into sterile tubes (preferably glass) with a syringe and cannula and measure protein at A_{280}. Repeat homogenization of the tissue pellet in water for 10 times total, but halve the final volume of water added each time, from the second to the fourth homogenization. Thereafter continue homogenization at 2–4 ml of water/gram of starting tissue. Keep fibril isolates (and residual tissue sediment) at 4° and analyze immediately by optical density (OD), Congo red staining, and sodium dodecyl sulfate–polyacrylamide gel electrophoresis (SDS–PAGE).

Modified Protocol for Isolation of ex Vivo Amyloid Fibrils (from ≤1-g Unfixed, Frozen Tissue)

Preparation of Tissue Homogenate in EDTA

1. Remove a portion of washed tissue from −70° and weigh: as little as 2 mg (wet weight) will yield fibrils sufficient for chemical analysis if ≥25% of the tissue section stained with Congo red is positive. Thaw tissue at 37°, dice into smaller pieces, and rinse well by placing in a conical bottom plastic tube (30 ml) filled with ice-cold TC buffer. Mix by rotation overnight at 4°. The supernatant should be clear and free of erythrocytes. Aspirate excess fluid, gently blot tissue dry on fiber-free filter paper, and reweigh. Transfer tissue to appropriate size centrifuge tube. (Typically a 2-ml polypropylene microcentrifuge tube is used for ~2–50 mg tissue and a 16-ml round-bottom polycarbonate tube is used for 50 mg to 1 g tissue.)

2. Add ice-cold TE buffer at 2 ml buffer/2–50 mg starting tissue or at 7 ml buffer/50 mg to 1 g tissue. Homogenize 2–50 mg tissue manually in the microcentrifuge tube for 2 min through a needle (21 gauge) and syringe (2 ml); homogenize 50 mg to 1 g tissue in the 16-ml centrifuge tube for 1 min at medium speed using a mechanical homogenizer (Ultra-Turrax T25, 8–24,000 rpm) mounted on a retort stand and the appropriate size dispersing tool. Remove a 15- to 20-μl aliquot of starting homogenate and keep aside to test by Congo red staining. Add a further 7 ml ice-cold TE buffer to the homogenate from 50 mg to 1 g tissue only and vortex vigorously for 2 min. Centrifuge all tubes at 15,000g for 30 min at 4–10°. Aspirate the supernatant into a clean tube, measure the protein concentration at A_{280}, and store at 4°. (Homogenization in EDTA removes tissue proteins and components soluble in EDTA, including the bulk of SAP from amyloid fibrils.)

Removal of Tissue Components Soluble in Water

3. Add ice-cold pure water to tissue pellet at 0.5–1.0 ml water/2–50 mg starting tissue; 1–2 ml water/50 mg to 1 g. Homogenize 2–50 mg manually and 50 mg to 1 g tissue mechanically, as described earlier. Centrifuge all tubes. Carefully aspirate supernatant into a sterile tube (preferably glass) and repeat homogenization of the tissue pellet in pure water for a total of 10 times. Keep residual tissue debris at 4° until analyzed by Congo red staining. (Homogenate may be left overnight at 4° if necessary. Homogenization in water extracts all components of the tissue that are soluble in water, including amyloid fibrils, SAP not removed by the single wash in EDTA, and other materials calcium dependently bound to the fibrils. These latter components are eluted from the fibrils in the absence of calcium, but remain in the supernatant.)

Precipitation with Salt to Recover Amyloid Fibrils in Suspension

4. Pool all pure water washes, remeasure A_{280}, add solid NaCl to 0.2 M and EDTA to 10 mM final, and vortex vigorously to dissolve. Leave at 4° for 2–3 days (longer if necessary) to selectively precipitate the amyloid fibrils. Centrifuge for 1 hr at 15,000g (or 20,000–27,000g if necessary), aspirate all of the saline supernatant carefully, remeasure A_{280}, and store at 4° until analyzed further. If the quantity of the starting tissue was small and/or the amyloid deposits in it were few, the saline pellet may be invisible. (The saline–EDTA supernatant contains residual SAP and all other components of the extracted tissue soluble in saline–EDTA and not precipitated under the condi-

tions described earlier, including some amyloid fibrils. The pellet contains the bulk of the amyloid fibrils precipitated by the hypertonic saline.)

5. Add ice-cold pure water to the saline pellet, visible or not, at 10–50 μl water/2–50 mg starting tissue: 50–100 μl water/50 mg to 1 g tissue and vortex to resuspend. Analyze immediately by Congo red staining and SDS–PAGE and N-terminal amino acid sequencing if required. (Dialyze the fibrils against pure water to remove salt, if necessary, using microdialysis cells).

Histochemical Identification and Immunohistochemical Typing of Amyloid Deposits

Histochemical Staining of Amyloid Deposits in Situ

Amyloid is identified by apple-green birefringence in unfixed cryostat sections (for speed and convenience) or in formalin-fixed deparaffinized sections of tissue stained with Congo red and viewed by polarized light. The intensity of Congo red staining is usually greater in frozen sections compared with wax-embedded sections. Congo red staining is performed on sections of the starting tissue prior to the extraction of amyloid fibrils. If only fragments of starting tissue are available, aliquots of the tissue homogenate in buffered saline are stained instead (see previous protocols and Congo red staining of isolated fibrils discussed later). Sections are cut at 6 μm thick, as serially as possible, onto microscope slides coated with chrome–alum–gelatin or poly(L-lysine) (Polysine slides, Merck Ltd) and dried under a stream of cool air for 1 hr at room temperature (frozen sections) or overnight at 37° (paraffin sections). Prior to staining, sections are rehydrated accordingly, counterstained in alum hematoxylin (10 min in Cole's or Mayer's hematoxylin solution, Pioneer Research Chemicals Ltd, Essex, UK), "blued" under running tap water, rinsed in pure water, and stained by the alkaline, alcoholic Congo red method according to Puchtler *et al.*[37] and as detailed elsewhere in this volume.[38] Working solutions of saturated salt and Congo red should be used within 20 min, and stock solutions of Congo red must be prepared fresh every 2 months. Section thickness of 5–10 μm and inclusion of known positive control tissue sections are also critical. Slides are viewed when dry by bright-field and

[37] H. Puchtler, F. Sweat, and M. Levine, *J. Histochem. Cytochem.* **10,** 355 (1962).

[38] G. T. Westermark, K. H. Johnson, and P. Westermark, *Methods Enzymol.* **309** [1] (1999) (this volume).

high-intensity polarized light microscopy and are examined for the severity and extent of amyloid deposition. In the absence of a dedicated polarizing microscope, most normal microscopes can be converted by placing one circular polarizer in the filter carrier of the substage condenser, where it can be rotated, and one rectangular polarizer (analyzer) between the objective and the eyepiece. The extent of amyloid deposition in the tissue used for extraction is determined by estimating the percentage of positive staining per section, either empirically, with the aid of a measuring graticule, or automatically using image analysis. This estimate is used to gauge the volumes of pure water to add when homogenizing tissue to recover the fibrils (standard protocol). GAGs are identified histochemically, if required, by sodium sulfate Alcian blue staining and by the Alcian blue technique involving magnesium chloride critical electrolyte concentration to test the amount of sulfation, as described elsewhere.[39,40]

Immunohistochemical Typing of Amyloid Deposits

Amyloid fibrils *in situ* are conveniently typed by the immunohistochemical staining of formalin-fixed, deparaffinized, and rehydrated tissue sections (cut as described earlier), using a screening panel of antibodies reactive with known amyloid fibril proteins in the most commonly occurring forms of amyloidosis. Ideally, immunostaining is performed on the tissue prior to the extraction of amyloid fibrils, but if not it should be done afterward to confirm the identity of the fibril protein. We use the peroxidase–antiperoxidase (PAP) technique, described elsewhere in this volume,[38] with purified IgG fractions of monospecific antibodies reactive with AA (polyclonal antibody produced in our laboratory or purchased from Calbiochem-Novabiochem Ltd, Nottingham, UK), AL (κ and λ), Aβ_2M, ALys, ATTR (all from Dako Ltd, Cambridge, UK), and AApoAI (Genzyme Diagnostics, Kent, UK) amyloid fibril proteins. There is often greater difficulty with immunohistochemical typing of AL amyloidosis, compared with other types, using standard antibodies to κ and λ immunoglobulin light chains, and in these cases antibodies directed against the amyloid fibril proteins themselves may be more sensitive.[41] The diluent and wash buffer is TBS pH 7.6, with the addition of 1% (w/v) bovine serum albumin (BSA) for primary and secondary antibody solutions. All incubation and washing steps are performed at room temperature unless stated otherwise. Endogenous

[39] H. C. Cook, *in* "Theory and Practise of Histological Techniques" (J. D. Bancroft and A. Stevens, eds.), p. 177. Churchill Livingstone, Edinburgh, 1990.

[40] N. A. Athanasou, B. Puddle, and B. Saillie, *Nephrol. Dial. Transplant.* **10,** 1672 (1995).

[41] R. J. Strege, W. Saeger, and R. P. Linke, *Virch. Arch.* **433,** 19 (1998).

peroxidase activity is usually quenched by incubation (15 min) in aqueous 0.5% (v/v) H_2O_2. Sections are pretreated (10–30 min at 37°) with 0.1% (w/v) trypsin (crude type, preferably containing chymotrypsin activity) in 0.1% (w/v) $CaCl_2$, pH 7.8 (adjusted with 0.1 *M* NaOH), where necessary to retrieve or to enhance immunoreactivity. TTR immunoreactivity is revealed by incubation in aqueous 1% (w/v) sodium meta-periodate (10 min, followed by two 5-min washes in water), aqueous 0.1% (w/v) sodium borohydride (10 min, followed by two 5-min washes in water), and 6 *M* guanidine hydrochloride freshly prepared in 0.9% (w/v) NaCl (overnight, followed by one 5-min rinse in saline). Prior to the application of primary antisera, nonspecific binding sites in the tissue sections are blocked for 30 min in 10% (v/v) normal nonimmune serum from species donating the secondary antibody. Sections are incubated with primary antisera overnight at 4°. Unlabeled secondary antibodies (specific for IgG of the host species of the primary antibody; usually purchased from Dako and diluted 1:50) and PAP complexes (reactive against peroxidase in the host species; usually purchased from Dako and diluted 1:100 in Protexidase, ICN Biomedicals Ltd., Oxfordshire, UK) are incubated for 30 min each. Antisera are centrifuged at 12,500*g* for 5 min at 4° prior to dilution, the optimum concentration of which is determined for every new batch by prior titration on positive control tissue sections. The antibody incubation and washing procedures (slides are washed once only for 5 min after each incubation with antibody) are performed manually, using Sequenza Coverplates (Life Sciences International Ltd, Hampshire, UK), which allow easy, and often economic, application of reagents and are a major time-saving device. Bound enzyme–antibody complexes are detected using metal-enhanced DAB solution (Pierce and Warriner Ltd, Chester, UK) according to the manufacturer's instructions, which results in a crisp, dark brown reaction product (the stock DAB concentrate is unstable and should be used within 2 months of purchase and storage at −20°). After a brief rinse in pure water, sections are counterstained lightly (30 sec) in hematoxylin, "blued" under running tap water, rinsed again in pure water, dehydrated through a series of ascending ethanol concentrations to xylene, and mounted (Histomount, National Diagnostics, Hessle, UK).

Separate, adjacent tissue sections are stained with Congo red, as mentioned previously, to confirm the presence of amyloid and to identify the precise distribution of the deposits. Alternatively, we often immunostain the sections first, counterstain with hematoxylin, and then overlay the same slides with alkaline, alcoholic Congo red.[42,43] Thus, amyloid deposits that

[42] R. P. Linke, H. V. Gärtner, and H. Michels, *J. Histochem. Cytochem.* **43,** 863 (1995).

[43] G. A. Tennent, K. D. Cafferty, and P. N. Hawkins, *in* "Amyloid and Amyloidosis 1998" (R. A. Kyle and M. A. Gertz. eds.). Parthenon Publishing, New York (in press).

display apple-green birefringence under polarized light are classified positive for the respective antibodies if identical structures display homogeneous, reddish-brown, immunospecific staining of the entire deposit (and the majority of amyloid deposits in the section) by bright-field microscopy. The sensitivity and optics of green birefringence are usually not impaired; however, we monitor the slides carefully during counterstaining and/or development with DAB to improve the discrimination where necessary.

Careful examination of appropriate positive control tissue sections and negative control sections is required. In our laboratory, the specificity of immunoreactivity is always controlled by demonstration that positive staining is abolished by prior absorption of the primary antiserum with its respective purified antigen or antigen-rich material. The following antigens to the antibodies used earlier are available commercially (Sigma-Aldrich): high-density lipoprotein (HDL) from acute-phase serum for anti-AA, HDL from normal serum for anti-apoAI, and pure lysozyme (after reconstitution this is stored immediately in small aliquots at $-70°$; each aliquot is thawed only once) and pure transthyretin for their respective antibodies. Pure β_2M is available from Calbiochem-Novabiochem Ltd. Whole pooled, normal serum is used for anti-k and λ immunoglobulin light chain antibodies. Affinity absorption is achieved simply by incubation of the primary antibody (undiluted or diluted 1:10–1:100 where necessary) with equal volumes of purified antigen or antigen-rich material (>0.2 mg/ml) for 1 hr at 37° (in a water bath) followed by incubation overnight at 4°. After separation by centrifugation at 12,500*g* for 5 min at 4°, the supernatant containing absorbed antibody is removed carefully (little or no precipitate is visible). It is diluted to the same optimal working dilution as the respective unabsorbed antiserum and is used in adjacent tissue sections in place of the primary antibody. If necessary, the supernatant is absorbed further in the same manner, or else solid-phase absorption is performed on antigen-coated Sepharose beads to remove antibody from the solution. When there is complete specific absorption of immunostaining, all that is visible under bright light in sections overlaid with Congo red is congophilia of the amyloid deposits. The effects of absorbing the primary antibody with whole normal serum in which the antigen of interest is depleted or is present at very low concentrations, and normal or acute-phase serum selected to contain high concentrations of antigen, can also be compared. If the staining is specific for the antigen of interest, it will be unaffected by normal serum absorption and abolished by the serum absorption containing high concentrations of antigen. In no case have we observed amyloid-specific, immunospecific staining of amyloid deposits with antibody against more than one type of fibril protein.

Immunohistochemical staining with antibodies to SAP is a valuable adjunct in the histological diagnosis of all forms of amyloidosis and may, on occasion, be a more sensitive method of detection in the laboratory than Congo red staining. However, human SAP antigenicity is frequently masked or destroyed completely by fixation in formalin and is not revealed easily by antigen retrieval techniques unless tissue is fixed and processed for short periods of time only.[32] Immunohistochemistry for human SAP is best demonstrated using unfixed, frozen tissue sections. Immunohistochemistry for GAGs, and for other proteins present in amyloid deposits, is described elsewhere.[40,44]

Fibril Identification and Characterization

We routinely analyze the fibrils during and after their isolation by optical density, Congo red staining, SDS–PAGE/immunoblotting, and SAP electroimmunoassay. Polypeptide analysis of the fibril isolates by N-terminal amino acid sequencing is also performed when necessary. All of these procedures should be carried out as soon as possible to minimize the risk of bacterial contamination in the aqueous isolates and before aggregation renders the fibrils difficult to analyze (this applies especially to fibrils prepared by the modified protocol).

Optical Density

In the standard protocol, the elution of soluble tissue proteins and amyloid fibril-associated proteins by homogenization in EDTA is monitored conveniently by measuring the absorbance of the TE buffer washes at 280 nm (A_{280}). Samples with A_{280} values greater than 1.0 are diluted to read between 0.3 and 1.0, and if necessary the A_{280} values are adjusted by subtraction of the A_{320} representing light scattering by particles or aggregates. If small particles of tissue are present in washes, these should be microcentrifuged and the OD repeated. The first two to three TE washes contain most of the eluted SAP and are kept at 4° for isolation of SAP if required (see later). Tissue is homogenized repeatedly in TE buffer until the A_{280} of the supernatant stabilizes at ≤ 0.05, a process that can take 15–30 cycles over 2 days or more, depending on the quantity of tissue processed. Subsequent homogenization of the tissue pellet in water yields amyloid fibrils in the supernatant as monitored by the increase in protein at A_{280}. Fibrils are recovered predominantly in the second to fourth water washes and their concentration is expressed as arbitrary A_{280} units/ml (A_{280}/

[44] C. Röcken, D. Paris, K. Steusloff, and W. Saeger, *Endocr. Pathol.* **8,** 205 (1997).

ml). Yields in the second water wash only range from 0.3 A_{280}/ml (100 ml) from ~10 g lightly laden amyloidotic tissue (spleen) to 2.0 A_{280}/ml (200 ml) from ~20 g moderately laden amyloidotic tissue (spleen). Water extraction with incrementally smaller volumes is continued 10 times. In our experience, this recovers almost all of the fibrils from the residual tissue sediment.

In the modified protocol, A_{280} values for the initial TE buffer wash are similar to those obtained for the standard protocol. A_{280} values for the second water wash are, not surprisingly, almost twice that for the standard protocol. This reflects the elution not only of all tissue components soluble in water (including amyloid fibrils, which are in solution–suspension), but also the elution of all components that are calcium dependently bound to the tissue and fibrils and removed by the extensive preliminary washes of the standard protocol. Yields of fibrils are nevertheless higher than those obtained after repeated homogenization in buffered saline and may be more representative of what is actually in the tissue. Using this rapid procedure, but without prior homogenization in EDTA, we have recovered analyzable native fibrils from tiny (≤2 mg) biopsies.[21,34]

Congo Red Staining of Isolated Fibrils

Aliquots of all fibril isolates (containing 0.03 A_{280}/ml minimum), homogenates of starting tissue and suspensions of remaining tissue debris (5-, 10-, and 20-μl volumes) are dried onto gelatin-coated microscope slides (10-well format) in a 37° incubator or a hot air oven. As mentioned earlier, sufficient amyloid density must be present to demonstrate apple-green birefringence, therefore aliquots of water washes are reapplied, if necessary (for three to four cycles), to the same spot after drying down. Suspensions of known amyloid fibrils are included as positive controls. The slides are fixed for 5 min with 10% (v/v) formalin in 90% (v/v) ethanol, rinsed in pure water, dried, and treated with hematoxylin, saturated alkaline alcoholic NaCl, and Congo red solutions before rapid dehydration and mounting as mentioned previously. Amyloid fibrils in solution–suspension in water (standard protocol) exhibit a sheet-like green birefringence in polarized light (little or no cellular debris is seen); individual fibrils are seen only by electron microscopy. When the fibrils in these isolates aggregate with time, especially if stored in the presence of salts or buffers, they are visible by light microscopy initially as a gauze-like mesh of fibrils (similar appearance to fibril suspensions prepared fresh by the modified protocol) and then progressively as more crystalline arrangements. If few fibrils are present in the water washes obtained with the standard protocol, these are pooled together, lyophilized, and resuspended in small volumes of water for restaining with Congo red. Fibrils concentrated by lyophilization and reconstituted in aqueous buffers or water form a suspension (opalescent) only, in which individual fibers are clearly visible by light microscopy. If amyloid fibrils are still present in the tissue sediment remaining after homogenization in water, they can be extracted in 6 *M* guanidine hydrochloride in TN buffer, with or without EDTA and dithiothreitol, and if necessary solubilized further in formic acid.[23,26,27] Fibrils are also present in the TE buffer washes (standard protocol), sometimes in significant amounts, which suggests the presence of a subpopulation of fibrils that are in solution–suspension in buffer.[45] Indeed, they are detected also in the saline supernatants of salt-precipitated fibrils (modified protocol), but only after their concentration by lyophilization.

SDS–PAGE and Immunoblotting

The purity and protein composition of all fibril isolates and buffered saline washes are analyzed by SDS–PAGE under denaturing, reducing conditions, with and without subsequent electroblotting and immunostaining to confirm the identify of the fibril protein subunits. Samples are boiled (2–5 min for buffer washes and 10 min for fibrils) in 10 m*M* Tris–HCl pH 8.0, containing 1 m*M* EDTA, 2.5% (w/v) SDS, 5% (v/v) 2-mercaptoethanol, 0.025% (w/v) bromphenol blue, and 10% (v/v) glycerol. The protein subunits of fibril isolates and buffer washes are separated routinely in our laboratory by horizontal SDS–PAGE using precast gradient (8–18%) or homogeneous gels (12.5 or 15%, depending on the molecular mass of the proteins of interest) (ExcelGels, Amersham Pharmacia Biotech, St. Albans, UK). Gradient gels are useful for initial screening and heavier loading of samples (typically 20- to 60-μl volumes; samples are applied with paper application pieces) to check for impurities; homogeneous gels are used for final sample analysis (10- to 18-μl volumes; samples are applied to preformed wells) and further characterization. It is important to ensure that fibril samples are solubilized sufficiently to enter the gels. The precast gels (~250 × 180 mm) are 0.5 mm thick, and electrophoresis is performed at 30–50 mA (constant/gel), 600 V, and 30 W at 15° until the bromphenol blue dye front migrates 8.5 cm from the point of sample application (~1.5–2.5 hr). Gels are calibrated with standard globular marker proteins of known relative molecular mass. Pure human SAP, pure preparations of fibril precursor proteins of interest (2–5 μg each), and other amyloid fibrils are included for reference. These large format gels accommodate 25 samples each, which allows for the analysis on the same gel of most of the fractions

[45] A. C. J. van Andel, T. A. Niewold, B. T. G. Lutz, M. W. J. Messing, and E. Gruys, *in* "Amyloidosis" (J. Marrink and M. H. van Rijswijk eds.), p. 169. Martinus Nijhoff, Dordrecht, 1986.

generated during fibril isolation. The entire gel is stained (30 min at 37°) for protein with Brilliant Blue R-350 (PhastGel Blue R, Amersham Pharmacia Biotech), a more sensitive dye (detects 50–100 ng protein/band) than the usual R-150 and R-250 forms, and destained [30% (v/v) methanol, 10% (v/v) acetic acid]. The fibril protein subunits are visualized as single major bands or as doublets of closely spaced bands of low molecular mass (<30 kDa). The presence of bands migrating directly ahead of the major stained band usually represent traces of proteolytically cleaved fibril protein in the fibril isolates, whereas the presence of bands migrating directly behind may be intact or minimally degraded fibril protein (confirmed by immunoblotting). We do not routinely centrifuge samples prior to loading, and diffuse staining, representing fibrils that are solubilized incompletely by SDS treatment, is often seen throughout the sample lanes containing the highest concentrations of fibrils. Trace, uncharacterized protein bands of higher molecular mass are seen in all *ex vivo* fibril isolates (especially after staining with silver), the numbers and intensities of which differ according to the type of tissue used for extraction. SAP, the major eluted protein in the TE buffer washes, aligns with the 30-kDa marker protein, because although the mass of the SAP protomer is 25,462 Da, it always migrates in such gels with an anomalously high apparent M_r. Traces of SAP are sometimes seen in the fibril isolates (especially after immunostaining). Wet gels are scanned densitometrically in two dimensions using a laser transmission densitometer at a spatial resolution (pixel size) of 100 μm (1000 data points/cm^2) and an optical resolution of 12 bits per pixel (4096 levels of resolution) to obtain optical density measurements of the fibril bands. The purity of the fibril protein in the isolates calculated by this analysis is usually ≥80%. Gels are then completely destained in 30% (v/v) ethanol overnight and restained with silver (detects 0.3 ng protein/band) (ExcelGel protocol, Amersham Pharmacia Biotech). The gels are preserved by soaking in 9% (v/v) glycerol containing 0.03% (w/v) sodium carbonate and drying onto cellophane sheets (24 hr in 37° incubator).

Separated proteins, including prestained marker proteins, are transferred from unstained gels (after removal of the support film) at 0.8 mA/cm^2 constant (50 V, 5 W, 15°) over 5–15 min onto a 0.2-μm pure nitrocellulose membrane by semidry blotting in 39 m*M* glycine, 48 m*M* Tris, 0.0375% (w/v) SDS, and 20% (v/v) methanol (NovaBlot protocol, Amersham Pharmacia Biotech). Nonspecific binding sites on the electroblotted membrane are quenched by incubation with 10% (w/v) milk powder in 20 m*M* Tris–HCl, 0.5 *M* NaCl, pH 7.5 (THS buffer), containing 0.1% (v/v) Tween 20 (THS-T buffer) for 2 hr at 37°. All incubation and washing steps are performed on an orbital shaking platform. After a brief rinse in THS-T buffer, the blot is probed (1 hr at room temperature or 4° overnight) with a monospecific IgG fraction of antiserum against the amyloid fibril protein of interest (as used in immunohistochemistry) diluted optimally (usually 1:2000–1:100,000 depending on the antigen concentration and the substrate used for detection) in 5% milk powder in THS-T buffer. The primary antibody must recognize the denatured, reduced form of the antigen. We also immunostain identical, separate blots with anti-SAP antibodies in the same manner. After extensive washing (5×5 min; 2×15 min), each blot is incubated at room temperature with an optimum dilution of affinity-purified, peroxidase-labeled antibody (usually 1:2000–1:4000 dilution in 5% milk powder in THS-T buffer) specific for the IgG of the species donating the primary antibody. One hour later, the blots are washed again extensively, overlaid with chemiluminescent substrate (SuperSignal, Pierce and Warriner Ltd) at 0.125 ml/cm^2 for 5–10 min, and exposed (1 min) to X-ray film (HyperFilm ECL, Amersham Pharmacia Biotech), which is developed automatically. Subsequent exposures are estimated on the basis of the first film. The specificity of primary antisera is confirmed by the absence of cross-reactivity with amyloid fibrils of different chemical types and by the absence of any reactivity with other proteins in normal human serum than the antigen in question. In the absence of the primary antibody, no staining should be seen with the secondary antibody alone. Immunostaining with anti-immunoglobulin light chain antisera is often positive in cases where inconclusive results are obtained by immunohistochemistry.

SAP Quantification, Isolation, and Binding Assays

Concentrations of SAP in TE buffer washes and fibril isolates (2 μl volumes) are quantified by electroimmunoassay (EIA), as described elsewhere in this series,[46] using monospecific polyclonal anti-SAP antibodies (produced in our laboratory or purchased from Calbiochem-Novabiochem) incorporated into 1-mm-thick 1% (w/v) agarose gel (IBF Indubiose A37, Universal Biologicals Ltd, Stroud, UK) in 70 m*M* barbitone buffer, pH 8.6, containing 10 m*M* EDTA. Gels are cast onto polyester support film (GelBond film, Flowgen Ltd, Lichfield, UK). Electrophoresis is performed in the barbitone–EDTA buffer at 200 V constant/gel at 15° for 6–8 hr with contact wicks of 2-mm-thick chromatography paper. The gels are subsequently washed overnight at 37° in 5% (w/v) NaCl with 0.1% (w/v) NaN_3, pressed, dried, stained with Brilliant Blue (Pharmacia), and destained as mentioned previously. The assays are standardized with dilutions of normal human serum spiked with isolated pure human SAP, for which the absolute mass units were determined spectrophotometrically after precise

[46] C. B. Laurell and E. J. McKay, *Methods Enzymol.* **73,** 339 (1981).

measurement of the extinction coefficient at 280 nm.[47] The total quantity of SAP eluted in TE buffer washes is proportional to the amount of amyloid deposits present in the tissue. The first two to three TE washes contain most of the eluted SAP. Yields of SAP in the first wash only (standard or modified protocols) are typically 25 mg from ~10 g lightly laden amyloidotic tissue (spleen) and 130 mg from ~20 g moderately laden amyloidotic tissue (spleen). The subsequent four to five TE washes contain only traces of SAP whereas no SAP should be detected in the final wash. Because ligand binding by SAP at physiological pH and ionic strength is absolutely dependent on the presence of calcium ions, SAP is eluted, although more gradually, in saline or other solutions containing no calcium. The TE washes containing the highest levels of SAP are pooled, if necessary, concentrated by ultrafiltration (30 kDa molecular mass cutoff membrane), and stored at −70° until use. They are dialyzed into TC buffer immediately prior to isolation of SAP, in pure form, by calcium-dependent affinity chromatography using *O*-phosphorylethanolamine coupled to CH-Sepharose 4B.[48]

Binding by SAP to amyloid fibrils is conveniently tested using radioiodinated SAP and fibrils that are immobilized on microtiter plates or in suspension, as described in detail elsewhere.[49,50] SAP must show reversible calcium-dependent binding to the fibrils that is inhibited by EDTA. Regardless of the chemical nature of the fibril protein, SAP binds with the same affinity to all types of *ex vivo* fibrils and to fibrils formed *in vitro* once they have adopted the fibrillar conformation.[50]

An important technical point to remember when performing functional assays with amyloid fibrils and SAP is that the physical properties of isolated amyloid fibrils on their own make it impossible to quantitatively pipette or transfer them (from uncoated plastic tubes or plastic tubes treated using a low adhesion process) under normal laboratory conditions. There are always significant losses during such procedures. In contrast, in the presence of SAP, and probably other proteins or glycans, such losses are largely attenuated. Thus, if a fixed volume sample is withdrawn from a tube containing fibrils alone and one containing the same amount of fibrils coated with SAP, the sample containing the SAP appears to have more fibrils (visualized in SDS–PAGE and immunoblotting and quantified by densitometry; quantified in ^{125}I-labeled fibril degradation–SAP protection assays using *ex vivo* fibrils, and in cytotoxicity assays, measuring cellular MTT

[47] S. R. Nelson, G. A. Tennent, D. Sethi, P. E. Gower, F. N. Ballardie, S. Amatayakul-Chantler, and M. B. Pepys, *Clin. Chem. Acta* **200,** 191 (1991).

[48] P. N. Hawkins, G. A. Tennent, P. Woo, and M. B. Pepys, *Clin. Exp. Immunol.* **84,** 308 (1991).

[49] G. A. Tennent, L. B. Lovat, and M. B. Pepys, *Proc. Natl. Acad. Sci. U.S.A.* **92,** 4299 (1995).

[50] L. B. Lovat, E. Hohenester, P. Westermark, S. P. Wood, and M. B. Pepys, submitted for publication.

reduction, using synthetic Aβ amyloid fibrils). This happens even when the entire liquid content of the tubes is transferred. We therefore design and perform all of these types of assays so that no transfer of aliquots of fibrils or SAP-coated fibrils takes place. In fibril degradation–SAP protection assays using *ex vivo* fibrils,[49] digestions are stopped in the assay tubes, and once the fibrils are denatured by boiling in SDS there is no further problem when removing samples for analysis.

Polypeptide Analysis

The ultimate identification of the amyloid fibril protein requires direct chemical characterization of the fibril protein subunits by N-terminal amino acid sequencing using fast cycle chemistry, either from SDS–PAGE directly (the thin precast ExcelGels are ideal for this) or after electroelution onto polyvinylidene difluoride (PVDF) membranes or in chromatographic fractions.[16,20,21,33,34,51] Molecular mass determinations using matrix-assisted laser desorption time-of-flight (MALDI-TOF) or electrospray mass spectrometry (ESMS) are also extremely useful in analyzing the dominant amyloid fibril subunit species and fragments thereof.[16,51]

Storage of Fibrils

The initial water wash obtained using the standard protocol contains few fibrils and is usually discarded. The second and third water washes usually contain the highest concentration of fibrils and are pooled if necessary. Separate pools are made of the two subsequent water washes and all of the remaining washes. Optical density measurements, Congo red staining SDS–PAGE, and EIA are performed again on all final pooled fractions as soon as possible. The final stock solutions of fibrils are kept at 4°, preferably in glass tubes, and are examined immediately by amino acid sequencing, mass spectrometry, and electron microscopy, if required. Provided that they are stored in water with no added salt or preservatives, the fibril isolates remain translucent for a week or two (depending on their protein concentration), but with time become increasingly opalescent as the fibrils aggregate (monitored by their absorbance spectrum) and eventually precipitate. Freezing and thawing of fibrils also cause their aggregation and precipitation. The fibrils are stored finally in 0.1% NaN_3 or in an equal volume of ×2 strength TC buffer for use in functional assays involving SAP. Alternatively, the fibrils are lyophilized and, when dry, are weighed

[51] D. R. Booth, S. Y. Tan, S. E. Booth, G. A. Tennent, W. L. Hutchinson, J. J. Hsuan, N. F. Totty, O. Nguyen, A. K. Soutar, P. N. Hawkins, M. Bruguera, J. Caballeria, M. Solé, J. M. Campistol, and M. B. Pepys, *J. Clin. Invest.* **97,** 2714 (1996).

and stored in a vacuum dessicator (designated for fibrils only) at room temperature. Fibrils reconstituted after lyophilization are disrupted ultrasonically on ice, using a probe sonicator (three 30-sec bursts at 18 μm), to obtain homogeneous suspensions.

We routinely use fibril isolates whole, without subsequent purification of the fibril proteins. However, this is achieved by extraction of the fibrils with high concentrations of a variety of protein denaturants, in the presence of EDTA and dithiothreitol to prevent covalent modification, and subsequent fractionation by gel filtration chromatography.[16,17,19,20,22,24] Fibril proteins purified in this manner have been used successfully in studies of protein refolding.[52]

Summary

Two simple protocols are described for the isolation of amyloid fibrils that consist of water extraction from homogenates of unfixed, frozen human amyloidotic tissues. Most of the contaminating plasma and fibril-associated proteins are removed to yield relatively pure amyloid fibrils, suitable for biochemical characterization, functional assays, and biophysical studies of their structure using many of the specialized techniques described elsewhere throughout this volume.

Acknowledgments

I thank Professor M. B. Pepys for reading the manuscript and for helpful comments. Thanks also to Ms. J. Gilbertson for advice on the use of Sequenza Coverplates for immunohistochemistry. This work was supported by a Medical Research Council (UK) Programme Grant to M.B.P.

[52] D. R. Booth, M. Sunde, V. Bellotti, C. V. Robinson, W. L. Hutchinson, P. E. Fraser, P. N. Hawkins, C. M. Dobson, S. E. Radford, C. C. F. Blake, and M. B. Pepys, *Nature* **385,** 787 (1997).

[3] Isolating Inclusion Bodies from Bacteria

By GEORGE GEORGIOU and PASCAL VALAX

Introduction

In 1975 Prouty *et al.*[1] first described the formation of dense, amorphous intracellular granules in *Escherichia coli* cells grown in the presence of the amino acid analog canavanine. These granules, comprised primarily of polypeptide chains, could be solubilized by sodium dodecyl sulfate (SDS) and were not surrounded by any sort of membrane. For several years this observation was considered an aberrant, and rather irrelevant, cellular response induced by growth under nonphysiological conditions. It was not until much later that it became apparent that protein aggregation *in vivo* is a widespread phenomenon, manifested in cells overexpressing heterologous proteins or native proteins beyond a certain level, and in cells exposed to thermal or other kinds of physiological stress. In addition, mutations resulting in amino acid substitutions, deletions, or insertions can interfere with the folding of a polypeptide to the native state, causing the formation of protein aggregates.[2]

Intracellular protein aggregates form dense, electron-refracting particles that can be distinguished readily from other cell components by electron microscopy. For this reason, protein aggregates, at least those observed in microorganisms, are usually called inclusion bodies or, less often, refractile bodies. However, it should be noted that protein misfolding and self-association may occur even when inclusion bodies cannot be detected by electron microscopy or following careful cell fractionation (H. G. Gilbert, personal communication, 1999). Nonetheless, for practical purposes, it is safe to assume that the expression of proteins susceptible to aggregation at a level 2% or greater of the total cell protein will be accompanied by the appearance of readily identifiable inclusion bodies.

Inclusion bodies have been used extensively as a source of relatively pure, albeit misfolded, polypeptide chains that can be renatured to give the biologically active soluble protein. The refolding of proteins from inclusion bodies has proven of great value for analytical and preparative purposes. As of 1998, there have been over 300 reports of mammalian, plant, and microbial proteins obtained and renatured from inclusion bodies formed in *E. coli.* Tens of kilograms of biologically active bovine somatotropin are

[1] W. Prouty, M. J. Karnovsky, and A. L. Goldberg, *J. Biol. Chem.* **250,** 1112 (1975).
[2] A. L. Fink, *Folding Design* **3,** R9 (1998).

0076-6879/99 $30.00

produced every year from inclusion bodies and even larger amounts of human hemoglobin may eventually be produced by renaturation. Irrespective of the scale, the recovery of purified active proteins from inclusion bodies involves the following steps[3]: (1) isolation of inclusion bodies, (2) solubilization, (3) removal of protein impurities, and (4) refolding.

Often the sequence of steps 3 and 4 is interchanged so that the desired protein is purified after refolding. Obviously, the success of steps 2–4 depends on the quality of material from step 1. Also, preparing high-quality inclusion bodies is important for structural studies.[4,5]

Inclusion bodies formed in cells overexpressing a certain polypeptide are generally expected to be comprised predominantly of that polypeptide. However, this is often not the case. We have found[6] that the composition of inclusion bodies is a complex function of the mode of expression and the growth conditions. For example, purified inclusion bodies from *E. coli* cells expressing β-lactamase contain between 35 and 95% intact β-lactamase polypeptides. The rest are composed of a variety of intracellular proteins, some lipids, and a small amount of nucleic acids. Homogeneous inclusion bodies [95% (w/w) β-lactamase] were obtained by expressing the protein without its leader peptide, in which case aggregation occurred within the bacterial cytoplasm. This gave rise to large, highly regular inclusion bodies that could be separated readily from other particulate matter in cell lysates (Fig. 1). As might be expected, the efficiency of β-lactamase refolding was inversely proportional to the level of contaminants present in the inclusion body preparation.[7]

"High-quality" inclusion bodies consist primarily of the overexpressed recombinant protein with as little contaminating material as possible. The quality of inclusion bodies obtained from bacteria overexpressing a desired protein depends on two parameters: (1) the degree to which extraneous polypeptides, and possibly other macromolecules, are incorporated within the aggregate and (2) the ability to separate inclusion bodies from other cellular particles having a similar sedimentation coefficient and from material, mostly membrane vesicles, that becomes adsorbed onto the surface of the protein particles following cell lysis. It is usually difficult to ascertain whether extraneous proteins form an integral part of the aggregates or represent copurifying contaminants. There is increasing evidence, discussed in other articles of this volume, that protein aggregation involves the associ-

[3] R. Ruboph and H. Lilie, *FASEB J.* **50,** 49 (1996).

[4] K. Oberg, B. A. Chrunyk, R. B. Wetzel, and A. L. Fink, *Biochemistry* **33,** 2628 (1994).

[5] T. M. Przybycien, J. P. Dunn, P. Valax, and G. Georgiou, *Prot. Engin.* **7,** 131 (1994).

[6] P. Valax and G. Georgiou, *Biotech. Progr.* **9,** 539 (1993).

[7] P. Valax, "*In Vivo* and *in Vitro* Folding and Aggregation of *Escherichia coli* β-Lactamase." Ph.D Dissertation, Univ. of Texas at Austin, 1993.

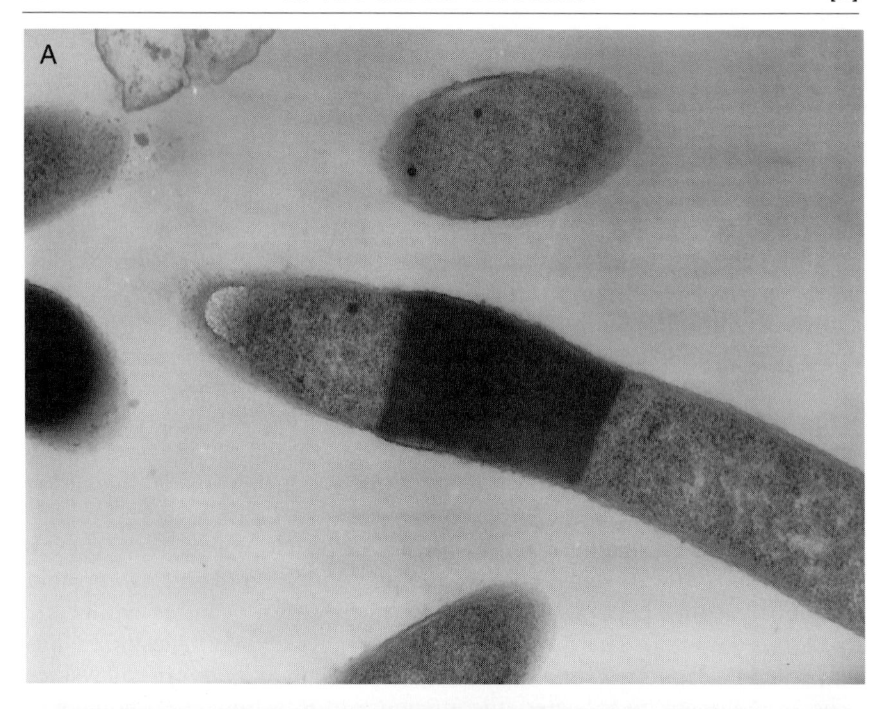

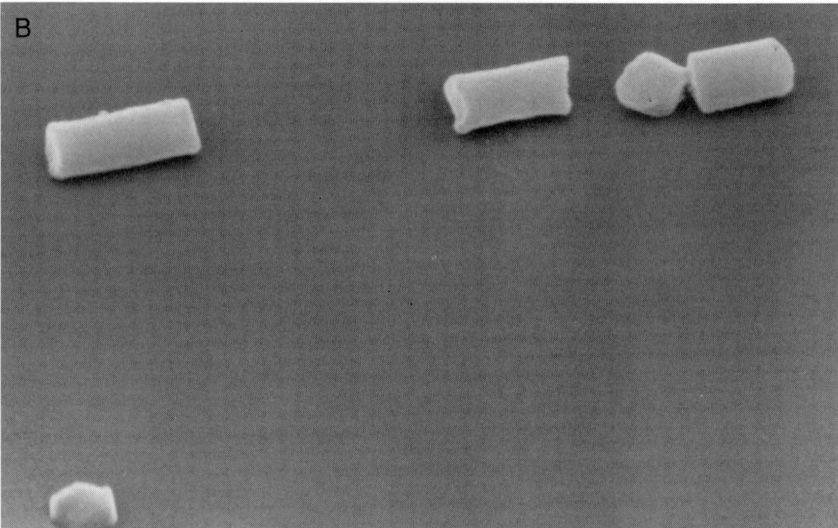

FIG. 1. (A) *E. coli* cells containing inclusion bodies and (B) inclusion bodies isolated by sucrose density centrifugation. Inclusion bodies were formed in the cytoplasm by expressing β-lactamase (a normally secreted protein) carrying a deletion of the first 20 amino acids (-20, -1) of the leader sequence.

ation of subdomains in partially folded intermediates.[2] The degree of specificity of the interactions that lead to protein aggregation varies from protein to protein. As a consequence, the extent of incorporation of cellular proteins and other extraneous macromolecules into inclusion bodies is protein dependent. Nonetheless, as was shown with β-lactamase,[6,8] it is often possible to modify the expression conditions to reduce the amount of extraneous material incorporated within the inclusion bodies.

Often the only step in the preparation of inclusion bodies is low-speed centrifugation of cell lysates. The effectiveness of this step depends to a large degree on the method of cell disruption. Low-speed centrifugation results in the sedimentation of membrane vesicles and cell wall fragments together with the inclusion bodies. In addition, other cellular components can become adsorbed nonspecifically onto the surface of inclusion bodies following cell lysis. The presence of contaminating substances causes a number of complications during subsequent solubilization and refolding steps: First of all, proteases may copurify with inclusion bodies during high-speed centrifugation.[9,10] Many proteases are active in the presence of high concentrations of denaturants used to solubilize the inclusion bodies and can rapidly cleave unfolded proteins under these conditions. For example, the presence of contaminating *E. coli* outer membrane protease OmpT was shown to cause significant reductions in the recovery of active porcine growth hormone[11] and creatine kinase[9] from inclusion bodies. Second, impurities found in inclusion bodies have to be removed eventually either prior to, or after, refolding. Third, protein as well as nonproteinaceous impurities interfere with refolding. In an interesting study, Maachupalli-Reddy and co-workers[12] examined the effects of typical inclusion body contaminants such as DNA, ribosomal RNA, lipids, and other proteins on the *in vitro* refolding of hen egg-white lysozyme. They found that the presence of other polypeptides prone to aggregation reduced the refolding yield significantly. The effect of RNA, DNA, and phospholipids at concentrations up to 30% of that of hen lysozyme did not have a significant effect on refolding yields. However, Darby and Creighton[13] found that nonproteinaceous contaminants had a dramatic effect on the refolding of bovine pancreatic trypsin inhibitor (BPTI) mutants from inclusion bodies.

[8] G. A. Bowden, A. M. Paredes, and G. Georgiou, *Bio/Technology* **9,** 725 (1991).

[9] P. C. Babbit, B. L. West, D. D. Buechter, I. D. Kuntz, and G. L. Kenyon, *Bio/Technology* **8,** 945 (1990).

[10] J.-M. Betton, N. Sasson, M. Hofnung, and M. Laurent, *J. Biol. Chem.* **273,** 8897 (1998).

[11] N. K. Puri, M. Cardamone, E. Crivelli, and J. C. Traeger, *Prot. Expr. Purif.* **4,** 164 (1993).

[12] J. Maachupalli-Reddy, B. D. Kelley, and E. De Bernardez-Clark, *Biotech. Prog.* **13,** 144 (1997).

[13] N. J. Darby and T. E. Creighton, *Nature* **344,** 715 (1990).

They suggested that even under the strongly denaturing conditions used to solubilize the inclusion bodies, BPTI is in a tight complex with a substance that affects its solubility and folding properties. This substance was found to be an acidic polymer, which unfortunately was not characterized further. Cardamone *et al.*[14] reported that the recovery yield of recombinant porcine growth hormone from inclusion bodies is lower than what can be achieved with the purified protein. They proposed that "morphopoietic factors" intrinsic to the inclusion bodies were responsible for preventing the protein from following the same folding pathway during renaturation, thus enhancing aggregation.

From these results it should be evident that the preparation of inclusion bodies suitable for protein recovery requires careful consideration of the expression conditions to minimize both the extent of nonspecific protein incorporation and the purification protocol. Because the formation of inclusion bodies is protein dependent, it is not possible to develop procedures that work for every case. Therefore, the remainder of this article is intended more as a set of recommendations that generally give good results in our experience.

Protein Expression

The bacterial strain, expression vector, and growth conditions have a pronounced effect on inclusion body formation. Expression from a T7 promoter[15] is accompanied by very high rates of protein synthesis, which enhance inclusion body formation. Genes placed downstream of a T7 promoter are transcribed by the T7 RNA polymerase. The gene for the latter is usually expressed from a relatively weak inducible promoter such as the lac promoter. Addition of the inducer IPTG turns on the synthesis of T7 RNA polymerase, which in turn transcribes the desired gene from the T7 promoter. For most applications, expression from a T7 promoter is the preferred way for inducing the formation of inclusion bodies. Ideally, the host strain should be a mutant defective in the gene responsible for the transport of the inducer into the cell. In such a host, induction with subsaturating concentrations of inducer can be employed to adjust the level of transcription per cell.[16] Another promoter that we have found to be favorable for inclusion body formation is the p_L of bacteriophage λ. With the p_L promoter, a very high rate of transcription is achieved by a temperature

[14] M. Cardamone, N. K. Puri, and M. R. Brandon, *Biochemistry* **34,** 5773 (1995).

[15] W. F. Studier, A. H. Rosenberg, J. J. Dunn, and J. W. Dubendorff, *Meth. Enzymol.* **85,** 61 (1990).

[16] D. A. Siegele and J. C. Hu, *Proc. Natl. Acad. Sci. U.S.A.* **94,** 8168 (1997).

upshift to 42°, although cold temperature induction can also be employed.[17] At 42° the combination of a higher temperature, which generally favors aggregation over folding, and the high rate of protein synthesis favor inclusion body formation. However, exposure of the bacteria to a supraoptimal temperature is also accompanied by the increased synthesis of heat shock proteases, which may cause degradation of the cloned product.

In *E. coli,* protein aggregation is enhanced by mutations in certain global regulatory pathways or in specific chaperone genes. The heat shock transcription factor σ^{32} is responsible for upregulating the expression of a number of *E. coli* chaperones under stressful conditions, including the DnaK-DnaJ-GrpE and GroEL-GroES systems. Mutations in the *rpoH* gene encoding σ^{32} have been shown to induce the complete aggregation of a normally soluble recombinant protein.[18,19] However, massive aggregation of host proteins also occurs in *rpoH* mutants and, as a result, the insoluble fraction contains, in addition to the desired polypeptide, a number of contaminating proteins. Mutations in the *dnaK, dnaJ, grpE, groEL,* or *groES* genes have also been found to favor the aggregation of recombinant proteins while having little effect on host protein solubility.[19] Hosts carrying mutations in *dnaK, dnaJ,* or *grpE* are particularly useful when it is desired to direct a protein into inclusion bodies. The DnaK-DnaJ-GrpE chaperone machinery binds newly synthesized polypeptides and mutations that interfere with its function are more likely to lead to rapid aggregation before proteolytic degradation can occur. Indeed, a number of unrelated overexpressed proteins have been found to aggregate extensively in a *grpE280* mutant, whereas the extensively studied *groES30* or *groEL140* alleles affected folding of only a limited number of proteins, often at the expense of reduced yields (J. Thomas, personal communication).[20,21]

Normally secreted heterologous proteins can be expressed either in the cytoplasm or with a prokaryotic leader peptide for targeting to the periplasmic space. Inclusion bodies can form in either cellular compartment. For secreted proteins containing two or more cysteine residues, the formation of protein aggregates in the periplasm may be increased by the coexpression of *E. coli* cysteine oxidoreductases (DsbC or DbA).[22] Periplasmic inclusion bodies are smaller and of irregular shape, most likely because

[17] S. C. Macrides, *Microbiol. Rev.* **60,** 512 (1996).

[18] A. I. Gragerov, E. S. Martin, M. A. Krupenko, M. V. Kashlev, and V. G. Nikiforov, *FEBS Lett.* **291**(2), 222 (1991).

[19] A. Gragerov, E. Nudler, N. Komissarova, G. A. Gaitanaris, M. E. Gottesman, and V. Nikiforov, *Proc. Natl. Acad. Sci. U.S.A.* **89,** 10341 (1992).

[20] J. G. Thomas and F. Baneyx, *Mol. Microbiol.* **21,** 1185 (1996).

[21] J. G. Thomas and F. Baneyx, *Prot. Express. Purif.* **11,** 289 (1997).

[22] J. C. Joly, W. S. Leung, and J. Swartz, *Proc. Natl. Acad. Sci. U.S.A.* **95,** 2773 (1998).

topological constraints limit the growth of the aggregate.[8] Inclusion bodies in the cytoplasm are larger and often highly regular (Fig. 1). For this reason, they can be separated more readily from cell debris and may contain lower amounts of extraneous polypeptides, phospholipids, and nucleic acids.[6] In some cases, high levels of expression of secreted proteins that are prone to aggregation are accompanied by the accumulation of the precursor form (the preprotein) with the leader peptide uncleaved.[6,23] The presence of the leader peptide renders preproteins highly insoluble and they are found either in the membrane fraction or are sequestered within cytoplasmic aggregates. These aggregates cofractionate with periplasmic inclusion bodies formed by the mature protein in the periplasm. The formation of a mixed aggregate population, consisting of preprotein and mature protein inclusion bodies, poses further complications during refolding and should be avoided.

The growth conditions can also be optimized to enhance inclusion body formation. As was mentioned earlier, growth at 42° results in decreased yields of soluble protein. Aeration, i.e., the concentration of dissolved oxygen, has been shown to affect protein aggregation.[24] The effect of dissolved oxygen on protein aggregation is complex and, in our experience at least, growth in a low dissolved oxygen environment can either enhance or reduce the formation of inclusion bodies, depending on the promoter and plasmid vector employed (unpublished data). The addition of ethanol to the growth media at concentrations around 3% (v/v) has been found to increase inclusion body formation with some proteins, such as human SPARC, but had the reverse effect with others.[21] Finally the culture pH, carbon source, and growth in minimal versus rich media also affect *in vivo* solubility in a protein-dependent manner.

Inclusion Body Isolation

Obtaining a homogeneous preparation of inclusion bodies requires a three-step process involving (1) cell lysis, (2) fractionation of the cell lysates to resolve the inclusion bodies from the cell debris by taking advantage of differences in size and density, and (3) removal of adsorbed contaminants.

Although incubation of intact *E. coli* with denaturants has been used for the *in situ* solubilization of aggregated proteins, such as IGF-I,[25] in general it is necessary to first lyse the bacteria. Laboratory methods for

[23] N, Sriubolmas, W. Panbangred, S. Sriurairatana, and V. Meevootisom, *Appl. Microbiol. Biotechnol.* **47,** 373 (1997).

[24] S. D. Betts, T. M. Hachigian, and E. Pichersky, *Plant. Mol. Biol.* **26,** 117 (1994).

[25] R. H. Hart, P. M. Lester, D. H. Reifsnyder, and J. R. Ogez, *Bio/Technology* **12,** 1113 (1994).

rupturing bacteria include repeated freeze–thaw cycles, lysozyme–EDTA treatment, sonication, and high-pressure homogenization using a French pressure cell. The method of lysis determines the size of cell debris that is generated. Because inclusion bodies are separated from the debris on the basis of size and density, it is an important parameter in recovery. High-pressure homogenization is the recommended method for obtaining a high degree of cell disruption (90–98% after one pass)[26] and smaller size debris. Multiple passes have been shown to further reduce the debris size down to a median diameter of around 0.3 μm.[26] For comparison, the size of inclusion bodies ranges between 0.3 and 1.2 μm. We have found that pretreatment with lysozyme–EDTA further improves inclusion body isolation.[6] In gram-negative bacteria, the outer membrane is linked covalently to the cell wall and therefore hydrolysis of the peptidoglycan by lysozyme may be necessary to facilitate the formation of smaller size outer membrane vesicles. As a rule, three passes through either a French press at 20,000 psi or equivalent conditions for industrial scale high-pressure homgenization equipment are adequate for good separation.[27,28]

Inclusion bodies are recovered from the cell lysate by centrifugation, usually at 15,000–30,000*g* in the laboratory (up to 20,000*g* for industrial scale). In addition to the aggregated protein, the pelleted fraction contains extraneous polypeptides and phospholipids. An appreciable amount of nucleic acids is also found.[29] The insoluble fraction obtained after centrifugation contains two prominent bands of molecular mass approximately 34 and 36 kDa (their migration may vary somewhat depending on the way the samples are treated prior to loading on the gel). These correspond to the major outer membrane proteins OmpA and OmpC/F, respectively, indicating that outer membrane fragments represent a major source of contaminating material. Outer membrane proteins are found in both cytoplasmic and periplasmic inclusion body preparations. Ribosomal proteins have also been found in inclusion bodies.[30]

Cosedimenting ribosomes and outer membrane vesicles can be separated from inclusion bodies by density gradient centrifugation.[8,31] Inclusion bodies have a density comparable to that of proteins (1.3–1.4 g/ml), whereas the density of outer membrane vesicles is 1.22 g/ml and that of ribosomes

[26] H. H. Wong, B. K. O'Neill, and A. P. J. Middleberg, *Biotech. Bioeng.* **55,** 556 (1997).

[27] E. A. Burks, and B. L.Iverson, *Biotech. Progr.* **11,** 112 (1995).

[28] M. E. Gustafson, K. D. Junger, B. A. Foy, J. A. Baez, B. F. Bishop, S. H. Rangwala, M. L. Michener, R. M. Leimgruber, K. A. Houseman, R. A. Mueller, B. K. Matthews, P. O. Olins, R. W. Grabner, and A. Hershman, *Prot. Expres. Purif.* **6,** 512 (1995).

[29] K. E. Langley, T. F. Berg, and T. W. Strickland, *Eur. J. Biochem.* **163,** 313 (1987).

[30] U. Rinas and J. Bailey, *Appl. Microbiol. Biotechnol.* **37,** 609 (1992).

[31] L. A. Classen, B. Ahn, H.-S. Koo, and L. Grossman, *J. Biol. Chem.* **266,** 11380 (1991).

around 1.5 g/ml. In a sucrose step gradient, inclusion bodies become focused into a relatively tight and visible band that can be distinguished clearly from membrane vesicles that are lighter and form a separate band. Good results have been obtained with a variety of inclusion body proteins using a 40–53–67% (w/w) sucrose step gradient, but slightly different conditions have also been used successfully.[30] The inclusion body band is found close to or at the interface of the 53 and 67% (w/w) sucrose layers and is collected with a Pasteur pipette. The inclusion bodies are then resuspended in buffer and precipitated by low-speed centrifugation several times to remove the sucrose. Analysis of the purified inclusion bodies by SDS–PAGE should show a clear reduction in the amount of outer membrane protein. The amount of lipid in the preparation is reduced by up to 8-fold, although in many cases a substantial amount of lipid still remains.[6] The nucleic acid content is also reduced significantly, typically by 10- to 20-fold over the amount in the inclusion body pellet prior to purification.

If the density of the inclusion bodies is low, the respective band is not well resolved from the cell membrane material. In that case, further purification can be obtained using a second sucrose density gradient centrifugation step.[6] Either an identical step gradient or a flotation gradient can be used at this stage.[32] For the latter, material is applied to the bottom of the tube and floats up a 60–40% (w/w) gradient until it reaches its buoyant density.

For large-scale protein recovery, cross-flow filtration has been used as an alternative to centrifugation. Cross-flow filtration is a separation technique in which the suspension is flowed parallel to the membrane in order to avoid the accumulation of material on the filter surface and to improve the filtration rate. Particles above a certain size cutoff are retained by a membrane and are collected continuously. With cross-flow filtration, inclusion body concentration and washing can be performed continuously. Protein recovery yields comparable to those obtained by centrifugation have been reported.[33]

After separation from cell debris, a series of wash/extraction steps is used to further remove contaminating material that had cosedimented or was adsorbed on the inclusion bodies.[7,34] Membrane material associated with the inclusion bodies is typically removed using mild detergents. An appropriate detergent should allow efficient extraction of impurities without solubilizing the inclusion bodies or irreversibly binding to the aggregated

[32] I. Poquet, M. G. Kormacher, and A. P. Pugsley, *Mol. Microbiol.* **9,** 1061 (1993).

[33] M. M. Meagher, R. T. Barlett, V. R. Rai, and F. R. Khan, *Biotech. Bioeng.* **43,** 969 (1994).

[34] F. A. O. Marston, *in* "DNA Cloning: A Practical Approach" (D. M. Glover, ed.), Vol. 3, p. 59. IRL Press, Oxford, 1987.

polypeptide chains. Price can also be a consideration as some detergents are very costly. Several detergents have been used in the literature, including Triton X-100, deoxycholate, Berol 185, and octylglucoside.[7,35] Nonspecifically adsorbed material can also be removed using low concentrations of chaotropic agents such as urea or guanidine hydrochloride. Because these agents are generally used for the solubilization of aggregated proteins, their concentration must be carefully optimized in order to efficiently remove contaminating material without significantly solubilizing the inclusion bodies. Alternatively, inclusion bodies may be treated with a combination of detergents and low concentrations of chaotropic agents. For example, human cathepsin B inclusion bodies were purified by extraction in 0.1% Triton X-100 once and 2 *M* urea twice,[36] whereas Belew *et al.*[35] used 0.1% Berol 185 and 0.5 *M* urea to remove impurities from recombinant human granulocyte–macrophage colony-stimulating factor.

The remainder of this article describes in detail a general procedure we have found useful for inclusion body isolation.

Cell Lysis

The conditions for *E. coli* growth and protein synthesis have to be optimized first, depending on the expression system. At the appropriate time after induction, the cells are harvested by centrifugation at 8000*g* for 10 min at 4° and washed once in buffer. The cell pellet is resuspended in 10 m*M* Tris–HCl, pH 7.5, containing 0.75 *M* sucrose and 0.2 mg/ml lysozyme. It is recommended to resuspend the cell pellet from a 50-ml culture of 1.0 OD_{600} into 1 ml of buffer. After a 10-min incubation at room temperature, a 3 m*M* EDTA solution is added at a 2 : 1 (v/v) ratio and transferred to ice for approximately 5 min. Subsequently, the cells are lysed by passing through a French press three times at 20,000 psi.

Inclusion Body Isolation

Following cell disruption, the lysate is centrifuged at 12,000*g* for 30 min at 4° and the pellet, which contains the inclusion bodies, is resuspended in 10 m*M* Tris–HCl buffer, pH 8.0, containing 0.25 *M* sucrose, 1 m*M* EDTA, and 0.1% sodium azide (1.25 ml of buffer per 50 ml of 1.0 OD_{600}). A tissue homogenizer may be used to resuspend the pellet thoroughly. The resuspended pellet is layered on the top of a sucrose step gradient [40, 53, and 67% (w/w)] in 1 m*M* Tris–HCl buffer, pH 8.0, containing 0.1% sodium azide and 1 m*M* EDTA. The sucrose gradient is prepared by carefully

[35] M. Belew, Y. Zhou, W. Wang, L.-E. Nyström, and J.-C. Janson, *Chromatogr. A* **679,** 67 (1994).
[36] R. Kuhelj, M. Dolinar, M. Pugercar, and V. Turk, *Eur. J. Biochem.* **229,** 533 (1995).

layering the sucrose solutions, with the more dense one at the bottom of the tube. Centrifugation is performed at 108,000*g* for 90 min at 4°.[8] The inclusion bodies become focused in a band at the interface between the 53 and 67% sucrose layers and are recovered and resuspended in either water or a suitable buffer, as required for further refolding steps. Several washes, followed by reprecipitation of the inclusion bodies by centrifugation at 12,000*g* for 30 min, are required to remove the sucrose. Alternatively, the sucrose can be removed by dialysis. At this point the aggregated protein should be examined by SDS–PAGE to evaluate the degree of purity. If the desired polypeptide constitutes less than 70% of the insoluble protein, then additional purification is recommended: The inclusion bodies are resuspended in 50 m*M* KH_2PO_4, pH 7.0, containing 50 m*M* octylglucoside and incubated for 15 min at room temperature. Following centrifugation at 12,000*g* for 20 min, the pellet is resuspended in 0.25 *M* sucrose solution and applied to a second sucrose gradient, as described previously.

[4] Isolation of Amyloid Deposits from Brain

By ALEX E. ROHER and YU-MIN KUO

Introduction

The profuse deposition of insoluble amyloid-β (Aβ) fibrils in the parenchymal and vascular extracellular spaces of the cerebral cortex and leptomeningeal vessels is one of the main histopathological lesions of Alzheimer's disease (AD). Fibrillar deposits of amyloid concentrate at the center of the senile plaques, usually surrounded by dystrophic neurites, or accumulate around cerebral blood vessels, leading to the death of vascular myocytes.[1] Nonfibrillar amyloid also deposits in the cerebral cortex in the form of diffuse plaques, which are apparently devoid of neuritic pathology.[2] The 40 to 42 amino acid Aβ peptides result from the proteolytic degradation of the transmembrane β-amyloid precursor protein.[3] Soluble monomeric and oligomeric forms of these peptides are normally present in the human

[1] H. M. Wisniewski, C. Bancher, M. Barcikowska, G. Y. Wen, and J. Currie, *Acta Neuropathol.* **78,** 337 (1989).

[2] H. Yamaguchi, Y. Nakazato, S. Hirai, M. Shoji, and Y. Harigaya, *Am. J. Pathol.* **135,** 593 (1989).

[3] D. J. Selkoe, *Annu. Rev. Neurosci.* **17,** 489 (1994).

0076-6879/99 $30.00

brain.[4] In AD, however, changes in the secondary structure of these peptides, probably induced by environmental factors or chaperone molecules, result in the polymerization of insoluble, protease-resistant, and cytotoxic amyloid fibrils that cannot be cleared from the brain.[5] The degradation of the amyloid of the brain is further hampered by the accumulation of irreversibly denatured Aβ dimers/tetramers and by the accumulation of posttranslational modifications,[5a] which increase within the fibrillar amyloid as a function of time.[6] In addition, the association of Aβ fibrils with glycoproteins and glycolipids increases their resistance to proteolytic degradation greatly. This article gives an account of techniques employed for the isolation and purification of amyloid from senile plaques, from cerebrovascular deposits, and from diffuse deposits found in the AD brain.

Isolation and Purification of Amyloid Cores from Senile Plaques

The enzymes collagenase CLS3 and DNase I were obtained from Worthington (Freehold, NJ), and sucrose and 98% formic acid were from Fluka (Ronkonkoma, NY). In order to remove impurities, all formic acid used in the following experiments was glass distilled in the laboratory. Enzyme inhibitors, Tris base, HCl, and thioflavin S were from Sigma (St. Louis, MO). Fast protein liquid chromatography (FPLC) chromatographic columns were from Pharmacia Biotechnology (Piscataway, NJ), and high-performance liquid chromatography (HPLC) columns were from Pharmacia-LKB (Uppsala, Sweden). Synthetic peptides Aβ1–40 and Aβ1–42 were from California Peptide Inc. (Napa, CA).

The cerebral cortex is dissected from the underlying white matter from the right hemispheres of patients who had been diagnosed histopathologically with AD. The following procedure applies to the cortex obtained from two hemispheres (approximately 400–450 g of tissue). However, it can be scaled up or down depending on the material available, considering that the final yield varies according to the extent of the pathology, which can differ greatly between AD patients.

The gray matter is minced finely with a razor blade and homogenized in a blender for 5 sec in 20 volumes of disaggregation buffer (DAB-S)

[4] Y.-M. Kuo, M. R. Emmerling, C. Vigo-Pelfrey, T. C. Kasunic, J. B. Kirkpatrick, M. J. Ball and A. E. Roher, *J. Biol. Chem.* **271,** 4077 (1996).

[5] A. E. Roher, D. Wolfe, M. Palutke, and D. KuKuruga, *Proc. Natl. Acad. Sci. U.S.A.* **83,** 2662 (1986).

[5a] J. D. Lowenstein, S. Clarke, and A. E. Roher, *Methods Enzymol.* **309** [7] (1999) (this volume).

[6] Y.-M. Kuo, S. Webster, M. R. Emmerling, N. De Lima, and A. E. Roher, *Biochim. Biophys. Acta* **1406,** 291 (1998).

made of 10 m*M* Tris–HCl, pH 7.4, 0.25 *M* sucrose, 3 m*M* EDTA, and protease inhibitors, including 200 μg/ml phenylmethylsulfonyl fluoride (PMSF), 0.5 μg/ml leupeptin, and 0.7 μg/ml pepstatin, as well as antibiotics gentamicin sulfate (50 μg/ml) and amphotericin B (0.25 μg/ml). The homogenate is stirred gently overnight at 4° and then filtered sequentially through a series of stainless-steel meshes (700, 350, 150, 75, and 45 μm) to disperse the tissue and remove blood vessels. Solid sucrose is added to the filtrate to a final concentration of 1.2 *M*, and this solution is then centrifuged at 25,000*g* in a Beckman type 19 rotor for 30 min at 4°. The top insoluble material containing myelin and cellular debris and intermediate supernatant is discarded. The pellet is resuspended in 12 volumes of DAB-S, filtered through a 45-μm mesh, adjusted to 1.9 *M* sucrose, and centrifuged at 125,000*g* in a Beckman SW28 rotor for 30 min at 4°. The solid material collected at the top of the tubes is removed with a spatula (the underlying supernatant and the pellet, which mainly contains free nuclei are discarded), dispersed in 200 ml of 50 m*M* Tris–HCl, pH 8.0 (Tris buffer), and centrifuged at 5000*g* for 15 min at 4°. This latter procedure is repeated twice to remove the sucrose and the components of DAB-S buffer. The recovered pellet is finally resuspended in 200 ml of Tris buffer containing 2 m*M* $CaCl_2$. Most of the contaminating collagen and DNA are hydrolyzed by the addition of 50 mg of collagenase CLS3 and 2 mg of DNase I, and the digestion is allowed to proceed for 14 hr at 37° with continuous shaking. The suspension is then centrifuged at 5000*g* for 15 min and the pellet is resuspended in Tris buffer. This step is repeated twice to remove the enzymes and $CaCl_2$. To the final suspension, solid sodium dodecyl sulfate (SDS) is added to make a final concentration of 5%. Once the SDS is dissolved, the sample is allowed to stand at room temperature for 2 hr and then centrifuged at 15,000*g* for 45 min at 20°. The supernatant is discarded and the pellet is submitted to sucrose gradient separation.

A discontinuous sucrose density gradient is prepared in 12-ml polycarbonate tubes by overlying 2-ml aliquots of 2.2, 2.0, 1.7, 1.4, and 1.3 *M* sucrose prepared in 50 m*M* Tris–HCl, pH 8.0, 1% SDS. The amyloid core-enriched material, resulting from the 5% SDS lysis, is resuspended in 12 ml of 1.1 *M* sucrose prepared in the same buffer, and aliquots of 2 ml are loaded at the top of each of the sucrose gradient tubes. The sucrose gradient tubes are centrifuged at 200,000*g* in a Beckman 41 Ti rotor for 1 hr at 20°. The amyloid cores, highly enriched at the 1.7/2.0 and 1.4/1.7 sucrose interphases, are collected carefully by aspiration using a peristaltic pump. The collected material is then diluted with 3 volumes of 50 m*M* Tris–HCl, pH 8.0, containing 0.1% SDS and centrifuged at 5000*g* for 15 min at 20°. To eliminate the sucrose the pellet is resuspended in the same Tris buffer and centrifuged again at 5000*g* for 15 min at 20°. This step is repeated two more times. The final pellet is dissolved in 8 ml of 80% formic acid. After standing at room temperature for 15 min to ensure that the pellet is dissolved completely, the suspension aliquots of 1 ml are centrifuged in 1.5-ml-thick wall polyallomer tubes at 435,000*g* for 20 min in a Beckman 100 TLA rotor at 4°. The pellets, consisting mainly of insoluble lipofuscin granules, are discarded and the supernatants containing the solubilized amyloid peptides are concentrated using vacuum centrifugation to a final volume of 4 ml and stored in aliquots of 500 μl at $-85°$.

To purify the Aβ peptides from glycoproteins and glycolipids associated with the amyloid filaments, we utilized a (FPLC) size-exclusion Superose 12 column (300 $\times$ 10 mm) equilibrated in 80% formic acid.[7] Aliquots of 500 μl of the formic acid-solubilized samples are loaded onto the column and the chromatography developed in 80% formic acid at a flow rate of 0.2 ml/min at room temperature and monitored at 280 nm. A representative chromatographic profile is shown in Fig. 1. The collected fractions, containing the different amyloid components, are concentrated by vacuum centrifugation and dialyzed (Spectrapor No. 6 membrane, molecular mass 1000 Da cutoff) against 4 liter of distilled water for 2 hr and then dialyzed against two changes of 4 liter each of 100 m*M* NH_4HCO_3 for a total of 16 hr. At this point the amyloid fractions can be stored at $-85°$, lyophilized, or digested directly by proteolytic enzymes.[5a]

Isolation and Purification of Amyloid from Cerebrovascular Deposits

Deposits of amyloid in the vasculature of the AD brain can be purified either from the cortical vessels or from the leptomeningeal vessels.

Isolation of Amyloid from Cortical Vessels

The initial amount of cerebral cortex to be invested in these experiments will depend on the amount of amyloid deposited in the vascular walls. This can be assessed from histological sections stained by thioflavin S. To achieve a high Aβ yield, the brains from AD patients carrying either apolipoprotein E genotype $\epsilon3/\epsilon4$ or $\epsilon4/\epsilon4$ will be the most desirable. The cerebral gray matter, from three 1-cm-thick coronal sections, is dissected carefully from the underlying white matter and then sectioned to yield cubes of approximately 7–9 mm^3. The gray matter cubes are stirred vigorously in 1500 ml of 50 m*M* Tris–HCl, pH 8.0, 5 m*M* EDTA containing 5% SDS, and 0.03%

[7] A. E. Roher, J. D. Lowenson, S. Clarke, C. Wolkow, R. Wang, R. J. Cotter, I. M. Reardon, H. A. Zurcher-Neely, R. L. Heinrikson, M. J. Ball, and B. D. Greenberg, *J. Biol. Chem.* **268,** 3072 (1993).

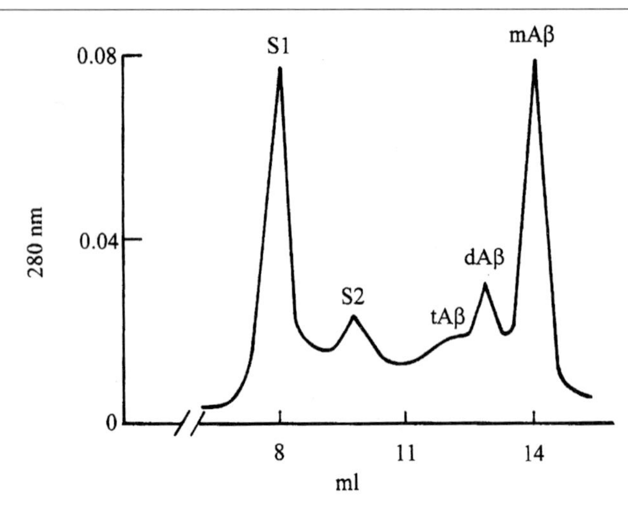

FIG. 1. FPLC Superose 12 elution pattern of amyloid peptides derived from AD senile plaques. Fractions S1 and S2 contain glycoproteins and glycolipids, representing between 20 and 25% of the total mass present in the cores of amyloid. The remaining 75–80% corresponds to Aβ peptides, of which irreversibly denatured tetrameric (tAβ) and dimeric (dAβ) Aβ peptides account for approximately 15 and 30%, respectively, of the total Aβ peptides. Dimeric and tetrameric Aβ peptides cannot be dissociated into monomeric forms.[6,10] Monomeric Aβ (mAβ), representing approximately 55% of the total Aβ, elutes last with a retention time corresponding to 4.5 kDa. Once dissociated from its ancillary molecules, the AD-purified mAβ can be degraded by proteolytic enzymes easily. The cores of amyloid purified by the present procedure contained approximately 85–90% Aβ terminating in residue 42, with the remaining 10–15% terminating at residue 40. A variable large proportion of the Aβ peptides in the senile plaques are N-terminal degraded and modified posttranslationally.[5a]

NaN_3 at room temperature. To achieve the complete removal of all SDS-soluble cellular material, the SDS solution should be changed six times during the course of 72 hr, using a nylon mesh (300 μm) to retain the vascular tufts. After this period, the only remaining SDS-insoluble components are the vascular extracellular matrix of blood vessels and the cores or sheets of amyloid attached firmly to the basal lamina. The vascular tufts are then rinsed six times with 2 liter each of 100 m*M* Tris–HCl, pH 8.0, until all the SDS in solution is totally eliminated. Collagenase CLS3 (50 mg) is added to the blood vessels suspended in 150 ml of Tris–HCl, pH 8.0, 1 m*M* $CaCl_2$ and the proteolysis is allowed to proceed for 16 hr at 37° with vigorous shaking. The vessels are filtered through a 150-μm mesh and the amyloid is recovered from the filtrate by centrifugation at 5000*g* for 15 min at 20°. The resulting pellet is dissolved in 12 ml of 80% formic acid with the aid of a Dounce homogenizer. After standing at room temperature for 15 min to dissolve the pellet, the specimen is centrifuged in 1-ml aliquots in 1.5-

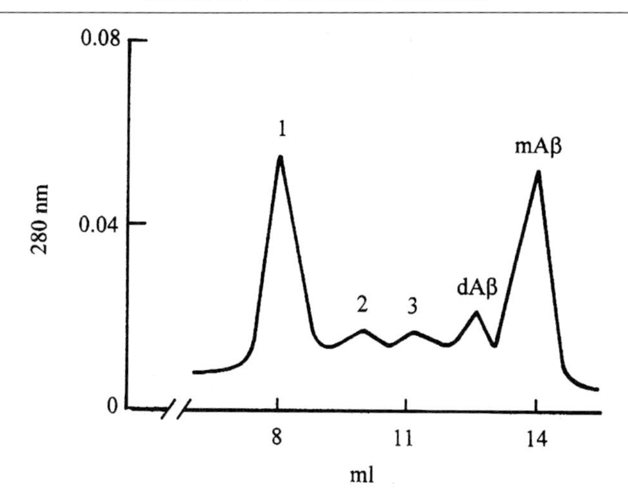

FIG. 2. FPLC Superose 12 elution profile of the AD amyloid proteins isolated from gray matter blood vessels. Most of this amyloid is localized in the walls of arteries rather than in veins.[11] Peaks 1, 2, and 3 are composed of water-insoluble glycoproteins and glycolipids. Approximately 70% of the Aβ is eluted as a monomer (mAβ) whereas the dimeric (dAβ) represents about 25%. There is up to twice as much Aβ terminating at residue 40 in the cortical vascular amyloid than in the neuritic plaque core amyloid.

ml-thick wall polyallomer tubes at 435,000*g* in a Beckman 100 TLA rotor for 20 min at 4°. The supernatant is concentrated by vacuum centrifugation and submitted to FPLC.

The FPLC separation of the Aβ peptides and ancillary components present in the amyloid of the cortical vessels is carried out under the same conditions as those described for the senile plaque amyloid. The resulting chromatographic profile of the vascular amyloid is shown in Fig. 2.

Isolation of Amyloid from Leptomeningeal Vessels

As in the case of cortical vascular Aβ, selection of tissue that is loaded with amyloid will provide a good yield of this material. The leptomeninges are teased carefully from the surface of four cerebral hemispheres and washed thoroughly with 10 changes of 100 m*M* Tris–HCl, pH 8.0 (Tris buffer), at 4° to remove entrapped blood. Vessels larger than 1 mm in diameter are dissected and discarded as they do not contain amyloid deposits.[8] The remaining tissue is finely cut down to 1- to 2-mm pieces, collected

[8] A. E. Roher, J. D. Lowenson, S. Clarke, A. S. Woods, R. J. Cotter, E. Gowing, and M. J. Ball, *Proc. Natl. Acad. Sci. U.S.A.* **90,** 10836 (1993).

by filtration through a 45-μm stainless-steel mesh, resuspended in 20 volumes of Tris buffer, 1 m*M* $CaCl_2$, containing 0.3 mg/ml of collagenase CLS3 and 10 μg/ml of DNase I, and incubated for 16 hr in a shaker bath at 37°. Insoluble large debris are removed by filtration using a 300-μm mesh, and the filtrate is centrifuged at 5000*g* for 15 min at 20°. The resulting pellet is resuspended with the aid of a Dounce homogenizer in 75 ml of Tris buffer followed by the addition of 75 ml of 6% SDS prepared in the same buffer. The specimen is maintained at room temperature for 2 hr, and the insoluble amyloid is separated from the solubilized material by centrifugation at 8000*g* for 30 min at 20°. To eliminate the excess amount of SDS, the recovered pellet is resuspended in 150 ml of Tris buffer and centrifuged as described earlier. The resulting pellet is dissolved in 10 ml of 80% formic acid, and the acid-soluble amyloid peptides are recovered by centrifugation by following the procedure described earlier for amyloid from the cortical vessels.

Fractionation of the different amyloid components is carried out by FPLC as outlined for the senile plaque amyloid. The elution pattern of the separated leptomeningeal amyloid components is given in Fig. 3.

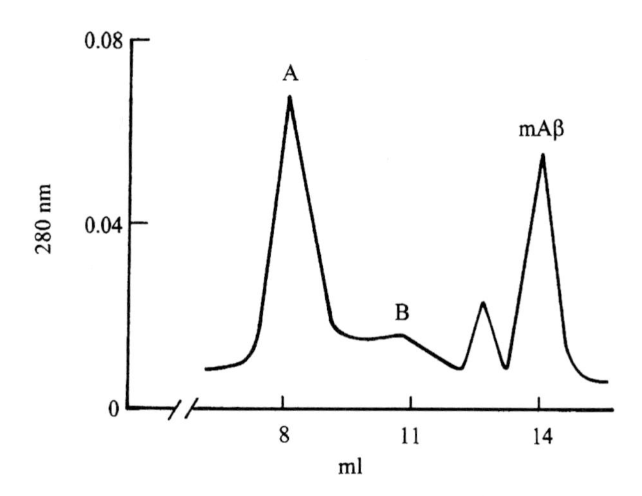

FIG. 3. FPLC Superose 12 elution profile of amyloid peptides from leptomeningeal vascular deposits. Peaks A and B represent insoluble glycoproteins and glycolipids. The most important difference between the gray matter vascular and the leptomeningeal vascular amyloid is that the amount of Aβ terminating at residue 40 is more abundant in the latter (65% average) than in the former. However, the ratio of vascular Aβ40 to Aβ42 varies widely from individual to individual depending on the vascular amyloid load, which reaches its maximum deposition in those individuals carrying the apolipoprotein E genotype $\varepsilon4/\varepsilon4$. The vascular A$\beta$ is also less posttranslationally modified and less N-terminally degraded.[5a] Monomeric Aβ (mAβ); dimeric Aβ (dAβ).

Isolation and Purification of Diffuse Amyloid from the Cerebral Cortex

The cerebral cortex, approximately 40 g, is minced finely and homogenized in a blender for 5 sec with 10 volumes of 100 m*M* Tris–HCl, pH 8.0, 3 m*M* EDTA, 200 μg/ml PMSF, 0.5 μg/ml leupeptin, 0.7 μg/ml pepstatin, 50 μg/ml gentamicin sulfate, and 0.25 μg/ml amphotericin B. An equal volume of 20% SDS solution, prepared in the same buffer, is added to give a final concentration of 10% SDS. After continuous stirring at room temperature for a period of 14 hr, the suspension is filtered through a 37-μm nylon mesh to eliminate most of the insoluble large blood vessels. The filtrate is centrifuged at 135,000 *g* at 20° for 2 hr in a Beckman SW28 rotor. The pellet is resuspended in 100 m*M* Tris–HCl, pH 8.0, and filtered through a 20-μm nylon mesh, and the filtrate is centrifuged at 1000*g* for 15 min at 20° to recover a pellet rich in compact cores of amyloid. The supernatant is centrifuged at 135,000*g* for 1 hr at 20° and the resulting pellet (P-135) is dissolved in 6 ml of 80% formic acid. To separate insoluble lipofuscin from the soluble amyloid components, 1.5 ml chloroform is added to the formic acid, mixed thoroughly, and centrifuged at 2000*g* for 5 min at 20°. The

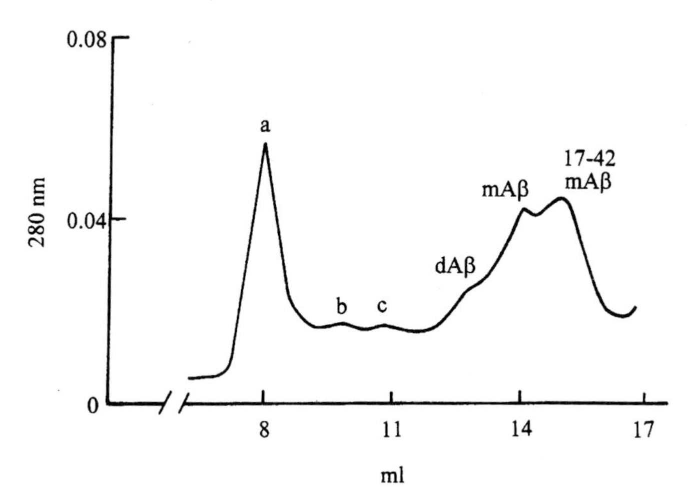

FIG. 4. FPLC Superose 12 elution pattern of peptides from diffuse amyloid deposits. Fractions a, b, and c contain a mixture of water-insoluble material. In addition to monomeric (mAβ) and dimeric Aβ (dAβ), an unresolved lower molecular weight fraction was eluted corresponding to a 3-kDa peptide. This component can be purified by HPLC on a Spherogel TSK 3000 SW size-exclusion column (60 × 0.75 cm) in 80% formic acid at a flow rate of 0.5 ml/min. Peptide mapping, amino acid sequencing, and mass spectrometry identified this peptide as Aβ residues 17–42.

upper phase containing the solubilized Aβ is divided into 1-ml aliquots in 12 × 75-mm glass tubes and the volume is reduced by vacuum centrifugation. This latter step prevents the sudden loss of the specimens due to the presence of small quantities of volatile chloroform in the acid phase during the vacuum centrifugation. Five hundred microliters of the reduced specimen is chromatographed on a Superose 12 column and equilibrated with 80% formic acid at a flow rate of 0.2 ml/min. The resulting elution profile is shown in Fig. 4.

The supernatant obtained from the first 135,000*g* centrifugation is further spun at 275,000*g* in a Beckman SW41 Ti rotor for 2 hr at 20°. The pellet (P-275) is resuspended in 100 m*M* Tris, pH 10.4, with the help of a Dounce homogenizer and centrifuged again at 275,000*g* for 2 hr at 20°, and the final pellet is resuspended in 80% formic acid. To eliminate contaminating lipids from the soluble Aβ peptides, the specimen is centrifuged first at 12,000*g* for 10 min at 4°, and the top layer of lipid is discarded. To remove insoluble lipofuscin, the specimen is further centrifuged at 435,000*g* in a Beckman 100 TLA rotor at 4° for 15 min. After concentration of the clear supernatant by vacuum centrifugation, the specimen is fractionated on a Superose 12 column under the conditions described for the amyloid

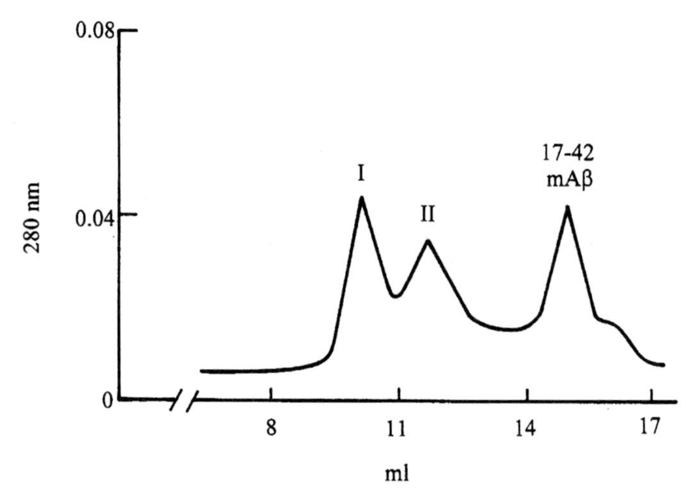

FIG. 5. FPLC Superose 12 fractionation of the P-275 pellet obtained from the cerebral cortex diffuse amyloid deposits. This pellet, which stains intensely for thioflavin S, yielded three components with molecular massers of 85 (I), 30 (II), and 3 kDa. The former two were immunochemically negative for Aβ and, on peptide mapping and amino acid analysis of tryptic peptides, these molecules demonstrated no relation to Aβ or its precursor protein. Analysis of the 3-kDa fraction by amino acid sequencing, peptide mapping, and mass spectrometry revealed the Aβ sequence of residues 17–42, free of contaminating peptides.

cores. The results of this chromatography are shown in Fig. 5. Amino acid analyses, amino acid sequencing, and mass spectrometry demonstrated that the Aβ peptides present in the diffuse amyloid deposits are mainly Aβ 1–42 and Aβ 17–42.[9]

[9] E. Gowing, A. E. Roher, A. S. Woods, R. J. Cotter, M. Chaney, S. P. Little, and M. J. Ball, *J. Biol. Chem.* **269,** 10987 (1994).

[10] A. E. Roher, M. O. Chaney, Y.-M. Kuo, S. D. Webster, W. B. Stine, L. J. Haverkamp, A. S. Woods, R. J. Cotter, J. M. Tuohy, G. A. Krafft, B. S. Bonelli, and M. R. Emmerling, *J. Biol. Chem.* **271,** 20631 (1996).

[11] R. O. Weller, A. Massey, T. A. Newman, M. Hutchings, Y. M. Kuo, and A. E. Roher, *Am. J. Pathol.* **153,** 725 (1998).

[5] Microextraction and Purification Techniques Applicable to Chemical Characterization of Amyloid Proteins in Minute Amounts of Tissue

By BATIA KAPLAN, RUDI HRNCIC, CHARLES L. MURPHY, GLORIA GALLO, DEBORAH T. WEISS, and ALAN SOLOMON

Introduction

Although amyloidosis has been recognized as a clinical entity for over 150 years, only recently has insight been gained into the heterogeneous nature of amyloid diseases. Knowledge in this area was advanced considerably when Pras *et al.*[1] reported in 1968 a water extraction technique to isolate amyloid fibrils from tissue and Glenner *et al.*[2] developed a method to purify the fibrillar components by gel permeation chromatography. To date, at least 17 different amyloid proteins have been described, many associated with a distinctive clinical disorder.[3] The proteins range in molecular mass from 3 to 30 kDa and are presumably derived from normal precursors that include monoclonal immunoglobulin light chains occurring in plasma cell dyscrasias (AL amyloidosis); serum amyloid A (SAA) in familial Mediterranean fever (FMF) and reactive amyloidosis (AA amyloidosis); transthyretin (TTR) in familial polyneuropathy, cardiomyopathy, or systemic senile amyloidosis; lysozyme in hereditary nonneuropathic systemic

[1] M. Pras, M. Schubert, D. Zucker-Franklin, A. Rimon, and E. C. Franklin, *J. Clin. Invest.* **47,** 924 (1968).

[2] G. G. Glenner, W. Terry, M. Harada, C. Isersky, and D. Page, *Science* **172,** 1150 (1971).

[3] J. Ghiso, R. Vidal, G. Gallo, and B. Frangione, *Rev. Bras. Rheumatol.* **35,** 93 (1995).

0076-6879/99 $30.00

amyloidosis (Ostertag type); and amyloid β precursor protein (AβPP) in Alzheimer's disease. Certain forms of amyloid are associated with point mutations or posttranslational modifications of the precursor protein, whereas others have no evident primary structural differences.[4] Further, the mechanisms of amyloid fibrillogenesis and deposition are not well understood, but seemingly involve the formation of unstable, partially unfolded intermediates.[5]

Because there are many types of amyloid-associated disease, elucidation of the chemical nature of the deposited pathologic components has obvious diagnostic, therapeutic, and prognostic importance. Although all amyloid proteins exhibit, after Congo red staining, a characteristic green birefringence, as seen with polarizing microscopy, there are no histologic or morphologic features that distinguish among the different chemical forms of this material. Historically, resistance to potassium permanganate treatment was used to differentiate AL from AA amyloid,[6] but this technique has only limited value. Through the development of antisera that recognize certain amyloids, e.g., AL, AA, ATTR, and Aβ_2M, immunohistochemical identification of the protein nature of the deposit has become possible. However, in some instances, the results of such analyses are inconclusive due either to the weak reactivity of the antibodies with amyloid or to their nonspecific binding with contaminating serum proteins, e.g., normal immunoglobulins. Thus, unequivocal identification of the type of amyloid present requires isolation of the extracted fibrillar components and determination of their primary structure.

Purification of amyloid fibrils by the classical methods used in studies of systemic amyloidosis[2] and Alzheimer's disease[7,8] generally requires gram amounts of starting tissue and is dependent on the availability for analysis of large surgical biopsies or autopsy-derived specimens. Unfortunately, these methods are not suitable for diagnostic, fine needle-derived samples where only minuscule amounts of tissue are typically obtained. Thus, the development of biochemical techniques applicable to the characterization of the amyloid protein contained in such small specimens has clinical

[4] G. G. Glenner, *in* "Amyloid and Amyloidosis 1993" (R. Kisilevski, M. D. Benson, B. Frangione, J. Gauldie, T. J. Muckle, and I. D. Young, eds.), p. 2. Parthenon, New York, 1993.

[5] M. R. Hurle, L. R. Helms, L. Li, W. Chan, and R. Wetzel, *Proc. Natl. Acad. Sci. U.S.A.* **91,** 5446 (1994).

[6] J. R. Wright, E. Calkins, and R. L. Humphrey, *Lab. Invest.* **36,** 274 (1977).

[7] G. G. Glenner and C. W. Wong, *Biochem. Biophys. Res. Commun.* **120,** 885 (1984).

[8] C. L. Masters, G. Simms, N. A. Weinman, G. Muthaup, B. L. McDonald, and K. Beyreuther, *Proc. Natl. Acad. Sci. U.S.A.* **82,** 4245 (1985).

value.[9] Further, such methodology would be useful in studies of the pathogenesis of amyloid deposition induced in small animal models.[10]

The identification of proteins of limited availability has been facilitated through technical advances in small-scale protein purification, including high-performance liquid chromatography (HPLC) and capillary zone electrophoresis. Sodium dodecyl sulfate–polyacrylamide gel electrophoresis (SDS–PAGE) has become increasingly important in amyloid research, especially since the introduction of polyvinylidene difluoride (PVDF) membranes as support media for electroblotting and microsequencing techniques.[11] Through this methodology, as well as mass spectroscopy, it is now possible to analyze picomole amounts of purified protein. This article describes the microtechniques used to extract, purify, and sequence amyloid proteins present in limited amounts of fresh or formalin-fixed tissue obtained from patients with systemic forms of amyloidosis.

Microextraction and Microsequencing Strategies

General

Several different methods have been reported regarding the isolation of amyloid proteins from relatively small amounts of tissue obtained by surgical biopsy or at autopsy. These include the use of guanidine hydrochloride,[12–14] acetonitrile–trifluoroacetic acid (TFA),[9,10,15–21] or formic

[9] B. Kaplan, S. Yakar, A. Kumar, M. Pras, and G. Gallo, *Amyloid Int. J. Exp. Clin. Invest.* **4,** 80 (1997).

[10] S. Yakar, B. Kaplan, A. Livneh, B. Martin, K. Miura, Z. Ali-Khan, S. Shtrasburg, and M. Pras, *Scand. J. Immunol.* **40,** 653 (1994).

[11] P. Matsudaira, *J. Biol. Chem.* **261,** 10035 (1987).

[12] P. Westermark, L. Benson, J. Juul, and K. Sletten, *J. Clin. Pathol.* **42,** 817 (1989).

[13] A. H. Forsberg, K. Sletten, L. Benson, J. Juul, and P. Westermark, *in* "Amyloid and Amyloidosis 1990" (J. B. Natvig, O. Forre, G. Husby, A. Husebekk, B. Skogen, K. Sletten, and P. Westermark, eds.), p. 797. Kluver Academic, Dordrecht/Norwell, MA, 1990.

[14] K. E. Olsen, K. Sletten, and P. Westermark, *Biochem. Biophys. Res. Commun.* 245, 713 (1998).

[15] B. Kaplan, R. Vidal, A. Kumar, J. Ghiso, B. Frangione, and G. Gallo, *Clin. Exp. Immunol.* **110,** 472 (1997).

[16] B. Kaplan, G. German, and M. Pras, *J. Liq. Chromatogr.* **16,** 2249 (1993)

[17] B. Kaplan, G. German, M. Ravid, and M. Pras, *Clin. Chim. Acta* **229,** 171 (1994)

[18] S. Yakar, B. Kaplan, G. German, A. Livneh, K. Miura, S. Shtrasburg, and M. Pras, *Amyloid Int. J. Exp. Clin. Invest.* **2,** 167 (1995).

[19] B. Kaplan, B. Martin, S. Yakar, M. Pras, T. Wisniewski, J. Ghiso, B. Frangione, and G. Gallo, *in* "Progress in Alzheimer's and Parkinson's Diseases" (A. Fisher, M. Yoshida, and I. Hanin, eds.), p. 823. Plenum Press, New York, 1998.

acid[22] for the extraction of fresh (nonfixed) specimens, or for formalin-fixed tissues, formic acid[22,23] or buffers containing 3% SDS[24–27] or 2.5% SDS/8 *M* urea.[28,29] Purification of amyloid protein from such extracts has been achieved through gel permeation chromatography[12–14] and/or HPLC.[12–14,22,30,31] Further, the purified material can be characterized through direct automated Edman degradation where N-terminal sequence data are obtained.[12–14] In those cases where a "blocked" N terminus is found, enzymatic or chemical cleavage of the protein results in peptides that can be analyzed successfully.[14,28] Sequence data also can be derived from an "SDS–PAGE-based microsequencing technique." Using this method, amyloid proteins are separated by SDS–PAGE[32–35] (17[10,15] or 10–20% polyacrylamide gradient[14,28,29] Tris–glycine gels, as well as 12.5% Tris–tricine gels[22]) followed by electrotransfer to PVDF membranes. The portion of the membrane containing the blotted protein of interest is excised, placed directly in the sequencer, and analyzed.

In general, the selection of the appropriate microextraction and purification technique depends, to a large extent, on the type of sample to be studied (fresh or fixed), as well as the amount of congophilic material

---

[20] B. Kaplan, V. Haroutunian, A. Koudinov, Y. Patael, M. Pras, and G. Gallo, *Neurobiol. Aging* (Suppl. 4S) (1998). [Abstract 993]

[21] B. Kaplan, V. Haroutunian, A. Koudinov, Y. Patael, M. Pras, and G. Gallo, *Clin. Chim. Acta* **280,** 147 (1999).

[22] E. M. Castano, F. Prelli, L. Morelli, A. Avagnina, A. Kahn, and B. Frangione, *Amyloid Int. Exp. Clin. Invest.* **4,** 253 (1997).

[23] F. Coria, F. Prelli, E. M. Castano, M. Larrondo-Lillo, J. Fernandez-Gonzalez, S. G. van Duinen, G. T. A. M. Bots, W. Luyendijk, M. L. Shelanski, and B. Frangione, *Brain Res.* **463,** 187 (1988).

[24] S. Shtrasburg, M. Pras, P. Langevitch, and R. Gal, *Am. J. Pathol.* **106,** 141 (1982).

[25] M. Picken, K. Pelton, B. Frangione, and G. Gallo, *Am. J. Pathol.* **129,** 536 (1987).

[26] R. Gal, S. Shtrasburg, M. Luria, B. Lifschitz Mercer, S. Viskin, S. Yakar, and M. Pras, *Amyloid Int. J. Exp. Clin. Invest.* **2,** 119 (1995).

[27] S. Shtrasburg, A. Livneh, M. Pras, T. Weinstein, and R. Gal, *Anat. Pathol.* **108,** 289 (1997).

[28] R. Layfield, K. Baily, J. Lowe, R. Allibone, R. J. Mayer, and M. Landon, *J. Pathol.* **180,** 455 (1996).

[29] R. Layfield, K. Baily, R. Dineen, P. Mehrotra, J. Lowe, R. Allibone, R. J. Mayer, and M. Landon, *Anal. Biochem.* **253,** 142 (1997).

[30] B. Kaplan and M. Pras, *J. Liq. Chromatogr.* **15,** 2486 (1992).

[31] B. Kaplan, *Anal. Chim. Acta* **372,** 161 (1998).

[32] U. K. Laemmli, *Nature* **227,** 680 (1970).

[33] H. Schagger and G. von Jagow, *Anal. Biochem.* **166,** 368 (1987).

[34] K. Ikeda, T. Monden, T. Kanoh, M. Tsujie, H. Izawa, A. Haba, T. Ohnishi, M. Sekimoto, N. Tomita, H. Shiozaki, and M. Monden, *J. Histochem. Cytochem.* **46,** 397 (1998).

[35] B. Kaplan, M. Pras, and M. Ravid, *J. Chromatogr.* **573,** 17 (1992).

present. A major limitation in acquiring biochemical information on microextracted proteins results from the often inadequate amounts of pathologic material that is available for analysis. Typically, the histochemical diagnosis of amyloidosis is made from fine needle biopsies of kidney, liver, heart, or other affected tissues that, in contrast to specimens of suspected malignant tumors, are obtained nonselectively, i.e., "blindly." The amyloid content of such samples can therefore vary widely and estimation of the amount of such material (as judged by the extent of green birefringence) must thus be considered when determining a purification strategy. Equally important is the fact that clinical specimens are subjected routinely to formalin fixation and the length of time in this fixative differs considerably among samples. Due to this process, the removal of blood contaminants and extraction of amyloid is more complicated and, in fact, conflicting data regarding the isolation of proteins from fixed tissues can be found in the literature. Specifically, some authors state that SDS-containing buffers are effective in the extraction of AA proteins only,[24,26,27] whereas others report that the heating of fixed tissue homogenates in SDS/urea-containing buffers results in the solubilization of AA, as well as AL and ATTR.[28,29] Data on the extraction of nonamyloid proteins from fixed tissues are also controversial.[34] The reasons for these discrepancies are unclear; however, certain factors must be considered, such as length of time of formalin fixation as well as the conditions of extraction, e.g. choice of buffer, incubation time, and temperature. Further, well-defined controls, both positive and negative, must be used to establish the validity of the methodology.

This article describes the procedures employed to extract, purify, and sequence amyloid proteins from milligram quantities of fresh or formalin-fixed tissue. Notably, through application of these techniques we have found it possible, in most cases, to identify unequivocally the chemical nature of amyloid deposits contained in patient specimens obtained by fine needle biopsy. The effectiveness of the procedures used has been verified by comparing results obtained from examination of (a) amyloid proteins isolated from tissues of the same patient using our microextraction techniques as well as the classical water extraction method[1]; (b) amyloid proteins microextracted from both fresh and formalin-fixed tissues of the same patient; and (c) proteins microextracted from both amyloid-containing and amyloid-free tissues.

Fresh (Nonfixed) Tissue

Fresh specimens of spleen, liver, kidney, heart, thyroid, brain, and others were snap frozen in liquid nitrogen and stored at $-78°$ until use (under these conditions, samples are stable indefinitely).

Extraction. The procedure used to extract fresh, frozen tissues is summarized in Scheme 1.[9,10,15–21] The specimen (5–50 mg, wet weight) is washed three times to remove soluble blood components by homogenizing the tissue manually with a glass rod in a 2-ml Eppendorf tube with 1.5 ml of saline or cold phosphate-buffered saline (PBS, pH 7.4), followed by centrifugation at 14,000*g* for 10 min at room temperature in an Eppendorf microfuge. The resulting pellet is rinsed with deionized water, centrifuged (as described earlier), and suspended in 1–1.5 ml of 20% acetonitrile containing 0.1% TFA. The mixture is incubated at room temperature for 1 hr with moderate mixing and centrifuged again. The incubation and centrifugation steps are repeated two more times and the supernatants are pooled and lyophilized.

Purification. Size-exclusion HPLC columns are used to purify microextracted amyloid proteins [Bio-Sil TSK-125, 600 × 7.5 mm, i.d. (Bio-Rad Labs., Richmond, CA)] and TSK-gel G 3000 SW_{XL}, 300 × 7.8 mm i.d. (TosoHaas, Stuttgart, Germany)].[16,17,30] Eluants consist of either an aqueous 20% acetonitrile–0.1% TFA solution (eluant "a") or a 0.1 *M* Tris–HCl buffer, pH 6.8, containing 0.1% SDS (eluant "b"). Lyophilized tissue extracts are dissolved to a final concentration of 6–8 mg/ml in 20% acetonitrile/0.1% TFA (to be used with eluant "a") or in 0.0625 *M* Tris–HCl buffer, pH 6.8, containing 3% SDS, 5% 2-mercaptoethanol (for eluant "b"). The samples are centrifuged at room temperature for 10 min at 14,000*g* (Eppendorf microfuge) and the supernatants are filtered in microfilterfuge tubes (Rainin, MA) and applied to the columns (60–100 μl per run). The elution is monitored at 220 nm and the collected peaks are analyzed by SDS–PAGE and Western blotting using a panel of antibodies specific for the most common types of amyloid protein deposited systemically, i.e., κ and λ immunoglobulin light chains and AA and ATTR. It should be noted that although elution with acetonitrile–TFA results in aberrant size-exclusion behavior (presumably due to the interaction of protein with unreacted silanol), the efficient purification of low molecular mass (<12 kDa) amyloid, e.g., AA and AL fragments, is achieved readily.[16,30] In contrast, use of the SDS-containing eluant results in classical size-exclusion separation of different types of protein, including amyloid components of higher molecular weight (>14 kDa); however, only partial purification results.[16,17,30,31]

Preparative slab PAGE gels also can be used to purify amyloid proteins microextracted with acetonitrile–TFA.[10,31] Lyophilized tissue extracts are dissolved in the SDS–PAGE sample buffer (final concentration, 15–20 μg/μl) and applied in multiple sample wells to 1.5-mm-thick 17% PAGE slab gels. To determine the location and approximate M_r of the separated proteins, the portion of the gel incorporating the standard molecular weight markers and one sample lane are stained with 0.1% Coomassie blue in 50%

TISSUE SPECIMEN
(NONFIXED 5-50 mg)

Homogenization in saline or PBS
Centrifugation (Eppendorf, 10 min, 14,000 *g*)
(Repeat X2)

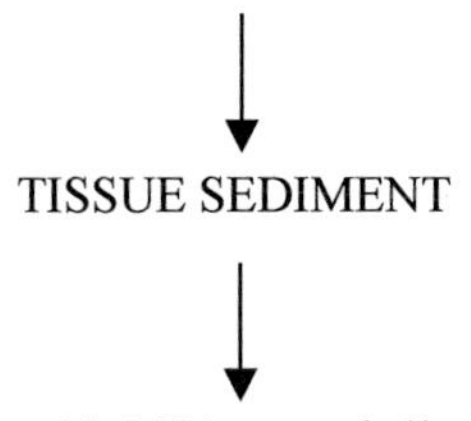

TISSUE SEDIMENT

Extraction with 20% acetonitrile-0.1% TFA
Centrifugation (Eppendorf, 10 min, 14,000 *g*)
(Repeat X2)

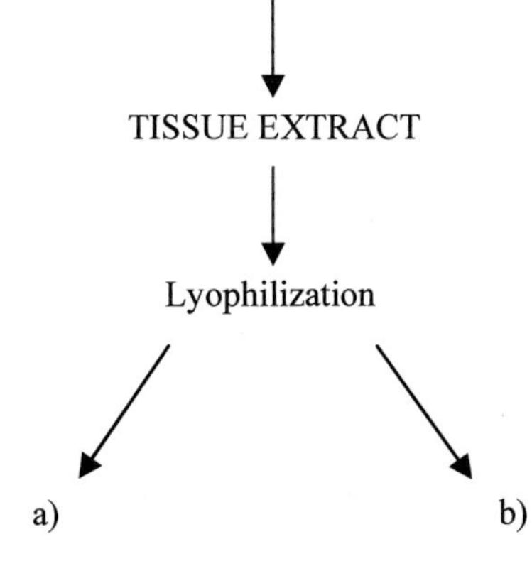

TISSUE EXTRACT

Lyophilization

a)

SDS-PAGE-based
microsequencing

b)

Small-scale purification
by HPLC or slab SDS-PAGE

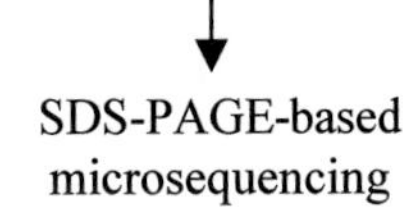

SDS-PAGE-based
microsequencing

SCHEME 1

methanol/10% acetic acid. Slices of the gel that contain the unstained proteins of interest are excised, placed in a dialysis bag (molecular weight cutoff 1000) filled with 0.025 *M* Tris, 0.192 *M* glycine buffer, pH 8.3, containing 0.1% SDS, and subjected to electroelution (120 V, 2 hr) in a horizontal electrophoresis cell filled with the same buffer. The contents of the dialysis bag are centrifuged and the supernatant containing the electroeluted proteins is dialyzed and lyophilized.[35,36] Alternatively, proteins separated by SDS–PAGE are stained and eluted by diffusion: The gel slices are minced and incubated at 37° for 2–3 h in 0.05 *M* Tris–HCl elution buffer, pH 7.0, containing 0.1% SDS, 0.1 m*M* EDTA, 5 m*M* dithiothreitol, and 0.2 *M* NaCl. Gel particles are removed by centrifugation. Two volumes of acetonitrile are added to the supernatant and the proteins are precipitated in the cold.[10,37]

Sequencing. The microextracted amyloid is either subjected directly to the SDS–PAGE-based microsequencing technique mentioned earlier (Scheme 1) or is first purified by HPLC or preparative slab SDS–PAGE (Scheme 2). The samples (10 μg/μl) are electrophoresed on 17% polyacrylamide gels and electrotransferred for 45 min at 400 mA to PVDF membranes (Millipore, Bedford, MA) using a 10 m*M* 3-cyclohexylamino-1-propanesulfonic acid buffer (pH 11) containing 10% methanol.[11] The electroblotted proteins are visualized by staining with 0.05% Coomassie blue in 40% methanol/1% acetic acid and the bands of interest are excised and sequenced on a 470A or 477A protein/peptide sequenator. The resulting phenylthiohydantoin derivatives are identified using an on-line phenylthiohydantoin derivative analyzer (Applied Biosystems, Foster City, CA).[10,15,38]

Comments. The acetonitrile–TFA microextraction technique is relatively simple and rapid and requires only milligram amounts of starting tissue as is typically obtained from diagnostic biopsy specimens. In comparative studies, we have analyzed the yield and electrophoretic profiles of amyloid extracted from the same tissue by both this method and the classical water extraction technique[9,16] and have obtained similar results (Fig. 1). Further, we have found that a 20% concentration of acetonitrile is optimal because increasing amounts of nonamyloid proteins are coextracted at higher levels. In contrast to the use of the formic acid method,[22] prolonged proteolytic digestion of the tissue homogenate is not required. This tech-

[36] B. Kaplan, S. Yakar, Y. Balta, M. Pras, and B. Martin, *J. Chromatogr.* **704,** 69 (1997).

[37] B. Kaplan and M. Pras, *Clin. Chim. Acta* **163,** 199 (1987).

[38] A. Rostango, B. Kaplan, R. Vidal, J. Chuba, A. Kumar, B. Frangione, G. Gallo, and J. Ghiso, *in* "VIII International Symposium on Amyloidosis," p. 46. Mayo Clinic, Rochester, MN, August 7–11, 1998. [Abstract 114]

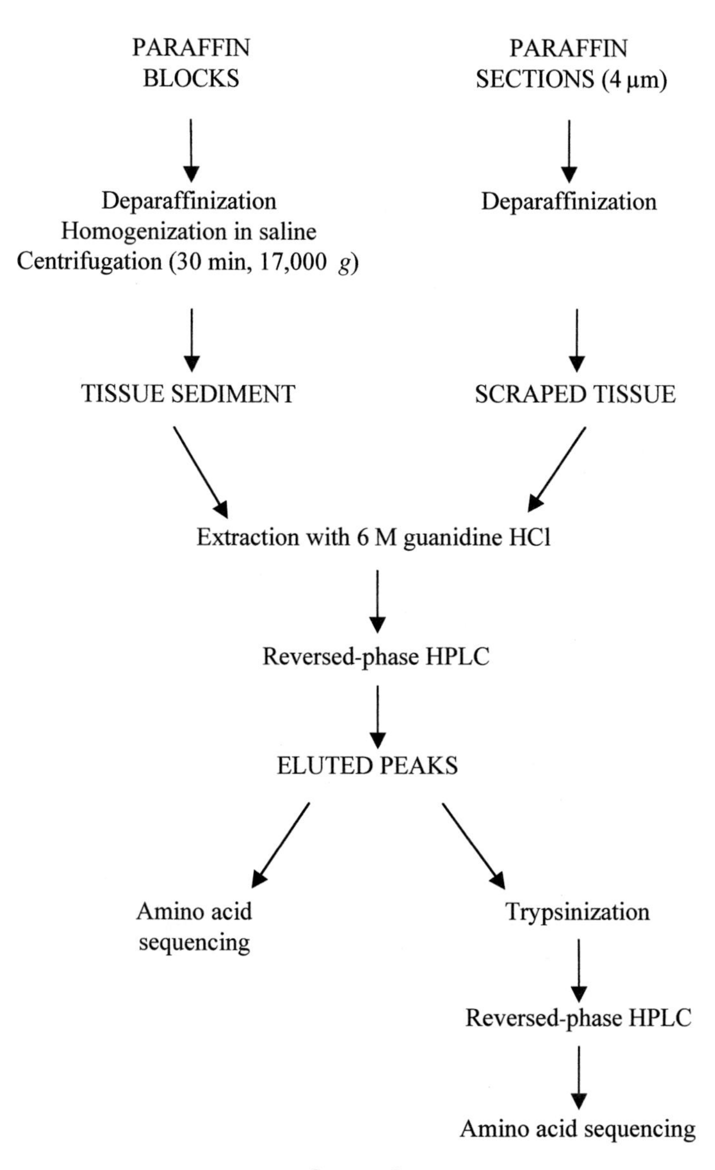
TISSUE SPECIMEN
(FORMALIN-FIXED)
PARAFFIN
BLOCKS
PARAFFIN
SECTIONS (4 μm)
Deparaffinization
Homogenization in saline
Centrifugation (30 min, 17,000 g)
Deparaffinization
TISSUE SEDIMENT
SCRAPED TISSUE
Extraction with 6 M guanidine HCl
Reversed-phase HPLC
ELUTED PEAKS
Amino acid
sequencing
Trypsinization
Reversed-phase HPLC
Amino acid sequencing

SCHEME 2

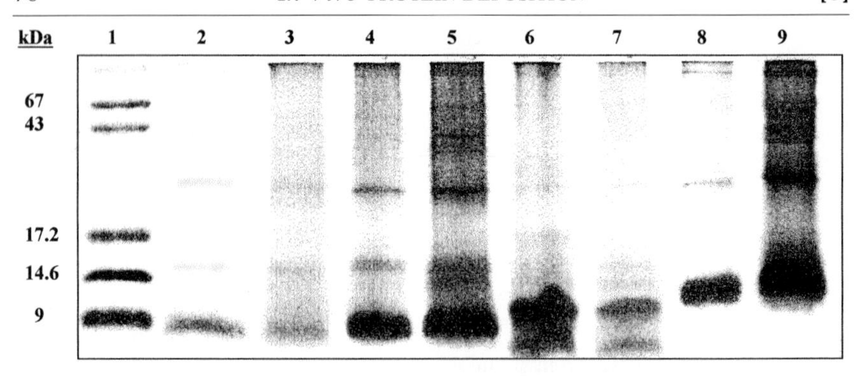

FIG. 1. Comparison of SDS–PAGE profiles of amyloid proteins microextracted with acetonitrile–TFA (lanes 3, 5, 7, and 9) and isolated by the classical water extraction method[1] (lanes 2, 4, 6, and 8). Lanes 2 and 3: spleen from patient COH with FMF; lanes 4 and 5: liver from patient GAM with idopathic AA amyloidosis; lanes 6 and 7: thyroid from patient NOR with FMF; and lanes 8 and 9: spleen from patient RAM with idopathic AL amyloidosis. Molecular masses of standard proteins (lane 1) are as indicated. From B. Kaplan, G. German, and M. Pras, *J. Liq. Chromatogr.* **16,** 2249 (1993), reprinted by courtesy of Marcel Dekker Inc., New York.

nique has other advantages as well. Protein recovery is efficient because the solvent is volatile and removed easily by lyophilization without need of dialysis. Acetonitrile–TFA is also compatible with reversed-phase HPLC as opposed to SDS-containing extraction buffers that cannot be applied directly to HPLC columns because SDS affects protein behavior adversely.[31] Notably, our microextraction method has proven to be an effective step in the immunochemical characterization of AA, AL, ATTR, and $A\beta$[9,15–21] type amyloid proteins by Western blotting and enzyme-linked immunosorbent assay (ELISA), as well as amino acid sequence analyses.[10,15,38]

As stated, the strategy to purify amyloid-related proteins from tissue depends largely on the tissue content and chemical nature of the fibril. If sufficient material is available, the tissue extract is analyzed by Western blotting and the proteins of interest are sequenced directly using the SDS–PAGE-based microsequencing technique.[10,15,38] However, in cases where the amyloid content is relatively limited or contaminating proteins that have a similar electrophoretic mobility are present, other small-scale separation methods are required. We have found that the combined use of slab gel electrophoresis and HPLC[31,35,36] can be useful due to the resolution power of analytical gels and the high-speed separation of HPLC.

Fixed (Formalin) Tissues

Extraction. The methods used to extract formalin-fixed tissues (spleen, kidney, heart, liver, abdominal fat, colon, lymph node, sternum) and to purify and sequence amyloid fibril proteins are summarized in Scheme 2. In those cases where the specimens are wax embedded, the block is deparaffinized prior to extraction[14,25–28] by melting at 65° followed by serial overnight washings at room temperature with xylene, absolute alcohol, and 95% alcohol. The resulting specimen is placed into an Eppendorf tube, homogenized in saline, and the insoluble sediment harvested by centrifugation for 30 min at 17,000*g* (Biofuge 17R, Baxter, Germany). Notably, we have found that we also can obtain sufficient amounts of amyloid-related protein for analysis from 4-μm sections of paraffin-embedded, formalin-fixed tissue biopsies. In this case, the tissue is scraped from the slides after deparaffinization, pooled, and transferred to an Eppendorf tube. The number of such sections required (~8 to 20) is dependent on the extent of amyloid present as judged by Congo red staining. Samples obtained from both blocks and sections are suspended in 0.2 to 1.0 ml of 0.15 *M* Tris–HCl buffer, pH 8.0, containing 6 *M* guanidine hydrochloride and 1 m*M* EDTA, incubated overnight at 37°, and stored at 4°.

Purification. Immediately prior to HPLC, the extracted protein is reduced with 2-mercaptoethanol (5 μl per 1 ml extract), vortex mixed, incubated for 1 to 2 hr at 37°, and alkylated by the addition of 4-vinylpyridine (10 μl/ml) for 0.5 to 1.0 hr at 37°. The samples are then centrifuged, passed through a 0.2-μm syringe filter (Millipore), and recentrifuged prior to reversed-phase HPLC (ABI Model, Applied Biosystems) on an Aquapore C_8 column [7 μm, 300 Å, 30 × 4.6 mm (Perkin Elmer, Norwalk, CT)]. Elution is performed over 45 min with a linear gradient of 7–70% acetonitrile in 0.1% TFA; the effluent is monitored at 220 nm and the peaks are collected manually. In general, amyloid proteins elute at an acetonitrile concentration of between 35 and 45% (peaks eluting before a concentration of 28% contain guanidine hydrochloride and other ingredients of the buffer necessary for sample solubilization, as well as chemicals employed in reduction and alkylation). To prevent cross-contamination, we change the small and relatively inexpensive Aquapore columns for each microsample. In addition, we found it useful to wash the injector and tubing system with a 6 *M* guanidine hydrochloride solution prior to replacing the column.

Sequencing. The collected peak volumes are reduced to 50–100 μl by vacuum centrifugation. Fifteen microliter aliquots are loaded onto a precycled glass fiber filter containing 1.5 mg of Polybrene (BioBrene, Applied Biosystems) and the N-terminal sequence is determined by automated Edman degradation (Procise Protein Sequencer, Applied Biosystems). In those cases where the N-terminal residue is blocked (e.g., λ type light chains of subgroups 1, 2, and 8) or possibly modified by formalin fixation, or if additional sequence data are required, the sample (~100 μl) is lyophilized, redissolved in 200–400 μl of 0.1 *M* methylmorpholine buffer (pH 8.0), and digested with trypsin (37°, 4 hr) at an enzyme : substrate ratio of 1 : 50

(weight/weight). The amount of amyloid is estimated from the peak area of the eluted protein trace as obtained using the Aquapore C_8 column described earlier. The volume of the digested sample is then reduced to 50 μl and the peptides are separated by HPLC (ABI Model 173 Capillary HPLC Microblotter, Applied Biosystems) on a Brownlee C_{18} column [5 μm, 300 Å, 150 × 0.5 mm (Perkin Elmer)] and sequenced.

The HPLC profiles of material microextracted from fixed biopsy specimens of patients with unknown forms of amyloidosis and sequence data on the purified amyloid proteins are illustrated in Figs. 2 and 3, respectively. In the case of a liver needle biopsy (seven 4-μm-thick sections), the major peaks eluting between acetonitrile concentrations of 42 to 45% (Fig. 2A) revealed sequences corresponding to that of the N-terminal portion of protein SAA, starting at positions 1, 2, 3, or 4 (Fig. 3, sample 1). In the case of a sternal mass biopsy (eight sections), the protein peak eluting at an acetonitrile concentration of 35% (Fig. 2B) contained κI light chain fragments (Fig. 3, sample 2). In another example, where the N-terminal residue was "blocked," the tryptic digest of material extracted from a formalin-fixed, paraffin-embedded heart biopsy specimen from a patient with λ-type AL amyloidosis was resolved by HPLC and the peak eluting at an acetonitrile concentration of 35% contained a peptide derived from the λ1 variable region (not illustrated).

Comments. We have applied this methodology to the chemical characterization of different types of amyloid proteins (e.g., AA, AL, A Lys) found in minute amounts of formalin-fixed tissues. Data were confirmed by sequence analyses of amyloid fibrils isolated classically[1] from fresh tissues of the same patients. We have determined that reduction and alkylation of the extracted protein facilitate HPLC purification of amyloid and also identification of cysteine residues. Most often, the microextracted amyloid protein elutes from Aquapore C_8 columns at an acetonitrile concentration of between 35 and 45%. In contrast, no detectable sequence signals are obtained in this concentration range from control, amyloid-free, formalin-fixed tissues extracted and analyzed under the same conditions. Using our chromatographic techniques, it is possible to remove blood-derived protein impurities (e.g., hemoglobin) that are coextracted with the amyloid (Fig. 2B).

Summary

This article described micromethods useful for the extraction, purification, and amino acid sequencing of amyloid proteins contained in minute specimens obtained from patients with systemic forms of amyloidosis. We

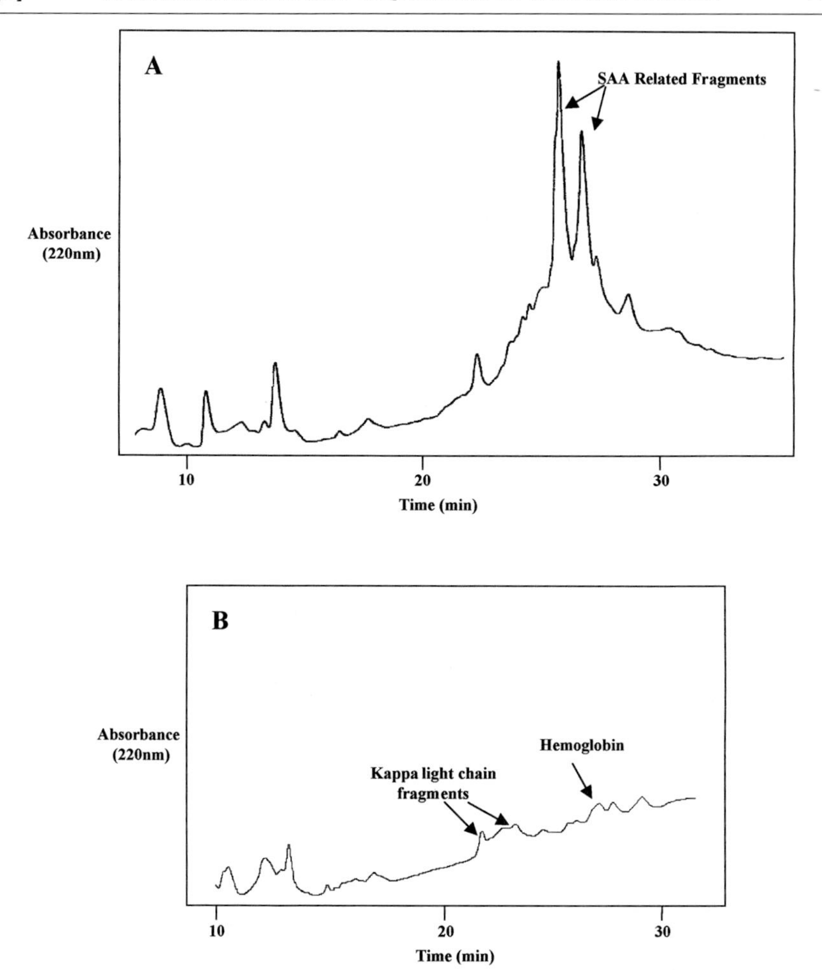

FIG. 2. Reversed-phase HPLC profiles of amyloid containing material microextracted from 4-μm-thick formalin-fixed, paraffin-embedded tissue sections from biopsy specimens: (A) liver needle biopsy, seven sections; and (B) sternal mass biopsy, eight sections. Samples were run for 45 min on an Aquapore C_8 column using a linear gradient of 7–70% acetonitrile in 0.1% TFA, as described in the text.

posit that these procedures can also be applied to the biochemical characterization of cerebral amyloid deposits. The selection of the techniques is dependent on the type of sample to be extracted (fresh or formalin fixed) as well as the amount of congophilic material present. Although amyloid proteins are isolated and purified more easily from fresh tissue, it must be noted that formalin-fixed specimens are available more readily for analysis

Sample 1: Liver needle biopsy, seven sections

AA
1 10 20 30
RSFFSFLGEAFDGARDMWRAYSDMREANYIGS
initial yield 79.2 pmoles,
repetitive yield 87.5%

Sample 2: Sternal mass biopsy, eight sections

AL κ1
1 10
DIQMTQSPSSLSASVGD
initial yield 17.6 pmole,
repetitive yield 85.3%

Sample 3: Kidney biopsy, 19 sections

ALys
1 10
KVFERCELARTLKRLGMD
initial yield 0.8 pmole,
repetitive yield 96%

FIG. 3. Amino acid sequence analysis of amyloid proteins microextracted and purified from formalin-fixed biopsy specimens. The specimens were extracted using 6 *M* guanidine hydrochloride, purified by HPLC, and microsequenced, as described in the text. Initial and repetitive yields of the sequenced material were calculated using 610A V2.1 software from Applied Biosystems.

due to the common diagnostic use of fine needle tissue biopsies and are, therefore, important for both current and retrospective studies. Remarkably, despite the expected difficulties associated with formalin treatment, we were able to extract and sequence amyloid proteins from fixed tissues, presumably due to the resistance of amyloid to formalin cross-linking. Through the continued development of techniques for small-scale protein separation and application of highly sensitive microsequencing and mass spectral methods, exact identification of the protein contained in fibrillar amyloid deposits can be determined. Such information has therapeutic and prognostic relevance and can increase our understanding of the pathogenesis of amyloidosis.

Acknowledgment

This work was supported in part by grants from the Chief Scientist's Office, Ministry of Health, Israel, USPHS Research Grant CA 10056 from the National Cancer Institute, and an American Cancer Society Harry and Elsa Jiler International Visiting Scientist Award. A.S. is an American Cancer Society Clinical Research Professor.

[6] Purification of Paired Helical Filament Tau and Normal Tau from Human Brain Tissue

By VIRGINIA M.-Y. LEE, JUN WANG, and JOHN Q. TROJANOWSKI

Introduction

Paired helical filaments (PHF) are the major structural component of the neurofibrillary tangles (NFTs) that form in neurons, which subsequently degenerate in the brains of patients with Alzheimer's disease (AD), Down syndrome, amyotrophic lateral sclerosis/parkinsonism–dementia complex of Guam (ALS/PDC), dementia pugilistica, and several other neurodegenerative disorders.[1,2] Paired helical filaments contain modified forms of central nervous system (CNS) tau proteins known collectively as PHF-tau (previously referred to as A68).[3,4] Tau proteins are a group of microtubule-associated proteins that are expressed predominantly in axons.[1] Some of the known functions of tau proteins are to bind to microtubules (MTs), to promote the assembly of tubulin monomers into MTs, and to stabilize polymerized MTs. The human tau gene is located on chromosome 17q21-22[5] and alternative splicing of this gene in the adult human CNS generates six different tau isoforms with M_r of 50,000 to 65,000.[1] The six tau isoforms differ with respect to the presence of three versus four MT binding repeats in the carboxy-terminal third of these proteins as well as by the presence of zero, one, or two amino-terminal inserts.[1] All six tau isoforms undergo

[1] M. Goedert, J. Q. Trojanowski, and V. M.-Y. Lee, "The Molecular and Genetic Basis of Neurological Disease" (R. N. Rosenberg, S. B. Prusiner, S. DiMauro and R. L. Barchi, eds.), p. 613. Butterworth Heinemann, 1997.

[2] M. B. Feany and D. W. Dickson, *Ann. Neurol.* **55,** 1051 (1996).

[3] V. M.-Y. Lee, B. J. Balin, L. Otvos, Jr., and J. Q. Trojanowski, *Science* **251,** 675 (1991).

[4] G. T. Bramblett, J. Q. Trojanowski, and V. M.-Y. Lee, *Lab. Invest.* **66,** 212 (1992).

[5] M. G. Spillantini, T. D. Bird, and B. Ghetti, *Brain Pathol.* **8,** 387 (1998).

0076-6879/99 $30.00

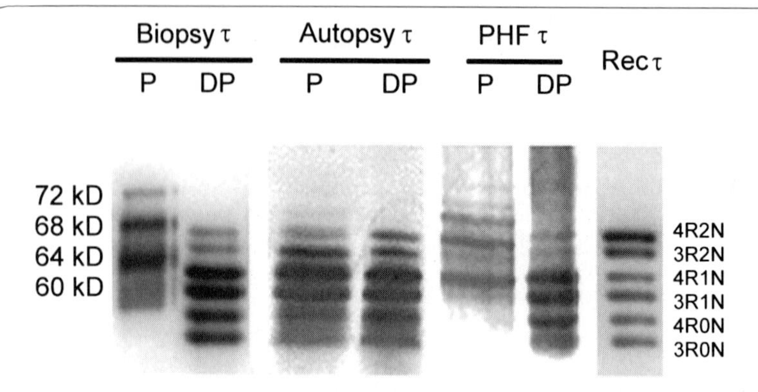

FIG. 1. Western blot analysis of biopsy-derived tau (biopsy τ), autopsy-derived tau (autopsy τ), and PHF-tau (PHF τ) before and after dephosphorylation with alkaline phosphatase. The Western blot is probed with a rabbit polyclonal antiserum raised to recombinant tau protein. To demonstrate complete enzymatic dephosphorylation, the recombinant tau protein corresponding to each of the tau isoforms is included here to mark the position of the isoforms (lane labeled Rec τ). The M_r of the biopsy-derived tau isoforms is also indicated on the left-hand side of the blot.

posttranslational modifications such as phosphorylation and glycosylation.[1,6] Although the biological significance of glycosylation remains to be elucidated, phosphorylation regulates the binding of tau to MT, i.e., increasing tau phosphorylation reduces MT binding.[1,7] Although previous studies have demonstrated that PHF-tau is hyperphosphorylated relative to normal human tau, normal human tau isolated from biopsy brain samples is much more phosphorylated than normal human tau isolated from autopsy brain samples.[1,8] Indeed, PHF-tau nearly comigrates with biopsy-derived normal tau and appears as three major polypeptide bands of 55, 64, and 69 KDa, whereas autopsy-derived normal human tau migrates as protein bands of 50–65 KDa as observed in sodium dodecyl sulfate–polyacrylamide gel electrophoresis (SDS–PAGE) gels (Fig. 1).[8] Significantly, PHF-tau does not bind to MTs unless it is enzymatically dephosphorylated.[7] Thus, phosphorylations regulate the binding of tau to MTs and could be one of the key events leading to the formation of PHFs in AD. While normal human tau isoforms are highly soluble proteins that are heat and acid stable, PHF-tau forms filamentous structures that are only soluble in strong detergents

[6] C. S. Arnold, G. V. W. Johnson, R. N. Cole, M. Lee, D. Y.-L. Dong, and G. W. Hart, *J. Biol. Chem.* **271,** 28741 (1996).

[7] G. T. Bramblett, M. Goedert, R. Jakes, S. E. Merrick, J. Q. Trojanowski, and V. M.-Y. Lee, *Neuron* **10,** 1089 (1993)

[8] E. S. Matsuo, R.-W. Shin, M. L. Billingsley, A. Van deVoorde, M. O'Connor, J. Q. Trojanowski, and V. M.-Y. Lee, *Neuron* **13,** 989 (1994).

such as SDS and guanidine isothiocyanate.[3,4] Although PHFs are formed from all six tau isoforms in the AD brain, the abnormal inclusions in a number of other neurodegenerative diseases contain insoluble straight, ribbon-like, or twisted filaments that are comprised predominantly of a subset of tau isoforms with either three or four MT binding repeats.[1,5] For example, abnormal filaments comprised primarily of four MT binding repeat tau isoforms have been found in the brains of patients with progressive supranuclear palsy (PSP), corticobasal degeneration (CBD), and several kindreds with familial frontal temporal dementia and parkinsonism (FTD) linked to chromosome 17 (FTD-17) caused by mutations in the tau gene.[2,5,9] Furthermore, Pick bodies of Pick's disease (PiD) are composed of only the three repeat tau isoforms.[5] Although the pathogenesis of PHFs and other tau filaments among the different neurodegenerative diseases is unclear, the analysis of these filaments may increase our understanding of the mechanisms leading to the abnormal transformation of soluble human brain tau into aggregated tau filaments. To begin to elucidate the processes involved in this abnormal transformation, the successful purification from human brain tissue of soluble normal tau proteins and insoluble filamentous tau aggregates is essential. This article describes a procedure to effectively purify PHF-tau and soluble normal tau preparations from human brain.[3,4]

Purification Procedure

Preparation of Highly Purified PHF-tau Proteins

Selection of Brain Tissues Enriched in PHF-tau. Source material for PHF-tau are brains of AD patients. However, because the amount of PHFs is highly variable among different AD brains and among different brain regions of the same AD brain, it is necessary to select brain tissues that contain abundant neurofibrillary pathology and consequently the highest amounts of PHF-tau. Because the diagnosis of AD is based on estimates of the abundance of NFTs and amyloid plaques in different brain regions,[10] this diagnostic screen will identify brains that should contain large amounts of PHFs. In general, PHF-rich neurofibrillary lesions accumulate preferentially in the hippocampus, amygdala, and neocortex of the AD brain, whereas other regions are more affected in other diseases.[1] A sebsequent screen for brain regions enriched in PHF-tau is semiquantitative Western blot analysis, which is performed on crude PHF-tau extracted from small

[9] M. Hong, V. Zhukareva, V. Volgelsberg-Ragaglia, Z. Wszolak, L. Reed, B. I. Miller, D. H. Geschwind, T. D. Bird, D. McKeel, A. Goake, J. C. Morris, K. C. Wilhelmson, G. D. Schellenberg, J. Q. Trojanowski, and V. M.-Y. Lee, *Science* **282,** 1914 (1998).

[10] B. T. Hyman and J. Q. Trojanowski, *J. Neuropathol. Exp. Neurol.* **56,** 1095 (1997).

pieces of tissue of different brain regions to estimate the amount of PHF-tau per gram wet tissue.[4] The following discussion describes a highly reproducible second screen for the estimation of PHF-tau in brain tissue.

Preparation of Crude PHF-tau. To estimate the amount of PHF-tau in different postmortem human AD brain regions, exactly 1 g of gray matter is dissected clean of meninges and blood vessels for homogenization in 1.5 ml of a buffer containing 0.75 *M* NaCl in RAB buffer [100 m*M* 2-(*N*-morpholino)ethanesulfonic acid (MES), 1 m*M* EGTA, 0.5 m*M* $MgSO_4$, 2 m*M* dithiothreitol (DTT), pH 6.8] supplemented with 0.5 m*M* phenylmethylsulfonyl fluoride (PMSF), and 1 μg/ml each of *N*-tosyllysine chloromethyl ketone, *N*-tosyl-L-phenylalanylchloromethyl ketone, pepstatin A, leupeptin, and soybean trypsin inhibitor to minimize proteolysis. Only gray matter is used for PHF-tau isolation as previous studies have shown that white matter contains little or no PHF-tau.[4] The homogenates are then incubated at 4° for 20 min to depolymerize any residual MTs and are centrifuged in an SS-34 rotor (Sorvall) at 11,000*g* for 20 min at 4°. The supernatant is then removed and recentrifuged at 100,000*g* for 60 min at 4° in a Beckman TL100 using a TLA45 rotor. The resulting supernatants contain primarily soluble normal tau, whereas PHF-tau remains in the pellets.

To obtain PHF-tau proteins, the pellets generated by the first and second cold centrifugations are combined and resuspended in 10 ml (1:10 w/v) of PHF extraction buffer (10 m*M* Tris, 10% sucrose, 0.85 *M* NaCl, 1 m*M* EGTA, pH 7.4) and spun at 15,000*g* for 20 min at 4°. In the presence of 10% sucrose and at low centrifugal forces, isolated PHFs or small PHF aggregates remain in the supernatant whereas intact or fragmented NFTs and larger PHF aggregates are pelleted. After this low-speed centrifugation, the pellets are reextracted in the same fashion, and the supernatants are pooled, made in 1% Sarkosyl, and stirred at room temperature for 1 hr. Sarkosyl treatment removes membranous material from the crude PHF preparations, thereby enriching for PHF-tau. The Sarkosyl extracts are then spun for 30 min at 4° in a Beckman 60 Ti rotor at 100,000*g* and the resulting pellets contain crude PHF-tau. To estimate the amount of PHF-tau in each gram of wet tissue, high-speed pellets are resuspended in 150 μl of Laemmli sample buffer and boiled for 5 min, and equal amounts of samples are loaded on a SDS–PAGE gel, transferred to nitrocellulose membranes, and probed with phosphorylation-dependent anti-tau antibodies.[4] The exact amount of PHF-tau in each gram (wet weight) of tissue can be estimated accurately using ^{125}I-labeled secondary antibodies together with a standard curve containing known amounts of tau proteins.[4]

Preparation of Highly Purified PHF-tau. Once AD brains enriched in

PHF-tau are identified, large quantities of highly purified PHF-tau can be obtained by scaling up the just-described protocol followed by further fractionation on a discontinous sucrose gradient and extraction with guanidine isothiocyanate as described later. To scale up, 20–100 g of gray matter from AD brains enriched in PHFs dissected clean of blood vessels and meninges is used as start material.

After Sarkosyl extraction and centrifugation, pellets containing PHF-tau are resuspended in 1–5 ml of RAB buffer (depending on the amount of starting material, a ratio of 1 ml to 25 g of gray matter is used), sonicated until smooth, and boiled for 5 min. Then 0.5-ml aliquots are loaded onto a stepwise 1.0–2.5 *M* discontinuous sucrose gradient and spun for 16 hr at 4° in a Beckman ultracentrifuge using a SW50.1 swinging bucket rotor at 175,000*g*. A white flocculent layer containing mostly membranes on the top of the gradient is discarded. A thick brown band is recovered between the 1.25 and 1.5 *M* sucrose interface, thick medium to light brown material is found at the 1.75 and 2.0 *M* interface, and a small amount of pale brown material is recovered at the 2.25 and 2.5 *M* sucrose interface. This last fraction contains PHF-tau proteins that are not contaminated with other proteins. By electron microscopy, only the heaviest fraction contains highly purified PHF-tau in the form of short but intact PHFs uncontaminated by amorphous material. However, this fraction represents less than 20% of the total PHF-tau recovered from all three fractions as determined by quantitative Western blotting. From 10 g of enriched gray matter, about 1 μg of highly enriched or nearly pure PHF-tau in the form of intact PHFs is generated from the heaviest fraction.

The remaining PHF-tau proteins in this preparation are concentrated in the dark (fraction 1) and light brown (fraction 2) bands of the sucrose gradient and are contaminated with other proteins. To purify this PHF-tau further, fractions 1 and 2 are collected and extracted with 2 *M* guanidine isothiocyanate (about 1 ml per gram wet tissue) at 37° for 1 hr. After centrifugation at 50,000*g* for 30 min at 25°, the supernatant is saved and the pellet discarded. Quantitative Western blotting indicates that >95% of the PHF-tau proteins are soluble in 2 *M* guanidine isothiocyanate. These soluble PHF-tau proteins are dialyzed against several changes of distilled water to remove the guanidine thiocyanate, and the dialyzed sample is spun at 50,000*g* for 30 min at 25° to remove any precipitates as a result of the dialysis. The PHF-tau proteins remaining in the supernatant after dialysis are lyophilized and resuspended in the appropriate buffer for subsequent studies. The recovery of PHF-tau proteins from extractions of material from all PHF-tau containing fractions is about 1–5 μg/g wet tissue. Highly purified PHF-tau proteins prepared as just described are soluble in aqueous buffer up to a concentration of about 100 μg/ml.

Purification of Normal Human Tau Preparations

This section summarizes the purification proctocol for biopsy- and autopsy-derived human adult tau isoforms (six are expressed in adult brain) as well as the fetal human tau protein (only the smallest tau isoform is expressed in the fetal brain).

Normal Human Adult Tau Isolated from Brain Biopsies. Requirements for the successful completion of the proposed experiments are reliable sources of fresh surgical human biopsy brain samples with little (<10 min) or no postsurgical delay between the excision of the tissue sample and the immersion of the tissue in cold buffer (see later for a detailed description) for the isolation of tau proteins in their native phosphorylation state. One source of human biopsy samples is residual tissue obtained from the lateral temporal lobe (occasionally frontal lobe) as a result of the therapeutic excision of an electroencephalographically characterized epileptogenic brain tissue focus in individuals with intractable epilepsy who are usually in the third to sixth decade of life. Another common source of biopsy tissue in the margin of samples obtained are residual tissue obtained during the removal of brain tumors. Thus, the sizes of the samples are highly variable and they frequently range from about 0.3 to 5 g. Notably, normal brain biopsy samples are from the margins around the epileptogenic focus or tumor that are removed to ensure complete resection of the lesion and minimize the need for a second surgery to remove lesional tissue left behind. Portions of the samples are examined to be certain that they are free of diagnostic abnormalities at the light microscopic level. This will ensure that residual tumor or other diagnostic abnormalities are not used here to prepare normal tau.

Two approaches have been used to isolate biopsy-derived human brain tau. For larger biopsy samples (i.e., 1 g or more), tau proteins are assembled using endogenous MTs because the integrity of the MTs is excellent in these biopsies. To do this, brain biopsies are homogenized in 1.5 volumes of cold RAB buffer [0.1 *M* MES buffer, pH 6.8, 0.5 m*M* $MgSO_4$, 1 m*M* EGTA, 2 m*M* DTT, and a cocktail of protease inhibitors (i.e., 1 m*M* PMSF, 1 μg/ml each of leupeptin, pepstatin, soybean trypsin inhibitor, TPCK, TLCK, TAME)] and 5 μM okadaic acid (OKA).[11] After leaving on ice for 10 min, the homogenate is centrifuged at 50,000*g* for 40 min at 4° and the supernatant is recentrifuged at 150,000*g* for 70 min at 4°. The high-speed brain extract containing most of the depolymerized tubulin subunits is supplemented with 4 *M* glycerol and 2 m*M* GTP and is incubated at 37°

[11] M. Mawal-Dewan, J. Henley, A. Van deVoorde, J. Q. Trojanowski, and V. M.-Y. Lee, *J. Biol. Chem.* **269,** 30981 (1994).

for 20 min to induce MT assembly. The mixture is then centrifuged at 150,000g for 20 min at 25° and tau bound to MTs is recovered in the pellet. Of the tau proteins from biopsy brain samples, 95% is recovered bound to MTs after one cycle of reassembly.[11] The bound tau is dissociated from the MTs by resuspending the pellet in RAB buffer containing 0.75 M NaCl, followed by incubation at 25° for 20 min. The mixture is centrifuged again, and the tau-containing supernatant is removed and boiled. Because tubulins are heat labile and tau is heat stable, the boiling step removes any residual tubulin in the supernatant. The purity of human tau recovered at this stage is about 60% with the major contaminants being other microtubule-associated proteins (MAPs). From 1 g of brain tissue, we recover 30–50 μg of enriched tau. Further, the inclusion of 5 μM of OKA completely inhibits the dephosphorylation of tau during processing. The enriched tau is purified further by 2.5% perchloric acid treatment to eliminate other MAPs (i.e., MAP2) followed by dialysis and (fast protein liquid chromotography) (FPLC) using a Mono S column (Pharmacia, Piscataway, NJ).[12] At this point, the human biopsy tau prepared by this method is >98% pure as monitored by silver-stained, overloaded SDS–PAGE gels.

Normal Adult Tau Purified from Small Brain Biopsies and Autopsy Brain. The second approach is designed to isolate tau from smaller brain biopsies and from normal postmortem brains. This method takes advantage of the ability of highly purified tubulin prepared from fresh bovine brains to bind to tau during the reassembly to tubulin into MTs.[13] The first step of this protocol involves extracting tau from either small biopsy samples or large amounts of postmortem brain with high salt RAB buffer containing the appropriate protease and phosphatase inhibitors as described earlier. The high salt extracts are then boiled and centrifuged, and the supernatant containing the heat-stable tau is then concentrated by 50% ammonium sulfate precipitation. Tau proteins recovered in the pellet after centrifugation are resuspended in a small volume of half concentrated RAB buffer and then dialyzed against the same buffer to remove excess salt. At this stage, tau only represent about 30–40% of the total protein.

Crude biopsy- or autopsy-derived tau after dialysis is treated with 2.5% perchloric acid to eliminate acid-insoluble contaminants, and acid-stable tau is recovered after centrifugation and dialysis. The recovery of autopsy-derived normal adult brain tau is about 1–2 mg/100 g tissue and the purity at this stage is about 70%. Human fetal tau has also been isolated successfully from postmortem brain to similar purity with this protocol. To purify

[12] M. Hasagewa, M. Morishima-Kawashima, T. K. Suzuki, L. Litani, and Y. Ihara, *J. Biol. Chem.* **267,** 17047 (1992).

[13] R. C. Williams, Jr., and H. W. Dietrich III, *Biochemistry* **18,** 2499 (1979).

adult autopsy tau even further, highly purified tubulin prepared from fresh bovine brain is added to autopsy-derived human adult tau to "fish out" tau during reassembly of bovine tubulin into MTs. Bovine tubulin is prepared by the reversible assembly/disassembly method[13] followed by chromatography through a phosphocellulose (PC) column. So-called "PC-tubulin" contains >99% tubulin with no detectable MAPs or other contaminating proteins as monitored in silver-stained overloaded SDS–PAGE gels. Exogenous bovine PC-tubulin is added to the human tau extract at a ratio of 10:1 (w:w), and MT assembly is conducted in the presence of taxol according to Vallee.[14] Tau bound to MTs is recovered as a pellet with the MT polymers after centrifugation. Low levels of residual tubulin in the tau preparation are eliminated after final boiling of the high salt supernatant as described earlier. Using the taxol method, about 1.0–2.0 mg of tau is recovered from 100 g of wet tissue from normal postmortem human brains. The purity of intact adult tau recovered by coassembly is about 90% of total protein as determined by silver-stained gels and the major contaminants are proteolytic fragments of tau itself as judged by Western blot analysis. However, further purification of the autopsy derived adult tau can be achieved by FPLC using a Mono S column.[12]

3. Purification of Human Fetal Tau. To purify fetal tau from postmortem brain tissues with either very short (< than 30 min) or normal (6–10 hr) postmortem intervals, a slightly different strategy is used because of the low affinity of fetal tau for MTs. Previous studies indicate that a ratio of 25:1 (w:w) of exogenous MTs to fetal tau is required for the optimal binding of tau to MTs. Because this requires large amounts of purified PC-tubulin, we purify fetal tau by taking advantage of the fact that fetal tau recovered from fetal brain after boiling and perchloric acid treatment contains relatively few contaminants. We showed that a 95% pure fetal tau preparation can be obtained when perchloric acid treated fetal tau is purified further using a PC column. Fetal tau is usually eluted from the column with about 0.3 *M* NaCl.

Dephosphorylation of Normal Human Tau and PHF-Tau Preparations

Because normal human tau is variably phosphorylated and PHF-tau is hyperphosphorylated, many of the studies would require the preparation of dephosphorylated tau proteins. Soluble heat-stable normal tau proteins can be dephosphorylated easily in a buffer containing 50 m*M* Tris-HCl, 1 m*M* $ZnSO_4$, pH 8.0, and *Escherichia coli* alkaline phosphatase (2–20 units/mg of tau protein). The enzyme reaction is allowed to proceed at 37°

[14] R. B. Vallee, *J. Cell Biol.* **92,** 435 (1982).

for 4 hr (up to 18 hr) and is terminated by the addition of excess sodium phosphate.[3,7] However, insoluble PHF-tau is dephosphorylated poorly at 37° and requires elevated temperature for complete dephosphorylation. PHF-tau can be dephosphorylated at 67° for 1 to 3 hr with *E. coli* alkaline phosphatase as described earlier.[15] Figure 1 shows a Western blot of biopsy-derived tau, autopsy-derived tau, and PHF-tau before and after dephosphorylation.

Acknowledgment

Work was supported by grants from the National Institute on Aging, NIH.

[15] M. Goedert, M. G. Spillantini, N. J. Cairns, and R. A. Crowther, *Neuron* **8,** 159 (1992).

[7] Chemical Modifications of Deposited Amyloid-β Peptides

By JONATHAN D. LOWENSON, STEVEN CLARKE, and ALEX E. ROHER

Introduction

Alzheimer's disease is a dementia characterized by the extracellular accumulation of amyloid in the cortical regions of the brain and in the walls of the cerebral and leptomeningeal blood vessels. This amyloid is composed primarily of insoluble fibrillar amyloid-β (Aβ), peptides of 40 to 42 amino acids derived by proteolysis of the type I transmembrane amyloid precursor protein.[1] The more insoluble 42 residue peptide is the major form of Aβ present in cortical neuritic deposits, or plaques, whereas the more soluble 40 residue peptide is predominant in the walls of the cerebral vasculature.[2,3] Deposition of amyloid fibrils precedes the onset of clinical manifestation of dementia by several decades, during which time the Aβ is subjected to a variety of degradation reactions. Aminopeptidases remove a variable number of N-terminal amino acid residues, yielding a heterogeneous mixture of shorter Aβ peptides, particularly in the amyloid

[1] D. J. Selkoe, *Annu. Rev. Neurosci.* **17,** 489 (1994).

[2] A. E. Roher, J. D. Lowenson, S. Clarke, C. Wolkow, R. Wang, R. J. Cotter, I. M. Reardon, H. A. Zurcher-Neely, R. L. Heinrikson, M. J. Ball, and B. D. Greenberg, *J. Biol. Chem.* **268,** 3072 (1993).

[3] A. E. Roher, J. D. Lowenson, S. Clarke, A. S. Woods, R. J. Cotter, E. Gowing, and M. J. Ball, *Proc. Natl. Acad. Sci. U.S.A.* **90,** 10836 (1993).

0076-6879/99 $30.00

associated with cortical neuritic plaques.[2,3] Spontaneous degradation reactions occurring over time include the isomerization of aspartyl residues,[2] the racemization of aspartyl and seryl residues,[2,4] and the cyclization of N-terminal glutamyl residues.[5–7] Accumulation of these chemical modifications may play a pathological role in Alzheimer's disease because they appear to contribute to the increased insolubility and resistance to enzymatic degradation observed in Aβ fibrils isolated from the brain.[4,8,9] This article describes the techniques utilized in our laboratory to characterize the chemical modifications observed in the Aβ peptides of Alzheimer's disease brains.

Characterization of Isomerized and Racemized Residues in Aβ Peptides

The purification of amyloid core protein from neuritic plaques and amyloid from vascular walls is described elsewhere in this volume.[9a]

Monomeric Aβ (2 mg) purified from the cerebral cortex or leptomeningeal vessels from Alzheimer's disease patients or the synthetic Aβ1–40 or Aβ1–42 (obtained from either California Peptide Inc., Napa Valley, CA, or Bachem, Torrance, CA) is dissolved in 1 ml of 80% (v/v) glass-distilled formic acid and dialyzed (Spectrapor No. 6 membrane, molecular weight cutoff 1000 flat width 18 mm, Spectrum Medical Industries, Inc., Los Angeles, CA) against 2 liter of distilled water for 3 hr, followed by 2 liter ($\times$ 2) of 100 m*M* ammonium bicarbonate under continuous stirring at room temperature. The dialyzed Aβ peptides are digested by trypsin (TPCK treated; Worthington Biochemical Corp., Freehold, NJ) at an approximately 1:50 enzyme:substrate ratio for 14 hr at 37°. Proteolysis is terminated by lyophilization. The tryptic peptides (1 mg) are dissolved in 500 μl of 10% (v/v) glass-distilled formic acid and are separated by high-performance liquid chromatography (HPLC) on an LKB C_{18} reversed-phase column (Spherisorb ODS2, 5-μm bead size, 4.6 $\times$ 250-mm column, Pharmacia-LKB, Sweden). Prior to loading onto the column, the insoluble tryptic

[4] T. Tomiyama, S. Asano, Y. Furiya, T. Shirasawa, N. Endo, and H. Mori, *J. Biol. Chem.* **269,** 10205 (1994).

[5] H. Mori, K. Takio, M. Ogawara, and D. J. Selkoe, *J. Biol. Chem.* **267,** 17082 (1992).

[6] D. L. Miller, I. A. Papayannopoulos, J. Styles, S. A. Bovin, Y. Y. Lin, K. Bieman, and K. Iqbl, *Arch. Biochem. Biophys.* **301,** 41 (1993).

[7] T. C. Saido, T. Iwatsubo, D. M. A. Mann, H. Shimada, Y. Ihara, and S. Kawashima, *Neuron* **14,** 457 (1995).

[8] Y.-M. Kuo, S. Webster, M. R. Emmerling, N. De Lima, and A. E. Roher, *Biochim. Biophys. Acta* **1406,** 291 (1998).

[9] S. B. Vyas and L. K. Duffy, *Biochem. Biophys Res. Commun.* **206,** 718 (1995).

[9a] A. E. Roher and Y.-M. Kuo, *Methods Enzymol.* **309** [4] (1999) (this volume).

core (particularly prevalent in the digests of AβN-42 peptides) is removed by centrifugation at 12,000*g* for 10 min in a 1.5-ml microcentrifuge tube. The column is eluted at a constant flow rate of 0.7 ml min^{-1} at room temperature with a linear gradient of 0–20% acetonitrile/0.1% trifluoroacetic acid over 90 min, followed by a linear gradient of 20–60% acetonitrile/0.1% trifluoroacetic acid for an additional 30 min. The column effluent is monitored at 214 nm and fractions are collected every 30 sec. The insoluble tryptic core, composed mainly of the hydrophobic peptide Aβ29–42, can be digested further by cyanogen bromide in formic acid.[2] The residues in this peptide, however, are not susceptible to the degradation reactions investigated in this article.

Only four peptides should be generated by tryptic digestion of Aβ at its one Arg and two Lys residues, and this is observed when synthetic Aβ is digested as described earlier (Fig. 1a). Peak C contains Aβ1–5, peak I corresponds to Aβ6–16, and peak J is Aβ17–28. When Alzheimer's disease brain-derived neuritic plaque Aβ is digested and the peptides are separated by HPLC, however, a more complicated chromatograph is obtained (Fig. 1b). Amino acid analysis (Table I) and mass spectrometry reveal that peaks A and B have the same mass and composition as peak C and that peaks G and H have the same mass and composition as peak I. The same new peaks are observed when vascular Aβ is subjected to this analysis, although there is less of them than in neuritic plaque Aβ. When these peptides are subjected to gas-phase automatic amino acid sequencing, only peptides in peaks C and I undergo complete Edman degradation. Peptides in peaks A and B are resistant to the Edman reaction, whereas peaks G and H release residue 6 (His) in the first cycle but no amino acids in the subsequent cycles. These results are consistent with the formation of isomerized aspartyl (isoaspartyl) residues in positions 1 and 7 via a spontaneous intramolecular reaction (Fig. 2). Isoaspartyl residues are bonded into the peptide backbone through their β-carboxyl group; although not released by Edman sequencing chemistry, they become normal aspartic acid upon acid hydrolysis. Aspartyl residues are also subject to racemization, resulting in D-aspartyl derivatives (Fig. 2). Both racemized and isomerized residues have the same mass as the L-aspartyl residue. Formation of the succinimide intermediate that produces these damaged forms from L-aspartyl (and in other proteins L-asparaginyl) residues is one of the most common spontaneous reactions affecting proteins as they age.[10–14] These residues are present at low levels

[10] T. Geiger and S. Clarke, *J. Biol. Chem.* **262,** 785 (1987).

[11] R. C. Stephenson and S. Clarke, *J. Biol. Chem.* **264,** 6164 (1989).

[12] I. M. Ota and S. Clarke, *J. Biol. Chem.* **264,** 54 (1989).

[13] J. Najbauer, J. Orpiszewski, and D. W. Aswad, *Biochemistry* **35,** 5183 (1996).

[14] H. T. Wright, *in* "Deamidation and Isoaspartyl Formation in Peptides and Proteins" (D. W. Aswad, ed.), p. 229. CRC Press, Boca Raton, FL, 1995.

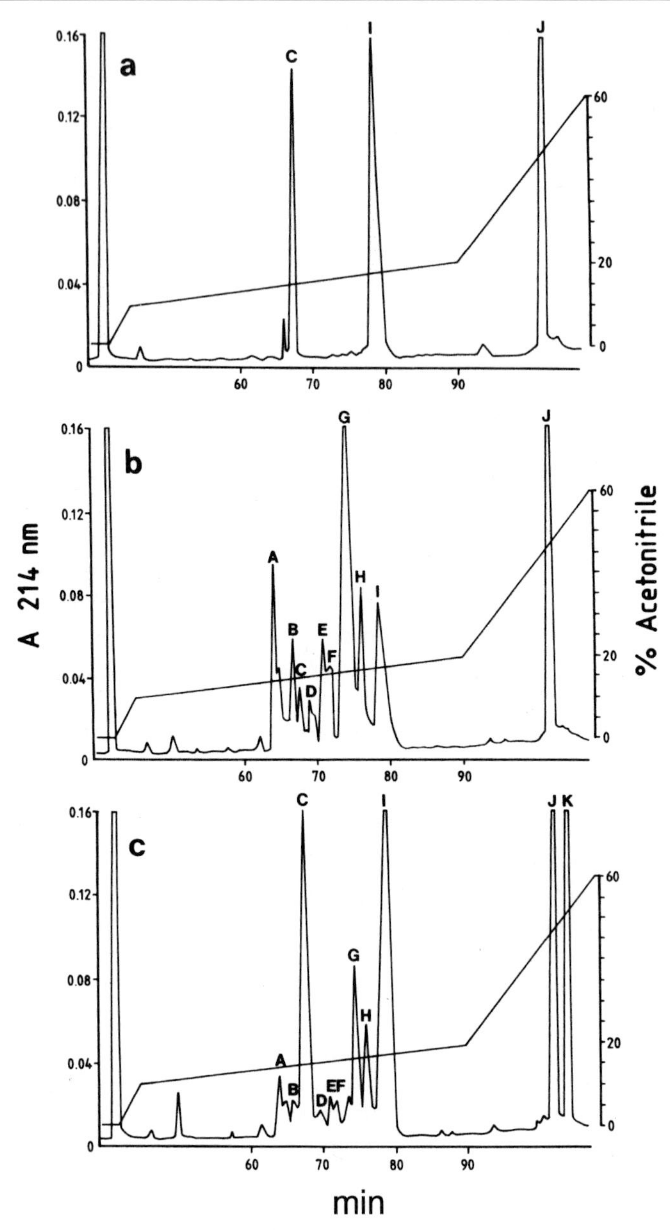

FIG. 1. Chromatographic profiles of neuritic plaque and vascular Aβ tryptic peptides separated on C_{18} reversed-phase HPLC. (a) Synthetic Aβ_{1-42} yields peaks C (residues 1–5), I (residues 6–16), and J (residues 17–28). (b) Neuritic plaque Aβ from Alzheimer's disease brain produces many additional tryptic peptides. (c) Vascular Aβ from leptomeningeal deposits yields lesser quantities of the modified peptides. Adapted with permission from A. E. Roher *et al., J. Biol. Chem.* **268,** 3072 (1993).

TABLE I
TRYPTIC FRAGMENTS OF PARENCHYMAL Aβ

Tryptic fragment	Predicted sequence	Peak	Observed sequence
Tp1	^{1}DAEFR^{5}	A	^{1}DAEFR^{5}
			^{2}AEFR^{5}
		B	^{1}DAEFR^{5}
			^{10}YEVHHQK^{16}
		C	^{1}DAEFR^{5}
Tp2	^{6}HDSGYEVHHQK^{16}	D	^{10}YEVHHQK^{16}
		E, F	^{9}GYEVHHQK^{16}
			^{8}SGYEVHHQK^{16}
		G	^{6}HDSGYEVHHQK^{16}
		H	^{6}HDSGYEVHHQK^{16}
		I	^{6}HDSGYEVHHQK^{16}
Tp3	^{17}LVFFAEDVGSNK^{28}	J	^{17}LVFFAEDVGSNK^{28}
Tp4	^{29}GAIIGLMVGGVVIA^{42}		^{29}GAIIGLMVGGVVIA^{42}

in most proteins, but are particularly abundant in cells with limited protein turnover, such as erythrocytes[15] and eye lens.[16] The presence of a single L-isoaspartyl residue has been observed to alter the structure and function of a number of proteins assayed *in vitro,* suggesting that the accumulation of this damage might be detrimental in living organisms.[17] Isoaspartyl residues are also resistant to enzymatic degradation[18–20] and appear to confer this resistance along with increased insolubility to Aβ.[8]

Determination of Aspartyl Isomerization

Several routine methods used in protein characterization, including those just described, can provide indirect evidence suggesting the presence of isoaspartyl residues. This evidence can be missed, however, if the normal and isoaspartyl forms of the peptide are not first isolated from each other. For example, the decrease in amino acid recovery with each cycle of Edman sequencing can make it difficult to detect the additional loss due to the presence of a small amount of an isoaspartyl residue at a particular position. Peptides that differ solely by the bond formed by a single amino acid can be difficult to separate. It is fortunate that the tryptic digestion of Aβ

[15] J. R. Barber and S. Clarke, *J. Biol. Chem.* **258,** 1189 (1983).
[16] P. N. McFadden and S. Clarke, *Mech. Ageing Dev.* **34,** 91 (1986).
[17] E. Kim, J. D. Lowenson, D. C. MacLaren, S. Clarke, and S. G. Young, *Proc. Natl. Acad. Sci. U.S.A.* **94,** 6132 (1997).
[18] E. D. Murray, Jr., and S. Clarke, *J. Biol. Chem.* **259,** 10722 (1984).
[19] B. A. Johnson and D. W. Aswad, *Biochemistry* **29,** 4373 (1990).
[20] J. D. Lowenson and S. Clarke, *J. Biol. Chem.* **267,** 5985 (1992).

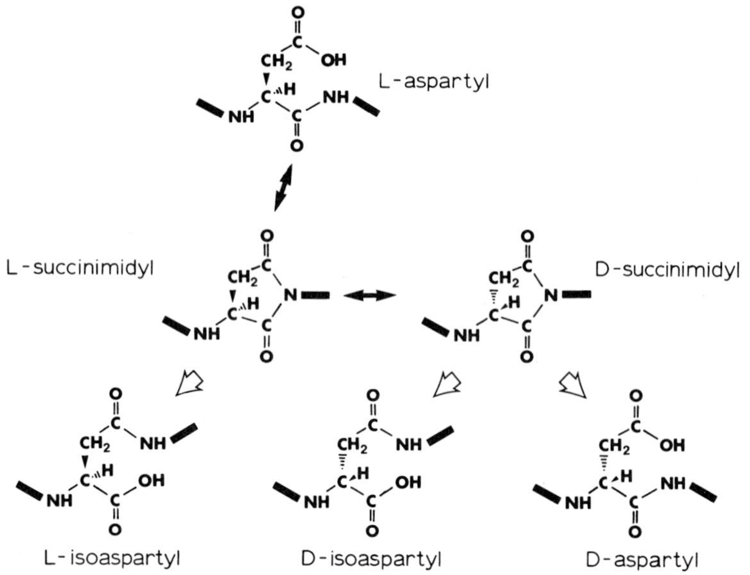

FIG. 2. Spontaneous reactions leading to the formation of altered aspartyl residues in polypeptides. At pH 7.4 and 37°, L-aspartyl residues in peptides and proteins can degrade to the L-succinimidyl form with $t_{1/2}$ values as short as 41 days for an Asp-Gly sequence and 168 days for an Asp-Ser sequence.[11] In similar peptides, $t_{1/2}$ values of 19.5 hr for racemization of the L-succinimide and 2.3 hr for hydrolysis of the succinimide to the normal and isoaspartyl forms have been observed.[18] Although not shown, each of these reactions is reversible. The attachment of the residue to the peptide backbone is represented by bold lines. Reproduced with permission from A. E. Roher *et al.*, *J. Biol. Chem.* **268,** 3072 (1993).

produces peptides that are small and balanced with hydrophobic and hydrophilic residues, allowing the aspartyl and isoaspartyl forms of the peptide to elute differently from a reversed phase HPLC column. The acidic pH of the 0.1% trifluoroacetic acid running buffer accentuates the pK_a difference between the free β-carboxyl group of the normal aspartyl residue and the more acidic free α-carboxyl of the isoaspartyl residue.

The best way to quantitate L-isoaspartyl residues directly is to use the L-isoaspartate (D-aspartate) *O*-methyltransferase (EC 2.1.1.77) as an analytical probe. This enzyme methylates the α-carboxyl group of L-isoaspartyl residues and, with approximately 1000-fold lower affinity, the β-carboxyl group of D-aspartyl residues, but does not recognize L-aspartyl residues in peptides or proteins.[20] Because methylation of D-aspartyl residues can, in some cases, be misinterpreted as L-isoaspartyl methylation,[20] additional characterization, including the racemization analysis described later, may be necessary. When *S*-adenosyl-L-[*methyl*-^{14}C]methionine ([^{14}C]AdoMet) or its [^{3}H]methyl analog is used as the methyl donor, L-isoaspartyl residues are radiolabeled by this methyltransferase. The methylation reaction is terminated by the addition of 50 μl of 0.2 *M* sodium hydroxide, 1% (w/v) sodium dodecyl sulfate (SDS), which also converts the base-labile methyl esters on the L-isoaspartyl residues to [^{14}C]methanol. In a "vapor diffusion assay," an aliquot (50–70 μl) of each base-treated incubation mixture is spotted on a 2 × 8-cm piece of thick filter paper (Bio-Rad, Richmond, CA) that had been prefolded in an accordion pleat.[21] This is wedged quickly into the neck of a 20-ml scintillation vial containing 10 ml of Safety-Solve II counting fluor (RPI), which is immediately capped tightly and allowed to equilibrate at room temperature for 2 hr. During this time, the volatile [^{14}C]methanol diffuses into the fluor, whereas the non-volatile [^{14}C]AdoMet remains on the filter paper. The filter paper is then removed and the vials are counted.

The volatile radioactivity obtained from a no-substrate blank containing methyltransferase, [^{14}C] AdoMet, and buffer is subtracted from each Aβ sample incubated for the same length of time. Each assay also includes a positive control in which a saturating quantity of a stock methyltransferase substrate [e.g., chicken ovalbumin (grade V from Sigma) or a synthetic L-isoaspartyl-containing peptide] is incubated under the same conditions as the Aβ samples to measure the maximum number of methyl groups that could be transferred during the assay.

In a typical assay, intact Aβ (150 pmol, as determined by amino acid analysis) or Aβ tryptic peptides (10 pmol) are incubated at 37° for various times in 20 μl of 0.1 *M* sodium citrate, pH 6, containing 3.2 units of methyltransferase and 10 μ*M* [^{14}C]AdoMet (ICN, Costa Mesa, CA, 52 mCi/mmol). The methyltransferase in this example was purified from human erythrocyte cytosol[21] to about 5000 units mg^{-1} [1 unit: 1 pmol methyl groups transferred/min to saturating (1 m*M*) chicken ovalbumin, a good substrate with an apparent K_m of 35 μ*M*]. The reaction is buffered to pH 6 because, although the methyltransferase is less active at this pH, the methyl esters produced are more stable and therefore less likely to spontaneously demethylate.[22] The concentration of [^{14}C]AdoMet, at 5-times its K_m for the methyltransferase, will not limit the rate of the reaction. As is illustrated in Fig. 3, solubilized Aβ1–42 from neuritic plaques accepts seven-fold more methyl esters than synthetic Aβ1–42, demonstrating that it contains at least one methylatable L-isoaspartyl residue. The stoichiometry of methylation is only 1 methyl per 21 Aβ molecules, but the continued increase in methylation with time indicates that all the available L-isoaspartyl residues have

[21] J. M. Gilbert, A. Fowler, J. Bleibaum, and S. Clarke, *Biochemistry* **27,** 5227 (1988).
[22] E. D. Murray, Jr., and S. Clarke, *J. Biol. Chem.* **261,** 306 (1986).

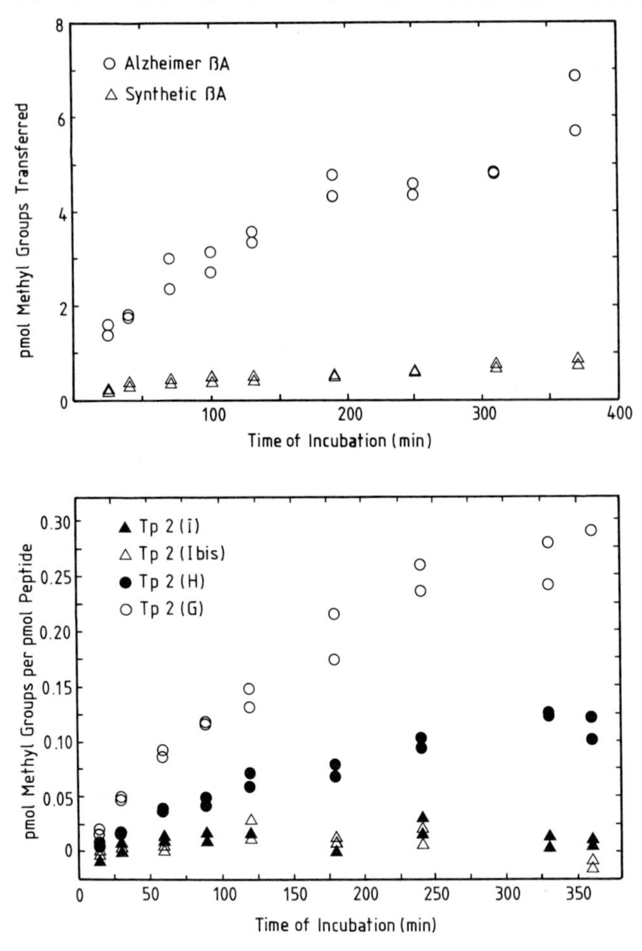

FIG. 3. Detection of altered aspartyl residues in Aβ and Aβ tryptic peptides by methylation with L-isoaspartate (D-aspartate) *O*-methyltransferase. Neuritic plaque or synthetic Aβ1–42 (150 pmol; top) and the tryptic peptides Aβ6–16 (10 pmol; bottom) are incubated with 3.2 units of methyltransferase and 10 μM [^{14}C]AdoMet at 37° for the times indicated. Reproduced with permission from A. E. Roher *et al., J. Biol. Chem.* **268,** 3072 (1993).

not been methylated (Fig. 3). An assay of the tryptic peptides in peaks G, H, I, and I-bis (a trailing shoulder of peak I) shows that G and H contain L-isoaspartyl residues, whereas I and I-bis do not (Fig. 3). As with intact Aβ, the absolute number of L-isoaspartyl residues cannot be determined from these results because even 360 min of incubation time is not long enough to fully methylate the peptide. This assay, however, combined with the mass spectrometry, sequencing, and composition analyses, indicates that peak G consists solely of the L-isoaspartyl-containing form of Aβ6–16, and thus the amount of this damage in the Aβ can be obtained from the UV absorbance on the HPLC chromatograph.

If an L-isoaspartyl-containing peptide or protein cannot be purified to homogeneity, the quantity of L-isoaspartyl residues can still be determined using the methylation assay as long as enough enzyme is present, the affinity of the substrate for the methyltransferase is high enough, and the reaction is allowed to proceed to completion. If the first two conditions are met, a time course of methylation should demonstrate saturation of the available substrate within 2 hr. If longer incubations are attempted, it should be determined whether the enzyme remains active and that the incubation is free from active proteases.

In the assay of Aβ described earlier, 3.2 units of methyltransferase is not enough to fully methylate the L-isoaspartyl residue in peak G within a time period during which the enzyme remains active. Other L-isoaspartyl-containing peptides of similar size and sequence to Aβ6–16 have K_m values of 1–16 μM,[23] but the concentration of peptide in this assay is just 0.33 μM, and therefore the reaction proceeds at a rate well below its maximal velocity. Simply adding more enzyme, the concentration of which is just 0.16 units μl^{-1}, dilutes the peptide further, counteracting the increase in enzyme activity. To circumvent this problem we have begun using recombinant human L-isoaspartyl (D-aspartyl) *O*-methyltransferase expressed in *Escherichia coli*,[24] which, when purified to a specific activity of more than 13,000 units mg^{-1} and a concentration of up to 58 units μl^{-1}, can provide as much as 350-fold higher activity in the same incubation volume, allowing faster methylation of even poor substrates. It should be noted, however, that while the recombinant methyltransferase provides high activity combined with a very low background of endogenous methylation due to its purity, even crude erythrocyte cytosol can be used to identify and, in some cases, quantitate L-isoaspartyl residues. Blood is drawn from a finger tip, disodium EDTA (1 mg ml^{-1} blood) is added to prevent clotting, and the erythrocytes are washed five times in 10 volumes of phosphate-buffered saline to remove L-isoaspartyl-containing plasma proteins. These cells are lysed by freezing in 5 volumes of 5 mM sodium phosphate, 1 mM EDTA, pH 7.4, to which 25 μM phenylmethylsulfonyl fluoride is added just prior to use. On thawing, the membrane ghosts are removed by centrifugation at 16,000*g* for 10 min at 4°. The cytosolic methyltransferase is only about 5–10 units mg^{-1} protein, but at 0.07–0.17 units μl^{-1} is suitable for some

[23] J. D. Lowenson and S. Clarke, *J. Biol. Chem.* **266,** 19396 (1991).
[24] D. C. MacLaren and S. Clarke, *Prot. Express. Purif.* **6,** 99 (1995).

analyses. For example, both recombinant methyltransferase (11.5 units, 0.2 μl) and crude erythrocyte cytosol (1.4 units, 18.68 μl) can stoichiometrically methylate 4.2 pmol of the high-affinity substrate VYP-isoD-HA, which has a K_m of 0.29 μM,[23] within 30 min (Fig. 4, top). The cytosolic methyltransferase, however, can only methylate half of the 3.2 pmol of L-isoaspartyl residues expected in 50 pmol of ovalbumin during a 40-min incubation (Fig. 4, bottom),[2] but still provides a good estimate of the amount of damaged residues. When using crude cytosol, care must be taken to assay and subtract endogenous methylation, ascertain that the substrate is not being degraded by erythrocyte proteases, and avoid the use of 2-mercaptoethanol, which is methylated by an erythrocyte thiol methyltransferase and can interfere with the vapor diffusion assay.

There are, however, some limitations in the use of methyltransferase to detect L-isoaspartyl residues. For example, the methyltransferase cannot be used to confirm that Aβ1–5 peak A (Fig. 1B) contains an L-isoaspartyl residue. Because this residue is at the N terminus, its affinity for the enzyme is too low to be measured given the amount of peptide available. In this case, the physical data presented earlier, as well as the racemization data given later, strongly support the presence of the L-isoaspartyl residue. As further support, the peptide L-isoAsp-Ala-Glu-Phe-Arg can be synthesized and shown to coelute with Aβ1–5 peak A. Furthermore, the soluble carbodiimide EDC could be used to attach a CBZ-derivatized amino acid to the amino terminus of the tryptic peptide, which would probably make it a much better methyltransferase substrate.[23,25] In fact, simply acetylating the N-terminal amino group of Aβ6–16 peak G, in which the L-isoaspartyl residue is penultimate to the N terminus, with acetic anhydride caused it to be methylated 2.5-fold faster than the unacetylated form (unpublished data, but see Ref. 15).

Peptides containing two aspartyl and/or asparaginyl residues provide another complication for the L-isoaspartyl methyltransferase assay. If two sites on a peptide have become damaged, a fraction of peptides in the population should contain two L-isoaspartyl residues, while many more will contain one such residue, with a different aspartyl derivative at the other site. These peptides can be distinguished by methylating them as described earlier and then separating them by reversed-phase HPLC. Methylated peptides are significantly more hydrophobic than their unmethylated forms. New peaks eluting later than the control peptide are collected and their radioactivity is quantitated in a scintillation counter. The amount of each peptide collected is calculated from its absorbance monitored with a UV

[25] P. Galletti, D. Ingrosso, C. Manna, F. Sica, S. Capasso, P. Pucci, and G. Marino, *Eur. J. Biochem.* **177,** 233 (1988).

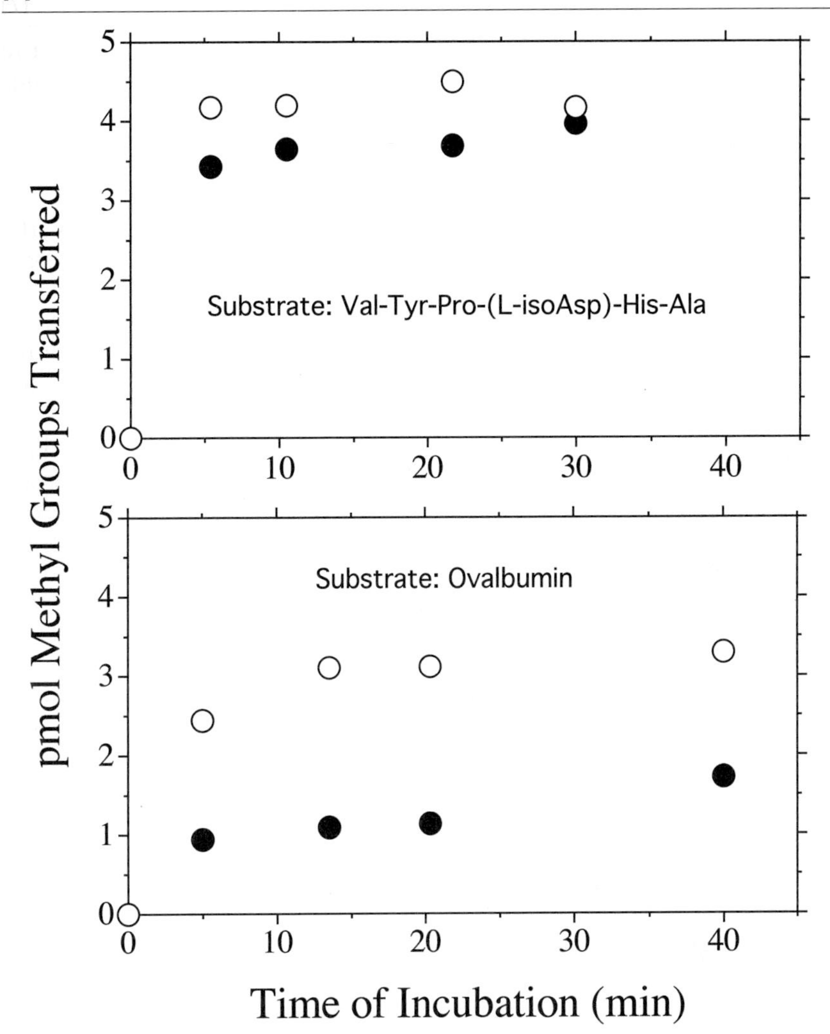

FIG. 4. Quantitation of L-isoaspartyl residues: Comparison of recombinant human L-isoaspartate (D-aspartate) *O*-methyltransferase, and crude human erythrocyte cytosol. The synthetic peptide Val-Tyr-Pro-(L-isoAsp)-His-Ala (4.2 pmol; top) or chicken ovalbumin (50 pmol; bottom) is methylated with 10 μM[^{14}C]AdoMet and 11.5 units of recombinant methyltransferase (○) or 1.4 units of crude erythrocyte cytosol methyltransferase (●). Just over 1 in 16 ovalbumin molecules contain a methylatable L-isoaspartyl residue, so about 3.2 pmol methyl esters are expected from 50 pmol ovalbumin.

detector. Peptides containing two L-isoaspartyl sites will show 2 pmol of methyl groups per pmol of peptide and should be the most hydrophobic peptides in the population.

Determination of Aspartyl Racemization

Although a link between isomerization and racemization of aspartyl residues had previously been observed *in vitro* with synthetic peptides,[10] analysis of Aβ from Alzheimer's disease brain provided the first evidence that these reactions also occurred at the same site *in vivo*. This has since been observed in aged eye lens crystallin,[26] and a mechanism by which the structure of the succinimide intermediate accelerates racemization[10] has been supported by *ab initio* theoretical calculations.[27]

To quantitate the abundance of D-aspartyl residues, Aβ tryptic peptides are dried down in 6 × 50-mm glass tubes that have been heated previously to 240 ° for 24 hr to destroy any contaminating amino acids. These peptides are hydrolyzed with vaporized HCl *in vacuo* at 108° for 3 hr using a PicoTag workstation (Waters, Milford, MA). Hydrolysis for 3 hr is long enough to release all aspartyl and asparaginyl residues, but short enough to limit the racemization resulting from these harsh conditions.[28] D-Aspartyl, D-isoaspartyl, and D-asparaginyl residues are each released as D-aspartate by hydrolysis. The hydrolysates are resuspended in double distilled water, and the amino acids are derivatized with *N*-acetyl-L-cysteine and ortho-phthalaldehyde.[29] The resulting fluorescent diastereomers are separated isocratically on a 3.9 × 150-mm reversed-phase Resolve C_{18} column (Waters) equilibrated in a mixture of 95% buffer A (50 m*M* sodium acetate, pH 5.4) and 5% buffer B (80% methanol, 20% buffer A)[28] and monitored with a Gilson Model 121 fluorometer (Middleton, WI. The excitation and emission filters used have windows of 305–395 and 430–470 nm, respectively. Under these conditions, the derivatives containing D-aspartate and L-aspartate elute at about 6.3 and 7.3 min, respectively. Their fluorescence color constants are determined with D- and L-aspartate standards. Comparison of the D-aspartate present in an L-aspartate standard before and after undergoing the acid hydrolysis used on the peptides, as well as the analysis of acid-hydrolyzed synthetic Aβ, indicates that no more than 2% of observed D-aspartate is due to racemization during the procedure. The method of

[26] N. Fujii, Y. Ishibashi, M. Fujino, and K. Harada, *Biochim. Biophys. Acta* **1204,** 157 (1994).
[27] J. L. Radkiewicz, H. Zipse, S. Clarke, and K. N. Houk, *J. Am. Chem. Soc.* **118,** 9148 (1996).
[28] L. S. Brunauer and S. Clarke, *J. Biol. Chem.* **261,** 12538 (1986).
[29] D. W. Aswad, *Anal. Biochem.* **137,** 405 (1984).

D'Aniello *et al.*[30] is reported to give a much lower racemization background but requires several enzymatic steps that may be difficult with very small samples.

The results of this analysis show that whereas peaks A and G contain primarily L-aspartate, peaks A-bis (the trailing shoulder of peak A) and H (the trailing shoulder of peak G) each contain more than 50% D-aspartate (Fig. 5). Combined with the blockage to sequencing, this indicates that these peptides contain D-isoaspartyl residues. As seen in Fig. 1B, neither peptide is well resolved from the more abundant L-isoaspartyl-containing analogs. In fact, the finding that half of the peptide in peak H contains an L-isoaspartyl residue explains why this material is methylated half as well as peak G (Fig. 3). The D-aspartate fraction in peak I-bis is twice that in peak I, indicating that this poorly resolved shoulder peak is the D-aspartyl-containing $A\beta_{6-16}$ contaminated with the L-aspartyl form (Fig. 5). Peak C from both plaque Aβ (Fig. 5 "C") and vascular Aβ (Fig. 5 "M-C") contains very little D-aspartate, confirming its identity as L-aspartyl-Aβ1–5. Interestingly, the elution pattern of L-isoaspartyl-, D-isoaspartyl-, L-aspartyl-, and D-aspartyl-Aβ6–16 peptides on reversed phase HPLC closely matches that observed with a synthetic tetrapeptide[31] and an 18 residue tryptic peptide[32] containing isomerized and racemized residues, suggesting that this pattern is independent of peptide length and sequence. The relative elution positions of the Aβ1–5 peptides are more difficult to interpret due to N-terminal heterogeneity of the Aβ. Although L-isoaspartyl-, D-isoaspartyl-, and L-aspartyl-Aβ1–5 elute in peaks A, A-bis, and C, respectively, peak B apparently contains both L-isoaspartyl and D-isoaspartyl residues, and D-aspartyl-Aβ1–5, the least abundant of the peptides, has not been identified.

Taken together, our analyses show that about 75% of the aspartyl residues in positions 1 and 7 in Aβ from dense neuritic plaques are in the isoaspartyl configuration and that about 25% of both the aspartyl and the isoaspartyl residues have racemized to the uncommon D form.[2] Aβ purified from vascular amyloid deposits has fewer damaged residues (Fig. 1C): only 20–25% of the residues in positions 1 and 7 are in the isoaspartyl configuration,[3] suggesting that this Aβ is younger than the dense neuritic plaque material or else is in a microenvironment that limits succinimide formation.

[30] A. D'Aniello, L. Petrucelli, C. Gardner, and G. Fisher, *Anal. Biochem.* **213,** 290 (1993).
[31] P. N. McFadden and S. Clarke, *Proc. Natl. Acad. Sci. U.S.A.* **84,** 2595 (1987).
[32] T. V. Brennan, J. W. Anderson, Z. Jia, E. B. Waygood, and S. Clarke, *J. Biol. Chem.* **269,** 24586 (1994).

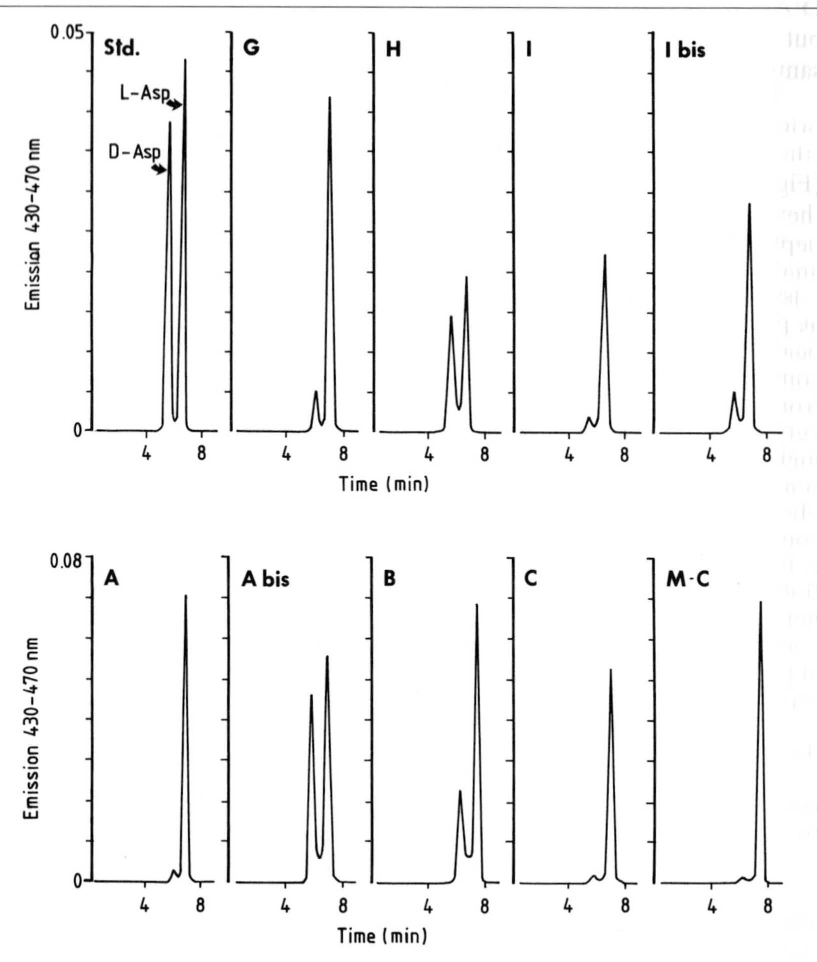

FIG. 5. Stereoconformational determination of Asp_1 and Asp_7 in $A\beta$ tryptic peptides. Aspartyl residues, released from the indicated neuritic plaque $A\beta$ tryptic peptides by mild acid hydrolysis, are derivatized with *N*-acetyl-L-cysteine and phthalaldehyde and resolved by reversed-phase HPLC. (Top) Analysis of Asp in neuritic plaque $A\beta$1–6 samples G, H, I, and I-bis and D L-aspartate standard. (Bottom) Analysis of Asp_1 in neuritic plaque $A\beta$1–5 samples A, A-bis, B, and C and meningeal $A\beta$1–5 sample C (M-C). Reproduced with permission from A. E. Roher *et al., J. Biol. Chem.* **268,** 3072 (1993).

Racemization of Seryl Residues in Aβ

Like aspartyl residues, racemized seryl residues have been observed in long-lived proteins.[33,34] Some synthetic Aβ derivatives containing a D-seryl residue are found to be more resistant to proteolytic degradation.[35] We measured the level of serine racemization in Aβ using the same method as for aspartyl residues. With the system described in this article, derivatized L-serine and D-serine elute at about 11 and 12 min, respectively. Hydrolysis of synthetic Aβ1–42 produces 1.6% D-serine, so this is the maximum D-serine that can be generated by the experimental procedures. Hydrolysis of Aβ1–42 from dense neuritic plaques, however, releases 6.2% D-serine, indicating that at least 4.6% of the two seryl residues in this Aβ is racemized. The serine in position 8 contains the same amount of the D form as does Aβ1–42, indicating that this residue and serine-26 are probably equivalently racemized.

Characterization of Aβ 3-Pyroglutamyl Residues

When glutamate-3 is exposed at the amino terminus of Aβ by proteolysis of the first two residues, it can undergo conversion into a pyroglutamyl residue (3pE). This modification prevents further proteolytic degradation of the Aβ-3pE by aminopeptidases, thus promoting the accumulation of this peptide. Immunocytochemical studies[7] and mass spectrometric analyses of complex mixtures of Aβ peptides[5,6,36] provided the first indication of pyroglutamyl in some of the Aβ. Tryptic digestion followed by separation of the resulting N-terminal peptides by HPLC permitted the quantitation of Aβ-3pE levels in Alzheimer's disease brains.[36]

Amyloid is purified from neuritic plaques and leptomeningeal vascular deposits,[2,36] and the Aβ isolated by FPLC and digested by trypsin as described earlier. The resulting tryptic peptides are separated by HPLC on an LKB C_{18} reversed-phase column (Spherisorb ODS2, 3-μm bead size, 4.6 × 100-mm column, Pharmacia-LKB, Sweden) using a linear gradient at a flow rate of 0.7 ml min^{-1} of 5–8.5% acetonitrile/0.05% trifluoroacetic acid over 70 min at room temperature. The column effluent is monitored at 214 nm and fractions are collected every 30 sec. This shallow gradient allows for a better separation of the N-terminal peptides ending at residue

[33] R. Shapira and C. H. Chou, *Biochem. Biophys. Res. Commun.* **146,** 1342 (1987).

[34] R. Shapira, G. E. Austin, and S. S. Mirra, *J. Neurochem.* **50,** 69 (1988).

[35] I. Kaneko, N. Yamada, Y. Sakuraba, M. Kamenosono, and S. Tutumi, *J. Neurochem.* **65,** 2585 (1995).

[36] Y.-M. Kuo, M. R. Emmerling, A. S. Woods, R. J. Cotter, and A. E. Roher, *Biochem. Biophys. Res. Commun.* **237,** 188 (1997).

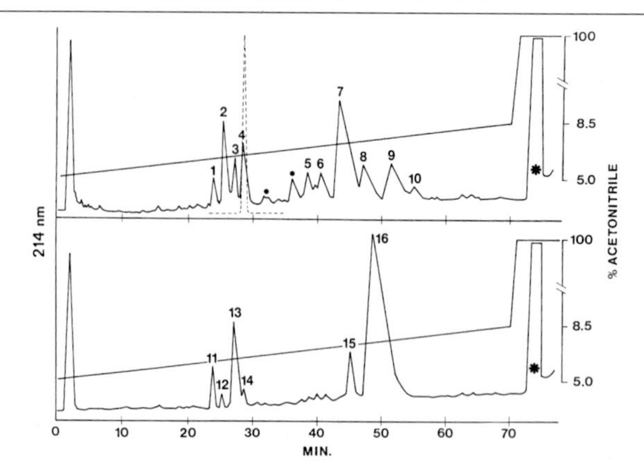

FIG. 6. Purification of Aβ 3-pyroglutamyl peptide on C_{18} reversed-phase HPLC. The N-terminal peptide 3pE-Aβ3–5 elutes as peak 4 in a tryptic digest of neuritic plaque core Aβ (top) and as peak 14 in a tryptic digest of vascular Aβ (bottom). The tryptic peptide in peak 4 coelutes exactly with the synthetic peptide pyroglutamylphenylalanylarginine (broken line). Reproduced with permission from Y.-M. Kuo *et al., Biochem. Biophys. Res. Commun.* **237,** 188 (1997).

TABLE II
Aβ N-TERMINAL TRYPTIC PEPTIDES FROM NEURITIC PLAQUE CORE AND VASCULAR AMYLOID

HPLC peak[c]	Sequence	Aβ residues	Neuritic plaque Aβ peptides (%)			
			A[a]	B[a]	C[a]	Average
1	AEFR	2–5	9	7	9	8
2	(isoD)AEFR	1–5	24	13	22	20
3	DAEFR	1–5	11	7	11	10
4	pEFR	3pE-5	46	62	45	51
5	GYEVHHQK	9–16	5	4	6	5
6	SGYEVHHQK	8–16	5	7	7	6
			Vascular Aβ peptides (%)			
		Aβ residues	D[b]	E[b]	F[b]	Average
11	AEFR	2–5	23	19	20	20
12	(isoD)AEFR	1–5	5	7	5	6
13	DAEFR	1–5	60	64	65	63
14	pEFR	3pE-5	12	10	10	11

[a] A, B, and C represent three independent neuritic plaque amyloid preparations.
[b] D, E, and F represent three independent vascular amyloid preparations.
[c] See Fig. 6.

Arg-5. Peptide 4 from neuritic plaque Aβ and an equivalent peptide from vascular Aβ (Fig. 6) elute with the same retention time as synthetic pyroglutamylphenylalanylarginine (obtained from California Peptide Inc.) at 28.7 min. Amino acid analysis confirmed that these peptides are tripeptides composed of glutamate, phenylalanine, and arginine (Aβ3–5). However, their molecular weight obtained by mass spectrometry, being 18 mass units lower than expected for this tripeptide, confirms the loss of water caused by cyclization of the N-terminal glutamyl residue.

Conclusion

The procedures described in this article demonstrate that few of the Aβ peptides present in dense neuritic plaques in Alzheimer disease brain are untouched by one or more of several spontaneous chemical modifications. Only 10% of the peptides possess the normal L-aspartyl residue expected at the N terminus, whereas 20% start with an L-isoaspartyl residue, 51% start with a pyroglutamyl residue, and the rest have suffered various amounts of proteolysis (Table II). Only 20% of the residues in position 7 are in the normal L-aspartyl configuration, with the rest being isomerized and/or racemized. Racemization also effects 4–6% of the serine residues. If these modifications arise randomly through the Aβ population in a stable amyloid deposit, fewer than 2% of the peptides should remain unaltered. Even if chemical modification is not random (i.e., younger peptides are undamaged, whereas older peptides are multiply modified), no more than 10% of the Aβ will be unaltered at the N terminus. Aβ from vascular amyloid deposits is less damaged, and thus perhaps younger, than that in neuritic deposits, but modification of 37% of these peptides at the N-terminus is still significant (Table II). It is therefore important to take these modifications into account when studying the role of Aβ in the development and progression of Alzheimer's disease.

[8] Monoclonal Antibodies Specific for the Native, Disease-Associated Isoform of the Prion Protein

By CARSTEN KORTH, PETER STREIT, and BRUNO OESCH

Introduction

Prion diseases are transmissible neurodegenerative diseases of humans and other mammals (see Ref. 1 for extensive review). The infectious agent, the prion, is thought to comprise a conformational isoform PrP^{Sc} of a membrane-anchored ubiquitous host protein PrP^{C} of unknown function. During the replication of prions it is hypothesized that disease-specific PrP^{Sc} converts host-resident PrP^{C}, possibly with the help of other cofactors, to new PrP^{Sc} molecules. The two isoforms of the prion protein, PrP^{C} and PrP^{Sc}, have the same amino acid sequence, but are folded differently.[2] On conversion to the disease-specific isoform, PrP^{Sc} aquires characteristic features such as partial protease resistance, increase in β-sheet structure, and insolubility as opposed to PrP^{C}, which is protease-sensitive, consists primarily of an α-helical secondary structure, and is soluble.[3,4]

The existence of a differently folded, conformational isoform of a normal host protein that is related intimately to a disease-causing process would predict the existence of epitopes on the surface of PrP^{Sc} that are unique to this isoform and can be detected by the humoral immune response on immunization. During natural or experimental prion infection in wild-type animals, no peripheral humoral immune response against PrP epitopes can be seen,[5,6] suggesting that the humoral immune response against the disease-specific isoform PrP^{Sc} is suppressed either by self-tolerance or by another mechanism.

Monoclonal antibodies against PrP have been produced using a protocol comprising immunization of wild-type mice with highly purified PrP^{Sc} mole-

[1] S. B. Prusiner, *in* "Field's Virology" (B. N. Fields, D. M. Knipe, and P. M. Howley, eds.), p. 2901. Lippincott Raven, Philadelphia, 1996.

[2] N. Stahl, M. A. Baldwin, D. B. Teplow, L. Hood, B. W. Gibson, A. L. Burlingame, and S. B. Prusiner, *Biochemistry* **32,** 1991 (1993).

[3] M. P. McKinley, D. C. Bolton, and S. B. Prusiner, *Cell* **35,** 57 (1983).

[4] K. M. Pan, M. Baldwin, J. Nguyen, M. Gasset, A. Serban, D. Groth, I. Mehlhorn, Z. Huang, R. J. Fletterick, F. E. Cohen, and S. B. Prusiner, *Proc. Natl. Acad. Sci. U.S.A.* **90,** 10962 (1993).

[5] R. A. Barry, M. P. McKinley, P. E. Bendheim, G. K. Lewis, S. J. DeArmond, and S. B. Prusiner. *J. Immunol.* **135,** 603 (1985).

[6] J. M. Bockman and D. T. Kingsbury, *J. Virol.* **62,** 3120 (1988).

0076-6879/99 $30.00

cules from hamster.[7,8] The resulting monoclonals, however, did not recognize native (i.e., folded), disease-associated PrP^{Sc}, but rather linear epitopes on either folded PrP^{C} or unfolded PrP of both isoforms.[8,9] The difficulties in producing monoclonal antibodies in wild-type mice able to bind exclusively to the native disease-specific isoform of the prion protein have been attributed to the high conservation of the PrP sequence between mammalian species.

When immunizing mice that are lacking a functional PrP gene[10] with highly purified PrP^{Sc}, recombinant antibodies derived from hybridoma cell mRNA could be produced that recognized both native PrP^{Sc} and PrP^{C}, but were unable to distinguish between these isoforms in their native conformation.[11] Furthermore, Williamson *et al.*[11] could not establish stable hybridoma cell lines secreting such antibodies, possibly because of an antibody-induced reaction to myeloma-resident PrP after the fusion of knockout splenocytes with myeloma cells. The same line of knockout mice has also been subjected to a DNA immunization protocol, but again, no conformation-specific monoclonal antibodies were obtained.[12]

Peptides comprising linear sequences of the prion protein have been used to immunize wild-type mice or rabbits.[13–18] The idea is, of course, that particular stretches of the linear PrP sequence are hidden in PrP^{C} but become exposed during conformational conversion to PrP^{Sc}. However, no antibodies specific for the disease-specific conformation have been reported with this approach either. It cannot be excluded that further attempts with peptide immunizations in knockout mice might yield interesting results

[7] R. A. Barry and S. B. Prusiner, *J. Infect. Dis.* **154,** 518 (1986).

[8] R. J. Kascsak, R. Rubenstein, P. A. Merz, M. Tonna DeMasi, R. Fersko, R. I. Carp, H. M. Wisniewski, and H. Diringer, *J. Virol.* **61,** 3688 (1987).

[9] M. Rogers, D. Serban, T. Gyuris, M. Scott, T. Torchia, and S. B. Prusiner, *J. Immunol.* **147,** 3568 (1991).

[10] H. Bueler, M. Fischer, Y. Lang, H. Bluethmann, H.P. Lipp, S. J. DeArmond, S. B. Prusiner, M. Aguet, and C. Weissmann, *Nature* **356,** 577 (1992).

[11] R. A. Williamson, D. Peretz, N. Smorodinsky, R. Bastidas, H. Serban, I. Mehlhorn, S. J. DeArmond, S. B. Prusiner, and D. R. Burton, *Proc. Natl. Acad. Sci. U.S.A.* **93,** 7279 (1996).

[12] S. Krasemann, M. Groschup, S. Harmeyer, G. Hunsmann, and W. Bodemer, *Mol. Med.* **2,** 725 (1996).

[13] R. A. Barry, S. B. Kent, M. P. McKinley, R. K. Meyer, S. J. DeArmond, L. E. Hood, and S. B. Prusiner, *J. Infect. Dis.* **153,** 848 (1986).

[14] M. Shinagawa, E. Munekata, S. Doi, K. Takahashi, H. Goto, and G. Sato, *J. Gen. Virol.* **67,** 1745 (1986).

[15] J. Safar, M. Ceroni, P. Piccardo, D. C. Gajdusek, and C. J. Gibbs, *Neurology* **40,** 513 (1990).

[16] A. Di Martino, E. Bigon, G. Corona, and L. Callegaro, *J. Mol. Recognit.* **4,** 85 (1991).

[17] T. Yokoyama, K. Kimura, Y. Tagawa, and N. Yuasa, *Clin. Diagn. Lab. Immunol.* **2,** 172 (1995).

[18] T. Yokoyama, S. Itohara, and N. Yuasa, *Arch. Virol.* **141,** 763 (1996).

once the design of suitable peptides can be based on novel insights into PrP structure.

Polyclonal antibodies have also been raised against PrP isoforms. These antisera from rabbits or knockout mice were consistently reported to bind to native PrP^{Sc} and PrP^{C}, as well as to unfolded PrP, but no antiserum would distinguish between PrP isoforms.[19–22]

This article describes in detail how a monoclonal antibody (MAb) has been raised that specifically recognizes only native, disease-associated PrP^{Sc} but not normal PrP^{C}.[22] The protocol involved immunization of knockout mice with recombinant bovine PrP produced in *Escherichia coli* and thorough screening of hybridoma for the production of monoclonal antibodies against recombinant bovine PrP as well as disease-associated PrP^{BSE} (i.e., native, bovine PrP^{Sc}).

Preparation of Immunogen

As an antigen to immunize knockout animals we have chosen to express recombinant PrP in inclusion bodies of *E. coli*. Because it was the initial goal to develop a test for the detection of bovine spongiform encephalopathy (BSE), we produced recombinant bovine PrP (rbPrP). cDNA coding for bovine PrP is obtained by polymerase chain reaction (PCR) of genomic bovine DNA (Promega, Madison, WI) with the following primers:

5′-GGGAATTCCATATGAAGAAGCGACCAAAACCTG-3′ (sense)

and

5′-CGGGATCCTATTAACTTGCCCCTCGTTGGTA-3′ (antisense)

Primers are designed such that a 5′*Nde*I and a 3′*Bam*HI restriction site are added to the cDNA of PrP that is deleted of the N-terminal signal sequence and the C-terminal GPI sequence. The added restiction sites allow cloning of bovine PrP cDNA into the pET11a vector (Novagen, USA). DNA sequencing confirms the sequence for bovine PrP containing six octarepeats.[23] The translated rbPrP would thus be expected to start with

[19] R. Gabizon, M. P. McKinley, D. Groth, and S. B. Prusiner, *Proc. Natl. Acad. Sci. U.S.A.* **85,** 6617 (1988).

[20] S. B. Prusiner, D. Groth, A. Serban, R. Koehler, D. Foster, M. Torchia, D. Burton, S. L. Yang, and S. J. DeArmond, *Proc. Natl. Acad. Sci. U.S.A.* **90,** 10608 (1993).

[21] S. Krasemann, M. Groschup, G. Hunsmann, and W. Bodemer, *J. Immunol. Methods* **199,** 109 (1996).

[22] C. Korth, B. Stierli, P. Streit, M. Moser, O. Schaller, R. Fischer, W. Schulzschaeffer, H. Kretzschmar, A. Raeber, U. Braun, F. Ehrensperger, S. Hornemann, R. Glockshuber, R. Riek, M. Billeter, K. Wuthrich, and B. Oesch, *Nature* **390,** 74 (1997).

[23] W. Goldmann, N. Hunter, T. Martin, M. Dawson, and J. Hope, *J. Gen. Virol.* **72,** 201 (1991).

an N-terminally added methionine ($NH_2MK^{25}KRPK$...; codon assignments in superscript according to Ref. 23) and end with the serine to which in mammalian cells the GPI anchor would be added (...$G^{240}ASCOOH$).

Escherichia coli bacteria (BL21/DE3, Novagen) are transfected, grown up to an OD of 0.8 at 37° with shaking at 250 rpm, then stimulated with 1 m*M* isopropylthiogalactoside (IPTG) and further grown for 3–4 hr at 30° with shaking at 250 rpm. Bacteria are pelleted and lyzed in one-tenth of the original volume TE buffer (50 m*M* Tris, 2 m*M* EDTA) with 100 mg/ml lysozyme, 1% Triton X-100 for 15 min at room temperature. DNA is digested by 10 μg/ml DNAse 1 (Sigma, St. Louis, MO) in 15 m*M* $MgCl_2$ for 30 min at 37°. Inclusion bodies are pelleted. The supernatant is discarded, and the pellet is solubilized by shaking overnight at room temperature in 8 *M* deionized urea/10 m*M* MOPS (UM buffer). Nonsolubilized inclusion bodies are pelleted again by centrifugation at 20,000 rpm at 4° in a Beckman SW-34 rotor and discarded.

The supernatant is loaded on a carboxymethyl (CM)-Sepharose column (Pharmacia, Sweden) that had been equilibrated previously with 0.5 *M* NaCl in UM buffer. The column is washed stepwise with UM buffer containing 0, 0.05, and 0.1 *M* NaCl. rbPrP is eluted with 0.5 *M* NaCl/UM buffer. This fraction is essentially free of other proteins as determined by sodium dodecyl sulfate–polyacrylamide gel electrophoresis (SDS–PAGE) and silver staining (data not shown).

rbPrP is then either oxidized or reduced with either 10 μM Cu_2SO_4 or 2% 2-mercaptoethanol, respectively, for at least 2 hr at room temperature or overnight at 4° by adding these reagents from stock solutions directly into the elution buffer.

We were particularly interested in obtaining reduced rbPrP, as it had been described previously that the reduced form of recombinant hamster PrP showed more β-sheet structure in CD spectroscopy than the oxidized form.[24] Although it is not probable that these two forms of rbPrP derived from *E. coli* correspond to the two PrP isoforms, as none was infectious,[22,24] we followed the idea that a maximum of β-sheet structure in recombinant PrP might represent β-sheet epitopes that also exist in PrP^{Sc}.

Finally, the oxidized or reduced CM-sepharose eluted fractions are purified further by reversed-phase high-performance liquid chromatography (HPLC) on a C4 column with a 0–85% gradient of acetonitrile in 0.1% trifluoroacetic acid (TFA). Typically, the oxidized or reduced rbPrP elutes at about 40 or 45% acetonitrile, respectively. Mass spectrometry indicates

[24] I. Mehlhorn, D. Groth, J. Stockel, B. Moffat, D. Reilly, D. Yansura, W. S. Willett, M. Baldwin, R. Fletterick, F. E. Cohen, R. Vandlen, D. Henner, and S. B. Prusiner, *Biochemistry* **35,** 5528 (1996).

that rbPrP is homogeneous and corresponds to the expected translation product with the N-terminal methionine uncleaved.

The eluted fractions are lyophilized and stored at −70°.

Immunization Procedure and Fusion

Four-to 6-week-old female mice without a functional PrP gene[10] are immunized with 100 μg rbPrP, diluted from a 3-μg/μl stock solution (lyophilized reduced or oxidized rbPrP in sterile water). Mice receive three subcutaneous injections (day 0 with Freund's complete adjuvant, days 21 and 42 with Freund's incomplete adjuvant) of the antigen (100 μg) in a total volume of 100 μl. On day 31, ca. 100 μl mouse blood (stabilized by 15 mM EDTA) is collected from the tail vein of anesthetized mice.

On day 49, mice are boosted with the antigen intraperitoneally and the next day intravenously with adjuvant Pertussi Berna (Berna, Switzerland; extract of *Bordetella pertussis* bacteria). On day 50, mice are anesthetized and decapitated. The spleens are removed, and splenocytes are recovered.

Mouse myeloma cells (cell line P3X63Ag8U.1, ATCC, Rockville, MD CRL 1597[25]) are mixed with the splenocytes at a ratio of 1 : 5 and fused by the addition of 50% polyethylene glycol (PEG) for 8 min at room temperature according to standard techniques.[26] Cells are then washed and grown overnight. The next day, cells are suspended in selective medium (HAT) and plated in 96-well microtiter plates.

Hybridoma cell lines are cloned three times and finally grown in serum-free medium (Turbodoma, Messi, Switzerland) in a Tecnomouse bioreactor device (integra Biosciences, Switzerland).

Screening of Hybridoma Cell Lines

Initial screening of supernatants from hybridomas is performed by testing for anti-PrP activity in an enzyme-linked immunosorbent assay (ELISA) against rbPrP and an enzyme-linked immunofiltration assay (ELIFA) against disease-associated PrP^{BSE}. Antibodies judged positive by either screening are tested further for anti-PrP reactivity on immunoblots, recognition of peptides comprised in the rbPrP amino acid sequence on a synthetic gridded array of cellulose-immobilized peptides, and immunoprecipitation.

Prior to the search for a conformation-specific MAb recognizing an epitope on the surface of native PrP^{Sc} that is composed of a unique three-

[25] M. D. Scharff, *Curr. Top. Microbiol. Immunol.* **81,** 1 (1978).

[26] R. H. Kennett, *in* "Monoclonal Antibodies. Hybridomas: A New Dimension in Biological Analysis" (R. H. Kennett, T. J. McKeam, and K. B. Bechtol, eds.), p. 365. Plenum Press, New York, 1980.

dimensional structure, the following assumptions were made: (1) such an antibody would yield a strong signal on the ELIFA against disease-associated PrP^{BSE} and (2) on Western blots there should be no detection of PrP with the respective MAb as on unfolding during boiling in SDS the BSE-specific conformation would get lost.

ELISA to rbPrP as Antigen

Ninety-six-well microtiter plates (Maxisorp, Nunc, Denmark) are coated overnight at 4° with rbPrP at a concentration of 0.5 μg/well in 50 μl phosphate buffered saline (PBS) and blocked with 5% bovine serum albumin (BSA) for 2 hr at room temperature. Supernatants of hybridoma cells are incubated in the wells for 2 hr at room temperature. After washing, the antimouse lgG H+L chain-specific antibody (Cappell, Switzerland) coupled to horseradish peroxidase is incubated in the wells for another hour, and the reaction is developed with TMB/H_2O_2 (Kierkegaard & Perry, USA). The reaction is stopped after 5–10 min with 2 *M* H_2PO_4, and the extinction is measured at 450 nm with a reference value at 620 nm in a conventional ELISA reader (Tecan, Switzerland).

ELIFA to PrP^{BSE} as Antigen

The ELIFA procedure is modified from Oesch *et al.*[27] Essentially, the ELIFA is a standardized dot-blot procedure using a particular apparatus (Pierce, Rockford, IL) that allows the quantification of samples. Blotting directly on nitrocellulose has the advantage over a conventional ELISA procedure that despite its poor solubility, disease-associated PrP^{BSE} can be immobilized and assessed immunologically.

Briefly, a 0.45-μm nitrocellulose membrane is put into an ELIFA apparatus. The membrane is covered with a plate containing 96 holes in the microtiter format. Equal amounts of antigen are pipetted and sucked through all of these holes with the help of a peristaltic vacuum pump. As antigen, 10% spinal cord homogenate (buffer: 10% sucrose, 20 m*M* HEPES, 2% sarcosyl, 5 m*M* EDTA, pH 7.5) of a Swiss cow with histologically confirmed BSE is digested with protease K (20 μg/ml) for 30 min at 37° and diluted 250-fold in PBS. The nitrocellulose membrane is removed from the apparatus and blocked for 30 min at room temperature with 5% low fat milk in TBST (Tris-buffered saline with Tween 20) and is reinserted into the apparatus. Fifty-microliter hybridoma supernatants are filled into each of the 96-well holes of the plate for incubation (2 hr at room temperature). The membrane is again removed from the apparatus, washed in

[27] B. Oesch, M. Jensen, P. Nilsson, and J. Fogh, *Biochemistry* **33,** 5926 (1994).

TBST, incubated for another hour with a horseradish peroxidase (HRP)-labeled antimouse IgG antibody (H + L chain; Cappell, Switzerland), and finally visualized with chemiluminescence (ECL, Amersham, USA).

Immunoblot

Supernatants of hybridoma cells are tested for their recognition of PrP from 10% brain homogenates from different species on Western blots using standard techniques.[28]

Peptide Library

A gridded array of peptides comprising 104 polypeptides of 13 amino acids, shifted by 2 amino acids and covering the entire mature bovine PrP sequence with six octarepeats,[23] are attached covalently at their C termini to a cellulose support as individual spots by the manufacturer (Jerini Biotools, Germany). After blocking in 5% BSA/TBST, the membrane is incubated for 2 hr at room temperature or at 4° overnight with the supernatant of a hybridoma clone, usually diluted in TBST to a final estimated antibody concentration of 1–10 μg/ml IgG. Subsequent washing and incubation with a HRP-coupled antimouse IgG (H + L chain) antibody reveals distinct dark spots as a positive reaction of a particular polypeptide with the antibody when processed with chemiluminescence as described earlier.

Regeneration of the membrane is achieved by stripping the antibody in a freshly prepared solution of 8 *M* deionized urea/2% SDS/0.5% 2-mercaptoethanol for 2 hr at room temperature.

Immunoprecipitation

For immunoprecipitation, 200 μl of 1% brain homogenate supernatants (precleared by centrifugation at 13,000 rpm for 15 min in a tabletop centrifuge) are incubated for 2 hr at room temperature with 200 μl of 0.25 mg/ml antibody-containing serum-free medium. After incubation with an additional 50 μl protein A or protein G-coupled agarose (Boehringer Mannheim, Germany) for 2 hr at room temperature, agarose beads are centrifuged at 13,000 rpm for 3 min, and the pellet is washed according to the manufacturer. Briefly, this involves two washing/centrifugation cycles of detergent buffer [50 m*M* Tris, 150 m*M* NaCl, 1% Nonidet P-40 (NP-40), 0.5% sodium deoxycholate] and high salt buffer (50 m*M* Tris, 500 m*M* NaCl, 0.1% NP-40, 0.05% sodium deoxycholate). Proteinase K (Sigma,

[28] J. Sambrook, E. F. Fritsch, and T. Maniatis, "Molecular Cloning: A Laboratory Manual." Cold Spring Harbor Laboratory Press, Cold Spring Harbor, NY, 1989.

USA) digestions of immunoprecipitates are done with 20 μg/ml for 30 min at 37° and are stopped with 2 m*M* phenylmethylsulfonyl fluoride (PMSF) (Sigma, USA). Immunoprecipitates are then analyzed by Western blotting. PrP on WB is detected with a polyclonal antiserum against rbPrP. The immunization procedure for this antiserum is basically the same as that used to immunize the knockout mice. Antisera from two rabbits (termed R26 and R29) are highly reactive against PrP on Western blots and are used in dilutions 1:1000 to 1:5000. As secondary antibodies, either HRP or alkaline phosphatase (AP) coupled antirabbit IgG (H + L chain) antibodies are applied. Bound enzymatic activity is visualized with chemiluminescent substrates (ECL, Amersham, or CSPD, Tropix, USA, respectively).

Results and Discussion

Peripheral Immune Response on Immunization with rbPrP

After immunization, the peripheral immune response of the knockout mice was examined at regular intervals. Four mice were immunized with oxidized rbPrP and another four mice were immunized with reduced rbPrP. As noted earlier, oxidized and reduced forms of rbPrP have different HPLC elution times and different characteristics in CD spectroscopy, suggesting a higher β-sheet content in reduced rbPrP.[24]

Peripheral immune response of the immunized mice was monitored by incubating a 1:1000 dilution of mouse serum (in TBST) with the described peptide library overnight at 4° and visualizing bound antibodies by chemiluminescence. In preimmune sera, no positive signals could be detected on the peptide library (data not shown). At day 31 after the initial immunization and 10 days after the first boost, serum reactivity against PrP-derived polypeptides on the peptide library was different between the group of mice that received oxidized rbPrP and that which received reduced rbPrP as antigen (Figs. 1 and 2). While serum reactivity against epitopes at the N terminus was about the same between groups (both with a strong signal against polypeptide sequence ^{38}RYPGQGSPGGNRY^{50} (sequence numbering in superscript adapted to human PrP[29]), sera from mice immunized with oxidized rbPrP reacted primarily with an epitope in the N-terminal region of α-helix 3 (residues 198–212; secondary structures according to Riek *et al.*[30]) and sera from mice immunized with reduced rbPrP reacted

[29] M. Billeter, R. Riek, G. Wider, S. Hornemann, R. Glockshuber, and K. Wüthrich, *Proc. Natl. Acad. Sci. U.S.A.* **94,** 7281 (1997).

[30] R. Riek, S. Hornemann, G. Wider, M. Billeter, R. Glockshuber, and K. Wüthrich, *Nature* **382,** 180 (1996).

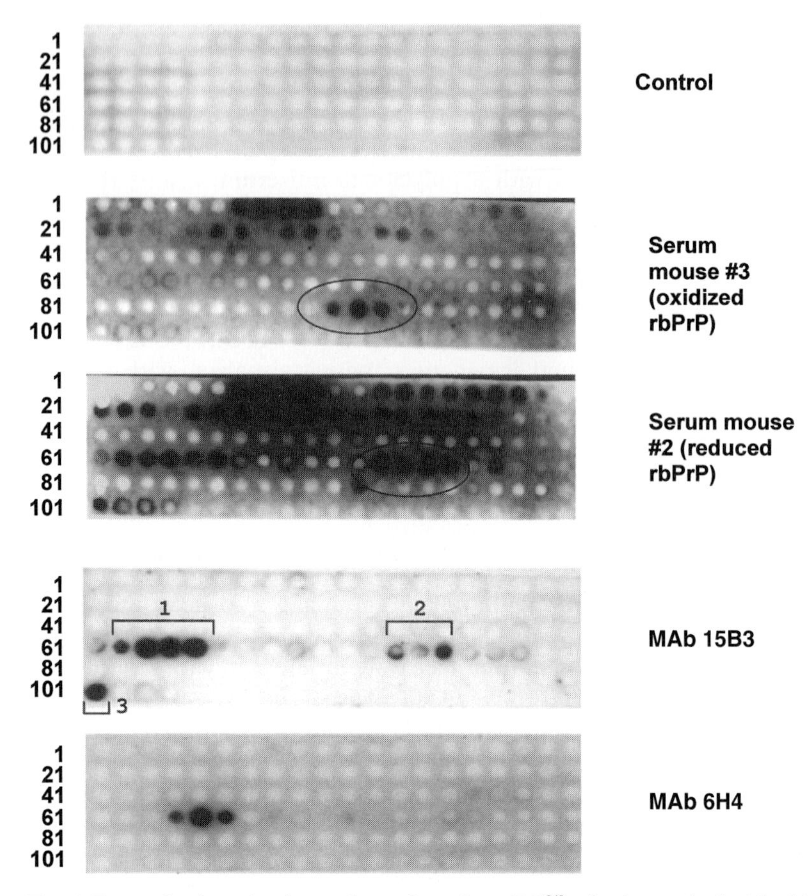

FIG. 1. Determination of epitopes for antisera from $PrP^{0/0}$ mice immunized with either oxidized rbPrP or reduced rbPrP and for MAbs 15B3 and 6H4. A gridded array of synthetic peptides ("peptide library" of 104 polypeptides) corresponding to rbPrP (13 amino acid peptides shifted by 2 amino acids along the bovine PrP sequence[23]) was incubated with the indicated antibodies and visualized with a secondary antibody (peroxidase-labeled goat antimouse lgG) and chemiluminescence. Control (secondary antibody only, above) was negative. Antisera from both mice (representative samples of those depicted in Fig. 2) show a strong immunological response against polypeptides from the N terminus, including the six octarepeats[23] of bovine PrP. At the C-terminal "core" of rbPrP the antisera recognized different polypeptides (encircled; see Fig. 2). Fusion of splenocytes from mouse #2 (immunized with reduced rbPrP) resulted in the 15B3 hybridoma cell line secreting a MAb specific against disease-associated PrP^{Sc}. MAb 15B3 recognizes three discontinous linear polypeptide segments (15B3-1, 2, and 3) that are hypothesized to form a conformational epitope on the surface of prions.[22] Note that an immunological reaction against these three polypeptide segments is already present in the antiserum of mouse #2 (reduced). Another MAb, termed 6H4, recognizes primarily native PrP^{C} and one distinct polypeptide segment corresponding to an epitope on the linear PrP amino acid sequence. Although the specificity of MAbs 15B3 and 6H4 is for native PrP^{Sc} and PrP^{C}, respectively, they have an overlapping (partial) epitope for polypeptides 64 and 65 (see also Fig. 2), indicating differential accessibility of this region in PrP isoforms.

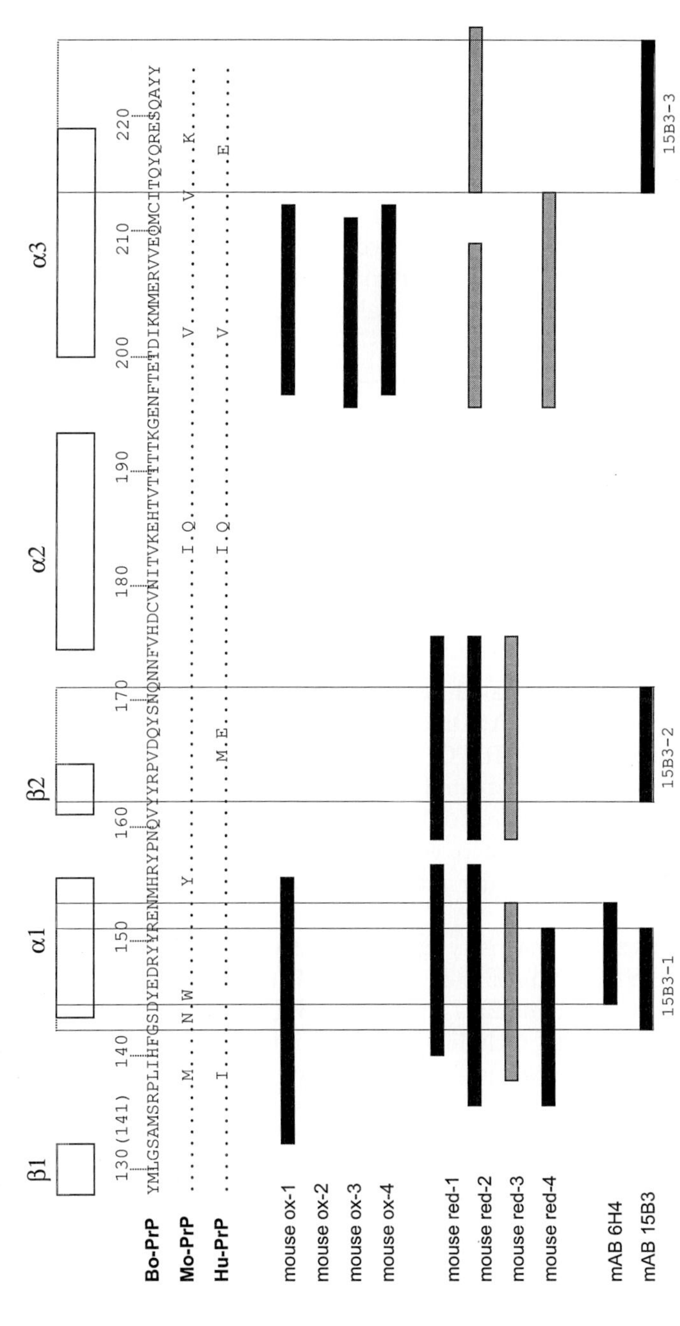

FIG. 2. Schematic representation of polypeptides recognized by antisera and MAbs in the C-terminal part of PrP. Numbers indicate the residue position relative to human PrP (Hu-PrP), and the relative residue position to bovine PrP is indicated in parentheses. Secondary structures as determined by nuclear magnetic resonance[30] of recombinant mouse PrP (Mo-PrP) are displayed (*β*1 and *β*2: *β*-sheets 1 and 2; *α*1, *α*2, and *α*3: *α*-helices 1, 2, and 3). Positive immunostaining of the peptide library by MAbs 6H4, 15B3, or antisera 1 to 4 of PrP$^{0/0}$ mice immunized with oxidized (mouse ox-1 to -4) and reduced (mouse red-1 to -4) rbPrP is indicated by bold (strong reaction) or gray (weak reaction) bars. For MAbs 6H4 and 15B3, the minimal binding sequence (smallest common denominator) is displayed. Note the different pattern of immunostaining between groups of PrP$^{0/0}$ mice immunized with oxidized and reduced rbPrP. Whereas antisera from mice ox-1 to -4 stain polypeptides located in *α*-helix 3 strongly, those of mice red-1 to -4 stain polypeptides comprising *β*-sheet 2 strongly. Note also that immunostaining of antiserum from mouse red-2 comprises polypeptides that are also stained by MAb 15B3, which is derived from splenocytes of mouse red-2.

primarily with an epitope comprising β-sheet 2 and the loop N-terminal of α-helix 2 (residues 160–172). Thus, oxidized and reduced rbPrP, in addition to differences in CD spectroscopy, elicit different immune responses in $PrP^{0/0}$ mice. Reduction of the disulfide bridge of rbPrP seems to hide a C-terminal epitope built by α-helix 3 and uncovers an epitope comprising β-sheet 2.

We consider the monitoring of the immune response on the peptide library as a valuable tool in assessing the success of the immunization. We also used this technique to map the epitopes of the described rabbit polyclonal antisera to rbPrP. Two immunized rabbits, R26 and R29, developed a strong peripheral immune response. As expected, mapping of the epitopes showed that they were localized at sequence patches around species-specific residues. One antiserum, R26, reacted with the linear polypeptide ^{115}KPKTNMKHV^{123} (underlined asparagine is serine in rabbit; sequence numbering adapted to human PrP[29]), and the other, R29, reacted with the linear polypeptides ^{36}TGGSRYPGQGSPG^{48} (underlined glycine is serine in rabbit) and ^{147}RPLIHFGSD^{155} (underlined serine is asparagine in rabbit).

Monoclonal Antibodies Specific for Native PrP^C

Screening of hybridomas revealed several antibodies that reacted strongly against rbPrP in the ELISA. One of them, termed 34C9 (Fig. 3), recognized the linear sequence ^{149}LIHFG^{153} in bovine PrP that is not well conserved between species and is situated in a turn region N-terminal to α-helix 1.[30] Accordingly, it recognizes only bovine and porcine PrP, but not PrP from hamsters, mice, and sheep, which differ in the corresponding peptide sequence. Antibody isotyping identified this MAb 34C9 as an IgG_{2b} isotype.

Another MAb with a strong reactivity against rbPrP, termed 6H4, recognized the epitope ^{144}DYEDRYYRE^{152} in bovine PrP that is highly conserved between species (Figs. 1 and 2). The MAb, an IgG_1, recognized PrP from all species examined so far, including mouse and hamster PrP, which have a tryptophan substituted for the first bovine tyrosine at residue 145 (Fig. 3B). Of note, this linear stretch of amino acids corresponds to the α-helical structure determined for recombinant mouse PrP.[30] Immunoprecipitation of normal and prion-infected brain homogenates showed that MAb 6H4 specifically recognizes the native PrP^c isoform and not the native PrP^{Sc} isoform, as the precipitated PrP from a BSE brain homogenate could be fully digested with protease K (Figure 4B).

In Western blotting, MAb 6H4 seems to recognize the double-glycosylated bovine PrP (Fig. 3, upper band) better than both the single-glycosylated

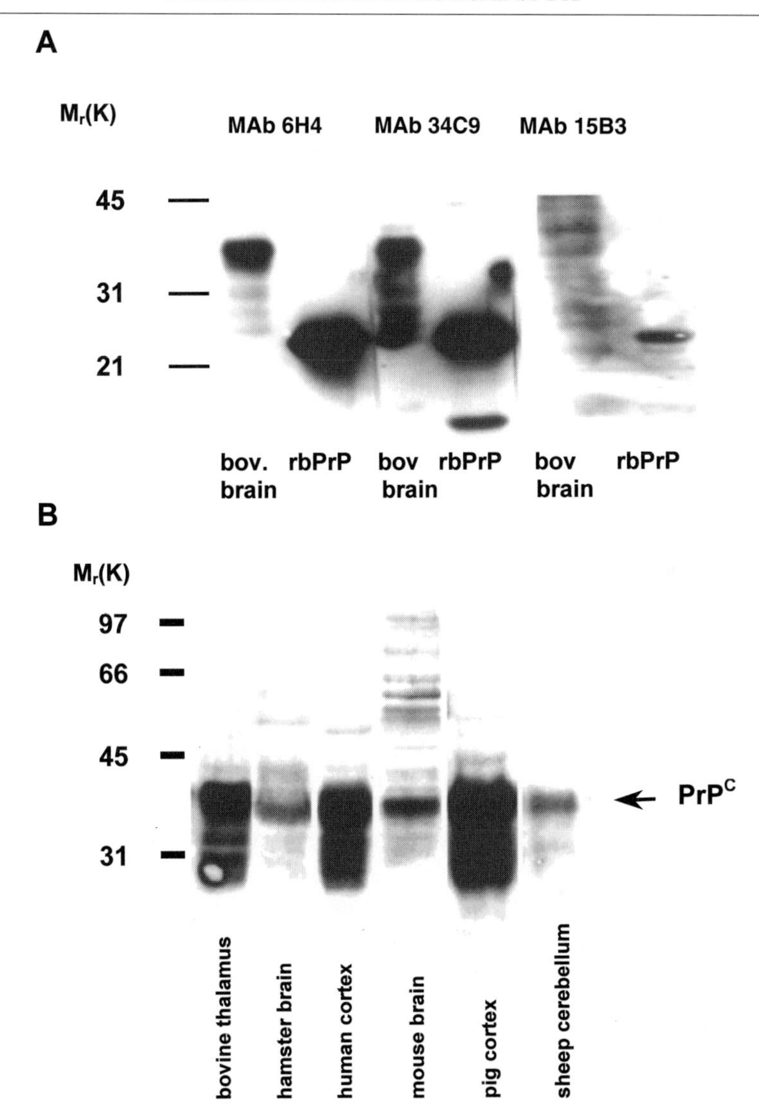

FIG. 3. Staining of PrP on Western blots. (A) Whereas MAbs 6H4 and 34C9 both recognize PrP from bovine brain homogenates and recombinant bovine PrP (rbPrP), MAb 15B3 does not recognize PrP from brain homogenates on Western blots and rbPrP is very weak (same amounts of brain homogenate and rbPrP for each MAb loaded). (B) MAb 6H4 recognizes a highly conserved epitope on PrP and can be expected to stain PrP on Western blots from all mammalia.

and the unglycosylated bovine PrP (Fig. 3, two lower bands), whereas MAb 34C9, for comparison, shows about an equal preference for either glycosylation form of bovine PrP (Fig. 3A).

Monoclonal Antibody Specific against Disease-Associated PrP^{Sc}

Screening for antibodies that recognize PrP^{Sc} conformational epitopes according to the described assumptions, i.e., strong signal on ELIFA and no signal on Western blot, led to the identification of a monoclonal antibody, named 15B3, that precipitated native PrP^{Sc}, but not native PrP^{C}, from BSE brain homogenates (Fig. 4). The antibody precipitated the disease-associated isoform of PrP from brain homogenates of BSE-infected cattle, scrapie-infected mice, and CJD-affected humans (Figs. 4A–4D). Antibody isotyping identified MAb 15B3 as an IgM.

Incubation of the peptide library with the MAb 15B3 demonstrated binding to three linear stretches of polypeptides ^{142}GSDYEDRYY^{150} (termed 15B3-1), ^{162}YYRPVDQYS^{170} (termed 15B3-2), and ^{214}CITQYQR ESQAYY^{226} (termed 15B3-3; see Figs. 1 and 2). Mapping of the three linear stretches on the nuclear magnetic resonance structure of mouse recombinant PrP[30] revealed spatial proximity of 15B3-2 and 15B3-3, whereas 15B3-1 was distant, as well as a partial overlapping of 15B3-1 with the linear epitope for MAb 6H4. Together with the fact that 15B3 did not recognize PrP on Western blots (Fig. 3A), it was assumed that the epitope was indeed conformational and that the three polypeptide segments represented partial epitopes thereof. 15B3 may recognize linear peptides of the PrP sequence on the peptide library but not on Western blots because the concentration of the peptides on the individual spots was much higher than that of denatured PrP on Western blots; rbPrP, when overloaded on a SDS–PAGE gel and blotted, was stained weakly by MAb 15B3 (Fig. 3A).

Of note, the surface region composed by partial epitopes 15B3-2 and -3 is thought to excert a major influence on the establishment of a species barrier and on the binding of conversion cofactors.[29,31] To complete the assumed conformational epitope of 15B3 on PrP^{Sc}, 15B3-1 should come into spatial proximity with partial epitopes 15B3-2 and -3. It is conceivable that either inter- or intramolecular rearrangement of rbPrP molecules or both contribute to the unity of a single continuous epitope patch on the surface of PrP^{Sc}. Possible models of these arrangements are discussed elsewhere.[22]

[31] K. Kaneko, L. Zulianello, M. Scott, C. M. Cooper, A. C. Wallace, T. L. James, F. E. Cohen, and S. B. Prusiner, *Proc. Natl. Acad. Sci. U.S.A.* **94,** 10069 (1997).

The production of antibodies specific for the disease-associated isoform of the prion protein when immunizing with recombinant PrP suggests that at least part of a disease-specific conformation of of PrP^{Sc} is present on rbPrP and can exist in mice without causing disease. We[22] and others[24] have shown that recombinant PrP, when injected intracerbrally into PrP-overexpressing mice[32] or hamsters, respectively, does not produce prion disease.

MAb 15B3 was obtained from a mouse immunized exclusively with the reduced form of rbPrP (mouse 2), and examination of the peripheral immune response shows that a binding pattern to the peptide library similar to that of mAB 15B3 is comprised in the binding pattern of the antiserum in this mouse (Figs. 1 and 2). However, we have not investigated systematically if the immunization of knockout mice with the reduced form of rbPrP is a necessary condition in obtaining PrP^{Sc}-specific monoclonal antibodies. The precise conditions under which recombinant PrP can mimic a PrP^{Sc} epitope, therefore, remain hypothetical. Factors other than reduction of rbPrP that may favor formation of a PrP^{Sc}-specific epitope are lypophilization and concentration effects that influence solubility and may favor formation of β-sheet structure and/or aggregation of PrP molecules.

Remarkably, MAb 15B3 is an IgM. However, we do not think that the particular qualities of IgM subtypes are a necessary condition for PrP^{Sc}-specificity. While in the short term the IgM subtype may complicate applications to certain immunological detection procedures, recombinant antibody technology will be able to refine the properties of MAb 15B3 and make the structural information about PrP^{Sc}-specific binding available for other applications.

Summary and Conclusion

Reviewing the circumstances that have led to the first monoclonal antibody against the disease-associated form of PrP, we consider the availability of PrP knockout mice and recombinant PrP, as well as a reliable conformational screening protocol as being important prerequisites for a successful immunization approach.

When considering presenting an antigen to a mouse with the goal of obtaining specific monoclonal antibodies against a misfolded or aggregated form of a host protein, it is desirable to increase the definition of a subtle conformational difference. This can be achieved by immunizing an antigen knockout mouse that has not developed self-tolerance against the respective

[32] M. Fischer, T. Rülicke, A. Raeber, A. Sailer, M. Moser, B. Oesch, S. Brandner, A. Aguzzi, and C. Weissmann, *EMBO J.* **15,** 1255 (1996).

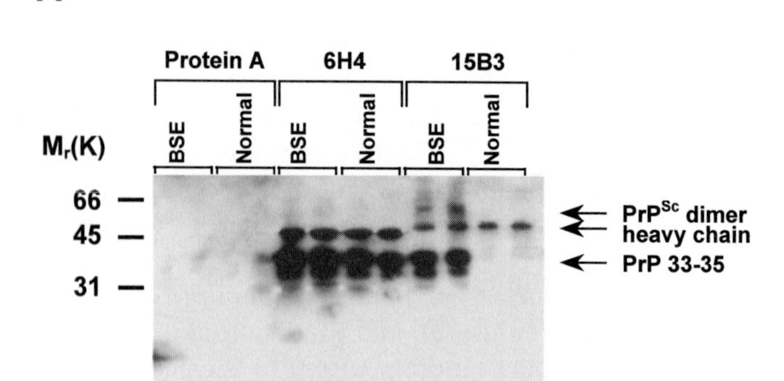

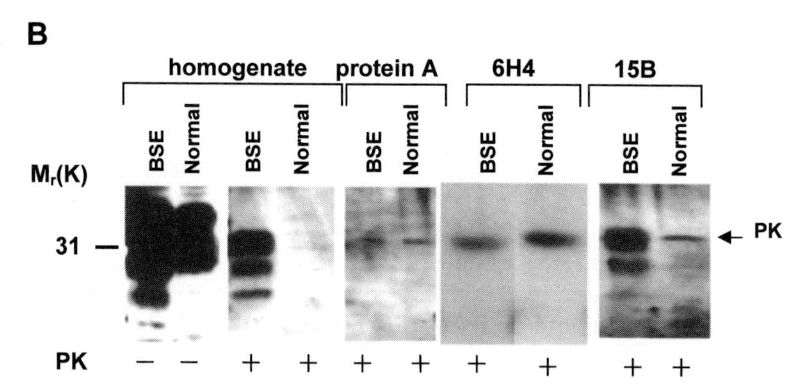

FIG. 4. Immunoprecipitation of bovine, mouse, and human PrP with MAbs 15B3 and 6H4. The supernatant of a precleared homogenate from brain of normal or prion-diseased subjects was incubated with antibodies 6H4 or 15B3. Antibodies were precipitated with protein A-agarose (15B3) or protein G-agarose (6H4). As a control, the same homogenates were incubated without antibodies and protein A only. Precipitates were analyzed on a Western blot for the presence of PrP using a polyclonal rabbit antiserum (R26) to PrP and goat antirabbit lg coupled to alkaline phosphatase. Signals were developed with chemiluminescence substrates. (A) Immunoprecipitation of normal or BSE-diseased bovine PrP. MAb 15B3, being specific for disease-associated PrP, precipitates PrP only from diseased but not normal bovine brain, whereas MAb 6H4 (specific for PrP^C) precipitates the normal isoform of both normal and diseased bovine brain. Cross-reaction of the secondary antibody with immunoprecipitated mouse immunoglobulins leads to the prominent band at about 50 kDa. Note the 60-kDa band characteristic for PrP^{Sc} in the 15B3 but not in the 6H4 immunoprecipitations. (B) Immunoprecipitates of normal and diseased bovine brain with MAbs 15B3 or 6H4 as shown in (A) were protease digested and compared to undigested and digested bovine brain homogenates. The sharp band at 31 kDa represents a cross-reactivity of the secondary antibody with proteinase K. Note that only the immunoprecipitate of 15B3 from BSE brain is protease resistant. (C) Immunoprecipitation of mouse PrP from normal and scrapie-infected mouse brains with MAbs 15B3 and 6H4. Homogenates from PrP null mice (0/0), wild-type mice

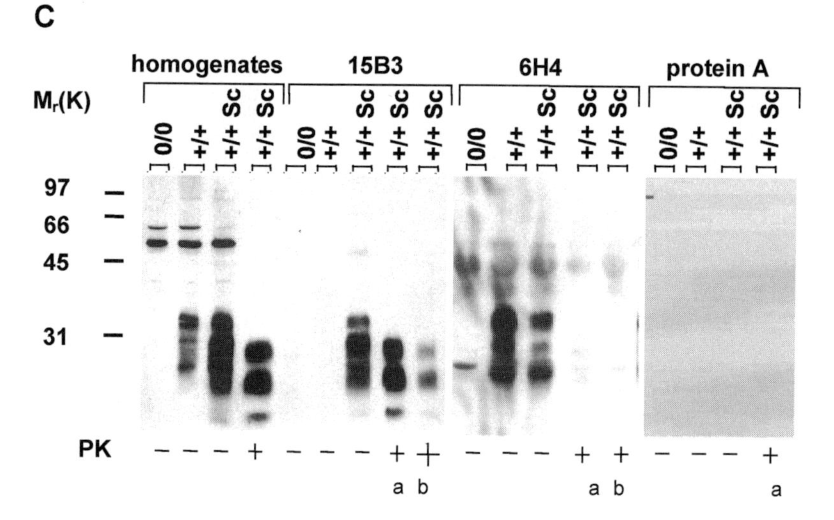

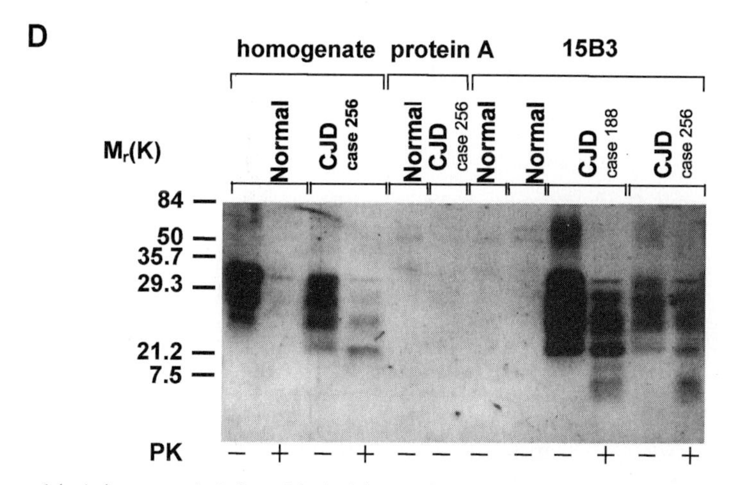

(normal (+/+), or scrapie-infected (+/+Sc) were immunoprecipitated with MAb 15B3, 6H4, or protein A-agarose only and analyzed by Western blotting as described earlier. Digestion with proteinase K after (a) or before (b) the immunoprecipitation is indicated. As for bovine brain homogenates, 15B3 precipitates PrP only from prion-diseased mouse brain homogenates, whereas 6H4 precipitates protease-sensitive PrP^C from both normal and scrapie-infected mouse brain homogenates. Note that a significant part of 15B3-precipitated full-length PrP from scrapie-infected mouse brain is protease sensitive, indicating binding of MAb 15B3 to several "subspecies" of disease-associated PrP. (D) Immunoprecipitation of human PrP^{CJD} with MAb 15B3. Brain homogenates from normal persons (normal) or CJD patients (CJD) were immunoprecipitated with 15B3 and analyzed as described. 15B3 precipitates PrP only from CJD-diseased human brains. (A) and (D) are reprinted by permission from C. Korth *et al., Nature* **390,** 74 (1997), copyright (1997) Macmillan Magazines Ltd.

antigen. Furthermore, if conformational isoforms and/or oligomeric forms of a protein sequence are understood to exist in an equilibrium, high and pure amounts of recombinant protein may increase the likelihood that a particular population of protein conformation passes an antigenic threshold necessary to start an immunogenic response.

Pulling out the monoclonal antibodies by correct screening is essential. Screening against the pure misfolded or aggregated protein is often complicated by its poor solubility and hence the ability to immobilize. In the present case, immobilization of disease-associated PrP on nitrocellulose had been established as a conformation-sensitive screening method, allowing to "freeze" PrP in its distinguishable, disease-associated conformation.[27,33]

We are cautious to generalize conclusions of how to assess the generation of monoclonal antibodies against these particular protein isoforms to other diseases of protein misfolding and/or aggregation, but ultimately the present approach may inspire respective experiments.

Acknowledgments

This work was supported by Grant 2BO from the SPP Biotech of the Swiss National Foundation to B.O.; P.S. is supported by Grant 31-49385-96 from the Swiss National Foundation.

[33] C. Korth and B. Oesch, *NeuroForum* **2S,** 229 (1996).

[9] Assays of Protease-Resistant Prion Protein and Its Formation

By BYRON CAUGHEY, MOTOHIRO HORIUCHI, RÉMI DEMAIMAY, and GREGORY J. RAYMOND

Introduction

The accumulation of abnormally protease-resistant prion protein (PrP-res) is common to transmissible spongiform encephalopathies (TSEs). PrP-res is formed posttranslationally from the host's normal, protease-sensitive prion protein isoform, PrP-sen.[1,2] Unlike wild-type PrP-sen, PrP-res is also

[1] D. R. Borchelt, M. Scott, A. Taraboulos, N. Stahl, and S. B. Prusiner, *J. Cell Biol.* **110** 743 (1990).

[2] B. Caughey and G. J. Raymond, *J. Biol. Chem.* **266,** 18217 (1991).

polymeric in most, if not all, circumstances.[3] This appears to be a central process in TSE pathogenesis and TSE agent replication. Detection of PrP-res in tissues and extracts of tissues and cells correlates widely with the presence of TSE infectivity, and treatments that inactivate or eliminate TSE infectivity also eliminate PrP-res. In the case of rare familial TSE diseases, it appears that PrP mutations can lead to aberrant behavior of PrP-sen[4,5] that predisposes it toward conversion to PrP-res. However, spontaneous conversion occurs rarely, if at all, in hosts with wild-type PrP-sen as evidenced by the 1 per million annual incidence of sporadic Creutzfeldt–Jakob disease in humans. Much more common in mammals is the induction of PrP-res formation from wild-type PrP-sen on infection of hosts with TSE agents. Thus, although the precise functional connections among PrP-res, the infectious agent, and TSE pathogenesis remain unclear, assays for PrP-res are used broadly and effectively as rapid surrogate tests for TSE infection.

Defining Protease-Resistant Prion Protein

The partial proteinase K (PK) resistance of the abnormal PrP that accumulates in TSE-infected hosts is the parameter used most widely to discriminate it from PrP-sen. Although PK completely digests PrP-sen, it typically removes only ~60–70 residues from the N terminus of PrP-res, resulting in a 6- to 8-kDa downward shift in its apparent molecular mass in sodium dodecyl sulfate–polyacrylamide gel electrophoresis (SDS–PAGE) gels.[6,7]

This difference is not subtle; the PK-resistant portion of the molecule can often resist concentrations of PK that are >1000-fold higher than the minimum required to eliminate PrP-sen or the PK-sensitive N-terminal residues of PrP-res. It is important to discriminate between this characteristic partial PK resistance, which is apparently indicative of a specific conformation and ordered polymeric state of PrP-res, and total PK resistance that would occur if the PrP molecules were buried in large amorphous aggregates, membrane bound vesicles, or any other physical state that would totally prevent the exposure of PrP molecules to PK. For instance, the latter has been seen with PrP incubated *in vitro* with vast stoichiometric

[3] B. Caughey and B. Chesebro, *Trends Cell Biol.* **7,** 56 (1997).

[4] S. Lehmann and D. A. Harris, *Proc. Natl. Acad. Sci. U. S. A.* **93,** 5610 (1996).

[5] S. A. Priola and B. Chesebro, *J. Biol. Chem.* **273,** 11980 (1998).

[6] B. Oesch, D. Westaway, M. Walchli, M. P. McKinley, S. B. H. Kent, R. Aebersold, R. A. Barry, P. Tempst, D. B. Teplow, L. E. Hood, S. B. Prusiner, and C. Weissmann, *Cell* **40,** 735 (1985).

[7] J. Hope, L. J. D. Morton, C. F. Farquhar, G. Multhaup, K. Beyreuther, and R. H. Kimberlin, *EMBO J.* **5,** 2591 (1986).

excesses of certain synthetic PrP peptides without any association with TSE infection.[8] Thus, in assaying for TSE-associated PrP-res, it is important to not only see PK resistance but also elimination of the full-length PrP in the sample along with the retention of bands that are $\geq \sim 6$ kDa smaller than the non-PK-treated full-length bands on SDS–PAGE.

Variations in Protease Resistance

The just-described description/definition is broadly applicable operationally for detecting TSE-associated forms of PrP-res such as those associated with scrapie (PrP^{Sc}), Creutzfeldt–Jakob disease (PrP^{CJD}), bovine spongiform encephalopathy (PrP^{BSE}), transmissible mink encephalopathy (PrP^{TME}), and chronic wasting disease (PrP^{CWD}). However, other TSE variants may stretch the limits of the previous characterization of TSE-associated PrP-res. Some scrapie and CJD strains give rise to PrP-res that is cleaved somewhat differently by PK, even in cases where the constituent PrP molecules have the same amino acid sequence.[9–16] Some cases of TSE transmissions into new species have been reported in which no PrP-res was detected.[17] Furthermore, degenerative diseases (but not necessarily TSE) have been caused in transgenic animals overexpressing wild-type or mutant PrP molecules without any apparent accumulation of PrP-res.[18,19] Thus,

[8] K. Kaneko, D. Peretz, K. Pan, T. C. Blockberger, H. Wille, R. Gabizon, O. H. Griffith, F. E. Cohen, M. A. Baldwin, and S. B. Prusiner, *Proc. Natl. Acad. Sci. U.S.A.* **92,** 11160 (1995).

[9] R. A. Bessen and R. F. Marsh, *J. Virol.* **66,** 2096 (1992).

[10] R. A. Bessen and R. F. Marsh. *J. Virol.* **68,** 7859 (1994).

[11] L. Monari, S. G. Chen, P. Brown, P. Parchi, R. B. Petersen, J. Mikol, F. Gray, P. Cortelli, P. Montagna, B. Ghetti, L. G. Goldfarb, D. C. Gadjusek, A. Lugaresi, P. Gambetti, and L. Autilio-Gambetti, *Proc. Natl. Acad. Sci. U.S.A.* **91,** 2939 (1994).

[12] G. C. Telling, P. Parchi, S. J. DeArmond, P. Cortelli, P. Montagna, R. Gabizon, J. Mastrianni, E. Lugaresi, P. Gambetti, and S. B. Prusiner, *Science* **274,** 2079 (1996).

[13] J. Collinge, K. C. L. Sidle, J. Meads, J. Ironside, and A. F. Hill, *Nature* **383,** 685 (1996).

[14] R. A. Somerville, A. Chong, O. U. Mulqueen, C. R. Birkett, J. Hope, J. Collinge, A. F. Hill, K. C. L. Sidle, and J. Ironside, *Nature* **386,** 564 (1997).

[15] P. Parchi, S. Capellari, S. G. Chen, R. B. Petersen, P. Gambetti, N. Kopp, P. Brown, T. Kitamoto, J. Tateishi, A. Giese, and H. Kretzschmar, *Nature* **386,** 232 (1997).

[16] A. F. Hill, M. Desbruslais, S. Joiner, K. C. L. Sidle, I. Gowland, J. Collinge, L. J. Doey, and P. Lantos, *Nature* **389,** 448 (1997).

[17] C. I. Lasmezas, J. P. Deslys, O. Robain, A. Jaegly, V. Beringue, J. M. Peyrin, J. G. Fournier, J. J. Hauw, J. Rossier, and D. Dormont, *Science* **275,** 402 (1997).

[18] K. Hsiao, H. F. Baker, T. J. Crow, M. Poulter, F. Owen, J. D. Terwilliger, D. Westaway, J. Ott, and S. B. Prusiner, *Nature* **338,** 342 (1989).

[19] D. Westaway, S. J. DeArmond, J. Cayetano-Canlas, D. Groth, D. Foster, S. Yang, M. Torchia, G. A. Carlson, and S. B. Prusiner, *Cell* **76,** 117 (1994).

there are some rare disease states caused by PrP abnormalities that may not be associated with typical PrP-res accumulation.

Types of PrP-res Assays

Several types of assays for PrP-res have been developed. Immunohistochemical staining[20,21] and histoblotting[22] of tissue sections have been described elsewhere. This article focuses on immunoblots, metabolic labeling, and assays based on the ability of PrP-res to induce the conversion of PrP-sen to PrP-res. Samples containing PrP-res are presumably infectious to at least some mammalian species and possibly humans. Thus, these biohazardous materials must be handled appropriately in the following procedures, although the biohazard containment and decontamination issues will not be considered here.

Reagent Solutions

Lysing buffer (LB): 0.5% Triton X-100, 0.5% (w/v) sodium deoxycholate, 5 m*M* Tris–Cl, pH 7.4, at 4°, 150 m*M* NaCl, 5 m*M* EDTA

Phosphate-buffered saline (PBS): 130 m*M* NaCl, 20 m*M* sodium phosphate, pH 6.9

SDS–PAGE sample buffer: 5% (w/v) SDS, 4% 2-mercaptoethanol, 4 *M* urea, 62.5 m*M* Tris–HCl, pH 6.8, at 20°, 5% glycerol, 3 m*M* EDTA, 0.02% bromphenol blue

Detergent, lipid, protein complex (DLPC) buffer: 0.84 ml of 100 mg/ml L-α-phosphatidylcholine (Sigma, St. Louis, MO)—dry in a 50-ml polypropylene tube under a stream of N_2 gas; then add 1 ml 1 *M* Tris–Cl, pH 7.5, at 20°, 17 ml 150 m*M* NaCl, and 2 ml of 20% sarkosyl (*N*-lauroylsarcosine). Sonicate in a cuphorn probe with occasional vortexing until suspended evenly, store at 4°, and make fresh monthly

TN: 50 m*M* Tris–HCl, pH 7.5, 150 m*M* NaCl

Proteinase K: PK aliquots are stored frozen at 1–10 mg/ml in 50 m*M* Tris–Cl, pH 8.0, at 4°, 1 m*M* $CaCl_2$, 50% glycerol.

[20] T. Kitamoto, R. W. Shin, K. Doh-ura, N. Tomokane, M. Miyazono, T. Muramoto, and J. Tateishi, *Am. J. Pathol.* **140,** 1285 (1992)

[21] M. E. Bruce, P. A. McBride and C. F. Farquhar, *Neurosci. Lett.* **102,** 1 (1989).

[22] A. Taraboulos, K. Jendroska, D. Serban, S. Yang, S. J. DeArmond, and S. B. Prusiner, *Proc. Natl. Acad. Sci. U.S.A.* **89,** 7620 (1992)

In situ reaction buffer (IRB): 25 m*M* HEPES, pH 7.2, 150 m*M* NaCl, 1 mg/ml bovine serum albumin

Rapid Immunoblotting Method to Screen Brain Tissue for PrP-res

This method may be used for frozen or fresh brains excised from animals euthanized or found dead. The method minimizes potential losses from extractions and fractionations. Place 0.2 g brain (e.g., approximately one-half of a mouse brain) in 1 ml LB and make a ~20% (w/v) homogenate/suspension by pipetting up and down several times using a micropipettor with a 1-ml aerosol-resistant tip. Avoid foaming. Vortex and let stand at room temperature for 30 min. Vortex and sonicate in a cuphorn probe at maximum power for 1 min. Centrifuge at 2000*g* at 4° for 10 min. Resuspend all but the hardest portion of the pellet by inverting the tube a few times. Transfer 20 μl of the suspension to a microtube and add 2 μl of 1 mg/ml PK, vortex, and incubate at 37° for 30 min. Note that amounts of PK to be used for different species may need to be adjusted for optimal detection of PrP-res. Inhibit PK with 1 μl of 0.1 *M* Pefabloc (Boehringer-Mannheim). Add 80 μl SDS–PAGE sample loading buffer, heat in boiling water bath, and load 10 μl onto SDS–PAGE gels. Analyze using standard immunoblotting techniques with appropriate antibodies.

Isolation of PrP-res from Cultured Cells for Immunoblotting Analysis

Scrapie-infected tissue cultures have been used to study the cell biology of PrP-res formation and to screen for compounds that can inhibit or enhance PrP-res formation and scrapie agent replication. Steady-state levels of PrP-res in such cultures can be assayed by the following extraction and immunoblotting procedure. Nearly confluent cell monolayers are lysed with 1 ml ice-cold LB per 25-cm^2 flask. The lysate is cleared of debris with a 5-min centrifugation at 1000*g*, and the supernatant is treated with 50 μg/ml PK for 30 min at 37°. After inactivation of the PK with 5 μl 0.1 *M* Pefabloc, the PrP-res is pelleted at 350,000*g* for 45 min and solubilized for analysis by SDS–PAGE and immunoblotting.[23]

Metabolic [^{35}S]Methionine Labeling and Immunoprecipitation of PrP-sen and PrP-res in Cultured Cells

Many types of studies of the biosynthesis and metabolism of PrP-sen or PrP-res begin with labeling different forms of PrP with [^{35}S]methionine

[23] B. Caughey and G. J. Raymond, *J. Virol.* **67,** 643 (1993)

in live cells followed by immunoprecipitation of the protein from cell lysates. Our usual procedure for attached mammalian cells involves preincubating the cells in methionine-deficient MEM (supplemented with 5% dialyzed fetal bovine serum and 1.5 mg/liter cold methionine) for 0.5 hr and incubating with the label (0.1–10 mCi per 25-cm^2 flask of 75–90% confluent cells) for various time periods (typically 2 hr) in this same medium. The cells can then be washed with physiological saline and lysed in ice-cold LB immediately or after a chase incubation in complete tissue culture medium. For maximal labeling of PrP-res, a chase of 16–48 hr is recommended.[2] The lysates are cleared by centrifugation for 5 min at 1000*g* at 4°. The proteins in the supernatant are precipitated with 4 volumes of methanol at −20° for ≥1 hr (or for PrP-res alternatives, see next paragraph). The precipitated proteins are collected by centrifugation in polypropylene tubes (4000*g* for 20 min at 4°) and resuspended into DLPC buffer[1] (1 ml/25-cm^2 flask equivalent of cell proteins) by sonication with a cuphorn probe. The appropriate PrP-specific antibody is incubated with the suspension and antibody–antigen complexes are collected by binding to protein A-Sepharose beads. The beads are washed once with DLPC buffer and twice with 0.5 *M* NaCl, 50 m*M* Tris–Cl, pH 7.0, 1% sarcosyl at 4° or other detergent/salt solutions, the strength of which can be adjusted for different antibodies to minimize nonspecific binding. Finally, the beads are washed once with water and the [^{35}S]PrP-sen is eluted in either 0.1 *M* acetic acid (for use in conversion reactions as described later) or SDS–PAGE sample buffer for SDS–PAGE/fluorography. In the former case, the eluant is two to three times the bed volume of the PAS beads and can be left with the beads during storage at 4°; this is often preferable to transferring the eluant to a new tube because relatively pure [^{35}S]PrP-sen can be lost due to binding to tube walls.

In Vitro Conversion Reactions for Detection and Formation of PrP-res

PrP-res preparations have what we have termed "PrP converting activity," i.e., the ability to induce the conversion of PrP-sen to PrP-res in a highly specific reaction *in vitro*.[24] Various permutations of the *in vitro* conversion reaction now exist that have been used to study (1) the relationship between PrP-res, converting activity, and infectivity,[25] (2) the conversion mechanism,[25–28] (3) the effects of PrP-res/PrP-sen sequence mis-

[24] D. A. Kocisko, J. H. Come, S. A. Priola, B. Chesebro, G. J. Raymond, P. T. Lansbury, and B. Caughey, *Nature* **370,** 471 (1994)

[25] B. Caughey, G. J. Raymond, D. A. Kocisko, and P. T. Lansbury, Jr., *J. Virol.* **71,** 4107 (1997)

[26] B. Caughey, D. A. Kocisko, G. J. Raymond, and P. T. Lansbury, *Chem. Biol.* **2,** 807 (1995).

[27] S. K. DebBurman, G. J. Raymond, B. Caughey, and S. Lindquist, *Proc. Natl. Acad. Sci. U.S.A.* **94,** 13938 (1997).

matches on converting activity[29–31] (4) the propagation of TSE strain-associated differences in PrP-res,[28] (5) inhibitors/stimulators of PrP-res formation,[32–35] the influence of chaperone proteins on the conversion reaction,[27] and (6) the colocalization of PrP-res and converting activity in tissue slices.[36] So far, converting activity has correlated with the presence of TSE infectivity.[25] All of the present manifestations of this reaction involve incubating labeled PrP-sen with a source of unlabeled PrP-res and then assaying for the formation of labeled PrP that acquired resistance to PK digestion. In its initial form, the cell-free conversion reaction was found to be stimulated greatly by pretreatments of the input PrP-res with 2–3 *M* guanidine hydrochloride.[24] More recent studies have determined that physiological salts (KCl or NaCl; see later) and chaperone proteins (GroEL and HSP104)[27] can at least partially substitute for the guanidine hydrochloride Gdn-HCL pretreatment to create more physiological conditions and that the quantity of PrP-res required to drive efficient conversion has dropped 60-fold.[31] Furthermore, the reaction has been shown to also be driven by intact brain slices from TSE-infected brain slices.[36]

Updated GdnHCl-Containing Cell-Free Conversion Reaction

1. PrP-res is prepared using the Bolton *et al.*[37] method or a modification thereof[38] and is stored in 0.5–1% sulfobetaine (SB, Calbiochem, La Jolla, CA, Zwittergent 3-14) in PBS in at 4°. Usually non-PK-treated

---

[28] R. A. Bessen, D. A. Kocisko, G. J. Raymond, S. Nandan, P. T. Lansbury, Jr., and B. Caughey, *Nature* **375,** 698 (1995).

[29] D. A. Kocisko, S. A. Priola, G. J. Raymond, B. Chesebro, P. T. Lansbury, Jr., and B. Caughey, *Proc. Natl. Acad. Sci. U.S.A.* **92,** 3923 (1995).

[30] A. Bossers, P. B. G. M. Belt, G. J. Raymond, B. Caughey, R. de Vries, and M. A. Smits, *Proc. Natl. Acad. Sci. U.S.A.* **94,** 4931 (1997).

[31] G. J. Raymond, J. Hope, D. A. Kocisko, S. A. Priola, L. D. Raymond, A. Bossers, J. Ironside, R. G. Will, S. G. Chen, R. B. Petersen, P. Gambetti, R. Rubenstein, M. A. Smits, P. T. Lansbury, Jr., and B. Caughey, *Nature* **388,** 285 (1997).

[32] J. Chabry, B. Caughey, and B. Chesebro, *J. Biol. Chem.* **273,** 13203 (1998).

[33] R. Demaimay, J. Harper, H. Gordon, D. Weaver, B. Chesebro, and B. Caughey, *J. Neurochem.* **71,** 2534 (1998).

[34] L. M. Herrmann and B. Caughey, *Neuroreport* **9,** 2425 (1998).

[35] W. S. Caughey, L. D. Raymond, M. Horiuchi, and B. Caughey, *Proc. Natl. Acad. Sci. U.S.A.* **95,** 12117 (1998).

[36] R. A. Bessen, G. J. Raymond, and B. Caughey, *J. Biol. Chem.* **272,** 15227 (1997).

[37] D. C. Bolton, P. E. Bendheim, A. D. Marmostein, and A. Potempska, *Arch. Biochem. Biophys.* **258,** 579 (1987).

[38] B. W. Caughey, A. Dong, K. S. Bhat, D. Ernst, S. F. Hayes, and W. S. Caughey, *Biochemistry* **30,** 7672 (1991).

PrP-res preparations (fraction P4) are used for the conversion reactions and have been stored at 0.1–5 mg/ml for up to 2 years (so far) without a noticeable loss of converting activity. PrP-res content and purity are determined by BCA (Pierce, Rockford, IL) protein assay in conjunction with analysis on silver-stained SDS–PAGE gels and immunoblots.

2. PrP-res has usually been pretreated in 2.0–2.5 *M* Gdn-HCl, although the optimal Gdn-HCl concentration can vary slightly among PrP-res preparations, TSE strains, and host species.[29,31] PrP-res is diluted to 0.1 μg/ml in its storage buffer, sonicated briefly, and transferred into a Gdn-HCl solution to achieve the desired final Gdn-HCl concentration and a total of 0.1–0.5 μg PrP-res per conversion reaction. Incubate at 37° for $\geq$30 min. When multiple conversion reactions are being prepared side by side it is helpful to make this suspension in bulk and aliquot it into separate tubes after this stage as the Gdn-HCl partially disaggregates the PrP-res and allows for more even distribution.
3. [^{35}S]PrP-sen is labeled, immunoprecipitated, and eluted from the protein A-Sepharose beads using 0.1 *M* acetic acid as described earlier. Approximately 10,000–50,000 cpm [^{35}S]PrP-sen is evaporated just to dryness in a vacuum centrifuge in a 0.5-ml siliconized screw-cap microtube. Do not overdry. It is not necessary to evaporate the acetic acid if the volume added to the reactions is less than ~20% of the total reaction volume. This usually contains ~1–10 ng total PrP-sen. If the [^{35}S]PrP-sen is dried, add 14 μl of a buffer (1.5×) containing 75 m*M* citrate/NaOH, pH 6.0; 7.5 m*M* cetylpyridinium chloride (CPC); 1.87% *N*-laurylsarkosine; and 0.14 *M* Gdn-HCl and vortex. If the [^{35}S]PrP-sen is not dried, adjust the volume of the buffer accordingly. Then add 8 μl of the Gdn-HCl-pretreated PrP-res from step 2 to each tube, mix, and incubate in a humidified chamber at 37° for 40–72 hr with occasional mixing. Include a negative control reaction without any PrP-res.
4. A 2.0-μl aliquot of the conversion reaction is analyzed without PK treatment by transferring it to a tube containing 2 μl of 10× TN, 11 μl of distilled H_2O, 4 μl of 5 mg/ml thyroglobulin (as carrier), and 1 μl 0.1 *M* Pefabloc. Precipitate with 80 μl of methanol (−20°).
5. Dilute the remaining 20 μl of the conversion reaction by adding 30 μl of TN and PK to a final concentration of 10–20 μg/ml PK. Incubate at 37° for 1 hr, add 1 μl of 0.1 *M* Pefabloc and 4 μl of 5 mg/ml thyroglobulin, and vortex. Place on ice for several minutes. Precipitate with 200 μl of methanol at −20° for $\geq$60 min. When varying the amounts of PrP-res and/or PrP-sen in the conversion reactions,

the amount of PK in the digestions is adjusted to assure that the [^{35}S]PrP from the no PrP-res control reaction is fully digested and that no full-length [^{35}S]PrP remains in the reactions containing PrP-res. Our rule of thumb is to maintain a PK : PrP-sen + PrP-res mass ratio of ~ 2 : 1; however, the relevant ^{35}S-PrP-res conversion products can usually tolerate considerably higher PK concentrations. The conversion reaction mixture should also be diluted so that the Gdn-HCl concentration is ≤0.4 *M* prior to PK digestion.

6. For SDS–PAGE analysis, the MeOH precipitates are pelleted, air dried for 5 min, and dissolved by vortexing into 15 μl SDS–PAGE sample buffer and heating in a boiling water bath for 5 min prior to loading on SDS–PAGE gels. The gels are dried and scanned for radioactive bands using a phosphor autoradiography instrument, e.g., a Storm PhosphorImager and Imagequant software (Molecular Dynamics, Sunnyvale, CA) (Fig. 1). The percentage conversion of [^{35}S]PrP-sen to PK-resistant forms is quantitated by comparing the intensities of the ^{35}S-PrP-sen bands in the non-PK-treated samples to the bands of interest in the PK-treated samples with appropriate factoring to compensate for the difference in reaction equivalents in the lanes compared (usually a 1 : 10 ratio of the −PK : +PK samples). Typical conversion efficiencies are 10–30% but can range from undetectable to >50%. At the low end, conversions of ≤~0.5% of the [^{35}S]PrP-sen are often difficult to detect above background.

Gdn-HCl-Free Conversion Reactions

For Gdn-HCl-free conversions (also without chaperone proteins), PrP-res is diluted to 50 ng/μl with water and sonicated briefly. Then 100 ng of PrP-res is mixed with [^{35}S]PrP-sen (e.g., 20,000 cpm) in a total volume of 20 μl, which also contains 200 m*M* KCl, 5 m*M* $MgCl_2$, 0.625% *N*-laurylsarcosine, and 50 m*M* sodium citrate (pH 6.0). Conversion reaction mixtures are incubated at 37° for 2 days. Nine-tenths of the reaction mixtures are diluted to 100 μl with TN and treated with PK (20 μg/ml) for 1 hr at 37°. Digestion by PK is stopped by adding 2 μl of 0.1 *M* Pefabloc and 20 μg thyroglobulin is added as a carrier. The remaining one-tenth of the reaction mixture is analyzed without PK treatment. Methanol precipitates of the proteins are subjected to SDS–PAGE and analyzed by phosphor audioradiography as described in the preceding paragraph. In our experience, the overall efficiencies of conversion reactions under these conditions are ~60–70% of those achieved in side-by-side reactions under the Gdn-HCl-containing reactions described earlier (Fig. 1). Optimal salt concentrations may vary with the types of PrP-sen and PrP-res used in the reactions. For

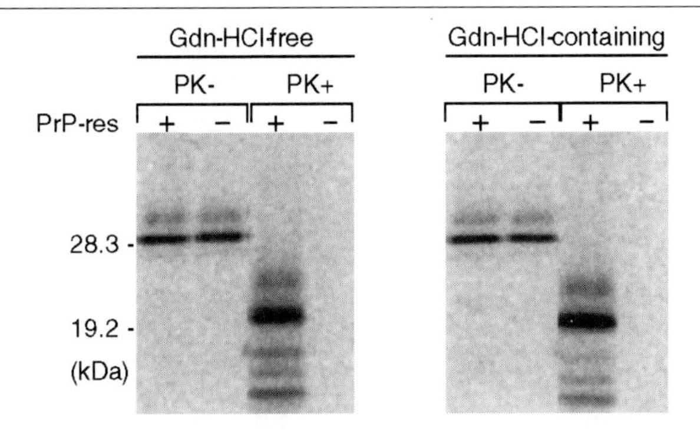

FIG. 1. Typical phosphor autoradiographic detection of ^{35}S-labeled PrP species from *in vitro* conversion reactions under the Gdn-HCl-containing and Gdn-HCl-free conditions described in Table I. Unlabeled hamster PrP-res (263K strain) was used to drive the conversion of recombinant hamster [^{35}S]PrP-sen lacking a glycosylphosphatidylinositol anchor. The major band of the latter is unglycosylated. Note that the PK-resistant [^{35}S]PrP products were generated only in those reactions containing PrP-res and that the major [^{35}S]PrP-res products were the characteristic 6–8 kDa lower in apparent molecular mass than the non-PK-digested [^{35}S]PrP-sen precursor bands. The thoroughness of the PK digestion is indicated by the lack of full-length [^{35}S]PrP-sen in the PK-digested samples. The ratio of reaction equivalents in non-PK-treated and PK-treated lanes is 1:10. Quantitation of the major [^{35}S]PrP-res (above the 19-kDa marker) relative to the precursor bands with appropriate factoring gave conversion efficiencies of 26 and 39%, respectively, in Gdn-HCl-free and Gdn-HCl-containing reactions.

instance, 200 m*M* KCl is optimal for the conversion of unglycosylated hamster GPI PrP-sen by hamster 263K strain PrP-res, but 100 m*M* NaCl works better for conversions with the wild-type, GPI-linked, glycosylated hamster PrP-sen (Table I).

TABLE I
REACTION CONDITIONS FOR CELL-FREE PrP CONVERSION REACTIONS

Step	Gdn-HCl containing	Gdn-HCl free
Pretreatment of PrP-res	2.0–2.5 *M* Gdn-HCl 37°, 1 h	Suspend in water Sonicate briefly
Reaction	[^{35}S]PrP-sen, 10–50,000 cpm PrP-res, 100–500 ng Gdn-HCl, 0.5–1 *M* Sodium citrate (pH 6.0), 50 m*M* CPC, 5 m*M* *N*-Laurylsarkosine, 1.25% 37°, 2 days	[^{35}S]PrP-sen, 10–50,000 cpm PrP-res, 100 ng KCl, 200 m*M* $MgCl_2$, 5 m*M* Sodium citrate (pH 6.0), 50 m*M* *N*-Laurylsarkosine, 0.62% 37°, 2 days

Chaperone proteins can also improve the efficiency of Gdn-HCl-free, PrP-res-induced conversion reactions under slightly different conditions, as has been described previously.[27]

In situ PrP Conversion Reactions in Tissue Slices

The PrP-sen to PrP-res conversion reaction can be induced not only with largely purified and concentrated reactants, but also in intact TSE-infected brain slices (Fig. 2).[36] Unfixed 10-μm-thick slices of frozen, unfixed brain tissue are air dried onto glass microscope slides and preincubated with IRB for 30 min at 37°. [^{35}S]PrP-sen (20,000–50,000 cpm/reaction) in IRB is then layered over the tissue and incubated at 37° for 18 hr in a humidified chamber. The supernatants are removed, and tissue sections are rinsed briefly and then treated with PK (150 μg/ml) in IRB lacking albumin followed by incubation at 37° for 1 hr. The sections are then washed in IRB supplemented with 2 m*M* phenylmethylsulfonyl fluoride for 15 min and dried. The slides are exposed directly to X-Omat AR film for autoradiography or, for higher resolution, dipped in NTB-2 liquid emulsion (Eastman Kodak Co., Rochester, NY) and developed according to manufacturer's instructions. The sizes of the [^{35}S]PrP-res conversion products in the tissue can be monitored by SDS–PAGE and fluorography after solubilizing the tissue off the slide in SDS–PAGE sample buffer. Adjacent tissue sections can be stained immunohistochemically for PrP-res[10] to allow comparison of the localization of preexisting PrP-res with the autoradiographic

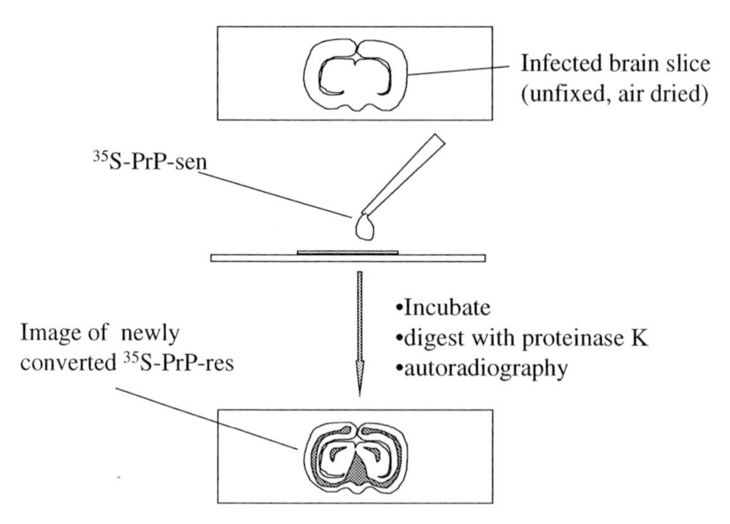

FIG. 2. Scheme for *in situ* conversion assay for PrP-res. For actual results, see Bessen *et al.*[36]

patterns of [^{35}S]PrP-res *in situ* conversion products in the intact tissue sections. In previous work, these patterns have been closely matched.[36] Furthermore, no *in situ* conversion pattern was observed when uninfected tissue or [^{35}S]PrP-sen from an incompatible species was used.

Current Limitations of PrP-res and Converting Activity Assays

A practical limitation to current assays for PrP-res and converting activity as surrogates for infectivity bioassays in animals is the relative lack of sensitivity of the former. The limit of detection for the immunoblot assays for PrP-res is usually 0.1–1 ng with good antibodies. The detection limit for present *in vitro* PrP conversion assays corresponds to ~10 ng of PrP-res. In contrast, samples containing 1 LD_{50} of 263K scrapie infectivity that is detectable by bioassay in hamsters are estimated to have only $\sim 10^5$ molecules (~5 fg) of PrP-res. Clearly, the *in vivo* bioassay for infectivity is time-consuming, animal intensive, and expensive, but it remains orders of magnitude more sensitive than biochemical assays. Thus, although the detection of PrP-res and/or converting activity provides strong evidence for the presence of TSE infectivity, the current *in vitro* assays cannot be used to establish that any given sample *lacks* TSE infectivity.

[10] *In Situ* Methods for Detection and Localization of Markers of Oxidative Stress: Application in Neurodegenerative Disorders

By LAWRENCE M. SAYRE, GEORGE PERRY, and MARK A. SMITH

Introduction

Neurodegenerative Disease, Pathological Deposition, and Oxidative Stress

Despite an extensive research effort aimed at characterizing the nature of protein abnormalities associated with age-related neurodegenerative conditions, the factors leading to the deposition of these protein aggregates remain elusive. Neurofibrillary tangles (NFT) and senile plaques (SP) in Alzheimer's disease (AD), intraneuronal filamentous inclusions in amyotrophic lateral sclerosis (ALS), Lewy bodies in Parkinson's disease (PD), and Pick bodies in Pick's disease are pathological hallmarks. These deposits

0076-6879/99 $30.00

are constituted of one or more proteins, including amyloid-β (Aβ), neurofilaments (NF), and microtubule-associated protein τ, which, in one or another form, are present in nonaggregated forms in normal cells. The various deposits also differ in terms of the degree to which aggregation is reversible and what conditions are needed to achieve their solubilization, thereby complicating efforts aimed at direct structural analysis.

Various factors have been proposed as the initiating events in pathological aggregation and include abnormal levels of protein synthesis, abnormal precursor protein processing, and abnormal phosphorylation or other posttranslational modification. In no case, however, is there a clear mechanistic consensus nor is there any direct evidence that connects protein deposition with the etiology of neuronal death in the respective neurodegenerative disease. Advances in molecular genetics and neurochemistry have associated neuronal death at least indirectly to oxidative stress, the condition when prooxidant events outweigh the cellular antioxidant defenses. In oxidative stress, a compromise in the normal physiological removal of reactive oxygen species (ROS) and/or an increase in their production results in irreversible damage to biomacromolecules.[1,2] Defects in mitochondrial energy metabolism[3,4] may be important and may explain in part the progressive oxidative modification of proteins in normal aging.[5,6]

It may be postulated that the pathological aggregation of cytoskeletal protein initially represents a reversible shift in the balance between monomeric and polymeric forms held together by noncovalent forces and that the ultimate insolubilization and proteolytic resistance seen for some pathological protein aggregates then arise, at least in part, from a covalent "cementing" of the aggregates by cross-linking reactions associated with oxidative stress. Chemical studies support the notion that oxidative covalent cross-linking of proteins is facilitated by their preassociation,[7,8] as protein-derived reactive moieties have a limited lifetime and will react preferentially with solvent or other small molecules if no suitably reactive and/or positioned protein group is nearby. Protein cross-linking lowers susceptibility

[1] M. A. Smith, L. M. Sayre, V. M. Monnier, and G. Perry, *Trends Neurosci.* **18,** 172 (1995).
[2] M. R. Markesbery, *Free Radic. Biol. Med.* **23,** 134 (1997).
[3] A. C. Bowling and M. F. Beal, *Life Sci.* **56,** 1151 (1995).
[4] R. Dawson, Jr., M. F. Beal, S. C. Bondy, D. A. Di Monte, and G. E. Isom, *Toxicol. Appl. Pharmacol.* **134,** 1 (1995).
[5] R. T. Dean, S. Fu, R. Stocker, and M. J. Davies, *Biochem. J.* **324,** 1 (1997).
[6] B. S. Berlett and E. R. Stadtman, *J. Biol. Chem.* **272,** 20313 (1997).
[7] K. C. Brown, S.-H. Yang, and T. Kodadek, *Biochemistry* **34,** 4733 (1995).
[8] M. L. Miller and G. V. W. Johnson, *J. Neurochem.* **65,** 1760 (1995).

to proteolytic degradation, thereby inhibiting turnover, and thus may itself represent a cytotoxic event.

Oxidative Stress Biochemistry

Protein damage that occurs under conditions of oxidative stress may represent direct protein oxidation, mediated by diffusible hydroxyl radical (HO·) or metal-catalyzed "site-specific" oxidation at the sites of metal ion binding in the protein.[9,10] This can result in backbone cleavage events or in the specific oxidation of side chain moieties, sometimes involving cross-linking reactions, such as coupling of two tyrosine radicals. Oxidative stress is also associated with the production of peroxynitrite, which is capable of nitration and/or oxidation of Tyr and Trp residues.[11,12]

The large majority of oxidative stress-dependent protein modification, however, involves the adduction of products of glycoxidation and/or lipoxidation. Modification of protein-based lysines and arginines by reducing sugars initiates a complex sequence of oxidation, rearrangement, condensation, and fragmentation reactions, known collectively as the Maillard reaction,[13] resulting ultimately in stable adducts called advanced glycation end products (AGEs). In the modification of proteins by lipoxidation products such as 4-hydroxy-2-nonenal (HNE) and malondialdehyde (MDA), several (especially lysine-based) adducts are also unstable and evolve over time to give stable advanced lipoxidation end products (ALEs).[14,15] Research over the last decade has elucidated the structure of many cross-link and non-cross-link AGE and ALE modifications.

Despite widespread recognition of the importance of independent mechanisms of protein oxidation and AGE and ALE modification, there is now substantial evidence that any one of these types of modification can promote the others, resulting in synergistically accelerated oxidative protein damage. Examples are the oxidative modification of proteins that occurs at loci of sugar attachment[16,17] and lipid peroxidation induced by glycation of

[9] E. R. Stadtman, *Science* **257,** 1220 (1992).
[10] E. R. Stadtman, *Annu Rev. Biochem.* **62,** 797 (1993).
[11] J. S. Beckman, *Chem. Res. Toxicol.* **9,** 836 (1996).
[12] B. Alvarez, H. Rubbo, M. Kirk, S. Barnes, B. A. Freeman, and R. Radi, *Chem. Res. Toxicol.* **9,** 390 (1996).
[13] V. M. Monnier and A. Cerami, *Science* **211,** 491 (1981).
[14] L. M. Sayre, W. Sha, G. Xu, K. Kaur, D. Nadkarni, G. Subbanagounder, and R. G. Salomon, *Chem. Res. Toxicol.* **9,** 1194 (1996).
[15] K. Itakura, K. Uchida, and T. Osawa, *Chem. Phys. Lipids* **84,** 75 (1996).
[16] J. V. Hunt, R. T. Dean, and S. P. Wolff, *Biochem. J.* **256,** 205 (1988).
[17] J. V. Hunt and S. P. Wolff, *Free Radic. Res. Commun.* **12-13,** 115 (1991).

membrane lipids[18] and cells in culture.[19] One chemical explanation is that the condensation of reducing sugars with amino groups leads to the equilibrium presence of redox-active enediol and enol-enamine moieties that can redox the cycle,[20] either directly or through the assistance of transition metals. Similar chemistry is expected for lipid-derived 4-hydroxy-2-alkenals.

Oxidative stress also leads to the oxidative modification of polynucleotides. This can again involve direct oxidation (such as the addition of HO· to purine and pyrimidine rings and the HO·-mediated cleavage of the sugar backbone) or the adduction of lipoxidation products such as MDA and HNE.

Specific Neuronal Vulnerability to Oxidative Stress Damage

The central nervous system seems to be especially vulnerable to oxidative stress on account of the high rate of oxygen utilization and the fact that neuronal membranes contain a high proportion of oxidation-prone polyunsaturated fatty acids compared to normal plasma membrane, making them more susceptible to peroxidative damage. At the same time, neurons contain low levels of the major cellular antioxidant, glutathione.[21] Furthermore, neurofilaments and cytoskeletal proteins such as τ have relatively long half-lives and contain high proportions of lysine residues. Although these lysine residues have been implicated in supramolecular cytoskeletal organization, their presence makes these proteins particularly susceptible to modification by carbonyl-containing products of glycoxidation and lipoxidation. Such modifications or direct oxidation results in neutralization of the charge and consequent alteration of protein–protein electrostatic interactions. The latter can provide a key noncovalent mechanism for aberrant protein aggregation/deposition even in the absence of covalent cross-linking.

In summary, although the complex interplay of biochemical mechanisms in force in neurodegenerative disease suggests several possible investigative avenues, oxidative stress appears to be a common contributor to both pathological protein deposition and irreversible cellular dysfunction leading to neuronal death. With the increasing availability of immunochemical tools

[18] R. Bucala, Z. Makita, T. Koschinsky, A. Cerami, and H. Vlassara, *Proc. Natl. Acad. Sci. U.S.A* **90,** 6434 (1993).

[19] S.-D. Yan, X. Chen, A.-M. Schmidt, J. Brett, G. Godman, Y.-S. Zou, C. W. Scott, C. Caputo, T. Frappier, M. A. Smith, G. Perry, S-H. Yen, and D. Stern, *Proc. Natl. Acad. Sci. U.S.A* **91,** 7787 (1994).

[20] T. Sakurai and S. Tsuchiya, *FEBS Lett.* **236,** 406 (1988).

[21] A. Slivka, C. Mytilineou, and G. Cohen, *Brain Res.* **409,** 275 (1987).

that recognize specific oxidative markers on the pathological deposits, the question of whether oxidative stress is contributory to, or merely a consequence of, the long-term presence of abnormal protein deposition might be answerable on the basis of spatiotemporal patterns of epitope appearance. Although the various neurodegenerative diseases have in common the presence of cytoskeletal abnormalities, it is unlikely that the vulnerable neurons (pyramidal neurons in the hippocampus/cortex in AD, pigmented neurons in the substantia nigra in PD, and motor neurons in ALS) respond in identical ways to the oxidative damage. Thus, while a certain degree of overlap among these diseases in associated oxidative stress biochemistry might be expected, each individual disease will likely show a distinct profile of specific oxidative modifications and therefore no single marker should be relied on as diagnostic of oxidative stress.

This article provides details of *in situ* detection methods that can be used for the qualitative and semiquantitative measurement of various oxidative stress markers in tissue sections obtained at autopsy. While oxidative stress damage detected no doubt implicates general or specific cytotoxicity, it is not the purpose of this article to suggest that the appearance of certain markers correlates with a particular mechanism of toxicity. Indeed, as discussed in the last section, any such correlation requires a consideration of whether the damage apparent represents a primary or secondary event in the cascade of biochemical processes leading to neuronal death.

Value of *in Situ* Methods

The importance of *in situ* methods over bulk analysis cannot be overplayed when considering the structural and cellular complexity of the nervous system. Not only do the affected neurons comprise a small percentage of the tissue, but we also do not know how other cells such as glia are affected, as they must be, by the disease. While neurons may show one type of damage or response, glia may show a compensatory response unrelated to, or opposite, of neurons. Although we do not have data on this point, it is enough of a concern to theoretically worry about the significance of bulk analysis studies in isolation from *in situ* studies. Further, bulk analysis of oxidative damage will always be plagued by the contribution of long-lived proteins modified during physiological aging. Extracellular matrix proteins of major vessels therefore provide a record of long-term oxidative insult, a facet that may be why vascular proteins are a sensitive marker of diabetic changes. However, the same properties that make vessels ideal monitors for aging limit their sensitivity to detect disease-specific changes. While the isolation of neurons or other cell fractions can address the issue of specificity, this is not always accomplished readily for poorly represented cell types, where there is always the concern of vascular contamination. Vascular contamination of insoluble proteins is a particular concern because even glass bead columns or collagenase treatments seldom quantitatively remove all vascular elements if the initial isolation procedure does not exclude them.

Experimental

Pathological Tissue

It is essential that the modifications analyzed be both stable and not produced as a result of postmortem or other time-dependent postexperimental changes. The latter is a particular concern if initial lipid oxidative adducts are analyzed. One way we have addressed this issue is to analyze and compare results of localizing early vs advanced lipoxidation end products. Tissue can be either flash frozen in liquid nitrogen/isopentane and cryostat sections prepared at 20 mm or fixed in methacarn (60% methanol, 30% chloroform, 10% acetic acid, v/v/v), paraffin embedded, and sections cut at 6 μm. Material fixed in glutaraldehyde or even paraformaldehyde is problematic because it precludes analysis of lipoxidative and other modifications that share the property of carbonyl condensation-dependent cross-links. The latter concern is the most significant limitation in practice as the majority of pathological tissue available for retrospective study is fixed in formalin. If formalin-fixed tissue must be used, it is best to focus on highly defined products, unlikely to share homology with formalin or nucleic acid modifications.

Direct Histochemical Methods

2,4-Dinitrophenylhydrazine Reactivity of Protein-Bound vs Protein-Based Carbonyls

Method Rationale. Two related methods for the detection of protein-based carbonyls are (i) derivatization with 2,4-dinitrophenylhydrazine (DNP-H) and detection of the protein-bound hydrazones using an enzyme-linked anti-2,4-dinitrophenyl antibody and (ii) derivatization with biotin hydrazide and detection of the protein-bound acyl hydrazone with enzyme-linked avidin or streptavidin. Although we have only utilized the DNP-H method *in situ*,[22,23] the biotin hydrazide method should perform similarly and may be preferable in cases where there appears to be excessive nonspecific binding of DNP-H. In either case, it is important to recognize that protein-bound carbonyls can represent either oxidized side chains or the univalent adduction of a *bi*functional glycoxidation or lipoxidation product such that the adduct still contains a free carbonyl group. Modification of unoxidized lysine amino groups by *mono*functional lipid- or sugar-derived carbonyl compounds will *not* interfere because DNP-H or biotin hydrazide will displace the carbonyl compound into bulk solution. In these cases, one can learn about the nature of the lysine modification through direct LC-MS analysis of the DNP derivatives in solution. If it is important to localize and/or quantify *total* protein-associated carbonyls (both protein-based carbonyls and carbonyl-modified lysines), one approach is to carry out an *in situ* determination of protein-bound tritium using [^{3}H]$NaBH_4$, with detection by fluorography.

Experimental. For all procedures involving carbonyl determination, it is essential that the tissue not be aldehyde fixed. After deparaffinization in xylene and rehydration through graded ethanol for methacarn-fixed material or brief fixation in ice-cold acetone for cryostat material, sections are covered with 0.1–0.001% (w/v) DNP-H in 2 *N* HCl. All incubations are conducted at room temperature unless otherwise indicated in a plastic box in which water-saturated paper towels are placed in the bottom over which the slides are placed on a platform. After a 1-hr incubation, sections are rinsed exhaustively in Tris-buffered saline (TBS) (50 m*M* Tris–HCl, 150 m*M* NaCl, pH 7.6) followed by a 30-min incubation in 10% normal goat serum (NGS) to block nonspecific binding sites. After rinsing with 1% NGS/TBS, a rat monoclonal antibody (LO-DNP-2; Zymed Laboratories, Inc., San Francisco, CA) to dinitrophenyl (DNP) is diluted 1:100 in 1% NGS/TBS and incubated with the sections at 4° for 16 hr. Sections are then rinsed with 1% NGS/TBS, followed by a 30-min incubation with goat antiserum to rat immunoglobulin G (IgG, Boehringer-Mannheim Corporation, Indianapolis, IN) diluted at 1:50 with 1% NGS/TBS. After rinsing in 1% NGS/TBS, the rat peroxidase–antiperoxidase complex (ICN Pharmaceuticals, Inc., Costa Mesa, CA) (1:250) in the same buffer is incubated with the section at room temperature for 1 hr, after which it is rinsed with 1% NGS/TBS. Peroxidase activity is localized by development for 5–10 min with 0.015% H_2O_2 in 50 m*M* Tris–HCl, pH 7.6, with 0.75 mg/ml 3,3′-

[22] M. A. Smith, G. Perry, P. L. Richey, L. M. Sayre, V. E. Anderson, M. F. Beal, and N. Kowall, *Nature* **382,** 120 (1996).

[23] M. A. Smith, L. M. Sayre, V. E. Anderson, P. L. R. Harris, M. F. Beal, N. Kowall, and G. Perry, *J. Histochem. Cytochem.* **46,** 731 (1998).

diaminobenzidine (DAB; Sigma, St. Louis, MO). The development of sections is monitored directly for maximum contrast under the 10× objective of a Zeiss Axioskop 20 microscope.

Chemical and immunochemical controls are used to define carbonyl-specific binding. Chemical reduction of free carbonyls and Schiff bases is performed by incubating sections with 25 m*M* $NaBH_4$ in 80% (v/v) methanol for 30 min before incubation with DNP-H. Selective reduction of Schiff bases as opposed to free carbonyls can be achieved by incubation with 50 m*M* sodium cyanoborohydride in 0.1 *M* phosphate buffer, pH 6.0, for 1 hr. Immunochemical specificity is demonstrated by omission of the DNP-H treatment or the antibody to DNP. Immunoabsorption of the antibody to DNP is performed by incubating the antibody (1 : 100) with 5 m*M* pyruvate 2,4-dinitrophenylhydrazone (stock 1 mg/ml in ethanol) at 4° for 16 hr and comparing the resulting immunoreactivity with an unabsorbed antibody that had been treated similarly with ethanol.

When the peroxidase–antiperoxidase method is used for visualization of bound DNP antibody, we have typically utilized a 20-min preincubation of tissue sections with 3% H_2O_2 in methanol to block artifactual staining arising from endogenous peroxidase activity in the tissue.[24] Although H_2O_2 might itself, or synergistically with tissue-bound transition metals, create oxidative modifications through hydroxyl radical formation (e.g., carbonyl residues), deletion of this step had no effect on subsequent DNP-H labeling conducted in our laboratory, although it did convert Fe(II) to Fe(III).[25,26] Therefore, a control that omits the H_2O_2 treatment step should be performed with each experimental paradigm. Alternatively, utilization of a secondary antibody coupled to alkaline phosphatase with *p*-nitophenyl phosphate as chromagen obviates the necessity of the prior H_2O_2 treatment.

Redox-Active Transition Metals

Method Rationale. The protein oxidation and modification observed in particular cells and subcellular compartments possibly reflect a prooxidant shift in biochemistry at an organismal level, perhaps reflective of present or impending degenerative disease, but arise ultimately from chemistry that is localized at the molecular level. For example, hydroxyl radicals generated from Fenton chemistry react with biomacromolecules at diffusion-controlled rates, and the reactive aldehydes generated during lipid oxidation are expected to be intercepted by circulating amino acids, detoxification

[24] L. A. Sternberger, ed., "Immunocytochemistry." Wiley, New York, 1986.

[25] M. A. Smith, P. L. R. Harris, L. M. Sayre, and G. Perry, *Proc. Natl. Acad. Sci. U.S.A* **94,** 9866 (1997).

[26] L. M. Sayre, G. Perry, and M. A. Smith, unpublished work (1999).

enzymes, and glutathione, if they do not first find a protein near their site of origin. This recognition has generated interest in determining whether it is possible to identify *sources* of oxidative stress in the same pathological structures that exhibit oxidative stress damage. The major culprits would be redox-active transition metals bound adventitiously to proteins or to protein sites with increased metal ion affinity due to modification by lipoxidation and glycoxidation products. Thus, we have developed methods for the *in situ* determination and localization of redox-active transition metals.[25,27]

Cytochemical Detection of iron(II)/(III). This technique incorporates three key modifications of the established histological method[28,29] that increase the detection sensitivity of this technique greatly: (i) increasing the concentrations of potassium ferrocyanide (7%, w/v) and hydrochloric acid (3%, v/v); (ii) lengthening the incubation time and/or increasing the temperature (15 hr at room temperature or 1 hr at 37°); and (iii) use of methacarn versus formalin-fixed tissue. The last modification is perhaps the most critical to iron detection in human tissue where formaldehyde-based fixation for extended periods is routine. Indeed, we found that even a brief postfixation with formaldehyde (3.7%, 5–60 min at room temperature) reduced labeling in a time-dependent manner.[25] Results with methacarn fixation are similar to those found in sections made from frozen blocks, which were neither fixed nor embedded, suggesting that the iron localization is not a result of tissue fixation or embedding.

This method takes advantage of forming the intensely blue mixed valence Fe(II)/Fe(III) cyanide complex, which can be observed directly as a pale blue stain on tissue sections or can be enhanced by taking advantage of the complex to catalyze the H_2O_2-dependent oxidation of 3,3′-diaminobenzidine (DAB), giving an insoluble brown precipitate localized to bound iron. For the latter case, the tissue sections are deparaffinized in xylene, rehydrated through graded ethanol, and then incubated for 15 hr in 7% potassium ferrocyanide [for iron(III) detection] or 7% potassium ferricyanide [for iron(II) detection] in aqueous hydrochloric acid (3%) and subsequently incubated with a solution containing 0.75 mg/ml DAB tetrahydrochloride in Tris–HCl, pH 7.6, and 0.015% H_2O_2 for 5–10 min.

In situ Oxidation. Protein-bound redox-active transition metals can be demonstrated directly by incubation of material with 3% H_2O_2 and 0.75 mg/ml DAB in 50 m*M* Tris–HCl, pH 7.6.[27] Because this protocol does not

[27] M. A. Smith, G. Perry, P. L. R. Harris, Y. Liu, K. A. Schubert, and L. M. Sayre, unpublished work (1999).

[28] M. Perls, *Virch. Arch. Pathol. Anat.* **39,** 42 (1867).

[29] R. Virchow, *Virch. Arch. Pathol. Anat.* **1,** 379 (1847).

involve preformation of a mixed valence iron complex, it does not directly identify the nature of the tissue-bound metal. However, iron and copper are the metals most likely to be bound adventitiously in redox-active form. In these experiments, it is important to recognize that DAB is supplied as the tetrahydrochloride and that DAB free base is only sparingly soluble in water. Thus, it is crucial that the tissue incubation solution remains at a low enough pH that the DAB stays in solution. In this manner, the insoluble brown DAB oxidation product formed at tissue sites of redox activity is visualized readily without interference from precipitated unoxidized DAB.

Metal Chelation/Binding. Redox-active metals bound adventitiously to proteins in tissue sections can be removed by exposure to up to 0.1 *M* deferoxamine (DFX) or diethylenetriaminepentaacetic acid (DTPA) at room temperature for a period of time dependent on the affinity of the tissue sites. For example, whereas for 0.1 *M* DFX, 20 min was sufficient to remove ferrocyanide-detectable iron from certain sites, a 15-hr incubation was required for the complete removal of iron bound to others, and use of 0.1 *M* EDTA for 15 hr was only partially effective in removing iron.[25] Observed differential efficiency of removal of the H_2O_2-dependent DAB-oxidizing activity by DFX vs DTPA provides information on the nature of endogenous metals present, as DFX and DTPA are somewhat selective for iron and copper, respectively. The fact that following chelation of iron, incubation of AD tissue sections with ferrocyanide or ferricyanide did not result in H_2O_2-dependent oxidation of DAB on the tissue indicated that formation of the mixed valence complex represents real tissue-bound iron and is not an artifact of the use of an iron-containing reagent.

Potential sites for binding of redox-active transition metals in tissue sections or attempted rebinding of endogenous metals following chelation-dependent removal can be carried out, for example, by incubation with 0.01 m*M* iron(III) citrate plus 0.01 m*M* iron(II) chloride or with 0.01 m*M* copper(II) chloride at room temperature for 3 hr followed by detection. By altering the parameters of iron chelation and/or rebinding, one can obviously obtain valuable information regarding the avidity and affinity of various iron sequestration sites.

Immunocytochemical Methods

General

Immunocytochemistry Using Peroxidase-Coupled Secondary Antibody. Following deparaffinization with xylene, sections are rehydrated through graded ethanol (70, 50, 30%). Endogenous peroxidase activity in the tissue is inactivated by a 20-min incubation at room temperature with 3% H_2O_2 in methanol (see earlier discussion for proper control needed) and nonspecific binding sites are blocked in a 30-min incubation at room temperature with 10% NGS in TBS. Immunostaining is accomplished by the peroxidase–antiperoxidase technique using DAB as cosubstrate chromogen.[24] Adjacent sections can be immunostained with antisera to known proteins and to the location of specific structures.

Antigen Retrieval on Tissue Sections. A great variety of protocols are available for the pretreatment of tissue sections to facilitate antigen retrieval (i.e., exposure of hidden epitopes at the protein–protein interface). For example, in previous studies we, and others, have used pretreatments with (i) NaOH (0.2 *M,* 15 min room temperature); (ii) trypsin (400 μg/ml in 0.05 *M* Tris, 0.3 *M* NaCl, 0.02 *M* $CaCl_2$, pH 7.6, for 10 min room temperature); (iii) microwave or pressure cooker treatment in citrate buffer (100 m*M*); (iv) guanidine (8 *M* for 15 min room temperature); (v) formic acid (70% for 15 min room temperature); (vi) antigen retrieval solution (Biogenix Laboratories); or (vii) proteinase K (Boehringer Mannheim, 10 μg/ml in pH 7.4 PBS, 40 min at 37°).

Nitrotyrosine Antibody for Determination of Protein Tyrosine Nitration by Peroxynitrite

Method Rationale. Because oxidative stress is associated with high local concentrations of both superoxide and nitric oxide, produced by the inducible isoform of nitric oxide synthase, the product of their combination, peroxynitrite, has become an important secondary oxidative stress marker. The complex chemistry of peroxynitrite has been the subject of intense study, and it has become clear that there are two principal pathways for protein modification: one involving homolytic hydroxyl radical-like chemistry that results in protein-based carbonyls and the other involving electrophilic nitration of vulnerable side chains, in particular the electron-rich aromatic rings of Tyr and Trp.[11,12] In the presence of buffering bicarbonate, peroxynitrite forms a CO_2 adduct, which augments its reactivity. Formation of 3-nitrotyrosine by this route has become the classical protein marker for the presence specifically of peroxynitrite.[30]

Experimental. Affinity-purified rabbit antiserum or mouse monoclonal antibodies (7A2 or 1A6), raised to nitrated keyhole limpet hemocyanin,[31]

[30] M. A. Smith, P. L. R. Harris, L. M. Sayre, J. S. Beckman, and G. Perry, *J. Neurosci.* **17,** 2653 (1997).

[31] J. S. Beckman, Y. Z. Ye, P. G. Anderson, J. Chen, M. A. Accavitti, M. M. Tarpey, and C. R. White, *Biol. Chem. Hoppe-Seyler* **375,** 81 (1994).

are used at a 1/100 dilution in 1% normal goat serum, 150 m*M* NaCl, 50 m*M* Tris–HCl, pH 7.6.

Controls consist of (i) omitting the primary antibody; (ii) absorption of the antibody with 50 μM nitrated bovine serum albumin (BSA) or 15 μM nitrated Gly-Tyr-Ala peptide at 4° overnight prior to application to the section; and (iii) chemical reduction of nitrotyrosine by treating sections with 15 m*M* sodium hydrosulfite in 50 m*M* carbonate buffer, pH 8.0, for 15 min at room temperature[32] prior to immunostaining. These procedures are performed in parallel with antisera to known markers as controls against artifactual inactivation of either primary or secondary antibodies from use of sodium hydrosulfite-reduced sections and against nonspecific adsorption with nitrated BSA or nitrated Gly-Tyr-Ala.

Advanced Glycation and Lipoxidation End Products

Method Rationale. Glycation of proteins by reducing sugars (Maillard reaction) is a key posttranslational event responsible for protein dysfunction in diabetes and kidney disease and results in a profile of time-dependent adduct evolution that becomes especially susceptible to oxidative elaboration. In addition, oxidative stress can result in oxidized sugar derivatives, which can subsequently modify protein through a process known as glycoxidation. Similarly, and of more general importance, oxidative stress results in lipid peroxidation and the production of a range of electrophilic and mostly bifunctional aldehydes that modify numerous proteins. Although there are a number of chemical events that occur early and rapidly, many of these are reversible and thus of less pathophysiologic significance. Thus, many workers in the field have focused their efforts on identifying the more stable protein modifications that form over time. These stable oxidative stress markers are referred to as advanced glycation end products and advanced lipoxidation end products. Protein modification can result in both non-cross-link and cross-link AGEs and ALEs, the latter arising from the potential bifunctional reactivity, such as of the lipid-derived modifiers 4-hydroxy-2-nonenal (HNE) and malondialdehyde. Examples of cross-link adducts are the fluorescent Arg-Lys AGE cross-link pentosidine[33] the fluorescent Lys-Lys ALE cross-link pyrrolidinone iminium[34] arising from 4-hydroxy-2-alkanals, the fluorescent 3-(1,4-dihydropyridin-4-yl)pyridinium and nonfluorescent 1-amino-3-iminopropene Lys-Lys cross-links arising from MDA,[15] and the imidazolium Lys-Lys cross-links GOLD and

[32] P. Cuatrecasas, S. Fuchs, and C. B. Anfinsen, *J. Biol. Chem.* **243,** 4787 (1968).
[33] D. R. Sell and V. M. Monnier, *J. Biol. Chem.* **264,** 21597 (1989).
[34] G. Xu and L. M. Sayre, *Chem. Res. Toxicol.* **11,** 247 (1998).

MOLD arising from glyoxal and methylglyoxal, respectively,[35] common to both AGE and ALE chemistry. Examples of non-cross-link modifications are the lysine-derived AGE pyrraline,[36] the HNE-derived 2-pentylpyrrole,[14,37] and carboxymethyllysine (CML)[38] resulting from glyoxal or from fragmentation of a glycated precursor, and the fluorescent arginine-based argpyrimidine arising from methylglyoxal.[39]

Antibodies to Advanced Glycation End Products. Polyclonal antibodies against pentosidine were raised in rabbits. Briefly, 250 μg pentosidine–keyhole limpet hemocyanin (KLH) conjugate prepared by the carbodiimide coupling technique[40] was emulsified with an equal volume of Freund's complete adjuvant and injected intradermally and intramuscularly into New Zealand Wistar rabbits. At 2- to 3-week intervals, booster injections of pentosidine–KLH emulsified in Freund's incomplete adjuvant were made. Serum was obtained 7–10 days postinjection and antibody titer assessed by ELISA. When antibody levels plateaued, the rabbits were exsanguinated by cardiac puncture and collected serum was stored at $-80°$ until required.

Labeling was by the peroxidase–antiperoxidase technique described earlier. Adsorption experiments can be performed in parallel by incubation of the primary antibody with either free pentosidine or pyrraline hapten (10 n*M*) for 3 hr at 37°. Cross-adsorption of pentosidine antisera with free pyrraline and pyrraline antibodies with free pentosidine was also performed as a control against artifactual adsorption. The preparation of pyrraline and pentosidine was described previously.[36,41]

Antibodies to HNE–Pyrrole. Two different antisera to HNE–pyrrole have been used, one raised to keyhole limpet hemocyanin modified by 4-oxononanal (ON–KLH),[14,42] which generates as the major product the same 2-pentylpyrrole on lysine ε-amino groups formed (in low yield) on direct exposure to HNE, but which also gives some side products. The other antiserum was raised to keyhole limpet hemocyanin conjugated with preformed 6-(2-pentylpyrrol-1-yl)caproic acid (PPC–KLH), which gener-

[35] K. J. Wells-Knecht, E. Brinkmann, and J. W. Baynes, *J. Org. Chem.* **60,** 6246 (1995).

[36] S. Miyata and V. M. Monnier, *J. Clin. Invest.* **89,** 1102 (1992).

[37] L. M. Sayre, P. K. Arora, R. S. Iyer, and R. G. Salomon, *Chem. Res. Toxicol.* **6,** 19 (1993).

[38] S. Reddy, J. Bichler, K. J. Wells-Knecht, S. R. Thorpe, and J. W. Baynes, *Biochemistry* **34,** 10872 (1995).

[39] I. N. Shipanova, M. A. Glomb, and R. H. Nagaraj, *Arch. Biochem. Biophys.* **344,** 29 (1997).

[40] H. Yamada, T. Imoto, K. Fujita, K. Okazaki, and M. Motomura, *Biochemistry* **20,** 4836 (1981).

[41] D. R. Sell and V. M. Monnier, *J. Clin. Invest.* **85,** 380 (1990).

[42] K. S. Montine, P. J. Kim, S. J. Olson, W. R. Markesbery, and T. J. Montine, *J. Neuropathol. Exp. Neurol.* **56,** 866 (1997).

ates the pure pyrrole epitope, but with a longer tether than when the pyrrole is formed from HNE.[14] Thus, the coincidence of labeling by both antisera was deemed to be most convincing of specific immunocytochemical recognition of the HNE-derived pyrrole.

HNE–pyrrole antisera were used at a dilution from 1:10–1000; we found a 1:100 dilution was optimal and this dilution was used throughout for immunostaining with the exception of the immunoadsorption experiment (see later). Immunostaining utilized the peroxidase–antiperoxidase procedure and was developed under a microscope using 3,3′-diaminobenzidine at 0.75 mg/ml in 0.015% H_2O_2, 50 m*M* Tris–HCl, pH 7.6. The sections were dehydrated through ethanol and xylene solutions and then mounted in Permount (Fisher). The location of aberrant protein aggregates such as NFT and senile plaques can be determined following immunostaining by counterstaining the sections with Congo red and viewing them under crossed polarized light. Antisera specific to the protein constituents of aberrant aggregates can be used on adjacent sections in order to determine the microscopic localization of the lipoxidation-dependent adduct.

The specificity of both HNE–pyrrole antibodies was confirmed by omitting the primary antibody or by performing an adsorption. The two antisera to HNE–pyrrole were diluted 1:150 and then incubated with 0.1 mg/ml of the bovine serum albumin (BSA)-derived antigens (ON–BSA or PPC–BSA)[43] for 16 hr at 4° prior to immunocytochemistry. The results of adsorbed immunostaining were compared with adjacent sections labeled with a 1:150 dilution of unabsorbed antibody.

DNA Damage

Method Rationale. Oxidative damage to nucleic acids results in base modification, substitutions, and deletions. Among the most common modifications, 8-hydroxyguanosine (8-OHG) is considered a signature of oxidative damage to nucleic acid.

Antibodies to 8-Hydroxyguanosine. Two mouse monoclonal antibodies to 8-OHG have been used[44]: anti-8-OHG (1:250, QED Bioscience Inc.,

[43] L. M. Sayre, D. A. Zelasko, P. L. R. Harris, G. Perry, R. G. Salomon, and M. A. Smith, *J. Neurochem.* **68,** 2092 (1997).

[44] A. Nunomura, G. Perry, M. A. Pappolla, R. Wade, K. Hirai, S. Chiba, and M. A. Smith, *J. Neurosci.* **19,** 1959 (1999).

San Diego) and 1F7 (1:30,[45] a gift of Regina Santella), which recognize RNA-as well as DNA-derived 8-OHG. For 1F7, sections were treated with proteinase K (see antigen retrieval section) prior to incubation with 10% NGS.

Immunostaining was developed by the peroxidase–antiperoxidase procedure[24] using 0.75 mg/ml 3,3′-diaminobenzidine cosubstrate in 0.015% H_2O_2, 50 mM Tris–HCl, pH 7.6.

The specificity of both antibodies to 8-OHG was confirmed by comparison with sections in which (i) the primary antibody was omitted or (ii) absorption with purified deoxyribo (8-OHG) (0.24 mg/ml) compared to deoxyribo (G).[44] Following proteinase K treatment, additional sections were pretreated with DNase I (10 U/μl, PBS, 1 hr at 37°, Boehringer Mannheim), S1 DNase (10 U/μl, PBS, 1 hr at 37°, Boehringer Mannheim), or RNase (5 μg/μl, PBS, 1 hr at 37°, Boehringer Mannheim) prior to incubation with 8-OHG antibody.

Cellular Response Factors

Method Rationale. Cells are not passive to increased oxygen radical production but rather upregulate protective responses. Failure to do so will initiate damage to mitochondria, leading to cell death. Whereas major features of antioxidant defense are determined by the basic building blocks of cells and diet, oxidation leads to the induction of specific changes. In neurodegenerative diseases, heme oxygenase-1 (HO-1) induction is coincident with the formation of neurofibrillary tangles.[46] This enzyme, which converts heme, a prooxidant, to biliverdin/bilirubin (antioxidants) and free iron, has been considered an antioxidant.[47] Seen in the context of arresting apoptosis, however, HO-1 and τ may coordinately play a role in maintaining the neurons free from the apoptotic signal cytochrome c, as τ has strong iron-binding sites.[48]

Given the importance of iron as a catalyst for the generation of reactive oxygen species, changes in proteins associated with iron homeostasis can be used as an index of a cellular response. One such class of proteins, the iron regulatory proteins (IRP), responds to cellular iron concentrations by

[45] B. Yin, R. M. Whyatt, F. P. Perera, M. C. Randall, T. B. Cooper, and R. M. Santella, *Free Radic. Biol. Med.* **18,** 1023 (1995).

[46] M. A. Smith, R. K. Kutty, P. L. Richey, S.-D. Yan, D. Stern, G. J. Chader, B. Wiggert, R. B. Petersen, and G. Perry, *Am. J. Pathol.* **145,** 42 (1994).

[47] S. M. Keyse and R. M. Tyrrell, *Proc. Natl. Acad. Sci. U.S.A* **86,** 99 (1989).

[48] M. Arrasate, M. Perez, J. M. Valpuesta, and J. Avila, *Am. J. Pathol.* **151,** 1115 (1997).

regulating the translation of proteins involved in iron uptake, storage, and utilization.[49,50] Therefore, IRPs are viewed as the central control of celular iron concentration.

Heme Oxygenase-1 Experimental. Rabbit antisera to heme oxygenase-1 are available from Stressgen Biotechnologies Corporation, Victoria, Canada. In addition, the multiple antigenic peptide (MAP) system described by Posnett *et al.*[51] was employed for the production of a rabbit antibody against a peptide corresponding to residues 1–30 from the sequence reported for rat HO-1.[52] Details of the preparation and characterization of the rabbit antibody to HO-1 peptide, as well as its ability to react with human HO-1, have been described.[53] Use of this antibody required a trypsin pretreatment of the section (see antigen retrieval section) prior to the immunocytochemical protocol described earlier. Additionally, for immunostaining with HO-1 antisera (Stressgen), we found that methacarn-fixed tissue was superior to formalin fixation, requiring shorter incubation times and lower titers of antisera.

Adsorption experiments were performed on anti-HO-1 to confirm the specificity of antibody binding. The immunostaining protocol was repeated, except using adsorbed antisera in parallel. Adsorbed antisera were generated by the incubation of primary antisera with purified HO-1 protein (Stressgen Biotechnologies Corporation) diluted to a final concentration of 10 μg/ml for 3 hr at 37°.

Iron Response Element Experimental. Rabbit polyclonal antibodies to iron response protein-2 (IRP-2) raised to a 75 amino acid peptide representing the N-terminal sequence unique to IRP-2[54] or IRP-1[55] can be used as described previously.

Adsorption experiments were performed with the antibody to IRP-2 to confirm the specificity of antibody binding.[50] The immunostaining protocol was repeated, except using adsorbed antisera in parallel. Adsorbed antiserum was generated by incubation overnight at 4° of primary antisera with purified IRP-2 protein[54] diluted to a final concentration of 100 μg/ml.

[49] Z. M. Qian and Q. Wang, *Brain Res. Rev.* **27,** 257 (1998).

[50] M. A. Smith, K. Wehr, P. L. R. Harris, S. L. Siedlak, J. R. Connor, and G. Perry, *Brain Res.* **788,** 232 (1998).

[51] D. N. Posnett, H. McGrath, and J. P. Tam, *J. Biol. Chem.* **263,** 1719 (1988).

[52] S. Shibahara, R. Muller, H. Taguchi, and T. Yoshida, *Proc. Natl. Acad. Sci. U.S.A* **82,** 7865 (1985).

[53] R. K. Kutty, C. N. Nagineni, G. Kutty, J. J. Hooks, G. J. Chader, and B. Wiggert, *J. Cell. Physiol.* **159,** 371 (1994).

[54] T. A. Rouault, K. C. Tang, S. Kaptain, W. H. Burgess, D. J. Haile, F. Samaniego, O. W. McBride, J. B. Harford, and R. D. Klausner, *Proc. Natl. Acad. Sci. U.S.A* **87,** 7958 (1990).

[55] K. Pantopoulos, N. K. Gray, and M. W. Hentze, *RNA* **1,** 155 (1995).

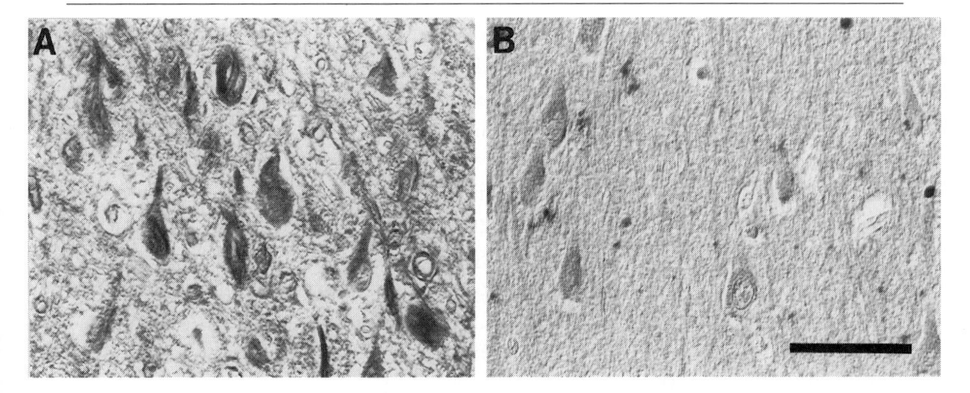

FIG. 1. Hydroxynonenal-derived pyrrole protein adduct in Alzheimer (A) and control brain (B). Only neurons, including those containing neurofibrillary tangles, in AD have markedly elevated levels of this lipid peroxidation-mediated protein modification. Bar: 50 μm.

Adsorption of antibodies to τ with IRP-2 protein was also performed as a control against artifactual absorption.

Application of *in Situ* Methods to AD Tissue Sections to Visualize Senile Plaques and NFT

The methods outlined in this article have been applied in an effort to localize markers of oxidative damage in AD. We and others[19,56,57] determined that AGE modifications are present on both neurofibrillary tangles and senile plaques in AD. These studies employed antibodies not only to the specific structures pentosidine and pyrraline, but also less defined AGE antibodies raised to carrier proteins treated with reducing sugars for long periods of time. AGE modification is likely an early *in vivo* event because both diffuse senile plaques[56] and PHF-τ,[19,57] considered two of the earliest pathological changes in AD,[58,59] were stained by the anti-AGE antibodies.

Immunostaining using antibodies to lipoxidation-derived modifications has exhibited more distinctive patterns (Fig. 1). Whereas some antibodies (e.g., to MDA adducts[19] or to ill-defined HNE adducts) display variable staining of both NFT and SP, antibodies to characterized HNE adducts are localized exclusively to neuronal cell bodies and neurofibrillary pathol-

[56] M. A. Smith, S. Taneda, P. L. Richey, S. Miyata, S.-D. Yan, D. Stern, L. M. Sayre, V. M. Monnier, and G. Perry, *Proc. Natl. Acad. Sci. U.S.A* **91,** 5710 (1994).

[57] M. D. Ledesma, P. Bonay, C. Colaco, and J. Avila, *J. Biol. Chem.* **269,** 21614 (1994).

[58] G. Perry and M. A. Smith, *Clin. Neurosci.* **1,** 199 (1993).

[59] J. Q. Trojanowski, M. L. Schmidt, R.-W. Shin, G. T. Bramblett, M. Goedert, and V. M.-Y. Lee, *Clin. Neurosci.* **1,** 184 (1993).

ogy.[42,43] In fact, antibodies to the HNE lysine-derived pyrrole, an advanced lipoxidation end product, were found to stain not only intraneuronal and extraneuronal NFT, but also apparently normal hippocampal neurons in AD but not in controls (Fig. 1).[43] Congo red counterstaining on the same sections or antisera to τ[60] on the adjacent serial sections were used to assess relationships to NFT. The same profile of staining seen for HNE–pyrrole was seen using antibodies to markers of direct protein oxidation, including nitrotyrosine and protein-based carbonyls.[22,23,30] Again, the heightened sensitivity of immunochemical follow-up subsequent to the derivatization of protein-based carbonyls with 2,4-dinitrophenylhydrazine permitted the *in situ* detection of carbonyl reactivity not only within NFT, but also within vulnerable neurons in AD.[22,23]

Two key objectives in the development of multiple *in situ* immunochemical markers for oxidative stress damage are (i) to obtain clues to the temporal aspects of pathogenesis and (ii) to help track down the sources of oxidative damage. In the former case, it is of interest to determine whether, and at what time point, there is a biochemical response to oxidative damage revealed by the induction of stress proteins such as HO-1. We found that HO-1 is associated with neurofibrillary pathology[46] at the same (early) stage of degeneration as is revealed by the antibody (Alz50) to abnormal conformation of tau. As far as sources of oxidative stress is concerned, the modified Perl's stain for protein-bound iron showed iron accumulation in both NFT and SP.[25] In the absence of the formation of the mixed valence iron complex through employment of hexacyanoiron(II/III), direct utilization of the H_2O_2-dependent DAB oxidation protocol localized endogenously bound redox-active transition metals at the same sites on NFT and SP.[27] Overall, the panel of *in situ* methods for localizing oxidative stress damage in AD is permitting an assessment of the spatiotemporal aspects to neuronal degeneration that define the disease process. One question that will ultimately be answered through better characterization of the cellular response to AD lesions is why certain oxidative stress markers show up only in neurofibrillary pathology whereas others are associated with both NFT and SP.

Quantification and Statistical Analyses

The intensity of immunoreaction on tissue sections can be evaluated by measuring the optical density (OD) with a suitable imaging system[44,61]

[60] G. Perry, M. Kawai, M. Tabaton, M. Onorato, P. Mulvihill, P. Richey, A. Morandi, J. A. Connolly, and P. Gambetti, *J. Neurosci.* **11,** 1748 (1991).

[61] E. Masliah, R. D. Terry, M. Alford, and R. DeTeresa, *J. Histochem. Cytochem.* **38,** 837 (1990).

[e.g., Quantimet 570C image processing and analysis system (Leica) linked to a COHU solid-state camera mounted on a Leitz Laborlux 12 ME ST microscope]. All measurements are done under the same optical and light conditions, as well as using an electronic shading correction to compensate for any unevenness that may be present in the illumination. Statistical analysis for the differences in the corrected OD value among subgroups is by analysis of variance (ANOVA) using the StatView 4.11 program (Abacus Concepts, Inc., Berkeley, CA). Fisher's protected least significant difference can be applied in the *post hoc* analysis.

Potential Complications and Artifacts

In assessing markers of oxidative damage, there are a number of potential complications that investigators should be aware of and, if possible, control against. For example, perhaps the most important aspect relates to processing artifacts, as oxidative processes continue after death and it is therefore imperative to minimize the time before fixation. In this respect, it is important to realize that iron, a potent catalyst of oxidative chemistry, is frequently liberated following death. Additionally, one needs to pay particular attention to the fixation protocol employed because oxidation-related modifications could either be destroyed or created by the fixative. In this regard, we routinely use fixation in methacarn (60% methanol, 30% chloroform, 10% acetic acid) that is relatively inert and appears to optimize labeling by immunocytochemical and histochemical techniques. The mutability of oxidative changes also serves as an important control: by destroying or altering the oxidative modification with specific reagents one can assert readily whether a particular technique is selective. For example, the reduction of free carbonyls with sodium borohydride, the reduction of nitrotyrosine with sodium dithionite, or the enzymatic removal of oxidized nucleic acids with DNase or RNase.

Finally, perhaps the most important aspect that one has to tackle is that oxidative modifications are a fundamental aspect of both aging and disease. Indeed, cell death, both by necrosis and by apoptosis, involves alterations in redox chemistry. Therefore, if one is looking at a pathological process, it is highly unlikely that one would not detect oxidative changes. In order to put such changes into context, it is important to understand the relevance of these changes with respect to other detrimental events. Thus, it is extremely important to empirically determine the conditions required for the detection of *selective* changes relative to the appropriate control. The definition of selective depends on the goal of the study. For example, if one is interested in observing immunocytochemical evidence for oxidative damage in a particular age-related neurodegenerative disease, then the antibody response should be titrated to a level such that age-matched control tissue exhibits changes at the background threshold level. However, for conclusions regarding the effect of aging itself, then the immunocytochemistry should be titrated such that young controls exhibit only background levels. Overall, it is important that the researcher take into consideration the various factors thought, at the time, to underlie the changes of interest in order to permit an informed evaluation of the experimental observations. Only then will they have the proper perspective to evaluate whether oxidative damage represents more of a primary as opposed to secondary phenomena.

Acknowledgments

Work in the authors' laboratories was funded by the National Institutes of Health, the Alzheimer's Association, and the American Health Assistance Foundation.

[11] Advanced Glycation End Products: Detection and Reversal

By YOUSEF AL-ABED, APHRODITE KAPURNIOTU, and RICHARD BUCALA

Protein modification, aggregation, and deposition feature prominently in many pathological processes and can play a direct, etiological role in tissue damage. The amyloidoses, for example, are characterized by the aggregation and progressive insolubilization of proteins or their fragments, leading to an accumulation of deposits that are resistant to proteolytic removal and normal tissue remodeling. The intra- and extracellular aggregates that comprise the senile plaques, the neurofibrillary tangles, and the amyloid deposits of the cerebrovasculature are considered the pathological hallmarks of Alzheimer's dementia, a degenerative disease of the central nervous system.[1] Similarly, amyloid deposition within the pancreatic islets of Langerhans occurs as a result of the aggregation and insolubilization of islet amyloid polypeptide (IAPP) and is associated with progressive β-cell dysfunction and the development of type II diabetes mellitus.[2] Amyloid-containing deposits also occur in the systemic amyloidoses associated with

[1] G. Perry, *in* "Neuroscience Year" (B. Smith and G. Edelman, eds.), p. 5.8. Birkhauser, Boston, 1992.

[2] P. Westermarck, L. Grimelius, J. M. Polak, L. I. Larsson, S. Van Noorden, E. Wilander, and A. G. E. Pearse, *Lab. Invest.* **37,** 212 (1977).

METHODS IN ENZYMOLOGY, VOL. 309
0076-6879/99 $30.00

hemodialysis, chronic inflammation, cancer, and aging.[3] Other forms of protein aggregation have also been described in association with degenerative conditions such as neurofilament conglomeration in amyotrophic lateral sclerosis.[4]

The pathological role of the nonenzymatic modification of proteins by glucose, or glycation, has become increasingly evident. It is now well established *in vivo* that early glycation products progress over time to irreversibly bound cross-links termed advanced glycation endproducts (AGEs), which in turn have been implicated in the development of many of the pathological sequelae of diabetes and aging, such as atherosclerosis and renal insufficiency.[5]

Evidence for a potential role of glycation products in protein deposition and amyloid formation was first provided by the immunochemical identification of AGEs in the brain tissue and lesions of Alzheimer's disease brains.[6,7] AGEs exist *in situ* within amyloid plaque deposits and also modify the insoluble, paired helical filament tau protein and the hemodialysis-associated amyloid protein, β_2-microglobulin.[6–11] The extreme insolubility and resistance to proteolytic degradation that are features of amyloid have led to the suggestion that covalent cross-linking may be an important mechanism for the stabilization of these structures *in vivo.*[12] Of even further significance is the finding that modification of the Alzheimer's β-amyloid peptide (β-AP) of IAPP by AGEs results in nucleation "seeds" that rapidly initiate the polymerization of β peptide into β-amyloid fibrils *in vitro.*[6,13]

The following discussion describes the present state of methodology for the detection of AGEs. While no single method may prove suitable for a particular experimental system or tissue-derived substance, the combination

[3] J. D. Sipe, *Annu. Rev. Biochem.* **61,** 947 (1992).

[4] S. M. Chou, *in* "Motor Neuron Disease: Biology and Management" (P. N. Leigh and M. Swash, eds.), p. 53. Springer-Verlag, New York, 1995.

[5] R. Bucala and A. Cerami, *Adv. Pharmacol.* **23,** 1 (1992).

[6] M. P. Vitek, K. Bhattacharya, J. M. Glendening, E. Stopa, H. Vlassara, R. Bucala, K. R. Manogue, and A. Cerami, *Proc. Natl. Acad. Sci. U.S.A.* **91,** 4766 (1994).

[7] M. A Smith, S. Taneda, P. L. Richey, S. Miyata, S-.D. Yan, D. Stern, L. M. Sayre, V. M. Monnier, and G. Perry, *Proc. Natl. Acad. Sci. U.S.A.* **91,** 5710 (1994).

[8] M. D. Ledesma, P. Bonay, C. Colaco, and J. Avila, *J. Biol. Chem.* **269,** 21614 (1994).

[9] T. Kimura, K. Ikeda, J. Takamatsu, T. Miyata, G. Sobue, T. Miyakawa, and S. Horiuchi, *Neurosci. Lett.* **219,** 95 (1996).

[10] T. Kimura, J. Takamatsu, N. Araki, M. Goto, A. Kondo, T. Miyakawa, and S. Horiuchi, *Neuroreport* **6,** 866 (1995).

[11] T. Miyata, S. Taneda, R. Kawai, Y. Ueda, S. Horiuchi, M. Hara, K. Maeda, and V. M. Monnier, *Proc. Natl. Acad. Sci. U.S.A.* **93,** 2353 (1996).

[12] G. Perry, P. Cras, M. Kawai, P. Mulvihill, and M. Tabaton, *Bull. Clin. Neurosci.* **56,** 107 (1991).

[13] A. Kapurniotu, J. Bernhagen, N. Greenfield, Y. Al-Abed, S. Teichberg, R. W. Frank, W. Voelter, and R. Bucala, *Eur. J. Biochem.* **251,** 208 (1998).

of immunochemical, high-performance liquid chromatography (HPLC), and mass spectroscopic methods has proven to be a powerful approach for the detection and quantitation of glycation-type covalent modifications in a variety of *in vitro* and *in vivo* settings. This article also discusses the experimental application of a new class of pharmacological agents capable of chemically dissociating AGE cross-links and physically "breaking" the covalent bond linking two polypeptide chains.

Pathway of AGE Formation

Glycation, or nonenzymatic glycosylation, begins with the covalent attachment of reducing sugars such as glucose to free amino groups.[5] In the first step of this pathway, the carbonyl group of the sugar attaches to a primary amino group to form a reversible Schiff base. The Schiff base then slowly undergoes an Amadori rearrangement to produce a more stable, but still slowly reversible adduct. Over a time period of days to weeks, additional dehydration, rearrangement, and oxidation reactions occur that lead ultimately to a structurally heterogeneous group of adducts that remain bound irreversibly to the protein. Still only partially characterized, this stage of glycation proceeds through the formation of reactive intermediates that have the capacity to cross-link proximate amino groups. It was realized many years ago that at least some of these advanced products, which comprise advanced glycation end products or AGEs, exhibit characteristic absorbance and fluorescence properties. While fluorescence-based methods have since been supplanted by more direct chemical and immunochemical techniques, these techniques nevertheless were useful in establishing that the glycation process occurs *in vivo* and contributes to much of the structural damage associated with "aged" proteins.

The structural identity of the many products that can form by glycation chemistry remains an area of active investigation. Studies performed largely *in vitro* have led to the elucidation of a number of defined AGE structures, and there is evidence that certain of these form *in vivo,* as assessed by immunochemical and chromatographic criteria (Fig. 1).[14–23] Many AGEs

[14] M. U. Ahmed, S. R. Thorpe, and J. W. Baynes, *J. Biol. Chem.* **261,** 4889 (1986).
[15] F. G. Njoroge, L. M. Sayre, and V. Monnier, *Carbohyd. Res.* **167,** 211 (1987).
[16] J. Farmer, P. Ulrich P, and A. Cerami, *J. Org. Chem.* **53,** 2346 (1988).
[17] D. R. Sell and V. M. Monnier, *J. Biol. Chem.* **264,** 21597 (1989).
[18] K. Nakamura, T. Hasegawa, Y. Fukunaga, and K. Ienaga, *J. Chem. Soc. Chem. Commun.* **14,** 992 (1992).
[19] F. Hayase, H. Hinuma, M. Asano, H. Akto, and S. Arai. *Biosci. Biotech. Biochem.* **58,** 1936 (1994).
[20] K. J. Wells-Knecht, E. Brinkmann, and J. W. Baynes, *J. Org. Chem.* **60,** 6246 (1995).
[21] Y. Al-Abed, T. Mitsuhashi, P. Ulrich, and R. Bucala, *Bioorg. Med. Chem. Lett.* **6,** 1577 (1996).

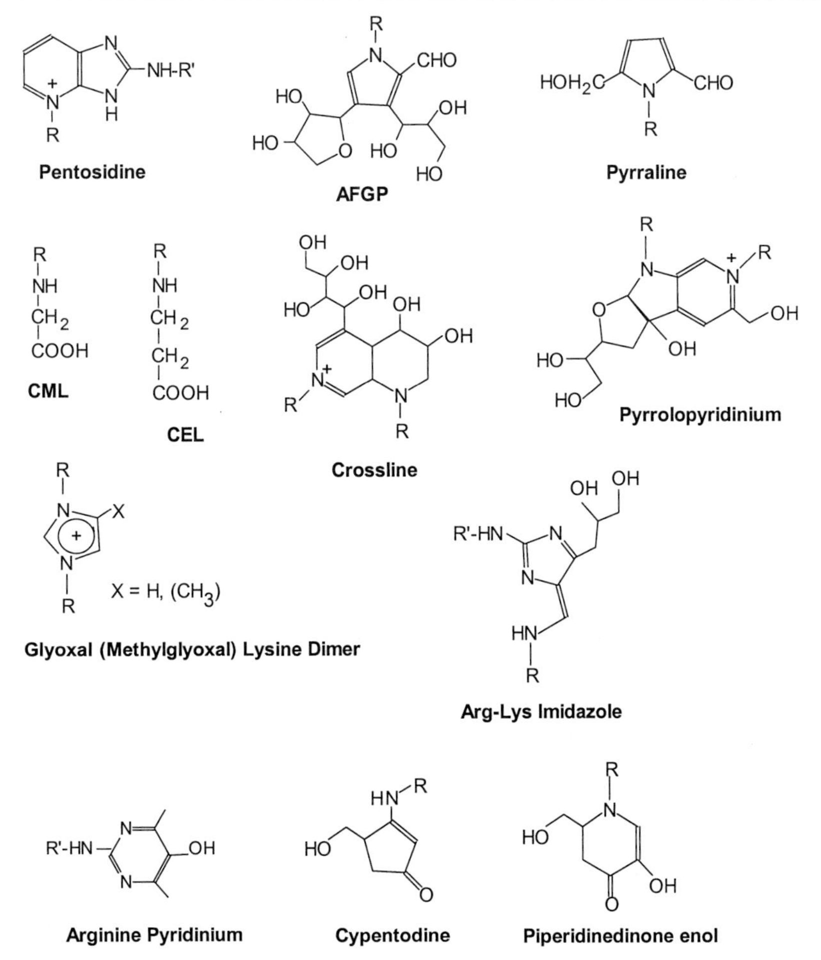

FIG. 1. Chemical structures of various AGEs along with their trivial names. "R" denotes the lysine and "R′" the arginine carbon backbone. Direct chemical or immunochemical evidence for the *in vivo* formation of pyrraline, pentosidine, carboxymethyllysine (CML), crossline, Arg-Lys imidazole (ALI), and cypentodine has been obtained.

appear to arise from Amadori product-derived dicarbonyl intermediates such as 1- and 3-deoxyglucosones, protein-bound dideoxyosones, and fragmentation products such as glyoxal and methylglyoxal. Of importance, however, there is increasing evidence to suggest that the pathologically

[22] Y. Al-Abed and R. Bucala, *in* "Maillard Reactions in Chemistry, Food, and Health," in press.
[23] X. Zhang and P. Ulrich, *Tet. Lett.* **37,** 4667 (1996).

important, cross-linking adducts are not fluorescent and that certain adducts may only represent biomarkers of the more "toxic" AGEs that form *in vivo*.

Immunochemical Detection of AGEs

While research into the development of sensitive and specific immunochemical methods for the detection of AGEs that form *in vivo* has been pursued for a number of years, it was only within the recent past that this approach has been validated in experimental animal systems and human clinical studies. Early immunization attempts aimed at developing specific antisera or monoclonal antibodies were frustrated by poor reactivity to the AGE epitope(s). This result was attributed to heterogeneity in the structure of AGEs, low epitope density, and self-tolerance. Over time, modifications of conventional immunization protocols resulted in the development of high-titer polyclonal and monoclonal anti-AGE antibodies that have been applied successfully to enzyme-linked immunosorbent assays (ELISA) and to immunohistochemical studies. In our laboratory, optimal antibody responses were obtained by hyperimmunization techniques that suppress the host response to common (peptide backbone) epitopes and enhance responsivity to the less commonly occurring epitopes such as AGEs.[24,25]

Detailed cross-reactivity studies of the first useful polyclonal antibody shown to be capable of detecting AGEs *in vivo* revealed a number of interesting features.[25] Of importance, this polyclonal antibody was prepared not against a structurally defined hapten, but against an AGE antigenic structure(s) that forms by the prolonged incubation of glucose with proteins *in vitro*. The first finding with respect to these antibodies was their lack of reactivity with each of the previously described, structurally defined AGEs such as AFGP, pyrraline, carboxymethyllysine (CML), and pentosidine; even when these compounds were tested at a several log-fold molar excess compared to native AGE antigen. These data suggested that the structurally defined AGEs that had been described until that point represented antigenically minor products when compared to the dominant class of AGEs that form *in vivo*.[25,26]

The second important finding was the identification of an apparently major class of cross-reactive AGE epitope(s) that are common to diverse proteins modified by advanced glycation.[25,26] The specific AGE epitope recognized by these antibodies was shown to increase as a consequence of

[24] R. Bucala and A. Cerami, *Mol. Immunol.* **20,** 1289 (1983).

[25] Z. Makita, H. Vlassara, A. Cerami, and R. Bucala, *J. Biol. Chem.* **267,** 5133 (1992).

[26] S. Horiuchi, N. Araki, and Y. Morino, *J. Biol. Chem.* **266,** 7329 (1991).

diabetes or protein age on various proteins such as collagen, hemoglobin, and LDL.[25,27,28] In addition, these immunoreactive AGEs were found to be inhibited from forming by administration of the pharmacological inhibitor, aminoguanidine, and to provide important prognostic information in diabetic renal disease.[25,27–29] Of importance, the use of one particular anti-AGE antibody, RU, as a molecular probe for AGE cross-links has led to the subsequent isolation and characterization of a nonfluorescent, acid-labile, imidazole-based arginine–lysine cross-link (ALI). ALI appears to account for the major AGE hapten that forms by incubating proteins with glucose *in vitro* and may represent the major class of the AGE cross-links that form *in vivo* (Fig. 1).[22]

Specific antibodies suitable for the detection of AGEs that form *in vivo* or *in vitro* are now available from several laboratories. Whereas some reagents have been raised to specific AGE structures, such as AFGP,[30] pyrraline,[31] pentosidine,[32] CML,[33] and crosslines,[34] others (as described earlier) have been produced by immunization with AGE antigens prepared by the incubation of glucose with protein substrates under "near" physiological conditions.[25,26,35] Cross-reactivity studies employing synthetic ligands are used to define the fine specificities present in the elicited antibodies. In our experience, the use of glucose-derived products as immunogens has proven to result in antibodies with the best reactivities to the AGE epitopes present within *in vivo* specimens. Nevertheless, it is important to recognize that the incubation conditions used to generate the AGE immunogens can bias the type of antibodies produced significantly. The use of metalloprotein carriers, such as keyhole limpet hemocyanin (KLH), for instance,

[27] Z. Makita, H. Vlassara, E. Rayfield, K. Cartwright, E. Friedman, R. Rodby, A. Cerami, and R. Bucala, *Science* **258,** 651 (1992).

[28] R. Bucala, Z. Makita, G. Vega, S. Grundy, T. Koschinsky, A. Cerami, and H. Vlassara, *Proc. Natl. Acad. Sci. U.S.A.* **91,** 9441 (1994).

[29] P. J. Beisswenger, Z. Makita, T. J. Curphey, L. L. Moore, S. Jean, T. Brink-Johnsen, R. Bucala, and H. Vlassara, *Diabetes* **44,** 824 (1995).

[30] W. Palinski, T. Koschinsky, S. W. Butler, E. Miller, H. Vlassara, A. Cerami, and J. L. Witztum, *Arterioscler. Thromb. Vasc. Biol.* **15,** 571 (1995).

[31] F. Hayase, R. H. Nagaraj, S. Miyata, F. G. Njoroge, and V. M. Monnier, *J. Biol. Chem.* **264,** 3758 (1989).

[32] S. Taneda and V. M. Monnier, *Clin Chem.* **40,** 1766 (1994).

[33] S. Reddy, J. Bichler, K. J. Wells-Knecht, S. R. Thorpe, and J. W. Baynes, *Biochemistry* **34,** 10872 (1995).

[34] K. Ienaga, K. Nakamura, T. Hochi, Y. Nakazawa, Y. Fukunaga, H. Kakita, and K. Nakano, *Contrib. Nephrol.* **112,** 42 (1995).

[35] T. Mitsuhashi, H. Nakayama, T. Itoh, S. Kuwajima, S. Aoki, T. Atsumi, and T. Koike, *Diabetes* **42,** 826 (1993).

leads to oxidative degradation and cleavage products such as CML, and the ensuing antibodies are necessarily directed toward epitopes bearing this or closely related structures.[33]

Immunohistochemical studies employing anti-AGE antibodies were among the first to establish a role for AGE modification in the protein deposition associated with the amyloidoses and other degenerative conditions.[6,7] The success of these techniques is dependent not only on the choice of antibody, but also on the often careful preparation of tissue, particularly in samples obtained from the central nervous system (CNS). Nevertheless, there is sufficient experience in different detection systems to ensure that immunohistochemical techniques will continue to occupy an important role in the analysis of AGE modification and protein deposition. ELISA methods employing anti-ALI and antipentosidine antibodies have also been successful in quantifying AGE modification in the CNS; however, these approaches remain limited by the necessity of introducing the protein components into a stable solution phase for presentation to the antibody. Thus, strict quantification of AGE-modified protein deposits by these methods has not yet become routine.

High-Performance Liquid Chromatography and Mass Spectroscopic Methods

The use of HPLC, particularly when combined with mass spectroscopy (MS), has proven invaluable in the structural elucidation of a number of the AGEs. Their application to *in vivo* samples has necessarily been limited to the detection and measurement of products that are stable to the acid hydrolysis methods required to liberate the AGE moieties from the peptide backbone. These techniques are useful for the detection of CML and pentosidine and, with appropriate standardization, can be made quantitative.[36,37] Moreover, when the exact structure of the protein substrate is defined, as in certain of the small, amyloidogenic peptides synthesized for *in vitro* studies, MS techniques can yield much useful information. Table I shows an example of the range of products that can be identified by matrix-assisted laser desorption ionization mass spectroscopy (MALDI-MS) on analysis of a model experimental system in which AGEs are produced by the reaction of glucose with islet amyloid polypeptide.

[36] D. G. Dyer, J. A. Dunn, S. R. Thorpe, K. E. Bailie, T. J. Lyons, D. R. McCance, and J. W. Baynes, *J. Clin. Invest.* **91,** 2463 (1993).

[37] P. Odetti, J. Fogarty, D. R. Sell, and V. M. Monnier, *Diabetes* **41,** 153 (1992).

TABLE I
MASS IONS PRESENT IN MALDI MASS SPECTRUM OF AGE–IAPP[a]

Mass		
Observed	Theoretical	Structure
4008	4014	IAPP + pyrraline
4070	4072	IAPP + pyrraline + CML
4070	4068	IAPP + Amadori product
4127	4127	IAPP + Amadori product + CML
4170	4176	IAPP + Amadori product + pyrraline
4190	4193	IAPP + Amadori product + pyrraline + Na^+
4230	4230	IAPP + Amadori product (2)
4286	4276	IAPP + Amadori product (2) + Na^+ (2)
4373	4373	IAPP + Amadori product (2) + imidazolone

[a] From A. Kapurniotu *et al., Eur. J. Biochem.* **251,** 208 (1998).

Reversal of AGE Cross-Links

Insights into the later stages of the glycation pathway have led to the concept that the Amadori double dehydration product, or AP-dione, is a critical reactive intermediate in the formation of protein–protein cross-links (Fig. 2). This realization has also prompted the development of a novel, pharmacological approach for the chemical "breakage," or removal

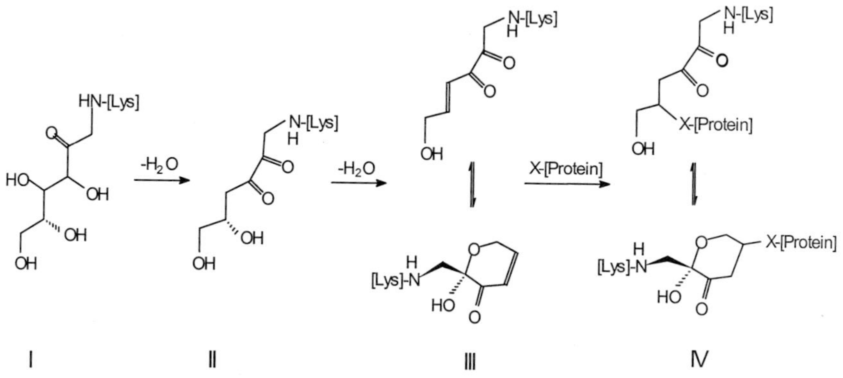

FIG. 2. Scheme for the formation of glucose-derived protein cross-links (AGEs) from Amadori products and cleavage by a thiazolium-based "AGE breaker." Successive dehydration via β elimination of protein-bound (lysine) Amadori products (**I**) to AP-dione (**II**), AP-ene-dione (**III**), and reaction with a protein nucleophile (X-[protein]) to form in a stable protein–protein cross-link (**IV**). **I** is known to exist predominantly in a pyranose form, and **II, IV,** and the *cis* form of **III** may also prefer pyranose- or furanose-like cyclic hemiacetal structures.

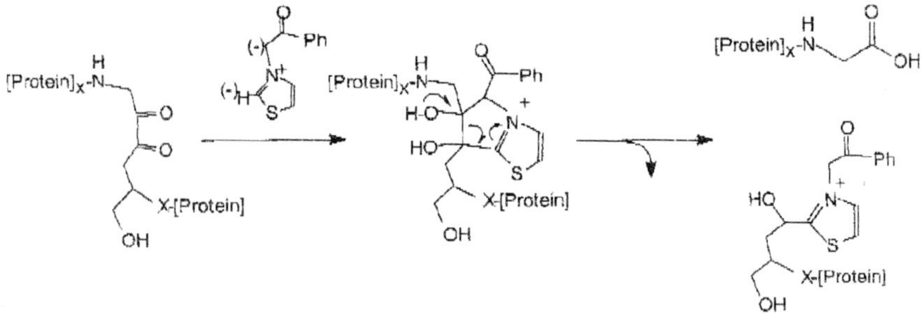

FIG. 3. Proposed reaction scheme for the cleavage of an AP-enedione-derived, protein–protein cross-link by *N*-phenacylthiazolium bromide.[38]

of the covalent, AGE cross-links. If cross-links in fact form by α-dicarbonyl intermediates, then specific dinucleophilic compounds that have the capacity to attach to, labilize, and cleave the covalent AGE cross-links may be envisioned (Fig. 3). In model studies, a prototypic compound, *N*-phenacylthiazoluim bromide (PTB), was found to break ~80% of the bovine serum albumin (BSA)–AGE–collagen cross-links that form *in vitro*. This agent also cleaved *in vitro* the AGE cross-links that formed *in vivo* in the tail tendon collagen isolated from rats that had been made diabetic by streptozotocin.[38] The ability of rationally designed compounds to remove established AGE cross-links suggests the first means to treat the adverse structural and functional consequences of glycation long after these adducts have formed and may prove to be of broad clinical utility in the treatment of advanced diabetic complications.

PTB has also been examined for its capacity to disaggregate the β-amyloid that forms by the AGE modification of native, monomeric β-amyloid peptide (β-AP).[38] In these studies, fibrillar aggregates of AGE-modified β-AP were radioiodinated, incubated with PTB, and then subjected to sodium dodecyl sulfate–polyacrylamide gel chromatography (SDS–PAGE) under reducing conditions. While the untreated preparation of β-AP migrated as stable, high-molecular weight aggregates near the very top of the polyacrylamide running gel, treatment of AGE–β-AP aggregates with PTB *in vitro* produced lower molecular mass β-AP forms with the same molecular mass region as that of the β-AP monomer (≈4400 Da). Electron microscopic examination also showed marked differences in the fibrillar structure of the β-AP aggregates before and after treatment with

[38] S. Vasan, X. Zhang, X. Zhang, A. Kapurniotu, J. Bernhagen, S. Teichberg, J. Basgen, D. Wagle, D. Shih, I. Terlecky, R. Bucala, A. Cerami, J. Egan, and P. Ulrich, *Nature* **382,** 275 (1996).

PTB. In control (untreated) preparations, the β-amyloid was organized into dense fibrillar aggregates. In contrast, the preparation of β-AP, which had been treated with PTB, showed fibrils that were less dense, less uniformly organized, and more filamentous in form.[38]

Methods

Preparation of AGE-Modified Proteins

Most studies of AGE modification (cross-linking, biochemical, and biological effects) have relied largely on AGE-modified proteins prepared by incubating test proteins with the high concentrations of glucose (or other reducing sugars) for periods of days to months. The rationale for these supraphysiological conditions has been to prepare substrates with appreciable amounts of biochemical modification that may then be investigated readily for their properties *in vitro* or *in vivo.* The initial Amadori rearrangement between the reducing sugar and the protein is generally slow, and the subsequent dehydrations and rearrangements that underlie cross-link formation occur over a time period of days. In general, heavily modified proteins have been found to be necessary for antigenicity and antibody production, recognizing that AGE modification is performed frequently on autologous proteins that are inherently poor immunogens.

To accelerate AGE formation *in vitro,* high concentrations of reactant are employed. While several hexoses other than glucose (such as glucose 6-phosphate, fructose) from AGEs faster than glucose (the rate of reaction is proportional to the anomerization rate of the sugar), in our own studies we have preferred to limit our investigations to glucose as the AGE substrate. Glucose is the major sugar present physiologically and is believed still to be the proximal reactant in the formation of AGEs *in vivo.* There is a concern that the use of other sugars may influence the type of AGEs that ultimately form and lead to structures that are not representative of the *in vivo* situation.

A major consideration in the preparation of AGE-modified proteins *in vitro* is buffer condition. Glycation proceeds with an increase in H^+ concentration, and unless the solution is buffered heavily, once the pH falls below 6.5, the course of the reaction slows markedly. While we employ phosphate buffers at concentrations of 200 m*M* or greater, interim monitoring of the pH is essential, and supplementation with OH^- is necessary on a weekly or biweekly schedule.

A second concern is the selection of buffer salts. Phosphate buffers are widely used and mimic somewhat physiological conditions. Nevertheless, commercial phosphate is generally prepared from diatomaceous earth, and trace metal contamination can be significant. By their redox properties, trace metals can affect Maillard processes significantly, skewing the types of adducts formed and favoring the production of "glycoxidation" products. The inclusion of a chelator such as EDTA can help obviate this concern, but recognizing that metal chelation itself can theoretically increase the solubility and reactivity of trace metals.[39] More pure, organic buffers such as HEPES can be utilized, as has been done in DNA glycation studies.[40] However, primary amine-containing buffers such as Tris should be avoided scrupulously.

The following sample recipe is used for the preparation of AGE-modified serum albumin that has been unitized for preparing competitor substrate for an AGE ELISA or for use in various cell- and animal-based investigations. A similar protocol has been employed in the generation of AGE–ribonuclease (AGE–RNase) for use as an immunogen for anti-AGE polyclonal and monoclonal antibody production.

Preparation of AGE-Modified Albumin

Bovine or human serum albumin (Sigma Chemical Co., St. Louis, MO) at 50 mg/ml is incubated with 0.5–1.0 *M* glucose in $NaPO_4$ buffer, pH 7.4, containing 1 m*M* EDTA for 60–90 days. The solution is filter-sterilized and incubated at 37° in the dark. At biweekly intervals, the solution is tested for pH and adjusted up to 7.4 with concentrated NaOH as needed. After incubation, unbound and low molecular weight material is removed by extensive dialysis against phosphate-buffered saline (supplemented with 1 m*M* EDTA) or by gel filtration over Sephadex G-10 (Pharmacia, Uppsala, Sweden). For many studies, a parallel incubation of albumin without glucose may be recommended as a control.

Preparation of High-Titer Anti-AGE Polyclonal Antibody

The first report of specific, high-titer anti-AGE antibodies recognizing *in vivo*-formed AGEs exploited the use of hyperimmunization techniques[24] to suppress the host response to common (peptide backbone) epitopes and to enhance responsivity to the less commonly occurring (and likely heterogeneous) AGE epitopes.[25]

Immunization is performed by intradermal injection into rabbits (female, New Zealand White). In our studies, the AGE immunogen has typically been AGE–RNase or AGE–BSA. While KLH is a very immunogenic carrier protein, we have observed that the use of metalloproteins leads to

[39] R. Bucala, *Redox Rep.* **2,** 291 (1996).
[40] R. Bucala, P. Model, and A. Cerami, *Proc. Natl. Acad. Sci. U.S.A.* **81,** 105 (1984).

adventitious oxidative phenomena and a preferential production of oxidation products such as CML.

One milliliter of a 2-mg/ml solution of AGE protein is emulsified with 1 ml of Freund's complete adjuvant (Miles Laboratories, Elkhart, IN). Each rabbit receives four intradermal injections over the back (0.2 mg each) and one injection in each hindquarter (0.1 mg each). This procedure is repeated at weekly intervals for 6 weeks. The rabbits are then rested for 4 weeks, following which a booster injection of 1 mg of AGE protein in Freund's incomplete adjuvant is administered. The animals are bled (50 ml) on the 10th day after this injection. Sera are prepared, cleared of debris by centrifugation (500*g*), and stored at −80°.

Glucose-derived AGE-modified proteins have also been found to be suitable for monoclonal antibody production. These procedures follow standard immunization, fusion, and screening protocols.[41]

Specific polyclonal and monoclonal antibodies directed against CML, pyrraline, AFGP, pentosidine, and crossline have been prepared by utilizing these structures as haptens in conventional immunization protocols.[30–34] While these antibodies have been useful in demonstrating the presence of immunochemically cross-reactive structures *in vivo,* they have been of somewhat lesser utility in pathological studies, as the molecular origin of some of these structures remains unclear. CML, for instance, does not crosslink polypeptides and has been shown to arise independently of glycation chemistry by lipid oxidation.[42]

AGE–ELISA

Polyclonal anti-AGE–RNase antibodies prepared as described earlier were instrumental in the development of the first quantitative techniques for the measurement of AGE-modified forms of collagen,[25] serum proteins including LDL,[28] and hemoglobin.[27] A standard ELISA procedure developed and utilized by several laboratories follows.

Preparation of AGE–BSA Plates

1. Prepare antigen (AGE–BSA) stock solution from dilution in coating buffer (0.1 *M* $NaHCO_3$, pH 9.6, 0.02% NaN_3), to give 1 mg/ml. Dilute the working solution in coating buffer such that 100 μl = 3 μg (30 μg/ml).

[41] E. Harlow and D. Lane, Eds., "Antibodies, a Laboratory Manual," p. 139. Cold Spring Harbor Laboratory, Cold Spring Harbor, NY, 1988.

[42] M.-X. Fu, J. R. Requena, A. J. Jenkins, T. J. Lyons, J. W. Baynes, and S. R. Thorpe, *J. Biol. Chem.* **271,** 9982 (1996).

2. Add 0.1 ml of antigen solution to the appropriate wells in a microtiter plate. Incubate for 2 hr to overnight at 4° after covering the plate with disposable sealing tape (Corning, Corning, NY).
3. Wash the plate in two orientations, three times each, by a plate washer (0.2 ml) with wash buffer (PBS, containing 0.05% Tween 20 and 0.02% azide).
4. Block the plate with blocking buffer (PBS with 2% goat serum, 0.2% BSA, 0.02% NaN_3), 0.15 ml, for 1 hr.
5. Wash three times as in step 3.

AGE Assay Procedure

1. To each well, add 50 μl of sample followed by 50 μl of the primary antibody of appropriate titer dissolved in blocking buffer (the RU anti-AGE–RNase antibody is typically diluted 1:1000). The primary antibody should be added with a multichannel pipetter and the plate rocked in order to mix the contents of wells prior to incubation.
2. Incubate at room temperature for 2 hr.
3. Wash in two orientations (three times in each orientation with wash buffer). The wash volume per well is 0.2 ml.
4. Add 0.1 ml of secondary antibody to the appropriate wells (goat antirabbit IgG, Cappel). Each new lot must be titered (1:3000–1:5000). Dilute with dilution buffer.
5. Incubate for 1 hr at 37°.
6. Wash as described earlier.
7. Add 0.1 ml of *p*-nitrophenyl phosphate (PNPP) substrate [1–15 mg PNPP tablet (Sigma) in 15 ml 1 *M* DEA buffer]. The DEA buffer (1 liter) is prepared by adding 105.1 g diethamolamine (liquid) to 800 ml doubly distilled H_2O. Sodium azide (0.2 g) and $MgCl_2 \cdot 6H_2O$ (0.1 g) are added and the pH adjusted to 9.8.
8. Read absorbance at a test wavelength of 410 nm and a reference wavelength of 570 nm. AGE "units" are typically calculated with reference to an AGE–BSA standard curve, with 1 U corresponding to 50% inhibition of antibody binding.[25]

Note: All solutions except the PNPP (substrate) should be brought to room temperature prior to use. The PNPP should be made and incubated at 37° for 30 to 60 min prior to use. The latter treatments will reduce but not eliminate the microtiter plate "edge effect." The mylar plate cover should be changed at each step to prevent cross-contamination of wells.

Measurement of Serum AGEs

AGEs modify plasma proteins by direct reaction of glucose (or other reducing sugars) and by the entry into the plasma compartment of reactive,

AGE peptides produced by the normal catabolism of AGE-modified tissue proteins.[43] High concentrations of circulating AGEs can occur under nonhyperglycemic conditions if plasma filtration is impaired by renal disease. The reactive nature of AGE peptides, together with their normal clearance by the kidneys, has led to the concept that circulating AGE peptides comprise an important component of the so-called uremic toxins or "middle molecules" that accumulate during renal insufficiency and contribute to the morbidity and mortality of chronic renal failure.

While serum protein AGEs are detected readily by ELISA of whole serum, the sensitivity of this analysis is enhanced markedly by predigesting serum with proteinase K. Given the cross-linking activity of AGEs, this technique is critical in adequately revealing AGEs for binding to specific antibodies.

Preparation of Serum Sample for AGE Assay

1. Prepare proteinase K solution (GIBCO, Grand Island, NY). The stock 8 mg/ml in Tris buffer (50 m*M*), pH 8, with 0.02% sodium azide; aliquot and store at 20° for up to 2 weeks. Working solution: dilute the stock 1:20 into 0.02 *M* sodium phosphate buffer.
2. Add 0.1 ml of serum to a microcentrifuge tube with 0.2 ml of the working proteinase K solution. Vortex to mix and place the solution on a rocker in a 37° incubator and incubate overnight.
3. Prepare an enzyme control (negative control) by adding 0.1 ml of sodium phosphate in place of the serum sample in step 2 and follow the procedure as for a sample. This control should not produce a signal in the assay $<95\%\ B/B_0$ where $>95\%\ B/B_0$ is considered no signal.
4. Prepare a fresh phenylmethylsulfonyl fluoride (PMSF) stock (100 m*M*) in ethanol. Make a 1:200 dilution in 0.02 *M* sodium phosphate buffer.
5. Centrifuge the digested sample for 10 min at 14,000 rpm (microfuge) to clear any debris.
6. Remove 0.2 ml of clear solution and add to PMSF (1:200 diluted) solution. Vortex to mix and add 50 μl of this preparation to the competitive ELISA. Also, add 50 μl of the negative control that has been treated in the same manner as the sample (see step 3). Further dilution of the sample may be necessary depending on where the sample falls on the standard curve.

[43] Z. Makita, R. Bucala, E. J. Rayfield, E. A. Friedman, A. M. Kaufman, S. M. Korbet, R. H. Barth, J. A. Winston, H. Fuh, K. Manogue, A. Cerami, and H. Vlassara, *Lancet* **343,** 1519 (1994).

Measurement of AGE–Hemoglobin

Hemoglobin was the first protein to be described to be modified by glycation products *in vivo,* and the clinical assessment of HbA_{1c} (the minor hemoglobin formed by the Amadori product modification of HbA) has become an important tool for assessing long-term glycemic control in diabetic patients. The development of an AGE-specific ELISA in turn led to the discovery of an AGE-modified form of hemoglobin in 1992.[27] Although Amadori adduct is slowly reversible (and reaches equilibrium in the blood over a period of 3–4 weeks), the AGE modification is irreversible and theoretically persists for the entire life of the red cell (120 days). This has been verified experimentally by kinetic comparisons of HbA_{1c} and Hb–AGE levels in diabetic patients introduced to insulin therapy. Hb–AGE levels decline much more slowly, indicating that the Hb–AGE measurement provides a longer-term index of ambient glucose levels than standard HbA_{1c} measurements.[44] The following standard method is used for Hb-AGE analysis.

AGE–Hemoglobin Isolation and Immunoassay: Hemolysate Preparation from Fresh Blood

1. Whole blood is collected in a heparinized tube.
2. The red blood cells (RBCs) are separated from the plasma by centrifugation at 2000–3000 rpm for 10 min, resuspended in sterile isotonic saline (0.85%) in a volume approximately equal to the plasma, and then repacked by centrifugation. If the cells are to be stored for up to 1 week before assay, the cells should soak at each of the two wash steps for 30 min and then stored in saline. For storage longer than 1 week, the cells should be stored at −20° as a pellet only after incubation in isotonic saline (4°) overnight to dialyze out the glucose contained within the RBCs.
3. To prepare hemolysate from fresh RBCs, pellet enough of the RBC suspension to give 1 ml of packed RBCs. To 1 ml of packed RBCs, add 3 ml of distilled water and 2 ml of toluene for delipidation.
4. Shake vigorously for a few minutes or vortex intermittently 6× to ensure complete extraction. Centrifuge the RBC preparation at 3000 rpm for 10 min to separate the preparation into two liquid phases and pellet the cellular debris.
5. Remove the toluene (top phase) above the lipid layer overlaying the hemolysate using a glass Pasteur pipette attached to a suction flask.

[44] B. H. R. Wolffenbuttel, D. Giordano, H. W. Founds, and R. Bucala, *Lancet* **347,** 513 (1996).

6. After removing the toluene, tip the tube carefully to expose the red aqueous phase below the white lipid layer and remove this phase using a glass Pasteur pipette taking care not to disturb either the white lipid layer or the pellet at the bottom of the tube.
7. The hemolysate may be stored at 4° if time does not allow completion of the assay.

Hemolysate Preparation from Frozen and Packed RBCs

1. Because the cells are hemolyzed after freezing, remove 1 ml of the packed RBCs without further treatment.
2. Follow the procedure from step 2 described earlier.
3. Add 50 μl of hemolysate to 3 ml of distilled water in disposable glass tubes capable of centrifugation. Add 1 ml of ice-cold 24% TCA, mix, and allow to stand 15 min. Centrifuge the TCA hemoglobin preparation at 3000 rpm for 30 min. Remove and discard the supernatant using a glass Pasteur pipette connected to a suction flask. Carefully remove all the remaining TCA from the pellet and tube inner wall by suction pipette.
4. Allow the pellets to sit for 10 min to drain. Remove any residual TCA as in the step described earlier as any residual TCA will affect the final pH and solubility of the pellet.
5. Dissolve the pellet in 75 μl of 1 *N* NaOH and vortex until all of the pellet is dissolved. If the pellet will not dissolve or has a flocculate in it, it should not be used. Add 0.5 ml of a 0.3 *M* KH_2PO_4 buffer, pH 7.4, and mix. This stock preparation should not have a precipitate and may be stored overnight at 4° if the assay cannot be completed.
6. A 1:3 dilution of the stock preparation must be made in 0.3 *M* KH_2PO_4, pH 7.4, just prior to running the Hb–AGE ELISA in order to adjust the pH of the preparation for immunoassay. Both AGE and total protein (Lowry) determinations are then made with this working preparation. In the preparation of the working preparation a precipitate will form. This precipitate must be vortexed prior to pipetting to assure accurate results in the Lowry and AGE assay.

Hb–AGE Assay Procedure

1. To each well of an AGE–BSA-coated microtiter plate, add 50 μl of the sample followed by 50 μl of primary antibody of the appropriate titer (as determined to provide a B_0 reading of approximately 1.5 optical density units after 2 hr) dissolved in dilution buffer (described earlier).
2. Incubate at room temperature for 2 hr.

3. Wash in two orientations, 3× in each orientation, with wash buffer on a plate washer. The wash volume per well is 200 μl. Blot excess liquid from the plate.
4. Add 100 μl of the secondary antibody (goat antirabbit antibody conjugate) to the appropriate wells.
5. Incubate for 1 hr at 37°.
6. Wash as described earlier; three times each in two orientations.
7. Add 100 μl of PNPP to the wells and incubate at 37° for 1–2 hr. The OD of the B_0 should read approximately 1.5 units within 2 hr.
8. Read wavelength absorbance as described earlier.

AGE–LDL ELISA

The lipid and lipoprotein components of LDL have also been identified as substrates for advanced glycation reactions,[45] and the presence of AGEs on apolipoprotein B has been shown to interfere with its normal uptake by high-affinity, tissue LDL receptors.[28,46] This can serve to delay the clearance of LDL from plasma and to promote its deposition into a vascular wall plaque. AGE modification of LDL has been proposed to contribute to the hyperlipidemia and accelerated atherosclerosis associated with diabetes, and this mechanism has been validated by the ability of aminoguanidine, a pharmacological inhibitor of advanced glycation, to lower LDL levels in human diabetic subjects.[47]

The AGE modification of LDL has been established to occur on both apolipoprotein (ApoB) and aminophospholipid components.[45] AGE–ApoB is measured readily by a sandwich AGE–ELISA employing an anti-AGE monoclonal antibody immobilized on microtiter plates and a polyclonal anti-ApoB–horseradish peroxidase (anti-ApoB-HRP)-conjugated antibody (1:500 dilution, Biodesign, Kennebunkport, ME). For this modification of the AGE–ELISA, wells are coated with anti-AGE MAb (1.5 g/well), followed by the addition of test serum (0.1 ml), 0.9 ml polyethylene glycol (PEG, 6.66%, Sigma), and 0.1 ml SDS (0.05% in PBS). Incubation is conducted for 10 hr at room temperature, after which anti-ApoB-HRP (0.1 ml) is added for 1 hr at 37°. The plates are then analyzed at an OD of 450 nm. The obtained values are then related to total serum ApoB, determined by the lncstar SPQ test system (lncstar Corp., Stillwater, MN).

[45] R. Bucala, R. Z. Makita, T. Koschinsky, A. Cerami, and H. Vlassara, *Proc. Natl. Acad. Sci. U.S.A.* **90,** 6434 (1993).

[46] R. Bucala, R. Mitchell, K. Arnold, T. Innerarity, H. Vlassara, and A. Cerami, *J. Biol. Chem.* **270,** 10828 (1995).

[47] R. Bucala and S. Rahbar, *in* "Endocrinology of Cardiovascular Function" (E. R. Levin and J. L. Nadler, eds.), p. 159. Kluwer Academic Publishers, Netherlands, 1998.

Immunohistochemical Staining for AGEs

Several laboratories have reported the successful use of anti-AGE antibodies, raised against either glucose-derived AGEs[29,48] or defined AGE haptens,[7,30,49–52] to identify the presence of cross-reactive epitopes *in situ.* Like any immunohistochemical study, tissue preparation, primary antibody concentrations, and the revealing methodology each require optimization. Also necessary are the judicious use of "control" tissues, a control antibody, and an antigen-blocking experiment in which tissue reactivity can be shown to be abolished by preincubating the primary anti-AGE antibody with an AGE-modified protein. The following procedure has been employed successfully for the detection of vascular wall AGEs with a polyclonal anti-AGE antibody.

Formalin-fixed and paraffin-embedded tissue sections are cut 3 μm thick and then deparaffinized by treatment with xylene three times for 5 min, followed by graded ethanol. Slides are incubated with 0.05% proteinase K (E. Merck AG, Germany) in 0.01 *M* phosphate-buffered saline (PBS), pH 7.4, for 30 min at 37°. This step serves to break down and disrupt connective tissue proteins and expose AGE epitopes, which as a result of their cross-linking properties may not be otherwise detectable. After washing with PBS, the slides are dipped in 0.3% peroxide/100% methanol for 30 min to block endogenous peroxidase. After washing with PBS three times, the slides are incubated with 20% normal goat serum (Vector Lab, Burlingame, CA) for 10 min at room temperature to block nonspecific binding. Slides are then incubated with a 1:30 dilution of rabbit anti-AGE for 24 hr at 4°. After washing with PBS three times, the slides are incubated with peroxidase-labeled second antibody: peroxidase-labeled goat antirabbit IgG (Cappel) for 2 hr at room temperature, rinsed with PBS, and incubated with 0.02% 3,3′-diaminobenzidine tetrahydrochloride (DAB) (Sigma Chemical, St. Louis MO) in 50 m*M* Tris–HCl buffer (pH 7.6) containing 0.003% peroxide for 3 min. After washing with PBS and DDW, the slides are finally counterstained with Mayer hematoxylin.

[48] Y. Nakamura, Y. Horii, T. Nishino, H. Shiiki, Y. Sakaguchi, T. Kagoshima, K. Dohi, Z. Makita, H. Vlassara, and R. Bucala, *Am. J. Pathol.* **143,** 1649 (1993).

[49] S. Miyata and V. Monnier, *J. Clin. Invest.* **89,** 1102 (1992).

[50] K. Horie, T. Miyata, K. Maeda, S. Miyata, S. Sugiyama, H. Sakai, C. Y. Strihou, V. M. Monnier, J. L. Witztum, and K. Kurokawa, *J. Clin. Invest.* **100,** 2995 (1997).

[51] T. Miyata, S. Taneda, R. Kawai, Y. Ueda, S. Horiuchi, M. Hara, K. Maeda, and V. M. Monnier, *Proc. Natl. Acad. Sci. U.S.A.* **93,** 2353 (1996).

[52] B. H. R. Wolffenbuttel, C. M. Boulanger, F. R. L. Crijns, M. S. P. Huijberts, P. Poitevin, G. N. M. Swennen, S. Vasan, J. J. Egan, P. Ulrich, A. Cerami, and B. I. Levy, *Proc. Natl. Acad. Sci. U.S.A.* **95,** 4630 (1998).

Mass Spectrometric Analyses of AGE Modification

In our experience, successful mass spectrometric analyses of AGE modification can be achieved by matrix-assisted laser desorption ionization. In a study of AGE-modified intraislet amyloid polypeptide, MS was performed with a VG Fisons TofSpec instrument (Altrincham, UK) and a Kratos Kompact MALDI 3 V4.0.0 instrument (Shimadzu, Duisburg, Germany) using α-cyanosinapic acid as the matrix.[13]

Human IAPP (836 g) is dissolved in 0.1 ml double-distilled water and 0.1 ml of 0.4 *M* sodium phosphate buffer (pH 7.4) containing 2 *M* D-(+)-glucose and 2 m*M* EDTA is added. This is followed by the addition of 1.1 ml of 0.2 *M* sodium phosphate buffer (pH 7.4) containing 1 *M* D-(+)-glucose and 1 m*M* EDTA. The final concentrations of IAPP and glucose are 178.4 μM and 1 *M*, respectively. The resulting gelatinous solution is incubated for 3 months at 37° in the dark and then dialyzed against double-distilled water using a Spectra/Por CE dialysis membrane (molecular mass cut off: 1000 Da, Spectrum Inc., Houston, TX). Control preparations of IAPP (control-IAPP) are incubated under the same conditions, but in the absence of D-(+)-glucose. For purpose of sample standardization, the amino acid content of AGE-IAPP and control-IAPP can be determined after acid hydrolysis (6 *N* HCl, 110°, 24 hr) by quantitative amino acid analysis using phenyl isothiocyanate derivatization. The masses obtained can then be compared to expected and published values for various AGE or AGE precursor structures (Table I).

Studies of AGE Cleavage by "AGE Breakers"

The description of dinuceophilic compounds with the capacity to cleave the carbon–carbon bonds of AGE cross-links has prompted significant interest in the pharmacological possibility of actually eliminating mature AGEs once formed *in vivo*.[38] The potential utility of these agents has been assessed in studies of diabetes-related increases in vascular wall collagen cross-linking,[51] amyloid fibril formation,[38] and, in an interesting application, the extraction of DNA from coprolites.[53]

Synthesis of PTB. The prototypic compound AGE breaker, *N*-phenacylthiazolium bromide (PTB) that has been described,[38] is prepared by heating a solution 1 *M* in both phenacyl bromide and thiazole in ethanol at reflux for 2 hr, cooling, filtering out the precipitated solid, and recrystallizing from aqueous 90% ethanol. mp: 222–223° [lit. mp: 223–223.5 (dec.)]. The yield is typically in the range of 74%.

[53] H. N. Poinar, M. Hofreiter, W. G. Spaulding, P. S. Martin, B. Artur Stankiewicz, H. Bland, R. P. Evershed, G. Possnert, and S. Pääbo, *Science* **281,** 402 (1998).

Assay for Cleavage of AGE Cross-Links That Form in Vitro. This microtiter plate-based assay is suitable for the rapid screening of cross-link formation and breakage. One gram of AGE–BSA (prepared as described earlier) is added to each well of a collagen-coated, 96-well microtiter plate (Biocoat, Collaborative Research, Bedford, MA). The AGE–BSA is allowed to react with the collagen for 4 hr at 37°, after which the wells are washed three times with PBS/0.05% Tween to remove the unattached AGE–BSA. Test concentrations of PTB (50 μl) dissolved in PBS are added to wells in triplicate and incubation is continued at 37° for the indicated times. The wells are then washed extensively with PBS/0.05% Tween, and the amount of BSA remaining attached to collagen is quantitated by direct ELISA employing a rabbit polyclonal anti-BSA antibody, a horseradish peroxidase-linked goat antirabbit antibody, and H_2O_2 substrate containing ABTS (2,2′-azino-di-3-ethylbenzthiazolinesulfonic acid) as chromogen. The percentage breaking activity is calculated by the following formula: $100 \times [(A_{410}$, PBS control)-(A_{410}, PTB)]/[A_{410}, PBS control]. The addition of PTB (0.003–3 mM) to wells for 7 hr at 37° following by washing and incubation with AGE–BSA does not affect the ability of AGE–BSA to subsequently attach to the collagen-coated surface, as detected by direct BSA ELISA. PTB is also known not to react with AGE–BSA to modify or obscure the BSA epitopes recognized by the anti-BSA antibody.

PTB Cleavage of AGE Cross-Links That Form in Vivo. PTB treatment *in vitro* can also decrease AGE cross-links that form *in situ* in diabetic, rat tail tendon collagen. For this study, diabetes is first induced in male Lewis rats (150–175 g) by the injection of steptozotocin (65 mg/kg, ip). Hyperglycemia is then confirmed 1 week later by plasma glucose measurement (≥250 mg/dl). Thirty-two weeks later, the rats are sacrificed and collagen is isolated from the tail tendon fibers by a standard protocol.[54] The insoluble collagen is treated with cyanogen bromide[55] and the hydroxyproline content is measured.[56] Aliquots containing 1 μg equivalent of hydroxyproline are subjected to SDS–PAGE under reducing conditions and are stained with Coomassie blue. Because diabetic collagen contains nonreducible, AGE cross-links, cyanogen bromide cleavage results in a pattern of higher molecular weight fragments than that observed in collagen obtained from nondiabetic animals. PTB treatment of diabetic collagen restores the electrophoretic pattern to that observed with nondiabetic collagen.

[54] J. Bochantin and L. I. Mays, *Exp. Gerontol.* **16,** 101 (1981).

[55] M. Itakura, H. Yoshikawa, C. Bannai, M. Kato, K. Kurikawa, Y. Kawakami, T. Yamaoka, and K. Yamashita, *Life Sci.* **49,** 889 (1991).

[56] H. Stegeman and K. Stadler, *Clin. Chim. Acta* **18,** 267 (1967).

PTB Disaggregation of β-Amyloid Fibrils in Vitro. AGE-modified β-amyloid peptide, prepared by the incubation of glucose with β-AP (amino acids 1–40, from Bachem, Torrance, CA), has been shown to efficiently initiate the aggregation and polymerization of β-AP into amyloid fibrils *in vitro.* PTB has been shown to disaggregate β-amyloid fibrils that have been aggregated in this manner. These assays are performed as follows.

AGE–β-AP is first radioiodinated using the Iodo-Beads iodination reagent (Pierce, Rockford, IL). Briefly, 50 μl of AGE–β-AP (25 μ*M*) is added to a solution of 215 μl PBS (pH 6.9) containing a washed Iodo-Bead and 15 μl of ^{125}I (100 mCi/ml, NEN Dupont, Wilmington, DE). The radioiodination reaction is stopped after 15 min by removing the bead, and the nonincorporated iodine is removed by extensive dialysis. For SDS–PAGE analysis, 40 μl of the ^{125}I-AGE–β-AP aggregate in 10 m*M* $NaPO_4$ buffer (pH 7.4) is incubated either alone or with 20 m*M* PTB for 24 hr at 37°. The samples are then mixed with an equal volume of 2× concentrated Laemmli sample buffer, heated to 100° for 10 min, and electrophoresed in a 16% gel under reducing conditions. The gel is fixed in 20% trichloroacetic acid for 16 hr and analyzed without further treatment by ^{125}I scanning with a Packard Instruments instant imager. The relative distribution of [^{125}I]β-AP is measured in each lane, which was gated and scanned individually.

Amyloid fibrils show a characteristic morphology by transmission electron microscopy. For this analysis, the AGE–β-AP aggregate in 10 m*M* $NaPO_4$ buffer (pH 7.4) is incubated either alone or with 20 m*M* PTB for 24 hr at 37°. A 10-μl drop of each sample is placed on a Formvar-coated, carbon-stabilized 200 mesh copper grid for 3 min and the excess liquid is removed slowly with filter paper. A drop of 2% (w/v) uranyl acetate is then placed on the grids for 1 min, the excess is removed with filter paper, and the sample is allowed to dry. The preparations are examined on a JEOL 100 CXII transmission electron microscope operated at 80 kV.

[12] Analysis of Transglutaminase-Catalyzed Isopeptide Bonds in Paired Helical Filaments and Neurofibrillary Tangles from Alzheimer's Disease

By BRIAN J. BALIN, ARIEL G. LOEWY, and DENAH M. APPELT

Introduction

Transglutaminases (TGases) are Ca^{2+}-activated enzymes that catalyze transamidation of available glutamine residues with the release of ammonia

METHODS IN ENZYMOLOGY, VOL. 309
0076-6879/99 $30.00

to form either N^{ϵ}-(γ-glutamyl)lysine (Glu–Lys) or (γ-glutamyl) polyamine covalent bonds between proteins.[1] This reaction results in cross-linked insoluble protein complexes that are stable and frequently resistant to most proteolytic enzymes. TGases are ubiquitous in nature and are found in plasma, blood cells, and tissues. They are involved in a number of normal fundamental biological processes, including maturation and differentiation of cells, the fibrin clotting cascade, cornification of the epidermis, hair, and nails,[2,3] and programmed cell death.[4] Additionally, TGases have been recognized and investigated in a number of disease states, in particular, neurodegenerative diseases such as Alzheimer's disease (AD).[5–12]

Isopeptidase (IPase) is an enzyme that can split the TGase-catalyzed Glu–Lys bonds cross-linking protein chains. Its activity can be assayed with ALMA casein, a novel radioactive substrate that is synthesized by cross-linking with tissue and/or extracellular TGase, such as factor XIIIa, the substrate *N*-α-[^{3}H]acetyl-L-lysine-*N*-methylamide (ALMA) to a carboxyl amide group of casein.[13] Isopeptidase is present in brain as well as in other cell types such as red blood cells (Loewy, unpublished observation, 1998).

A characteristic of the amyloid and tau proteins that accumulate in Alzheimer's disease brains is their deposition as insoluble neuritic senile plaques and neurofibrillary tangles (NFTs), both refractory to proteolysis, harsh solvents, denaturants, and reducing agents.[6] The first suggestion that TGase-catalyzed bonds might be involved in the neuropathology of Alzheimer's disease arose from early work investigating the formation of NFTs thought to be derived from neurofilament proteins.[5,14] These studies stimulated other investigations into the assembly of the microtubule-associated protein tau by TGases into insoluble paired helical filaments (PHFs) present in neurofibrillary tangles.[7,9,12] Increased levels of TGase enzymatic activity

[1] J. E. Folk and J. S. Finlayson, *Adv. Prot. Chem.* **31,** 1 (1977).
[2] C. S. Greenberg, P. J. Birckbichler, and R. H. Rice, *FASEB J.* **5,** 3071 (1991).
[3] D. Hand, M. J. M. Perry, and L. W. Haynes, *Int. J. Dev. Neurosci.* **11,** 709 (1993).
[4] S. el Alaoui, S. Mian, J. Lawry, G. Quash, and M. Griffin, *FEBS Lett.* **311,** 174 (1992).
[5] D. J. Selkoe, C. Abraham, and Y. Ihara, *Proc. Natl. Acad. Sci. U.S.A.* **79,** 6070 (1982).
[6] D. J. Selkoe, Y. Ihara, and F. J. Salazar, *Science* **215,** 1243 (1982).
[7] S. M. Dudek and G. V. Johnson, *J. Neurochem.* **61,** 1159 (1993).
[8] S. M. Dudek and G. V. Johnson, *Brain Res.* **651,** 129 (1994).
[9] M. L. Miller and G. W. Johnson, *J. Neurochem.* **65,** 1760 (1995).
[10] G. V. W. Johnson, T. M. Cox, J. P. Lockhart, M. D. Zinnerman, and M. L. Miller, *Brain Res.* **751,** 323 (1997).
[11] D. M. Appelt, G. C. Kopen, L. J. Boyne, and B. J. Balin, *J. Histochem. Cytochem.* **44,** 1421 (1996).
[12] D. M. Appelt and B. J. Balin, *Brain Res.* **745,** 21 (1997).
[13] A. G. Loewy, J. K. Blodgett, F. R. Blase, and M. H. May, *Anal. Biochem.* **246,** 111 (1997).
[14] C. J. Miller and B. H. Anderton, *J. Neurochem.* **46,** 1912 (1986).

have been reported in AD brains as compared to brains of normal aged individuals.[10] These findings are intriguing because the AD proteins tau,[7,9,12] β-amyloid (Aβ),[15,16] and synuclein (the non-Aβ component of amyloid plaques)[17] have been shown to act as good substrates for TGases *in vitro.*

Detection of the Glu–Lys bond historically has involved exhaustive enzyme digestion of the cross-linked protein with a number of proteases in the presence of an isodipeptide marker in which the lysine is labeled radioactively. The isodipeptide is subsequently purified with high-performance liquid chromatography (HPLC) anion-exchange after which it is acid hydrolyzed with the resulting lysine purified chromatographically. By measuring the specific radioactivity of the purified lysine and comparing it with that of the marker, one can calculate how much nonradioactive isodipeptide was contributed by the protein digest (for a review, see Ref. 18).

With the advent of monoclonal antibodies to both TGase and the Glu–Lys isodipeptide (CovalAb; Lyon, France), we can now utilize combinations of immunological and biochemical techniques to identify the TGase enzyme, measure its activity levels in tissues, and establish the presence of the isodipeptide *in situ* as well as in isolated/purified proteins and protein complexes. In addition, using the newly identified isopeptidase specific for TGase-catalyzed Glu–Lys bonds, we can now reverse *in vitro* the cross-linking of proteins heretofore refractory to proteolytic digestion.

This article discusses the deposition of neuritic senile plaques and NFTs in relation to TGase, the enzyme cross-linking them, and IPase, the enzyme reversing the cross-links. We will examine the hypothesis that the formation of these abnormal intracellular structures occurs when the normal balance of TGase and IPase is altered in the direction of an increase in the TGase/IPase ratio. We begin by discussing methods for measuring and identifying TGase in tissues and protein complexes, followed by methods to evaluate IPase activities and reversal of Glu–Lys cross-links in protein assemblies such as the PHFs/NFTs in Alzheimer's disease.

Detection of Transglutaminase Activity

TGase activity has been reported to be higher in AD brain tissues than in those of aged matched controls[10] and has been localized in tangle bearing

[15] K. Ikura, K. Takahata, and R. Sasaki, *FEBS Lett.* **326,** 109 (1993).

[16] L. K. Rasmussen, E. S. Sorensen, T. E. Petersen, J. Gliemann, and P. H. Jensen, *FEBS Lett.* **338,** 161 (1994).

[17] P. H. Jensen, E. S. Sorensen, T. E. Petersen, J. Gliemann, and L. K. Rasmussen, *Biochem. J.* **310,** 91 (1995).

[18] A. G. Loewy, *Methods Enzymol.* **107,** 241 (1984).

neurons from AD brains.[11] Numerous assays are available for measuring TGase activity, but this discussion focuses on the incorporation of [^{3}H]putrescine into protein substrates.[10] The method discussed herein is used to determine TGase activity in AD brain tissues and is a modification from an earlier protocol described by Hand *et al.*[19]

Autopsy tissues from human brain collected at postmortem intervals of less than 8 hr are typically used for TGase and IPase analyses. These tissues are obtained from aged normal individuals and from AD individuals confirmed using standard accepted criteria (NINDS/CERAD)[20] for the histological diagnosis of AD. Samples taken for immunohistochemistry are fixed in 10% formalin; those for enzyme and biochemical analyses are used fresh or frozen at −70°.

For the TGase activity assay,[10] 100 mg of fresh or frozen tissue samples is homogenized in a 1:10 (w/v) ratio in 10 m*M* Tris–HCl, pH 7.5, 1 m*M* EDTA, 1 m*M* phenylmethylsulfonyl fluoride (PMSF), 10 μg/ml of leupeptin, 10 μg/ml aprotinin, and 10 μg/ml pepstatin A. The homogenate is centrifuged at 16,000*g* for 10 min at 4° to obtain a supernatant to be assayed for TGase. To 50 μl of the supernatant, add 50 μl of assay buffer consisting of 10 m*M* Tris–HCl, pH 7.5, 0.5 m*M* EDTA, 5 m*M* dithiothreitol (DTT), 15 m*M* NaCl, 0.2 m*M* unlabeled putrescine, 1 μCi [1,4(*n*)-^{3}H]putrescine dihydrochloride (Amersham International, Inc.), 5 m*M* $CaCl_2$, 3 mg/ml *N,N*-dimethylated casein (DMC), and protease inhibitors (1 m*M* PMSF, 10 μg/ml leupeptin, 10 μg/ml aprotinin, 10 μg/ml pepstatin A). Baseline values are obtained by incubating the tissue supernatant in buffer containing 5 m*M* EGTA with no $CaCl_2$ or DMC. The reaction mixtures are incubated for 1 hr at 37° and are terminated with the addition of trichloroacetic acid (TCA) to 8.5% final concentration, followed by incubation on ice for an additional hour and centrifugation for 20 min at 16,000*g* at 4° to obtain a pellet. The supernatant is removed and the pellet is rinsed twice with 5% TCA followed by resuspension in 250 μl of 0.25 *N* NaOH. The samples are then placed in boiling water for 5 min, cooled, and mixed. [^{3}H]Putrescine incorporation into the precipitated proteins can be determined by liquid scintillation counting. Following protein determination,[21] TGase activity is determined after baseline subtraction and is expressed as picomoles of putrescine incorporated per hour per milligram of protein.

Using this protocol, TGase activity was shown to be significantly greater in AD brain samples, particularly from the prefrontal cortex, as compared to similar areas of the brain from non-AD controls.[10]

[19] D. Hand, F. J. Campoy, S. Clark, A. Fisher, and L. W. Haynes, *J. Neurochem.* **61,** 1064 (1993).
[20] S. S. Mirra, A. Heyman, and D. McKeel, *Neurology* **41,** 479 (1991).
[21] D. H. Lowry, N. J. Rosebrough, A. L. Farr, and R. J. Randall, *J. Biol. Chem.* **193,** 265 (1951).

Detection of Transglutaminase *in Situ*

To detect the TGase enzyme and the isodipeptide Glu–Lys *in situ,* immunohistochemistry protocols using specific monoclonal and polyclonal antibodies to the TGase enzyme and to the isodipeptide are utilized. Antibodies that can be used are CUB7402 from Dr. Paul Birckbichler and DAKO (Carpinteria, CA), a monoclonal antibody 4C1 made against liver TGase from Dr Gail Johnson, antifactor XIIIa from Calbiochem (LaJolla, CA), and a series of monoclonal anti-isodipeptide antibodies, including 81D1C2 from CovalAb (Lyon, France).[22]

Sections from paraffin-embedded formalin-fixed human brains are deparaffinized in xylenes followed by graded alcohols 100%–70%. After sections are rehydrated in phosphate-buffered saline (PBS), pH 7.4, endogenous peroxidase activity is blocked with 3% H_2O_2 in PBS for 15 min at 22°. Sections are rinsed in 10 m*M* sodium citrate at pH 6.0 and microwaved for 1–3 min for antigen retrieval. These sections remain in sodium citrate buffer for an additional 15 min, followed by rinsing in warm doubly distilled H_2O, and finally rinsing in 4° doubly distilled H_2O. Nonspecific background staining is reduced by blocking in 0.1% cold water fish gelatin (CWFG, Sigma) in PBS or 2% bovine serum albumin (BSA), pH 7.4, for 30–60 min at 4°. The sections are then incubated in primary antibodies diluted in PBS for 12–18 hr at 4° in a humid chamber. Following this incubation, sections are rinsed three times for 15 min each in 0.1% CWFG in PBS or 2% BSA in PBS at 22° and then incubated in secondary antibodies conjugated to horseradish peroxidase (HRP) (Amersham) diluted in PBS for 2–4 hr at 22°. The sections are rinsed in PBS three to five times for 5 min each prior to reacting the sections with diaminobenzidine (DAB) for 8 min. After the DAB reaction, they are rinsed in doubly distilled H_2O, counterstained with hematoxylin, rinsed with PBS, dehydrated in graded alcohols and xylenes, and mounted. For negative controls, sections are treated as described previously except that primary antibodies are omitted and substituted with nonimmune serum. Positive controls should include sections from human brain tissue immunoreacted with antibodies of known specificity.

Biochemical and Immunological Analyses of Transglutaminase

Quantitative immunoblotting can be used to measure the amount of TGase in tissues.[10] Using an internal standard, densitometry can be performed to obtain these quantitative measures. Immunoblotting with various

[22] S. el Alaoui, S. Legastelois, R. Am, J. Chantepie, and G. Quash, *Int. J. Cancer* **48,** 221 (1991).

antibodies and comparing protein band migrations provide evidence for the direct association of TGase with substrates such as the tau protein. In addition, anti-isodipeptide antibodies immunolabel dipeptide moieties within proteins separated by sodium dodecyl sulfate (SDS)–polyacrylamide gel electrophoresis that contain the TGase-catalyzed Glu–Lys bond. Using this method of detection (Fig. 1), we are able to determine that TGase associates with its substrate, in this instance, the tau proteins from AD brain, that the Glu–Lys isodipeptides are present in PHF/NFT complexes,[11]

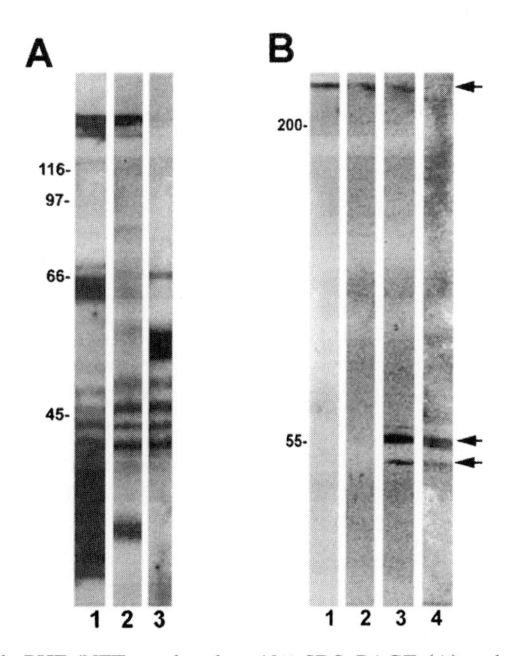

FIG. 1. Insoluble PHFs/NFTs analyzed on 10% SDS–PAGE (A) and on a 4–20% Tris–glycine SDS–PAGE (B) followed by Western blotting. (A) A heterogeneous population of soluble and insoluble PHFs was analyzed by Western analysis for the presence of tau (polyclonal antibody B-19; courtesy of Dr. J. Nunez) in lane 1, tissue transglutaminase (4C1; courtesy of Dr. Gail Johnson) in lane 2, and the isopeptide N^{ε}-(γ-glutamyl)lysine (CovalAb; 81D1C2) in lane 3. (B) A pure fraction of insoluble PHFs after they have been exposed to 4 *M* urea, DTT, SDS, and a series of enzymatic digestions with trypsin, chymotrypsin, and collagenase to ensure removal of any residual nonspecific proteins that were not covalently cross-linked to these structures. These structures were analyzed by Coomassie staining in lane 1 and Western analysis with the polyclonal B-19 tau antibody in lane 2. Note that the extremely insoluble PHFs remain at the top of the gel and blot and are refractory to electrophoresis. Lanes 3 and 4 represent insoluble PHFs after they were incubated at 37° for 18 hr with *B. cereus* isopeptidase. PHFs were analyzed for breakdown products using B-19 to determine the presence of tau in lane 3 and 81D1C2 for the presence of the isopeptide in lane 4. Note that in both lanes 3 and 4 the banding pattern is similar in the 45- to 55-kDa range.

and that the prefrontal cortex of AD brains as compared to controls contains threefold higher amounts of tissue TGase.[10]

The hippocampal region (~0.8–1.4 g wet weight) or another region of common AD neuropathology from autopsy AD brains and from brains of normal aged individuals is homogenized separately in 1–2.5 ml of 1× RAB buffer composed of 0.1 *M* MES, 1.0 m*M* EGTA, 0.75 *M* NaCl, 0.5 m*M* $MgCl_2$, 0.5 m*M* PMSF, 10 μg/ml leupeptin, and 0.01% Triton X-100, pH 6.9, at 4°. The samples are stirred at 4° for 1 hr and centrifuged at 1000*g* for 30 min at 4°. The resultant supernatants are centrifuged at 200,000*g* for 1 hr at 4°. The supernatant obtained from this step contains TGase(s). Lammeli sample buffer is added to the supernatants, followed by boiling for 5 min and analysis by SDS–polyacrylamide gel electrophoresis.

For Western blotting, proteins are transferred onto nitrocellulose membranes and blocked for 30 min with 5% PBS–nonfat milk prior to incubation for 18 hr at 4° with primary antibodies diluted in the blocking solution. Antibodies that have been useful for the identification of TGase, tau proteins, and the Glu–Lys isodipeptides include (a) anti-factor XIIIa, CUB 7402, 4C1; (b) tau 1, tau 2, tau 46, B19, AT8, and PHF1; and (c) 81D1C2, respectively. Controls to include are incubations of nitrocellulose transfers of purified human recombinant tau (courtesy of Dr. Michel Goedert) and bovine tau with the anti-TGase and anti-isodipeptide antibodies to demonstrate specificity of the antibodies for their proteins. To detect primary antibody immunolabeling, anti-mouse and anti-rabbit secondary antibodies conjugated to HRP (Amersham) are typically used at dilutions of 1:300–1:500 in the blocking solution for 2 hr at 22°, rinsed in PBS, and reacted for 10–20 min with 4-chloronaphthol.

Colocalization of Transglutaminase and Insoluble Protein Complexes

To verify that TGase is expressed in the cells/tissues in which the Glu–Lys bonds are formed, double immunolabeling techniques for TGase and the insoluble protein complex are used. The following colocalization protocol will demonstrate that cells in the AD brain contain deposited PHFs/NFTs directly associated with tissue TGase.[11]

Sections of hippocampi from paraffin-embedded formalin-fixed AD brains are rinsed in xylenes followed by graded alcohols 100%–70%. They are then placed in 10 m*M* sodium citrate at pH 6.0 and microwaved for 15 min. The sections remain in the citrate buffer for an additional 15 min after which they are rinsed in 4° doubly distilled H_2O and placed in PBS with normal goat serum at pH 7.4 for 5 min. These sections are blocked further with normal swine serum (DAKO) for 10 min followed by incubation for 30 min with the monoclonal antibody PHF1 (1:250 dilution), which recognizes phosphorylated tau proteins in PHFs/NFTs in AD.[23] The sections are then rinsed in PBS, blocked a second time with normal goat serum in PBS for 5 min, and incubated for 30 min in TRITC (rhodamine)-labeled goat antimouse secondary antibody (1:50) (Southern Biotechnology Assoc., Inc., Birmingham, AL), which is preabsorbed with normal rat serum. Following a rinse in PBS alone and PBS with normal goat serum for 5 min, the sections are incubated for 30 min with rabbit polyclonal antifactor XIIIa (an extracellular TGase) (1:50), which recognizes tissue TGase. Subsequently, sections are rinsed in PBS and blocked in PBS with normal goat serum for 5 min followed by incubation for 30 min in FITC (fluorescein isothiocyanate)-labeled goat antirabbit secondary antibody (1:50) (Southern Biotech Assoc.) preabsorbed with mouse and human serum. The sections are rinsed repeatedly in PBS, mounted, and photographed.

NFT/PHF Isolation

In order to determine that TGase is involved biochemically and ultrastructurally in the cross-linking of NFTs/PHFs, a protocol was developed to enrich for a homogeneous population of the insoluble NFTs/PHFs that would presumably contain the covalently bound Glu–Lys isodipeptides. Therefore, PHFs/NFTs are isolated from AD brains using the following methods. Brain tissue (100 g) is minced in a blender at 22° in 300 ml of buffer A containing 50 m*M* Tris–HCl, 0.1% Triton X-100, 0.75 *M* NaCl, pH 7.4. The tissue is homogenized and stirred for 2 hr at 22°. The homogenized brain tissue is then centrifuged at 110,000*g* at 22° in a TI45 rotor (Beckman) for 30 min. The pellet obtained from this centrifugation is homogenized in 150 ml of buffer B containing 50 m*M* Tris–HCl, 8 *M* urea, 1% SDS, 10 m*M* DTT, pH 7.4. This homogenate is layered onto a 0.34 *M*/1.0 *M* sucrose pad that is prepared in buffer A prior to centrifugation at 190,000*g* at 22° for 1.5 hr. The pellets are collected and resuspended in buffer A. This resuspended mixture is centrifuged for 30 min at 190,000*g* at 22° and the resulting pellet is homogenized in 150 ml of buffer A and then layered onto a discontinuous sucrose gradient (0.34 *M*/1.2 *M*/1.8 *M* sucrose) prepared in buffer A. This suspension is centrifuged for 1.5 hr at 22° at 190,000*g*. The sucrose interface at 1.2 *M*/1.8 *M* and pellets from this centrifugation are subsequently collected and homogenized in 200 ml of buffer B and stirred for 18 hr at 22°. Following this step, the mixture is centrifuged at 190,000*g* for 30 min at 22°. The pellet is collected, resuspended in 20 ml of buffer B, and dialyzed in 300,000 molecular weight cutoff dialyzers at 4° for 2–4 hr in 50 m*M* Tris–HCl, 0.1 *M* NaCl, pH 7.4. The

[23] S. G. Greenberg and P. Davies, *Proc. Natl. Acad. Sci. U.S.A.* **87,** 5827 (1990).

dialyzed material containing the PHFs/NFTs is centrifuged at 110,000*g* for 30 min at 22°. The pellets are homogenized and sonicated for 40 sec in the following buffer: 0.1 *M* morpholinoethanesulfonic acid (MES), 0.5 m*M* $MgCl_2$, 1 m*M* EGTA, 0.5 m*M* PMSF, 1 μg/ml leupeptin, 10 m*M* benzamidine, and 1 μg/ml trypsin at pH 6.4. The homogenate is finally layered onto a 30% sucrose pad and centrifuged at 190,000*g* for 1 hr at 4° resulting in a pellet consisting of an enriched population of soluble and insoluble PHFs/NFTs. In order to obtain solely insoluble PHFs/NFTs and to eliminate collagen contamination, an additional protocol is utilized as outlined in the following section.

Purification of Insoluble NFT/PHF

To obtain a pure population of insoluble PHFs/NFTs, the pellet is resuspended and homogenized in 4 *M* urea, 10 m*M* DTT, and 1% SDS and, following sonication for ~40 sec, is electrophoresed through an agarose plug composed of 1.5% agarose, 4 *M* urea, and 5 m*M* DTT that had been poured to a thickness of 4 mm onto the stacking gel of a 4–20% Tris–glycine gradient gel or onto a 10% Tris–glycine gel. The PHFs are electrophoresed and the remaining band at the top of the agarose plug is excised. The plug is melted at 90°, resuspended in (HPLC) H_2O, and centrifuged for 10 min at 13,000*g* at 22°. This step should be repeated several times to ensure that all of the agarose is removed. Following agarose removal, the pellet is homogenized in a 50 m*M* Tris–HCl buffer containing 0.1 *M* NaCl, pH 7.4. To remove any remaining collagen contaminants, this suspension is heated to 50° for 10 min, allowed to cool to 42°, and incubated for 10 min in 100 μg/ml of trypsin (Sigma), 200 μg/ml of chymotrypsin (Sigma), and 100 μg/ml of collagenase (Sigma). Subsequently, 500 μg/ml of soybean trypsin inhibitor (Sigma) should be added to the mixture and incubated for an additional 10 min to quench the previous enzymatic reaction. Following this incubation, the solution is centrifuged at 200,000*g* in a TL 100 ultracentrifuge (Beckman). The pellet obtained from this centrifugation is resuspended in 50 m*M* Tris–HCl buffer, layered onto a 1.5% agarose plug, and electrophoresed. The protein band remaining at the top of the agarose plug following electrophoresis is once again excised, melted, centrifuged, and resuspended in a 50 m*M* Tris–HCl buffer or a 0.1 *M* MES buffer as described previously. These steps are necessary to ensure that one obtains extremely clean fractions of insoluble PHFs to be analyzed for the isodipeptide bond.

Ultrastructural Analysis of NFTs/PHFs

Analysis of the PHFs/NFTs isolated using the preceding method initially involves negative staining and immunolabeling with colloidal gold to identify component proteins of these insoluble structures. This approach provides visual evidence that PHFs/NFTs have been isolated, are indeed in a complex, and contain epitopes reactive with anti-tau antibodies, anti-TGase antibodies, and the anti-isodipeptide antibody (Fig. 2). In this manner, we

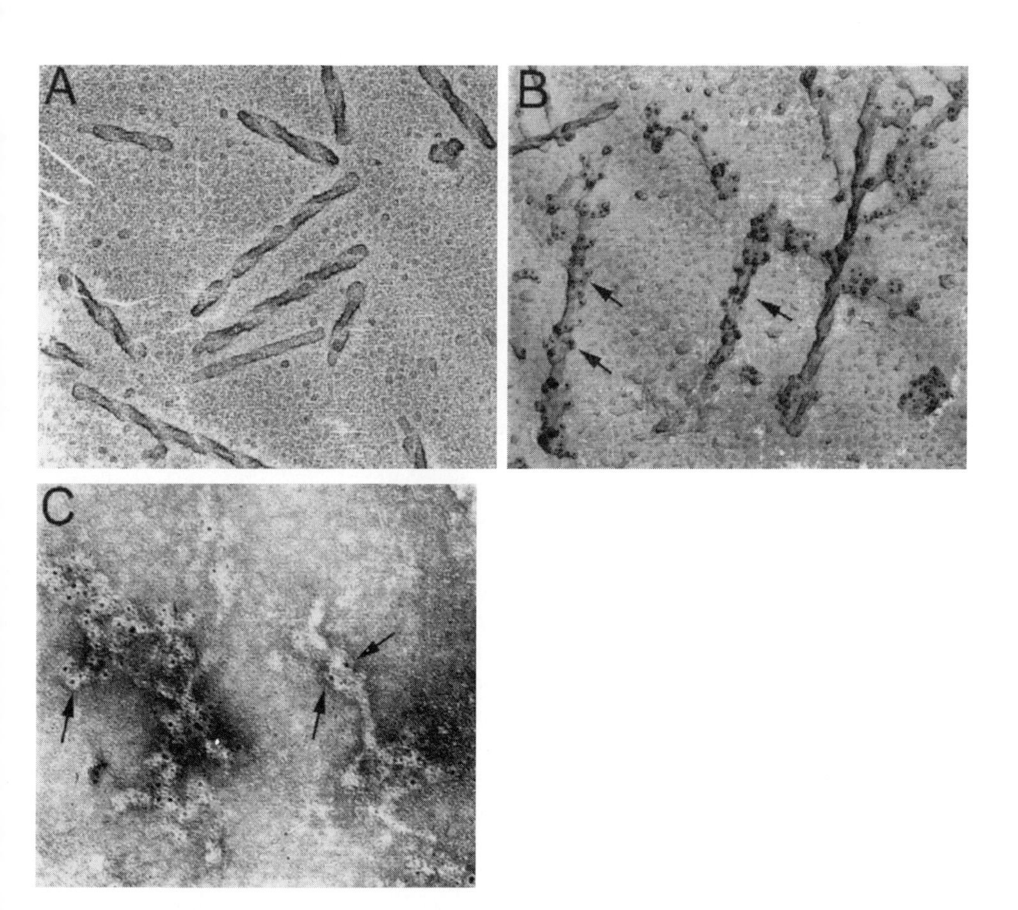

FIG. 2. Electron micrographs of insoluble paired helical filaments (PHFs) isolated from AD brains. (A) PHFs that have been freeze-dried/rotary shadowed reveal the characteristic twisted structure. (B) Immunoelectron microscopy in tandem with freeze-dried/rotary shadowing provides a unique view of tau epitopes on the PHFs as revealed by the 5-nm gold particles (arrows). These filaments have been immunolabeled with the monoclonal tau46 antibody (Courtesy of Dr. Virginia Lee). (C) Negatively stained filaments, which appear to be unraveled, are immunolabeled with the monoclonal antibody that reacts with the isopeptide N^{ε}-(γ-glutamyl)lysine (CovalAb; 81D1C2), which is detected with a 5-nm gold-conjugated secondary antibody (arrows) (Amersham). Magnification: (A) ×79,000; (B) ×72,000; (C) ×85,000; (A and B) 15° angle of platinum/carbon rotary shadow.

are able to demonstrate that TGase is involved in catalyzing Glu–Lys cross-links in the PHF and NFT complexes.

Isolated PHFs/NFTs are stained negatively with a 1% aqueous uranyl acetate solution. In brief, 5–10 μl of the sample is layered onto a carbon-coated copper electron microscopy (EM) grid, washed with 10 drops of 100 m*M* ammonium acetate followed by 10 drops of 1% aqueous uranyl acetate, and air dried.

For immunoelectron microscopy, 5–10 μl of the sample is layered onto the carbon-coated copper EM grid followed by washing with PBS, pH 7.4. The grids are blocked with 0.1% CWFG in PBS for 10 min prior to incubating the grid with the primary antibody diluted in PBS for 15–30 min at 22°. Subsequently, the grids are rinsed in PBS and blocked in 0.1% CWFG in PBS for 10 min and then incubated in the appropriate secondary antibody conjugated to 5- to 10-nm colloidal gold particles at 1.0 mg/ml (Amersham) for 30–60 min at 22°. Prior to negative staining the grids in 1% aqueous uranyl acetate, the grids are rinsed in doubly distilled H_2O. Grids can be examined at 80 kV on an electron microscope.

Antigenic sites on PHFs isolated from AD NFTs are analyzed further by freeze-drying/rotary shadowing with a greater degree of certainty as compared to traditional microscopy techniques. To avoid artifacts introduced by traditional freezing methods, samples are processed using a modified glass sandwich technique.[24] Subsequently, this protocol was modified for its application in immunoelectron microscopy.[25] Purified suspensions of PHFs are incubated for 2 min on glass coverslips and blocked with 0.1% CWFG in 10 m*M* Tris–HCl, pH 7.2, for 10 min prior to a 20-min incubation with the primary antibody solution containing antibodies such as anti-factor XIIIa and AT8 (Innogenetics). The samples are blocked a second time for 20 min followed by a 20-min incubation with a secondary antibody conjugated to 5- to 10-nm colloidal gold (Amersham). The coverslips on which the proteins were adsorbed are rinsed with 100 m*M* ammonium acetate and are incubated for 1–2 min in 0.5% aqueous uranyl acetate. Following this incubation, the glass coverslip is rinsed with 100 m*M* ammonium acetate and incubated for 1–2 min in 0.25% tannic acid. Finally, the coverslip is rinsed with 30% methanol and overlaid with a second coverslip to form the glass sandwich, which is flash frozen by immersion in liquid nitrogen. While immersed in liquid nitrogen, the sandwich is split and mounted sample side up onto a specimen holder and then inserted at −150° into a Balzers 400 freeze-fracture machine. The samples are etched for ~45 min at −90° at 2×10^{-6} mbar and rotary shadowed with platinum at 15° and

[24] K. Loesser and C. Franzini-Armstrong, *J. Struct. Biol.* **103,** 48 (1990).
[25] D. M. Appelt and B. J. Balin, *J. Struct. Biol.* **111,** 85 (1993).

25° angles to a thickness of ~2 nm, followed by a carbon shadow. The glass coverslip is dissolved in full-strength hydrofluoric acid under a fume hood and the replica is recovered on 600 mesh fine-bar EM grids in doubly distilled H_2O for examination.

This method provides a visual verification that tau and TGase are both present in isolated PHFs/NFTs. In addition, immunolabeling with the anti-isodipeptide antibody clarifies that the specific cross-links catalyzed by TGase are present in PHFs/NFTs. This confirms that TGase must be utilizing the glutamine and lysine amino acid residues in the tau proteins as substrates for the formation of the covalent Glu–Lys bond.

Reversal of Cross-Links in Protein Assemblies with Isopeptidase

Isopeptidase Mechanism for Cleavage of [^{3}H]ALMA–Casein

The determination of isopeptidase activity either in homogenates from eukaryotic tissues or excreted by the bacterium *Bacillus cereus* uses a substrate synthesized for this purpose. It consists of ALMA, synthesized and linked by its ε amino group to a γ-carboxyl amide group of casein with either guinea pig liver TGase or factor XIIIa.[13] Isopeptidase cleaves the Glu–Lys cross-link of [^{3}H]ALMA–casein. The substrate is able to reliably detect specific activities as low as 10^{-9} μmol of [^{3}H]ALMA/min/μg protein.[13] The isopeptidase assay utilizes thin-layer chromatography to measure the cleavage of [^{3}H]ALMA from [^{3}H]ALMA–casein.

Enzymes capable of breaking the ALMA–casein isopeptide bond are distributed broadly in prokaryotes such as *B. cereus* and eukaryotes such as human erythrocytes and embryonic mouse brains.[13] In addition, isopeptidase activity has been measured from mature human brains.[26] To determine that insoluble protein assemblies are cross-linked with Glu–Lys isopeptide bonds and that the level of these assemblies depends, in part, on the level of isopeptidase activity, we again used [^{3}H]ALMA–casein to assay for enzyme activity.

To confirm that ALMA–casein is split by an isopeptidase mechanism, alternative mechanisms are ruled out.[13] These mechanisms are γ-glutamyl transpeptidase(γ-glutamyltransferase), γ-glutamyl cyclic transferase, and the reversal of TGase. To rule out the γ-glutamyl transpeptidase mechanism, we use 10 m*M* acivicin, which does not inhibit [^{3}H]ALMA release, although it inhibits γ-glutamyl transpeptidase activity. The reaction is not inhibited competitively by 50 m*M* γ-glutamyl-*p*-nitroamalide. Furthermore,

[26] D. M. Appelt, M. P. Howard, A. G. Loewy, and B. J. Balin, *Soc. Neurosc. Abstr.* **22,** 559.9 (1996).

γ-glutamyl transpeptidase does not release [^{3}H]ALMA from [^{3}H]ALMA–casein. The γ-glutamyl cyclic transferase mechanism[27] is ruled out because the reaction is not inhibited by its isodipeptide substrate, 50 m*M* N^{ε}-(γ-glutamyl)lysine. In addition, the reversal of the TGase mechanism is ruled out because exhaustive dialysis to remove even traces of ammonia (a product of the TGase reaction) does not inhibit the release of [^{3}H]ALMA; 50 m*M* lysine, a product of TGase reversal, does not inhibit the rate of release of [^{3}H]ALMA and Mg^{2+} can substitute for Ca^{2+} in the [^{3}H]ALMA casein splitting reaction, whereas Ca^{2+} and not Mg^{2+} is required for the TGase reaction.

It has been shown that the erythrocyte Glu–Lys isopeptidase is not the ubiquitin–protein isopeptidase found in erythrocytes[28] because ubiquitin aldehyde does not inhibit the isopeptidase while being a strong inhibitor of the ubiquitin–protein isopeptidase. Despite evidence for an isopeptidase-based mechanism of ALMA–casein cleavage presented earlier, cytosolic TGases and coagulation factor XIIIa are able, under very specific conditions *in vitro,* to hydrolyze synthesized γ-branched peptides,[29] thereby suggesting that TGases could act as isopeptidases as well; K_m values for the isopeptidase activity of these TGase enzymes ranged from 10^{-4} to 10^{-5} *M*. However, no evidence has been presented nor has it been determined that TGase and factor XIII can proteolyze the Glu–Lys cross-links of NFTs/PHFs isolated from the Alzheimer brain. Therefore, it is not clear whether, unlike isopeptidase, these enzymes could play an intracellular role in both the assembly and the disassembly of the cross-linked structures.

Digestion of NFTs/PHFs with Bacillus cereus Isopeptidase

If NFTs/PHFs are cross-linked with Glu–Lys isopeptide bonds, it should be possible to use isopeptidase preparations to digest these structures, which have hitherto resisted a wide range of proteases. A preparation of an isopeptidase secreted by *B. cereus,* which was high in isopeptidase activity (7×10^{-8} μmol/min/μg protein) and low in protease activity, was shown to partially digest NFTs/PHFs.[26] The isopeptidase is prepared by growing *B. cereus* in an LB medium (10 g Bacto-tryptone, 5 g Bacto yeast extract, 10 g NaCl, pH 7.5) plus 1 m*M* $CaCl_2$, pelleting the cells at 4500*g* for 15 min at 22°, resuspending the cells in 1/10 original culture volume of doubly distilled H_2O at 22° for 3 hr, repelleting the cells, and finally precipitating the isopeptidase from the supernatant at 40% saturation ammonium sulfate.

[27] M. I. Fink and J. E. Folk, *Methods Enzymol.* **94,** 347 (1983).

[28] J. Moskovitz, *Biochem. Biophys. Res. Commun.* **205,** 354 (1994).

[29] K. N. Parameswaran, X. F. Cheng, E. C. Chen, P. T. Velasco, J. H. Wilson, and L. Lorand, *J. Biol. Chem.* **272,** 10311 (1997).

Isopeptidase preparations obtained from *B. cereus* can be used on NFTs/PHFs that have been purified by removing electrophoretically protein chains not covalently cross-linked into these structures.[26] Concentrations of isopeptidase ranged from 1.07 to 3.85 mg/ml (specific activities of 3.75×10^{-9} to 1.0×10^{-8} μmol/min/μg protein). PHFs/NFTs are suspended in a Tris buffer, pH 7.5 (50 m*M* Tris–HCl + 0.1 *M* NaCl). PHFs/NFTs are incubated at 37° for 18 hr with the *B. cereus* isopeptidase (v/v) in the presence of 33–50 m*M* $CaCl_2$. In this regard, $MgCl_2$ could be substituted for $CaCl_2$ with equal efficacy. These preparations are capable of partially digesting NFTs/PHFs, a process that can be analyzed by immunoblotting with anti-tau and anti-isopeptide antibodies and electron microscopy. Because of the specificity of our isopeptidase preparation, we can conclude that this digestion of PHFs/NFTs proceeds by cleavage of Glu–Lys isopeptide bonds. Further digestion may have to be achieved by including additional proteases, which may make the remaining cross-links more accessible to the isopeptidase.

In the cell, we believe that an increase in the TGase/IPase ratio would cause the steady-state level of covalently cross-linked protein structures to shift in the direction of increasing the amount of insoluble material. Tau, a protein shown to be capable of being cross-linked by TGase, would be expected to be a major component of the insoluble structures, but other proteins are also likely to be involved. Thus, one aspect of the molecular changes taking place in Alzheimer's disease may be an increase in TGase activity and a decrease in IPase activity occurring either at the level of enzyme expression and/or at the level of regulation of enzyme activity.

Conclusion

Recognition of transglutaminase-catalyzed cross-links in deposited proteins found in Alzheimer's disease brains is important to our understanding of protein modifications in neurodegenerative diseases. While numerous protein modifications have become apparent in AD, few are as difficult to evaluate and reverse as is the transamidation of susceptible proteins. Identification of specific isopeptidases capable of cleaving the TGase-catalyzed Glu–Lys bonds in protein assemblies provides insight into a potential steady-state mechanism important to maintaining the stability and viability of eukaryotic cells. Future studies will focus on delineating how TGase/IPase are regulated normally in the nervous system to provide the proper homeostatic environment in which nerve cells can function properly. It is hoped that this article will enable further investigation into the intriguing nature of the formation and reversal of transglutaminase-catalyzed cross-links found in proteins in neurodegenerative diseases of the human nervous system.

Acknowledgments

We are grateful to the following individuals for the use of their antibody probes. Dr. Virginia Lee for tau 46; Dr. Lester Binder for tau 1 and tau 2; Dr. Jacques Nunez for B19; Dr. Peter Davies for PHF1, Dr. E. Vanmechelen and Innogenetics for AT8; Dr. Gail Johnson for the anti-TGase 4C1; Dr. Paul Birckbichler for the anti-TGase CUB7402; and Dr. Said El Alaoui and CovalAb for the anti-isopeptide 81D1C2. This work was supported in part by NIH Grant AG10160 and by the Department of Pathology and Laboratory Medicine of the Allegheny University of the Health Sciences.

Section II

Characterization of *in Vitro* Protein Deposition

A. Managing the Aggregation Process
Articles 13 through 16

B. Aggregation Theory
Article 17

C. Monitoring Aggregate Growth by Dye Binding
Articles 18 through 20

D. Measurement and Characterization of Assembly Intermediates
Articles 21 and 22

E. Monitoring Aggregate Growth by Measuring Solid-Phase Accumulation
Articles 23 through 26

F. Monitoring Aggregate Growth and Structure Using Light Scattering
Articles 27 and 28

G. Aggregation Inhibitors
Articles 29 and 30

[13] Methodological and Chemical Factors Affecting Amyloid β Peptide Amyloidogenicity

By MICHAEL G. ZAGORSKI, JING YANG, HAIYAN SHAO, KAN MA, HONG ZENG, and ANITA HONG

Introduction

The phrase "the peptide from hell" has been used frequently to describe amyloid-β, particularly by the bench chemist or biologist who must endeavor on a day-to-day basis with its ever-changing demeanor. Biophysical and biological studies of the Aβ peptides,[1] especially studies with the less soluble and more pathogenic 42 residue Aβ(1–42), are plagued by many difficulties. The major difficulty relates to the ease in which the Aβ peptides aggregate and precipitate. Problems related to the inability to reproduce amyloid inhibitor or neurotoxicity assays are particularly prevalent [for a earlier review, see *Neurobiology of Aging* **13,** 535–623 (1992)]. Depending on the commercial source, peptide batch, and the particular aggregation conditions, considerable discrepancies exist across different laboratories as well as within the same laboratory over time.

An intense area of research in Alzheimer's disease (AD) involves unraveling the mechanisms of Aβ-induced neurotoxicity and aggregation.[2,3] One of the earliest studies reported direct toxicity on rat neurons when the Aβ peptide is fully soluble in acetonitrile, whereas subsequent reports showed that aggregation or aging is a prerequisite for neurotoxicity. Other laboratories showed that Aβ is not directly toxic to cortical and hippocampal neurons. Despite these and other discrepancies, there is now general agreement that aging of the Aβ to achieve particular assembly states is required for neurotoxicity.[2,4,5] During the aging process, the Aβ undergoes

[1] The primary amino acid sequence for the Aβ(1–42) peptide is H_3N^+-D[1]A-E-F-R[5]-H-D-S-G-Y[10]-E-V-H-H-Q[15]-K-L-V-F-F[20]-A-E-D-V-G[25]-S-N-K-G-A[30]-I-I-G-L-M[35]-V-G-G-V-V[40]-I-A[42]-COO^-. The Aβ(1–40) peptide has the identical sequence, except that the last two C-terminal residues (Ile-41 and Ala-42) are absent.

[2] L. L. Iversen, R. J. Mortishire-Smith, S. J. Pollack, and M. S. Shearman, *Biochem. J.* **311,** 1 (1995).

[3] D. B. Teplow, *Amyloid Int. J. Exp. Clin. Invest.* **5,** 121 (1998).

[4] L. M. Sayre, M. G. Zagorski, W. K. Surewicz, G. A. Krafft, and G. Perry, *Chem. Res. Toxicol.* **10,** 518 (1997).

[5] M. P. Lambert, A. K. Barlow, B. A. Chromy, C. Edwards, R. Freed, M. Liosatos, T. E. Morgan, I. Rozovsky, B. Trommer, K. L. Viola, P. Wals, C. Zhang, C. E. Finch, G. A. Krafft, and W. L. Klein, *Proc. Natl. Acad. Sci. U.S.A.* **95,** 6448 (1998).

0076-6879/99 $30.00

conformational conversions, from soluble, monomeric random coil or α-helix conformations[6] into aggregated β-sheet structures.[7] These conversions are affected by several factors, including the length of the Aβ peptide [the Aβ(1–42) aggregates faster than the Aβ(1–40)],[8] solvent hydrophobicity,[9] ionic strength,[10,11] pH,[6,12,13] peptide concentration, initial aggregation state,[14,15] buffer type, peptide counterions (i.e., CF_3COO^- vs Cl^-),[16] and the presence of partially oxidized or preaggregated forms (seeds) of the peptide.[8,11]

In view of the general agreement that Aβ aging and aggregation are required for toxicity, the most serious problem arises from the different starting conformational and aggregational states of the various lyophilized peptide batches. This situation comes about from the rapid Aβ aggregation in the acetonitrile–water solutions used during high-performance liquid chromatography (HPLC) purification[9]; the greater the length of time in the acetonitrile–water, the more likely the Aβ will aggregate into β-sheet structures (Fig. 1). Consequently, peptide batches containing mostly monomeric or smaller sized aggregates will be more soluble, less neurotoxic, and less prone to aggregation, whereas other less soluble batches with higher ordered aggregates will show contradictory results. Additional problems result from the presence of minor impurities, as well as differences in the types and quantities of salts that can accelerate aggregation,[10,11,16] small quantities of peptides with partial racemization,[17] and/or the presence of oxidation at Met[35]-CH_3S, in which the latter can affect aggregation rates.[11] These complications undoubtedly contribute to the poor reproducibility and the lack of consensus regarding the mechanisms of neurotoxicity and

[6] C. J. Barrow, A. Yasuda, P. T. M. Kenny, and M. G. Zagorski, *J. Mol. Biol.* **225,** 1075 (1992).
[7] S. B. Malinchik, H. Inouye, K. E. Szumowski, and D. A. Kirschner, *Biophys. J.* **74,** 537 (1998).
[8] J. D. Harper and P. T. Lansbury, *Annu. Rev. Biochem.* **66,** 385 (1997).
[9] C.-L. Shen and R. M. Murphy, *Biophys. J.* **69,** 640 (1995).
[10] C. Hilbich, B. Kisters-Woike, J. Reed, C. L. Masters, and K. Beyreuther, *J. Mol. Biol.* **218,** 149 (1991).
[11] S. W. Snyder, U. S. Ladror, W. S. Wade, G. T. Wang, L. W. Barrett, E. D. Matayoshi, H. J. Huffaker, G. A. Krafft, and T. F. Holzman, *Biophys. J.* **67,** 1216 (1994).
[12] P. E. Fraser, J. T. Nguyen, W. K. Surewicz, and D. A. Kirschner, *Biophys. J.* **60,** 1190 (1991).
[13] D. Burdick, B. Soreghan, M. Kwon, J. Kosmoski, M. Knauer, A. Henschen, J. Yates, C. Cotman, and C. Glabe, *J. Biol. Chem.* **267,** 546 (1992).
[14] C. Soto, E. M. Castaño, R. A. Kumar, R. C. Beavis, and B. Frangione, *Neurosci. Lett.* **200,** 105 (1995).
[15] D. R. Howlett, K. H. Jennings, D. C. Lee, M. S. G. Clark, F. Brown, R. Wetzel, S. J. Wood, P. Camillri, and G. W. Robert, *Neurodegeneration* **4,** 23 (1995).
[16] I. Kaneko and S. Tutumi, *J. Neurochem.* **68,** 438 (1997).
[17] T. Tomiyama, S. Asano, Y. Furiya, T. Shirasawa, N. Endo, and H. Mori, *J. Biol. Chem.* **269,** 10205 (1994).

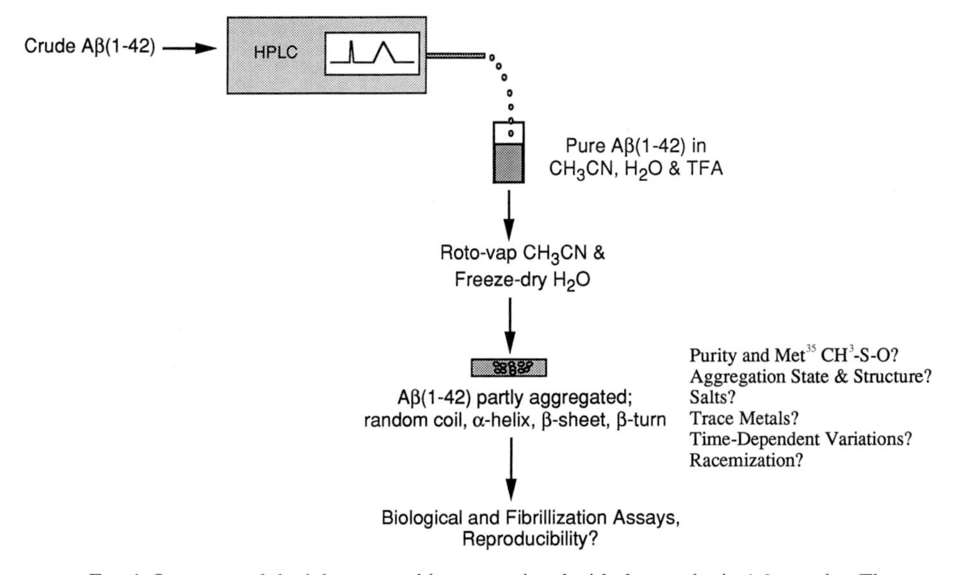

FIG. 1. Summary of the inherent problems associated with dry synthetic Aβ samples. The major difficulties are the presence of minor impurities, as well as the mixed structures and aggregation states.[14] These factors result from poor peak resolution plus time- and concentration-dependent aggregation in the acetonitrile–water solution during the HPLC purification.[9] Other variables include potential metal impurities, salts, and partial racemization, with the latter event possibly occurring during the peptide synthesis.

thus suggest an urgent need for developing generally accepted standards regarding the purity and handling of synthetic Aβ peptides.

The following sections discuss in greater detail the issues and approaches for dealing with them. Any experienced Aβ investigator will agree that these factors are critically important and can save enormous amounts of time and money.

Aβ Sample Preparation Variables

Purity and Met-35 Oxidation State

The presence of minor impurities in Aβ peptide batches is often an overlooked, yet important factor that can contribute to the inconsistencies and the batch-to-batch variability. It is important to recognize that the starting aggregation state and purity levels are completely independent factors. Many laboratories incorrectly assume that when a particular peptide batch is water soluble and neurotoxic that the purity level is high or if a peptide batch is less water soluble and shows modest neurotoxicity that the purity level is low. These factors are totally unrelated, and it is quite conceivable that a very pure Aβ batch could have poor water solubility and low neurotoxicity or vice versa. Low water solubility usually indicates that the particular Aβ batch is highly preaggregated and needs to be disaggregated (discussed in the next section).

Most studies utilize synthetic Aβ peptides that are purified by conventional reversed-phase high-performance liquid chromatography (HPLC) methods. Synthetic Aβ is available from many commercial sources, many of which proclaim high (>95%) purity levels. To assess these purities, standard mass spectrometry (MS), HPLC, and occasionally amino acid analysis (AAA) are employed, which in our opinion are not sufficient to determine the Aβ purity properly. Standard AAA and MS are incapable of providing quantitative purity levels. For AAA, it is critically important that the peptide is completely hydrolyzed, and problems related to incomplete hydrolysis or decomposition are quite common. It is well known that when valine and isoleucine residues are adjacent in the primary sequence, complete hydrolysis can take up to 3–4 days, during which time the serine and threonine residues can decompose.[18]

With MS analysis, it is imperative that a single, clean molecular ion peak is obtained at the expected mass. We routinely obtain MS spectra before and after HPLC purification. Standard fast atom bombardment (FAB) or matrix-assisted laser desorption/ionization (MALDI) analysis are suitable.[19] For FAB, an important point is that sample should be disaggregated before applying to the matrix, which can be accomplished easily by predissolving the Aβ in trifluoroacetic acid (TFA).[6] We normally use a bench-top, Voyager BioSpectrometry MS instrument (PerSeptive Biosystems) for routine MALDI analysis. Only small quantities of sample are required, usually microgram amounts, and a α-cyano-4-hydroxycinnamic acid matrix is appropriate. The detection of impurities is usually limited to 5–10%.

The purity content of most commercial Aβ is determined with HPLC. Under standard reversed-phase HPLC conditions, the Aβ(1–42) peptide elutes as an extremely broad peak with poor resolution due to its aggregation in the acetonitrile–water solution.[9] Therefore, it is possible that impurities will coelute and overlap with the pure Aβ peptide. We have also found that the collection of one peak corresponding to pure Aβ(1–42), followed by reinjection, produces several peaks, establishing that the Aβ(1–42) aggregates under HPLC conditions. Additional problems result from the

[18] M. Bodanszky and A. Bodanszky, "The Practice of Peptide Synthesis." Springer-Verlag, Berlin, 1994.

[19] M. W. Senko and F. W. McLafferty, *Annu. Rev. Biophys. Biomol. Struct.* **23,** 763 (1994).

Aβ adhering to the HPLC column. Newer HPLC purification methods, such as those employing elevated temperatures and Zorbax bonded-phase or Vydac silica columns, can sometimes provide better resolution and purity levels.[20,21] However, impurities can often coelute with the Aβ at higher temperatures. Work from our laboratory has shown that a collection of single sharp HPLC peaks for Aβ(1–40) at 60°, followed by reinjection at 25°, produced four peaks, many being smaller peptide impurities. As a result, standard HPLC methods are questionable for determining the purity, particularly with the more aggregation-prone Aβ(1–42) peptide.

We showed that the inclusion of standard ^1H nuclear magnetic resonance (NMR) spectroscopy can assist in providing a more realistic assessment of the Aβ purity.[22] Despite its lower sensitivity, the NMR method is noninvasive, and after analysis the sample can be recovered. Usually about 0.5- to 1-mg quantities of the Aβ are required for one-dimensional NMR analysis. The usefulness of the NMR approach is demonstrated with the spectra shown in Fig. 2, which display ^1H NMR spectra of two Aβ(1–42) samples. Both samples were delineated pure by HPLC (eluted as sharp peaks at 60°) and MS [single molecular ion peaks, FAB-MS (MH^+) *m/z* 4515 or MALDI-MS (M) *m/z* 4514]. However, as shown in the NMR spectra, one of the batches (supplied by a collaborator) contains many small molecular weight impurities that came from the solvents and plasticizers in the tubing and plastic vials (Fig. 2B). These impurities are nonvolatile and can adhere to the surface of the Aβ. In addition, because of their low molecular weight and lack of strong ultraviolet absorbance, they were not detected by either MS or HPLC.

To provide a numerical value of purity, peak integration is used. The integral tracings are normalized to the well-separated His-2H signal (three protons, 8.56 ppm). As shown in the spectra (Fig. 3), this normalization causes the integral height for the His-2H to be 3.000 with the remaining heights scaled accordingly. Similar scaled integral values were obtained when the Tyr-10 3,5-H (6.85 ppm) was normalized to 2.000, thus establishing that the His-2H is a reliable integration marker for pure Aβ(1–42) peptide. The purity was determined using the αH region (3.87–5.25 ppm), which is perfect for detecting peptide impurities, as virtually all peptides have αH NMR peaks that will contribute to the integral height. For pure Aβ(1–42)

[20] B. E. Boyes, "Efficient High Yield Reversed-Phase HPLC Separations of Amyloid Precursor Polypeptide C-Terminal Fragments," 14th American Peptide Symposium, Columbus, OH, 1995.

[21] N. K. Menon, A. E. Przybyla, E. B. Neuhaus, and R. A. Makula, *in* "Vydac Advances," Vol. Winter, 1998.

[22] S.-C. Jao, K. Ma, J. Talafous, R. Orlando, and M. G. Zagorski, *Amyloid Int. J. Exp. Clin. Invest.* **4,** 240 (1997).

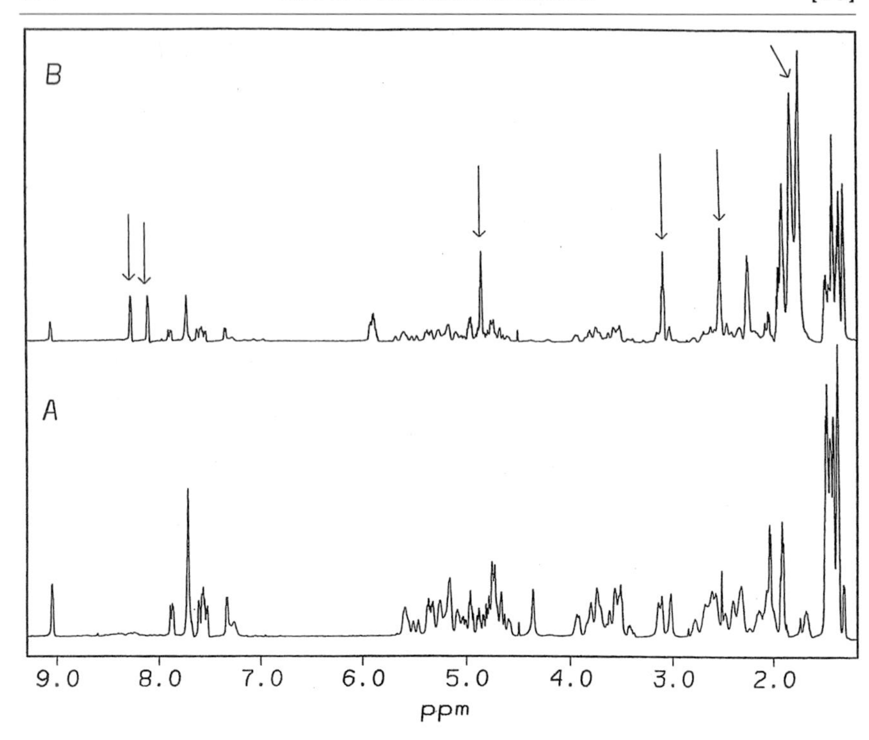

FIG. 2. ^{1}H NMR spectra (600 MHz) of two Aβ(1–42) samples that were pure by HPLC (single peak, 60°) and MS (single molecular ion peak). The presence of one, degenerate His-2H signal for His-6, His-13, and His-14 is consistent with the Aβ(1–42) peptide adopting a predominantly random coil structure. (A) The upper spectrum shows the presence of small molecular weight impurities (peaks labeled with arrows) that come from solvents or plasticizers of the hoses or tubing. (B) The lower spectrum corresponds to another peptide batch, in which the impurities are absent. The solvent was TFA-*d* and the chemical shifts are reference to internal TSP.

peptide, there are 12 protons for the six Gly residues, 4 β-CH_2-OD protons for Ser-8 and Ser-26, plus one αH signal for each of the remaining residues. This corresponds to a total of 52 protons. Because the observed height in the αH region is 62.254 (rounded to 62, Fig. 3), the percentage purity is 52/62 (100) = 84%. It is important to recognize that the integration procedure is susceptible to spectral processing variations, particularly with the phasing, integral slope, and integral bias that are applied in the integration procedure. Thus, it is important that flat, baseline integrals are present before and after each peak.

With the NMR method, we normally obtain purities ranging from 80 to 95% for different batches of Aβ(1–42) peptide that have been purified

Purified β-(1-42) in CF_3COOD purity 52/62 (100) = 84%

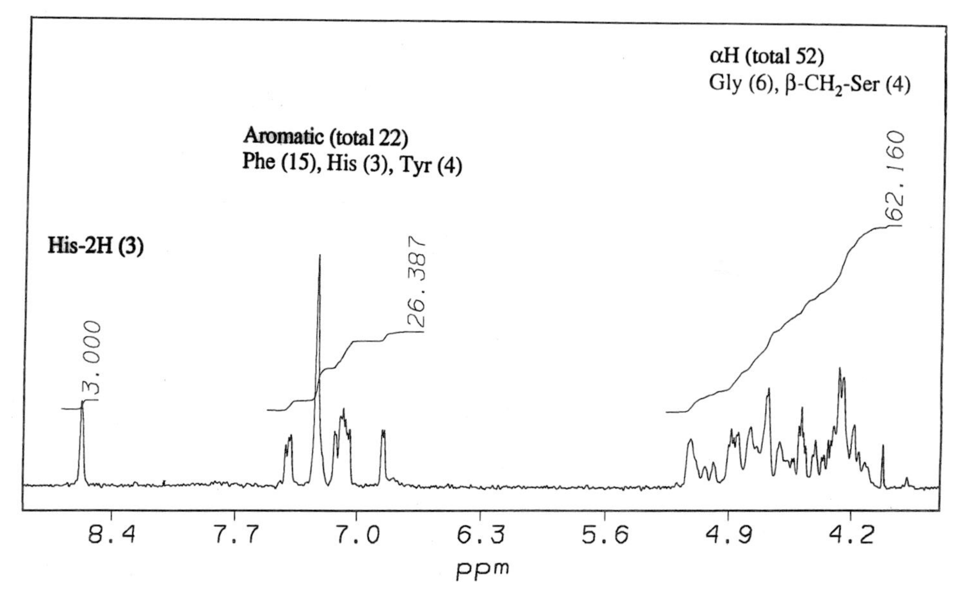

FIG. 3. ^{1}H NMR spectra (600 MHz) of an Aβ(1–42) sample (0.40 m*M*) in TFA-*d* solution (0.5 ml), which represents an expanded region of the spectrum in Fig. 2B. Integrals were scaled to the three His-2H protons at 8.56 pm, and the total number of protons expected for pure Aβ(1–42) in the His-2H (8.56 ppm), aromatic (7.02–7.44 ppm), and the αH (3.87–5.25 ppm) regions are indicated in parenthesis. The percentage purity was based on the integral heights in the αH region (see text). To ensure complete relaxation of all protons between scans, the pulse width was set to 45° rather than 90°, and the total recycle delay was 8.2 sec.

by standard (acetonitrile–water–TFA) reversed-phase HPLC conditions. For our biophysical experiments, purities greater than 90% are usually acceptable and generally will provide reproducible experimental data. As described earlier, the major reason for the purity not exceeding 90% results from the difficulties associated with the standard reversed-phase HPLC purification method, particularly the Aβ(1–42) that elutes as a relatively broad peak.

An additional advantage of performing a final purity check by NMR is that detection of oxidation of the Met-35 methyl [$CH_3S(O)$] is possible. The Met^{35}-$CH_3S(O)$ signal appears at 2.20 ppm, whereas the reduced Met^{35}-CH_3S signal appears further upfield at 2.02 ppm.[22] In our experience, the Met^{35}-CH_3S is extremely sensitive to air oxidation and can be avoided with distilled solvents that are stored in the dark under nitrogen or argon. Additionally, the methionine sulfoxide can also be reverted back easily to the reduced form using reagents such as 2-mercaptoethanol.[23] The detection of oxidation at Met[35]-CH_3S is important, as this can affect aggregation rates and toxicity. However, as discussed earlier,[23] MS can also identify oxidation at Met-35.

In conclusion, the NMR method can provide a precise quantity for Aβ purity. The NMR technique alone cannot be used to characterize the peptide, but that together with HPLC, MS, and NMR are probably sufficient techniques for the appraisal of the Aβ purity state.

Starting Aggregation State and Structure

As mentioned earlier, a significant issue with the Aβ batch-dependent variations are the different starting aggregation states and structures.[14,15] Numerous studies have established that the neurotoxicity and the kinetics of β aggregation are directly related to the assembly state in solution. Thus, direct solubilization of the Aβ peptides in aqueous media should always be avoided, as it generates batch-dependent mixtures of aggregates and structures.

To overcome these problems, many laboratories prepare concentrated Aβ stock solutions in organic solvents in which, compared to water, the Aβ is more soluble and time-dependent aggregation is presumably avoided. Some of these solvents include formic acid and trifluoroethanol, which reportedly prevented the loss of Aβ immunoreactivity by aggregation,[24] hexafluoro-2-propranol (HFIP), aqueous acetonitrile with TFA, dimethyl sulfoxide (DMSO), dilute aqueous TFA, neat TFA, or mixtures of both HFIP and DMSO. Aliquots of these stock solutions are usually diluted directly into an excess volume of aqueous solution. However, there are many potential pitfalls with these approaches, as different solvents can promote different structures and aggregation states, as well as different amyloid fibril sizes and dimensions.[23] Additionally, the variant solution compositions may have different behavior when diluted into aqueous buffered solution. For example, even though formic acid is very effective in dissolving Aβ in amyloid plaques from AD brain,[25] it promotes dimer formation of Aβ(1–42)[26] and also reacts with the Ser-βCH_2OH side chains

[23] C.-L. Shen, M. C. Fitzgerald, and R. M. Murphy, *Biophys. J.* **67,** 1238 (1994).

[24] P. C. Southwick, S. K. Yamagata, C. L. Echols, G. J. Higson, S. A. Neynaber, R. E. Parson, and W. A. Munroe, *J. Neurochem.* **66,** 259 (1996).

[25] C. L. Masters, G. Simms, N. A. Weinman, G. Multhaup, B. L. McDonald, and K. Beyreuther, *Proc. Natl. Acad. Sci. U.S.A.* **82,** 4245 (1985).

[26] A. E. Roher, M. O. Chaney, Y.-M. Kuo, S. D. Webster, W. B. Stine, L. J. Haverkamp, A. S. Woods, R. J. Cotter, J. M. Tuohy, G. A. Krafft, B. S. Bonnell, and M. R. Emmerling, *J. Biol. Chem.* **271,** 20631 (1996).

to produce relatively stable formate esters (βCH$_2$OOCH).[27,28] In addition, the toxicity of Aβ to mixed cortical neuronal cultures was greatest, intermediate, and weak for stock solutions prepared in aqueous acetonitrile, water, and DMSO, respectively.[29] More recent results showed that Aβ(1–40) stock solutions in DMSO produced fibrils[3] (H. Levine, personal communication) and cortical culture cells showed 100-fold greater toxicity to glutamate with Aβ stock solutions in DMSO rather than stock solutions in water.[30] Because the kinetics of Aβ assembly is slower in DMSO than water,[9] these problems are clearly related to the different structures and aggregation states of the Aβ peptide.

The organic solvent should completely disaggregate the Aβ and promote monomeric, unstructured peptide. The solutions should be clear, with residual aggregated peptide removed by filtration or centrifugation, along with complete removal of the organic solvent before dissolution in aqueous solution. In this way, the dry, unstructured, and monomeric Aβ will fold into its native-like conformation in water without potential interferences by trace organic cosolvents.

Observations from our laboratory established that dissolution with neat TFA efficiently breaks up any preexisting aggregates (seeds) and ensures that the dry Aβ peptide is monomeric and unstructured.[22] An outline of this protocol is shown in Fig. 4. The TFA pretreatment method is relatively simple and involves dissolving the Aβ(1–40) and Aβ(1–42) peptides (after HPLC purification) in distilled TFA. TFA is added to the peptide in a glass tube at an approximate 1:1 ratio (mg:ml), followed by sonication at 25° with a Sonicor instrument (Model SC-101TH) for about 15 min, during which additional portions (1 ml) of TFA are added until the peptide dissolves completely. Complete solubilization is judged by the absence of precipitate after centrifugation. It is critically important that all the Aβ peptide dissolves in the TFA, as any remaining precipitated peptide can later act as a seed for initiating amyloid formation.[8] The TFA is removed with dry N_2 gas, after which the peptide coats the glass walls of the tube as a waxy layer. Trace amounts of TFA can be removed by the addition of distilled HFIP, sonication, and removal of the HFIP with N_2 gas. The latter is repeated three times and the trace amounts of HFIP are removed under vacuum (0.5 mm Hg, 2 hr). After the removal of TFA, a concentrated stock solution prepared in HFIP is stable for several months with no pep-

[27] W. E. Klunk and J. W. Pettegrew, *J. Neurochem.* **54,** 2050 (1990).

[28] R. Orlando, P. T. M. Kenny, and M. G. Zagorski, *Biochem. Biophys. Res. Commun.* **184,** 686 (1992).

[29] J. Busciglio, A. Lorenzo, and B. A. Yankner, *Neurobiol. Aging* **13,** 609 (1992).

[30] M. P. Mattson, B. Cheng, D. Davis, K. Bryant, I. Lieberburg, and R. E. Rydel, *J. Neurosci. Res.* **12,** 376 (1992).

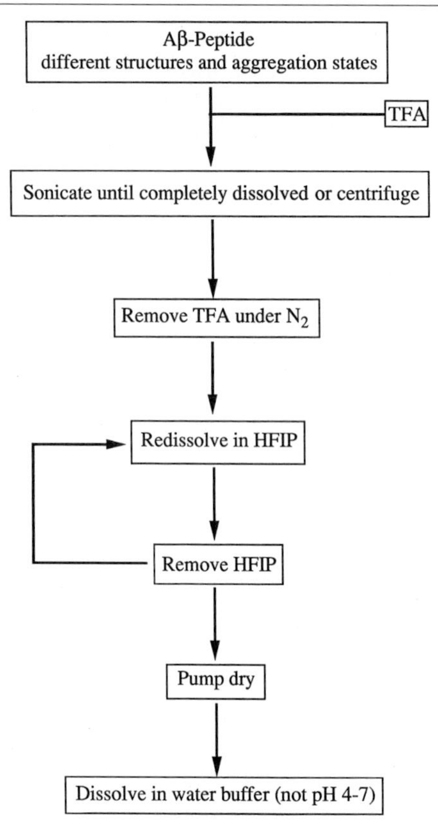

FIG. 4. Protocol for disaggregation of Aβ samples using neat TFA solvent.[22] Trace amounts of TFA are removed by redissolution in HFIP, followed by evaporation under N_2 and vacuum. The solutions should be filtered or centrifuged to remove residual sediment that could later seed Aβ aggregation.[8] The final dry peptide will be a monomeric, α-helical structure[6] and readily soluble in water. The solvents TFA and HFIP are sensitive to light and oxygen, plus commercial sources frequently contain metal impurities that can promote aggregation of the Aβ peptide. For these reasons, it is best to first distill the TFA (bp 72°) and HFIP (bp 59°) under an inert atmosphere of nitrogen and followed by storage in an opaque bottle at 5°.

tide precipitation. The TFA-pretreated Aβ(1–42) is still capable of producing β-amyloid-like fibrils that are neurotoxic. Studies of TFA-pretreated Aβ(1–42) showed binding to plasma lipids and eventually with aging become neurotoxic *in vivo* (G. Cole, personal communication). However, it should be kept in mind that because the TFA-pretreated Aβ(1–42) is more soluble, it will take longer to aggregate and become toxic.

Control experiments showed that TFA is more effective than either DMSO or HFIP in dissolving aged and preaggregated Aβ(1–42).[22] In fact, despite repeated sonication, HFIP was incapable of completely dissolving preaggregated Aβ(1–42). Although previous hydrodynamic measurements showed that Aβ(1–40) and Aβ(1–42) are predominantly monomeric in DMSO solvent,[11] previous work showed that analysis of the Aβ(1–42) peptide by ^{1}H NMR in DMSO-d_6 showed broader line widths than those seen for an identical sample that was first predissolved in TFA solution.[22] The solvents DMSO and HFIP may be suitable for dissolving the more soluble Aβ(1–40) peptide, but they are less effective in breaking up preaggregated Aβ(1–42). Another drawback with DMSO relates to its high freezing point (19°) and boiling point (180°), which make the solvent difficult to remove from the peptide. The solvent TFA is volatile (bp 72°) and thus removed easily from the peptide.

Salts and Trace Metals

The general effect of the presence of salts[10,11] and metal ions such as aluminum, iron, and zinc is to accelerate β aggregation.[31] Divalent cations such as Zn^{2+} presumably bind to the His residues of the Aβ. The best method of dealing with these problems is to not introduce any metal ions into the solution. It is best to avoid contact of the sample with so-called stainless-steel spatulas, so only Teflon-coated spatulas should be used. All glassware and plasticware should be treated with 1 m*M* ethylenediaminetetraacetic acid (EDTA) prior to use, followed by rinsing thoroughly with deionized water. In addition, it is useful to pretreat and filter all buffer solutions through a column containing the metal-chelating agent Chelex. Because of its lack of specificity, Chelex is very effective in removing metal ions at neutral pH.

Time-Dependent Variations

One of the biggest problems related to the Aβ is the time-dependent aggregation in aqueous solution, which is particularly serious for experiments requiring higher (m*M*) concentrations. During the aggregation, the more soluble random coil and α-helix structures rearrange into the less soluble oligomeric β sheet (random coil → β sheet and α helix → β sheet), which eventually precipitates as an amyloid-like deposit. This dilemma has clearly contributed to the inconsistencies with studies of the Aβ(1–40) and Aβ(1–42) peptides in water solution.

The best approach to avoid these complications is to always prepare fresh Aβ solutions, i.e., after the peptide is pure and completely disaggregated according to the procedures described earlier. Because stock solutions

[31] R. Balakrishnan, R. Parthasarathy, and E. Sulkowski, *J. Peptide Res.* **51,** 91 (1998).

in water exhibit Aβ concentration-dependent changes, more dilute solutions will be less likely to show the undesirable time-dependent variations. Therefore, stock solutions should be kept as dilute as possible, preferably below 50 μ*M*.

Other important factors are the temperature and pH. Work from our laboratory established that structure-dependent variations are considerably slower at 0–5°. Thus, Aβ(1–40) and Aβ(1–42) stock solutions at 100–200 μ*M* concentrations in water at pH 7.2 are relatively stable for several days at 0°, with no precipitation or other conformational changes noted by NMR or circular dichroism (CD) spectroscopy. It is also critical that the pH never fall within the 4–7 range,[6,12,13,32] which is a region where β aggregation proceeds very rapidly and reconversion back into the more soluble extended chain structures may not be possible. The pH of stock solutions should be checked periodically to ensure that the buffer is keeping the pH above 7. Buffer concentrations at approximately 10–50 m*M* are usually sufficient to maintain physiological pH 7.2–7.4 and still not increase the ionic strength to levels that promote β aggregation.

Racemization

Despite addressing issues related to the purity and starting aggregation states using the TFA pretreatment method, we were unable to completely reproduce our previous nicotine inhibition to Aβ(1–42) aggregation.[33] Subsequent results with another Aβ(1–42) batch showed that nicotine still slowed down the aggregation, but unlike the past results, complete inhibition was not observed. In light of previous work showing that Aβ peptides with D-aspartate (D-Asp, rather than the natural L-Asp) displayed enhanced aggregational properties,[17] we decided to determine whether our inability to reproduce these results may be due to partial recemization or trace amounts of D-amino acids in the different synthetic Aβ peptide batches. It is well known that different coupling reagents employed during peptide synthesis can alter the extent of racemization[34,35] and that the Asp residue is usually the most susceptible. For example, the more expensive reagent *N*-[(dimethylamino)-1*H*-1,2,3-triazolo[4,5-*b*]pyridino-1-ylmethylene]-*N*-methylmethanaminimum hexafluorophosphate *N*-oxide (HATU) has a maximum D/L isomer ratio of 5%, whereas less expensive reagents such as *N*-[(1*H*-benzotriazol-1-yl)(dimethylamino)methylene]-*N*-methyl-

[32] S. J. Wood, B. Maleeff, T. Hart, and R. Wetzel, *J. Mol. Biol.* **256,** 870 (1996).

[33] A. R. Salomon, K. J. Marcinowski, R. P. Friedland, and M. G. Zagorski, *Biochemistry* **35,** 13568 (1996).

[34] G. B. Fields, *Methods Enzymol.* **289** (1997).

[35] PerSeptive Biosystems Catalog (1998).

TABLE I
Aβ PEPTIDES AND PERCENTAGE D-Asp CONTENT

Aβ Peptide sequence	Source	D-Asp/total Asp (%)
1–28	Synthesized by standard t-BOC methods[a]	1.24
1–40	AnaSpec, Lot 3763	0.70
1–40	BACHEM, Lot ZM365	0.97
1–40	BACHEM, Lot ZN571	1.60
1–40	QCB, Lot 13	1.36
1–42	AnaSpec, Lot 3764	2.92
1–42	AnaSpec, Lot 2031	2.16
1–42	Synthesized by standard t-BOC methods[a,b]	11.9
1–42	U.S. Peptide, Lot 12	4.03
1–42	BACHEM, Lot 508780	3.44

[a] From C. J. Barrow *et al., J. Mol. Biol.* **225,** 1075 (1992).
[b] This batch was used in previous nicotine-inhibition studies.[33]

methanaminimum hexafluorophosphate *N*-oxide (HBTU) can have D/L isomer ratios up to 15%.

To assess the amount of D-Asp, procedures similar to those of Aswad and Roher were followed with minor modifications in the HPLC analysis.[36–38] Approximately 1.0 mg of purified Aβ peptide was hydrolyzed at 100° in 5 ml of 6 *M* HCl, and the chiral fluorogen derivatives were prepared according to standard procedures. Before analysis of the Aβ samples, the HPLC column was calibrated using a series of standard DL-Asp mixtures to obtain a correlation plot of the D-Asp signal intensity ratio versus concentration. The correlation plots usually had good linear coefficients over the 2–70% D-Asp concentration range. The percentage of D-Asp in each Aβ batch was read from the correlation plot obtained on the same day. A Waters gradient HPLC system equipped with a 616 pump (50 I pump head), a 600S controller and a 474 scanning fluorescence detector connected with a linear chart recorder was used. D-Asp contents for different Aβ batches are summarized in Table 1.

Significantly, the Aβ(1–42) batch with a relatively high D-Asp content (11.9%) was used in the previous studies in which nicotine completely blocked the Aβ(1–42) precipitation.[33] These results are intriguing and indi-

[36] D. W. Aswad, *Anal. Biochem.* **137,** 405 (1984).
[37] A. E. Roher, J. D. Lowenson, S. Clarke, C. Wolkow, R. Wang, R. J. Cotter, I. M. Reardon, H. A. Zürcher-Neely, R. L. Heinrikson, M. J. Ball, and B. D. Greenberg, *J. Biol. Chem.* **268,** 3072 (1993).
[38] A. E. Roher, *Methods Enzymol.,* in press.

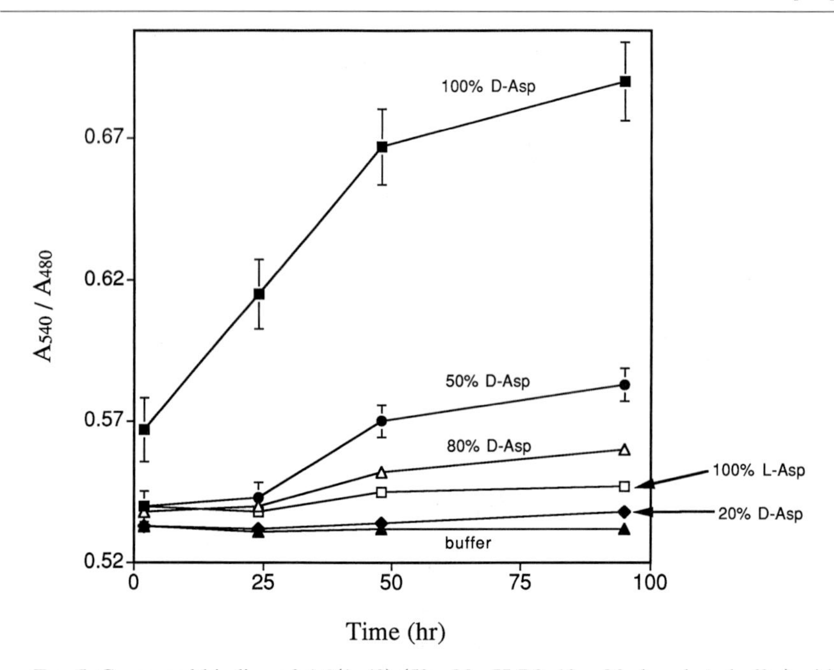

FIG. 5. Congo red binding of Aβ(1–42) (50 μ*M*, pH 7.2, 10 m*M* phosphate buffer) with peptides containing α-L-Asp or α-D-Asp at residues 1, 7, and 23. The samples were mixed at time 0 hr and incubated at 37°. Aliquots were removed and analyzed periodically according to well-established procedures.[32] For mixed samples, the percentage (by weight) of the 1,7,23-α-D-Asp-Aβ(1–42) is provided. The percentage α-D-Asp in the 100% L-Asp sample was less than 1%.

cate that there may be some stereospecificity of the nicotine binding and inhibition processes with the Aβ peptide. Although studies with an all L- and D-Aβ(1–42) peptide demonstrated identical aggregation rates and neurotoxicities, mixtures of the D and L peptides were not tested.[39] As it happens, the percentage of D-amino acid residues, particularly D-Asp, is higher in aging tissues and in the amyloid plaque cores of patients with AD.[37] We are currently examining mixtures of D- and L-Aβ(1–42) peptides for aggregation rates and secondary structures, which required us to synthesize an Aβ(1–42) peptide containing α-D-Asp at residues 1, 7, 23, and a brief report of these results is shown here.

Preliminary studies monitoring the progress of β aggregation with a standard Congo red binding assay[32] showed that the 1,7,23-α-D-Asp-Aβ(1–42) peptide aggregates faster than the natural, L-Asp-Aβ(1–42) (Fig. 5).

[39] D. H. Cribbs, C. J. Pike, S. L. Weinstein, P. Velazquez, and C. W. Cotman, *J. Biol. Chem.* **272,** 7431 (1997).

Other experiments monitoring time-dependent random coil → β-sheet transitions by CD also established that the 1,7,23-α-D-Asp-Aβ(1–42) has a greater tendency to form β-sheet structure (data not shown). The increased β aggregation seen with the 1,7,23-α-D-Asp-Aβ(1–42) is consistent with previous studies in which a 1,7,23-α-D-Asp-Aβ(1–35) peptide showed greater aggregation than the α-L-Asp-Aβ(1–35).[17] As also shown in Fig. 5, mixtures of the 1,7,23-α-D-Asp-Aβ(1–42) and L-Asp-Aβ(1–42) peptides have noticeably different aggregation rates, with the 20 : 80 mixture showing the lowest β-aggregation rate. Because differences in the observed β-aggregation rates between the all L-Asp and 20% D-Asp peptides are very similar, it is likely that commercial Aβ batches with D/L ratios less than 4% do not cause significant inconsistencies. However, differences in amyloid fibril-inhibition assays may be seen if an inhibitor is optically active. Further experiments using mixtures of 1–4% 1,7,23-α-D-Asp-Aβ(1–42), with and without chiral inhibitors, are currently underway and a manuscript describing these results is forthcoming.

Conclusion

The immense difficulties and unexpected questions presented by hundreds of amyloid toxicity studies beg for careful studies of Aβ assembly processes, evaluating kinetics and structures of different aggregation pathways that convert the soluble monomer into the higher order Aβ assemblies that harbor neurotoxic properties. A large array of sophisticated instrumental techniques, including analytical ultracentrifugation, atomic force microscopy, solid-state and solution NMR, and fluorescence tracer techniques, will help elucidate the bioactive Aβ structure(s), whereas specific cellular and biochemical techniques will elucidate the specific steps involved in the aftermath of Aβ cellular interactions. Nonetheless, all these potentially useful high technology methods are ineffective unless proper care regarding Aβ sample preparation is taken into account.

We have briefly summarized potential problems involved with Aβ sample preparation and have also described methods to deal with these issues. The most critical complications are related to the starting structures and aggregation states of the Aβ. Our suggestions include the following.

1. Ensure that the Aβ peptide batch is pure by MS and NMR. The purity should be at least 90% as determined by NMR and particular care must be exercised with HPLC analysis.
2. Use an efficient protocol such as pretreatment with TFA to dissaggregate the Aβ, particularly one that generates monomeric and unstructured peptide. Storage of the pretreated peptide in HFIP (not

DMSO) is recommended at 0–5° for brief 1- or 2-week intervals. Complete removal of the organic solvent should be done prior to dissolution in aqueous solution.

3. All solvents should be of the highest purity level and virtually free of all metal contaminants. If possible, all glassware should be rinsed with EDTA solution and deionized water before use.
4. Analysis of the D-Asp content of the Aβ may not be required. Instead, before purchasing the Aβ from a commercial source, request information about the coupling reagents that were employed during the peptide synthesis. If more reliable reagents such as HATU were used, then the possibility of racemization is not serious.

With a greater appreciation of the purity content and consistency with handling the Aβ, the end result is an enormous savings in time and money, which in turn may hasten the development of a useful therapeutic inhibitor of Aβ-amyloid formation for the treatment of AD patients.

Acknowledgments

Supported in part by grants from the National Institutes of Health (AG-08992-06 and AG-14363-01), Philip Morris, Inc., the Smokeless Tobacco Research Council, and a Faculty Scholars Award from the Alzheimer's Association to M.G.Z (FSA-94-040). The 600-MHz NMR spectrometer was purchased with funds provided by the National Science Foundation, the National Institutes of Health, and the state of Ohio. We also thank Witold Surewicz (CWRU), Nanda Menon (University of Georgia), Beth Neuhaus (University of Georgia), and Mark Smith (CWRU) for useful comments, Shu Guang Chen (CWRU) for analysis of the Aβ by MS, Harry LeVine (Parke-Davis, Inc.) and Gregory Cole (UCSD) for providing preliminary data about the Aβ fibril deposition in DMSO solution, and finally Russel Rydel (Athena Neurosciences, Inc.) for suggesting the phrase "the peptide from hell."

[14] *In Vitro* Immunoglobulin Light Chain Fibrillogenesis

By JONATHAN WALL, CHARLES L. MURPHY, and ALAN SOLOMON

Amyloid: An Introduction

Amyloidosis is a disease state resulting from the aggregation and deposition of normally soluble, innocuous proteins as insoluble fibrils. At present, more than 17 precursor proteins have been associated with amyloid syndromes. Irrespective of the type of precursor protein, however, amyloid deposits *in vivo* share a remarkable number of common structural and

0076-6879/99 $30.00

compositional similarities.[1] The precise mechanism of amyloidogenesis *in vivo* remains enigmatic. When resolved, the process will undoubtedly involve a number of accessory macromolecules in addition to the amyloid precursor protein itself, as well as the occurrence of favorable physiological factors. The assistance of chaperones or accessory molecules during amyloid formation *in vivo* is inferred by the ubiquitous presence of proteins such as P-component, glycosaminoglycans, and apolipoproteins.[2,3] How these accessory molecules initiate, perpetuate, or stabilize fibril formation, however, remains unclear. What is certain is that they are not mandatory requirements for fibril formation, implying that the amyloidogenic potential of a precursor protein resides in part within the protein itself.[4]

All amyloid deposits are clinically identified according to the following three criteria: (1) The appearance of unbranching, linear fibrils of approximately 7–10 mm in diameter and of variable length when viewed by electron microscopy; (2) binding of Congo red, resulting in a hyperchromic shift in the absorbance spectrum and blue-green birefringence when viewed microscopically using cross-polarizing filters[5]; and (3) green fluorescence on addition of the benzothiazole dye, thioflavin T (ThT), using excitation and emission wavelengths of 450 and ~485 nm respectively.[6,7] These characteristics are also used for the experimental investigation of *in vitro* fibrillogenesis.[8,9]

Immunoglobulin Light Chain Aggregation

Immunoglobulin light chain (IgLC), primary amyloidosis (AL) is an invariably fatal, systemic syndrome resulting from the overproduction and fibrillogenesis of light chain proteins, generally resulting in multiple organ dysfunction.[10,11] The major form IgLC protein associated with AL amyloid is the N-terminal 110 amino acids comprising the light chain variable region,

[1] J. D. Sipe, *Crit. Rev. Clin. Lab. Sci.* **31,** 325 (1994).

[2] M. B. Pepys, D. R. Booth, W. L. Hutchinson, J. R. Gallimore, P. M. Collins, and E. Hohenester, *Amyloid Int. J. Exp. Clin. Invest.* **4,** 274 (1997).

[3] J. H. Magnus and T. Stenstad, *Amyloid Int. J. Exp. Clin. Invest.* **4,** 121 (1997).

[4] M. Schiffer, *Am. J. Pathol.* **148,** 1339 (1996).

[5] H. Puchtler, F. Sweat, and M. Levine, *J. Histochem. Cytochem.* **10,** 355 (1962).

[6] H. Naiki, K. Higuchi, M. Hosokawa, and T. Takeda, *Anal. Biochem.* **177,** 244 (1989).

[7] H. Levine, *Amyloid Int. J. Exp. Clin. Invest.* **2,** 1 (1995).

[8] H. Naiki, K. Higuchi, K. Nakakuki, and T. Takeda, *Lab. Invest.* **65**(1), 104 (1991).

[9] L. R. Helms and R. Wetzel, *J. Mol. Biol.* **257,** 77 (1996).

[10] A. Solomon and D. T. Weiss, *Amyloid Int. J. Exp. Clin. Invest.* **2,** 269 (1995).

[11] C. A. Vaamonde, G. O. Perez, and V. Pardo, *in* "Diseases of the Kidney" (R. W. Schrier and C. W. Gottschalk, eds.), p. 2189. Little, Brown and Co., Boston, 1992.

V_L. In addition, nonfibrillar, "amorphous" aggregates may also form in patients with monoclonal gammopathies. Amorphous aggregates of IgLCs often deposit in the kidney and exhibit neither microscopic nor spectroscopic properties of amyloid fibrils. Two pathologies are distinguished based on the distribution of these kidney deposits: Light chain deposition disease characterized by glomerular precipitates and cast nephropathy associated with renal tubular deposits. Furthermore, crystalline deposits have also been observed in the renal tubules and, rarely, within the urine itself.[12] No clinical evidence exists to suggest that all three types of aggregate can be formed from the same protein in an individual patient. However, they appear not to be mutually exclusive, as both casts and amyloid have been documented in a single patient. The same protein was implicated in both aggregates by N-terminal, amino acid sequence analysis.[13,14] We suggest that the three aggregate types represent competitive pathways down which a susceptible protein may travel. Ultimately, the resultant aggregation will depend on kinetic factors, favorable associations with host macromolecules, and the microenvironment of the proteins, as well as the inherent proclivity of the protein itself, related to its primary structure.[10]

This article describes techniques specifically for managing and monitoring the production of synthetic fibrils derived from IgLC peptides V_L, proteins, and whole Bence Jones proteins (BJps). These *in vitro* techniques have been developed in order to minimize complications arising from amorphous aggregation, thus allowing the investigation of the underlying biochemical and physical factors governing light chain fibrillogenesis.

Agitation-Stimulated Fibrillogenesis

This approach has been applied successfully to the fibrillogenesis of peptides, recombinant V_L fragments, and whole BJps. Initial experiments used a 24-mer peptide as the fibril precursor, originally isolated as fibrils from a tryptic digest of the κ1 Bence Jones protein isolated from patient WAT. The peptide constitutes the N-terminal 24 amino acid residues in framework region one (FR I), which was synthesized by standard F-moc procedures. The sequence of two peptides used extensively as a fibril precursor in the agitation-stimulated fibrillogenesis (ASF) system follows:

[12] A. Solomon, D. T. Weiss, and A. A. Kattine, *N. Engl. J. Med.* **324,** 1845 (1991).
[13] M. B. Stokes, J. Jagirdar, O. Burchstin, S. Kornacki, A. Kumar, and G. Gallo, *Mod. Pathol.* **10,** 1059 (1997).
[14] B. Kaplan, R. Vidal, A. Kumar, J. Ghiso, B. Frangione, and G. Gallo, *Clin. Exp. Immunol.* **110,** 472 (1997).

	1 10 20
WAT(1–24)	DIQMTQSPSSLSASVGDRVTITCR-
WAT(13–24)	ASVGDRVTITCR-

The 12-mer designated WAT(13–24) has also been tested thoroughly and found to form fibrillar aggregates readily using this technique (Fig. 1).

The standard protocol employed in our laboratory for the production of fibrils is as follows. A 1-mg/ml^{-1} solution of protein/peptide (used interchangeably hereafter) is prepared in phosphate-buffered saline (PBS) using high-performance liquid chromatography (HPLC)-grade water and adjusted to pH 7.5. The solution is passed through a 0.2-μm pore-sized filter to remove preformed aggregates. A 1-ml volume of this solution is placed in a 10-ml volume glass test tube and placed in a thermostatted orbital shaker (Queue Orbital Shaker, Parkersburg, WV). The protein solution is agitated at 37° and 225 revolutions per minute. Depending on the type of protein substrate used, solution turbidity can be detected within 1–10 days (Fig. 2). Peptides WAT(1–24) and WAT(13–24) generally form precipitates within 24 hr, V_L fragments may require up to 5 days, and whole BJps aggregate within 3–10 days (Fig. 2). There are, however, exceptions to these generalities, as certain V_L proteins may require extensive shaking; 21 days is the longest time observed to date.

ASF has been used to monitor both the kinetics and the equilibrium concentration of fibrillar material in solution. The kinetics may be accessed by routine sampling of the mixture over the course of the experiment. Samples are removed from the shaker and a 200-μl aliquot is placed in

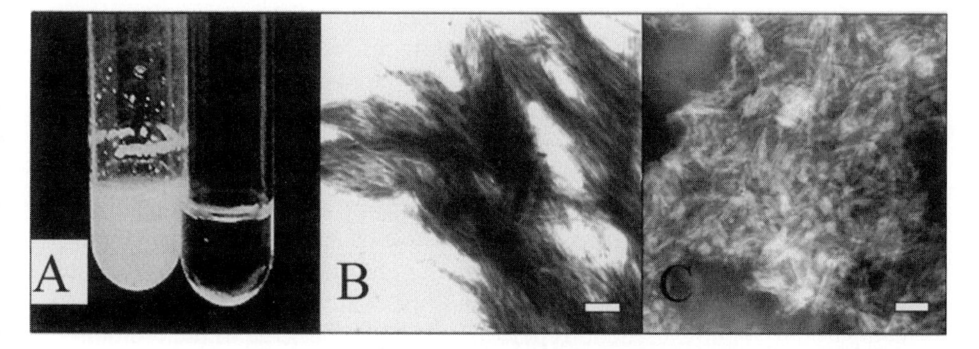

FIG. 1. Production of amyloid fibrils by ASF in an orbital shaker. (A) Test tubes showing the formation of fibrils by ASF using 1 mg/ml of V_L (left) and control protein (right). (B) Negatively stained, electron microscopy of WAT(1–24) fibrils formed by ASF. Bar: 120 nm. (C) Negatively stained, electron microscopy of V_L fibrils formed by ASF. Bar: 90 nm.

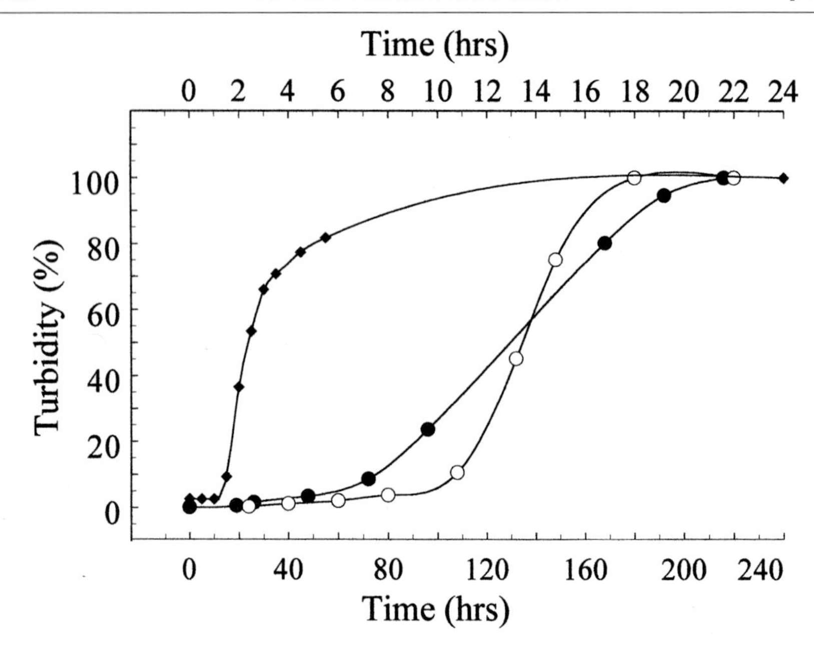

FIG. 2. Agitation-stimulated fibrillogenesis. Changes in the turbidity (A_{405}) of WAT(1–24) (◆, upper abscissa), a V_L protein (●, lower abscissa), and a BJp (○, lower abscissa) all at 1 mg/ml^{-1} in PBS, pH 7.5, during ASF.

a minicuvette (250 μl maximum volume; Shimadzu, Norcross, GA) and interrogated using a standard bench spectrophotometer (Shimadzu UV160U). The turbidity (optical density) is assessed by measuring the absorbance at 400 nm (Fig. 2). Other wavelengths may be chosen, but the same wavelength should be used consistently throughout the course of the reaction due to the wavelength dependence of scattered light intensity.[15] Generally, fibrillogenesis occurs within 10 days and, considering that a small volume of sample will be lost at every measurement point, the experimenter must determine how often to screen the samples in order to minimize changes in the sample volume over the course of the experiment. A plot of the measured absorbance versus time yields kinetic information for further analysis.[16] The plateau phase at the end point of fibril formation provides only a qualitative value of the fibril content that should only be compared to identical reactions performed in parallel. Due to the relationship between optical density and aggregate structure, this measurement

[15] I. D. Campbell and R. A. Dwek, "Biological Spectroscopy." Benjamin-Cummings, Redwood City, CA, 1984.

[16] M. F. Bishop and F. A. Ferrone, *Biophys. J.* **46,** 631 (1984).

cannot be used as a direct measure of fibril content. Furthermore, other methods, such as ThT fluorescence and electron microscopy, should be employed to confirm that the aggregates are fibrillar.

In addition, quantification of the amount of precursor protein incorporated into fibrils can be achieved indirectly by measuring the free protein concentration and subtracting from the initial. This is accomplished simply by sedimentating fibrillar material at 17,000*g* and 25° for 30 min. Remaining polymers are removed from the resulting supernatant by filtration through a 0.2-μm pore-size, nonadsorbing filter (e.g., Acrodisc, Pall Corp., East Hills, NY). Soluble protein can then be quantified using standard techniques, e.g., HPLC, ε_{280} or colorimetric/fluorescent assays, as appropriate.[17] Centrifugation at 100,000*g* and 25° for 30 min did not significantly alter the concentration of soluble protein in the supernatant when compared to lower speed separations.

Based on the premise that perturbation of a solution of WAT(1–24) peptide is required to induce fibrillogenesis, several other methods have been assessed. Although rigorous in certain cases, and applicable only to peptide fibrillogenesis, they warrant a brief description. Protein concentrations tested using these techniques range between 0.4 and 2 mg/ml suspended in PBS, pH 7.5. The following procedures result in the confirmed fibril formation of WAT peptides within 24 hr: (1) sonication of the solution, delivering 3 W for a 4-sec period every 15 min; (2) continuous stirring of the solution using a magnetic stir bar (kinetics are dependent on the rate of stirring); and (3) manual agitation by repeated pipetting using a p1000 Gilson (Rainin, Woburn, MA). Somewhat intuitively, differences in fibril morphology were observed by electron microscopy; sonication yielding shorter dispersed material and the latter techniques generally result in longer, bundled fibril aggregates.

ASF has proved to be an ideal tool for monitoring fibril formation in systems that require many days to attain equilibrium, such as with whole BJPs; however, the sampling technique can be cumbersome and inconvenient. ASF is well suited for the production of milligram quantities of fibrillar material for use in seeding experiments and structural analyses. Because this method does not rely on extremes of pH or the presence of chaotropic ions to induce fibrillogenesis, it may be considered a better approximation of the physiological paradigm.

Monitoring ASF with in Situ Thioflavin T

Ideally, an *in vitro* fibrillogenesis model should be amenable to rapid and reproducible production of data for subsequent analysis, adhere to

[17] C. M. Stoscheck, *Methods Enzymol.* **182,** 50 (1990).

physiological constraints wherever possible, and be applicable to a wide range of precursor proteins. Methods have been developed from ASF that are capable of monitoring the fibrillogenesis of peptides, IgLC V_L fragments, and BJps using ThT as a reporter of fibril production. This method differs from previously employed ThT-based assays in that the dye is added to the precursor solution *ab initio.* This permits changes in fluorescence intensity at fibril-bound wavelengths to be measured in real time, thus circumventing the need for continuous sampling of the reaction mixture (Fig. 3A). The spectroscopic properties of ThT and its application for detecting and estimating fibril concentration are dealt with in detail elsewhere in this volume.[17a] Suffice to say is that in the presence of amyloid fibrils, ThT fluorescence is enhanced greatly using an excitation wavelength of 450 nm and emission at 490 nm. Furthermore, unbound ThT is essentially nonfluorescent at these wavelengths, a feature that permits its inclusion in the reaction solution without interfering with the measured fluorescence signal. Care should be taken to exclude sulfated macromolecules and other biological polymers, as well as lipid membranes from the solution, as they compete for ThT binding.[18] In addition, nucleic acids and solvents, such as chloroform and dichloromethane, should be excluded as their presence results in ThT fluorescent properties equivalent to the fibril-bound dye.

Before undertaking a systematic investigation of protein fibrillogenesis using this method, important preliminary data must be gathered. Inherent in the application of this technique is the assumption that the inclusion of ThT in the sample mixture will not influence the kinetic or thermodynamic properties of the reaction. This assumption can be validated by performing identical, parallel fibrillogenesis experiments: one reaction followed using *in situ* ThT fluorescence and the second in the absence of ThT, monitored by changes in solution turbidity (Fig. 3B).

Stock solutions of ThT are prepared in HPLC-grade water to a final concentration of 1 m*M*; this solution is then cleared by filtration through a 0.2-μm pore-sized filter. The final concentration may be determined using the Beer–Lambert equation; add 10 μl of stock solution to 0.99 ml of ethanol and measure the absorbance at 416 nm, concentration is calculated using a molar extinction coefficient, ε_{416}, of 26,620 (Sigma product information). The concentration should be determined routinely to compensate for changes due to evaporation or loss of material in subsequent filtrations. The stock solution can be kept for over 12 months without loss of amyloid-bound fluorescence intensity.

[17a] H. LeVine, *Methods Enzymol.* **309** [18] (1999) (this volume).
[18] M. T. Elghetany and A. Saleem, *Stain Technol.* **63,** 201 (1988).

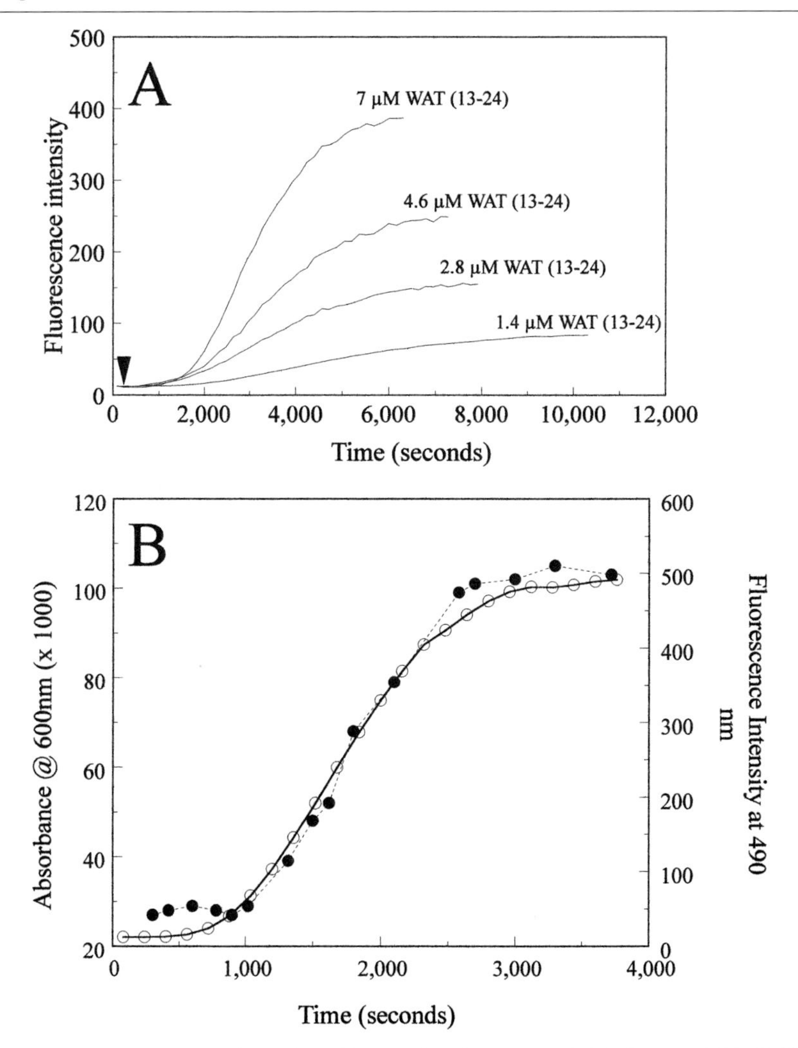

FIG. 3. Analysis of WAT(13–24) fibrillogenesis using *in situ* ThT. (A) *In situ* ThT fibrillogenesis of increasing concentrations of WAT(13–24). Arrow indicates addition of protein. (B) Comparison of WAT(13–24) fibril formation using *in situ* ThT (○, right ordinate) and turbidity (●, left ordinate). Fluorescence measurements were taken using an SLM Aminco Bowman Series II as described.

ASF is monitored using *in situ* ThT using an SLM Aminco-Bowman Series II spectrofluorometer with thermostatted cuvette housing and stirring capabilities. Excitation and emission wavelengths are 450 and 490 nm, respectively, with corresponding slit widths of 4 and 8 nm. Data are collected at 25°, over 8 hr, in "time trace" mode and a frequency of one data point per second. The fluorescence signal is corrected internally for lamp drift and pulsing by division by a reference PMT signal. Precursor proteins are prepared as described in the ASF technique at 1 mg/ml^{-1} in PBS, pH 7.5. Immediately prior to the start of the reaction, stocks of precursor and ThT are diluted into filtered PBS, pH 7.5, in a 4-ml volume fluorimetric cuvette. The solution is mixed by inversion and analyzed. Agitation is provided by inclusion of a microstir bar (7 × 2 mm). The importance of the mixing speed is discussed later. The precise micromolar concentrations of protein and ThT vary depending on the nature of the protein to be studied, generally 1–10 μM protein and 2–20 μM ThT are employed. A twofold molar excess of ThT is used to ensure saturation of the binding sites. The precise quantities should be determined by the experimenter. Care should be taken not to use much greater than saturating concentrations of the dye as it significantly self-quenches at these levels and sensitivity will be compromised.

The inclusion of ThT in the reaction mixture has several advantages over previous sampling methodologies. Most notable is the enhanced fluorescence intensity of the sample with respect to the addition of ThT to an equivalent sample, postfibril formation. This phenomenon consistently results in a two- to eightfold increase in the fluorescence intensity of a synthetic fibril preparation at equilibrium (end point), formed in the presence of ThT, relative to a parallel solution into which ThT is titrated to saturation postfibril formation (Fig. 4). Preliminary data suggest that this results from the inaccessibility of many ThT-binding sites at the reaction end point, perhaps due to the organization of ThT-positive filaments into fibers and higher order macrocomplexes.

Agitation-Independent Fibrillogenesis of Recombinant V_L Proteins

An attempt to construct a unified theory to rationalize the fibrillogenic potential of IgLCs and their fragments has led to an evaluation of the thermodynamic stability of proteins and its correlation with the propensity to form aggregates. This has initiated a search for destabilizing amino acid substitutions that may confer to the protein, an inherent propensity to aggregate.[19,20] Based on this postulate, a rapid reproducible method of fibrillogenesis has been developed and applied, as in previous systems, to

[19] R. Wetzel, *Adv. Prot. Chem.* **50,** 183 (1997).

[20] M. R. Hurle, L. R. Helms, L. Li, W. Chan, and R. Wetzel, *Proc. Natl. Acad. Sci. U.S.A.* **91,** 5446 (1994).

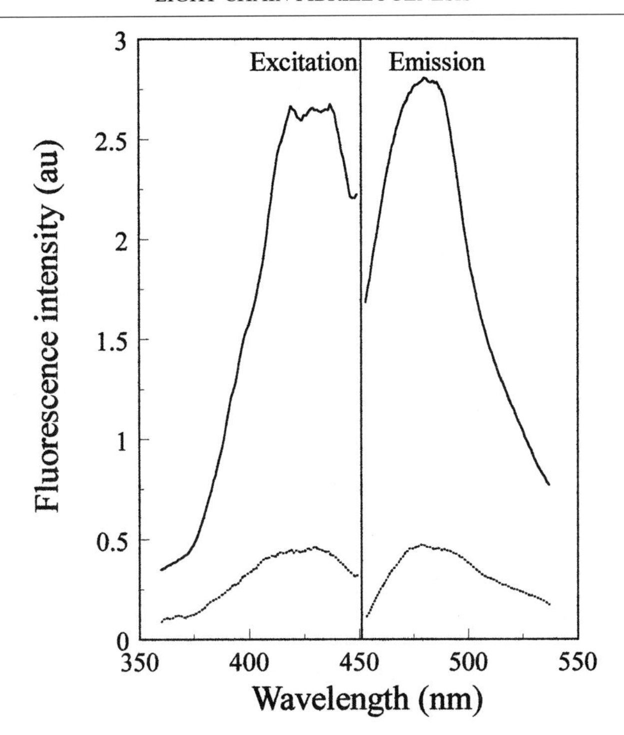

FIG. 4. Comparison of ThT fluorescence sensitivity. Excitation and emission spectra of ThT when added *ab initio* to the fibrillogenesis of WAT(13–24) (solid) and titrated to saturation postfibril formation (dotted).

recombinant V_L fragments.[20,21] The principle underlying this fibrillogenesis assay is that temperature-induced destabilization of the precursor protein results in fibril formation. The assay is performed essentially as that described for agitation-induced fibrillogenesis in the fluorimeter. The major difference is that agitation is no longer required. A stock solution of V_L protein at 1 mg/ml^{-1} (approximately 90 μM) is filtered through a 0.2-μm pore-sized filter immediately prior to polymerization. The A_{280} is measured and the protein concentration is reconfirmed by application of the Beer–Lambert equation using the appropriate molar extinction coefficient. The fibrillogenesis reaction mixture consists of 3 μM V_L protein and 6 μM filtered ThT. This is prepared by dilution of the protein stock solution into PBS, pH 7.5, at 37° to yield a final volume of 3 ml. A microstir bar is included at a constant stirring speed of one to two revolutions per second

[21] P. W. Stevens, R. Raffen, D. K. Hanson, Y.-L. Deng, M. Berrios-Hammond, F. A. Westholm, C. Murphy, M. Eulitz, R. Wetzel, M. Schiffer, and F. J. Stevens, *Prot. Sci.* **4,** 421 (1995).

to prevent the settling of precipitating material. Fibrillogenesis is performed at 37° and is modeled using the heterogeneous nucleation model of Bishop and Ferrone[16] (Fig. 5). Higher temperatures result in a decrease in lag phase concomitant with an increase in the rate of hyperbolic fibril growth. This technique has been applied successfully to distinguish an amyloidogenic λ VI V_L from a cast-forming λ VI V_L based on the rates of fibrillogenesis at 37°.

Phenomenology

When ASF is performed in an orbital shaker, certain technical nuances need to be considered. The first is precautionary; namely, that the reaction mixtures should contain 0.02% NaN_3 as a preservative. Its addition, however, results in a high absorbance at 230 nm that will prevent turbidity measurements using UV wavelengths and preclude the use of spectrophotometric protein determination assays using this wavelength. Second, because the samples are incubated over periods of days at 37°, the tubes should be stoppered securely and held in place with Parafilm to prevent evaporation that would result in changes in protein concentration. Although not as widely available as bench-top spectrophotometers, instruments are available commercially to measure absorbance in 1-cm-diameter test tubes; the Spectronic 20 Genesys is one example (Spectronic, Rochester, NY). This

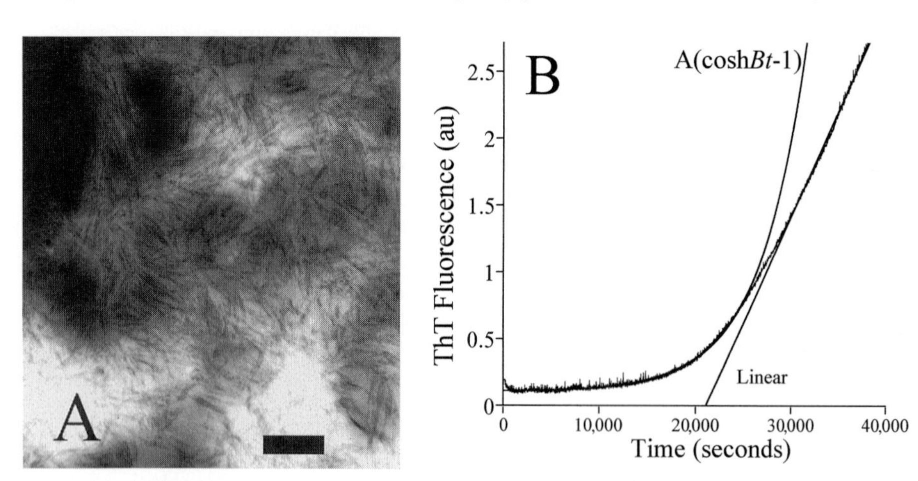

FIG. 5. Fibrillogenesis of V_L proteins at 37°. (A) Negatively stained, electron microscopy of V_L fibrils formed by fibrillogenesis at 37°. Bar: 300 nm. (B) Kinetics of V_L fibril formation. Least-squares analysis of data using A(cosh *Bt*-1) according to Bishop and Ferrone[16] provides excellent agreement with the first 10% of the data, as predicted, indicative of nucleation-dependent polymerization. Linear extrapolation of the later aggregation curve yields a value for the lag phase, i.e., 22,000 sec.

instrument would preclude the need for sampling the reaction mixture and obviate many of the aforementioned problems. Finally, the orientation of the tubes in the orbital shaker is critically important. This is believed to relate to the extent of agitation achieved. Samples placed upright mix vortically, which generally results in little or no fibrillogenesis. When placed at a 45° angle, however, mixing is erratic; this motion results in optimal fibril formation.

At room temperature without agitation, IgLC peptides, V_L, and Bence Jones proteins do not form fibrils. The process is initiated in the fluorimetric cuvette by inclusion of a microstir bar, and the sample is agitated rather than vortex-mixed, somewhat analogous to the importance of tube angle in ASF. This is achieved by setting the stirrer speed such that the stir bar "jumps" in the cuvette. Although somewhat anecdotal, the importance of the motion of the stir bar cannot be understated. When V_L fragments and BJPs are mixed vortically, the predominant aggregates are amorphous with none of the microscopic or tinctorial properties of fibrils. However, when the same protein is agitated, fibrils dominate (Fig. 6).

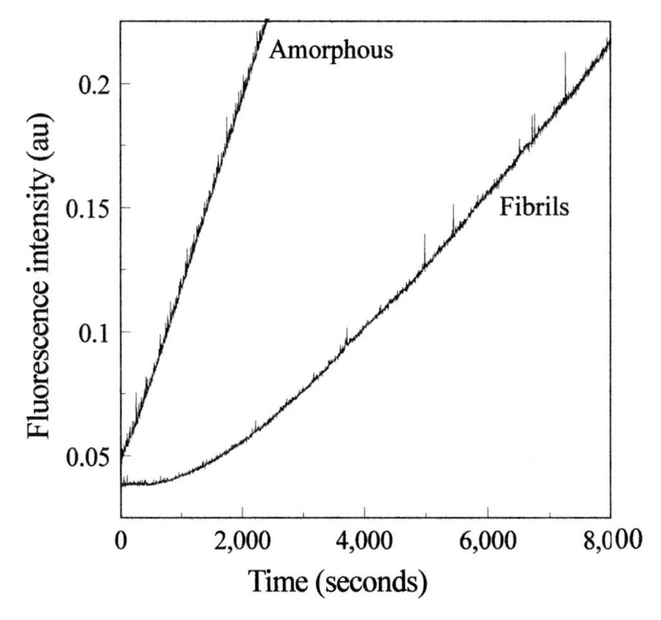

FIG. 6. Fibrillogenesis in the cuvette. Vortex mixing vs agitation. Amorphous precipitates arise when the solution is mixed vortically, which results in a linear increase in the intensity of scattered light. Fibrils are formed when the stir bar motion is erratic and the solution agitated. Both samples contain 3 μM BJp and 10 μM ThT in 3 ml of PBS, pH 7.5, and 25°. Emission data were collected using an excitation of 450 nm. The PMT voltage and slit widths were identical for both samples.

Quartz cuvettes should not be used to monitor fibril formation in the fluorimeter. Fibrillogenesis of IgLC proteins has not been observed within 72 hr using a quartz cuvette. Polystyrene cuvettes (Starstedt, Newton, NC) yield rapid reproducible results and are used routinely in our laboratory. Polymethacrylate cuvettes (Sigma) are less amenable to fibril formation, observed as an increase in the lag time with respect to fibrillogenesis performed in polystyrene cuvettes. This result remains enigmatic, but probably reflects the role that surface adhesion plays in the aggregation of these proteins.

Sampling versus in Situ ThT Monitoring

In many instances when sampling reaction mixtures, an aliquot is diluted into a larger volume followed by titration with ThT to measure fibril content. This approach results in a rapid, exponential decrease in the fluorescence intensity of the ThT within 5 min of dilution and has been witnessed with synthetic fibrils prepared from IgLC peptides, V_L proteins, Aβ(25–35), and Aβ(1–40) alike. The phenomenon is under investigation, but may be related to reequilibration of the monomer, polymer, or polymeric aggregates.

When synthetic V_L and Aβ(25–35) fibrils, formed in the presence of 10 μM ThT, are diluted into a solution of 10 μM ThT, the decrease in the fluorescence intensity is again observed, suggesting that ThT dissociation from the fibrils is not a factor. Whatever the mechanism, there is clearly a decrease in the fluorescence intensity. The inclusion of ThT within the polymerization reaction, therefore, precludes the need for reaction mixture sampling and circumvents the sampling problems associated with this phenomenon.

Summary

These techniques permit the production of bulk quantities of fibrils and provide methods for monitoring the kinetics of fibrillogenesis. Experiments performed in the fluorimeter require low protein concentrations, sampling is not necessary (with ThT *in situ*), and the measured fluorescence signal is indicative of fibril content and is not complicated by the presence of amorphous aggregates. However, ASF using the orbital shaker is a simple, rapid, initial procedure, adequate for screening for fibrillogenic potential, in which multiple experiments can be performed simultaneously and over long periods of incubation. These methods may be used to investigate the fibrillogenesis of V_L proteins and BJps as a means of predicting pathogenicity, as well as providing information on the basic biophysical principles underlying light chain aggregation.

Acknowledgments

We thank Valerie Brestel and Rich McCoig for assistance in the preparation of this manuscript, Bradley Hamilton who provided early WAT(1–24) data, Dick Williams for the electron microscopy, and Maria Schell for invaluable assistance on the V_L fibril project.

[15] Inhibition of Aggregation Side Reactions during *in Vitro* Protein Folding

By ELIANA DE BERNARDEZ CLARK, ELISABETH SCHWARZ, and RAINER RUDOLPH

When synthetic or natural genes or cDNAs are overexpressed in the cytosol of microbial host cells such as *Escherichia coli,* recombinant proteins can be produced in large amounts. Although high-level expression is usually achieved using standard recombinant DNA techniques, the polypeptides are often sequestered in the form of insoluble, inactive inclusion bodies. These large, dense particles often span the whole diameter of the host cell. After proper isolation they consist primarily of the recombinant protein. Active protein can be recovered from the inclusion bodies by solubilization in chaotrophic buffer systems and subsequent *in vitro* folding. However, unproductive side reactions (predominantly aggregation) often compete with correct folding during *in vitro* folding. Various techniques, some of which are summarized in this article, have been developed to inhibit aggregation side reactions and to ensure efficient *in vitro* protein folding. These methods, which can now be considered as standard laboratory techniques, allow the refolding of many inclusion body proteins on the laboratory scale or even in industrial production processes.

Isolation of Inclusion Bodies

Many different protocols for inclusion body isolation are described in the literature. Generally, inclusion bodies are harvested by centrifugation after cell lysis.[1] The quality of the inclusion bodies can be improved considerably before centrifugation by maximum cell lysis and dissociation of all particulate matter. Maximum cell disruption can be achieved by combining

[1] R. Rudolph, G. Böhm, H. Lilie, and R. Jaenicke, *in* "Protein Function" (T. E. Creighton, ed.), p. 57. IRL Oxford Univ. Press, Oxford, 1997.

0076-6879/99 $30.00

enzymatic and mechanical cell lysis. Membrane fragments are then disintegrated by adding detergents at a high salt concentration. Using the following protocol we have been able to obtain almost homogeneous inclusion body preparations of many different recombinant proteins.

Protocol

One gram *E. coli* cells (wet weight) is suspended in 5 ml of 0.1 *M* Tris–HCl, pH 7.0, 1 m*M* EDTA at 4°. After addition of 7.5 mg lysozyme, the mixture is incubated for 30 min at 4°, followed by high-pressure dispersion to completely disrupt the cells. For small-scale preparations, cells can be disrupted by sonication instead of high-pressure dispersion. In this case care should be taken to maintain the sample at low temperature. After cell disruption, cellular DNA is digested by the addition of DNase and $MgCl_2$ to a final concentration of 10 μg/ml and 3 m*M,* respectively. After 30 min incubation at 25°, EDTA, Triton X-100, and NaCl are added to a final concentration of 20 m*M,* 2% and 0.5 *M*, respectively, working with a three times concentrated stock solution of pH 7.0 containing 60 m*M* EDTA, 6% Triton X-100, and 1.5 *M* NaCl. After 30 min incubation at 4°, inclusion bodies are harvested by centrifugation. In order to remove residual detergent, the pellet is resuspended in 40 ml of 0.1 *M* Tris–HCl, pH 7, 20 m*M* EDTA and the inclusion bodies are finally isolated by a second centrifugation step.

This vigorous inclusion body isolation procedure guarantees maximum cell disruption even in those cases where the *E. coli* cells are difficult to disintegrate because of changes in cell wall morphology as a response of the cells to the physiological stress caused by the overexpression of a heterologous protein.

Solubilization of Inclusion Bodies

Various reagents have been used for the release of recombinant proteins from inclusion bodies.[2] Strong denaturants such as 6 *M* guanidinium chloride (Gdm/Cl) or 8 *M* urea are generally employed for inclusion body solubilization.[1] Gdm/Cl is preferred over urea for two reasons: (1) Gdm/Cl is a stronger denaturant, which allows disintegration of extremely sturdy inclusion bodies that are resilient to solubilization by urea. (2) Under the given solvent conditions, urea decomposes to a significant extent into cyanate, which may cause chemical modification of lysine and cysteine side chains.[3,4] Inclusion bodies often contain disulfide bonds that reduce the solubility in the absence of reducing agents such as dithiothreitol (DTT), dithioerythritol (DTE), or 2-mercaptoethanol. Addition of these low molecular weight thiol reagents in combination with a chaotroph allows reduction of disulfide bonds by thiol disulfide exchange. Because the reactive species in thiol disulfide exchange is the thiolate anion, inclusion body solubilization in the presence of reducing agents should be performed under mildly alkaline conditions. Inclusion body isolates can be solubilized by the following method.

Protocol

Suspend the inclusion body pellet in 6 *M* Gdm/Cl, 0.1 *M* Tris–Cl, pH 8.0, in the presence of 100 m*M* DTE, 1 m*M* EDTA at a protein concentration of ca. 10 mg/ml. The concentration of the solubilized protein can be determined by the Bradford assay[5] with bovine serum albumin (BSA) in solubilization buffer as a standard. After a 2-hr incubation at 25°, the pH is lowered to pH 3–4 by the dropwise addition of 0.5 *M* HCl. Residual insoluble cell debris is then removed by centrifugation. In the case of disulfide-bonded proteins, where oxidation must occur in the subsequent refolding step (see later), DTE (or DTT) has to be removed by extensive dialysis against 4 *M* Gdm/Cl in 10 m*M* HCl. If the respective protein does not contain disulfide bonds, refolding can be performed without prior removal of the reductant.

Transfer of Unfolded Protein in Refolding Buffer

The buffer exchange from unfolding to folding conditions is usually performed by dialysis or dilution. In dialysis the denaturant is removed gradually. Because many proteins form structured intermediates at intermediate denaturant concentrations, which are prone to irreversible precipitation, aggregation often predominates when refolding by dialysis. In this case, rapid removal of the denaturant by dilution reduces loss by aggregate formation, especially if the propensity for aggregation is reduced further

[2] F. A. O. Marston and D. I. Hartley, *Methods Enzymol.* **182,** 264 (1990).

[3] G. R. Stark, W. H. Stein, and S. Moore, *J. Biol. Chem.* **235,** 3177 (1960).

[4] Over a 8-hr period, more than 0.1 m*M* cyanate accumulates in an aqueous solution of 8 *M* urea at 25° at pH values above pH 5.0 [P. Hagel, J. J. T. Gerding, W. Fieggen, and H. Bloemendal, *Biochim. Biophys. Acta.* **243,** 366 (1971)], which increases to 20 m*M* at equilibrium.

[5] M. M. Bradford, *Anal. Biochem.* **72,** 248 (1976).

by the addition of low molecular weight folding enhancers, molecular chaperones, or foldases in the refolding buffer. For other proteins the yield of refolding is, however, higher with gradual denaturant removal by dialysis.[6,7]

Successful *in vitro* folding has also been achieved by applying size-exclusion chromatography.[8] In this case, the solubilized protein is loaded on a gel-filtration column equilibrated with refolding buffer. One explanation for the increased refolding yield obtained by this technique is that high molecular weight polypeptides escape the denaturing buffer front. They then either fold to the native state and continue to elute or precipitate to large aggregates that stop migrating. These aggregates are subsequently resolubilized by the following denaturant front, which gives them a second chance for escape and folding. If this explanation holds true, *in vitro* folding by gel filtration should, however, only work in those cases where folding is much faster than the gel-filtration process time. As an alternative mechanism by which refolding via chromatography works, a general reduced diffusion of proteins during chromatography has been suggested.[9]

Because of its good performance in many examples, dilution is considered the preferred method for transfer of the unfolded protein into refolding conditions in the following sections. Because dilution is simple compared to dialysis or gel filtration, it allows extensive optimization of the folding conditions with respect to protein concentration, temperature, time, pH, low molecular weight additives, and, in the case of disulfide-bonded proteins, redox conditions.

Folding with Concomitant Disulfide Bond Formation

Whereas the cysteine residues of native cytosolic proteins are almost exclusively in the reduced form, they are mostly oxidized, forming disulfide bonds in the case of most extracellular proteins.[10] Protein folding with concomitant disulfide bond formation includes both the regeneration of native, noncovalent interactions *and* the formation of correct covalent chemical bonds. The most common methods used to promote disulfide bond formation during refolding are (1) air oxidation in the presence of trace amounts of metal ions,[11] (2) the oxido-shuffling system developed by

[6] J. London, C. Skrzyna, and M. E. Goldberg, *Eur. J. Biochem.* **47,** 409 (1974).
[7] Y. Maeda, T. Ueda, and T. Imoto, *Prot. Engin.* **9,** 95 (1996).
[8] M. H. Werner, G. M. Clore, A. M. Gronenborn, A. Kondoh, and R. J. Fisher, *FEBS Lett.* **345,** 125 (1994).
[9] B. Batas and J. B. Chaudhuri, *Biotechnol. Bioengin.* **50,** 16 (1996).
[10] R. C. Fahey, J. S. Hunt, and G. C. Windham, *J. Mol. Evol.* **10,** 155 (1977).
[11] A. K. Ahmed, S. W. Schaffer, and D. B. Wetlaufer, *J. Biol. Chem.* **250,** 8477 (1975).

Ahmed and co-workers,[11] and (3) the use of mixed disulfides.[12] Protocols describing how to implement each of these methods can be found in Rudolph *et al.*[1] Although air oxidation in the presence of trace amounts of metal ions is simple and inexpensive, renaturation rates and yields are generally low due to the mass transfer limitations of oxygen in aqueous solutions and incorrect disulfide bonding, respectively. Furthermore, proteins may be modified chemically by prolonged exposure to molecular oxygen. Therefore, air oxidation is not described in this article. Experimental details of this procedure have been described elsewhere.[1,11]

The oxido-shuffling system utilizes low molecular weight thiols in reduced and oxidized form to promote both formation and reshuffling of disulfide bonds. This method usually results in much higher rates and yields than those obtained through air oxidation and has been the most widely applied oxidation method. The most common oxido-shuffling reagents are reduced and oxidized glutathione (GSH/GSSG), but the pairs cysteine/cystine, cysteamine/cystamine, 2-mercaptoethanol, and 2-hydroxyethyl disulfide have also been employed. Typically, 5 m*M* reduced thiol and a 10 : 1 to 5 : 1 ratio of reduced and oxidized thiol are used to promote proper disulfide bonding.[13] Because, however, other ratios of reduced to oxidized thiol components may be optimal for a given protein, this ratio has to be optimized on a case-by-case basis. The following general method may be applied for oxidative folding via oxido shuffling.

Protocol

Dilute the solubilized inclusion bodies rapidly (with all DTT or DTE removed by dialysis) in 0.1 *M* Tris–HCl, pH 8.5, 15°, 1 m*M* EDTA in the presence of 5 m*M* GSH and 0.5 m*M* GSSG or other low molecular weight monothiol reagents in reduced and oxidized form. The protein concentration should be between 20 and 50 μg/ml after the dilution in refolding buffer. If necessary, low molecular weight folding enhancers (see later) should be included in the refolding buffer.

A third method to promote oxidation to disulfides during folding is the formation of mixed disulfides between glutathione and protein before renaturation. The introduction of mixed disulfides increases the solubility of the unfolded protein by enhancing the hydrophilic character of the polypeptide chain.[14] Disulfide bond formation is then triggered by adding

[12] T. W. Odorzynski and A. Light, *J. Biol. Chem.* **254,** 4291 (1979).

[13] R. Rudolph and H. Lilie, *FASEB J.* **10,** 49 (1996).

[14] R. Rudolph, *in* "Modern Methods in Protein and Nucleic Acid Research" (H. Tschesche, ed.), p. 149. Walter deGruyter, New York, 1990.

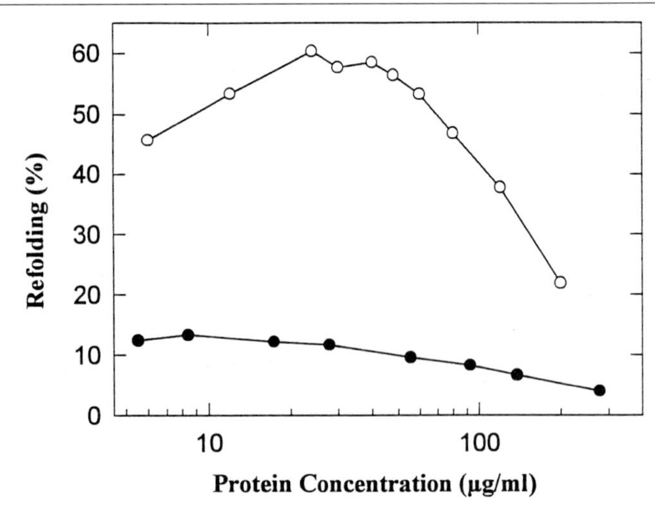

FIG. 1. Comparison of the concentration dependence of reactivation of the reduced (●) and mixed disulfide derivative (○) of recombinant human tissue-type plasminogen activator (rh-tPA) (U. Opitz, Ph.D. Thesis, University of Regensburg, Germany, 1988). The solubilized reduced protein was refolded by dilution in 0.1 *M* Tris–Cl, pH 10.5, 1 m*M* EDTA, 0.5 *M* L-arginine, 1 m*M* glutathione (GSH), and 0.2 m*M* GSSG at the indicated protein concentration. In the case of the mixed disulfide derivative, refolding was performed by dilution in 0.1 *M* Tris–Cl, pH 8.5, 1 m*M* EDTA, 0.8 *M* L-arginine, and 2 m*M* GSH. The folding yield as percentage of the total amount of protein was determined after 48 hr of renaturation at 20°.

catalytic amounts of a reducing agent in the final renaturation step. As shown in Fig. 1, refolding of the mixed disulfide derivative of recombinant human tissue-type plasminogen activator (rh-tPA) results in a more efficient folding at higher protein concentrations than refolding of the reduced protein in oxido-shuffling systems. The following protocol can be followed for mixed disulfide formation and subsequent folding.

Protocol

First, mixed disulfides are formed by adding 100 m*M* GSSG to the solubilized inclusion bodies (with DTE or DTT removed by dialysis). The pH is then adjusted to pH 9.3 by the dropwise addition of 0.5 *M* NaOH. After a 2-hr incubation at 20°, the pH is lowered to pH 3–4 by the dropwise addition of 0.5 *M* HCl. Excess GSSG is removed by exhaustive dialysis against 4 *M* Gdm/Cl in 10 m*M* HCl. Refolding is initiated by rapid dilution of the mixed disulfide derivative in 0.1 *M* Tris–Cl, pH 8.5, 15°, 1 m*M* EDTA in the presence of 1 m*M* GSH or similar concentrations of other low molecular weight monothiol reagents in the reduced form. The protein concentration after dilution into the refolding buffer should again be in the range of 20–50 μg/ml. As described later, low molecular weight folding enhancers can be added to improve folding.

Folding of Nondisulfide-Bonded Proteins

When folding nondisulfide-bonded proteins, the removal of DTT/DTE during solubilization by dialysis and the addition of a redox buffer, which is essential for the oxidative folding of disulfide bonded proteins, can be omitted. After removal of cell debris by centrifugation, the inclusion body proteins solubilized by Gdm/Cl in the presence of DTT or DTE can be diluted directly into the refolding buffer supplemented further with 5 m*M* DTT or DTE to prevent disulfide bond formation. If the native protein contains prosthetic groups such as cofactors or metal ions, these compounds should be added to the refolding buffer.

Aggregation during Folding

Features of Protein Aggregates

Overall yields of the refolding process are usually low due to the formation of inactive misfolded species, particularly aggregates. Very little is known about the mechanism of aggregate formation. Intermediates with hydrophobic patches exposed to the solvent play a crucial role in the partition between native and aggregated conformations. Folding intermediates possess significant secondary structure elements of the native protein but only little tertiary structures characteristic of the native conformation. Oberg *et al.*[15] showed that the secondary structure content of aggregates obtained when refolding the K97V mutant of interleukin-1β was very similar to that of the native protein,[15] indicating that aggregation is due to the interactions of folding intermediates with native-like secondary structure elements.

Pioneer work by Goldberg *et al.*[16] shed light on the nature of interactions leading to aggregation. This group showed that incorrect disulfide bonding may not be the major cause of aggregation because aggregates were formed even when a carboxymethylated protein, which can neither fold nor form disulfide bonds, was transferred to folding conditions. During the simultaneous renaturation of turkey lysozyme and excess BSA, lysozyme molecules were trapped in the heterologous aggregates formed by BSA.[16] These experiments indicate that aggregation can be a nonspecific phenomenon. Simi-

[15] K. Oberg, B. A. Chrunyk, R. Wetzel, and A. L. Fink, *Biochemistry* **33,** 2628 (1994).
[16] M. E. Goldberg, R. Rudolph, and R. Jaenicke, *Biochemistry* **30,** 2790 (1991).

larly, Maachupalli-Reddy *et al.*[17] performed mixed folding experiments with hen egg-white lysozyme and three other proteins to investigate further the specificity of the aggregation reaction. These experiments showed that aggregation during *in vitro* folding is due to nonspecific interactions between polypeptide chains. Corefolding of hen egg-white lysozyme was analyzed with a set of acidic and basic proteins.[18] In this study, coaggregation of the basic protein lysozyme was especially pronounced with acidic proteins, whereas the yield of lysozyme refolding increased when corefolding with basic proteins. These data suggest that electrostatic interactions play a major role in coaggregation.

Analysis of soluble aggregates in mixed folding experiments with the tailspike and coat proteins of bacteriophage P22 led Speed *et al.*[19] to conclude that aggregation is a specific phenomenon, as they were only able to observe self-association of the individual proteins instead of coaggregation. However, because they only analyzed soluble aggregates, it is possible that larger aggregates may grow by a mechanism involving nonspecific interactions, such as intermolecular disulfide bonding between unpaired cysteines[20] or nonspecific electrostatic interactions.[18]

Kinetic Competition between Folding and Aggregation

Analysis of the kinetics of aggregation showed that the process exhibits an apparent reaction order ≥ 2,[21,22] whereas the correct folding steps are generally first-order reactions. As a consequence the rate of aggregate formation increases when increasing the initial concentration of unfolded protein. Because of this kinetic competition, the yield of correctly folded protein decreases when refolding at rising protein concentrations (Fig. 1). Thus, the most direct means of preventing aggregation is to work at low protein concentrations. Concentrations in the range of 10 to 50 μg/ml are typically used,[13] but renaturation at such low protein concentrations requires large volumes of refolding buffers.

Preventing Aggregation by "Pulse Renaturation"

In order to achieve high refolding yields at a relatively high final protein concentration, a method was developed that reduces renaturation volumes

[17] J. Maachupalli-Reddy, B. D. Kelley, and E. De Bernardez Clark, *Biotechnol. Prog.* **13,** 144 (1997).

[18] V. D. Trivedi, B. Raman, Ch. Mohan Rao, and T. Ramakrishna, *FEBS Lett.* **418,** 363 (1997).

[19] M. A. Speed, D. I. C. Wang, and J. King, *Nature Biotechnol.* **14,** 47 (1996).

[20] E. De Bernardez Clark, D. Hevehan, S. Szela, and J. Maachupalli-Reddy, *Biotechnol. Prog.* **14,** 47 (1998).

[21] G. Zettlmeissl, R. Rudolph, and R. Jaenicke, *Biochemistry* **18,** 5567 (1979).

[22] D. Hevehan and E. De Bernadez-Clark, *Biotechnol. Bioengin.* **54,** 221 (1997).

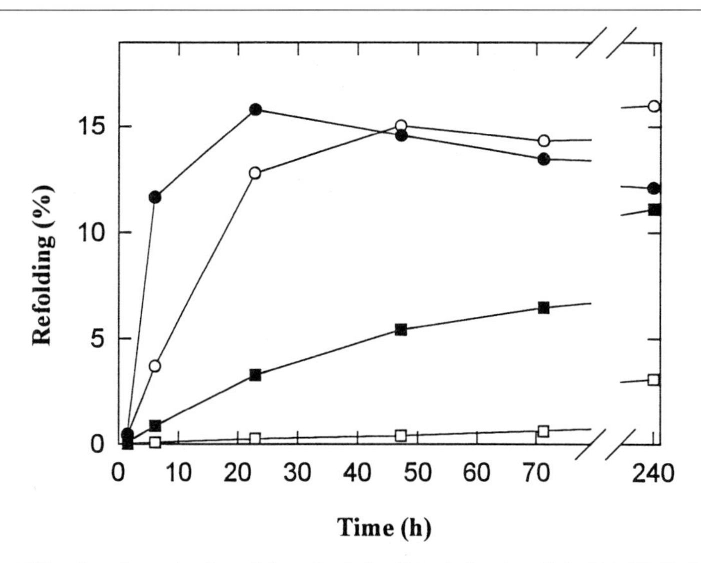

FIG. 2. Kinetics of reactivation of the mixed disulfide derivative of rh-tPA (U. Opitz, Ph.D. Thesis, University of Regensburg, Germany, 1988). The mixed disulfide derivative of rh-tPA with glutathione was refolded by rapid dilution in 0.1 *M* Tris–Cl, pH 8.5, 1 mg/ml EDTA, 1 mg/ml BSA, 0.5 *M* L-arginine, and 2 m*M* reduced glutathione. Refolding was performed at 0° (□), 10° (■), 20°(○), and 30° (●). The folding yield as percentage of the total amount of protein was determined after 17 hr of renaturation at 25°.

by adding the denatured protein into renaturation buffer in a stepwise fashion.[23] By this stepwise addition of denatured protein to the refolding vessel, the concentration of unfolded polypeptides is kept low throughout the refolding process. Naturally, enough time has to be allowed between the additions to enable the polypeptides to fold past the early stages in the folding pathway where they are still susceptible to aggregation.

Pulse renaturation for a given protein is optimized by the following strategy. (1) Determine the critical concentration during refolding above which the yield decreases due to an increase in aggregation. (2) Monitor the kinetics of refolding. Note that folding may be extremely slow in the case of complex proteins containing multiple disulfide bonds. As shown in Fig. 2, refolding of rh-tPA, which requires the correct formation of 17 disulfide bonds, is extremely slow. Further note that in the case of oligomeric proteins, the rate of renaturation often depends on the protein concentra-

[23] R. Rudolph and S. Fischer, U.S. Patent 4,933,434 (1990).

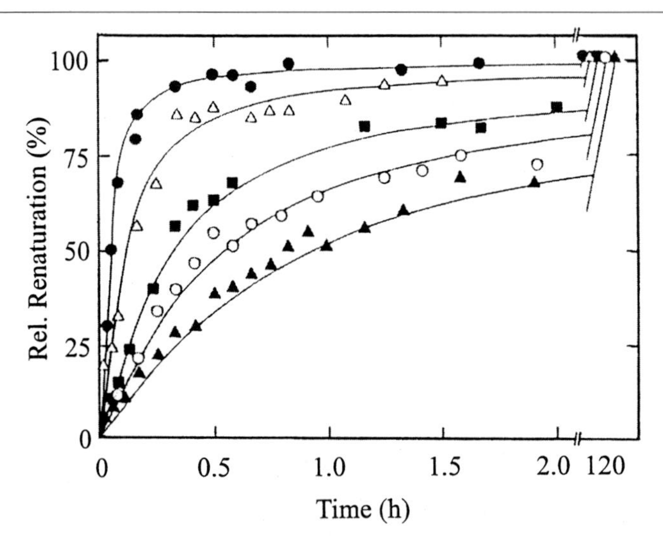

FIG. 3. Kinetics of reactivation of dimeric rabbit muscle triose-phosphate isomerase [S. Zabori, R. Rudolph, and R. Jaenicke, *Zeitschr. Naturforsch.* **25c,** 999 (1980)]. After 5 min denaturation in 6.5 *M* Gdm/Cl at 25°, the protein was renatured at 0° by dilution in 0.1 *M* triethanolamine buffer, pH 7.5, containing 1 m*M* EDTA and 1 m*M* DTT at varying protein concentrations: (●) 39.5 nmol; (△) 10.5 nmol; (■) 3.0 nmol; (○) 1.8 nmol; and (▲) 1.0 nmol. Reactivation was calculated relative to the final yield of ca. 95% determined after 120 hr of reactivation according to an irreversible Uni/Bi-molecular mechanism with $k_1 = 1.9 \times 10^{-2}$ sec^{-1} and $k_2 = 3 \times 10^5$ M^{-1} sec^{-1}.

tion. As an example, Fig. 3 shows the reactivation of dimeric triose-phosphate isomerase, which depends on first-order folding and second-order subunit association.[24] (3) Determine the maximum concentration of residual denaturant that is still compatible with high-yield folding. Careful experimentation through points 1–3 is necessary before setting up a protocol for the actual pulse renaturation. (4) Depending on the conditions worked out as just described (points 1–3), pulse renaturation is subsequently performed by the repetitive dilution of unfolded protein into the refolding buffer. In the first step, a low concentration of unfolded protein, which still allows maximum refolding, is diluted into the refolding buffer. After refolding ca. 80% of this initial batch, another aliquot of denatured protein (with denaturant!) is added to the same refolding solution and the procedure is repeated until the residual denaturant concentration in the refolding solution, which increases with each addition of the unfolded protein, interferes with structure formation.

[24] S. Zabori, R. Rudolph, and R. Jaenicke, *Zeitschr. Naturforsch.* **35c,** 999 (1980).

Preventing Aggregate Formation by Low Molecular Weight Folding Enhancers

A very efficient strategy to suppress aggregation is the inhibition of intermolecular hydrophobic interactions leading to precipitation by the use of special additives. These small molecules are relatively easy to remove when refolding is complete. Numerous additives have been shown to prevent aggregation. The mechanism of action of additives on the folding process is still unclear. They may influence both the solubility and the stability of the native, denatured, and intermediate state(s) or they may act by changing the ratio of the rates of proper folding and aggregate formation. Due to the lack of knowledge about the influence of additives on structure formation, the beneficial effect of a given additive on the folding of a particular protein cannot be predicted. Specific additives that work well for a certain protein have proven to be disadvantageous during the refolding of other proteins (see Table I).

The function of L-arginine as a suppressor of protein aggregation during *in vitro* folding was first described in a patent application in 1985.[25] Meanwhile, L-arginine is the most widely used additive today. L-Arginine is added to the refolding buffer at concentrations in the range of 0.4 to 0.8 *M*.[26–30] As shown in Fig. 4, the yield of correct folding increases tremendously if relatively high molar concentrations of L-arginine are included in the refolding buffer. This effect is caused by a decrease in aggregate formation, as determined by light scattering (Fig. 5). Although containing a guanidinio group, L-arginine has, in contrast to Gdm/Cl, only a minor effect on protein stability. However, this additive enhances the solubility of partially structured folding intermediates, leading to a decrease in aggregate formation. A similar mechanism may be responsible for the beneficial effect on folding observed when including polyethylene glycol (PEG) in the renaturation buffer.[31,32] Solvent–protein interactions studies indicate that PEG interacts preferentially with the denatured state of the protein, but is excluded from the native state.[33] Because PEG is partially hydrophobic, it interacts with

[25] R. Rudolph, S. Fischer, and R. Mattes, U.S. Patent 5,593,865 (1997).
[26] J. Buchner and R. Rudolph, *Bio/Technology* **9,** 157 (1991).
[27] J. Buchner, I. Pastan, and U. Brinkmann, *Anal. Biochem.* **205,** 263 (1992).
[28] W.-J. Lin and J. A. Traugh, *Prot. Express. Purif.* **4,** 256 (1993).
[29] D. Arora and N. Khanna, *J. Biotechnol.* **52,** 127 (1996).
[30] W.-J. Lin, R. Jakobi, and J. A. Traugh, *Genet. Engin.* **18,** 101 (1996).
[31] J. L. Cleland, S. E. Builder, J.-R. Swartz, M. Winkler, J. Y. Chang, and D. I. C. Wang, *Bio/Technology* **10,** 1013 (1992).
[32] J. L. Cleland, C. Hedgepeth, and D. I. C. Wang, *J. Biol. Chem.* **267,** 13327 (1992).
[33] L. L.-Y. Lee and J. C. Lee, *Biochemistry* **26,** 7813 (1987).

TABLE I
ADDITIVES USED TO INCREASE YIELD OF CORRECT FOLDING AND PREVENT AGGREGATION

Additive	Recommended concentration	References[a]
L-Arginine hydrochloride	0.4–0.8 *M*	1, 2
Polyethylene glycol (PEG, 3350 MW)	0.1–0.4 g/liter	3, 4
Nondenaturing concentrations of		5, 6
Urea	≤2.0 *M*	
Gdm/Cl	≤1.0 *M*	
Methylurea	1.5–2.5 *M*	1
Ethylurea	1.0–2.0 *M*	1
Formamide	2.5–4.0 *M*	1
Methylfomamide	2.0–4.0 *M*	1
Acetamide	1.5–2.5 *M*	1
Ethanol	Up to 25% (v/v)	7
n-Pentanol	1.0–10.0 m*M*	8
n-Hexanol	0.1–10.0 m*M*	8
Cyclohexanol	0.01–10.0 m*M*	8
Tris	≥0.4 *M*	9
Na_2SO_4 or K_2SO_4	0.4–0.6 *M*	10
Glycerol	20–40% (v/v)	10
Sorbitol	20–30% (v/v)	10
α-Cyclodextrin	20.0–100.0 m*M*	11
Lauryl maltoside	ca. 0.06 mg/ml	12
Cetyltrimethylammonium bromide	ca. 0.6 mg/ml	13
CHAPS	10.0–60.0 m*M*	8, 14
Triton X-100	ca. 10.0 m*M*	8
Dodecyl maltoside	2.0–5.0 m*M*	15
Mixed micelles	Depending on compounds used	16, 17

[a] (1) R. Rudolph, S. Fischer, and R. Mattes, U.S. Patent 5,593,865 (1997); (2) J. Buchner and R. Rudolph, *Bio/Technology* **9,** 157 (1991); (3) J. L. Cleland, S. E. Builder, J.-R. Swartz, M. Winkler, Y. Chang, and D. I. C. Wang, *Bio/Technology* **10,** 1013 (1992); (4) J. L. Cleland, C. Hedgepeth, and D. I. C. Wang, *J. Biol. Chem.* **267,** 13327 (1992); (5) G. Orsini and M. E. Goldberg, *J. Biol. Chem.* **253,** 3453 (1978); (6) S. E. Builder and J. R. Ogez, U.S. Patent, 4,620,948 (1986); (7) K. R. Hejnaes, S. Bayne, L. Nørskov, H. H. Sørensen, J. Thomsen, L. Schäffer, A. Wollmer, and L. Skriver, *Prot. Engin.* **5,** 797 (1992); (8) D. B. Wetlaufer and Y. Xie, *Prot. Sci.* **4,** 1535 (1995); (9) D. Ambrosius and R. Rudolph, U.S. Patent 5,618,927 (1997); (10) U. Michaelis, R. Rudolph, M. Jarsch, E. Kopetzki, H. Burtscher, and G. Schumacher, U.S. Patent 5,434,067 (1995); (11) N. Karuppiah and A. Sharma, *Biochem. Biophys. Res. Commun.* **211,** 60 (1995); (12) S. Tandon and P. Horowitz, *Biochim. Biophys. Acta.* **955,** 19 (1988); (13) S. Tandon and P. Horowitz, *J. Biol. Chem.* **262,** 4486 (1987); (14) N. Cerletti, G. K. McMaster, D. Cox, A. Schmitz, and B. Meyhack, U.S. Patent 5,650, 494 (1997); (15) J. Stockel, K. Doring, J. Malotka, F. Jahnig, and K. Dornmair, *Eur. J. Biochem.* **248,** 684 (1997); (16) G. Zardeneta and P. M. Horowitz, *J. Biol. Chem.* **267,** 5811 (1992); (17) G. Zardeneta and P. M. Horowitz, *Anal. Biochem.* **218,** 392 (1994).

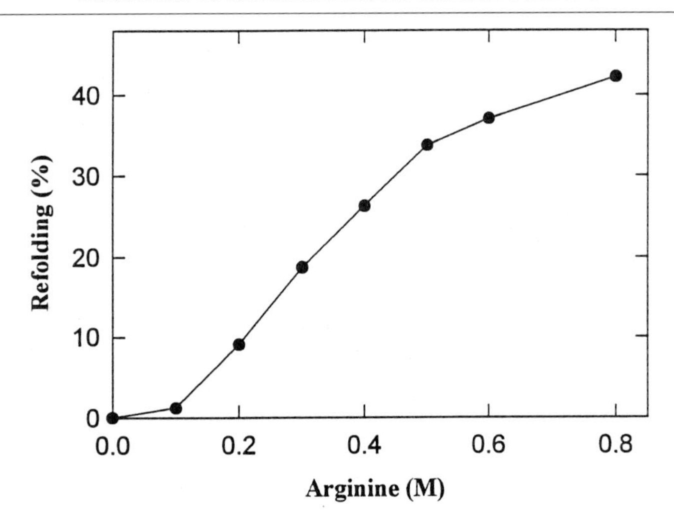

FIG. 4. Effect of L-arginine on the refolding of rh-tPA (U. Opitz, Ph.D. Thesis, University of Regensburg, Germany, 1988). rh-tPA as the mixed disulfide derivative with glutathione was refolded by dilution in 0.1 *M* Tris–Cl, 1 m*M* EDTA, 1 mg/ml BSA, 1 m*M* GSH, pH 8.5, at a protein concentration of 3 mg/ml. The folding yield as percentage of the total amount of protein was determined after 17 hr of renaturation at 25°.

hydrophobic side chains exposed in unfolded polypeptides or partially folded intermediates.[34]

In early studies on the effect of additives, denaturants such as Gdm/Cl or urea at nondenaturing concentrations were found to improve the yield of correctly folded protein.[35,36] Using lysozyme as a model protein, Hevehan and De Bernardez-Clark[22] demonstrated that this effect is due to a strong deceleration of aggregation as compared to folding. Improving refolding by the addition of nondenaturing concentrations of denaturants is, however, only possible if the native state is not destabilized under these solvent conditions.

Among denaturing substances, alkylureas, carbonic acid amides,[13] or alcohols[37,38] have also been used to improve *in vitro* folding. The *in vitro* structure formation of rh-tPA, human granulocyte colony-stimulating factor, and antibody fragments could be stimulated by refolding in concen-

[34] J. L. Cleland and T. W. Randolph, *J. Biol. Chem.* **267,** 3147 (1992).

[35] G. Orsini and M. E. Goldberg, *J. Biol. Chem.* **253,** 34 (1978).

[36] E. Builder and J. R. Ogez, U.S. Patent 4,620,948 (1986).

[37] K. R. Hejnaes, S. Bayne, L. Nørskov, H. H. Sørensen, J. Thomsen, L. Schäffer, A. Wollmer, and L. Skriver, *Prot. Engin.* **5,** 797 (1992).

[38] D. B. Wetlaufer and Y. Xie, *Prot. Sci.* **4,** 1535 (1995).

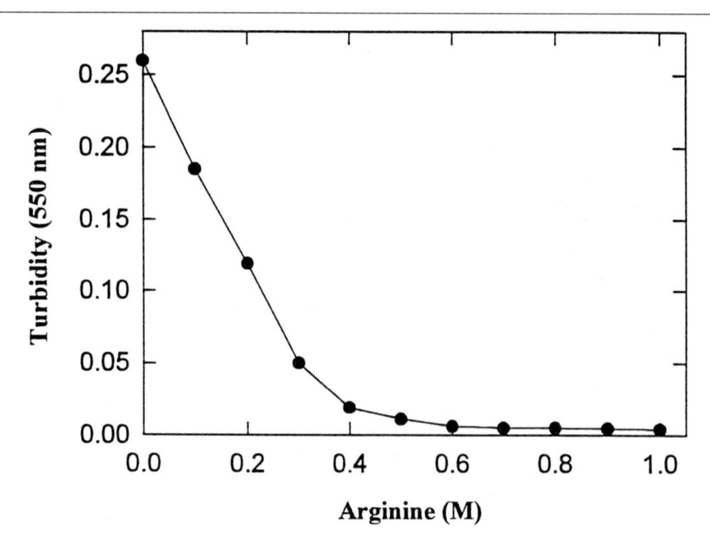

FIG. 5. Effect of L-arginine on the aggregation competing with correct folding of recombinant human interferon-γ [D. Arora and N. Khama, *J. Biotechnol.* **52,** 127 (1996)]. After solubilization of the inclusion bodies in 6 *M* Gdm/Cl, refolding was initiated by rapid 1:100 dilution in refolding buffer containing 0.1 *M* Tris, pH 8.0, 2 m*M* EDTA in the presence of increasing concentrations of L-arginine. After 1 hr the turbidity of the samples was measured at 550 nm.

trated Tris buffer.[39] For some proteins, stabilizing additives are essential for efficient *in vitro* folding.[40] In these rare cases, the stabilization of folding intermediates is obviously essential for proper structure formation.

For some model proteins, refolding could be enhanced by adding cyclodextrin to the refolding buffer.[41] This additive may contribute to the solubility of folding intermediates by interchelating hydrophobic, aromatic amino acid side chains.[42]

Other low molecular weight additives, such as detergents and surfactants or mixed micelles, have been found to promote correct folding.[38,43–48] These

[39] D. Ambrosius and R. Rudolph, U.S. Patent 5,618,927 (1997).

[40] U. Michaelis, R. Rudolph, M. Jarsch, E. Kopetzki, H. Burtscher, and G. Schumacher, U.S. Patent 5,434,067 (1995).

[41] N. Karuppiah and A. Sharma, *Biochem. Biophys. Res. Commun.* **211,** 60 (1995).

[42] A. Cooper, *J. Am. Chem. Soc.* **114,** 9208 (1992).

[43] S. Tandon and P. Horowitz, *Biochim. Biophys. Acta* **955,** 19 (1988).

[44] S. Tandon and P. M. Horowitz, *J. Biol. Chem.* **262,** 4486 (1987).

[45] T. Zardeneta and P. M. Horowitz, *J. Biol. Chem.* **267,** 5811 (1992).

[46] T. Zardeneta and P. M. Horowitz, *Anal. Biochem.* **218,** 392 (1994).

[47] J. Stockel, K. Doring, J. Malotka, F. Jahnig, and K. Dornmair, *Eur. J. Biochem.* **248,** 684 (1997).

additives act by binding to folding intermediates, reducing self-association, which leads to aggregation. The binding of detergents to hydrophobic intermediates results in changes in their tertiary structure[49] and a dampening of their amino acid side chain fluctuations.[47] Proper folding commences on the release of detergent molecules from folding intermediates, which is facilitated by the use of weakly binding detergents[47] or by extraction with cyclodextrin.[50] The optimal range of concentrations for improved folding varies with detergents, with some being more effective above their critical micelle concentration (cmc),[43,44] whereas others either do not form micelles or are equally effective below and above their cmc,[38] a fact that agrees with the hypothesis that folding intermediates associate with individual detergent molecules rather than with micelles.[47]

In order to optimize the refolding of a given protein, the proper additive for improving correct structure formation can be identified in a first screen by diluting the denatured protein in refolding buffer containing the amounts of additives provided in Table I. If the protein of interest has a distinct biological activity, e.g., enzymatic activity, which is easy to measure, the yield of refolding can be determined directly in the refolding solution, provided that the additives do not affect the assay. If correct folding can only be analyzed by complex assays, e.g., cell culture assay, care must be taken to remove misfolded species that may partially refold during the activity test. In this case the additives should be removed by dialysis before further analysis, as some additives, such as L-arginine, increase the solubility of misfolded species. Following removal of the additives, misfolded proteins may form a visible precipitate. Despite this precipitation of misfolded species during dialysis, the yield of correctly folded protein can still be high.

Preventing Aggregate Formation by Antibodies

Aggregation can also be reduced by adding monclonal antibodies that bind preferentially to hydrophobic patches of the refolding protein, thus protecting it from incorrect intermolecular interactions. Katzav-Gozansky *et al.*[51] used this strategy successfully to prevent aggregation during the folding of carboxypeptidase A. Tight binding of antibodies to folding intermediates may, however, inhibit native structure formation. Antibodies may also affect folding by stabilizing the native state of inherently unstable

---

[48] N. Cerletti, G. K. McMaster, D. Cox, A. Schmitz, and B. Meyhack, U.S. Patent 5,650,494 (1997).

[49] N. B. Bam, J. L. Cleland, and T. W. Randolph, *Biotechnol. Prog.* **12,** 801 (1996).

[50] D. Rozema and S. H. Gellman, *Biochemistry* **35,** 15760 (1996).

[51] T. Katzav-Gozansky, E. Hanan, and B. Solomon, *Biotechnol. Appl. Biochem.* **23,** 227 (1996).

proteins. As an example, reduced S-protein (a proteolytic fragment of ribonuclease A), which is difficult to refold due to its low stability, could be renatured in the presence of monoclonal anitbodies.[52]

Preventing Aggregate Formation by Site-Directed Mutagenesis

Because interactions among folding intermediates with exposed hydrophobic patches are the major cause of aggregation, aggregate formation could be prevented by introducing amino acid mutations that disrupt such hydrophobic patches. This method was tested by Knappik and Plückthun[53] and Nieba and co-workers,[54] who demonstrated by *in vivo* and *in vitro* folding experiments that mutations in turns and hydrophobic patches of a protein result in decreased aggregation.

Preventing Aggregate Formation by High Hydrostatic Pressure

Hydrostatic pressure in the order of 1–3 kbar can be used to dissociate oligomeric protein complexes without significant changes in protein structure.[55] Higher hydrostatic pressures are needed to induce conformational changes leading to denaturation. Thus, high hydrostatic pressures can potentially be applied during renaturation to prevent aggregate formation. Gorovits and Horowitz[56] demonstrated that hydrostatic pressures up to 2 kbar are able to reverse the aggregation of rhodanese folding intermediates, resulting in increased renaturation yields. Combinations of pressure and glycerol resulted in a 10-fold increase in rhodanese folding yields.

Preventing Aggregate Formation by Transient Binding of Folding Polypeptide to Solid Support

As first demonstrated by Creighton,[57] *in vitro* folding can be improved by noncovalent binding of the unfolded polypeptide to an ion-exchange matrix. In this case, transient binding of the folding polypeptide to the support reduces misfolding and aggregation. However, matrix-assisted folding using commercially available chromatography supports often fails to improve *in vitro* folding. The reason for this is mainly a too strong interaction of the folding intermediates by hydrophobic or ionic forces with the matrix material leading to an inhibition of structure formation.

[52] J. D. Carlson and M. L. Yarmush, *Bio/Technology* **10,** 86 (1992).
[53] A. Knappik and A. Plückthun, *Prot. Engin.* **8,** 81 (1995).
[54] L. Nieba, A Honegger, C. Krebber, and A. Plückthun, *Prot. Engin.* **10,** 435 (1997).
[55] B. C. Schade, R. Rudolph, H.-D. Lüdemann, and R. Jaenicke, *Biochemistry* **19,** 1121 (1980).
[56] B. M. Gorovits and P. M. Horowitz, *Biochemistry* **37,** 6132 (1998).
[57] T. E. Creighton, Patent U.S. Patent 4,977,248 (1990).

While guaranteeing maximal freedom for folding, specific binding of the unfolded protein to a matrix has been achieved using histidine (His) tags or polyionic fusion constructs.[58–61] Aggregation can be completely prevented by this specific, noncovalent immobilization of the folding polypeptides via the affinity tags at their N or C terminus. This method must, however, be further optimized to allow its routine application for *in vitro* folding.

Suppressing Aggregate Formation Using Fusion Constructs with Hydrophilic Proteins or Peptides

Due to the exposure of hydrophobic patches, folding intermediates are often prone to aggregation. These unproductive side reactions can be reduced using fusion constructs with hydrophilic proteins or peptides. If proteins or protein domains serve as fusion partners, they should refold fast and quantitatively and not interfere with the folding of the target protein. As an example, these requirements are fulfilled by the immunoglobulin G (IgG)-binding domain from staphylococcal protein A. Fusion of this IgG-binding domain to insulin-like growth factor I improved the oxidative folding of the growth factor substantially.[62] Similar results were obtained on the oxidative folding of human granulocyte colony-stimulating factor using small hydrophilic peptides as fusion partners.[63] As for folding on solid supports, folding with the help of fusions has not yet been established to a level that could be resorted to as a standard laboratory operation.

Preventing Aggregate Formation by Molecular Chaperones and Foldases

Protein folding is facilitated by molecular chaperones and protein folding catalysts. Of the latter, prolyl isomerases and disulfide oxidoreductases catalyze the *cis/trans* isomerization of prolyl residues and formation and/or the correction of disulfide bonds, respectively, reactions that are usually rate limiting during folding. In contrast, to the very defined action of folding catalysts, molecular chaperones promote native structure formation by a very nonspecific mode of interaction with unfolded proteins. By associating transiently with hydrophobic sequences, which in the native conformation are buried in the core of the protein, but are solvent-exposed in unfolded

[58] A. Holzinger, K. S. Phillips, and T. E. Weaver, *Biotechniques* **20,** 804 (1996).
[59] G. Stempfer, B. Höll-Neugebauer, and R. Rudolph, *Nature Biotechnol.* **14,** 329 (1996).
[60] A. Negro, M. Onisto, L. Grassato, C. Caenazzo, and S. Garbisa, *Prot. Engin.* **10,** 593 (1997).
[61] H. Rogl, K. Kosemund, W. Kühlbrandt, and I. Collison, *FEBS Lett.* **432,** 21 (1998).
[62] E. Samuelsson, H. Wadensten, M. Hartmanis, T. Moks, and M. Uhlen, *Bio/Technology* **9,** 363 (1991).
[63] D. Ambrosius, C. Dony, and R. Rudolph, Patent U.S. Patent 5,578,710 (1996).

proteins, molecular chaperones suppress aggregation and contribute to folding.[64] Although a wealth of data obtained from *in vitro* and *in vivo* studies demonstrates the power of chaperones during protein structure formation, their featuring during the *in vitro* refolding of large amounts of proteins has not been documented so far. Neither have folding catalysts been reported as helpful tools during the *in vitro* folding of quantitative amounts of protein. To date, the use of specific molecular chaperones is limited to their application as stabilizers of recombinant proteins in solution during storage.[65]

A number of proteins such as zymogens are synthesized as proforms. After attainment of the native structure the propeptide is removed proteolytically. In the cases studied so far (subtilisin, α-lytic protease, carboxypeptidase, etc.), the propeptides are required for the folding of the mature protein.[66] Accordingly, the propeptides have been referred to as "intramolecular chaperones." In contrast to "real" chaperones, these propeptides do not act promiscuously and are dedicated to their specific covalently linked substrate. Structure formation facilitated by propeptides is likely to involve a positioning effect of the propeptide on the polypeptide chain in its folding intermediate state(s). Because folding yields are extremely poor when the mature protein without propeptide is renatured, folding of those proteins as proforms should, in any case, be worth a trial.

Optimization of Folding Conditions

Because the rates of folding, misfolding, and aggregation as well as the relative stability of the native protein, correctly folded intermediates, misfolds, and aggregates depend on the folding conditions, these have to be optimized carefully in order to obtain a maximum yield of correctly folded protein. Usually, folding conditions are optimized in a rational, step-by-step fashion. First, the proper additives that may be necessary for folding of a given protein are identified by diluting the unfolded protein into refolding buffer at a protein concentration of 20–50 μg/ml. Different refolding buffers are prepared containing the amounts of additives proposed in Table I. Because a temperature of 15° is often optimal for folding, this temperature is chosen for this initial screen. Furthermore, EDTA is added during refolding of nondisulfide-bonded proteins as well as during the refolding of disulfide bond-containing proteins. EDTA complexes trace

[64] F. U. Hartl, *Nature* **381,** 571 (1996).

[65] U. Jakob, J. Buchner, M. Gaestel, D. Ambrosius, and R. Rudolph, Patent 0,599,344 A1 (1994).

[66] J. Eder and A. R. Fersht, *Mol. Microbiol.* **16,** 609 (1995).

amounts of metal ions, especially Cu^{2+}, which catalyze cysteine oxidation by molecular oxygen.[11] If, however, the given protein contains metal ions in its native state, inclusion of these metal ions in the refolding buffer is essential (fivefold molar excess to the respective protein) and EDTA has to be omitted. When refolding nondisulfide-bonded proteins, cysteine oxidation is reduced further by adding 5 m*M* DTT or DTE. In the case of disulfide-bonded proteins, a redox system consisting of 5 m*M* reduced glutathione and 0.5 m*M* oxidized glutathione is added to facilitate protein disulfide bond formation by oxido shuffling (see earlier discussion). Because thiolate anions are the reactive species in thiol disulfide exchange reactions, a slightly alkaline pH value (pH 8.5) is used in the initial screen. In the case of nondisulfide-bonded proteins, a more physiological pH value (pH 7.5) is preferred. In the initial screen, the yield of correct folding is determined after 2 and 24 hr of incubation under refolding conditions. If the protein refolds under certain conditions identified in these preliminary experiments, the solvent parameters affecting the kinetic partitioning between folding and aggregation have to be optimized step by step in a second screening round. In addition to the concentration of unfolded protein, critical parameters for refolding are temperature, folding time, pH, concentration of low molecular weight folding enhancers, ligands, and, in the case of disulfide-bonded proteins, redox systems consisting of low molecular weight thiol reagents in oxidized and reduced forms. In the case of disulfide-bonded proteins, the yield also depends on the relative amount of reduced to oxidized low molecular weight thiol reagents as well as on the chemical nature of the redox buffer compounds. This rational, step-by-step optimization usually allows rapid identification of the proper folding conditions. If necessary, fine tuning of the folding conditions can be performed in a third round of optimization.

Fractional factorial folding screens have been proposed for the identification of optimum folding conditions.[67] This approach is, however, not yet used as a standard laboratory protocol.

Prospect

In the past, mechanistic aspects of protein folding have been studied in minute detail.[68] In most of these studies, small single domain proteins have been analyzed, which can be refolded quantitatively *in vitro*. In the case of more complex proteins (large multidomain, oligomeric, multiple

[67] G.-Q. Chen and E. Gouaux, *Proc. Natl. Acad. Sci. U.S.A.* **94,** 13431 (1997).

[68] R. Jaenicke, *in* "Current Topics in Cellular Regulation" (E. R. Stadtman and P. B. Chock, eds.), p. 209. Academic Press, New York, 1996.

disulfide bonds), aggregation processes often compete with correct folding. Aggregation has attracted far less attention than protein folding. To date, little is known about the nature of these aggregation processes competing with correct folding or about the structure of the aggregates. Despite this lack of fundamental knowledge, powerful methods are available to inhibit aggregation side reactions in protein refolding. More detailed analyses of the refolding processes competing with correct folding will improve protein folding further.

Acknowledgments

This work was supported by the Federal German Ministry for Bildung Wissenschaft, Forschung, und Technologie (BMBF). We thank Susanne Richter for critically reading the manuscript and Hauke Lilie for the artwork.

[16] Inhibition of Stress-Induced Aggregation of Protein Therapeutics

By JOHN F. CARPENTER, BRENT S. KENDRICK, BYEONG S. CHANG, MARK C. MANNING, and THEODORE W. RANDOLPH

Introduction

There are numerous unique, critical applications for proteins in human healthcare. Dozens of protein therapeutics are already on the market or are in clinical trials.[1] However, even the most promising and effective protein therapeutic will not be beneficial to human health if its stability cannot be maintained during packaging, shipping, storage, and administration. Instability can lead to protein aggregation, which is a major problem in the biopharmaceutics field. The aggregates are usually nonnative in structure and may remain soluble and/or precipitate from solution. Typically, the aggregated molecules cannot be dissociated without the addition of a strong chaotropic agent (e.g., several molar guanidine hydrochloride). In some cases, the aggregates have nonnative intermolecular covalent linkages, e.g., disulfide bonds.

In addition to reducing efficacy, aggregates in parenterally administered proteins can cause adverse patient reactions such as immune response,

[1] Y. J. Wang and R. Pearlman, *in* "Pharmaceutical Biotechnology," Vol. 5. Plenum Press, New York, 1993.

0076-6879/99 $30.00

sensitization, or even anaphylactic shock.[2,3] Therefore, if even a small percentage of the protein molecules are aggregated, a product can be rendered unacceptable. Thus, what can be simply a nuisance—to be removed by centrifugation—to the protein chemist doing basic laboratory research can completely derail the product development process for a biotechnology company. Unless costly human clinical trials are conducted to prove that a product containing aggregates is both safe and effective, it is essential that aggregate formation is completely prevented during all stages of product processing, shipping, and storage.

This goal is difficult to achieve because the free energy barrier between native and unfolded states is only on the order of 50 kJ/mol.[4] Furthermore, it appears that proteins can form nonnative aggregates from protein molecules that are only slightly perturbed from the native structure (e.g., molten globules). These protein molecules can even be a part of the ensemble of species that encompass the native conformation. Thus, even minor stresses (e.g., subdenaturing concentrations of chaotropic agents, small changes in pH, or temperatures well below the melting point of the protein) can foster protein aggregation. Aggregation is greatly favored if the protein is at relatively high concentrations (e.g., >10 mg/ml), which is often the case for protein therapeutic products. The situation becomes even more complex if one considers that, in addition to resistance to acute stresses encountered during product processing, therapeutic proteins must maintain stability during storage for up to 18–24 months. Even under conditions greatly favoring the native state (e.g., optimal pH, buffer, and ionic strength at 30°), there can be slow accumulation of aggregated molecules, which can be significant on the time scale for the shelf-life of pharmaceutical products.

This article describes the types of stresses and conditions that are routinely found to cause aggregation of purified therapeutic proteins. Stresses encountered during processing include short-term exposure to high temperature during pasteurization of aqueous solutions, freeze-thawing, freeze-drying, and exposure to denaturing interfaces due to agitation, filtration, air bubble entrainment during filling, and so on. It also considers the effects of long-term storage on protein stability in aqueous solutions and dried solids. Each section describes how the rational choice of stabilizing additives (Table I) can be used to inhibit protein aggregation. These choices are based on a clear understanding of the mechanisms by which different additives succeed or fail as protein stabilizers under different conditions. The mechanisms are considered in detail and are illustrated by selected exam-

[2] R. E. Ratner, T. M. Phillips, and M. Steiner, *Diabetes* **39,** 728 (1990).

[3] C. A. Thornton and M. Ballow, *Arch. Neurol.* **50,** 135 (1993).

[4] C. N. Pace, *Trends Biochem. Sci.* **15,** 41 (1990).

TABLE I
CLASSES OF PROTEIN STABILIZERS EFFECTIVE AT INHIBITING STRESS-INDUCED AGGREGATION[a]

Type of stress condition	Effective stabilizers
Short-term thermal stress	Sugars; polyols; salting-out salts; methylamines; polymers
Long-term storage in aqueous solution	Same
Freezing/freeze-thawing stress	Same; surfactants
Dehydration stress	Disaccharides; surfactants
Long-term storage in dried solid	Disaccharides; carbohydrate polymers in combination with disaccharides; surfactants
Interfacial stresses	Surfactants

[a] See text for example compounds and their mechanisms of action and for definitions of stresses.

ples. Finally, the article briefly describes a model that allows the calculation of degree of expansion of the native state needed to form an aggregate-fostering species in aqueous solution and the utility of infrared spectroscopy to characterize the structure of proteins in precipitates.

Short-Term Thermal Stress in Aqueous Solution

It is well recognized that aggregation can be triggered by the exposure of proteins to elevated temperatures in aqueous solution. Although this stress may seem simple to avoid, there are certain pharmaceutical processes that must employ heat treatments. Regulatory agencies require that any protein derived from blood or circulatory organs must undergo some type of viral inactivation. The most widely accepted method is exposure of aqueous solutions to 60° for 10 hr. However, many protein products exhibit some level of thermally induced aggregation during heat treatment.

In general, thermally induced aggregation appears to follow a simple scheme. In a homogeneous protein preparation, perturbation of the native protein structure during heating can foster sufficient unfolding to promote aggregation. Alternatively, a population of protein molecules may contain a fraction that is especially sensitive to thermal stress. The fraction may be a misfolded form of the active protein, a chemically modified (e.g., proteolytically clipped) version of the active protein or a less stable protein impurity. With either type of sample, during heating the unfolded protein molecules then associate, often with native protein as well as themselves, forming soluble, nonnative aggregates. Finally, soluble aggregates accumulate and eventually associate into insoluble protein precipitates.

Protein aggregation during thermal treatment can be minimized by increasing the free energy of protein unfolding (i.e., increase the melting point of the native conformation), which is often accomplished by employing stabilizing additives. The most potent stabilization can usually be obtained if a specific ligand, which binds strongly to the native state (e.g., an enzyme substrate or cofactor) and interacts minimally with the denatured state, can be identified.[5] These additives are highly effective at relatively low bulk concentration, in stoichiometric ratios with the protein. Also, nonspecific stabilizers can be used alone or in combination with specific stabilizers.[6–10] Low molecular, nonspecific compounds usually only confer useful stabilization at concentrations of several hundred millimolar or higher. These stabilizers include sugars (e.g., sucrose), salting-out salts (e.g., ammonium sulfate), polyols (e.g., glycerol), and certain methylamines (e.g., trimethylamine *N*-oxide). High molecular weight polymers such as polyethylene glycol (PEG), gelatin, and hydroxyethyl starch are also effective stabilizers, often at concentrations above a few percent (w/v). However, low molecular polyethylene glycol (e.g., <3 kDa) or polyvinylpyrrolidone should be avoided because they can reduce the melting temperature of the native state and increase protein aggregation.[7] Finally, combinations of stabilizers from one or more of these classes can be used to enhance the inhibition of thermally induced protein unfolding and the resultant protein aggregation.[8]

Mechanism for Thermodynamic Stabilization by Additives

The effects of both specific and nonspecific stabilizers can be explained by a single, straightforward thermodynamic mechanism. It is important to emphasize that the following arguments are applicable to many different reversible equilibria between protein states or species; even those that would be encompassed within the native state ensemble, but only have small differences in surface area (see later).

For simplicity, we will consider a two-state model, in which there is an equilibrium between native and denatured states of the protein ($N \leftrightarrow D$). At room temperature and in nonperturbing solvent environments, the na-

[5] J. Brandt and L. O. Andersson, *Int. J. Peptide Protein Res.* **8,** 33 (1976).

[6] W. R. Porter, H. Staack, K. Brandt, and M. C. Manning, *Thromb. Res.* **71,** 265 (1993).

[7] M. Vrkljan, M. E. Powers, T. M. Foster, J. Henkin, W. R. Porter, J. F. Carpenter, and M. C. Manning, *Pharm. Res.* **11,** 1004 (1994).

[8] T. M. Foster, J. J. Dormish, U. Narahari, J. D. Meyer, M. Vrkljan, J. Henkin, W. R. Porter, J. F. Carpenter, and M. C. Manning, *Int. J. Pharm.* **134,** 193 (1996).

[9] L. Bjerring-Jensen, J. Dam, and B. Teisner, *Vox Sang.* **67,** 125 (1994).

[10] P. K. Ng and M. B. Dobkin, *Thromb. Res.* **39,** 439 (1985).

tive state is favored because it has a lower free energy than the denatured state. The magnitude of the difference in free energy between the two states (i.e., the free energy of denaturation) dictates the relative stability of the native state. Any alteration in a system that decreases this difference will reduce stability, e.g., reduce the melting temperature of the protein. Conversely, increasing this free energy difference will increase the stability of the native state.

The effects of additives on the relative stabilities of each state can be described by an application of the Wyman linkage function, which in this case can be defined as the link between ligand binding and stability of protein states binding the ligand. More detailed explanations can be found in publications by Timasheff,[11] Wyman,[12] and Wyman and Gill.[13] Binding of a ligand to a given state will reduce the free energy (chemical potential) of that state. The effect of ligand binding on protein stability depends on the difference in binding between the two states. If more ligand binds to the native state than to the denatured state, then the free energy of denaturation will be increased and the native state will be stabilized. This is the case for potent specific protein stabilizers (e.g., enzyme cofactors and other ligands that interact strongly with the native protein conformation). The opposite effect will be seen if more ligand binds to the denatured state.

This general ligand-binding argument can also explain the mechanism for nonspecific protein stabilization and destabilization by additives. Detailed reviews of this mechanism, which has been developed by Timasheff and colleagues, can be found elsewhere.[11] For the nonspecific additives, relatively high concentrations (ca. $>0.3M$) are needed to affect protein stability. This is because the interactions of the solute with the protein are relatively weak. Binding of these additives is actually a measure of the relative affinities of the protein for water and ligand. Therefore, the ligand interaction is referred to as preferential.

Timasheff and colleagues have found that denaturants (e.g., urea and guanidine hydrochloride) are bound preferentially to proteins and that the degree of binding is greatest for the denatured state.[11] The free energy (chemical potential) of the denatured state is decreased more than that for the native state because more surface area for binding is exposed to solvent as the protein unfolds. Therefore, the free energy barrier between the two states is reduced, and the resistance to stress in the native state is reduced

[11] S. N. Timasheff, *Adv. Protein Chem.* **51,** 355 (1998).

[12] J. Wyman, *Adv. Protein Chem.* **19,** 223 (1964).

[13] J. Wyman and S. J. Gill, *in* "Functional Chemistry of Biological Molecules." University Science Books, Mill Valley, CA, 1990.

(e.g., the melting point of the protein in lowered). If this effect is great enough, the protein will be denatured at room temperature.

Conversely, Timasheff and colleagues have observed experimentally that there is a deficiency of stabilizing solutes (e.g., sugars and polyols) in the presence of the protein, relative to that seen in the bulk solution.[11] That is, the solutes are preferentially excluded from contact with the surface of the protein. In a thermodynamic sense the solute (ligand) has negative binding to the protein. Thus, there is an increase in the free energy (chemical potential) of the protein. The native state is stabilized because denaturation leads to a greater surface area of contact between the protein and the solvent and to greater preferential solute exclusion. Thus, even though there is an increase in the free energy of the native state, this effect is offset by the greater increase in the free energy of the denatured state.

Long-Term Storage in Aqueous Solution

During the time frame (i.e., several months) of storage that is needed for a therapeutic protein product, the formation of soluble aggregates and/or protein precipitates is often a major problem. These degradation products can arise under conditions greatly favoring the native state and in the absence of agitation (which is treated separately as a stress later). For example, the product may be prepared at the optimum pH and ionic strength, stored at temperatures well below the melting point of the protein, and still exhibit protein aggregation.

Under these conditions, any spectroscopic measurement of the protein sample prior to storage indicates a "fully native" protein. In addition, in some cases, only minimal changes in secondary structure are detected in the aggregated species. Thus, a far more minor structural fluctuation than the unfolding of secondary structure is sufficient to cause problematic levels of aggregation during months of storage of aqueous therapeutic protein formulations.

There are a number of reasons why the minor structural changes responsible for undesirable aggregations have been difficult to determine. The degree of structural change is probably too small for many analytical methods to resolve. The change is also presumably a reversible process, unless the molecules involved are kinetically trapped by aggregation. The aggregate-competent species in the molecular population are most likely relatively scarce when compared to that of the most compact, native protein molecules. In addition, minor structural changes are constantly occurring so that what is defined as a "native" structure is actually an average of a continuous ensemble of molecules with fluctuating conformations.

An example of aggregation of a therapeutic protein during storage under minimally stressful conditions is provided by recombinant interleukin-1 receptor antagonist (IL-1ra).[14] Native IL-1ra is monomeric and does not associate into higher order states. However, during storage at 30°, which is well below the melting point of 56° for the solution conditions used, the protein irreversibly forms a soluble dimer. The soluble dimeric IL-1ra is biologically active and is almost structurally indistinguishable from native monomeric IL-1ra, based on infrared and circular dichroism spectroscopies. These results are consistent with the hypothesis that the dimer forms from a species in the native state ensemble, which is only minimally different from the most compact species.

Dimer formation is inhibited greatly by sucrose.[14] This finding can be explained by the sucrose preferential exclusion mechanism presented earlier. In the presence of sucrose, protein species having an increase in surface area (or volume) are thermodynamically less favorable than the more compact species. Thus, in the presence of sucrose, the equilibrium is shifted away from the aggregate-competent species and, hence, with time there is less accumulation of aggregates. Furthermore, with rhIL-1ra, the rates of both H–D exchange and reaction of side chains, which are chemically modified on exposure to the surface, are reduced in the presence of sucrose.[15] Thus, with short-term measurements that are indicative of the conformational mobility of the native state, it can be seen that sucrose makes structural expansion less favorable. In equilibrium terms, the population is shifted toward the more compact species. In general, for proteins that form nonnative aggregates during long-term storage in aqueous solution, the effect of thermodynamic stabilizers such as sucrose should be useful for increasing kinetic stability and minimizing aggregation.

Pathway for Protein Aggregation and Calculation of Structural Expansion to Form Aggregation-Fostering Species

Kinetic analyses of many protein aggregation pathways has resulted in the well-known Lumry–Eyring model.[16,17] The model, shown in Scheme 1, involves a first-order reversible unfolding of the protein and subsequent aggregation of nonnative species in a higher order process.[16,17] In Scheme

[14] B. S. Chang, R. M. Beauvais, T. Arakawa, L. N. Narhi, A. Dong, D. I. Aparisio, and J. F. Carpenter, *Biophys. J.* **71,** 3399 (1996).

[15] B. S. Kendrick, B. S. Chang, T. Arakawa, B. Peterson, T. W. Randolph, M. C. Manning, and J. F. Carpenter, *Proc. Natl. Acad. Sci. U.S.A.* **94,** 11917 (1997).

[16] K. W. Minton, P. Karmin, G. M. Hahn, and A. P. Minton, *Proc. Natl. Acad. Sci. U.S.A.* **79,** 7107 (1982).

[17] R. Lumry and H. Eyring, *J. Phys. Chem.* **58,** 110 (1954).

$$\text{(a)} \quad N \underset{k_{-1}}{\overset{k_1}{\rightleftharpoons}} A$$

$$\text{(b)} \quad A_m + A \xrightarrow[k_m]{} A_{m+1} \quad \text{(Aggregate)}$$

Scheme 1

SCHEME 1

1, N refers to native protein and A to an intermediate conformational state preceding aggregation. A_m refers to an aggregated form composed of m protein molecules. The rate constants for each reaction, i, are represented by k_i. If the first step is in equilibrium, the model predicts that aggregation should follow second- or higher-order kinetics.

However, with some proteins aggregation has been found to follow first-order kinetics.[18] This is an unexpected result for a bimolecular reaction and shows that the process is not rate limited by protein–protein collisions, but rather by the preceding unimolecular step shown in Scheme 2. In Scheme 2, N* is a transiently expanded conformational species within the native state ensemble, which is in equilibrium with N. K_{eq} is the equilibrium constant for the reaction N to N*. N* is transformed irreversibly to an aggregation-competent state, A. State A undergoes further reaction to form insoluble aggregates, A_m, composed of m monomer units. The irreversible, unimolecular isomerization reaction of N* to A is the rate-limiting step in the formation of aggregates and has a rate constant denoted as k_c. For unimolecular isomerizations, k_c is not expected to depend on solvent viscosity.

For proteins that aggregate via this pathway, a decrease in the aggregation rate by sucrose, which is excluded preferentially from the surface of proteins, can be explained readily by the Timasheff mechanism outlined earlier.[11,18] The degree of preferential exclusion and the increase in chemical potential are directly proportional to the surface area of protein exposed to solvent. By the LeChatelier principle, the system will minimize the thermodynamically unfavorable effect of preferential sucrose exclusion by favoring the state with the smallest surface area. This corresponding shift in the equilibrium (K_{eq} in Scheme 2) toward compact species (N in Scheme 2), which can be explained by the Wyman linkage relationship, will lead to an overall decrease in the rate of protein aggregation.

[18] B. S. Kendrick, J. L. Cleland, J. F. Carpenter, and T. W. Randolph, *Proc. Natl. Acad. Sci. U.S.A.* **95,** 14142 (1998).

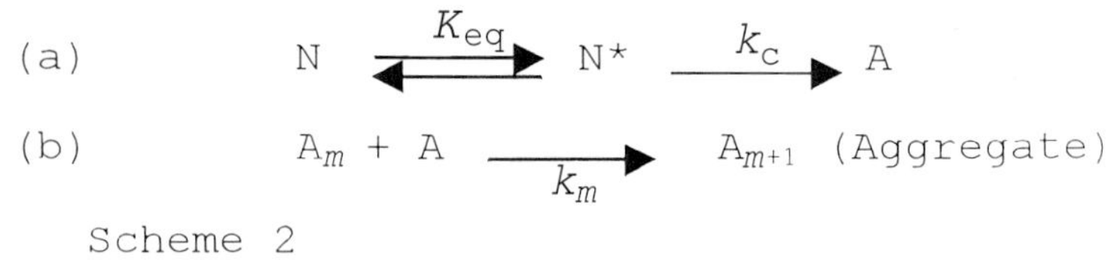

(a) $N \underset{}{\overset{K_{eq}}{\rightleftharpoons}} N^* \xrightarrow{k_c} A$

(b) $A_m + A \xrightarrow[k_m]{} A_{m+1}$ (Aggregate)

Scheme 2

SCHEME 2

Based on the aggregation pathway presented in Scheme 2 and insight into protein–sucrose thermodynamic interactions, a model has been developed to calculate the expansion in protein surface area needed to form the intermediate state preceding aggregation.[18] A detailed development of this model and an example using first-order aggregation of interferon-γ can be found in Kendrick *et al.*[18] Application of the model requires determination of initial rates of loss of native protein as a function of sucrose concentration and use of thermodynamic parameters available from the literature (e.g., surface tension increment of sucrose in water). With interferon-γ, it was found that only a 9% expansion in surface area of the most compact species of the native state was needed to foster protein aggregation. Such a minor change in conformation is consistent with the finding that proteins can aggregate under conditions that favor the native state greatly.

Finally, these results indicate that for the many protein systems that are prone to aggregation both *in vitro* and *in vivo,* only modest or even undetectable applied stress may be sufficient to drive aggregation. Similarly, mutations that have apparently minimal effect on properties, such as free energy of unfolding, may cause aggregation because they promote a slight increase in the fraction of the native state population that is structurally expanded at any instant in time. Methods that probe for these minor changes (e.g., measurement of H–D exchange kinetics) may be needed to discern mechanistically why some systems aggregate, whereas others that are very similar do not.

Freezing and Freeze-Thawing Stresses

There are several points in the processing, shipping, and storage of proteins at which a solution may be frozen, either by design or accident. For example, large volumes of therapeutic protein solutions are stored frozen as an intermediate holding step. Freezing stress can also contribute to protein damage during the freeze-drying process. Liquid aqueous formulations that may be adequately stable during storage under controlled conditions (e.g., at 4°) can be damaged severely if frozen accidently during shipping. Often the consequence of freeze-thawing is the formation of protein aggregates.

During freezing a protein is exposed to low temperatures and the formation of ice. Cold denaturation has been documented for many proteins and by itself may be sufficient to account for at least some of the damage noted during freezing.[19] Also, the protein, which partitions into the non-ice liquid phase, is exposed to extremely high solute concentrations as the sample is frozen. If solutes that are destabilizing to the protein are present or if the protein is perturbed by high ionic strength, then this concentrating effect can contribute to protein denaturation. Also, there can be dramatic pH changes during freezing. For example, the dibasic form of sodium phosphate crystallizes in frozen solution, which results in a system that contains essentially solely the monobasic salt and has a very low pH, e.g., pH 4.[20] To minimize problems associated with pH changes, sodium phosphate should be avoided whenever possible. Other buffers such as Tris, which do not have large freezing-induced changes in pH, should be used.[20]

Fortunately, to prevent freezing-induced damage to proteins, often it is not necessary to discern which stresses are responsible for the damage. The first step is to choose specific conditions (e.g., pH or ligand) that maximize the thermodynamic stability of the protein.[21] Next, nonspecific stabilizers can be added, if needed. A wide variety of compounds have been found to provide nonspecific protection to proteins by the preferential exclusion mechanism described earlier.[21] These include sugars, amino acids, polyols, certain salts, methylamines, alcohols, other proteins, and synthetic polymers.[21] Compared to sugars, polymers such as PEG, polyvinylpyrrolidone, and other proteins (e.g., albumins) are much more potent cryoprotectants and can be effective at concentrations $<1.0\%$ (w/v).[21] However, for most proteins, sufficient freezing protection can be obtained by using a disaccharide (e.g., sucrose), which has the added benefit of also protecting the protein during subsequent drying (see later). Relatively high sugar concentrations [e.g., $>30\%$ (w/v)] may be needed to confer adequate protection during freezing.

Often, increasing the initial protein concentration increases the percentage recovery of native protein during freeze-thawing.[20,21] At least in some

[19] P. L. Privalov, *Crit. Rev. Biochem. Mol. Biol.* **25,** 281 (1990).

[20] T. J. Anchordoquy and J. F. Carpenter, *Arch. Biochem. Biophys.* **332,** 231 (1996).

[21] J. F. Carpenter and B. S. Chang, *in* "Biotechnology and Biopharmaceutical Manufacturing, Processing and Preservation" (K. E. Avis and V. L. Wu, eds.), p. 199. Interpharm Press, Buffalo Grove, IL, 1996.

cases, in a given sample volume, there is the same mass of protein aggregated, independent of the initial protein concentration.[22] These results suggest that protein denaturation and aggregation may occur at the ice–water interface, which has a finite area under a given set of freezing conditions (e.g., sample volume and cooling rate). Other observations support this suggestion. Increasing the rate of cooling, which fosters the formation of smaller ice crystals and greater overall ice surface area, often leads to increased aggregation during freeze-thawing.[23] Strambini and Gabellieri[24] have directly documented with phosphorescence lifetime studies that perturbation of protein tertiary structure in the frozen state was almost twofold greater for protein frozen by cooling at 100°/min than that for samples cooled at 1°/min.

Consistent with the suggestion that the ice–water interface can denature proteins is the finding that many different nonionic surfactants, at very low concentration [ca. <1% (w/v)], can inhibit protein aggregation greatly during freeze-thawing.[23] This protection has been attributed to the competition of the surfactant with the protein for the ice–water interface.[22,23] This mechanism and other aspects of protection of proteins against interfacial stress will be considered further below.

Freeze-Drying Stress

To obtain the requisite storage stability for a pharmaceutical protein, it is often necessary to prepare a freeze-dried formulation.[21] The problem is that the stresses encountered during this process can cause protein denaturation and aggregation, the latter of which is usually noted after the protein sample has been rehydrated.[21,25] The effects of freezing, and how protein damage can be minimized during this stress, have already been addressed. This section focuses primarily on the dehydration stress. However, it is important to realize that to recover a native, functional protein after freeze-drying and rehydration, it is essential that the protein be protected during both freezing and drying streps.[21]

The functional, three-dimensional structure of a protein is determined to a large degree by the interaction of the protein residues and backbone with water.[26] Thus, it is not surprising that, in the absence of stabilizing additives, protein unfolding is often induced by dehydration, which can

[22] L. Kreilgard, J. L. Flink, S. Frojaer, T. W. Randolph, and J. F. Carpenter, *J. Pharm. Sci.* **88,** 281 (1999).

[23] B. S. Chang, B. S. Kendrick, and J. F. Carpenter, *J. Pharm. Sci.* **85,** 1325 (1996).

[24] G. B. Strambini and E. Gabellieri, *Biophys. J.* **70,** 971 (1996).

[25] S. J. Prestrelski, N. Tedeschi, T. Arakawa, and J. F. Carpenter, *Biophys. J.* **65,** 661 (1993).

[26] I. D. Kuntz and W. Kauzman, *Adv. Protein Chem.* **28,** 239 (1974).

be brought about by freeze-drying or other drying methods (e.g., spray-drying).[21,25,27] Infrared spectroscopy has proven invaluable in studying these structural alterations because protein samples in both liquid and dried states can be examined. For an unprotected protein in the dried solid, the infrared spectrum in the conformationally sensitive amide I region is often altered greatly relative to that for the native protein in aqueous solution.[21,25,27] Broadening and shifting of the component bands indicate that unfolded protein molecules are present in the dried solid.[21,25,27]

Unprotected lyophilized proteins often form nonnative aggregates on rehydration.[21,25,27] The degree of aggregation correlates with the degree to which the secondary structure of the protein has been perturbed during freeze-drying.[21,25,27] In the dried solid there are unfolded protein molecules at essentially infinite concentration. When water is added, intermolecular contacts are favored over refolding and a substantial fraction of the molecular population can form aggregates.

Reducing protein unfolding during freeze-drying also serves to reduce the degree of aggregation noted after rehydration. Sometimes, simply altering a specific solution condition, such as initial pH, can have dramatic effects. For example, with infrared spectroscopy, Prestrelski and colleagues[28] found that interleukin-2, lyophilized from a solution of pH 7.0, was unfolded. On rehydration, a large fraction of the sample was aggregated. Furthermore, the appearance of new infrared bands at 1617 and 1690 cm^{-1} suggested that the protein sample was at least partially aggregated in the dried solid.[27] When pH 5.0 was employed, the spectrum of the dried protein was almost identical to that for the native aqueous protein, and aggregation after rehydration was minimal. The resistance of many proteins to stresses imposed in solution is known to show a strong pH dependence. For lyophilization, it is not known whether the influence of pH manifests itself during freezing, drying, or rehydration, where charge–charge repulsion may decrease intermolecular interactions.

Optimizing specific solution conditions for a given protein often is not sufficient to inhibit lyophilization-induced unfolding and the resulting aggregation.[21,25,27] A nonspecific stabilizer, such as sucrose or trehalose, is then needed. To date, these two disaccharides have been found to be the most effective stabilizers during lyophilization.[21,25,27] Importantly, they provide protection during both freezing and dehydration, such that the native secondary structure can be recovered in the dried solid. As a consequence, aggregation is reduced in the rehydrated sample.

[27] J. F. Carpenter, S. J. Prestrelski, and A. Dong, *Eur. J. Pharm. Biopharm.* **45,** 231 (1998).
[28] S. J. Prestrelski, K. A. Pikal, and T. Arakawa, *Pharm. Res.* **12,** 1250 (1995).

Sometimes, a nonionic surfactant, such as polyoxyethylene sorbitan monolaurate (Tween 20), can be used, in addition to the sugar, to inhibit aggregation further. The surfactant may contribute to the inhibition of protein unfolding during freezing and drying.[23] In addition, the surfactant may serve a "chaperone" function and foster refolding over aggregation during rehydration.[23]

The mechanism by which disaccharides protect proteins during dehydration has been debated in the literature. One mechanism states that the sugar forms a highly viscous glass around the protein molecules that prevents protein unfolding and aggregation.[29] The second mechanism contends that sugar molecules hydrogen bond to the surface of the dried protein in place of the removed water and, hence, inhibit unfolding.[21,25] Hydrogen bonding between dried protein and sugars has been measured directly with infrared spectroscopy and shown to correlate directly with the inhibition of dehydration-induced protein unfolding.[21,25] For effective hydrogen bonding between the sugar and the dried protein to occur, they both must be in the same amorphous phase. If the sugar, or other potential stabilizer, crystallizes during freeze-drying, protection of the protein is abolished.[21,25]

However, the majority of data support that glass formation alone is not sufficient to stabilize proteins. For example, the large carbohydrate dextran has been shown to remain amorphous during drying, but does not prevent protein unfolding.[28,30] This is because steric hindrance prevents effective hydrogen bonding to the protein.[28,30]

Storage in Dried Solid

The main reason to prepare a lyophilized protein formulation is to obtain a product in which the protein is stable during shipping and long-term storage (e.g., 18–24 months at room temperature). The first step in the formulation development process is to stabilize the protein adequately during freezing and drying, such that protein aggregation does not occur if the product is rehydrated immediately. Just meeting this criterion does not assure that the protein will be stable during subsequent storage in the dried solid. Unless certain critical physical properties are obtained in the dried product, unacceptable levels of protein degradation, which is often manifested as aggregation on rehydration, can occur during storage.

[29] F. Franks, R. H. M. Hatley, and S. F. Mathias, *BioPharm.* **4**(9), 38 (1991).

[30] L. Kreilgaard, S. Frokjaer, J. M. Flink, T. W. Randolph, and J. F. Carpenter, *Arch. Biochem. Biophys.* **360,** 121 (1998).

First the protein formulation must be maintained below its glass transition temperature, T_g.[21,28–32] In the dried powder, the protein is a component of a glassy phase that includes amorphous additives and residual water. If this phase is held below T_g, the rate of diffusion-controlled reactions, including protein unfolding and chemical degradative processes, should be reduced greatly, relative to rates noted above the transition temperature.[21,28–32] T_g can be determined with differential scanning calorimetry or other thermal scanning methods.[21]

Water is a potent plasticizer for glasses.[29,32] Thus, in addition to playing a direct role in certain degradation reactions,[33] water in the dried formulation can reduce T_g greatly. For example, increasing the residual moisture of pure sucrose from 1 to about 3–4% (g H_2O/100 g dried powder) is sufficient to reduce the T_g to below room temperature.[34] It is critical to achieve a sufficiently low residual moisture for a given formulation such that T_g exceeds the planned storage temperature.

To obtain long-term storage stability of a dried protein formulation, it is also necessary to minimize protein unfolding during lyophilization, which can be accomplished by adding the appropriate stabilizer (e.g., sucrose or trehalose).[21,28,30,31,34] To date there have been published studies on four different proteins (interleukin-2, interleukin-1 receptor antagonist, factor XIII, and a lipase) that document this effect.[21,28,30,31,34] Thus, a disaccharide additive increases protein storage stability in the dried solid by preventing protein unfolding during the freeze-drying process and by providing a glassy matrix, which has a T_g above the storage temperature. Just meeting the latter criterion alone is not sufficient to optimize storage stability. Proteins that have been lyophilized in dextran, which provides a glassy matrix with a relatively high T_g for a given moisture content (e.g., >100° with 1% water), are nonnative.[28,30,34] These proteins degrade in the dried solid, even though the formulation T_g is much greater than the storage temperature (e.g., 60°).[30]

As a corollary to the conclusion that formulation T_g must be greater than the storage temperature, stability of a lyophilized protein also depends on avoiding a phase transition of the stabilizer during storage. For example, in a study with factor XIII, a sucrose formulation degraded rapidly during storage at 60°.[30] This was due to the crystallization of sucrose during storage. The transfer of moisture from the vial stopper to the dried formulation

[31] B. S. Chang, R. M. Beauvais, A. Dong, and J. F. Carpenter, *Arch. Biochem. Biophys.* **331,** 249 (1996).

[32] H. Levine and L. Slade, *J. Chem. Soc. Faraday Trans.* **84,** 2619 (1988).

[33] M. C. Manning, K. Patel, and R. T. Borchardt, *Pharm. Res.* **6,** 903 (1989).

[34] J. H. Crowe, J. F. Carpenter, and L. M. Crowe, *Annu. Rev. Physiol.* **60,** 73 (1998).

caused formulation T_g and sucrose crystallization temperature to be reduced to less than 60°. Sucrose forms an anhydrous crystal.[34,35] As a result, the residual moisture of the remaining amorphous phase containing the protein would be increased, causing T_g to be reduced to well below the storage temperature.

In practice, the risk of sucrose crystallization can be minimized by using vial stoppers that are treated to reduce moisture transfer to the product and avoiding storage temperatures approaching the formulation T_g. Sucrose crystallization can also be inhibited if the ratio of protein : sugar in the dried solid is high enough.[35] For example, Sarciaux and Hageman[35] found that 29 mass percent bovine somatotropin was sufficient to prevent the crystallization of sucrose during exposure to up to 80% relative humidity at 23°. Thus, if sucrose is the stabilizing additive, it is important to use a sufficient amount to protect the protein during lyophilization, but not to have an excessive amount that is prone to crystallization during storage. Fortunately, usually this criterion can be met by using the minimal amount necessary to inhibit maximally lyophilization-induced unfolding.

In addition to having a higher T_g for a given moisture content than sucrose, trehalose can also form a dihydrate crystal.[34] Thus, if trehalose did crystallize due to storage conditions (e.g., relatively high storage temperature and moisture transfer into product), the remaining amorphous phase would actually have a lower moisture content and a higher T_g. Thus, trehalose crystallization might be expected to increase the storage stability of the protein, assuming that no other adverse effects were induced. Potential adverse effects include perturbation of protein structure due to the formation of new interfaces or reduction of the mass of amorphous phase sufficiently to eliminate the "dilution" of protein molecules in the dried solid.

Finally, Allison and colleagues[36] have compared the capacities of dextran, sucrose, and trehalose alone and disaccharide–dextran mixtures to stabilize actin during lyophilization and storage in the dried solid. They found that dextran alone failed to confer optimum stability, as protein unfolding was not inhibited during lyophilization. Optimal stability was found with mixtures of sucrose or trehalose and dextran. In these systems, the disaccharide protected protein structure during lyophilization. The presence of dextran increased the T_g of the dried powder relative to that measured for formulations lyophilized with just the disaccharides. In addi-

[35] J. M. Sarciaux and M. J. Hageman, *J. Pharm. Sci.* **86,** 365 (1997).

[36] S. D. Allison, A. Davis, K. Middleton, T. W. Randolph, and J. F. Carpenter, submitted for publication.

tion, dextran provided an amorphous "bulking agent," which fostered the formation of strong, elegant, dried cakes.

Dextran is not always acceptable for parenteral products. However, other polymers have the potential to offer the same benefits as dextran to the physical properties of a lyophilized product. For example, hydroxyethyl starch also has a high T_g and generally should be more acceptable than dextran for parenteral administration. It is hoped that the rational use of polymers as "Tg modifiers" will make formulations more robust and much easier to lyophilize rapidly.

Interfacial Stresses

During the processing, shipping, storage, and delivery of proteins, it is inevitable that they come in contact with interfaces such as air–water interfaces and vial–water interfaces. Amphiphilic molecules are generally highly surface active, and proteins are no exception. Proteins tend to adsorb to surfaces, often nearly irreversibly,[37] due in part to the multipoint nature of protein–surface contacts.[38] This adsorption often results in alteration of the protein structure, which, coupled with the high protein concentrations found at the interface, may in turn lead to aggregation.[39] Processes where surface-induced protein aggregation may be particularly problematic include mechanical mixing,[40,41] spray draying,[42] agitation during shipping, and filtration.

Surfactants are commonly added to inhibit protein aggregation, both for liquid and lyophilized preparations.[23,43] The mechanism(s) by which surfactants inhibit protein aggregation is not always clear. However, several studies suggest that a mechanism that is frequently operative is competition by the surfactant against the protein for adsorption on various inter-

[37] W. H. Nord and C. A. Haynes, *in* "Proteins at Interfaces II: Fundamentals and Applications" (T. A. Horbett and J. L. Brash, eds.), p. 26. Am. Chem. Soc., Washington, DC, 1995.
[38] B. W. Morrisey and R. R. Stromberg, *J. Colloid Interface Sci.* **46,** 152 (1974).
[39] J. R. Lefebvre and P. Relkin, *in* "Surface Activity of Proteins" (S. Magdassi, ed.), p. 181. Dekker, New York, 1996.
[40] H. L. Levine, T. C. Ranashoff, R. T. Kawahata, and W. C. McGregor, *J. Parent. Sci. Tech.* **45,** 160 (1991).
[41] S. A. Charman, K. L. Mason, and W. N. Charman, *Pharm. Res.* **10,** 954 (1993).
[42] Y. N. Maa, P. Nguyen, and S. Hsu, *J. Pharm. Sci.* **87,** 152 (1998).
[43] L. S. Jones, N. S. Bam, and T. W. Randolph, *in* "Therapeutic Protein and Peptide Formulation and Delivery" (Z. Shahrokh, V. Sluzky, J. L. Cleland, S. J. Shire and T. W. Randolph, eds.), p. 206. Am. Chem. Soc., Washington, DC, 1997.

faces.[43–48] For example, insulin adsorption to solid–liquid interfaces was implicated in the agitation-induced denaturation and aggregation of insulin.[49] Competition for these interfaces by polyoxyethylene sorbitan monolaurate (Tween 20) reduced the susceptibility of insulin to denaturation and aggregation. Tween surfactants have proved effective in reducing or eliminating aggregation of proteins under a number of conditions and process steps, including guanidine hydrochloride-induced unfolding,[50] freezing,[23] freeze-drying,[23,51] agitation,[52] reconstitution,[53] nebulization,[54] and spray drying.[55] In each of these examples (with the possible exception of the guanidine hydrochloride-induced aggregation study) a surface was present that could have contributed to interfacial stresses that caused the aggregation.

If surfactants inhibit aggregation by competing with the protein for the interface, then the maximum degree of protection should be noted at or above the critical micelle concentration of the surfactant. Micelles form once sufficient surfactant is present to saturate the interface. Also, the degree of protection afforded by the surfactant should be dictated by the concentration of surfactant, independent of the protein concentration.[56]

For some proteins (e.g., recombinant human growth hormone) it has been found that the optimal degree of inhibition of aggregation during agitation was dependent of the mol ratio of surfactant to protein.[57] The mol ratio for optimal stability matched the binding stoichiometry between protein and surfactant, as determined by electron paramagnetic spectroscopy.[58] It was proposed that interaction of the surfactant with hydrophobic

[44] D. Andrade and V. Hlady, *Adv. Polymer Sci.* **70,** 1 (1986).

[45] H. Elwing, A. Askendal, and I. Lundström, *J. Colloid Interface Sci.* **128,** 296 (1989).

[46] T. Arnebrandt and M. C. Wahlgren, *in* "Proteins at Interfaces II: Fundamentals and Applications" (T. A. Horbett and J. L. Brash, eds.), p. 239. Am. Chem. Soc., Washington, DC, 1995.

[47] P. M. Claesson, E. Blomberg, J. C. Fröberg, T. Nylander, and T. Arnebrandt, *Adv. Colloid Interface Sci.* **57,** 161 (1995).

[48] M. M. Feng, A. Poot, T. Beugeling, and A. Bantjes, *J. Biomaterials Sci.* **7,** 415 (1995).

[49] V. Sluzky, J. A. Tamada, A. M. Klibanov, and R. Langer, *Proc. Natl. Acad. Sci. U.S.A.* **88,** 9377 (1991).

[50] D. B. Wetlaufer and Y. Xie, *Prot. Sci.* **4,** 1535 (1995).

[51] T. F. Osterberg, A. Fatouros, and M. Mikaelsson, *Pharm. Res.* **14,** 892 (1997).

[52] H. G. Thurow and K. Geisen, *Diabetologia* **27,** 212 (1984).

[53] M. Z. Zhang, K. Pikal, T. Nguyen, T. Arakawa, and S. J. Prestrelski, *Pharm. Res.* **13,** 643 (1996).

[54] A. Ip, T. Arakawa, H. Silvers, C. M. Ransone, and R. W. Niven, *J. Pharm. Sci.* **84,** 1210 (1995).

[55] M. H. Mumenthaler, C. C. Hsu, and R. Pearlman, *Pharm. Res.* **11,** 12 (1994).

[56] L. Kreilgaard, L. S. Jones, T. W. Randolph, S. Frokjaer, J. M. Flink, M. M. Manning, and J. F. Carpenter, *J. Pharm. Sci.* **87,** 1597 (1998).

[57] N. B. Bam, J. L. Cleland, J. Yang, M. C. Manning, J. F. Carpenter, R. F. Kelly, and T. W. Randolph, *J. Pharm. Sci.* **87,** 1554 (1998).

[58] N. B. Bam, T. W. Randolph, and J. L. Cleland, *Pharm. Res.* **12,** 2 (1995).

patches on the surface of the native protein sterically hindered the intermolecular interactions needed to form protein aggregates. It is important to note that this mechanism is only applicable to proteins for which the surfactant interacts with the surface of the native state. For other proteins that do not have this interaction, the competition of surfactant with the denaturing interface will be more important for inhibiting aggregation.

Finally, surfactants can also minimize aggregation by fostering refolding. This has been documented during the rehydration of lyophilized protein formulations.[23,53] For aqueous solutions, aggregation is also often a problem when the refolding of proteins, which have been fully denatured chemically, is attempted by diluting the denaturant.[50] At least for the case of recombi-

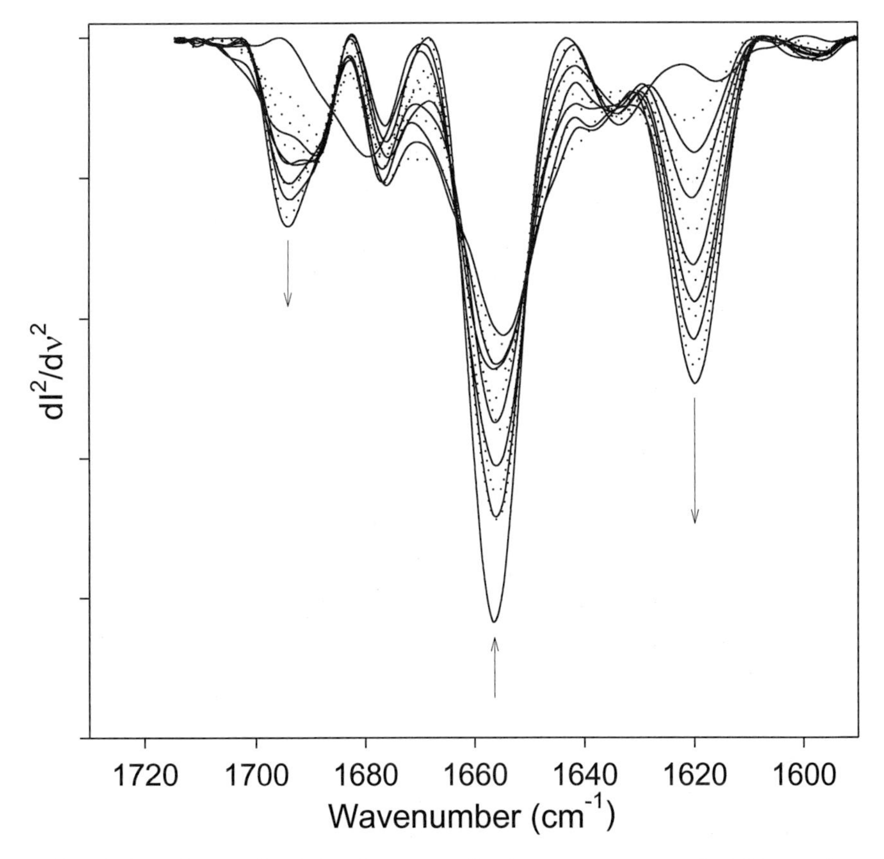

FIG. 1. Second derivative infrared spectra of inteferon-γ aggregation in 1 *M* guanidine hydrochloride at various time points up to 90 min in the order of solid–dot–solid–dot plots. Modified from B. S. Kendrick *et al., J. Pharm. Sci.* **87,** 1069 (1998).

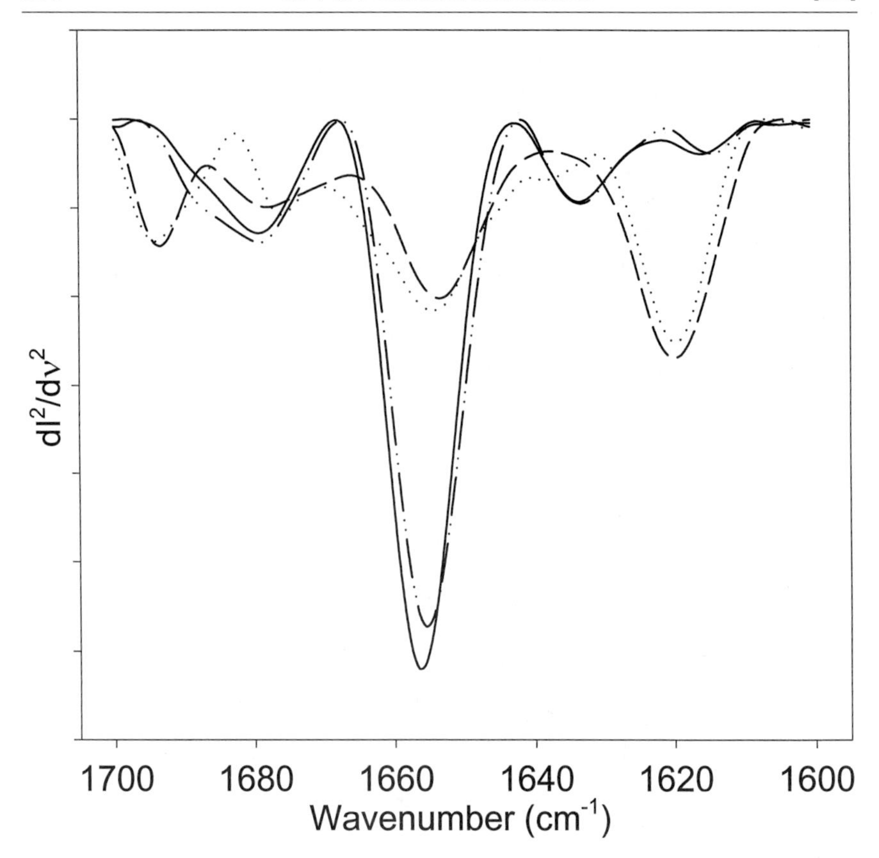

FIG. 2. Second derivative infrared spectra of interferon-γ after aggregation in 1 *M* guanidine hydrochloride (dotted line) or 300 m*M* NaSCN (dashed line) and salted out in 25% (w/v) polyethylene glycol 8000 (dash–dot–dot line). Solid line indicates native protein. Modified from B. S. Kendrick *et al., J. Pharm. Sci.* **87,** 1069 (1998).

nant growth hormone, inhibition of aggregation by surfactants has been attributed to the interaction of the surfactant with a folding intermediate (molten globule).[59]

Structural Characterization of Aggregates with Infrared Spectroscopy

Infrared spectroscopy is not a sensitive method for detecting aggregation of a small fraction (e.g., a few percent) of molecules in a protein solution. However, it is a uniquely powerful tool for studying the structure of purified

[59] N. B. Bam, J. L. Cleland, and T. W. Randolph, *Biotechnol. Prog.* **12,** 801 (1996).

aggregated and precipitated protein, and for monitoring the aggregation process in situations where large fractions of the molecular population are forming aggregates. Details of the method are presented in other chapters in this volume.

An example of the results from a study of protein aggregation is presented in Fig. 1, which shows infrared spectra of interferon-γ as a function of time of exposure to 1 *M* guanidine hydrochloride,[60] a mildly perturbing environment. Compared to the spectrum for the native protein, there is a progressive loss of the dominant helix band and a concomitant increase in absorbance of bands at 1620 and 1692 cm^{-1}. The latter bands are from intermolecular β sheets, which are the result of protein aggregation.[60] There is still a substantial fraction of helix band area remaining, even after complete aggregation. The results of this study document that the increased β sheet content noted in aggregates is a result of the aggregation process and not the cause of aggregation.

Salting-out agents cause protein precipitation in a manner distinct from protein destabilizing conditions. These agents are excluded preferentially from the surface of proteins, which increases protein chemical potential and favors the native state of the protein.[11] Thus, the structure of proteins in precipitates induced by salting-out agents is fully native. For example, the infrared spectrum for interferon-γ, which has been salted out by 25% (w/v) PEG 8000, is presented in Fig. 2.[60] In contrast, spectra of the protein aggregated in either 1 *M* guanidine hydrochloride or 0.3 *M* NaSCN have a substantial loss of helix band area and the presence of large intermolecular β sheet bands.

Infrared spectroscopy provides a valuable tool not only to follow perturbation-induced protein aggregation processes, but also to understand protein aggregation when it is induced by not readily apparent factors, which can be a problem in the processing of protein products.

Acknowledgments

We gratefully acknowledge support by grants from the National Science Foundation (BES9505301), Amgen, Genentech, Genetics Institute, Genencor International, and Boehringer Mannheim Therapeutics and by predoctoral fellowships to our students from the NSF, NIH, the American Foundation for Pharmaceutical Education, the Pharmaceutical Manufacturers and Research Association, and the Colorado Institute for Research in Biotechnology.

[60] B. S. Kendrick, J. L. Cleland, X. Lam, T. Nguyen, T. W. Randolph, M. C. Manning, and J. F. Carpenter, *J. Pharm. Sci.* **87,** 1069 (1998).

[17] Analysis of Protein Aggregation Kinetics

By FRANK FERRONE

This article takes up the question of describing the formation of large aggregates of proteins ordered by specific contacts. There are two goals in any modeling of protein aggregation. The first is to validate a possible mechanism by reducing a given proposal to a kinetic scheme whose predictions (time course, concentration dependence, etc.) can be compared with experiments. The second goal is to determine the molecular ingredients once a viable scheme has been established: what the rate constants are and what determines their observed values.

In following a kinetic process, modeling the initiation of the process poses the greatest challenges. This is because the latter stages are usually more amenable to direct observation, whereas the initial phases are more likely to be controlled by intermediates that are difficult to observe directly. Although protein aggregation has been studied for quite a long time, a number of fallacies persist. Probably the most notable is the assumption that a lag time in the kinetics represents a nucleation phase and that the end of such a lag corresponds to cessation of nucleation. This article first develops a basic description of the association process using a set of fairly reasonable assumptions and then turns to more advanced topics in the sections that follow. While it is impossible to cover all the possible mechanisms or their ramifications, the goal of this article is to provide a robust method of attacking all types of assembly process, keeping in mind the two goals of establishing mechanism and determining rates.

This article concentrates on the description of net properties, such as total mass polymerized, for aggregations at times near their initiation, e.g., during the first 10% or so of the reaction. There are a number of other fascinating topics such as polymer length redistribution, which will not be discussed here. The reader is referred to the discussion of other relevant work throughout the article. We will also omit any systems for which the polymerization is coupled to energy sources, such as the hydrolysis of ATP or GTP.

Basic Description of Aggregation

Protein aggregation processes are formally represented by the addition reaction

$$A + A_n \rightleftharpoons A_{n+1} \tag{1}$$

METHODS IN ENZYMOLOGY, VOL. 309
0076-6879/99 $30.00

This reaction is, in principle, characterized by the knowledge of all rate constants for elongation and depolymerization for species of all size, i.e., the knowledge of all values of k_n^+ and k_n^-. A plausible goal therefore might be to determine the individual reaction step rates for all n values, and given the ready accessibility of computers that can perform numerical integration of rate equations, such a brute force model is easy to construct and solve numerically. If a comprehensive table of rate constants were to be constructed, one would immediately need to winnow the set down to an understandable pattern, or, equivalently, in building a model, one would need a rationale for a particular pattern of constants. In short, some rule is required to make sense of the various rates, if they were known, or to simplify the model if one is being constructed. This is especially true for the testing phase, where one is faced with the problem of varying $2n$ parameters to obtain their optimum values for an n step model.

One immediate simplification occurs for long polymers. It is generally found that once n gets to be large, elongation and shortening become size independent, and one can describe the net elongation rate of polymers, denoted by J, in terms of the elementary rate constants as

$$J = k_+c - k_- \tag{2}$$

in which c is the concentration of monomers [A]. From Eq. (2) it is clear that a basic experiment consists in the simple measurement of elongation or shrinkage of polymers, so as to measure J. If c is known, or ideally if the measurement is performed as a function of c, the elementary rate constants can be inferred.

Most interesting assembly reactions do not begin with the same rate constants as they conclude. In a great many cases of interest, the initial reaction steps are slower than the later ones. In evolved systems this allows for control of the reaction because the initiation can be spatially localized, despite the general presence of favorable growth conditions. In some pathologic systems this initial inhibition has the advantage of allowing the organism to survive longer.[1] Whatever the reason, a fundamental question for any kinetic mechanism must be whether it shows such an initial inhibition.

It is a great help in understanding such a system if the initial reaction steps may be considered close to equilibrium, as the aggregates may be considered as thermodynamic species rather than kinetic intermediates. In terms of the preceding comments on inhibition, this equilibrium represents a series of unfavorable equilibria, at least up until some point is reached, i.e.,

[1] A. Mozzarelli, J. Hofrichter, and W. A. Eaton, *Science* **237,** 500 (1987).

$$\begin{array}{ll} A + A_n \overset{\rightarrow}{\longleftarrow} A_{n+1} & n < n^* \\ A + A_n \underset{\leftarrow}{\longrightarrow} A_{n+1} & n > n^* \end{array} \tag{3}$$

The equilibrium probability of finding a given species A_n can be related to a Gibbs free energy (relative to some standard state) as

$$[A_n] = [A_{\text{standard}}] \exp[-\Delta G(n)/RT] \tag{4}$$

in which R is the gas constant, T the temperature in Kelvins, and $[A_{\text{standard}}]$ is the standard state concentration. One logical choice for this "standard state" is the initial monomer concentration. Then the energy of each aggregate is measured relative to the initial concentration, although differences between curves as the initial concentration is varied would then be masked. Another common choice is some arbitrary concentration, say, 1 m*M*. Then the free energies are measured relative to a fixed concentration. Such fixed standard states can produce a paradoxical result on occasion, namely that the aggregate may be favored relative to the standard state. This simply means that the arbitrarily chosen standard is too high. In any case, the decrease in probability in finding A_n relative to $[A_{\text{standard}}]$ corresponds to an increase in energy. When a sequence of steps involves the increase in energy, the steps in the reaction can be viewed as climbing an energetic barrier that must be overcome for the aggregation process to proceed (Fig. 1). At equilibrium,

$$k_n^+ c[A_n] = k_{n+1}^-[A_{n+1}] \tag{5}$$

from which it follows that $[A_{n+1}]/[A_n] = ck_n^+/k_{n+1}^-$. However, it is also clear that

$$\begin{aligned} \frac{[A_{n+1}]}{[A_n]} &= \exp\left\{-\left[\frac{\Delta G(n+1) - \Delta G(n)}{(n+1) - n}\right] \Big/ RT\right\} \\ &= \exp\left(-\left\{\frac{d\Delta G}{dn}\right\} \Big/ RT\right) \end{aligned} \tag{6}$$

In other words, the slope of the free energy plot, $d\Delta G/dn$, is related to the ratio of the rate constants into and out of a state for a given monomer concentration c. The changeover in rates is therefore related to the change in slope of the free energy barrier, and a barrier that is linear with size gives a constant rate ratio. When the turning point is sufficiently sharp, the implication is that there is one state with a particularly small population that will represent the rate-limiting step for the reaction. This bottleneck is a thermodynamic nucleus, a necessary but very scarce species in the reaction path. This is quite distinct from a structural nucleus, in which a

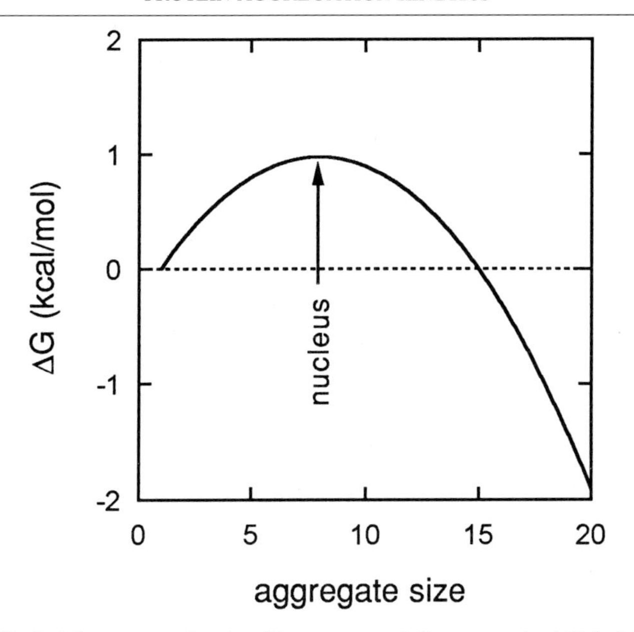

FIG. 1. Typical free energy barrier. Free energy of the aggregate ΔG (relative to the monomer) is shown on the vertical axis, whereas size of the aggregate is shown on the horizontal axis. The nucleus is the species whose size corresponds to the peak of the energy curve, and thus for which the population is smallest. Polymerization requires that the aggregate size pass through this maximum, which equates the reaction to a barrier crossing. The slope of the curve at any size n is controlled by the concentration times the ratio of rate constants, ck_n^+/k_{n+1}^-. To assume that the rate constants are independent of n on either side of the nucleus would imply that the slope is the same for different sizes n, in turn implying that the free energy is linear with n in that range. At large values of n, this assumption is reasonable.

specific stable structure fosters further growth. A thermodynamic nucleus, by its nature, is the least stable and hence least prevalent species in the reaction. While stable structural nuclei may also be scarce, nothing in the mechanism intrinsically requires their scarcity, and their stability permits a variety of strategies for their capture and study that are simply infeasible for unstable nuclei.

When the small concentration of nuclei effectively form a barrier to further growth, then the rate of the formation of polymers is set by the population of nuclei and the rate of elongation of the nuclei themselves, i.e., the rate of crossing this effective barrier. (For a more detailed treatment, see the section on Generalized Nucleation.) If c^* is the concentration of nuclei and J^* is the rate of elongation of the nucleus, then polymers at concentration c_p are formed at a rate

$$dc_p/dt = J^*c^* \tag{7}$$

Note that polymers may formally be counted by their ends. Once polymers are formed, they add mass by accretion to their ends. If the assumption is made that polymer addition is all by the same rates J which do not depend on size n, then if we call Δ the total concentration of monomers that have gone into polymers,

$$d\Delta/dt = Jc_p \tag{8}$$

If all molecules must be classed as either polymers or monomers, then the original concentration c_o becomes split into these, and we can write,

$$\Delta(t) = c_o - c(t) \tag{9}$$

The accuracy of this separation into polymers and monomers depends on the rarity of intermediate species, and this should be a good approximation.

J and J^* clearly depend on c as well. The solution of this set of Eqs. (7)–(9) is not simple, although given values for the various parameters, it is straightforward to construct a numerically integrated solution (see Fig. 2). It is possible to obtain an analytic solution for such equations when the

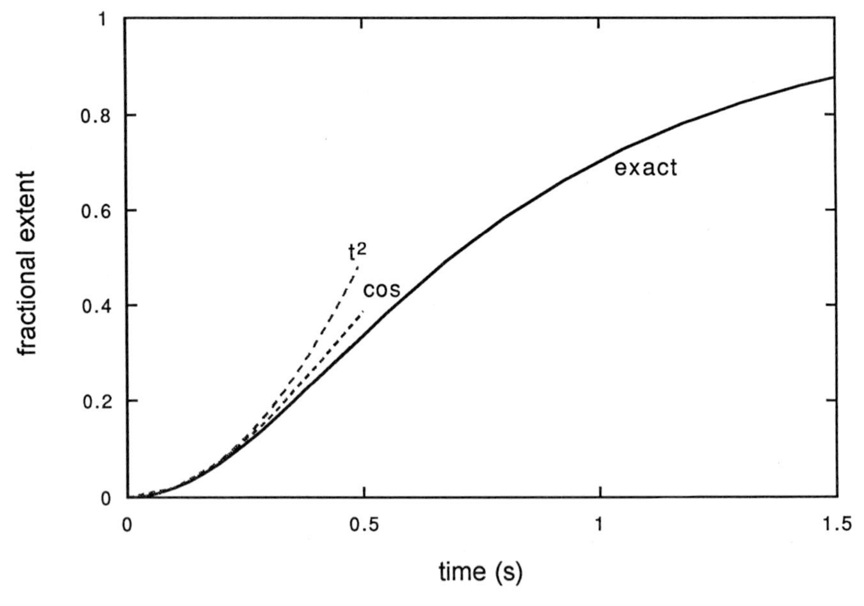

FIG. 2. Nucleation-controlled aggregation kinetics. The solid curve is an exact, numerically integrated solution to Eqs. (7)–(9), in which it is assumed that $J^* = k_+c$. In the numerical solution, it is not assumed that the forward rates are much greater than the reverse rates. At long times, the exact solution goes to 1. The long dashed curve, labeled t^2, shows the result of simply treating the monomer concentration as a constant, as described in Eq. (18). The short dashed curve, labeled cos, shows the result of using the linearized equations, as given by Eq. (12). Note that the cosine solution also begins as t^2 but is closer to the exact solution.

forward rates significantly exceed the reverse,[2] as well as to deduce the concentration dependence from scaling arguments independent of the actual solution of the equations.[3,4] We would like to introduce a different approach because of its general utility. The approach we take is known as a perturbation approach, whose central idea is to expand various quantities about their initial values in such a way that the resulting equations become linear and soluble.[5] In this approach, one formally expands all the equations about their initial values. For example,

$$J(c) = J(c_o) + (dJ/dc)_o(c - c_o) + \dots = J_o - (dJ/dc)_o\Delta + \dots \quad (10)$$

where J_o is defined as $J(c_o)$ and $(dJ/dc)_o$ means the derivative is evaluated at $c = c_o$. Only lowest order terms are retained. Smallness is formally defined relative to c_o, the initial concentration.

Then Eqs. (7) and (8) become

$$dc_p/dt = J_o{}^*c_o{}^* - [d(J^*c^*)/dc]\,\Delta \quad (11a)$$
$$d\Delta/dt = J_o c_{po} \quad (11b)$$

Equation (11b) contains no higher terms because c_{po} begins as a small term intrinsically, in contrast to c_o, which is the initial value of c and which is not small at all.

The solution to the just-described set of equations has the form

$$\Delta = A[1 - \cos(Bt)] \quad (12)$$

Before examining this further, the reader should note that the cosine is only employed near the initial time, so that Bt never becomes large and the oscillatory behavior of the cosine function is not seen.

In terms of parameters that appear in the original rate equations,

$$A = \frac{J_o^* c_o^*}{\dfrac{d}{dc}(J^*c^*)} \quad (13a)$$

and

$$B^2 = J_o[d(J^*c^*)/dc] \quad (13b)$$

[2] F. Oosawa and S. Asakura, "Thermodynamics of the Polymerization of Protein." Academic Press, New York, 1975.

[3] R. F. Goldstein and L. Stryer, *Biophys. J.* **50,** 583 (1986).

[4] H. Flyvbjerg, E. Jobs, and S. Leibler, *Proc. Natl. Acad. Sci. U.S.A.* **93,** 5975 (1996).

[5] M. F. Bishop and F. A. Ferrone, *Biophys. J.* **46,** 631 (1984).

The product of these two is also interesting, giving

$$B^2A = J_o J^*_o c^*_o \tag{13c}$$

In Eq. (12), B is clearly the effective rate constant and $1/B$ is the time constant for the reaction. So we ask how the initial rate constant depends on the concentration. Consider two cases. First suppose the nucleus is an aggregate of some size n, so that

$$c^* = K_{n^*} c^{n^*} \tag{14}$$

where K_{n^*} is used to indicate the equilibrium constant for association of n^* monomers. Let us also assume that the reverse rate k^*_- can be neglected so that

$$B^2 = (n^* + 1)k^*_+ J K_{n^*} c^{n^*} \tag{15}$$

If, in fact, the forward rates are all greater than the reverse rates (i.e., $k^*_+ \gg k^*_-$ and $k_+ \gg k_-$), then $B \sim c^{(n^*+1)/2}$, or the characteristic time $(1/B) \sim c^{-(n^*+1)/2}$. This is a familiar result[2]; i.e., a plot of log B vs log c has a slope of $(n^* + 1)/2$, or perhaps slightly less if the depolymerization rates cannot be ignored. However, the salient feature of an equilibrium nucleus in a simple linear elongation reaction will be rates with higher than unity dependence on concentration.

The other extreme occurs when the concentration of nuclei is fixed, as when they are provided by using preformed seeds. Returning to Eq. (11a), the concentration of nuclei is fixed in this case and therefore is not given by a monomer equilibrium. Mathematically, this means c^* does not depend on c and hence $dc^*/dc = 0$, and then

$$B^2 = Jc^* dJ^*/dc = k^*_+ c^*(k_+ c - k_-) \tag{16}$$

Now the concentration dependence of B is sublinear, and at best B will go as the square root of c! For either of these extremes, however, the time course of the reaction is similar. Its leading term is parabolic (i.e, t^2). The effect of this parabolic initiation is to give a weak delay at the start of the reaction.

Note that the shape of the curve is the same regardless of whether nuclei are formed. Such a curve is sometimes taken as indicative of nucleus formation, but as we have seen just now, even with a supply of preformed nuclei and no new nuclei created, the time dependence will be the same as the case when nuclei are in equilibrium with monomers. For both preformed nuclei and thermodynamic nuclei, all nuclei appear at the start of the reaction, and the upward curvature has nothing to do with their formation. As we will see later, it is possible to have a much more abrupt curve than

t^2 (equivalent to a more pronounced delay). It is also evident that nuclei do not disappear after the lag period, even in the case of thermodynamic nucleation. To verify this comment, consider that the concentration of polymers may be obtained using Eq. (12) in Eq. (11b) to give $c_p = (BA/J_o) \sin(Bt)$. For nucleation to abate after the lag requires the concentration of polymers to fall, but from the foregoing it is clear that c_p only abates slightly during the weak "lag phase."

If one observes such a shape of the time course (t^2 or $\cos Bt$), what strategies are appropriate? Clearly it is helpful to know J, so that some method of following the growth of polymers is desirable. A very effective way to do this is by differential interference contrast (DIC) microscopy.[6] Next, the parameters A and B that describe the growth are determined by curve fitting the initial growth phase. The concentration dependence of B^2A [using Eq. (13c)] will give n^*, the nucleus size. K_{n^*} and J^* remain to be determined and are more problematic as they may apply to the properties of the least populous species, the nuclei. (Of course, if nuclei are preformed, then the issue is quite different). It may be possible to relate K_{n^*} to other known equilibria of the system. As described later, k_-^* can be neglected. Even so, one is left with a product of k_+^* and K_{n^*}, which cannot be separated empirically. One approach is to assume that $k_+ = k_+^*$, thereby placing the entire difference of J^* and J in the off rates. This is not likely to be a bad approximation, as discussed later.

One might ask if the expansion and linearization process represents mathematical overkill. For example, it might appear intuitive to assume that near the beginning of the reaction all variables take their original values, i.e., $c = c_o$, $J = J_o$, and $c^* = c_o^*$. Then direct integration of Eq. (7) gives

$$c_p = J^* c^* t \tag{17}$$

from which

$$\Delta = \tfrac{1}{2} J J^* c^* t^2 \tag{18}$$

If our previous expression [Eq. (12)] for Δ were to be expanded, this would be the leading term, so this simple idea is not far off. However, it is not at all apparent from this equation if the t^2 is a lower or upper limit; as we shall see later, augmented pathway polymerization can give way to exponential growth after beginning with a t^2 time dependence. It is also not so clear what should be viewed as the rate constant in Eq. (18). For example, one might falsely conclude that the product JJ^* provides the effective rate constant, as it has the units of reciprocal time squared, leaving

[6] R. E. Samuel, E. D. Salmon, and R. W. Briehl, *Nature* **345,** 833 (1990).

c^* to provide the amplitude. That would suggest incorrectly a rate constant with very low concentration dependence. While the simple process of equating of all terms to initial values without expansion is very convenient, it must be viewed as a kind of "quick and dirty" approach, rather than a rigorous one. The linearization approach described earlier, although limited to the initial reaction, is rigorous if applied consistently. Moreover, it can be used in other ways, such as constructing a solution of equations with an activation step for which the initial values of some parameters would be zero.

Two important points for the analysis of assembly must be made here. First, the t^2 dependence alone is not that unique, and other reactions may lead to such a relationship. Rather it is the concentration dependences of the rate constants that serve to provide true diagnostics. Second, t^2 represents the maximal time dependence of the initial course, with further time "flattening" the curve. As we shall see later, other initial t^2 results can have the opposite effect, namely that their time dependence exceeds the initial value. The reason for this is that in Eq. (11a) the coefficient of Δ is negative, i.e., the further the reaction proceeds, the slower the rate. However, the opposite can happen and this is discussed next.

Polymerization with Secondary Pathway

It is possible to have a term in the rate of growth of the polymer concentration [i.e., Eq. (11a)] that is positive so that the reaction accelerates rather than decelerates. Three possibilities readily come to mind: fragmentation, branching, and heterogeneous nucleation. We distinguish these in the following way: fragmentation is the result of breaking polymers to produce new polymer ends onto which growth may occur. The simplest model for this process is that it occurs with a rate proportional to Δ, i.e., that breaks are possible anywhere, and because polymers are long and linear, Δ is a good measure of the net length of all polymers. (Fancier models are possible in which breakage is proportional to higher orders of Δ, but that will only be manifest as higher order terms in the expansions discussed earlier.[7] Fragmentation appears to be operative for some condition of actin filament growth.[8]

Branching is almost as simple as fragmentation and represents the beginning of a new polymer from an existing site by the addition of the first monomer to that site. (In other words, a polymer is not branched until a branch site begins the new polymer.) This is given by a term $\phi k_{\text{branch}} Jc$ in

[7] T. L. Hill, *Biophys. J.* **44,** 285 (1983).

[8] A. Wegner and P. Savko, *Biochemistry* **21,** 1909 (1982).

which ϕ represents the frequency of branch sites and $k_{\text{branch}}J$ is the rate constant for addition of a monomer to such a site.

Finally, it is possible to nucleate on the surface of a polymer. This resembles branching in structural terms, but thermodynamically it is distinct in the same way downhill polymerization and nucleated polymerization are distinct, namely the new polymer is incapable of growth until a minimum number of monomers are present that form a heterogeneous nucleus. This effect is described by a term $\phi K^{**}J^{**}c^{**}$, where ϕ is the fraction of sites that can support heterogeneous nucleation, K^{**} is the equilibrium constant for attaching a heterogeneous nucleus to that site, J^{**} is the rate of elongation of the heterogeneous nucleus, and c^{**} is the concentration of heterogeneous nuclei. Naturally, c^{**} expected to be related to the monomer concentration by the same type relationship as Eq. (14), namely $c^{**} = K_{m*}c^{m^*}$, where in general the nucleus sizes n^* and m^* are not equal. Heterogeneous nucleation has been observed in sickle hemoglobin polymerization.[6,9,10]

If we denote these various processes by Q, then Eq. (11a) becomes

$$dc_p/dt = J_o^*c_o^* + [Q - (dJ^*c^*/dc)]\,\Delta \tag{19}$$

with no change to Eq. (11b).

The solution of Eqs. (11b) and (19) is then

$$\Delta = A(\cosh Bt - 1) \tag{20}$$

The cosh function begins as t^2 and then (for large Bt) becomes an exponential, $\exp(Bt)$ (Fig. 3). (Note that it is now possible to have Bt large but still remain within the consistent limits for the solution. However, in the $\cos Bt$, the solution is limited to small Bt.)

Now we have

$$A = \frac{J^*c^*}{Q - \dfrac{d}{dc}(J^*c^*)} \tag{21a}$$

and

$$B^2 = J\left[Q - \frac{d}{dc}(J^*c^*)\right] \tag{21b}$$

The secondary process Q has affected both the rate parameter B and the amplitude parameter A. Remarkably, however, the product B^2A remains unaffected by the presence of the secondary process and is still given by

[9] F. A. Ferrone, J. Hofrichter, H. Sunshine, and W. A. Eaton, *Biophys. J.* **32,** 361 (1980).
[10] F. A. Ferrone, J. Hofrichter, and W. A. Eaton, *J. Mol. Biol.* **183,** 591 (1985).

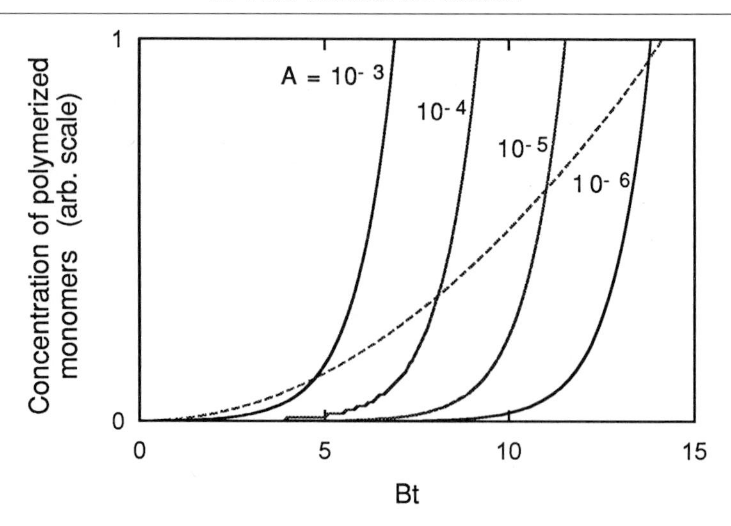

FIG. 3. Polymerization kinetics with a secondary pathway. The concentration of polymerized monomers, Δ, is shown as a function of *Bt,* as given by Eq. (20). All curves are in the exponential limit of the function. *A* is varied from 10^{-3} to 10^{-6} as labeled. Note that all curves start from time 0, but the exponential time course gives rise to the apparent delay. A curve of the form $\Delta = A(Bt)^2$ is shown as the dashed line for comparison [cf. Fig. 2 and Eq. (18)]. Contrast the delay or lag with that in Fig. 2. Only an exponential (or high power of time) will give the abruptness shown here.

Eq. (13c) and permits a means of deducing the concentration of homogeneous nuclei c^* without a precise specification of the process Q.

The strategy for analyzing assembly now becomes the following. When an abrupt time course for assembly is seen, the curve should be fit to Eq. (20). If this succeeds, the analysis of B^2A gives J^*c^* and the concentration dependence of B^2A gives the nucleus size. By observing the growth of polymers it is possible to determine J as before. Then from either B or A (B being preferred) one can isolate Q and study its concentration dependence to identify which of the just-described types of secondary process it might be.

Once again a word of caution is in order. The apparent lag or delay is the consequence of the exponential time dependences and is not a phenomenon of nucleation at all! For example, it is entirely possible for a downhill polymerization to have no nucleus, i.e., $c^* = c$, but yet possess a secondary process that then gives the reaction an exponential time course and a distinct lag time. The lack of nucleation in that example would then be revealed in the concentration dependence of B^2A rather than in the shape of the curve itself. Again, as described earlier, the lag time cannot be associated with a unique period during which nuclei are being formed.

Inserting Eq. (20) into Eq. (11b) reveals that the concentration of polymers is almost exponential, i.e., most nuclei are formed after the lag!

With a secondary process, the reaction assumes a spatial character, as the presence of a given polymer affects the likelihood of forming others. This has been explored only slightly, but the interested reader may wish to consult Zhou and Ferrone[11] or Dou and Ferrone[12] for some approaches to this issue.

An apparent lag time, sharper than seen in the t^2 dependence described earlier, is also found in cascade-type reactions, i.e., downhill polymerizations. If there are no reverse rates it is easy to show by direct integration that the reaction moves forward with a power law dependence. What is not immediately obvious, but is demonstrated readily, is that the concentration dependence of the characteristic rate is simply linear or, if the characteristic time is denoted by τ, then

$$\Delta \sim (t/\tau)^n \tag{22}$$

and

$$\tau \sim 1/c \tag{23}$$

so that

$$d \log \tau / d \log c = -1 \tag{24}$$

Thus, without a secondary pathway, either the time dependence or the concentration dependence can be high, but not both.

This concludes the description of basic nucleation–elongation kinetics. We now turn to a series of more detailed topics.

Generalized Nucleation: Effect of Near-Nuclear Species

So far our approach to describing nucleation has been based on the notion that a single species creates the bottleneck for growth. What if there are a few species in the pathway that have very small concentrations? This section provides a more rigorous way to deal with this problem following the treatment by Burton[13] (which is based on the venerable treatment by Becker and Döring[14]). While the exercise may appear somewhat academic, its importance lies in showing how the assumption of a single species in

[11] H.-X. Zhou and F. A. Ferrone, *Biophys. J.* **58,** 695 (1990).

[12] Q. Dou and F. A. Ferrone, *Biophys. J.* **65,** 2068 (1993).

[13] J. J. Burton, *in* "Nucleation Theory" (B. J. Berne, ed.). Plenum Press, New York, 1977.

[14] R. Becker and W. Döring, *Ann. Physik* **24,** 719 (1935).

the reaction path affects the final results, as well as the justification for ignoring the back reaction in nucleation theories.

We consider the equations for the change in c_i, the concentration of the ith species, $i \geq 2$

$$\frac{dc_i}{dt} = k_{i-1}^{+} c c_{i-1} + k_{i+1}^{-} c_{i+1} - (k_i^{+} c - k_i^{-}) c_i \tag{25}$$

Now define a flux, F_i

$$F_i \equiv k_i^{+} c c_i - k_{i+1}^{-} c_{i+1} \tag{26}$$

so that

$$\frac{dc_i}{dt} = F_{i-1} - F_i \tag{27}$$

At steady state the flux through all the states is the same, i.e.,

$$f \equiv k_i^{+} c c_i - k_{i+1}^{-} c_{i+1} \tag{28}$$

The equilibrium populations will be denoted here by uppercase letters, i.e., C_i, and are such that no flux exists. In other words,

$$k_i^{+} C C_i = k_{i+1}^{-} C_{i+1} \tag{29}$$

Using this relationship to eliminate the back rates, we can write

$$f = k_i^{+} C C_i \left(\frac{c c_i}{C C_i} - \frac{c_{i+1}}{C_{i+1}} \right) \tag{30}$$

Define $s(t) \equiv c/C$. Then

$$\frac{f}{k_i^{+} C C_i} = s(t) \frac{c_i}{C_i} - \frac{c_{i+1}}{C_{i+1}} \tag{31}$$

and then if we sum over i and assume that, past some size N, $c_{N+1} \sim 0$, we get

$$\sum_{i=1}^{N} \frac{f}{k_i^{+} C C_i} = s(t)^2 + \left[s(t) - 1 \right] \sum_{i=2}^{N} \frac{c_i}{C_i} \tag{32}$$

Initially there are so few aggregates in the system that $c \approx C$, or $s(t \approx 0) \approx 1$, from which we get the relation

$$f_o = \left[\sum_{i=1}^{N} \frac{1}{k_i^{+} C C_i} \right]^{-1} \tag{33}$$

Another way to view this relationship is that the time for nucleation, $(1/f_o)$, is the sum of the times of all the preceding steps. If there is one species whose concentration C^* is much lower than all others, its reciprocal will dominate the sum and we have

$$f_o = k_*^{\ddagger} C C^* \tag{34}$$

Because the flux through these states is the rate of formation of polymers, dc_p/dt, Eq. (34) is essentially Eq. (7) (in which we had not distinguished between equilibrium values C and instantaneous steady-state values c).

Note that the reverse rate has disappeared, i.e., that $J^* = k_*^{\ddagger} C$. The validity of this was verified in an independent way by Goldstein and Stryer.[3] Equally important, it is also evident how one is to include species of size near that of the critical nucleus, if the concentration of nuclei is not significantly less than the species of similar size. If all the monomer addition rates are equal, the result is particularly simple, namely

$$f_o = k_*^{\ddagger} C \left[\sum_{i=1}^{N} \frac{1}{C_i} \right]^{-1} \tag{35}$$

This therefore provides a systematic way to include species of size similar to the nucleus. Conversely, if one assumes a single rate-limiting species, of concentration C^*, what one is actually determining is an effective concentration whose reciprocal $(1/C^*)$ is equal to the sum of the reciprocals of the concentrations of aggregates near the nucleus. Hence the concentration of nuclei in the single-species assumption underestimates the actual concentration of nuclei; this approximation will be improved by including more species, as is evident from Eq. (35).

The time required to establish the steady-state flux is also a matter of interest. Roughly it goes as $n^*\tau$, where n^* is the size of the nucleus and τ is the step time, approximately $1/J^*$ (see Firestone *et al.*[15]). Then $n^*\tau$ needs to be compared to $1/B$. In most cases, the time to establish the flux is short, i.e., $n^*\tau \ll 1/B$, as generally $n^3 \ll c/c^*$ where the latter inequality is based on the relative smallness of c^*.

Some Physical Issues

Given the correct phenomenology, one would wish to rationalize or deduce from first principles the nature of the rate constants and the nucleation barrier. It becomes logical to think in terms of an energetic barrier rather than in terms of the rate constants themselves as the nucleus is being treated as being in equilibrium with the monomers.

[15] M. P. Firestone, S. K. Rangarajan, and R. de Levie, *J. Theor. Biol.* **104,** 535 (1983).

Early approaches to nucleation imagined an abrupt transition such as might occur in the closure of a ring in the formation of helical polymers.[2,15] This was based on the need to justify the change in the free energy by invoking a greater number of contacts once the ring closed. It is important to ensure that such models are faithful to the assumptions of thermodynamic nucleation and that the contacts along the initial chain are not weaker than those up and down, as otherwise the structure will form in short double layers in preference to the single strand that later wraps around itself. As pointed out earlier, it is not necessary to have a change in structure such as this for a nucleation barrier. If the nucleus is the result of simple thermodynamics of small clusters, its size will be a function of the initial concentration, and the nucleation barrier will be curved rather than having a cusp as for a helix closure. In the case of a thermodynamic barrier, it is possible to construct models based on simple thermodynamic principles that reduce the nucleus calculation to energetic parameters. These models are beyond the scope of this article, but suffice it to say that the guiding principle is the competition between the free energy increase due to contacts within the nucleus versus the loss in entropy due to immobilization of the monomers in aggregates. In terms of entropic considerations, it is also important to include the redemption of entropy arising from vibration of the monomers themselves within the framework of the aggregate.[9,16–18] This is easily overlooked and can be very significant.

An important feature of the thermodynamic models is that the nucleus size depends on the initial concentration.[17] This is easy to rationalize in physical terms. The loss in translational entropy is the result of relative immobilization of the monomers in solution. The initial concentration essentially determines the volume each molecule has to "wander around in." In a higher concentration solution, the entropy loss is not so dire as in a solution of lower concentration. This can easily create confusion if different experiments are performed at drastically different concentrations, which can give rise to different size nuclei.

The nucleus elongates with a rate we have denoted as J^*, whereas long polymers elongate at a rate J. In the absence of other specific information, should one plan to model the difference in J^* and J in addition rates or in off rates? To gain some insight into this question, it is worth observing that on rates are controlled by long-range forces whereas off rates are controlled by short-range forces. Therefore, the association rates will be limited by diffusion control, with some possible additional effects of electro-

[16] H. P. Erickson and D. Pantaloni, *Biophys. J.* **34,** 293 (1981).
[17] F. A. Ferrone, J. Hofrichter, and W. A. Eaton, *J. Mol. Biol.* **183,** 611 (1985).
[18] Z. Cao and F. A. Ferrone, *Biophys. J.* **72,** 343 (1996).

statics. Conformational changes that are required for the assembly can slow the association further, but there is not much to speed them up. However, the dissociation rates are basically governed by the same things that affect the equilibrium constant and possess a far wider range of variability. In short, variation of the dissociation rates is more likely than the association rates.

Stochastic Methods

Polymerization reactions with a secondary pathway show exponential proliferation of polymers, meaning a single polymer, and hence a single nucleation event, can lead to a large and observable polymer mass. Because the origin of these polymers is one molecular event, the time of the event following initiation must be intrinsically random. That is, the first nucleation represents the random walk process that leads to crossing the nucleation barrier. This idea was first described in detail by Hofrichter,[19] and a concise formula was subsequently developed by Szabo.[20] In this case, it is possible to collect an ensemble of times $T(t)$ and thereby deduce the nucleation rate, denoted by ζ. The rate ζ is related to nucleation rate constant f by Avogardo's number N and the volume V in which the reaction is observed, i.e.,

$$f = \zeta / NV \tag{36}$$

If the 10th time (i.e., time for the reaction to reach 1/10 its final value) occurs at a time $\langle t \rangle$ for a volume sufficiently large that fluctuations are not significant (i.e., that many nucleations occur), then the constant ν is defined as

$$\nu = e^{B\langle t \rangle} \tag{37}$$

Usually, $\nu \gg 1$. With these definitions, Szabo[20] showed that the distribution of 10th times can be written as

$$T(t) = \frac{B\Gamma(\nu + \zeta/B)}{\Gamma(\nu)\Gamma(\zeta/B)} e^{-\zeta t}(1 - e^{-Bt})^{\nu-1} \tag{38}$$

where Γ is the gamma function. This has been implemented successfully to determine nucleation rates in sickle hemoglobin assembly.[18,21,22] Note, however, that the formalism is applicable regardless of the nature of the

[19] J. Hofrichter, *J. Mol. Biol.* **189,** 553 (1986).

[20] A. Szabo, *J. Mol. Biol.* **199,** 539 (1988).

[21] Z. Cao and F. A. Ferrone, *J. Mol. Biol.* **256,** 219 (1996).

[22] Z. Cao, D. Liao, R. Mirchev, J. J. Martin de Llano, J.-P. Himanen, J. M. Manning, and F. A. Ferrone, *J. Mol. Biol.* **265,** 580 (1997).

secondary pathway. It is only necessary that the reaction display exponential growth from a single nucleation event.

Other Treatments

A variety of treatments of nucleation kinetics exist in the literature. One school arises from the droplet condensation/crystallization approach, which has advanced the art of nucleus calculation to a high order (see, e.g., the book by Abraham[23] or the more concise treatment by Burton[13]). However, the geometry of droplet or crystal growth is intrinsically three dimensional, not one dimensional, so that the utility of the work in that field does not extend very far in terms of analysis of polymerization.

The classic work of polymerization analysis is that of Oosawa and Asakura,[2] which contains a number of valuable insights regarding relevant time scales for growth and length distribution, for example. More rigorous and mathematical approaches to the basic problem of nucleated linear polymerization are given by de Levie and co-workers[15,24] and by Goldstein and Stryer.[3] The papers by de Levie's group provide a detailed look at the distribution of species of various sizes under the assumption that monomer concentration does not change. As discussed earlier, this is an intuitive assumption but not as rigorous as the perturbation expansion approach used here.

Goldstein and Stryer[3] adopt an interesting approach to the problem of assembly by producing dimensionless variables, thereby allowing critical pieces of the physics to emerge simply by observing how the parameters scale as concentration is changed, without requiring solutions to the equations involved (although a number of solutions are produced). A pitfall inherent in that type of approach is that physical variables become transformed into dimensionless quantities whose meaning is sometimes obscure. For example, they criticize the assumption that nuclei (which they refer to as "seeds") are in equilibrium when the monomer concentration is high enough to neglect depolymerization rates and nuclei are small. However, the central construct of nucleation is that depolymerization rates initially are greater than forward rates. Therefore, the incompatability of equilibrium nucleation with a high concentration and small nuclei is a problem built in by their assumptions, as the physical parameters themselves do not permit nucleation, regardless of the numerical or analytic solutions. In addition, Goldstein and Stryer[3] vary their model by the adjustment of on rates rather than off rates, which, as discussed earlier, is less likely than the reverse.

[23] F. F. Abraham, "Homogeneous Nucleation Theory." Academic Press, New York, 1974.
[24] S. K. Rangarajan and R. de Levie, *J. Theor. Biol.* **104,** 553 (1983).

Both of these weaknesses are disguised inadvertently by the use of the dimensionless approach and are a signal that this methodology should be employed with caution, although it can be useful in certain cases (especially as mentioned later).

With the accessibility of good integration and differential equation-solving routines and reasonably quick desktop computers to run them, there are an increasing number of investigators who prefer to construct a model and simply integrate the rate equations with no attempt at analytic solution. This approach has merit when the broad outlines of the problem are in place. However, the large number of variables makes it difficult for this approach to be exhaustive. For example, one must choose a nucleus size and then specify all rate constants before and after nucleation, in addition to whatever ancillary properties are being introduced (alternate pathways, activation steps, etc.) This makes it very tedious to demonstrate that a given description is necessary, rather than sufficient, to describe the observations. When numerical approaches are employed, the dimensionless approach of Goldstein and Stryer[3] can be extremely useful to ensure that various patterns of parameters are not explored needlessly.

Summary

Given a set of kinetic data, then, the preceding discussions suggest the following approach to its analysis.

1. For purposes of establishing the reaction, ignore the final stages and concentrate on the initial 10–20% of the reaction at first. A globally optimized model may be based on a faulty assumption for the initial steps. Thus, although the whole data set may look reasonably well fit, the reaction could be misrepresented, and thus the fit unhelpful if accuracy at the later stages has come at the expense of the initial phase of the reaction.
2. What is the time course of the initial reaction? (A) Is the reaction exponential? Exponential growth gives dramatic lag times (see Fig. 3), whereas nonexponential "lag times" have a visible signal from time 0 (i.e., Fig. 2). If the data set shows the abrupt appearance of signals after a period of quiescence, the chances are excellent that the time course is exponential. High sensitivity measurement of the signal at times during the lag phase should be used to confirm the exponential nature quantitatively. Exponential reactions mean a secondary pathway is operative. (a) A cascade (t^n) can look similar to an exponential, but may proceed from a multistep single-path reaction. Thus the exponential needs to be ascertained with some accu-

racy. (b) It is possible that some or all of the lag results from a stochastic process, i.e., formation of a single nucleus being observed. This, however, is likely to be accompanied by a secondary process, as few techniques are sensitive enough to detect a single polymer at a time, and having one nucleus form many polymers is a hallmark of a secondary process. Thus, the reproducibility of the kinetics must be established to rule out stochastics. If data show wide variation, stochastic methods as described earlier may be employed. (c) Given a secondary process, one must separate the primary nucleation process from the secondary process (by stochastic means or by use of the product B^2A, as described earlier). (B) If the reaction does not begin with an exponential, is it parabolic? If so, it falls in the general class of linear polymerizations.

3. What is the concentration dependence of the reaction(s)? This will separate nucleation processes from growth, and so on.
4. If the initial reaction is neither exponential nor parabolic, a reaction mechanism needs to be proposed and evaluated. Solving the resulting equations is best done by linearization, which has the best chance of giving equations whose solutions and their sensitivity to parameters are readily understood. If this proves fruitful, full numeric solutions may be useful.
5. At this point, the full reaction may be considered to completion.
6. The physical basis of the description (sizes of parameters and their dependencies) needs to be finally considered to ensure that the mathematical success of the description rests on tenable physical grounds.

Acknowledgment

The author is pleased to acknowledge helpful discussions with Professor H.-X. Zhou.

[18] Quantification of β-Sheet Amyloid Fibril Structures with Thioflavin T

By HARRY LEVINE III

Background

Fibrous tissue deposits staining blue with iodine after treatment with acid were called amyloid by Virchow in 1855 to describe their "starch-like"

0076-6879/99 $30.00

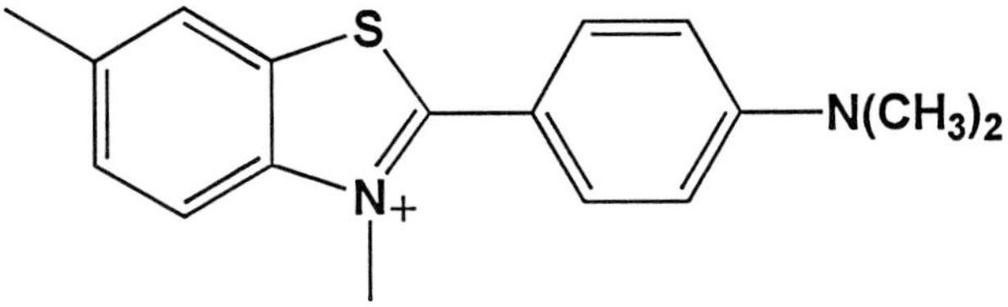

FIG. 1. Structure of thioflavin T.

appearance. Despite the presence of a significant amount of carbohydrate in these fibrils, the staining reaction was eventually shown to be due to the protein component. The histologic benzothiazole dyes thioflavin S (ThS) and thioflavin T (ThT) (Fig. 1) under the appropriate conditions selectively stain amyloid structures in a number of pathological settings,[1,2] as does the diazobenzidine sulfonate dye, Congo red, which is also birefringent when bound to fibrils. Phorwhite BBU, Sirius Red, and several other fluorescent and nonfluorescent aromatic molecules[3] also show this property.

Investigation of the amyloid fibril formation process requires not only the ability to distinguish the characteristic amyloid β-sheet structure from amorphous aggregates of the monomer or nonamyloid fibril forms of the precursor protein, but quantitation of the amyloid form as well. Congo red and thioflavin T undergo characteristic spectral alterations on binding to a variety of amyloid fibrils that do not occur on binding to the precursor polypeptides, monomers, or amorphous aggregates of peptide. Both dyes have been adapted to *in vitro* measurements of amyloid fibril formation. The metachromatic spectral shift of Congo red on binding to amyloid fibrils formed of various subunits has been used as the basis for the quantitation of amyloid fibril formation[4] and is the subject of another article in this volume.[4a] Radiolabeled Congo red derivatives bind to Aβ amyloid fibrils,[5] as does radiolabeled chrysamine G[6] and some derivatives.[7] The radiolabeled materials are problematic to synthesize and are not available commercially. Binding assays with these compounds require the inclusion of an organic cosolvent to demonstrate binding selectivity.[6]

[1] P. S. Vassar and C. F. A. Culling, *Arch. Pathol.* **68,** 487 (1959).
[2] G. Kelenyi, *J. Histochem. Cytochem.* **15,** 172 (1967).
[3] H. Puchtler, F. Sweat-Waldrop, and S. N. Meloan, *Histochemistry* **77,** 431 (1983).
[4] W. E. Klunk, J. W. Pettegrew, and D. J. Abraham, *J. Histochem Cytochem.* **37,** 1273 (1989).
[4a] W. E. Klunk, R. F. Jacob, and R. P. Mason, *Methods Enzymol.* **309** [19] (1999) (this volume).
[5] T. T. Ashburn, H. Han, B. F. McGuinness, and P. T. Lansbury, Jr., *Chem Biol.* **3,** 351 (1996).
[6] W. E. Klunk, M. L. Debnath, and J. W. Pettegrew, *Neurobiol. Aging* **15,** 691 (1994).
[7] W. E. Klunk, J. W. Pettigrew, and C. A. Mathis, Jr., International Patent WO 96/34853 (1996).

Naiki *et al.*[8–10] showed that apolipoprotein AII (ApoAII) and amyloid A fibril suspensions, but not monomers, bound ThT and ThS with distinct fluorescence spectral changes. ThS is a methylated, sulfonated polymerized primulin preparation whose structure has not been characterized. Although commonly used for the histological demonstration of amyloid fibrils where long wavelength UV excitation can be used and the free dye can be washed away, commercial preparations of ThS are complex mixtures of molecules. Binding to amyloid fibrils enhances the emission intensity of ThS severalfold with no change in the excitation or emission spectra, resulting in a high background fluorescence in solution, which makes this dye unsuitable for quantitative analysis in solution.

With ThT, amyloidogenic Aβ synthetic peptides derived from the major protein of the Alzheimer's disease senile plaque protein give rise to a large fluorescence excitation spectral shift that allows selective excitation of the amyloid fibril-bound ThT.[11] Figure 2 shows excitation and emission spectra of the dye in the absence and presence of β(1–42). Precursor monomers and low multimer (oligomer) complexes,[11] high β-sheet content, and $\alpha\beta$ proteins fail to elicit the fluorescence changes induced by amyloid fibrils.[12] The fibrillar β-sheet synthetic homopolymeric amino acids poly(L-serine) and poly(L-lysine) (β form) do not react with ThT, nor do amyloid fibrils of a truncated human amylin peptide (20–29).[12] Many other types of amyloid or β-sheet fibrils, fortunately, are reactive, including polyglycine (β form), insulin,[12] transthyretin,[13] β_2-microglobulin,[12,14] ApoAII,[8,14] amyloid A,[10] C410Y presenilin 1 mutant, N141I mutant presenilin 2,[15] immunoglobulin light chain AL,[16] tau parahelical filaments,[17] gelsolin,[18] and dialysis-related amyloid (DRA)[19] fibrils. This selective response forms the basis for

[8] H. Naiki, K. Higuchi, M. Hosokawa, and T. Takeda, *Anal. Biochem.* **177,** 244 (1989).
[9] H. Naiki, K. Higuchi, K. Matsushima, A. Shimada, W. H. Chen, M. Hosokawa, and T. Takeda, *Lab. Invest.* **62,** 768 (1990).
[10] H. Naiki, K. Higuchi, K. Nakakuki, and T. Takeda, *Lab. Invest.* **65,** 104 (1991).
[11] H. LeVine III, *Prot. Sci.* **2,** 404 (1993).
[12] H. LeVine III, *Amyloid Int. J. Exp. Clin. Invest.* **2,** 1 (1995).
[13] M. J. Bonifacio, Y. Sakaki, and M. J. Saraiva, *Biochim. Biophys. Acta* **1316,** 35 (1996).
[14] H. Naiki, K. Higuchi, A. Shimada, T. Takeda, and K. Nakakuki, *Lab. Invest.* **68,** 332 (1993).
[15] C. P. Maury, E. L. Nurmiaho-Lassila, and M. Liljestrom, *Biochem. Biophys. Res. Commun.* **235,** 249 (1997).
[16] Y. M. Tagouri, P. W. Sanders, M. M. Picken, G. P. Siegal, J. D. Kerby, and G. A. Herrera, *Lab. Invest.* **74,** 290 (1996).
[17] O. Schweers, E. M. Mandelkow, J. Biernat, and E. Mandelkow, *Proc. Natl. Acad. Sci. U.S.A.* **92,** 8463 (1995).
[18] C. P. Maury, E. L. Nurmiaho-Lassila, and H. Rossi, *Lab. Invest.* **70,** 558 (1994).
[19] D. Brancaccio, G. Ghiggeri, A. Gaberi, A. Anelli, F. Ginevri, G. Loggi, and R. Gusmano, *Biochem. Biophys. Res. Commun.* **176,** 1037 (1991).

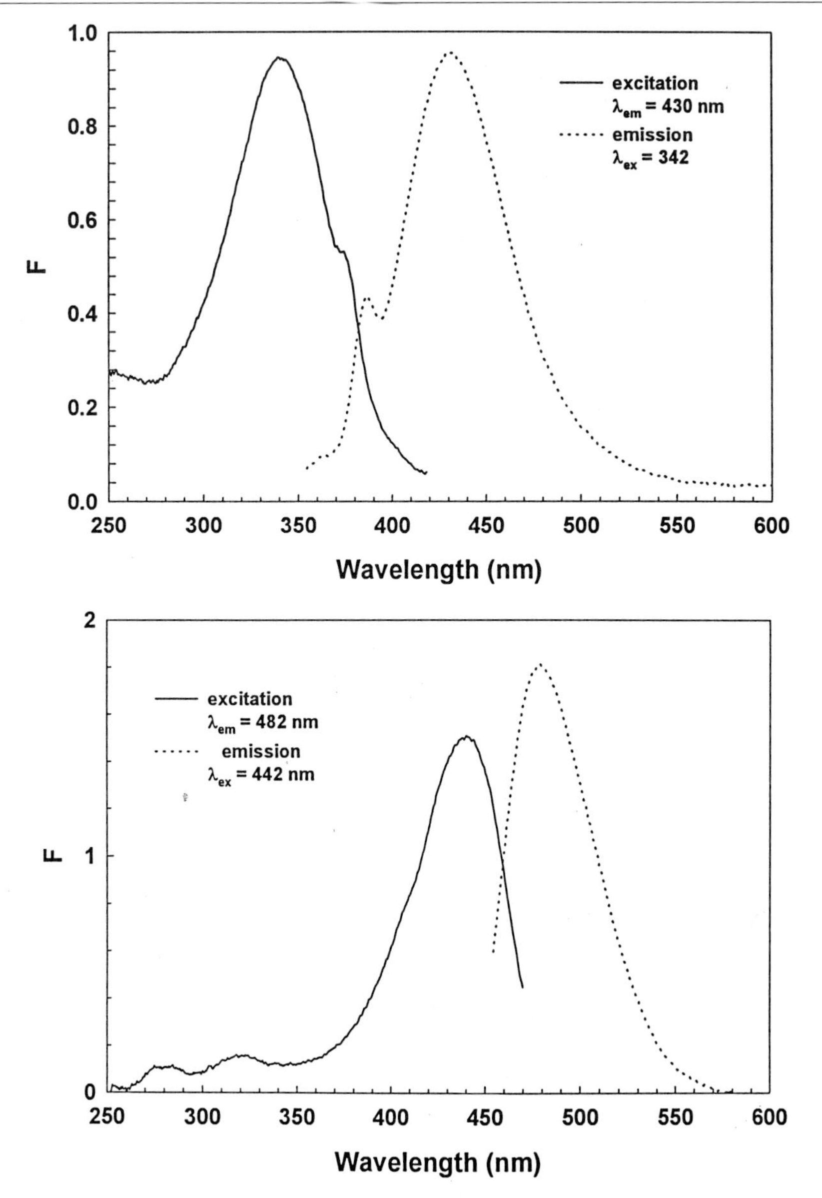

FIG. 2. Fluorescence spectra of free and $\beta(1\text{–}42)$-bound thioflavin (ThT): 5 μM ThT in 50 mM glycine–NaOH, pH 8.5; 2 μM peptide when present. Excitation/emission fluorimeter slits = 3/10 nm. Solid line, excitation; dotted line, emission spectra. (*Top*) 5 μM ThT alone. (*Bottom*) 5 μM ThT + 2 μM $\beta(1\text{–}42)$ amyloid fibrils.

an assay utilizing ThT fluorescence changes as an indicator of amyloid fibrillar structure.

Basis of ThT Spectral Alteration on Binding to β-Peptide Amyloid Fibrils

The biophysical mechanism by which amyloid fibrils induce the notable 115-nm hypochromic ThT spectral red shift is presently unknown. The pronounced excitation spectral changes, coupled with the lack of an effect on the emission spectrum, suggest that the ground state of the chromophore is altered by binding to amyloid fibrils rather than the excited state, which is usually associated with the emission spectrum. The difference absorption spectrum for ThT in the presence of $\beta(1-40)$ fibrils shown in Fig. 3 displays a maximum around 440 nm, clearly shifted from both the free dye absorbance (400 nm) and the fluorescence excitation maximum (335 nm). $\beta(1-42)$ fibrils induce the same spectral changes. A classical metachromatic transition (blue shift) of chromophores such as Congo red is usually caused by dye stacking. There is no spectral evidence for ThT dimerization in solution at concentrations up to 10 m*M* using short path length cells. Explanations invoking dye rotational restriction or environmental dielectric constant are ruled out by a lack of a solvent viscosity spectral effect on the

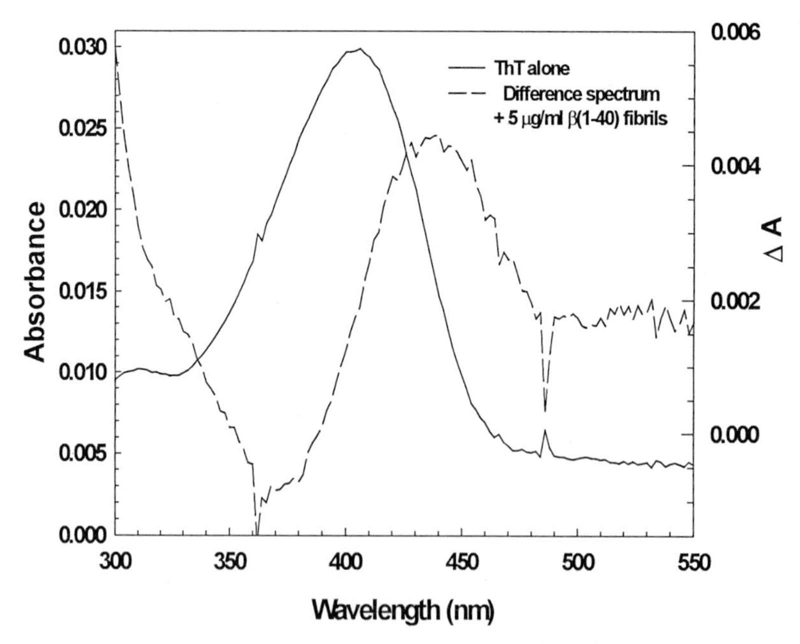

FIG. 3. Absorption spectrum of free and fibril-bound ThT: 5 μM ThT in 50 m*M* glycine–NaOH, pH 8.5; 5 μg/ml $\beta(1-40)$ peptide when present. Solid line, ThT alone; dashed line, difference spectrum of ThT [(ThT + fibrils)-ThT].

excitation spectrum, and low dielectric solvents enhance only the emission intensity. The characteristic excitation maximum of amyloid fibril-bound ThT appears in the presence of high concentrations of polyhydroxy solvents such as ethylene glycol and glycerol, suggesting that the dye-binding site may possess similar hydrogen-bonding characteristics.[12]

The principle of the ThT determination of amyloid fibril formation depends on the unique fluorescence of the ThT : amyloid fibril complex. Scatchard plot analysis of dye absorbance indicates that β(1–40) amyloid fibrils bind substoichiometric amounts of ThT. Job plot analysis of binding site heterogeneity indicates that several sites of similar apparent affinity contribute to the fluorescence signal, although not all bound dye molecules may be spectroscopically visible.[20] Some ThT molecules therefore can apparently bind to a fibril without eliciting the signal characteristic of the amyloid state. Because fluorescence depends on binding site number, affinity, and quantum yield, it is difficult to obtain an exact relationship between ThT fluorescence and absolute amyloid fibril concentration. Standard curves are constructed with the particular peptide under study to normalize the determinations. Comparable selectivity for amyloid fibrils is reflected in Congo red metachromasy but probably not in the binding of radiolabeled dye molecules. Direct comparisons of binding and metachromasy have not been published. Thus, the ThT and Congo red spectroscopic techniques are favored for studies of complex systems where the structure of the complexes is unknown. It is useful to be able to infer something about the repeating β-sheet structure of the fibrils and to distinguish them from irregular aggregates.

The apparent affinity of ThT for different amyloid fibrils depends on the identity of the fibril building block, ranging from 0.033 μM (amyloid A protein) to 11 μM [Aβ(1–40)].[12] Because apparent affinities for ThT usually increase with higher pH due to the quaternary positively charged benzothiazole ring nitrogen, measurements are often made between pH 8 and 9.[12] The stability of the fibrils also depends on the identity of the peptide subunits as well as the pH, so measurements must often be made rapidly, less than 1 min, after dilution of the samples into the high pH solution that optimize the ThT signal but which may depolymerize the fibrils.

Materials and Methods

Synthetic β(1–40) peptide, TFA salt (BACHEM, Torrance, CA) is dissolved in ice-cold HPLC-grade water at 0.5 mM (pH 1.6) and frozen immediately in 50- to 100-μl aliquots at either $-20°$ or $-80°$. On storage

[20] H. LeVine III, *Arch. Biochem. Biophys.* **342,** 306 (1997).

at either temperature the peptide seeds spontaneously over the period of weeks to several months. Fresh unopened lyophilized peptide or peptide disaggregated by treatments involving neat trifluoroacetic acid,[21] concentrated formic acid,[22] and/or 1,1,1,3,3,3-hexafluoro-2-propanol[23,24] (HFIP) exhibits a prolonged lag phase before fibril formation is apparent. Different batches of peptide from the same supplier and material from various suppliers will display altered kinetics of fibril formation. Similar irreproducibility is observed for other methods following fibril formation and is generally ascribed either to the conformational ambiguity of Aβ peptides, which leads to its amyloidogenicity, or to the presence of variable contamination by a partially deblocked peptide. Any batch of Aβ peptide should be characterized under standard conditions before experimental use. Matrix-assisted laser dissociation mass spectroscopy (MALDI-MS) and capillary electrophoresis are particularly useful in detecting impurities. Fibrils formed from other amyloidogenic Aβ peptides such as 1–28, 1–42, 12–28, and 25–35 yield different ThT fluorescence signal intensities without a drastically altered stoichiometry of ThT binding, suggesting that the fluorescent yield of ThT depends dramatically on the nature of the peptide forming the fibril structure.

Assay Method

The fluorescent yield of ThT is relatively low for most amyloid fibrils, necessitating the use of microgram quantities of fibrils for detection, although the use of a 96-well microtiter plate reader decreases the assay volume (100–200 μl), thus reducing the consumption of fibrils (1–3 μg per assay). Fibril formation is frequently measured in aliquots of a fibril-forming reaction. Light scattering at the wavelengths and peptide concentrations used rarely contributes substantially to the observed fibril-specific ThT signal because of the excitation/emission filter cutoffs and sample dilution.

Steady-state fluorescence measurements for Aβ(1–40) can be performed with a Cytofluor 2350 fluorescence plate reader or equivalent instrument (filters: excitation 440/20 nm and emission 485/20 nm for ThT) in round-bottom polystyrene 96-well microtiter plates (Corning, NY) containing 250 μl/well of 5 μM ThT in 50 mM glycine–NaOH, pH 8.5.[12] Any standard fluorimeter can be used, preferably with a cuvette volume of 0.5

[21] S. C. Jao, K. Ma, J. Talafous, R. Orlando, and M. G. Zagorski, *Amyloid Int. J. Exp. Clin. Invest.* **4,** 240 (1997).

[22] A. E. Roher, K. C. Palmer, V. Chau, and M. J. Ball, *J. Cell Biol.* **107,** 2703 (1988).

[23] S. J. Wood, W. Chan, and R. Wetzel, *Biochemistry* **35,** 12623 (1996).

[24] C. J. Barrow, A. Yasuda, P. T. Kenny, and M. G. Zagorski, *J. Mol Biol.* **225,** 1075 (1992).

ml or less; λ_{ex} 450 nm (bandwidth 5 nm) and λ_{em} 482 nm (bandwidth 10 nm). The inner filter effect is also negligible with ThT because the free dye ($\lambda_{max} \approx$ 335 nm) absorbs very little at 450 nm or 482 nm, and the fibril-induced absorbance is also small in the diluted sample.

Effects on Fibril Formation

The formation of Aβ(1–40) fibrils can be measured either by taking aliquots of a reaction mixture incubated under appropriate conditions and reading the fluorescence after dilution into ThT-containing buffer or by running multiple small volume reactions and quenching with ThT-containing buffer. Higher throughput may be achieved using a fluorescence plate reader and small volume reactions. The basic conditions used here are essentially those of Naiki and Nakakuki.[25] Other buffer constituents, temperature, and shaking may be employed. Assay kinetics should be established as factors such as pH and ionic strength can have drastic effects on the time course.[25]

To measure the effect of an added component on the rate of Aβ(1–40) fibril formation, 15 μl of an ice-cold reaction mix composed of 33 μM Aβ(1–40) peptide in 50 mM sodium phosphate, pH 7.5, 100 mM NaCl, and 0.02% NaN_3 is pipetted into the wells of a 96-well plate on ice wells containing the additive and mixed. If a seeded assay is desired, an appropriate concentration (determined empirically) of sonicated seed fibrils in buffer is added, and the plate is sealed with an aluminum film (Beckman, Fullerton, CA). Assays can also be run in individual 0.5-ml microfuge tubes or reactions run in bulk with samples removed at selected times for assay. The reaction is initiated by bringing the plate or tubes to 37° and incubating unshaken for the desired time. The reaction time required will depend on the state of the peptide used, temperature, and stirring and must be determined by experiment. Usually, a detectable reaction is obtained between 3 and 6 hr under the indicated conditions and reaches equilibrium overnight. Aβ(1–40) fibril formation is highly temperature dependent, with the rate increasing dramatically between 25° and 37°. A more rapid rate of reaction can be obtained by shaking the samples gently.

The reaction is terminated after removal of the sealing film by the addition of 250 μl/well of 5 μM ThT in 50 mM glycine–NaOH, pH 8.5, and the amyloid-specific fluorescence read within 1–2 min in the plate reader. For reactions to be read in a regular fluorimeter, the samples are removed into a cuvette containing 0.5 ml of ThT solution or the microfuge tubes are filled with 0.5 ml of ThT solution and transferred to a cuvette.

[25] H. Naiki and K. Nakakuki, *Lab. Invest.* **74,** 374 (1996).

Immediate quenching of a reaction and reading gives the starting point of the reaction. Reaction mixture stored on ice is a less reliable measure of the starting point. Figure 4 shows amyloid-specific fluorescence for several Aβ fragments as a function of peptide concentration. Note that although β(25–35) is known to form modified amyloid fibrils, it is weakly fluorescent with ThT. The error bars give an indication of intraassay variation.

Quantification of Amyloid Fibrils

Although for most purposes only a relative determination of the amount of amyloid fibrils in several samples is required, a semiquantitative measurement can be made by reference to a standard curve constructed with completely fibrillized protein composed of the same material being assayed. An excess of ThT over fibril-binding sites is required to maintain a linear relationship between fibril added and signal. Because fluorescence intensity is a relative measurement, a standard curve should be run at the same time as the samples unless the instrument is run in a quantum-counting mode and standardized for lamp intensity and photodetector response. Sample temperature control is also required for consistent standard intensities. The

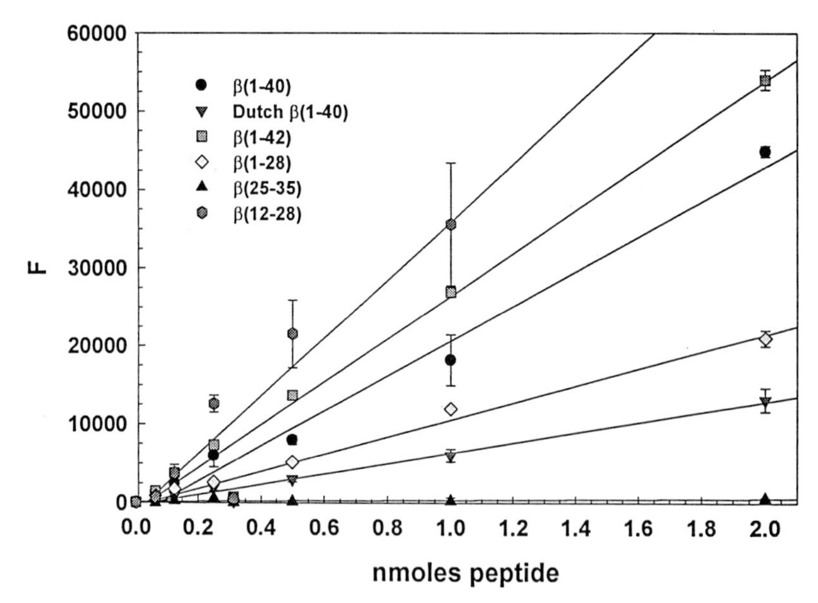

FIG. 4. ThT fluorescence with different Aβ fibrils. Data obtained with Cytofluor 2350 plate reader in 250-μl volumes of 5 μM ThT in 50 mM glycine–NaOH, pH 8.5. Fibrils formed for 2 days at 37° in 50 mM phosphate buffer at pH 7.5 β(1–40), Dutch mutant (E22Q) β(1–40), β(1–42), or pH 6.0 β(1–28), β(12–28), and β(25–35). Lines are linear least-square fits of data.

measurement being made is the fraction of the total amount of fibrils that can be assembled as defined by dye binding.

Caveats of Using ThT Fluorescence to Measure Amyloid Fibril Formation

ThT measurement of Aβ fibril formation can be used to assess whether an added component will interfere or potentiate amyloid fibril formation. In such applications, the investigator must determine whether the effect is on the maximal extent of fibril formation or on the rate of fibril formation. Although this corresponds superficially with K_m and V_{max} effects with enzymes, for amyloid fibrils the question is whether the effect is kinetic or thermodynamic. In looking for inhibitors this is an important distinction as kinetic inhibition of an intermediate step in fibril formation will eventually be overcome by the overwhelming thermodynamic stability of the fibril as was demonstrated for ApoE and Aβ.[26] Of further concern is that as with all fluorescent probes, apparent inhibitors or potentiators could affect either the quantum yield or the number of ThT molecules bound to the amyloid fibril without actually changing the amount of fibril formation. Colored compounds may also affect determinations by interfering with either the excitation or the emission from the fluorophore. Light scattering due to the precipitation of compound/peptide complexes or compound insolubility may be reduced by dilution. Further verification of the quantity of amyloid fibrils formed should be done as with the fluorescent ApoE : ThT : Aβ multimer complex isolated by Chan *et al.*[27] Methods such as negative staining or Pd/C rotary shadowing electron microscopy or atomic force microscopy that visualize fibril structure should be used to assess fibril formation independently.

Comparison with Other Methods of Fibril Quantification

Regardless of the precursor protein or peptide, amyloid fibril formation is defined operationally as the assembly of monomeric units into fibrils with a high content of β-sheet conformation of defined dimensions [7–10 nm for Aβ(1–40) and Aβ(1–42)]. The resulting fibrils are resistant to proteolytic attack, bind amyloid P protein in the presence of calcium, form a birefringent metachromatic green complex with Congo red, and many form a fluorescent ThT complex under standard conditions. Fibril sedimentation[28] or filtration[29] can be used with labeled or unlabeled peptide as surro-

[26] K. C. Evans, E. P. Berger, C. G. Cho, K. H. Weisgraber, and P. T. Lansbury, Jr., *Proc. Natl. Acad. Sci. U.S.A.* **92,** 763 (1995).

[27] W. Chan, J. Fornwald, M. Brawner, and R. Wetzel, *Biochemistry* **35,** 7123 (1996).

[28] D. Burdick, J. Kosmoski, M. F. Knauer, and C. G. Glabe, *Brain Res.* **746,** 275 (1997).

[29] S. J. Hayes, H. LeVine III, and J. D. Scholten, International Patent WO 97/16194 (1997).

gates for the optical methods, but they only infer the presence of fibrils by size. Nonamyloid fibril amorphous aggregates would give similar results with these methods.[30] Electron microscopy or atomic force microscopy unequivocally identifies amyloid fibrils but they require expensive equipment, technically demanding analysis, and can process only small numbers of samples. Quantifying fibril formation by microscopy is difficult as well. For many applications, optical methods are the choice for sensitivity and high sample throughput. A series of fluorescently labeled Aβ(1–40) peptides minimally perturbed with respect to amyloid fibril formation[31] promise to increase sensitivity and are commercially available (Amersham). Assay methods using these modified peptides are under development and promise to be highly sensitive, although the proper interpretation of results brings its own share of caveats.

Congo red difference spectroscopy and ThT fluorescence can be considered complementary methods for amyloid fibril quantitation. The two probe molecules appear to bind to different sites on the Aβ fibril as their opposite charges would indicate. Neither is particularly sensitive, although radiolabeled derivatives of Congo red[5] and chrysamine G[6] improve the detection of small quantities of amyloid fibrils, assuming that their binding selectivity for the amyloid fibril vs the amorphous aggregate can be demonstrated. Syntheses of the labeled compounds, however, are not trivial and the binding conditions must be optimized. The primary advantage of the dye-binding approaches lies in the relative simplicity of their application. Potential caveats for both Congo red and ThT detection arising from optical artifacts with some colored compounds or possible displacement of the probe molecules continue to plague these systems. Because both of these extrinsic probes recognize the peculiar characteristics of most amyloid fibrils, no special derivitization is required for different amyloidogenic proteins.

[30] S. J. Wood, B. Maleeff, T. Hart, and R. Wetzel, *J. Mol. Biol.* **256,** 870 (1996).

[31] W. Garzon-Rodriguez, M. Sepulveda-Becerra, S. Milton, and C. G. Glabe, *J. Biol. Chem.* **272,** 21037 (1997).

[19] Quantifying Amyloid by Congo Red Spectral Shift Assay

By WILLIAM E. KLUNK, ROBERT F. JACOB, and R. PRESTON MASON

Introduction

Congo red (CR)* is a diazo dye that is widely used as a postmortem histological indicator of amyloid β-peptide deposition in Alzheimer's disease brain tissue.[1] Its usefulness as an indicator of amyloid in pathological specimens was substantiated by early recognition that CR binds amyloid proteins with relatively high specificity.[2] Binding interactions of CR and amyloid have since been characterized further and shown to depend on the secondary conformation of the amyloid—specifically, the β-pleated sheet conformation.[3] In fact, this β-sheet conformation of amyloid appears to be *the* crucial factor in CR binding.[4–6] Other proteins that lack or contain only a minor proportion of β-sheet secondary structure do not stain with CR.[4,7] The exact structure of a CR–amyloid protein complex has never been determined by rigorous techniques such as X-ray crystallography or two-dimensional nuclear magnetic resonance. The crystal structure of a nonfibrillar insulin complex with CR has been published.[8] However, this is not a good surrogate representation of a CR–amyloid complex, as the CR "interaction" with nonfibrillar (i.e., native) insulin is of much lower affinity and fails to induce the spectral changes that are so characteristic of CR "binding" to β-sheet fibril proteins.[7] Several mutually exclusive models have been proposed to explain the β-sheet dependency of CR binding to amyloid proteins. In one, CR lies perpendicular to the peptide backbone of the amyloid protein and spans five β sheets, binding to posi-

* Based largely on W. E. Klunk, R. F. Jacob, and R. P. Mason, Quantifying amyloid β-peptide (Aβ) aggregation using the Congo Red-Aβ (CR–Aβ) spectrophotometric method. *Anal. Biochem.* **266,** 66 (1999).

[1] P. Ladewig, *Nature* **156,** 81 (1945).

[2] H. Puchtler, F. Sweat, and M. Levine, *J. Histochem. Cytochem.* **10,** 355 (1962).

[3] R. A. DeLellis, G. G. Glenner, and J. S. Ram, *J. Histochem. Cytochem.* **16,** 663 (1968).

[4] G. G. Glenner, E. D. Eanes, and D. L. Page, *J. Histochem. Cytochem.* **20,** 821 (1972).

[5] G. G. Glenner, E. D. Eanes, J. D. Termine, H. A. Bladen, and R. P. Linke, *J. Histochem. Cytochem.* **21,** 406 (1973).

[6] J. H. Cooper, *Lab. Invest.* **31,** 232 (1974).

[7] W. E. Klunk, J. W. Pettegrew, and D. J. Abraham, *J. Histochem. Cytochem.* **37,** 1273 (1989).

[8] W. G. Turnell and J. T. Finch, *J. Mol. Biol.* **227,** 1205 (1992).

0076-6879/99 $30.00

tively charged amino acid side chains.[7,9] In another, CR lies parallel to the peptide backbone and intercalates between two β sheets.[10] A third model predicts that CR binds directly to the side chains of a single β sheet.[11] The key component common to all of these models is the idea that there is a stoichiometric and saturable interaction between CR and the amyloid fibril. This interaction leads to a change in the spectral characteristics of the dye for reasons that are not well understood. However, the spectral shift induced by this stoichiometric/saturable interaction allows accurate and absolute quantitation of the concentration of the CR–amyloid complex.

Our laboratory previously developed a method for studying the interaction of Congo red with amyloid proteins using fibrillar β-sheet insulin as an amyloid prototype.[12] This method was not designed to quantify the amyloid protein itself, but several investigators have recognized the potential to measure Aβ aggregation using the equations developed for the CR–insulin method.[13–17] While this is valid in concept, application of the CR–insulin equations to the quantitation of other amyloids suffers from potentially invalid assumptions (i.e., that the CR–amyloid complex will have exactly the same absorption characteristics as the CR–insulin complex) and improper application of the technique, depending on the physical characteristics of the amyloid studied. The intended application of the CR–insulin method was to quantitatively understand the kinetics and stoichiometry of CR binding to a representative amyloid, namely insulin fibrils. That is, the emphasis was on the dye, not the amyloid. In many cases in the development of the CR–insulin method, inferences were made from situations in which the amyloid was in great excess. Furthermore, no attempt was made to develop this technique to quantify the amyloid, although this certainly could be done.

Our original method (developed using a model amyloid system, i.e., insulin fibrils)[12] has been optimized for application in the study of a specific

[9] W. E. Klunk, M. L. Debnath, and J. W. Pettegrew, *Neurobiol. Aging* **15,** 691 (1994).

[10] D. B. Carter and K. C. Chou, *Neurobiol. Aging* **19,** 37 (1998).

[11] F. Cavillon, A. Elhaddaoui, A. J. P. Alix, S. Turrell, and M. Dauchez, *J. Mol. Struct.* **408/409,** 185 (1997).

[12] W. E. Klunk, J. W. Pettegrew, and D. J. Abraham, *J. Histochem. Cytochem.* **37,** 1293 (1989).

[13] A. M. Brown, D. M. Tummolo, K. J. Rhodes, J. R. Hofmann, J. S. Jacobsen, and J. Sonnenberg-Reines, *J. Neurochem.* **69,** 1204 (1997).

[14] S. J. Wood, B. Maleeff, T. Hart, and R. Wetzel, *J. Mol. Biol.* **256,** 870 (1996).

[15] D. R. Howlett, K. H. Jennings, D. C. Lee, M. S. Clark, F. Brown, R. Wetzel, S. J. Wood, P. Camilleri, and G. W. Roberts, *Neurodeg.* **4,** 23 (1995).

[16] S. K. Brining, *Neurobiol. Aging* **18,** 581 (1997).

[17] S. J. Wood, L. MacKenzie, B. Maleeff, M. R. Hurle, and R. Wetzel, *J. Biol. Chem.* **271,** 4086 (1996).

amyloid, the amyloid β (Aβ) peptide.[18] The importance of the secondary structure of amyloid β-peptide (Aβ) in Alzheimer's disease neuropathology has received significant emphasis in recent years. Numerous studies have clearly demonstrated that Aβ-induced neurotoxicity is dependent on the aggregation of Aβ into β-sheet fibrils.[19–23] The importance of Aβ fibril formation has led to the search for efficient and broadly applicable techniques for quantifying this aggregation and the effect of agents that may prevent it.

The specific protocols discussed in this article refer to the use of the CR spectral shift assay to quantify Aβ aggregation. However, this method has the potential to be very general and the strategy for optimizing the method for other amyloid proteins is discussed in detail later. The CR spectral shift assay has several advantages over other approaches. First, this method requires only a simple spectrophotometer, making it broadly applicable. Second, it requires no radioisotopes or other expensive reagents. It is rapid and technically simple because there is no need to separate amyloid-bound from free indicator. Perhaps the greatest advantage of this technique is its ability to quantitate dye–amyloid interactions in absolute terms, allowing the direct comparison of results from day to day and from one laboratory to another. Initial discussion of the mathematical derivation of the equations used in this method incurs the risk of obscuring the relative ease and straightforward nature of this technique. To avoid this, we will first present a basic protocol that allows rapid and direct application of the assay to the quantitation of Aβ fibril test samples prior to discussion of the theoretical basis of the method.

CR–Aβ Spectrophotometric Assay: Basic Protocol

1. Prepare a 100–300 μM stock solution of Congo red (C.I. 22120, Direct Red 28; Aldrich, Milwaukee, WI) in a solution of 90% filtered phosphate-buffered saline (PBS, Sigma, St. Louis, MO, 0.01 M phosphate buffer, 0.0027 M KCl, and 0.137 M NaCl; pH 7.4) and 10% (v/v) ethanol.

[18] W. E. Klunk, R. F. Jacob, and R. P. Mason, *Anal. Biochem.* **266,** 66 (1999).

[19] A. Lorenzo and B. A. Yankner, *Proc. Natl. Acad. Sci. U.S.A.* **91,** 12243 (1994).

[20] B. Seilheimer, B. Bohrmann, L. Bondolfi, F. Muller, D. Stuber, and H. Dobeli, *J. Struct. Biol.* **119,** 59 (1997).

[21] L. K. Simmons, P. C. May, K. J. Tomaselli, R. E. Rydel, K. S. Fuson, E. F. Brigham, S. Wright, I. Lieberburg, G. W. Becker, D. N. Brems *et al., Mol. Pharmacol.* **45,** 373 (1994).

[22] C. J. Pike, A. J. Walencewicz, C. G. Glabe, and C. W. Cotman, *Brain Res.* **563,** 311 (1991).

[23] C. J. Pike, D. Burdick, A. J. Walencewicz, C. G. Glabe, and C. W. Cotman, *J. Neurosci.* **13,** 1676 (1993).

2. Filter the CR stock solution three times using Gelman extra-thick glass fiber filters (~0.3-μm nominal pore retention) to remove any CR micelles (microaggregates).
3. Determine the concentration of the CR stock solution by measuring the absorbance of a diluted aliquot of the filtrate prepared in a solution of sodium phosphate (1 m*M*, pH 7.0) and 40% ethanol at 505 nm [$\varepsilon = 5.93 \times 10^4$ AU (absorbance unit)/(cm · *M*)]. Maximum accuracy will be achieved by attempting to keep the concentration of the diluted aliquot between 10 and 15 μ*M*.
4. Mix the Aβ test sample with a solution of CR in PBS to yield a final concentration of 2–20 μ*M* CR. The ratio of CR (in μ*M*) to Aβ fibrils (in μg/ml, determined by the total, initial concentration of soluble Aβ) should not fall below 1 : 5. Therefore, the total Aβ concentration should not exceed 100 μg/ml. Incubate the CR–Aβ fibril test mixture at room temperature for 15 min prior to spectral analysis.
5. Prepare controls samples: (a) a CR sample without Aβ in which the concentration of CR matches that used in the CR–Aβ test sample and (b) a preaggregated Aβ test sample prepared in the absence of CR (necessary only if determining the *r* value, see later).
6. After blanking the spectrophotometer on PBS, record the absorbances of all test samples and controls at 541 and 403 nm.
7. To determine the concentration of CR bound (i.e., CR–Aβ), enter the results of these spectrophotometric analyses into one of the following equations.
   a. For simplicity, one can use the average value for *r* (average $r = 0.7$, see later) and use the following equation [see Eq. (11)]:

$$[\text{CR–A}\beta] = ({}^{541}A_t/47{,}800) - ({}^{403}A_t/68{,}300) - ({}^{403}A_{\text{CR}}/86{,}200)$$

   where ${}^{541}A_t$ equals the absorbance of the CR + Aβ test sample at 541 nm, ${}^{403}A_t$ equals the absorbance of the CR + Aβ test sample at 403 nm, and ${}^{403}A_{\text{CR}}$ is the absorbance of the CR-only sample at 403 nm. To have results expressed in terms of μg/ml Aβ fibrils, use the following equation [Eq. (15)]:

$$\text{A}\beta_{\text{fib}} = ({}^{541}A_t/4780) - ({}^{403}A_t/6830) - ({}^{403}A_{\text{CR}}/8620)$$

   b. For added accuracy, *r* can be determined from the ratio of the light scattering of the Aβ-alone control sample at 541 and 403 nm:

$$r = {}^{541}S_{\text{A}\beta}/{}^{403}S_{\text{A}\beta}$$

where S represents the scattering (as measured in absorbance units) of the Aβ control sample. This r value can then be entered along with data from the CR–Aβ test sample into the following equation [Eq. (9)]:

$$[\text{CR–A}\beta] = ({}^{541}A_t/47{,}800) - [{}^{403}A_t/(47{,}800/r)] + {}^{403}A_{CR}[(r/47{,}800) - (1/38{,}100)]$$

or, for results in terms of μg/ml Aβ, use the following equation [Eq. (13)]:

$$A\beta_{fib} = ({}^{541}A_t/4780) - [{}^{403}A_t/(4780/r)] + {}^{403}A_{CR}[(r/4780) - (1/3810)]$$

Spectral Characteristics of CR and Aβ

When CR binds to excess fibrillar Aβ(1–40), a change in color from orange-red to rose is induced that corresponds to a shift in the characteristic absorbance spectrum of CR. This change in CR spectral properties has been noted with its interactions with other fibrillar proteins such as insulin.[12] The ability of amyloid proteins to induce CR spectral shifts is dependent on protein aggregation state.[7] In addition to inducing changes in CR spectra, many amyloid fibrils also possess inherent light-scattering properties.[24–28]

Figure 1 demonstrates the individual and combined spectral properties of fibrillar Aβ(1–40) and CR. The concentration-dependent properties of free CR are depicted in Fig. 1A. The point of maximal absorbance remains at ~488 nm at all tested concentrations (in PBS). The effect of Aβ light scattering is shown in Fig. 1B. The Aβ + CR (uncorrected) absorbance spectrum is appreciably greater than the spectral curve that is corrected for Aβ scattering (Aβ + CR, corrected); this discrepancy increases with decreasing wavelength due to increased Aβ scattering at lower wavelengths. Figure 1C demonstrates Aβ + CR (corrected) spectra at a variety of CR concentrations. In Figure 1D, an "isosbestic" point is defined at 403 nm, as both bound and free CR have equal molar absorptivity values at this wavelength. Thus, the difference between Aβ + CR (uncorrected) and CR alone is equal to the light scattering of Aβ alone at 403 nm and is designated as ${}^{403}S_{A\beta}$ (Fig. 1B). ${}^{541}S_{A\beta}$ is equivalent to ${}^{403}S_{A\beta}$, representing the light scattering due to Aβ fibrils at this wavelength. However, ${}^{541}S_{A\beta}$ cannot be

[24] E. M. Castano, J. Ghiso, F. Prelli, P. D. Gorevic, A. Migheli, and B. Frangione, *Biochem. Biophys. Res. Commun.* **141,** 782 (1986).

[25] J. T. Jarrett and P. T. Lansbury, Jr., *Biochemistry* **31,** 12345 (1992).

[26] J. H. Come, P. E. Fraser, and P. T. Lansbury, Jr., *Proc. Natl. Acad. Sci. U.S.A.* **90,** 5959 (1993).

[27] C. L. Shen, G. L. Scott, F. Merchant, and R. M. Murphy, *Biophys. J.* **65,** 2383 (1993).

[28] J. Ma, A. Yee, H. B. Brewer, Jr., S. Das, and H. Potter, *Nature* **372,** 92 (1994).

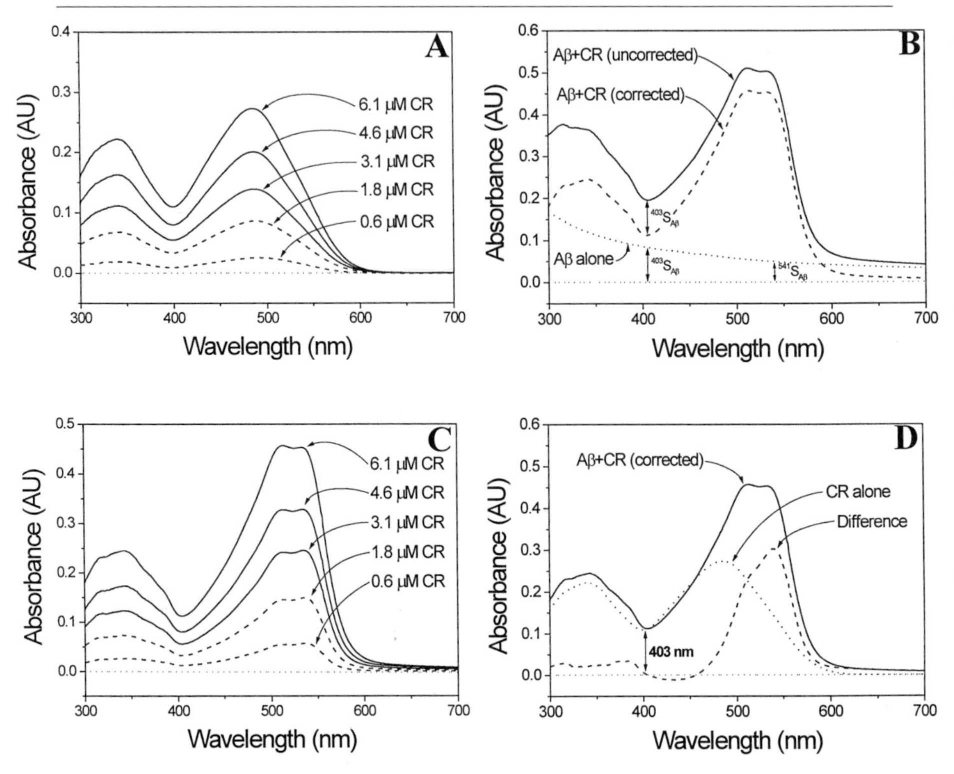

FIG. 1. Spectral characteristics of CR and fibrillar Aβ(1–40). (A) Absorbance spectra of free CR at five concentrations. (B) Absorbance spectra of a suspension of 97.2 μg/ml Aβ fibrils in the absence and presence of 6.1 μM CR. Aβ + CR (corrected) results from the subtraction of Aβ scattering (Aβ alone) from the Aβ + CR (uncorrected) spectral curve. ${}^{403}S_{A\beta}$ represents the degree of light scattering due to Aβ fibrils at the isosbestic point; ${}^{541}S_{A\beta}$ is defined similarly at the point of maximal spectral difference. (C) Aβ + CR (corrected) spectra at five concentrations of CR. (D) Absorbance spectra of a 6.1 μM CR solution in the absence and presence of 97.2 μg/ml Aβ fibrils, corrected for Aβ scattering. The difference spectrum is obtained by subtracting the CR spectrum in the absence of Aβ fibrils from the corrected CR spectrum in the presence of Aβ fibrils.

measured directly due to the variable contribution of the CR–Aβ complex to the total absorbance at this wavelength. Therefore, ${}^{541}S_{A\beta}$ is derived from ${}^{403}S_{A\beta}$ as follows.

The light scattering of Aβ fibrils at both 403 and 541 nm is linear with concentration (Fig. 2), and the ratio between the degree of scattering at these two wavelengths remains constant (Fig. 2, inset). This ratio (termed *r* in the equations) varies slightly from preparation to preparation but is typically near 0.7 (range 0.65–0.75). Therefore, knowing the absorbance of

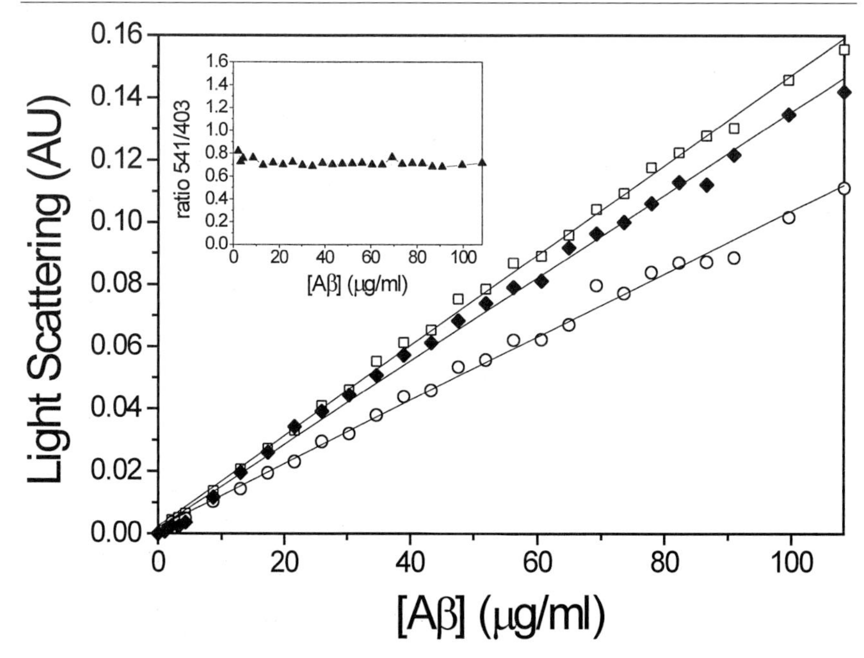

FIG. 2. Light-scattering effects of $A\beta$(1–40) fibrils. Scattering of $A\beta$ fibrils at 403 (□) and 541 (○) nm was measured at various peptide concentrations in the absence of CR. Light scattering remained linear with $A\beta$ concentration at wavelengths of both 403 nm (Pearson correlation coefficient = 0.9992) and 541 nm (Pearson correlation coefficient = 0.9973). In another experiment using a separately prepared sample of $A\beta$(1–40) fibrils, light scattering of $A\beta$(1–40) was measured indirectly by calculating the difference between the absorbance of CR + $A\beta$ fibrils and the absorbance of CR alone at 403 (${}^{403}S_{A\beta}$, ◆). Scattering determined in the presence of CR also remained linear with concentration (Pearson correlation coefficient = 0.9983) and was very similar to that determined in the absence of CR. Differences between the two curves are partially due to inherent differences in separately prepared samples. The ratio of light scattering at 541 and 403 nm remained relatively constant at all tested $A\beta$ concentrations (inset), yielding a ratio (r) value of 0.7 (where $r = {}^{541}S_{A\beta}/{}^{403}S_{A\beta}$).

a CR–$A\beta$ complex (at an unknown concentration of $A\beta$) and CR alone at 403 nm will allow calculation of both ${}^{403}S_{A\beta}$ and ${}^{541}S_{A\beta}$, where ${}^{541}S_{A\beta} = 0.7 \times {}^{403}S_{A\beta}$. ${}^{403}S_{A\beta}$ calculated in this manner (Fig. 2, closed diamonds) closely approximates direct measurement in an $A\beta$-alone control sample (Fig. 2, open squares). These ${}^{403}S_{A\beta}$ and ${}^{541}S_{A\beta}$ values will be used to correct for light scattering (if present) in subsequent equations.

The point of maximal spectral difference between bound and free CR occurs at 541 nm (Figs. 1D and 3). Difference spectra at five separate CR concentrations are depicted in Fig. 3. The isosbestic point and the point of maximal difference remain constant at all concentrations of CR used, and

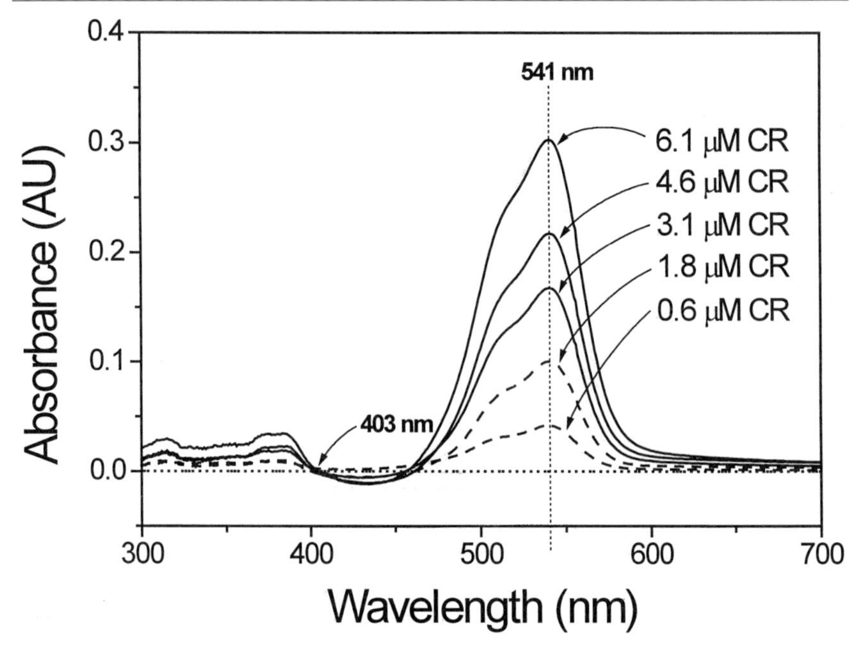

FIG. 3. Difference spectra obtained at various concentrations of CR in the presence of 97.2 μg/ml Aβ fibrils. All curves are approximately equal to zero at the isosbestic point (403 nm) and 541 nm defines the point of maximum difference at all concentrations of CR.

the maximal spectral difference remains linear with CR concentration (Pearson correlation coefficient = 0.9971).

Derivation of CR–Aβ Equations

Congo red spectral changes induced by binding to Aβ fibrils as seen in Fig. 1 can be used to determine the concentrations of bound and free CR. According to the Beer–Lambert law, $A = abc$, where A is the absorbance in absorbance units (AU), a is the molar absorptivity in units of AU/(cm $\cdot$ M), b is the path length in cm, and c is the concentration in moles/liter. The total absorbance at any wavelength (w), ${}^{w}A_t$, equals the sum of the absorbances of bound CR (A_b) and free CR (A_f):

$$ {}^{w}A_t = {}^{w}A_b + {}^{w}A_f = {}^{w}a_b c_b + A_f c_f \qquad (1) $$

Path length (b) is omitted because it is asumed to be held constant at 1 cm. Measuring the total absorbance at two wavelengths, w1 and w2, yields two equations with two unknowns (c_b and c_f):

$$^{w1}A_t = {}^{w1}a_b c_b + {}^{w1}a_f c_f \tag{2}$$
$$^{w2}A_t = {}^{w2}a_b c_b + {}^{w2}a_f c_f \tag{3}$$

Equating Eq. (2) to Eq. (3) and solving for c_b yields

$$c_b = \frac{({}^{w1}A_t/{}^{w1}a_f) - ({}^{w2}A_t/{}^{w2}a_f)}{({}^{w1}a_b/{}^{w1}a_f) - ({}^{w2}a_b/{}^{w2}a_f)} \tag{4}$$

In the spectrum of a mixture containing both bound and free CR, the highest signal to noise is obtained at the point of maximal spectral difference, 541 nm (Fig. 1B). Using this as one of the wavelengths maximizes the accuracy of the calculations at low ratios of c_b/c_f. Additionally, if w2 is an isosbestic point, then, by definition, ${}^{w2}a_b = {}^{w2}a_f$ and $({}^{w2}a_b/{}^{w2}a_f) = 1$. The most consistent isosbestic point was found to be at 403 nm (Fig. 3, Table I). Based on these substitutions, and substituting [CR–Aβ] for c_b, Eq. (4) can be altered to yield

$$[\text{CR–A}\beta] = \frac{({}^{541}A_t/{}^{541}a_f) - ({}^{403}A_t/{}^{403}a)}{({}^{541}a_b/{}^{541}a_f) - 1} \tag{5}$$

where ${}^{403}a$ represents the average molar absorptivity of bound and free CR (${}^{403}a_f$ and ${}^{403}a_b$) at the isosbestic point of 403 nm (i.e., 18,100 after rounding, see Table I).

At this point, the concentration of CR bound to Aβ (CR–Aβ) can be calculated simply by measuring the absorbance at 541 and 403 nm, once the molar absorptivities are known. These absorptivity values are determined easily for the free ligand by traditional methods using Aβ fibril-free solutions of CR (Fig. 1A). For the bound ligand, molar absorptivity values

TABLE I
MOLAR ABSORPTIVITIES OF CONGO RED SOLUTIONS[a]

Wavelength (nm)	a_b[b,c]	a_f[c,d]	Difference[e] (%)
541	70,500	22,700	211
403	18,070	18,040	0.17

[a] Reproduced with permission from W. E. Klunk *et al., Anal. Biochem.* **266,** 66 (1999).
[b] Molar absorptivity of bound CR, determined in the presence of 97.2 μg/ml Aβ fibrils.
[c] Units, AU/(cm · *M*).
[d] Molar absorptivity of free CR, determined in the absence of Aβ fibrils.
[e] Percentage difference between a_b and a_f.

can be determined by taking spectrophotometric measurements under conditions that assure essentially complete binding, i.e., $c_f = 0$. This was accomplished by using a concentration of 97.2 μg/ml Aβ fibrils (Fig. 1C). At this concentration of Aβ fibrils, the absorbance of even the most concentrated CR solution (6.1 μM) was decreased by 97% with filtration to remove the fibril–dye complex from suspension (data not shown). Table I lists molar absorptivity values obtained at 541 and 403 nm for both free and bound ligand (derived from data in Figs. 1A and 1C). Substituting the values from Table I into Eq. (5) gives

$$[\text{CR–A}\beta] = \frac{({}^{541}A_t/22{,}700) - ({}^{403}A_t/18{,}100)}{(70{,}500/22{,}700) - 1}$$

This can be simplified to

$$[\text{CR–A}\beta] = ({}^{541}A_t/47{,}800) - ({}^{403}A_t/38{,}100) \tag{6}$$

Equation (6) is analogous to that described previously for CR binding to insulin fibrils.[12] However, this equation assumes that the absorbances at 541 and 403 nm have already been corrected for light scattering (if present) by the use of a reference cell containing an equal (i.e., known) concentration of fibrils. Since the purpose of the current method is to determine the concentration of unknown preparations of fibrils, this approach is not possible. Therefore, a mathematical correction for scattering, based on the determination of ${}^{403}S_{A\beta}$ as described earlier, must be incorporated into Eq. (6). The importance of this correction can be seen in Fig. 4.

Figure 4 shows the calculation of the concentration of the CR–Aβ complex with and without correction for the light scattering of Aβ fibrils. This particular sample of Aβ fibrils was prepared with stirring according to the method of Evans *et al.*,[29] resulting in a visibly turbid suspension. At very low concentrations of Aβ fibrils (<5 μg/ml), this correction may be unnecessary for certain noncritical applications of this technique as the percentage error is only 5–10%. As the ratio of CR : Aβ reaches typical assay conditions at about 15 μg/ml Aβ fibrils, the correction becomes necessary as the percentage error ranges from 20 to $>$50%. The omission of this critical correction makes the CR–insulin equation invalid for the purpose of quantitating aggregation in Aβ suspensions, which possess significant light-scattering properties. Other methods of aggregating Aβ without stirring (e.g., Wood *et al.*[14] and Howlett *et al.*[15]) may result in suspensions having minimal turbidity, decreasing the necessity for this correction (see caveats later). The correction for scattering can be accomplished as

[29] K. C. Evans, E. P. Berger, C. G. Cho, K. H. Weisgraber, and P. T. Lansbury, Jr., *Proc. Natl. Acad. Sci. U.S.A.* **92,** 763 (1995).

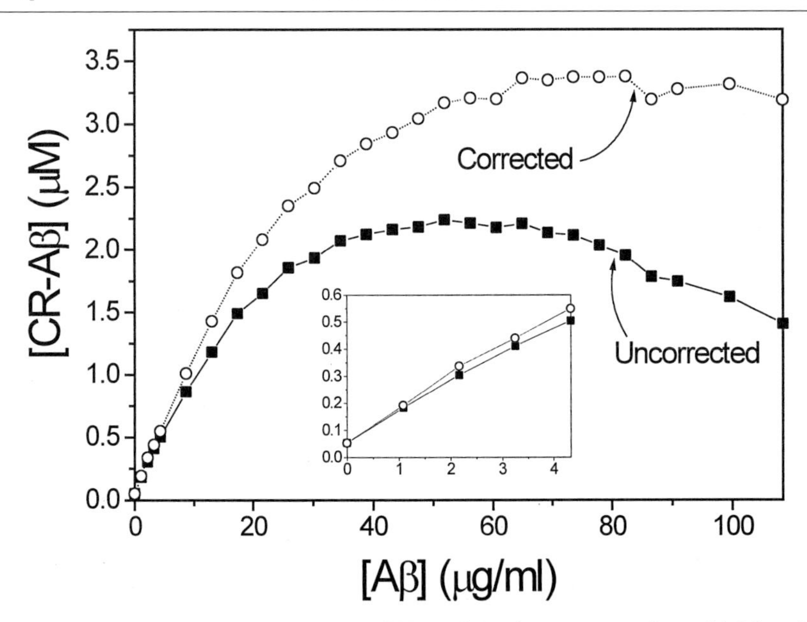

FIG. 4. $A\beta$ fibril concentration effects on [CR–$A\beta$]. Various concentrations of $A\beta$ (ranging from 0 to 108 μg/ml) were incubated with 3.3 μM CR for 15 min. CR–$A\beta$ was determined using Eq. (9), which corrects for $A\beta$ scattering, indicated by the open circles and dotted line. To demonstrate the necessity of correcting for $A\beta$ light scattering, CR–$A\beta$ was calculated using Eq. (6), which was uncorrected for $A\beta$ scattering (closed squares and solid line). (Inset) Expansion of the initial portion of the curve (axes of the inset have the same units as the larger graph).

$$[\text{CR–A}\beta] = [({}^{541}A_t - {}^{541}S_{A\beta})/47{,}800] - [({}^{403}A_t - {}^{403}S_{A\beta})/38{,}100] \quad (7)$$

where ${}^{403}S_{A\beta}$ and ${}^{541}S_{A\beta}$ represent light scattering due to $A\beta$ fibrils at the respective wavelengths. Because ${}^{541}S_{A\beta} = r \times {}^{403}S_{A\beta}$ (where $r = {}^{541}S_{A\beta}/{}^{403}S_{A\beta}$):

$$[\text{CR–A}\beta] = \{[{}^{541}A_t - ({}^{403}S_{A\beta} \times r)]/47{,}800\} - [({}^{403}A_t - {}^{403}S_{A\beta})/38{,}100] \quad (8)$$

Knowing that ${}^{403}S_{A\beta} = {}^{403}A_t - {}^{403}A_{CR}$ (where ${}^{403}A_{CR}$ equals the absorbance of CR at its isosbestic point):

$$[\text{CR–A}\beta] = ({}^{541}A_t/47{,}800) - [({}^{403}A_t - {}^{403}A_{CR})r/47{,}800] - \{[{}^{403}A_t - ({}^{403}A_t - {}^{403}A_{CR})]/38{,}100\}$$

which reduces to

$$[\text{CR–A}\beta] = ({}^{541}A_t/47{,}800) - [{}^{403}A_t/(47{,}800/r)] + {}^{403}A_{CR}[(r/47{,}800) - (1/38{,}100)] \quad (9)$$

$^{403}A_{CR}$ can be determined in one of two ways. First, the CR-alone sample absorbance at 403 nm can be used. This can be done in typical aggregation experiments in which [CR] is kept constant and the scattering due to the unknown Aβ samples varies among the samples. This practice also assures accurate definition of the final concentration of CR in the assay. The second method of determining $^{403}A_{CR}$ applies only when the concentration of Aβ fibrils is held constant and the concentration of CR is varied (e.g., Fig. 5). In this situation, the $^{403}S_{A\beta}$ is held constant and $^{403}A_{CR} = {}^{403}A_t - {}^{403}S_{A\beta}$, where $^{403}S_{A\beta}$ is determined from a control sample containing only Aβ. Under these conditions, the concentration of CR in each individual sample is actually measured, correcting for any variability between samples. This is not useful for typical Aβ aggregation experiments, however, and only applies when studying the kinetics and stoichiometry of CR binding to known samples of Aβ fibrils.

It can also be recognized that $^{403}A_{CR} = {}^{403}a_{CR} \times [CR] = 18{,}040 \times [CR]$, where [CR] is the total concentration of CR added to the assay and 18,040 is the molar absorptivity of CR at 403 nm (Table I). When substituted into Eq. (9), Eq. (10) is obtained:

$$[\text{CR–A}\beta] = ({}^{541}A_t/47{,}800) - [{}^{403}A_t/(47{,}800/r)] + ([\text{CR}])[(r/2.64) - 0.475] \quad (10)$$

Equation (10) is also adequate for determining Aβ aggregation in the presence of a single concentration of CR.

Substituting the typical value of r into the equation ($r \cong 0.7$) simplifies Eq. (9) to

$$[\text{CR–A}\beta] = ({}^{541}A_t/47{,}800) - ({}^{403}A_t/68{,}300) - ({}^{403}A_{CR}/86{,}200) \quad (11)$$

and Eq. (10) to

$$[\text{CR–A}\beta] = ({}^{541}A_t/47{,}800) - ({}^{403}A_t/68{,}300) - ([\text{CR}]/4.77) \quad (12)$$

Derivation of Aβ_{fib} Equations

Using these equations, the relationship of [CR–Aβ] to the amount of added Aβ fibrils was determined. Figure 4 demonstrates that, at the concentration of CR used in this experiment (3.3 μM), the concentration of CR–Aβ determined by Eqs. (6) and (9) is linear up to approximately 16 μg/ml Aβ fibrils. Beyond this ratio of 3.3 μM CR to 16 μg/ml Aβ fibrils (or approximately 1 : 5), the CR-binding sites on the Aβ fibrils are no longer saturated and the assay is not linear. Furthermore, Fig. 4 clearly demonstrates the necessity to correct for light scattering by Aβ fibrils, particularly at higher concentrations of Aβ fibrils. Of note, Eq. (9) yielded a maximum value for [CR–Aβ] of 3.38 μM, which is in excellent agreement with the total CR concentration of 3.3 μM.

Figure 5 shows the effect of increasing CR concentration on the value of CR–Aβ determined by Eqs. (9) and (10). Again, there is a clear concentration dependency with a maximum value (i.e., saturation of all CR-binding sites on the Aβ fibrils) obtained at a concentration of about 8 μM CR in the presence of 37 μg/ml Aβ fibrils. This yields a CR : Aβ ratio of approximately 1 : 5, consistent with that derived previously from data in Fig. 4.

It can be seen from Fig. 5 that under the conditions of excess CR (conditions essential to this assay), there is approximately 1 μg/ml of aggregated, fibrillar Aβ (Aβ_{fib}) for every 0.1 μM CR–Aβ. This factor of 10 can be used to convert Eqs. (9) and (10) to yield the following equations [Eqs. (13) and (14), respectively], which give the results in terms of the concentration of Aβ_{fib} rather than the concentration of the CR–Aβ complex:

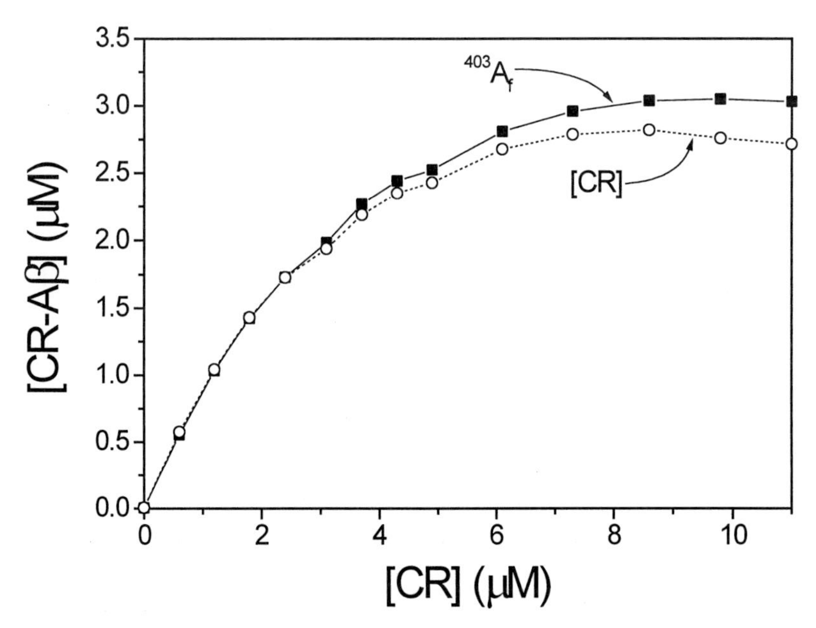

FIG. 5. Concentration effect of CR on [CR–Aβ]. Various concentrations of CR were incubated with 37 μg/ml Aβ fibrils (determined accurately by protein analysis) for 15 min prior to spectrophotometric measurements. CR–Aβ was calculated using both Eq. (9), which is based on direct measurement of the absorbance of free CR ($^{403}A_{CR}$), and Eq. (10), which assumes a known [CR] that is equal in all samples.

$$A\beta_{fib} = (^{541}A_t/4780) - [^{403}A_t/(4780/r)] + {^{403}A_{CR}}[(r/4780) - (1/3810)] \quad (13)$$

$$A\beta_{fib} = (^{541}A_t/4780) - [^{403}A_t/(4780/r)] + ([CR])[(r/0.264) - 4.75] \quad (14)$$

Likewise, Eqs. (11) and (12) can be revised to yield Eqs. (15) and (16) by assuming $r = 0.7$:

$$A\beta_{fib} = (^{541}A_t/4780) - (^{403}A_t/6830) - (^{403}A_{CR}/8620) \quad (15)$$

$$A\beta_{fib} = (^{541}A_t/4780) - (^{403}A_t/6830) - ([CR]/0.477) \quad (16)$$

Theoretical Considerations and Caveats

Limiting Concentration Ranges of Assay

A significant advantage of this technique is its ability to quantitate dye–amyloid interactions in absolute terms. For this absolute quantitation to be accurate, the limiting concentration ranges of the assay must be kept in mind. The primary limiting factor is the concentration of CR. In our experience, absorbance measurements above 1.0 AU run the risk of nonlinear artifacts. Therefore, we prefer to keep the total absorbance of each sample below 1.0 AU. At the wavelength maximum of the CR–Aβ complex (541 nm), this means that no more than 14 μM bound CR or 44 μM free CR be present. Because some of the CR will exist in both bound and free states in the actual assay, the maximum concentration of CR will be somewhere between these extremes. At the limiting CR : Aβ ratio of 1 : 5, the concentration of bound and free CR is approximately equal (Figs. 4 and 5), and approximately half of the CR exists as a CR–Aβ complex, which has a molar absorptivity at 541 nm of 70,500 AU/(cm $\cdot$ M) (Table I). The remaining free CR has a molar absorptivity of 22,700 AU/(cm $\cdot$ M) (Table I). Thus, in order to stay below a total absorbance of 1.0, the total CR concentration should stay below 21 μM. At this concentration of CR, the CR : Aβ ratio of 1 : 5 dictates that the concentration of Aβ fibrils remains below 100 μg/ml (~23 μM). At the other end of the scale, we can reproducibly detect less than 5 μg/ml Aβ (Fig. 4, inset). However, at these very low concentrations of Aβ fibrils, the signal to noise is better using concentrations of CR between 2.5 and 5 μM.

Adaptation of Method to 96-Well Plates

An important caveat in these calculations is the definition of the path length as 1 cm. We are aware that some investigators will prefer to adapt the CR method to 96-well plates. Under these conditions, the path length for the plate reader will depend on the volume in each well (being approximately 0.1 cm/100 μl). This allows for considerably more CR to be present before reaching the limiting absorbance of 1.0 AU. However, new artifacts are possible as levels of CR surpass 50 μM. The most important is micelle formation,[30] which will add to the light scattering and may affect the CR–Aβ interaction. Because of this we do not recommend exceeding 25 μM under any assay conditions.

The second effect of the 96-well plate is that the concentration of the CR–Aβ complex determined by the equations given earlier will be an underestimate by a factor of (1/path length). This simple correction factor must be applied in order to obtain accurate absolute concentrations. However, meniscus effects can cause variation in the actual path lengths measured from well to well. This will decrease the accuracy of the technique compared to the use of a standard cuvette. This tradeoff of convenience/speed versus accuracy may be desirable depending on the particular application. It is worth noting that some 96-well plate readers can be programmed to correct for path length, giving readouts that can be used directly in the equations. For those that do not, the equations need to be modified to account for the shorter path length. For example, Eqs. (11) and (15) would be modified as

$$[\mathrm{CR{-}A}\beta]/b = ({}^{541}A_t/47{,}800) - ({}^{403}A_t/68{,}300) - ({}^{403}A_{\mathrm{CR}}/86{,}200) \quad (11)$$

$$\mathrm{A}\beta_{\mathrm{fib}}/b = ({}^{541}A_t/4780) - ({}^{403}A_t/6830) - ({}^{403}A_{\mathrm{CR}}/8620) \quad (15)$$

where b is the path length in the well. For samples not containing Aβ or for samples with negligible light scattering, the actual path length for a given sample can be determined by measuring the absorbance at 403 nm and calculating the path length according to

$$b = A/ac$$

or

$$b = A/(18{,}100 \times [\mathrm{CR}])$$

Potential Competition with CR for Binding to Aβ

A potential disadvantage of the CR spectral shift assay in general is the possibility that a test compound could compete with CR for binding to Aβ. This would give the false appearance of inhibition of aggregation. Furthermore, a compound with strong absorbance at 403 or 541 nm will interfere with the assay. Correction can be made for the absorbance of

[30] F. C. Bedaux, P. J. O. Kerssemakers, and C. A. M. Meijers, *Pharm. Weekbl.* [*Sci.*] **98,** 189 (1963).

interfering compounds in the same way that correction is made for fibril light scattering. That is, a blank with the same concentration of the test compound but no CR or Aβ can be used to subtract out the interference at 403 and 541 from the compound. Similar sorts of corrections cannot be made for compounds that quench the fluorescence of amyloid probes such as thioflavin T, which represents a significant limitation of the thioflavin T method.[31]

Correction for competitive binding inhibitors is more difficult. First, it can be minimized by using a high concentration of CR in the assay. The K_i of Congo red binding to Aβ(1–40) is ~50 nM (unpublished results). In an assay in which the [CR] is 20 μM, the apparent IC_{50} of a potential competitive binding compound is equal to

$$IC_{50} = K_i\ (1 + 20{,}000\ \text{n}M/50\ \text{n}M) \qquad \text{or} \qquad IC_{50} = 401 \times K_i$$

where K_i is the actual inhibitory constant of that compound.[32] That is to say, at this relatively high concentration of CR, it would take a concentration of the potential competitor 400 times the actual K_i of that competitor to displace 50% of CR from Aβ. In typical assays,[33] compounds are tested at 10–50 μM and then diluted by a factor of 10 before addition of the CR. In order to cause 50% inhibition of CR binding, the K_i of these potential competitors would have to be 2.5–12.5 nM. For 10% inhibition, the K_i would need to be 12.4–113 nM. These are attainable values and, in such cases, controls must be included in which the potential compound is added *after* the period of aggregation. In this way, the decrease in CR–Aβ due solely to inhibition of CR binding to Aβ (and not to inhibition of aggregation) can be determined.

Choice of Equations

Finally, some consideration should be given to which of Eqs. (9)–(16) should be chosen as the actual "working equation," although all are acceptable. Obviously, if the emphasis is on the concentration of the CR bound, then one of Eqs. (9)–(12) should be chosen. However, when the goal is quantitation of Aβ aggregation, one of Eqs. (13)–(16) should be used. We have provided Eqs. (11) and (12) and (15) and (16) mainly for reasons of simplification. These equations presume a value of 0.7 for r. Our observa-

[31] H. LeVine, *Protein Sci.* **2,** 404 (1993).

[32] J. P. Bennett and H. I. Yamamura, *in* "Neurotransmitter Receptor Binding" (H. I. Yamamura, S. J. Enna, and M. J. Kuhar, eds.), p. 61. Raven Press, New York, 1985.

[33] S. J. Pollack, I. I. J. Sadler, S. R. Hawtin, V. J. Tailor, and M. S. Shearman, *Neurosci. Lett.* **197,** 211 (1995).

tions suggest that this is reasonably accurate. Because some preparations of aggregated Aβ yield slightly different values for r, we recommend measurement of r using a control sample of Aβ and then recalculation of Eqs. (11), (12), (15), or (16) with the appropriate value for r. Values of r less than 0.65 or greater than 0.75 should be held suspect. When the degree of light scattering falls below 0.02 AU (1 cm path length) for the Aβ sample, the determination of r becomes inaccurate. However, the correction for light scattering is less critical at these low concentrations of Aβ.

Equations (9) and (13) can offer the advantage of actually measuring the concentration of CR in each individual sample. In order to do this, however, the contribution of light scattering at 403 nm must be either known or negligible. To be negligible (i.e., $<5\%$), the CR : Aβ ratio must be above 1.5 : 1. Otherwise, the scattering of a known sample of Aβ fibrils must be measured and used to correct the total absorbance at 403 nm. This is only useful when the focus is on the amount of CR–Aβ formed by varying the [CR] in the presence of a known and consistent amount of aggregated Aβ, such as the experiment shown in Fig. 5. For this purpose, we favor Eq. (9) (see Fig. 5). In order to use Eqs. (9) and (13) to quantify unknown degrees of aggregation in Aβ samples, the absorbance at 403 nm of the CR-alone control sample is used as ${}^{403}A_{CR}$ in the equation. While this will not correct each individual sample, it will correct for inaccuracies in dilution from a concentrated stock of CR. These inaccuracies usually stem from the difficulty in precisely defining the concentration of CR in the very concentrated stock solution.

Therefore, if r can be determined accurately, either Eq. (9) or (13) should be used. If r cannot be determined accurately, Eqs. (11) and (15) should be sufficiently accurate for the quantitation of most fine suspensions of aggregated Aβ. We have observed some very large, easily precipitable aggregates of Aβ under aggregation conditions different from those used in this study. The CR–Aβ method would not be expected to work well with such large aggregates. Equations (13) and (15) give the concentration of Aβ_{fib} in units of micrograms per milliliter, but, if desired, these units can be converted to micromolar by division by 4.33 [for Aβ(1–40)]. These units can be useful in determining the stoichiometry of the CR–Aβ interaction. However, in a rigorous sense, it is not appropriate to refer to a suspension of fibrillar aggregates of unknown size in terms of moles per liter. We, therefore, prefer the more general units of mass per liter.

Generalization to Nonturbid Samples of Aβ

The degree of turbidity in samples of aggregated Aβ depends closely on the method of aggregation. Methods that employ stirring or agitation (e.g., Evans *et al.*[29]), such as employed in our preparation, result in turbid samples that have significant light scattering. Methods that employ static conditions (no stirring/agitation; e.g., Wood *et al.*[14] and Howlett *et al.*[15]) result in Aβ fibril preparations that have minimal turbidity (which does not necessarily equate to zero light scattering). This decreases the necessity for light-scattering corrections, which constitute the major distinction between our original CR–insulin equations and the optimized equations[18] presented here. However, as seen in the following example, equations that have been corrected for light scattering are still valid when applied to samples with no light scattering.

As an example, consider two 70-μg/ml Aβ fibril preparations, one of which has significant turbidity/light-scattering properties and another which causes no light scattering. Each is combined with CR to yield a final CR concentration of 10 μM and each preparation of Aβ binds 50% (5 μM) of the CR. Based on the molar absorptivity values in Table I and obtaining typical Aβ light scattering from Fig. 2, we obtain the data in Table II.

The transparency of why these two samples should yield equivalent results in Eq. (11) is obscured by the numerous mathematical simplifications used to derive this equation. Therefore, we will begin our comparison with Eq. (7).

For the turbid sample:

$$
\begin{aligned}
[\text{CR–A}\beta] &= [({}^{541}A_t - {}^{541}S_{A\beta})/47{,}800] - [({}^{403}A_t - {}^{403}S_{A\beta})/38{,}100] \\
&= [(0.5380 - 0.0720)/47{,}800] - [(0.2836 - 0.1030)/38{,}100] \\
&= 9.75 \times 10^{-6} - 4.74 \times 10^{-6} \\
&= 5.01 \times 10^{-6}\,M = 5.0\ \mu M
\end{aligned}
$$

TABLE II

COMPONENTS OF TWO THEORETICAL CR–Aβ SAMPLES

Sample component	${}^{541}A$	${}^{403}A$
5 μM CR–Aβ	0.3525[a]	0.0904[a]
5 μM CR free	0.1135	0.0902
70 μg/ml Aβ	0,[b] 0.0720[c]	0,[b] 0.1030[c]
Total absorbance	0.4660,[b] 0.5380[c]	0.1806,[b] 0.2836[c]

[a] CR–Aβ values do not include the Aβ-scattering components, which are listed separately in the third row.

[b] Values for a theoretical Aβ sample with *no* inherent light-scattering characteristics.

[c] Values for an Aβ sample with inherent light-scattering characteristics typical of the samples used in this study.

For the sample lacking turbidity/light scattering:

$$
\begin{aligned}
[\text{CR–A}\beta] &= [({}^{541}A_t - {}^{541}S_{\text{A}\beta})/47{,}800] - [({}^{403}A_t - {}^{403}S_{\text{A}\beta})/38{,}100] \\
&= [(0.4660 - 0)/47{,}800] - [(0.1806 - 0)/38{,}100] \\
&= 9.75 \times 10^{-6} - 4.74 \times 10^{-6} \\
&= 5.01 \times 10^{-6}\,M = 5.0\,\mu M
\end{aligned}
$$

Completing the same exercise with Eq. (11):

For the turbid sample:

$$
\begin{aligned}
[\text{CR–A}\beta] &= ({}^{541}A_t / 47{,}800) - ({}^{403}A_t/68{,}300) - ({}^{403}A_{\text{CR}}/86{,}200) \\
&= (0.5380/47{,}800) - (0.2836/68{,}300) - [(0.0904 + 0.0902)/86{,}200] \\
&= 11.25 \times 10^{-6} - 4.15 \times 10^{-6} - 2.10 \times 10^{-6} \\
&= 5.01 \times 10^{-6}\,M = 5.0\,\mu M
\end{aligned}
$$

For the sample lacking turbidity/light scattering:

$$
\begin{aligned}
[\text{CR–A}\beta] &= ({}^{541}A_t/47{,}800) - ({}^{403}A_t/68{,}300) - ({}^{403}A_{\text{CR}}/86{,}200) \\
&= (0.4660/47{,}800) - (0.1806/68{,}300) - [(0.0904 + 0.0902)/86{,}200] \\
&= 9.75 \times 10^{-6} - 2.64 \times 10^{-6} - 2.10 \times 10^{-6} \\
&= 5.01 \times 10^{-6}\,M = 5.0\,\mu M
\end{aligned}
$$

Clearly, and as expected, both equations yield the same result whether the particular Aβ preparation is turbid or not. Because of the general nature of these corrected equations and the unreliability of visual inspection of samples for assessing the presence of light scattering, we recommend that equations corrected for light scattering always be used.

Modification of Method for Other Amyloid Fibrils

The CR spectral shift assay has been applied to two different amyloid proteins: insulin fibrils[12] and Aβ.[18] In addition, data have been presented to suggest it could be applied to β-poly(L-lysine) as well.[7] It is likely that the method described earlier can be generalized to a wide variety of amyloid proteins. This is predicted by our model[7] of CR binding to amyloid, which is independent of the primary amino acid sequence and dependent only on the presence of a β-pleated sheet secondary structure (a characteristic of amyloid proteins in general) and the presence of at least one positively charged amino acid residue (a very high likelihood in any peptide over 15 amino acids in length). In order to quantitate CR binding to another amyloid protein using the CR spectral shift assay, the following steps will need to be carried out.

1. Verify the CR molar absorptivity values in the CR stock prepared for the assay by comparison to Fig. 1A and Table I.

2. Find the concentration of the particular amyloid fibril (AM_{fib}) preparation that binds over 95% of the highest concentration of CR (usually 5–10 μM) used in determining the molar absorptivity of the CR–AM_{fib} complex. This can be determined by filtering or centrifuging the CR–AM_{fib} suspensions and measuring the absorbance of the filtrate/supernatant at 488 nm (in PBS).
3. Determine the need for correction for light scattering by running a spectrum of AM_{fib} in the absence of CR as shown in Fig. 1B for Aβ. This data can also be used to determine the r value of AM_{fib}.
4. The molar absorptivity of the CR–AM_{fib} complex can be determined by measuring spectra of a series of CR concentrations (in the presence of the excess concentration of AM_{fib} determined in step 2) as shown in Fig. 1C. CR–AM_{fib} spectra will first need to be corrected for light scattering as shown in Fig. 1B.
5. Determine the point of maximal spectral difference as shown in Fig. 1D and check the linearity of the difference with [CR] as shown in Fig. 3.
6. Determine the best isosbestic point as shown in Fig. 3 by checking difference spectra at a variety of CR concentrations. The best isosbestic point is the wavelength at which all difference spectra most closely equal zero. It should be noted that the existence of an isosbestic point is not absolutely necessary. If one cannot be found, then the equations will need to be derived from Eq. (4), omitting the simplification introduced by the use of an isosbestic point in Eq. (5).
7. Insert the new molar absorptivities into Eq. (5), using the appropriate wavelength notations. This yields a new Eq. (6).
8. Derive the light-scattering-corrected equations by following the derivations shown in Eqs. (7)–(9).
9. Utilize the specific molar absorptivities and r values to derive Eqs. (10)–(12).
10. An accurately quantified standard sample of AM_{fib} is necessary to determine the stoichiometry of the CR–AM_{fib} complex and the correction factor (which is 10 in the case of CR–Aβ), which can be used to determine Eqs. (13)–(16).

Conclusions

In summary, this study describes the modification of the CR–insulin method for use in the quantitation of Aβ aggregation. The CR–Aβ method is simple, rapid, and requires no specialized equipment or expensive/radioactive reagents. By simply measuring the absorbance of a sample at two wavelengths, the absolute amount of aggregated Aβ can be determined by using the equations provided. This absolute quantitation of Aβ is important in the study of the effect of peptide alterations on aggregation. Quantifying aggregation is also essential in the evaluation of potential drug candidates designed to prevent Aβ fibril formation. Furthermore, the neurotoxicity of various Aβ preparations in cell culture has been shown in a qualitative sense to be dependent on the ability of Aβ to bind CR.[16] The CR–Aβ assay has the potential to quantitatively define the neurotoxic potential of various Aβ preparations. This could help standardize a variety of considerably difficult and variable Aβ neurotoxicity assays.

[20] Kinetic Analysis of Amyloid Fibril Formation

By HIRONOBU NAIKI and FUMITAKE GEJYO

Introduction

A nucleation-dependent polymerization model has been proposed to explain the mechanisms of amyloid fibril formation *in vitro*.[1–4] This model consists of two phases, i.e., nucleation and extension phases. Nucleus formation requires a series of association steps of monomers that are thermodynamically unfavorable, representing the rate-limiting step in amyloid fibril formation. Once the nucleus (n-mer) has been formed, the further addition of monomers becomes thermodynamically favorable, resulting in a rapid extension of amyloid fibrils. We have developed a first-order kinetic model of amyloid fibril extension *in vitro* and confirmed that the extension of amyloid fibrils proceeds via the consecutive association of monomeric precursor proteins onto the ends of existing fibrils.[2,5,6] A characteristic sigmoidal time–course curve of amyloid fibril formation from monomeric precursor proteins at a physiological pH is widely believed to represent the essence

[1] J. T. Jarrett and P. T. Lansbury, Jr., *Cell* **73,** 1055 (1993).
[2] H. Naiki and K. Nakakuki, *Lab. Invest.* **74,** 374 (1996).
[3] A. Lomakin, D. S. Chung, G. B. Benedek, D. A. Kirschner, and D. B. Teplow, *Proc. Natl. Acad. Sci. U.S.A.* **93,** 1125 (1996).
[4] A. Lomakin, D. B. Teplow, D. A. Kirschner, and G. B. Benedek, *Proc. Natl. Acad. Sci. U.S.A.* **94,** 7942 (1997).
[5] H. Naiki, K. Higuchi, K. Nakakuki, and T. Takeda, *Lab. Invest.* **65,** 104 (1991).
[6] H. Naiki, N. Hashimoto, S. Suzuki, H. Kimura, K. Nakakuki, and F. Gejyo, *Amyloid Int. J. Exp. Clin. Invest.* **4,** 223 (1997).

METHODS IN ENZYMOLOGY, VOL. 309
0076-6879/99 $30.00

of a nucleation-dependent polymerization model, i.e., an initial lag phase represents the thermodynamically unfavorable nucleus formation.[1,7] However, no convincing kinetic models to explain the sigmoidality of the curve have been reported.

This article describes our paradigm to analyze the kinetic aspect of amyloid fibril formation *in vitro*. During the investigation of murine senile amyloidosis observed in the senescence-accelerated mouse (SAM),[8] we developed a novel fluorometric method to determine amyloid fibrils *in vitro* based on the unique characteristics of thioflavin T (ThT).[9] In recent years, this method has been applied to the analysis of several human amyloidoses, including gelsolin-derived amyloidosis, dialysis-related amyloidosis, and Alzheimer's β-amyloidosis.[10] With this fluorometric method, we successfully measured the polymerization velocity of three types of amyloid fibrils, i.e., murine senile amyloid fibrils (fAApoAII),[5] Alzheimer's β-amyloid fibrils (fAβ),[2,7] and dialysis-related amyloid fibrils (fAβ_2-m),[6] and performed kinetic analysis of amyloid fibril polymerization *in vitro*.

Preparation of Amyloid Fibrils

Purification of fAApoAII and fAβ_2-m from Amyloid-Deposited Tissue

Amyloid fibrils can be purified from amyloid-deposited tissues by the following three-step procedure. First, crude amyloid fibrils are isolated as a water suspension from the tissues according to the method of Pras *et al.*[11] fAApoAII are isolated from the livers of 16- to 18-month-old SAMP1 mice. About 1 g of liver tissue and ice-cold 150 m*M* NaCl are put into a Beckman polycarbonate tube (size; 1 × 3.5 inch) and then homogenized on ice with five intermittent pulses (dispersion, 30 sec; interval, 30 sec; output level, 10,000 rpm) using an ultradisperser (Ultra-Turrax T25, Janke & Kunkel IKA-Labortechnik, Staufen, Germany). This homogenate is then ultracentrifuged at $2.84 \times 10^4 g$ for 20 min at 4° using a Beckman L8-60M ultracentrifuge and a Beckman 60Ti rotor. After ultracentrifugation, the supernatant is discarded and the just-described procedure is repeated until the absorbance of the supernatant drops to 0.05. The homogenizing solution is then switched to ice-cold distilled water and the just-described procedure is repeated until the absorbance of the supernatant drops to 0.5. This time,

[7] H. Naiki, F. Gejyo, and K. Nakakuki, *Biochemistry* **36,** 6243 (1997).
[8] K. Higuchi, M. Hosokawa, and T. Takeda, *Methods Enzymol.* **309** [42] 1999 (this volume).
[9] H. Naiki, K. Higuchi, M. Hosokawa, and T. Takeda, *Anal. Biochem.* **177,** 244 (1989).
[10] H. LeVine III, *Methods Enzymol.* **309** [18] 1999 (this volume).
[11] M. Pras, D. Zucker-Franklin, A. Rimon, and E. C. Franklin, *J. Exp. Med.* **130,** 777 (1969).

the supernatant, except for the first fraction, is collected and pooled (crude fAApoAII). f$A\beta_2$-m are isolated from the Baker's cyst wall excised from the popliteal fossa or the soft tissues excised from the carpal tunnel of patients suffering from dialysis-related amyloidosis (DRA). Crude f$A\beta_2$-m are isolated as described earlier with minor modifications. As f$A\beta_2$-m-deposited tissues are composed mainly of collagenous connective tissue, 2–3 g is first diced into small pieces on ice using a pair of scissors, then made into a paste on ice using a ceramic mortar after the addition of 2–3 ml of ice-cold 150 m*M* NaCl. To reduce the fragmentation of f$A\beta_2$-m, samples are homogenized with a Teflon homogenizer (size: 50 ml) equipped with a low-speed electronic motor.

A water suspension of crude amyloid fibrils is then ultracentrifuged at 10^5g for 2 hr at 4°. Practically all of the amyloid fibrils are collected to the pellet fraction, leaving fatty debris in the supernatant. Pellets are resuspended in ice-cold distilled water by mixing thoroughly on ice with a Teflon homogenizer (size: 10 ml). These suspensions are then applied on a discontinuous sucrose density gradient prepared as follows. First, 1.5 ml of 60% (w/v) sucrose is placed on the bottom of a Beckman Ultra-Clear tube (size: 0.5 × 2 inch). On this bed, 2.5 ml of 50% (w/v) sucrose is placed and, finally, 1.0 ml of amyloid suspension is applied. This preparation is ultracentrifuged at 10^5g for 24 hr at 16° using a Beckman SW55Ti swing rotor. Both pellets and fine aggregates formed around a 50–60% interface are collected by careful aspiration and dialyzed against 0.05% (w/v) NaN_3 at 4°. After dialysis, these aggregates are resuspended by vortexing in an Eppendorf tube, sonicated on ice with five intermittent pulses (pulse, 0.6 sec; interval, 0.4 sec; output level, 2) using an ultrasonic disruptor (UD-201, Tomy, Tokyo, Japan) equipped with a microtip (TP-030, Tomy, Tokyo, Japan) and stored at 4° until assay (purified fAApoAII and f$A\beta_2$-m). Electron microscopically, although there may be some contaminants, typical amyloid fibrils can be purified.[6,9]

Formation of fAβ from Synthetic Amyloid β-Peptides

fAβ(1–42) are formed from the fresh β-amyloid(1–42) [Aβ(1–42)] solution described in the following section. The reaction mixture in an Eppendorf tube is 950 μl and contains 25 μ*M* Aβ(1–42), 50 m*M* phosphate buffer, pH 7.5, and 100 m*M* NaCl. After vortexing briefly, the mixture is incubated at 37° for 6 hr for polymerization reactions. The reaction tubes are not agitated during the reaction. After incubation, the mixture is centrifuged at 4° for 3 hr at 1.5×10^4 rpm using a high-speed refrigerated microcentrifuge (MRX-150, Tomy, Tokyo, Japan). More than 95% of fAβ(1–42) precipitates as measured by the fluorescence of ThT. The pellet is resuspended in 300–400 μl of 50 m*M* phosphate buffer, pH 7.5, 100 m*M* NaCl, and 0.05% NaN_3 in an Eppendorf tube, sonicated on ice with 15 intermittent pulses (pulse, 0.6 sec; interval, 0.4 sec; output level, 2) using an ultrasonic disruptor and stored at 4° until assay. Another type of fAβ [fAβ(1–40)] is formed from the fresh β-amyloid(1–40) [Aβ(1–40)] solution described in the following section. The reaction mixture is 600 μl and contains 50 μ*M* Aβ(1–40), 50 m*M* phosphate buffer, pH 7.5, and 100 m*M* NaCl. After incubation at 37° for 24 hr, the mixture is centrifuged at 4° for 3 hr at 1.5×10^4 rpm. fAβ(1–40) precipitates completely as measured by the fluorescence of ThT. The pellet is resuspended in 300–400 μl of 50 m*M* phosphate buffer, pH 7.5, 100 m*M* NaCl, and 0.05% NaN_3, sonicated as described earlier, and stored at 4° until assay.

Preparation of Precursor Proteins

Preparation of Fresh Aβ Solutions

Synthetic Aβ(1–42) (Bachem AG, Bubendorf, Switzerland) is dissolved by vortexing briefly in an ice-cold 0.02% ammonia solution at a concentration of about 250 μ*M* (1.1 mg/ml) in a 4° room. Although the solution is clear, short fibrils are visualized by electron microscopy. Significant ThT fluorescence is also detected by fluorescence spectroscopy. To remove these fibrils, the solution (0.8 ml) is applied to a Beckman polycarbonate tube (size, 11 × 34 mm,) and ultracentrifuged at $2 \times 10^5 g$ for 3 hr at 4° using a Beckman Optima TLX tabletop ultracentrifuge and a Beckman TLA-120.2 fixed angle rotor. Although no visible pellets are formed after ultracentrifugation, the ThT fluorescence is collected to the bottom one-quarter fraction and no significant fluorescence is detected in the upper three-quarter fraction. Electron microscopically, no fibrillar components are observed in the upper three-quarter fraction. Therefore, the upper three-quarter fraction is collected by careful aspiration, aliquoted into 50- to 100-μl fractions and stored at −80° until assay [fresh Aβ(1–42) solution]. Synthetic Aβ(1–40) (Bachem AG) is dissolved by vortexing briefly in ice-cold distilled water at a concentration of about 500 μ*M* (2.2 mg/ml) in a 4° room. No significant ThT fluorescence is detected in this solution. Electron microscopically, no fibrillar components are observed either. Therefore, the solution is aliquoted into 50- to 100-μl fractions and stored at −80° until assay [fresh Aβ(1–40) solution].

Purification of Monomeric Constituents of fAApoAII and fAβ_2-m

The monomeric constituent of fAApoAII (AApoAII) is purified from a pellet of the just-described ultracentrifugation ($10^5 g$) of crude fAApoAII.

Pellets are resuspended in ice-cold distilled water on ice using an ultradisperser and lyophilized. The following procedure is performed in a 4° room. Six molar urea polyacrylamide gel electrophoresis is performed by the method of Davis[12] on a 16.2 × 13 × 0.4-cm slab gel with 7.5% acrylamide. The lyophilized sample (about 2.5 mg of protein) is dissolved in 2.0 ml of sample buffer containing 6 *M* urea and incubated for 8 hr to depolymerize fAApoAII into AApoAII. It is then applied onto a gel and electrophoresed overnight at 12.5 mA. AApoAII-containing bands located by stained reference gels are excised, diced into small pieces, and suspended in 10 m*M* NH_4HCO_3. AApoAII is eluted for 24 hr with gentle stirring. The elution buffer is exchanged twice, pooled, and dialyzed against the same elution buffer in Spectrapor 3 tubes (Spectrum Medical Industries Inc., Terminal Annex, LA) for 24 hr. Fine gel fragments are removed by centrifugation at 3000*g* for 5 min at 4°. The solution is aliquoted into fractions containing 400 μg of AApoAII and lyophilized. The lyophilized samples are stored at −20° until assay. The polymerization activity of these samples remains intact for a few months. AApoAII is dissolved in ice-cold distilled water at a concentration of 400 μM (3.48 mg/ml).

The monomeric constituent of fAβ_2-m(β_2-m, β_2-microglobulin) can be purified from the urine of patients with renal insufficiency who are negative for hepatitis B virus antigen and antibodies to human immunodeficiency virus (HIV). The following procedure is performed in a 4° room. The urine is first concentrated by ultrafilteration through cuprophane membrane dialyzers. The concentrated urine is then fractionated by gel chromatography on Sephadex G-75 (Pharmacia Fine Chemicals, Uppsala, Sweden). Three milliliters of the concentrated urine is applied to a column (1.4 × 120 cm), and elution is carried out with 100 m*M* ammonium acetate buffer, pH 8.0, at a flow rate of 6.0 ml/hr. The β_2-m-containing fraction is purified further by ion-exchange chromatography on DEAE-Sephacel (Pharmacia Fine Chemicals). The column (1.2 × 44 cm) is equilibrated with 10 m*M* phosphate buffer, pH 7.1 (buffer A). After application of the β_2-m-containing fraction, the column is washed with 150 ml of buffer A and eluted with a linear gradient of NaCl using 250 ml each of buffer A and buffer B (10 m*M* phosphate buffer, pH 7.1, and 200 m*M* NaCl) at a flow rate of 8 ml/hr. After dialyzed against distilled water at 4° and lyophilized, the β_2-m-enriched fraction is dissolved in phosphate-buffered saline (PBS) and purified further by high-performance liquid chromatography (HPLC) at room temperature using the gel filtration column (TSK G2000 SW, Toyo Soda Manufacturing Co., Ltd., Tokyo, Japan). Elution is carried out with PBS at a flow rate of 1.0 ml/min. Following further purification by the

[12] B. J. Davis, *Ann. N.Y. Acad. Sci.* **121,** 404 (1964).

ion-exchange chromatography described earlier, the final purified β_2-m is obtained by repeated HPLC. After dialyzed against distilled water at 4°, the samples are lyophilized and stored at −20° until assay. The polymerization activity of these samples remains intact for a few years. β_2-m is dissolved in 50 m*M* NaCl at a concentration of 400–500 μM (4.7–5.9 mg/ml).

Determination of Protein Concentrations

Protein concentrations of the AApoAII and fAApoAII solutions are determined by the method of Lowry *et al.*[13] using bovine serum albumin as the standard. Protein concentrations of the β_2-m and fAβ_2-m solutions are determined by the method of Bradford[14] using a protein assay kit (500-0001, Bio-Rad Laboratories, Inc., CA) and bovine γ-globulin as the standard. Protein concentrations of the Aβ and fAβ solutions are determined by the method of Bradford[14] using the Aβ(1–40) solution quantified by amino acid analysis as the standard.

Fluorescence Spectroscopy

A 100 μM ThT (Wako Pure Chemical Industries, Ltd., Osaka, Japan) solution and buffer solutions are prepared and stored at room temperature until assay. The ThT solution should be kept in a brown bottle to avoid quenching by light and mixed appropriately with buffer solutions and distilled water before use. Immediately after making the mixture, fluorescence is measured on a Hitachi F-3010 fluorescence spectrophotometer in the ratio mode at 25°, with an assay volume of 1.0 ml, and is averaged for the initial 5 sec. Excitation and emission slits are set at 5 and 10 nm, respectively.

In the presence of amyloid fibrils, ThT emits a novel fluorescence at 482–490 nm with an excitation maximum at 446–455 nm [fAApoAII: 450–482 nm,[9] fAβ(1–40) and (1–42): 446–490 nm,[2] and fAβ_2-m: 455–485 nm[6]]. Under the same experimental conditions, fluorescence spectra for ThT in the presence of freshly prepared monomeric constituents of amyloid fibrils completely overlap spectra for ThT in the absence of amyloid fibrils. Thus, this system detects only the polymeric form of precursor proteins (i.e., amyloid fibrils) present in the solution.

The fluorescence of ThT in the presence of amyloid fibrils is pH dependent, reaching a maximum at approximately pH 8.5.[2,6,9] The Scatchard plot of ThT binding to amyloid fibrils reveals a single population of binding sites. In the cases of fAApoAII, fAβ(1–40), and fAβ_2-m, individual K_d

[13] O. H. Lowry, N. J. Rosebrough, A. L. Farr, and R. J. Randall, *J. Biol. Chem.* **193,** 265 (1951).
[14] M. M. Bradford, *Anal. Biochem.* **72,** 248 (1976).

values are about 0.04, 0.86, and 1.1 μM, respectively.[2,6,9] Thus, optimum florescence measurements of amyloid fibrils are obtained with the reaction mixture containing 0.25 (fAApoAII), 5(fAβ), or 3 μM (fAβ_2-m) ThT and 50 mM of glycine–NaOH buffer, pH 8.5.

The fluorescence intensity of ThT is used as a measure of the total amount of amyloid fibrils (i.e., the concentration of monomeric proteins constituting the whole amyloid fibrils), based on the following: (1) the fluorescence intensity is linear with the increase in the protein concentration of amyloid fibrils[2,6,9]; (2) the fluorescence intensity is independent of the number concentration or the mean length of amyloid fibrils if the protein concentration of amyloid fibrils is constant[2,5]; and (3) in the extension kinetics study, where the number concentration of amyloid fibrils is constant, the increase in the average length of amyloid fibrils corresponds to the increase in the fluorescence intensity.[5] At present, there is no proof that the signal response of ThT is constant among the different fibrillar structures formed from soluble precursor proteins (e.g., Aβ) during fibrillogenesis (e.g., prenuclei, nuclei, short fibrils, long fibrils, and fibril bundles). In fact, some morphological heterogeneity of fAβ(1–42) is observed in the reaction mixture after proceeding to equilibrium.[15] However, for the simplicity of the model, we assume that the fluorescence intensity of ThT is proportional to the total amount of amyloid fibrils.

Morphological Analysis

Electron Microscopy and Morphometry

A carbon-coated 150-mesh copper grid (150-A, Nisshin EM Co., Ltd., Tokyo, Japan) is placed for 30 sec on 30 μl of a reaction mixture dropped on a Parafilm and diluted appropriately with distilled water. After excess liquid is removed with filter paper, the grid is dried by gently blowing with a blower, then placed for 30 sec on a 20-μl drop of fresh 1% phosphotungstic acid, pH 7.0. After removing excess liquid and drying, specimens are examined under a Hitachi H-7000 electron microscope with an acceleration voltage of 75 kV.

After taking photographs on which both ends of amyloid fibrils can be identified, appropriately, the length of each amyloid fibril on the prints is measured using a pair of dividers and a scale. The average length of amyloid fibrils of the reaction mixture is obtained and compared with the fluorescence of ThT in the extension reaction described later.

[15] H. Naiki, K. Hasagawa, I. Yamaguchi, H. Nakamura, F. Gejyo, and K. Nakakuki, *Biochemistry* **37,** 17882 (1998).

Polarized Light Microscopy

Part of the reaction mixtures is centrifuged at 4° for 90 min at 1.5×10^4 rpm. Pellets are spread on glass slides, dried overnight in an incubator set at 37°, stained with Congo red, and examined under a polarized light microscope to check for orange–green birefringence (one of the classical criteria of amyloid fibrils).

Secondary Structure Analysis

The secondary structure of precursor proteins is analyzed by circular dichroism (CD). The mixture contains 25 μM of precursor proteins and 10 mM phosphate buffer, pH 7.5. The CD spectrum is recorded at room temperature using a 1.0-mm path length cell on a Jasco 720WI spectropolarimeter (Jasco Corporation, Hachioji, Japan). Ten cumulative readings at a 1-nm bandwidth, a resolution of 0.1 nm, a sensitivity of 10 mdeg, a response time of 2 sec, and a scan speed of 50 nm/min are taken from each sample, averaged, and base line subtracted. Results are expressed in terms of molar ellipticity (ϑ), ranging from 260 to 183 nm. The content of β-sheet structure is estimated by curve fitting using the algorithm of Yang *et al.*[16]

Polymerization Reaction

Reaction Mixtures

Reaction mixtures are prepared on ice at 4°, with neither polymerization nor depolymerization of amyloid fibrils being observed by fluorometric analysis. Distilled water is put into a tube. Then 500 mM phosphate buffer, pH 7.5 (AApoAII, Aβs), or 500 mM citrate buffer, pH 2.5 (β_2-m), is added to yield a final buffer concentration of 50 mM, and 5 M NaCl is added to a final concentration of 100 mM. Fresh AApoAII, Aβs, or β_2-m solutions are added to yield a final concentration of 0–100 (AApoAII), 0–30 [Aβ(1–42)], 0–50 [Aβ(1–40)], or 0–100 μM (β_2-m). Finally, the amyloid fibril solution is added to yield a final concentration of 0–50 (fAApoAII), 0–15 [fAβ(1–42)], 0–50 [fAβ(1–40)], or 0–150 ng protein/μl (fAβ_2-m).

To analyze the effects of apolipoprotein E (apoE) on amyloid fibril formation *in vitro,* recombinant human apoE2, E3, or E4 [lot numbers: CAF7252, LEP 7841 (E2), SKK7520, LEQ7625 (E3), and SKP7961, LEL 7273 (E4), respectively, Wako Pure Chemical Industries, Ltd.] dissolved

[16] J. T. Yang, C. S. C. Wu, and H. M. Martinez, *Methods Enzymol.* **130,** 208 (1986).

in 50 mM phosphate buffer, pH 7.5, 100 mM NaCl is added before adding amyloid fibrils to yield a final apoE concentration of 0–4 μM. Recombinant apoE lots are supplied at 0.42–0.65 mg/ml in 0.7 M ammonium bicarbonate. To remove ammonium bicarbonate, they are lyophilized and dissolved in ice-cold 6 M urea at a concentration of 12 μM to obtain all molecules in monomeric form. To remove urea, they are subsequently dialyzed against 50 mM phosphate buffer, pH 7.5, 100 mM NaCl at 4°. In some experiments, human α_1-microglobulin (α_1-MG) [lot number: 065(501), DAKO A/S, Glostrup, Denmark] dissolved in 15 mM NaN_3 at a concentration of 554 mg/liter (20.7 μM) is added as a negative control of apoE, to a final α_1-MG concentration of 0–3 μM.

Initiation and Termination of Reaction

After a brief vortexing of the mixture, 30-μl aliquots are put into oil-free polymerase chain reaction tubes (size, 0.5 ml; code number, 9046; Takara Shuzo Co. Ltd., Otsu, Japan). The reaction tubes are then transferred into a DNA thermal cycler (PJ480, Perkin Elmer Cetus, Emeryville, CA). Starting at 4°, the plate temperature is elevated at maximal speed to 37°. Incubation time ranges from 0 to 15 hr, and the reaction is stopped by placing the tubes on ice. The reaction tubes are not agitated during the reaction. From each reaction tube, triplicate 5-μl aliquots are removed and subjected to fluorescence spectroscopy, and the mean of each triplicate is determined.

Reaction Temperature

The velocity of amyloid fibril formation *in vitro* is greatly dependent on the incubation temperature.[2,6] For example, the velocity of fAβ(1–40) extension at 25° is about 10% of that at 37°.[2] Therefore, the kinetic experiment should be performed at a constant temperature (e.g., 37°).

First-Order Kinetic Model of Amyloid Fibril Extension *in Vitro*

Analysis of Time–Course Curves

When amyloid fibrils are incubated with their monomeric constituents at 37°, the fluorescence of ThT increases without a lag phase and proceeds to equilibrium (Fig. 1A).[2,5,6] A semilogarithmic plot shown in Fig. 1B represents a good linearity. Interpretation of this plot yields the following equation:

$$\log[A - F(t)] = a - bt \qquad (1)$$

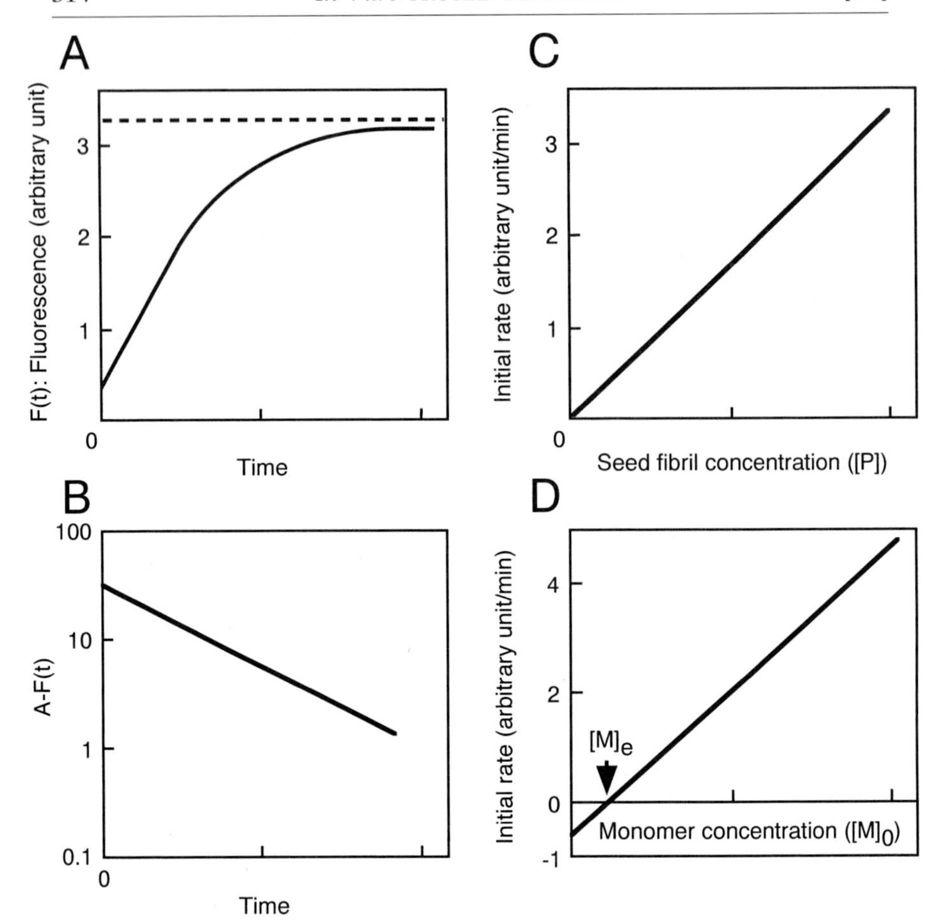

FIG. 1. Kinetics of amyloid fibril extension *in vitro*. (A) Time course of fluorescence after the initiation of the polymerization reaction. $F(t)$ represents fluorescence as a function of time. The reaction mixture contains both amyloid fibrils and their monomeric constituents. (B) The semilogarithmic plot of the difference: $F(\infty) - F(t)$ versus incubation time. A is tentatively determined as $F(\infty)$ and is shown as a broken line in A. (C) Effect of the seed fibril concentration on the initial rate of amyloid fibril extension. The initial concentration of the monomeric constituents in the reaction mixture ($[M]_0$) is constant. (D) Effect of the monomer concentration on the initial rate of amyloid fibril extension. The seed fibril concentration in the reaction mixture ([P]) is constant. Note that at $[M]_0 = 0$, the negative initial rate, i.e., the depolymerization of amyloid fibrils, is observed. Note also that at $[M]_0 = [M]_e$, the net rate of extension is 0.

where t is the reaction time, $F(t)$ is the fluorescence as a function of time, A is tentatively determined as $F(\infty)$, a and $-b$ are the y intercept and the slope of the straight line, respectively. Differentiating Eq. (1) by t yields

$$\frac{1}{\ln 10} \times \frac{-F'(t)}{A - F(t)} = -b \tag{2}$$

Rearranging Eq. (2) yields

$$F'(t) = B - CF(t) \tag{3}$$

where $B = bA \ln 10$, $C = b \ln 10$, and $F'(t)$ represent the rate of fluorescence increase at a given time.

We now assume that the kinetic properties of amyloid fibril extension can be described as

$$[\mathrm{P}] + [\mathrm{M}] \underset{k_{-1}}{\overset{k_2}{\rightleftharpoons}} [\mathrm{P}] \tag{4}$$

where [P] is the number concentration of amyloid fibrils, [M] is the concentration of the monomeric constituents, and k_2 and k_{-1} are the apparent rate constants for polymerization and depolymerization, respectively. [P] is constant throughout the reaction.

If t is the reaction time, $f(t)$ is the concentration of the monomeric constituents that have newly polymerized into amyloid fibrils during the reaction, and $[\mathrm{M}]_0$ is the initial monomer concentration, then Eq. (4) can be written as

$$f'(t) = k_2[\mathrm{P}][\mathrm{M}] - k_{-1}[\mathrm{P}] \tag{5}$$

$$[\mathrm{M}] = [\mathrm{M}]_0 - f(t) \tag{6}$$

where $f'(t)$ represents the rate of amyloid fibril extension at a given time and $k_2[\mathrm{P}][\mathrm{M}]$, and $-k_{-1}[\mathrm{P}]$ denote the rate of polymerization and depolymerization, respectively.

The insertion of Eq. (6) into Eq. (5) and subsequent rearrangement yields the following differential equation:

$$f'(t) = D - Ef(t) \tag{7}$$

where $D = (k_2[\mathrm{M}]_0 - k_{-1})[\mathrm{P}]$ and $E = k_2[\mathrm{P}]$.

Equation (7) is the same as Eq. (3). Therefore, Fig. 1B shows that the kinetics of amyloid fibril extension follows a first-order kinetic model as described by Eq. (4).

We now obtain the equilibrium monomer concentration (critical concentration) $[M]_e$ by setting $f'(t)$ in Eq. (5) equal to zero to obtain

$$[M]_e = \frac{k_{-1}}{k_2} = \frac{1}{K} \tag{8}$$

where K is the equilibrium association constant.

Measurement of Initial Rate of Amyloid Fibril Extension

In the initial phase of amyloid fibril extension, a linear increase is observed (see Fig. 1A). Therefore, a fluorescence increase within the linear phase, $F(t_1) - F(0)$, can be taken as the initial rate of extension.

If $t = 0$, then $f(0) = 0$. Therefore, from Eqs. (5) and (6)

$$f'(0) = k_2[P][M]_0 - k_{-1}[P] \tag{9}$$

Equation (9) explains the following results. First, when $[M]_0$ is constant, the initial rate of extension is proportional to the number concentration of amyloid fibrils ([P]) (Fig. 1C). Second, when [P] is constant, the initial rate of polymerization is found to be proportional to the initial concentration of monomeric constituents ($[M]_0$) (Fig. 1D). Finally, at each monomer concentration, the net rate of extension is the sum of the rates of polymerization and depolymerization (Fig. 1D).

Kinetics of fAβ Formation from Fresh Aβ

Analysis of Time–Course Curves

When fresh Aβ are incubated at 37°, the fluorescence of ThT follows a characteristic sigmoidal curve (Fig. 2A).[7,15] Plotting fluorescence data as common logarithms, shown in Fig. 2B, gives a linear semilogarithmic plot.[15] Interpretation of this plot yields the following equation:

$$\log\left[\frac{F(t)}{A - F(t)}\right] = at + b \tag{10}$$

where t is the reaction time, $F(t)$ is the fluorescence as a function of time, A is tentatively determined as $F(\infty)$, a and b are the slope and the y intercept of the straight line, respectively. Differentiating Eq. (10) by t and subsequent rearrangement yields the following differential equation:

$$F'(t) = BF(t)[A - F(t)] \tag{11}$$

where $B = \ln 10\,(a)/A$, and $F'(t)$ represents the rate of fluorescence increase at a given time.

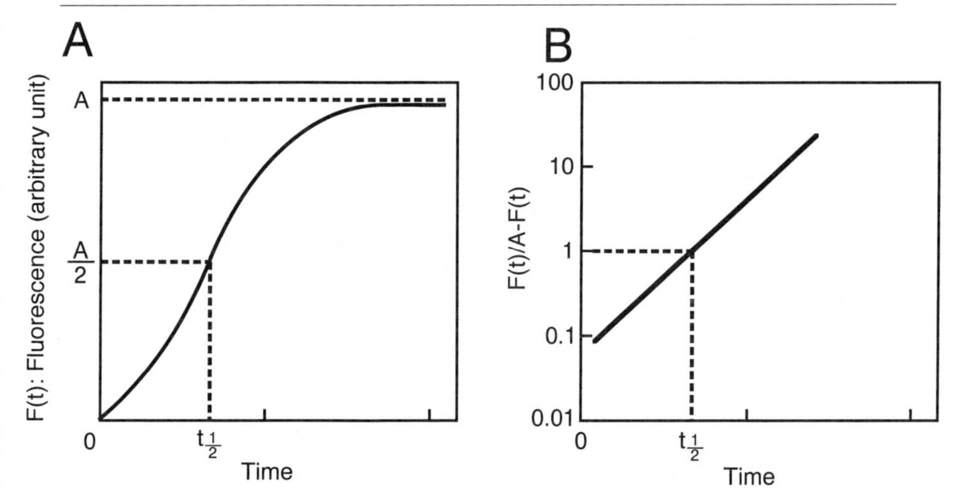

FIG. 2. Kinetics of fAβ formation from fresh Aβ. (A) Time course of fluorescence after the initiation of the reaction. $F(t)$ represents fluorescence as a function of time. (B) The semilogarithmic plots of the value: $F(t)/A - F(t)$ versus incubation time. A is tentatively determined as $F(\infty)$ and is shown as a broken line in A. Note that at $t = t_{1/2}$, $F(t) = A/2$ and $F'(t)$ reaches its maximum, $\ln 10 \cdot a \cdot A/4$.

Equation (11) is a logistic differential equation and clearly shows that the sigmoidal curve described previously is a logistic curve.[17] Various biological phenomena can be explained by a logistic equation; e.g., growth processes of bacteria, yeast, and plants, aging/survival data for various organisms, and autocatalytic reaction of trypsin that converts trypsinogen into trypsin.[17] When $F(t) = A/2$, $F(t)/A - F(t) = 1$. Moreover, from Eq. (11), F(t) reaches its maximum, $\ln 10 \cdot a \cdot A/4$. Therefore, from the straight lines shown in Fig. 2B, we obtain the time when $F'(t)$ is maximum. We describe this time point as $t_{1/2}$.

Concluding Remarks

The effects of various biological molecules and chemical compounds on amyloid fibril formation *in vitro* can be examined utilizing the paradigm and kinetic models presented here. We have shown that apoE inhibits fAβ formation *in vitro* by making a complex with Aβ, thus eliminating free Aβ from the reaction mixture.[7] The experimental system described here may

[17] L. C. Cerny, D. M. Stasiw, and W. Zuk, *Physiol. Chem. Phys.* **13,** 221 (1981).

prove useful in the development of nonpeptide inhibitors of amyloid fibril formation *in vivo*.

Acknowledgments

This research was supported in part by Grant-in-Aid 10670198 for Scientific Research (C) from the Ministry of Education, Science, Sports, and Culture of Japan.

[21] Small Zone, High-Speed Gel Filtration Chromatography to Detect Protein Aggregation Associated with Light Chain Pathologies

By ROSEMARIE RAFFEN and FRED J. STEVENS

Introduction

Proteins may self-associate by well-defined mechanisms that have functional relevance or may aggregate through unknown interactions that occur between partially unfolded molecules.[1] Either case results in variation of the particle size distribution in solution. This variation provides the basis for ultracentrifugational and light-scattering approaches to analyze aggregation. Protein self-association or aggregation can also be readily revealed by gel filtration chromatography. The chromatographic approach is attractive from the perspective that suitable experimental systems are available to most biochemistry laboratories. In comparison to light scattering and ultracentrifugation, the chromatographic phenomenon is defined less rigorously at the physical level, which restricts the development of quantitative analyses that might be desired. However, the chromatographic approach readily identifies proteins that tend to aggregate in an abnormal manner and provides a convenient method by which to survey the influence of solution conditions on aggregation. Furthermore, the development of computer models that simulate the chromatographic behavior of interacting proteins has made it possible to extract kinetic and stoichiometric parameters from experimental data.

Chromatographic separation of macromolecules on the basis of size is variously referred to as gel filtration or size-exclusion chromatography and has been used traditionally to purify proteins and to estimate approximate

[1] R. Wetzel, *Adv. Prot. Chem.* **50,** 183 (1997).

0076-6879/99 $30.00

molecular weights.[2] Separations are obtained (ideally) in a mechanical manner rather than through physicochemical interactions between proteins and ionic or hydrophobic moieties on the surface of the matrix. Chromatographic media generally consist of a porous matrix of cross-linked polymers. Separation of proteins according to size is accomplished by partitioning between the mobile phase, consisting of the solution that is exterior to the chromatographic particles, and the stationary phase, consisting of the solution within the particles themselves.[2] The effective transport velocity of any molecule is determined by the relative amount of time it spends in the mobile and stationary phases, which is a function of the size-restricted ability of the molecule to diffuse into the chromatographic particles. Consequently, molecules that are too large to penetrate the matrix are said to be totally excluded and elute from the chromatographic column at a time determined directly by the flow rate of solution. Smaller molecules that can pass into the matrix have a larger volume within which to diffuse and elute later.

Although size-exclusion chromatography has been used in most laboratories primarily for preparative purposes, analytical applications were developed soon after introduction of this technology.[3] Hummel and Dryer[4] introduced a method to estimate the equilibrium constant governing the interaction between proteins and small ligands. This technique involves saturating the chromatographic column with buffer that contains the ligand. A small sample of the protein containing ligand at a concentration equivalent to that in the column buffer is injected onto the column. Binding of the ligand to the protein creates an elution zone in which the ligand is depleted. The stoichiometry and equilibrium constant for the protein–ligand interaction is obtained by measurement of the depleted zone under several different ligand concentrations.

The technique of large zone gel filtration chromatography was developed for analysis of affinity, kinetics, and stoichiometry of protein–protein interactions (for reviews, see Ackers[2] and Winzor[5]). A large volume of sample, which may occupy 80% of the total column volume, is loaded onto the column. This creates an elution zone with a constant protein concentration and constant elution velocity, allowing a mathematically rigorous representation of the effects of protein–protein interaction on the observed leading and trailing boundaries of the solute zone to be devel-

[2] G. K. Ackers, *in* "The Proteins" (H. Neurath and R. L. Hill, eds.), Vol. 1, p. 1. Academic Press, New York, 1975.

[3] J. Porath and P. Flodin, *Nature* (*London*) **183,** 1657 (1959).

[4] J. P. Hummel and W. J. Dryer, *Biochim. Biophys. Acta* **63,** 530 (1962).

[5] D. J. Winzor, *in* "Protein–Protein Interactions" (C. Frieden and L. W. Nichol, eds.), p. 129. Wiley, New York, 1981.

oped.[6] Although large sample requirements and long column run times have been a major limitation to this method, the large zone technique has been adapted to a smaller column format.[7]

Small zone size-exclusion chromatography involves sample volumes that are on the order of 10% or less of the column volume. Minimal sample expenditure and the potential to use analytical scale columns and fast run times are desirable features of this mode of chromatography. Small zone chromatography can be used to study the stoichiometry, kinetics, and affinity of protein interactions and aggregation. From a qualitative perspective, much can be learned about protein–protein interactions simply by varying the concentrations of the sample components. Interactions characterized by high affinity and slow equilibration, as might be observed for antibody–antigen interactions, yield discrete peaks for the complexed and uncomplexed species and a concentration-dependent change in the ratio of the peak areas of the different species. In the case of kinetically controlled interactions, where dissociation and reassociation occur during the column run, albeit at rates too slow to allow complete equilibration, partial intermixing of the complexed and uncomplexed species occurs. Depending on the kinetics of the interaction, individual peaks will begin to merge.[8]

Rapidly equilibrating systems are characterized by a single peak and concentration-dependent changes in the peak shape and position. Figure 1A shows a series of elution profiles generated from different concentrations of a dimerizing protein, an immunoglobulin light chain variable domain (V_L). The rapid kinetics of association and dissociation exhibited by the interacting V_L's produces single peaks that are notably asymmetric; only the peak for the sample with the lowest concentration shows near symmetry. Rapid dissociation accounts for the sharp leading edge of the peaks. Dimeric complexes that migrate in front of the bulk band have entered into a dilute protein zone. On dissociation, the now monomeric proteins are quickly overtaken by the faster migrating, dimer-rich zone that follows. In contrast, dissociation at the dilute trailing edge of the band generates monomers that have less chance to redimerize and that, even as dimers, could not be transported at a rate sufficient to rejoin the bulk band. Therefore, a continuous trail of protein follows the predominant band. Within the bulk band itself, rapid dissociation and reassociation ensure that each V_L alternates between monomeric and dimeric forms. Thus, the average transport velocity of the peak is intermediate between that of the monomer and dimer.

[6] G. K. Ackers and T. E. Thompson, *Proc. Natl. Acad Sci. U.S.A.* **53,** 342 (1965).

[7] E. Nentoras and D. Beckett, *Anal. Biochem.* **222,** 366 (1994).

[8] F. J. Stevens, *Methods Enzymol.* **178,** 107 (1989).

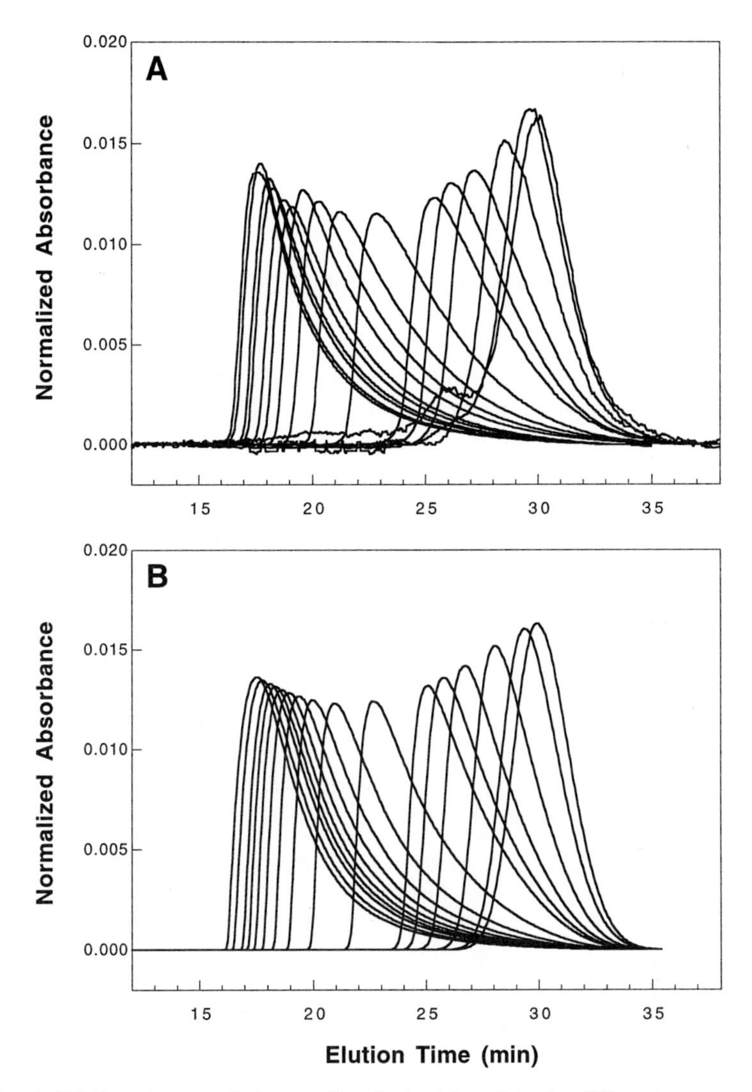

FIG. 1. (A) Experimental elution profiles obtained from injecting different concentrations of a V_L dimer. The initial V_L monomer concentration for each chromatogram was (from left to right) 16.4, 13.7, 10.9, 9.6, 8.2, 6.8, 5.5, 4.1, 2.7, 1.4, 0.55, 0.14, 0.27, 0.41, 0.055, and 0.027 mg/ml. Chromatography was performed as described in the text on a 3 mm $\times$ 25 cm Superdex 75 column operating at a flow rate of 0.06 ml/min. The buffer was 20 m*M* potassium phosphate, 100 m*M* NaCl, pH 7.0 (B) Simulated elution profiles. V_L monomer concentrations of the simulated profiles correspond with the experimental profiles. The simulation was performed as described in the text, using the instantaneous equilibration assumption. The best fit between experimental and simulated data was obtained by assigning monomer and dimer dispersion parameters of 0.500 and 3.500 (cells), respectively, and monomer and dimer diffusion parameters of 0.500 and 0.250, respectively. This yielded an association constant of $1.0 \times 10^5 M^{-1}$. Used with permission from R. Raffen *et al., Prot. Eng.* **11,** 303 (1998).

In the small zone chromatography of interacting proteins (under rapid equilibrium or kinetic control), dilution of the sample peak and the consequent shift of the association equilibrium to lower molecular weight species result in a continuous decrease in the velocity of the peak as it moves through the column. It has been demonstrated that small zone elution behavior is not subject to the same analysis used for large zone experiments,[9] and a mathematical solution for small zone behavior has not been found. However, a mathematical description of the chromatographic process can be substituted with an iterative simulation. Computer simulations have been developed by many researchers to study protein–protein interactions in the context of a variety of transport or mass migration processes,[5] which in addition to gel filtration chromatography[2,10–15] include ultracentrifugation[16] and electrophoresis.[17] The simulation developed in this laboratory[18,19] is conceptually similar to many of the examples just cited, incorporating an iterative sequence of transport, equilibration, and diffusion steps. We have used computer simulation for the quantitative interpretation of homogeneous (dimerization) interactions[18,20–22] and heterogeneous (antibody:antigen)[19,23–25] interactions. Improvements to the simulation extend to the analysis of kinetically controlled interactions[26,27] and higher order oligomerization.[27]

[9] J. K. Zimmerman and G. K. Ackers, *J. Biol. Chem.* **246,** 7289 (1971).
[10] J. K. Zimmerman, *Biochemistry* **13,** 384 (1974).
[11] J. K. Zimmerman, *Biophys. Chem.* **3,** 339 (1975).
[12] J. K. Zimmerman, D. J. Cox, and G. K. Ackers, *J. Biol. Chem.* **246,** 4242 (1971).
[13] J. R. Cann and D. J. Winzor, *Arch. Biochem. Biophys.* **256,** 78 (1987).
[14] J. R. Cann, E. J. York, J. M. Stewart, J. C. Vera, and Ricardo B. Maccioni, *Anal. Biochem.* **175,** 462 (1988).
[15] T. W. Patapoff, R. J. Mrsny, and W. A. Lee, *Anal. Biochem.* **212,** 71 (1993).
[16] D. J. Cox, *Methods Enzymol.* **48,** 212 (1978).
[17] J. R. Cann, *Anal. Biochem.* **237,** 1 (1996).
[18] F. J. Stevens and M. Schiffer, *Biochem. J.* **195,** 213 (1981).
[19] F. J. Stevens, *Biochemistry* **25,** 981 (1986).
[20] F. J. Stevens, F. Westholm, A. Solomon, and M. Schiffer, *Proc. Natl. Acad. Sci. U.S.A.* **77,** 1144 (1980).
[21] E. A. Myatt, F. J. Stevens, and P. B. Sigler, *J. Biol. Chem.* **266,** 16331 (1991).
[22] H. Kolmar, C. Frisch, G. Kleemann, K. Goetz, F. J. Stevens, and H.-J. Fritz, *Biol. Chem. Hoppe Seyler's Z.* **375,** 61 (1994).
[23] F. J. Stevens, W. Carperos, W. J. Monafo, and N. S. Greenspan, *J. Immunol. Methods* **108,** 271 (1988).
[24] E. A. Myatt, P. T. P. Kaumaya, and F. J. Stevens, *BioChromatography* **4,** 282 (1989).
[25] E. A. Myatt, F. J. Stevens, and C. Benjamin, *J. Immunol. Methods* **177,** 35 (1994).
[26] F. J. Stevens, *Biophys. J.* **55,** 1155 (1989).
[27] F. J. Stevens, unpublished results (1998).

Dimerization of Immunoglobulin Light Chain Variable Domains

The immunoglobulin light chain variable domain is a model system for understanding processes involved in light chain amyloidosis and light chain deposition disease. Variations in the self-association properties of V_L's are a consequence of amyloid-associated amino acid substitutions[28,29] and structural variations that alter domain association.[30,31] Computer simulation methods have been used to determine V_L association constants.[28,30] V_L domains form rapidly equilibrating homodimers and therefore provide a simple system for illustrating the utility of the simulation method for determining association constants. Using the following experimental conditions, we have measured association constants for V_L's ranging between 10^3 and 10^7 M^{-1}.

Experimental

Column Preparation

Silanized 0.3 × 25-cm glass columns (Alltech Associates, Deerfield, IL) are packed with Superose 12 HR or Superdex 75 HR resin (Pharmacia, Piscataway, NJ). Suitable glass columns are also available from Omnifit (Cambridge, England) and Pharmacia. The gel filtration resin can be varied, if necessary, to obtain a separation range suitable for the protein under study. To pack the column, a larger, donor column containing a suspension of the gel filtration media is connected to the high-performance liquid chromatography (HPLC) pump. The recipient column is attached to the other end of the donor column, and frits in the joining end fittings are removed. The resin is pumped into the recipient column at low flow rates. The back pressure is monitored as the resin packs in the recipient column, and the flow rate is adjusted accordingly. Because we store packed columns and chromatographic media in 20% (v/v) ethanol, we find it convenient to undergo the packing procedure in this solvent, and later equilibrate the columns with the desired running buffer. After the recipient column is

[28] P. Wilkins Stevens, R. Raffen, D. K. Hanson, Y.-L. Deng, M. Berrios-Hammond, F. A. Westholm, C. Murphy, M. Eulitz, R. Wetzel, A. Solomon, M. Schiffer, and F. J. Stevens, *Prot. Sci.* **4,** 421 (1995).

[29] R. Raffen, L. J. Dieckman, M. Szpunar, C. Wunschl, P. R. Pokkuluri, P. Dave, P. Wilkins Stevens, X. Cai, M. Schiffer, and F. J. Stevens, *Prot. Sci.* **8,** 509 (1999).

[30] R. Raffen, C. Boogaard, P. Wilkins Stevens, M. Schiffer, and F. J. Stevens, *Prot. Eng.* **11,** 303 (1998).

[31] P. R. Pokkuluri, D.-B. Huang, R. Raffen, X. Cai, G. Johnson, P. Wilkins Stevens, F. J. Stevens, and M. Schiffer, *Structure* **6,** 1067 (1998).

filled, the top frit is replaced, and the column is connected to the HPLC system, keeping lengths of all connections to a minimum. At this stage, the column is equilibrated with the appropriate buffer and is allowed to run overnight. Occasionally, a small void space may form at the top of the column after overnight packing. A Pasteur pipette can be used to fill the space, if present, with resin. The column is tested with gel filtration calibration standards (Pharmacia), and only columns that produce symmetrical peaks are used. With occasional replacement of resin within the top few millimeters of the column, columns produced in this manner can be used for many months with no noticeable reduction in performance.

Generation of Experimental Elution Profiles

The identification of authentic self-association (as opposed to high molecular weight contaminants or cross-linked components) is confirmed by demonstrating the concentration dependence of the elution profile. Ideally, one would like to obtain sample elution profiles over a wide concentration range that allows observation of the monomer and dimer or multimer positions on the gel filtration column. This is not always possible, especially when very high or very low association constants are under study; however, reasonable simulations of data can be obtained even if these end points are not achievable experimentally.[30] We generally prepare V_L at 10 to 15 different sample concentrations; a span of 25 mg/ml to about 0.005 mg/ml (the latter concentration approaches the limit of our detector) provides a good cross section of V_L peak profiles in most cases. A sample volume of 5 μl is injected into the column with a Perkin Elmer Series 200 autosampler. Solvent is delivered at 0.06 ml/min by a Pharmacia 2248 HPLC pump. Considerations in selecting a pump include the precision of flow rate (particularly at low flow rates) and its ability to function under conditions of low back pressure (which may affect the operation of check valves). The size of piston stroke volumes should be small, and tolerance of the materials in contact with buffer to saline solutions used for protein studies is also important. Typical run times are on the order of 30 min; variations in the flow rate allow runs of 15 min to 1 h as appropriate. The slow flow rate used in the V_L experiments ensures that near equilibrium conditions are maintained throughout the run. Sample elution is measured at 214 nm with a Hewlett Packard 1040A diode array detector. High concentration sample data are collected at 280 or 254 nm. Data are recorded with Hewlett Packard ChemStation software. Injections of standard protein are interspersed with the samples to monitor the operation of the system.

The experimental peak areas are normalized with program KRUNCH.[27] KRUNCH scales data so that the integrated area under the elution profile is equal to 1. This enhances visualization of the concentration-dependent changes in the peak shape. Additionally, the KRUNCH program reduces the number of data points to coincide with the number of cycles in the simulation (see later). This is convenient for the direct comparison of simulated and experimental data.

Simulation of V_L Dimerization

The simulation, SCIMZ, consists of an interative sequence of transport, equilibration, and diffusion steps. The simulated "column" consists of a linear array of cells onto which the V_L sample of defined volume (in cells) and concentration is "loaded." This is followed by execution of a five-step iteration cycle. The first step is an equilibration step during which the concentration of each species is calculated for each cell in the column. For example, in the case of a dimerizing system

$$2A \underset{k_r}{\overset{k_f}{\rightleftharpoons}} B \tag{1}$$

the rate of formation of the dimer, B, is given by

$$\frac{db}{dt} = k_f a^2 - k_r b \tag{2}$$

where a is the concentration of monomer, b is the concentration of dimer, and k_f and k_r are the forward and reverse rate constants. This is evaluated in the simulation by the expression[27]

$$\Delta b = (k_f a^2 - k_r b)\, \Delta t \tag{3}$$

For rapidly equilibrating systems, such as V_L dimers, a simplifying assumption of "instantaneous equilibration" can be made. This means that complete equilibration is obtained at each cycle in the simulation. Equation (3) therefore simplifies to

$$K_a = \frac{a^2}{b} \tag{4}$$

where K_a is the association constant. This is not an unreasonable assumption; for although gel filtration chromatography is a transport process, and migration of the interacting proteins through the column results in a continual departure from equilibrium,[5] the rapid association and dissociation of V_L's combined with a relatively slow column flow rate ensure that rapid reequilibration occurs throughout the chromatographic run. The equilibration step is followed by a transport step (step 2) during which monomers and dimers are advanced through the column according to a Gaussian distribution. The movement of each species is characterized by a velocity (in cells/cycle) that defines its mean displacement and a dispersion factor (in cells) that represents the standard deviation, σ, of the displacement about the mean position. The Gaussian transport strategy creates a device to represent the axial dispersion or band spreading that occurs during chromatography due to incomplete chromatographic partitioning, flow irregularities, edge effects, and other physical effects.[2] This is followed by another equilibration step (step 3). Next, monomers and dimers are individually subjected to a diffusion cycle (step 4) according to Eq. (5)[18]

$$c_i^* = c_i - f_d(c_i - c_{i-1}) + f_d(c_{i+1} - c_i) \tag{5}$$

where c_i^* represents the concentration of monomer or dimer in cell i following the diffusion step, c_i represents the initial monomer or dimer concentration of protein in cell i, and f_d is the fractional diffusion coefficient assigned to the monomeric or dimeric species. For $f_d = 0.5$, diffusion equilibrium is reached, and c_i is the average of c_{i-1} and c_{i+1}. In step 5, the protein concentration in a postcolumn "monitor queue," analogous to the detector in the experimental system, is recorded, and the cycle is repeated.

User-defined parameters include the number of cells in the column and cell volume, which are adjusted to emulate the experimental column volume. For example, our 1.77-ml experimental column is simulated with 708 cells with a volume of 2.5 μl/cell. A smaller cell volume improves the resolution of the simulated chromatogram, but this is balanced by increased computational requirements. The user also defines the number of iteration cycles and the time per cycle. We typically use 708 cycles and 3 sec/cycle to simulate a run time of approximately 35 min. The simulation for dimerizing systems using the instantaneous equilibration assumption has been partially automated. It determines the monomer and dimer velocities and the association constant, K_a, by fitting the peak positions of simulated chromatograms with those of experimental chromatograms. However, the monomer and dimer dispersion and diffusion parameters are defined by the user. These parameters affect the shape of the simulated peaks; they are adjusted by trial and error until the simulated chromatogram matches the experimental chromatogram. Diffusion and dispersion parameters are oppositely related to the molecular weight of the molecule; thus, monomers diffuse rapidly but have small dispersion parameters, whereas dimers diffuse slower but have larger dispersive properties.[2] Figure 2 illustrates the effects of varying dispersion and diffusion parameters on the simulated elution profiles. The final set of simulated data, normalized for comparison with experimental data, is shown in Fig. 1B.

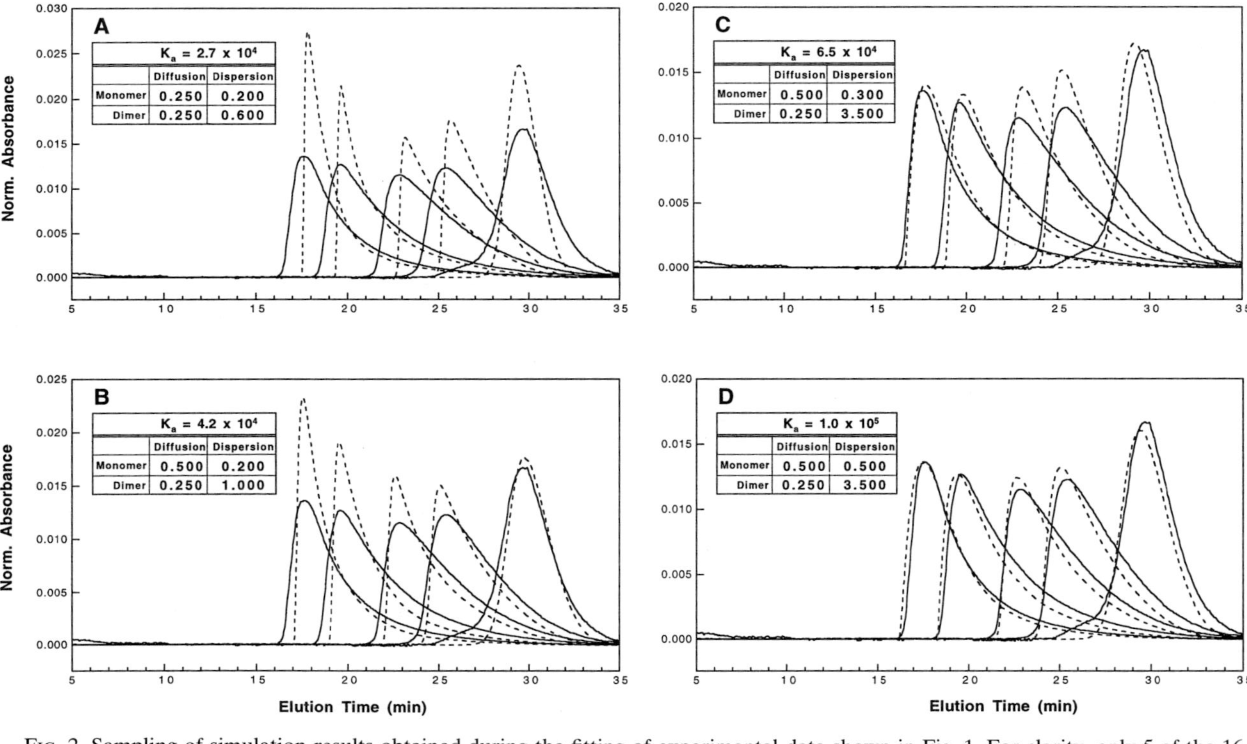

FIG. 2. Sampling of simulation results obtained during the fitting of experimental data shown in Fig. 1. For clarity, only 5 of the 16 elution profiles are displayed in each panel. Experimental chromatograms are shown as solid lines and simulated results are shown as dashed lines. The tabulation in each panel gives the monomer and dimer diffusion and dispersion parameters used in the simulation and the association constant determined by the simulation. Note that the absolute values of the dispersion parameters that give the best fit will vary with different simulated column cell sizes. (D) The best possible fit to experimental data.

Estimation of Affinities from Gel Filtration Data without Computer Simulation

It is also possible to obtain reasonable estimates of absolute affinity and relative affinities for homologous proteins without computer analysis. For a dimerizing system, the molar concentration of monomer at equilibrium may be written as $m = [-1 + (8m_oK + 1)^{1/2}]/4K$[18] in which m_o is the total molar concentration and K is the equilibrium constant for dimerization. The fractional monomer concentration is therefore $\phi_m = [-1 + (8m_oK + 1)^{1/2}]/4m_oK$. Thus, the monomer/dimer composition is determined solely by the product of protein concentration and equilibrium constant. As a result, relative affinity constants of two proteins can be estimated from the ratio of concentrations that result in equivalent elution positions. Variations in the shapes of elution profiles obtained at equivalent positions may indicate differences in the kinetics of dimerization, the presence of low levels of higher order aggregates, or differences in nonspecific interactions with the chromatographic matrix.

To illustrate, an inspection of Fig. 1A indicates that a V_L concentration of 1.4 mg/ml (1.1×10^{-4} M) gives an elution peak near the midpoint between dimer and monomer elution positions. This elution profile will serve as our reference peak; it is chosen near the midpoint of the monomer and dimer elution positions, as it is here that the elution position is most sensitive to changes in concentration. The reciprocal concentration of 9×10^3 M^{-1}, as discussed earlier, is proportional to the dimerization constant (1×10^5 M^{-1}). Thus, we can calibrate this particular column with a scaling factor of 11, i.e., 11 multiplied by the inverse concentration resulting in elution at the reference peak position is a reasonable estimate of affinity. Note that the absolute scaling factor depends on the column volume, flow rate, ratio of sample volume to column volume, and fractionation properties of the resin in use. By this method, one can "calibrate" a column with a protein of known dimerization constant (determined by any of a number of means) and evaluate dimerization constants for homologous proteins without the use of computer simulation.

Aggregation of Bence Jones Proteins

Small zone size-exclusion chromatography is also useful for the qualitative analysis of more complex protein aggregation processes. This technique was used to demonstrate extensive aggregation in a recombinant V_L that incorporated an extremely destabilizing mutation found in a light chain associated with light chain deposition disease.[32] Myatt *et al.*[33] used small zone size-exclusion chromatography to compare the aggregation properties of immunoglobulin light chain proteins (Bence Jones proteins[34]) obtained from patients with clinically characterized light chain pathologies. This study found a correlation between the formation of high molecular weight aggregates and the existence of protein deposition in the patient (or inferred when tested in an *in vivo* animal model[35]).

Bence Jones proteins are typically found as mixtures of light chain monomer (22,500 Da) and dimer (45,000 Da). A varying proportion of the light chain protein is usually in the form of a disulfide-bonded covalent dimer. The remainder consists of noncovalent dimer that is in dynamic equilibrium with free monomer. Chromatography was performed as described earlier, but under buffer conditions designed to mimic the physiological conditions encountered by the light chains during filtration in the kidney.[33] The following examples illustrate the type of qualitative information on kinetics, type of aggregates, and so on that can be extracted from gel filtration experiments with more complicated aggregating systems.

Figure 3 displays chromatograms obtained for the $\kappa 1$ light chain protein Wms at three concentrations (1.5, 0.15, and 0.015 mg/ml). Protein Wms produced crystals in the kidney tubules of the patient. The chromatograms were obtained in a phosphate-buffered saline (PBS). The elution patterns observed in Fig. 3 demonstrate the characteristics of high molecular weight aggregation characterized by slow kinetics. If high molecular weight aggregates are of moderately low affinity but are characterized by dissociation rates on the order of 0.01 sec^{-1} or slower, then column volumes and flow rates that ensure that the excluded buffer volume passes through the system within a few minutes allow the method to detect the presence of aggregates in the original sample. Elution data shown in Fig. 3 have been scaled to elution units of V_e/V_t (i.e., ratio of actual elution volume to total column volume) to facilitate comparison, if desired, of chromatograms obtained at different flow rates, different columns, etc. In this experimental system, the covalent dimer component has an expected elution position of approximately $V_e/V_t = 0.6$, and the free monomer has an elution position of 0.7. At the highest concentration, elution is found near the expected total excluded volume ($V_e/V_t \sim 0.36$). This finding indicates that some aggregates

[32] L. R. Helms and R. Wetzel, *J. Mol. Biol.* **257,** 77 (1996).

[33] E. A. Myatt, F. A. Westholm, D. T. Weiss, A. Solomon, M. Schiffer, and F. J. Stevens, *Proc. Natl. Acad. Sci. U.S.A.* **91,** 3034 (1994).

[34] H. B. Jones, *Philos. Trans. R. Soc. London* 55 (1848).

[35] A. Solomon, D. T. Weiss, and A. A. Kattine, *N. Engl. J. Med.* **324,** 1845 (1991).

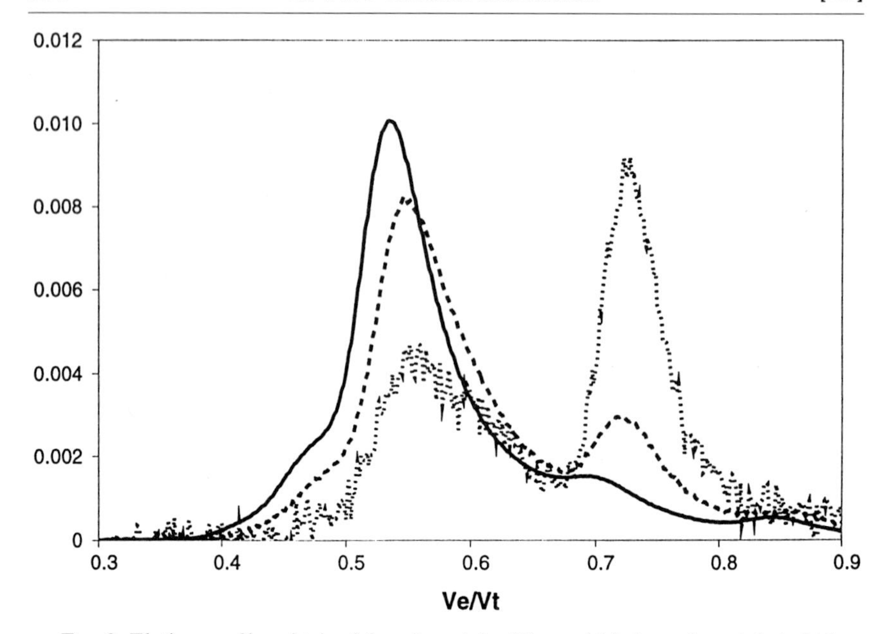

FIG. 3. Elution profiles obtained for κ1 protein, Wms, which formed crystals in kidney tubules. Protein concentrations were 1.5 (solid line), 0.15 (dashed line), and 0.015 (dotted line) mg/ml. Buffer was PBS (50 m*M* sodium phosphate, 100 m*M* NaCl, pH 7.2). In this and the subsequent figure, areas under the elution profiles have been normalized to facilitate comparison of data obtained over a wide range of initial sample concentrations. Note that Bence Jones proteins from individual patients are identified by an encrypted form of the patient's name.

survive up to 10 min after they have been introduced into the column medium. Note the discrete shoulders at positions 0.4 and 0.45, particularly on the profile of the sample with the highest concentration. This finding also indicates slow kinetics of turnover of components between different size aggregates. As anticipated, with decreasing concentration of the initial sample, elution shifts to later times (larger V_e/V_t) that correspond to the increased proportion of smaller aggregates and monomer. Wms appears to form a tetramer and larger aggregates and shows little intermixing with free monomer.

Figure 4 compares elution profiles for the λ2 cast-forming light chain protein, Mora, in two buffers. Figure 4 (top) depicts profiles obtained in PBS buffer whereas Fig. 4 (bottom) shows interaction behavior in a lower pH acetate buffer. When one compares the elution behavior seen in Fig. 4 (top) to that of Wms, it is possible to conclude that the effective "aggregation affinity" of Mora is less than that of protein Wms but that its dissocia-

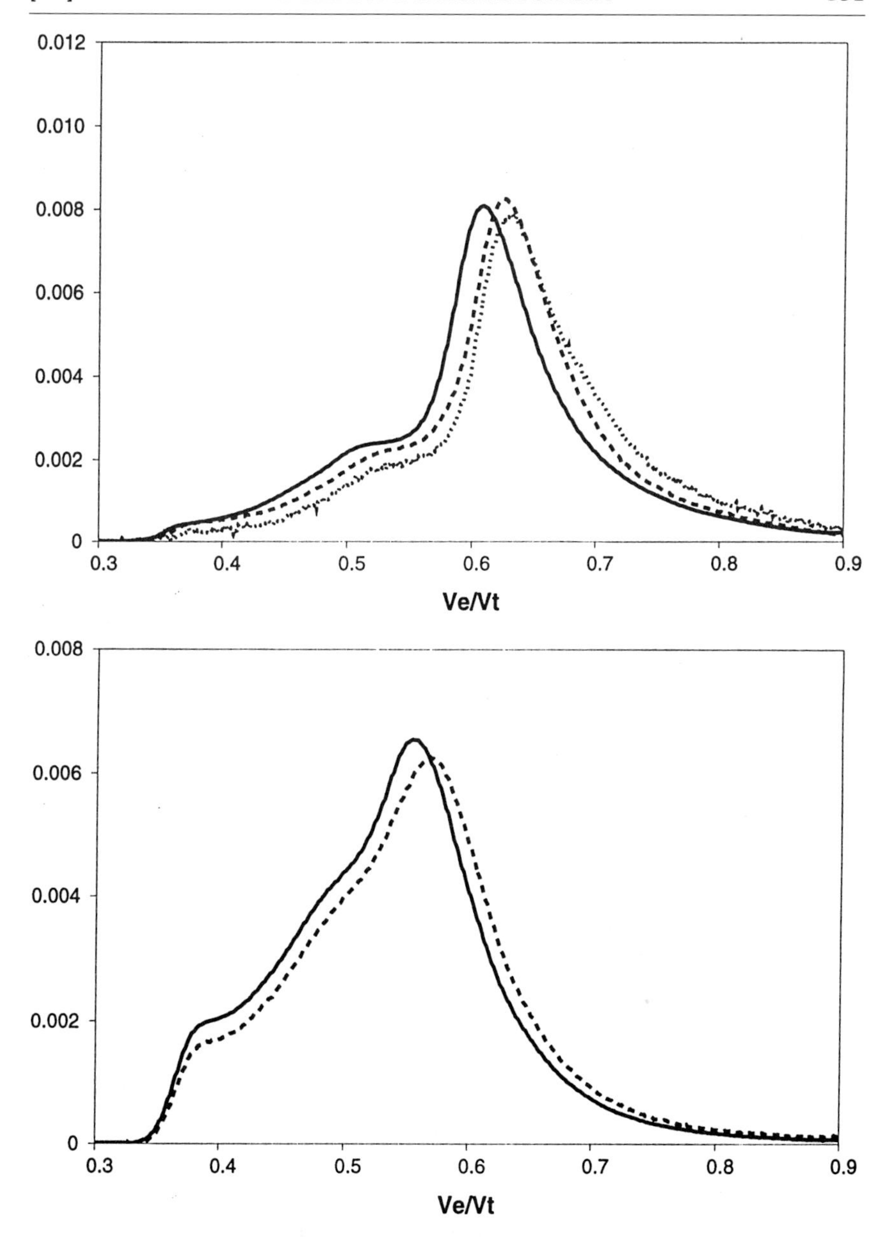

FIG. 4. Elution profiles obtained for λ2 protein, Mora, which caused cast nephropathy in the patient. (Top) Elution profiles in PBS; protein concentrations were 1.0 (solid line) 0.4 (dashed line), and 0.04 (dotted line) mg/ml. (Bottom) Elution profiles in acetate buffer (pH 4.5, 0.245 *M* NaCl): 2.0 (solid line), and 0.5 (dashed line) mg/ml.

tion rate is slower. These inferences from the shapes of the profiles implicitly assume that the mechanisms of aggregation for the two proteins are the same. Given the validity of that presumption, the inference that Mora has a lower affinity is simply based on the apparently lower average molecular weight exhibited by this sample. The predominant peak is at position 0.6, which corresponds to that of a dimer; the corresponding Wms peak eluted at ~0.53, which suggests tetramer formation. The conclusion that the dissociation rate of Mora light chain aggregates is slower than that of Wms is based on the increased proportion of the highest molecular weight aggregates eluting at or near the void volume. If the aggregation (equilibrium) affinity of Mora is less than that of Wms, it is reasonable to conclude that its rate of aggregate formation is also slower as affinity is the ratio of assembly rate to dissociation rate.

Figure 4 (bottom) shows elution patterns of the protein Mora in acetate buffer. The elution behavior is very different at low pH on the basis of a comparison with the interaction pattern seen in PBS buffer (top, Fig. 4). It is clear that even at a 2-mg/ml concentration, most of the protein elutes as a monomer ($V_e/V_t \sim 0.7$). However, it is also evident that a continuum of protein elution, extending from the excluded volume position to the monomer position, is present. This implies that in the 2-mg/ml sample, a significant fraction (~10%) was in an aggregated form (of indeterminate molecular weight) and that dissociation of this aggregate in response to dilution during chromatography was a slow process. Because the initial aggregate was a relatively small fraction of the original sample, the rate of aggregate formation must also have been slow.

Summary

Small zone gel filtration chromatography can be used for qualitative and quantitative analysis of protein interactions and aggregation phenomena. The technique is fast, accessible to most laboratories, and can be combined with computer simulation to extract quantitative information from experimental data. The programs KRUNCH and SCIMZ will be furnished on written request to the authors.

Acknowledgments

We are grateful to Dr. E. Myatt for the generation of Bence Jones protein data. This work was supported in part by the U.S. Department of Energy, Office of Biological and Environmental Research, under contract W-31-109-ENG, and by U.S. Public Health Service Grant DK43757.

[22] Detection of Early Aggregation Intermediates by Native Gel Electrophoresis and Native Western Blotting

By SCOTT BETTS, MARGARET SPEED, and JONATHAN KING

Introduction

Protein misfolding and aggregation have been implicated in the molecular basis of several human diseases.[1-3] The failure of correct folding and assembly processes also limits yields of recombinant proteins in research laboratories and in the biotechnology industry.[4] Understanding the earliest stages in the formation of protein aggregation precursors is essential to controlling and suppressing nonproductive chain-association reactions. Given the apparent stability of the aggregated state, the ability to detect early misfolded intermediates and their association into off-pathway multimers will aid in the targeting and design of therapeutics and in the optimization of strategies for both *in vivo* protein expression and *in vitro* protein folding.

The tendency of partially folded intermediates and aggregation precursors to self-associate limits the methods that can be used for fractionation and analysis of these species. Transient intermediates are frequently difficult species to detect and identify, and a similar situation holds for aggregation intermediates derived from these species. Thus the presence of protein-folding intermediates is usually deduced by some optical signal, e.g., fluorescence, or by the shape of an equilibrium denaturation curve. They are rarely directly isolated from the reaction mixture. Efforts to fractionate aggregation intermediates have met with similar difficulties.

In our initial studies of P22 tailspike aggregation intermediates, many standard column matrices, whether ion exchange, gel filtration, high-performance (HPLC), or fast protein (FPLC) liquid chromatography, failed to reveal intermediate species that must have been present in the reaction mixture.[5] Our conclusion was that these species were adsorbing to the matrices of the columns or were polymerizing further during fractionation,

[1] G. Taubes, *Science* **271,** 1493 (1996).
[2] R. Wetzel, *in* "Protein Misassembly" (R. Wetzel, ed.), Vol. 50. Academic Press, San Diego, 1997.
[3] R. W. Carrell and D. A. Lomas, *Lancet* **350,** 134 (1997).
[4] J. L. Cleland, *in* "Protein Folding: In Vivo and In Vitro" (J. L. Cleland, ed.), Vol. 526, p. 1. Am. Chem. Soc., Washington, DC, 1993.
[5] M. A. Speed, D. I. C. Wang, and J. King, *Prot. Sci.* **4,** 900 (1995).

METHODS IN ENZYMOLOGY, VOL. 309
0076-6879/99 $30.00

preventing recovery of the aggregation intermediates. As a result we turned to methods in which protein adsorption was minimal. One of the factors in Davis'[6] original choice of polyacrylamide was the lack of adsorption of blood and urine proteins to the matrix. The *in vivo* tailspike folding intermediates had originally been identified through the use of native gel electrophoresis[7] and it seemed reasonable that other species might be identifiable by the same method.

Another direction was capillary zone electrophoresis (CZE), as this proceeds without a solid matrix. In fact, tailspike folding intermediates are resolved from the native trimer by CZE,[8,9] and aggregation intermediates can be detected. However, they are not resolved into discrete species. Thus we have continued to employ native gel electrophoresis in the cold as a method of choice.

Optical methods do provide some useful information. Quasi-elastic light scattering avoids the difficulty of fractionation and allows the estimation of submicron multimers on a time scale of seconds.[10,11] Multimers can be fractionated and quantitated by size-exclusion chromatography on a time scale of several minutes.[12] However, gel filtration and light-scattering methods are not sensitive enough to distinguish conformational differences between multimeric species of the same molecular weight. This is the primary advantage of native gel electrophoresis. For example, the early dimeric and trimeric intermediates on the folding pathway to the P22 tailspike can be resolved from the corresponding intermediates on the aggregation pathway.[5,13]

Polyacrylamide gel electrophoresis (PAGE) was developed as an analytical method to fractionate native proteins under nondenaturing conditions.[6,14] The migration rate of polypeptide chains moving through polyacrylamide depends both on the viscosity of the gel and on the charge density on the migrating chain. The charge density depends on both the isoelectric point of the polypeptide chain and the pH of the gel buffer (see later). The apparent charge can vary depending on the conformation of

[6] B. J. Davis, *Ann. N.Y. Acad. Sci.* **121,** 404 (1964).

[7] D. Goldenberg and J. King, *Proc. Natl. Acad. Sci. U.S.A.* **79,** 3403 (1982).

[8] Z. H. Fan, P. K. Jensen, C. S. Lee, and J. King, *J. Chromatogr.* **669,** 315 (1997).

[9] P. K. Jensen, A. K. Harrata, and C. S. Lee, *Anal. Chem.* **70,** 2044 (1998).

[10] J. L. Cleland and D. I. C. Wang, *Biochemistry* **29,** 11072 (1990).

[11] A. Lomakin, G. B. Benedek, and D. B. Teplow, *Method. Enzymol.* **309** [27] 1999 (this volume).

[12] J. L. Cleland and D. I. C. Wang, *in* "Protein Refolding" (E. G. Georgiou and E. De Bernardez-Clark, eds.), Vol. 470, p. 169. Am. Chem. Soc., Washington, DC, 1991.

[13] S. D. Betts and J. King, *Prot. Sci.* **7,** 1516 (1998).

[14] L. Ornstein, *Ann. N.Y. Acad. Sci.* **121,** 321 (1964).

the migrating chain, thus allowing resolution of different conformations of the same polypeptide chain. Partially folded intermediates may be less compact than their corresponding native states. Such an increased hydrodynamic volume would likely decrease electrophoretic mobility.

Another advantage of native gel electrophoresis over alternative methods is the ability to analyze multiple samples simultaneously and side by side. The technique is restricted, however, to systems in which the multimerization reaction is quenched or slowed in the cold. Multimers must also resist both dissociation in the cold and further aggregation during the concentration of proteins into discrete bands, which occurs during migration through the upper, stacking gel.

Even though electrophoretic mobility in native gels is quite sensitive to conformation, interpretation of the mobility differences is often difficult.[15] Monoclonal antibodies provide more reliable information, at least with respect to particular epitopes. Unfortunately, in the normal Western blot procedure, the denaturing step during transfer to the blot loses conformational information. The second part of this article describes a Western blotting procedure that preserves and detects some of the conformational differences that underlie electrophoretic separation.

Native Gel Electrophoresis of Aggregation Intermediates

Early multimeric intermediates on the *in vitro* aggregation pathways of several proteins have been identified directly by nondenaturing gel electrophoresis. These include the P22 tailspike and coat proteins, bovine carbonic anhydrase II, and α_1-antitrypsin. Aggregation intermediates of the *Salmonella* phage P22 tailspike protein, a trimer of 72-kDa subunits, were generated *in vitro* following the dilution of unfolded chains to nondenaturing conditions.[5,13,16] Aggregation intermediates of recombinant human α_1-antitrypsin, a 52-kDa monomer, were generated during the *in vitro* refolding of denatured polypeptide chains[17] and also by heating native protein under otherwise nondenaturing conditions.[18]

Denaturation and Refolding/Aggregation

Aggregation intermediates can be generated during the *in vitro* refolding of denatured protein or by thermal unfolding of the native protein. Urea

[15] K. A. Ferguson, *Metabolism* **13,** 985 (1964).

[16] M. A. Speed, D. I. C. Wang, and J. King, *Nat. Biotech.* **14,** 1283 (1996).

[17] K. S. Kwon, S. Lee, and M. H. Yu, *Biochim. Biophys. Acta* **1247,** 179 (1995).

[18] J. Konz, "Oxidative Damage to Recombinant Proteins during Production," Ph.D. thesis, Massachusetts Institute of Technology, 1998.

is the denaturant of choice over guanidinium because it is nonionic. Guanidinium precipitates sodium dodecyl sulfate (SDS) and must be removed from samples intended for analysis by SDS–PAGE. At sufficiently low concentrations of residual guanidinium (millimolar range), electrophoretic separation is possible under nondenaturing conditions (no SDS).

The rate of aggregation can be controlled by adjusting various parameters, including temperature, ionic strength, residual denaturant concentration, and protein concentration both before and after dilution. The protein concentration must be sufficiently high to detect polypeptide chains distributed among a variety of multimeric species. Also, the concentration of early aggregation intermediates will decrease with time as they associate into higher order aggregates that are excluded from the gel. The extent of aggregation as well as the rate of aggregation can be adjusted by supplementing the dilution buffer with urea.

We typically denature protein in 50 μl of acid urea within a 1.5-ml Eppendorf tube. Refolding and aggregation are initiated by rapid dilution, usually to a final volume of 1 ml. This is done using a Pipetman P1000 pipettor. The diluted sample is mixed immediately on dilution by pumping it into and out of the pipettor a few times. Temperature is controlled using a circulating water bath. At intervals, portions of the reaction are mixed with chilled 3× electrophoresis sample buffer. These samples are stored on ice during the collection of additional samples at later time points.

For the set of proteins examined using this method, chilling in an ice-water bath appears to effectively stop the off-pathway multimerization reaction. The efficiency of cold quenching can be determined by assaying samples at various times after chilling. The sample buffer contains glycerol to stabilize multimers when it is desirable to freeze samples. Aggregation intermediates of the P22 tailspike protein do not dissociate when frozen in liquid nitrogen and thawed at 4°. However, the stability of multimers to freezing and thawing should be demonstrated for each protein.

Stock Solutions and Buffers for Nondenaturing Gel Electrophoresis

50× running buffer: 248 m*M* Tris base, 1.918 *M* glycine. Store at 4°.

3× electrophoresis sample buffer: 0.6 ml 50× running buffer, 30 ml glycerol, add H_2O to 10 ml, 0.05% (w/v) bromphenol blue. Store at −20°.

4× resolving gel buffer: 1.5 *M* Tris base. Adjust to pH 8.9 with hydrochloric acid. Store at 4°.

7× stacking gel buffer: 0.47 *M* Tris base. Adjust to pH 6.7 with phosphoric acid. Store at 4°.
30% acrylamide/bisacrylamide: 29.2% (w/v) acrylamide, 0.8% (w/v) bisacrylamide. We use premixed and preweighed acrylamide/bisacrylamide (30%T/2.7%) from Bio-Rad Laboratories (Hercules, CA). Store at 4°.
10% (w/v) ammonium persulfate (prepare fresh weekly). Store at 4°.
TEMED (Bio-Rad Laboratories). Store at room temperature.

Nondenaturing Polyacrylamide Gel Electrophoresis

The purpose of the stacking gel in disc electrophoresis is to concentrate proteins into discrete bands in order to increase the sharpness of bands resolved in the lower, separating gel. One concern is that stacking of aggregation intermediates will promote further aggregation in the gel. Early experiments demonstrated that multimers of tailspike polypeptides did not aggregate in the stacking gel. This was demonstrated simply by comparing the distribution of multimers in the same sample separated by electrophoresis with and without a stacking gel.

The concentration of acrylamide/bisacrylamide in the resolving gel will vary depending on the molecular weight of the protein. The stacking gel is usually 4.3% acrylamide/bisacrylamide. Gels are cast at room temperature. We use gels of the following dimensions. The stacking gel is 1.8 cm long and the resolving gel is 8 cm long. The volume of the stacking gel is about 9 ml. Spacers and combs are 0.75 mm thick. We have achieved excellent resolution of tailspike multimers using precast gradient gels from Bio-Rad (4–15% Tris–glycine Ready Gels). However, we often observe significant background staining with minigels. Although it does not interfere with qualitative analysis of the gels, the decreased signal-to-noise ratio does limit quantitative analysis by densitometry. For this reason, we prefer custom gels for the quantitative analysis of multimerization reactions.

Gels and running buffer (1×) are equilibrated to 4° before loading chilled samples. Sample loading and electrophoresis are performed in a cold room (4°). Electrophoresis is started at a constant current of 10 mA (500 V maximum). After the samples have entered the gel, usually within 1 hr, the current is increased to 15 mA. The duration of electrophoresis will depend on the acrylamide concentration and the degree of separation desired. Gels are stained at room temperature or prepared in the cold for electrotransfer to polyvinylidene difluoride (PVDF) membranes as described later.

Examples

Quantitative Analysis of Tailspike Aggregation

The intracellular folding intermediates to the P22 tailspike protein were first identified using nondenaturing polyacrylamide gel electrophoresis.[7] Speed *et al.*[5] adapted this technique to study the competition *in vitro* between the productive refolding of tailspike polypeptides and their nonproductive self-association and aggregation. This method allowed the direct identification of the early multimeric intermediates on the pathway to the aggregated inclusion body state.[5]

Under the electrophoretic conditions described earlier, tailspike intermediates containing from one- to six polypeptide chains (72–432 kDa) can be resolved. Figure 1 shows the electrophoretic profiles of tailspike folding and aggregation intermediates collected from a single reaction over a 24-hr period. Refolded, native tailspikes migrate as a sharp band in the middle of the gel. Monomeric and multimeric intermediates on the competing folding and aggregation pathways can be seen in the earliest sample (left lane). These species form a ladder of bands, with monomers migrating fastest, followed by dimers, nonnative trimers, tetramers, etc.

Bands labeled as monomers, dimers, and trimers include both productive and off-pathway species. The stoichiometries of these nonnative tailspike species were determined by Speed *et al.*[5] using the method of Ferguson. This method involves measuring the relative mobilities of each species in a series of gels containing different concentrations of polyacrylamide.[15]

Over a period of several hours, intermediates are depleted either due to further folding and assembly into native tailspikes or association of off-pathway multimers to form higher order aggregates (Fig. 1). This can be seen most clearly when comparing the 4- and 24-hr samples in Fig. 1 (right lanes). Lower polyacrylamide concentrations and prolonged electrophoresis (overnight at 5 mA) resolve even larger multimers but have the disadvantage of decreasing the overall sharpness of bands.

The kinetics of folding and aggregation can be quantified using laser densitometry. We use the Personal Densitometer SI (Molecular Dynamics, Sunnyvale, CA) equipped with a helium–neon laser. Gels are scanned wet at room temperature; complete scans are obtained in 1–3 min. The staining intensity of individual bands is determined by integrating the absorbance of the entire band using the volume integration function in ImageQuant software (Molecular Dynamics). A standard curve is generated for each gel from three to four lanes of native tailspike protein (0.1–2.5 μg). Staining is linear over this range, which corresponds to 3–76% refolded tailspike.

The plots in Figs. 1B and 1C show the concentrations of native tailspikes and of intermediates containing one to six chains during the first 6 hr of

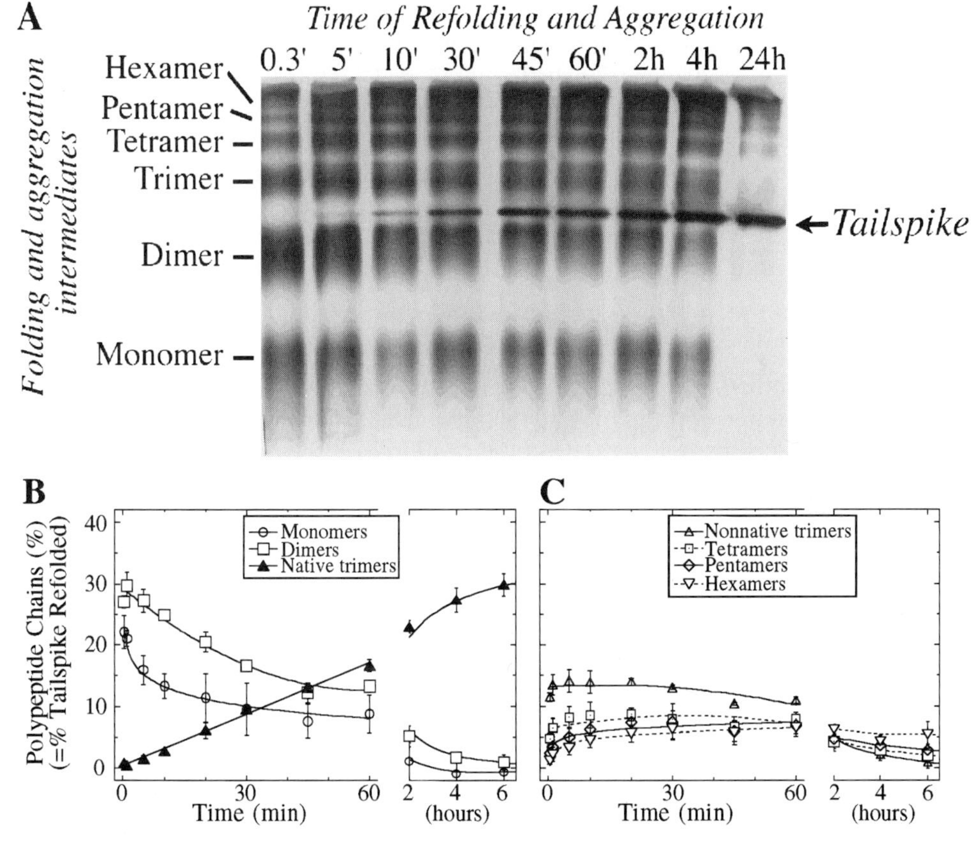

FIG. 1. Kinetics of *in vitro* folding and multimerization of the P22 tailspike protein. (A) Native polyacrylamide gel showing the time course of multimerization at 20°. Tailspike trimers were denatured in 5 *M* urea at pH 3.0 (urea–citrate). Refolding and aggregation were initiated by 20-fold dilution to 100 μg/ml with 100 m*M* Tris–HCl (pH 7.6). The dilution buffer included 0.58 *M* urea to give a final denaturant concentration of 0.8 *M*. The reaction was maintained at 20° for 24 hr. Samples were quenched at intervals by mixing with ice-cold sample buffer and stored at 0–4° until electrophoresis. The stacking gel is not pictured. (B and C) Quantitative analysis of refolding and multimerization. Samples collected up to 60 min were fractionated immediately by electrophoresis. Samples collected between 2 and 6 hr were kept on ice overnight before electrophoresis. Gels were stained with Coomassie blue and analyzed by laser densitometry. Relative intensities were determined by the volume integration of entire bands. Each data point is the average of three independent experiments (error bars ± 1 standard deviation).

the reaction. Data in these panels were obtained by densitometric analysis of native gels from three independent experiments. The yield of refolded native tailspikes (solid triangles) was 33 ± 3% after 24 hr, with a half-time of 62 min. Under these conditions, only one of every three tailspike polypeptide chains refolded into native tailspikes. Nonnative species containing one to six polypeptide chains (open symbols) reached maximum concentrations in order of size, with monomers peaking before 0.3 min, dimers at 1 min, nonnative trimers at 5 min, tetramers at 10 min, and pentamers and hexamers between 30 and 60 min.

Resolution of Conformational Isomers

The method just described was used to compare the *in vitro* refolding and aggregation of wild-type and mutant tailspike polypeptide chains. Temperature-sensitive folding (tsf) mutations increase the partitioning of partially folded tailspike polypeptide chains onto the aggregation pathway.[19,20] Mutations of this class destabilize a partially folded monomeric intermediate at the junction between competing folding and aggregation pathways.[21,22] Destabilization of the productive, junctional folding intermediate, called [I], favors formation of an alternative, nonproductive conformation, called [I*].[13,23] Analysis of the *in vitro* folding and aggregation of the tsf mutant Trp202Gln provides the first direct evidence that [I] and [I*] have different conformations.

Figure 2 shows the side-by-side profiles of intermediates formed during the refolding and aggregation of wild-type and tsf Trp202Gln polypeptide chains. Native trimers accumulated with time in the wild-type reaction (+ lanes), whereas the tsf mutation completely blocked their formation (− lanes). The mutant polypeptide chains appear to proceed down the aggregation pathway quantitatively.

Close inspection of the gel reveals that wild-type monomers migrate as a doublet band (Fig. 2, + lanes). In the mutant refolding reaction, only the faster of these two bands is present (− lanes). A simple interpretation of this result is that the low-mobility monomer, which is abundant only in wild-type lanes, is the productive intermediate [I]. The tsf mutation destabilizes this species and prevents its accumulation, thus blocking refolding and

[19] J. King, C. Haase-Pettingell, A. S. Robinson, M. Speed, and A. Mitraki, *FASEB J.* **10,** 57 (1996).

[20] A. Mitraki, C. Haase-Pettingell, and J. King, *in* "Protein Refolding" (G. Georgiou and E. De Bernardez-Clark, eds.), p. 35. Am. Chem. Soc., Washington, DC, 1991.

[21] M. Danner and R. Seckler, *Prot. Sci.* **2,** 1869 (1993).

[22] B. Schuler and R. Seckler, *J. Mol. Biol.* **281,** 227 (1998).

[23] C. A. Haase-Pettingell and J. King, *J. Biol. Chem.* **263,** 4977 (1988).

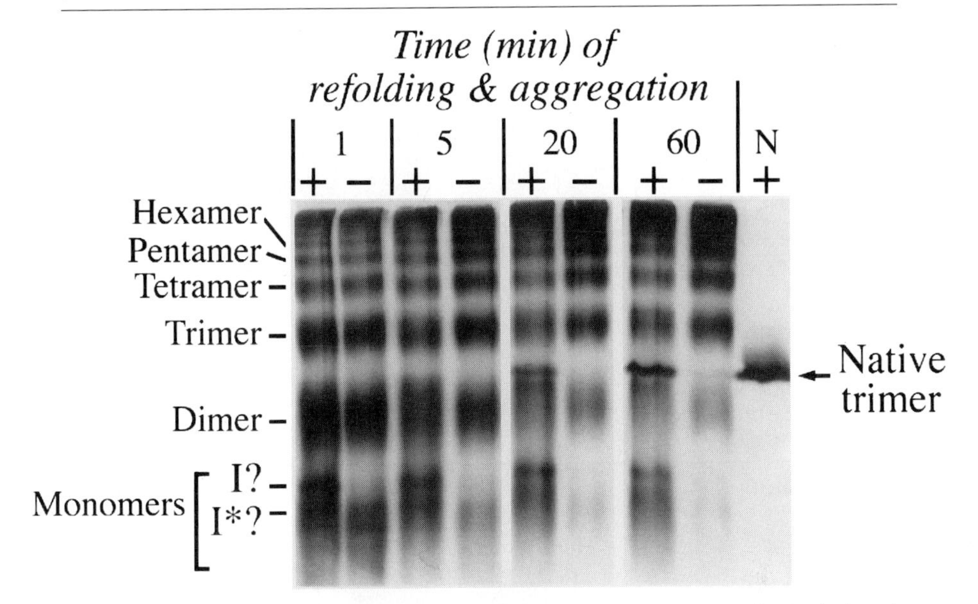

FIG. 2. Electrophoretic resolution of conformational isomers. Wild-type tailspike (+ lanes) and the tsf mutant Trp202Gln (− lanes) were denatured and diluted as in Fig. 1. Samples were taken at intervals for up to 60 min and analyzed by native gel electrophoresis (9% polyacrylamide). Proteins were visualized by silver staining. I, productive partially folded monomer; I*, nonproductive aggregation precursor; N, native trimer control.

assembly at the level of a single-chain intermediate. In the mutant-refolding reaction, the predominant monomeric species is the high-mobility monomer, a good candidate for the aggregation precursor [I*].

The two conformations of monomeric intermediates in Fig. 2 are clearly sensitive to a single amino acid substitution. By adapting the methods described here, it should be possible to resolve conformational effects of single amino acid substitutions on early aggregation intermediates of other proteins.

Carbonic Anhydrase, P22 Coat Protein, and α_1-Antitrypsin

Figure 3A shows the off-pathway multimerization of recombinant human α_1-antitrypsin during heat denaturation at 57°.[18] The fastest migrating band at $t = 0$ is the predominant band in fresh preparations and is most likely the native monomer. Self-association of partially unfolded chains generates a ladder of multimers in a time-dependent manner. The band migrating just behind the native monomer band appears only after 15 min at 57° and may be the partially unfolded aggregation precursor.

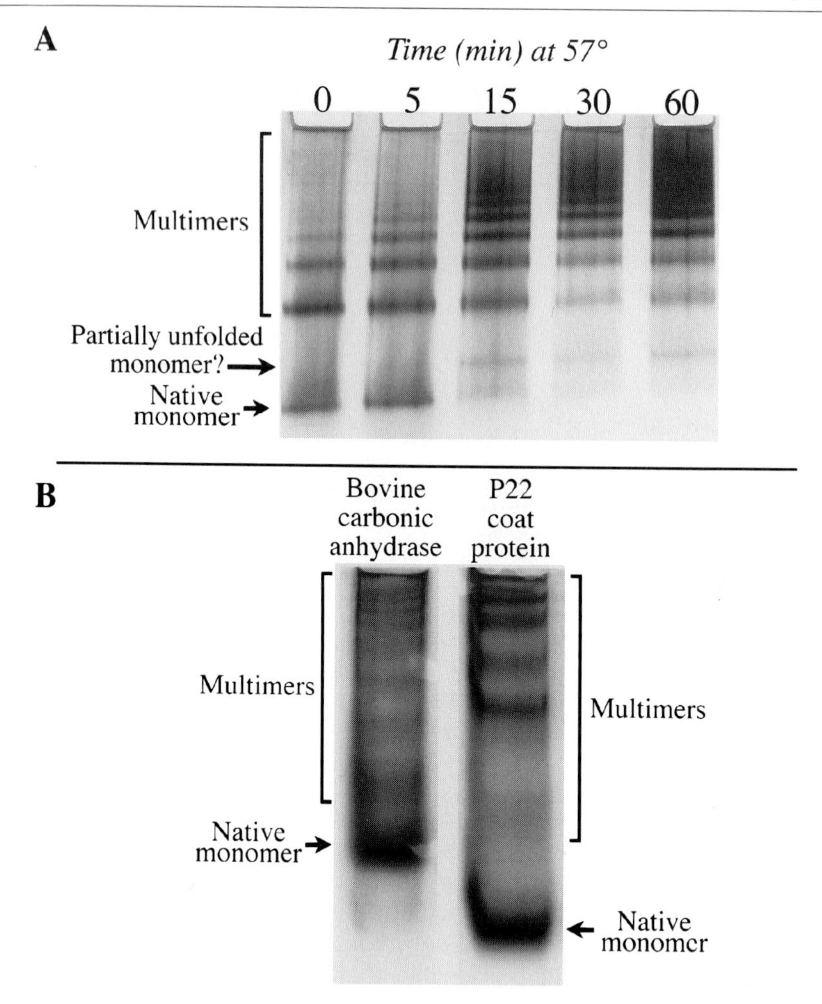

FIG. 3. Off-pathway multimerization of α_1-antitrypsin, carbonic anhydrase II, and P22 coat protein. (A) Recombinant human α_1-antitrypsin in buffer was incubated at 57°. Portions of the reaction were chilled on ice at the indicated times and then fractionated on a 10–20% polyacrylamide gel at 4°.[18] (B) Denatured carbonic anhydrase was diluted into buffer (pH 7.6) at 1 mg/ml and 0.4 *M* urea and incubated for 0.3 min at 20°. Denatured P22 coat protein was diluted into buffer (pH 7.6) at 100 μg/ml and 0.4 *M* urea and incubated for 0.3 min at 30°. Adapted from M. A. Speed, D. I. C. Wang, and J. King, *Prot. Sci.* **4,** 900 (1995). Reprinted with permission of Cambridge University Press.

Bovine carbonic anhydrase (carbonate dehydratase), a monomeric protein, and the coat subunit of phage P22 were denatured *in vitro* and refolding/aggregation was initiated by dilution.[5] The gel lanes in Fig. 3B show the native monomers and off-pathway multimers that accumulated in these reactions. The examples presented here demonstrate the effectiveness of the method discussed earlier for monitoring both conformational and kinetic properties of the earliest intermediates on protein aggregation pathways.

Nondenaturing Western Blotting of Aggregation Intermediates

Introduction and Basic Principles

Western blotting is a standard method used to identify and characterize submicrogram amounts of protein displaying distinct epitopes characteristic of a specific protein. In conjunction with nondenaturing polyacrylamide gel electrophoresis, nondenaturing Western blotting can be used to characterize structural epitopes of metastable folding intermediates and multimeric aggregation intermediates. Previous work has utilized the specificity of monoclonal antibodies to probe the structure of productive folding intermediates in solution during *in vitro* or *in vivo* folding, including intermediates of bovine serum albumin,[24] the β_2 subunit of native tryptophan synthase,[25] influenza hemagglutinin,[26] and the P22 tailspike.[27–29] In addition, the specificity of antibodies to certain folding conformers can be used to identify and characterize folding and aggregation intermediates trapped by nondenaturing gel electrophoresis.[30] Significant modifications must be incorporated into the standard Western blot protocol to eliminate the denaturing elements of the procedure, which would cause conformational changes to the thermolabile folding and aggregation intermediates.

Standard Denaturing Western Blotting

In the standard Western blotting procedure, the conditions used to transfer polypeptide chains from the separating gel to the surface of the

[24] L. G. Chavez, Jr., and D. C. Benjamin, *J. Biol. Chem.* **253,** 8081 (1978).

[25] S. A. G. Blond, *Proteins* **1,** 247 (1987).

[26] I. Braakman, H. Hoover-Litty, K. R. Wagner, and A. Helenius, *J. Cell Biol.* **114,** 401 (1991).

[27] B. Friguet, L. Djavadi-Ohaniance, C. A. Haase-Pettingell, J. King, and M. E. Goldberg, *J. Biol. Chem.* **265,** 10347 (1990).

[28] B. Friguet, L. Djavadi-Ohaniance, J. King, and M. E. Goldberg, *J. Biol. Chem.* **269,** 15945 (1994).

[29] K. Tokatlidis, B. Friguet, D. Deville-Bonne, F. Baleux, A. N. Fedorov, A. Navon, L. Djavadi-Ohaniance, and M. E. Goldberg, *Philos. Trans. R. Soc. Lond. B Biol. Sci.* **348,** 89 (1995).

[30] M. A. Speed, T. Morshead, D. I. C. Wang, and J. King, *Prot. Sci.* **6,** 99 (1997).

membrane are denaturing for many proteins. The electrotransfer solution usually contains 20% methanol and 0.1% (w/v) sodium dodecyl sulfate (SDS) to promote the efficient transfer of polypeptide chains from the gel to the surface of the membrane. The addition of SDS in the electrotransfer solution ensures the efficient transfer of macromolecules out of the gel, which is often necessary to counteract the effect of methanol tightening the gel pores. The binding of negatively charged SDS to the polypeptide chains ensures migration toward the anode (positive electrode) regardless of the p*I* of the protein. Methanol in the transfer solution improves the binding efficiency of the protein to the membrane through hydrophobic interactions. Without SDS in the electrotransfer solution, methanol further improves binding efficiency to the transfer membrane by stripping residual SDS from the polypeptide chain.

After electrotransfer using the standard denaturing buffer, partial renaturation during the initial steps of the Western blot is often required for the chains to reacquire the structural epitopes that the antibodies recognize in the Western blot. Under standard Western blot conditions, incubation of the blots in reconstituted milk and buffer solutions at room temperature allows the polypeptide chains to renature on the surface of the membrane and form distinct native and nonnative epitopes. The structural epitopes recognized by the primary antibodies of the Western blot procedure may not be indicative of the structural characteristics of the protein conformer that was present during nondenaturing gel electrophoresis.

Nondenaturing Western Blotting

The Western blot protocol can be modified to avoid the exposure of polypeptide chains to denaturing conditions to ensure that the protein does not undergo a conformational change during the procedure. Multimeric species are first separated by native gel electrophoresis in the cold as described earlier. Methanol and SDS are omitted from the Tris–glycine transfer buffer. Electrotransfer and all immunoblotting steps through the final membrane wash are performed at 4° to maintain the conformation of the intermediates.

In both standard and nondenaturing electrotransfer procedures, the transfer membrane must be prewetted with methanol in order to enable macromolecules in the aqueous phase to bind to the hydrophobic surface of the membrane. For nondenaturing applications, the membrane is then rinsed repeatedly with Tris–glycine buffer to remove all the methanol before use.

Although nondenaturing buffers and low temperatures used in the gel electrophoresis and Western blotting procedures are designed to minimize any conformational changes of the folding and aggregation intermediates, certain minor structural changes may occur during these procedures. Additional controls can be included to determine the temperature dependence of potential structural changes, test the effects of additional denaturing elements to the Western blot procedure, and correlate the results to data obtained by traditional *in vitro* refolding studies.

For the Western blotting method, monoclonal antibodies against the particular protein of interest are isolated from hybridomas derived from mice or rabbits immunized with native protein. Because of the dynamic structure of proteins, animals immunized with native protein may develop antibodies against distinct nonnative structural epitopes. The purified monoclonal antibodies are screened by enzyme-linked immunosorbent assay (ELISA) for binding to native protein and/or thermally denatured protein. The set of monoclonal antibodies against nonnative chains may recognize structural epitopes found on early folding intermediates and aggregates. These antibodies may serve as tools to follow the *in vitro* refolding of purified polypeptide chains and to characterize the structural epitopes of folding and aggregation intermediates by nondenaturing Western blotting.

Considerations

The choice of membrane type depends on the protein properties and its binding efficiency to the transfer membrane. Polyvinylidene difluoride (PVDF) membranes have a more hydrophobic surface than traditional hydrophilic nitrocellulose membranes, and therefore the proteins generally display greater binding efficiencies to PVDF. Immobilon-P low-retention membrane (Millipore, Bedford, MA) is recommended over other PVDF membranes for Western blotting of proteins to obtain low background. Note that to quantify populations of different protein conformers of varying hydrophobicity, proper controls must be included to ensure consistent transfer efficiency of species with possible differences in mobility out of the gel, binding to the membrane, and reactivity to the antibodies.

The electrotransfer procedure must be performed under conditions to minimize conformational changes of the polypeptide chains. In addition to removing the denaturing elements from transfer buffer, the entire procedure should be done in the cold (4°).

Stock Solutions and Buffers for Nondenaturing Western Blotting

Transfer buffer: 20 m*M* Tris base, 150 m*M* glycine. Store at 4°.
10× Tris-buffered saline (TBS): 200 m*M* Tris base, 1.37 *M* NaCl. Adjust to pH 7.5 with hydrochloric acid.

Blocking solution: 10% (w/v) nonfat dry milk in 1× TBS. Prepare 1 day in advance. Store at 4°.

5% milk-TBS: 5% (w/v) nonfat dry milk in 1× TBS ± 0.05% (v/v) Tween 20. Prepare 1 day in advance. Store at 4°.

Nondenaturing Western Blot Procedure

Prewet PVDF membrane with methanol (10 min) and rinse with transfer buffer to remove methanol (3 × 10 min). All subsequent steps should be performed in a cold room (4°) with chilled buffers. Following electrophoresis under nondenaturing conditions as described earlier, place gel directly into transfer buffer and incubate for at least 15 min. The following instructions were developed for full-size custom gels described earlier. We have not yet adapted the procedure for use with precast minigels. Layer gel and membrane between filter paper and sponges according to standard procedures for assembling transfer cassettes. Take care to exclude air bubbles from between the layers.

Submerse the cassette in chilled transfer buffer in the transfer tank. The orientation of the gel/membrane sandwich is determined by the p*I* of the protein and the pH of the transfer buffer. The Tris–glycine buffer has a pH of 8.6 ± 0.1 at room temperature. Proteins with p*I* values lower than the pH of the transfer buffer will migrate toward the positive electrode, whereas proteins with p*I* values >8.6 will migrate toward the negative electrode. Electrotransfer overnight at a constant current of 100 mA (20V maximum) at 4° with stirring.

On disassembly of the transfer apparatus, the polypeptide chains on the surface of the transfer membrane are exposed to antibodies that recognize distinct structural epitopes displayed by the protein. This antibody-binding procedure must be performed in the cold to minimize structural changes of the protein. Carefully lift the membrane off of the gel and immediately submerse it in chilled blocking solution. The protein side of the membrane should be facing up throughout all subsequent steps to ensure the uniform exposure of adsorbed proteins to the antibodies. Wear nonpowdered gloves or use forceps to handle the membrane. Incubate the membrane in blocking solution with gentle rocking for 30 min. This step is essential to prevent nonspecific antibody binding to the membrane.

Decant blocking solution and add 1× TBS containing the primary antibody. The membrane will adhere to the glass or plastic tray if the solutions are decanted slowly, thus avoiding any need for further handling of the membrane prior to preparation for enhanced chemiluminescence (ECL). Add primary antibody diluted in 1× TBS (±Tween 20, see below). Incubate for 1 hr with gentle rocking. The optimal concentration of primary antibody is usually in the nanomolar range and will vary according to binding affinity. Tween 20 is included to prevent nonspecific binding of the antibody to the transferred polypeptide chains. However, we have observed that exposure to Tween 20 can either release primary antibody from native tailspikes or release native tailspikes from the PVDF membrane. This effect may vary depending on the source of the PVDF membrane. In contrast, binding and/or detection of aggregation and folding intermediates does not appear to be affected by exposure to the detergent. Excess primary antibody is removed by washing the membrane with 5% milk-TBS (±Tween 20) (3 × 10 min).

Bound primary antibodies are next labeled with enzyme-linked secondary antibodies. Dilute the secondary antibody in 1× TBS and incubate for 1 hr in the cold. We use a horseradish peroxidase-linked sheep antimouse antibody from Amersham (Arlington Heights, IL). The dilution of the secondary antibody depends on the sensitivity of the ECL reaction. We currently dilute the secondary antibody 100,000-fold for use with the SuperSignal Ultra Chemiluminescent Substrate from Pierce (Rockford, IL). A 10,000-fold dilution for use with Amersham's ECL kit also gives excellent signal and very low background. Remove the unbound secondary antibody by washing the membrane with 5% milk-TBS (±Tween 20) (3 × 10 min).

The blots are developed using enhanced chemiluminescence according to the manufacturer's instructions. The reactions proceed efficiently at 4° and require only 1–10 min. This is the same amount of time recommended in the instructions for reactions performed at room temperature. Following draining of the substrate solution from the membranes (about 1 min), they are sealed with tape inside standard overhead transparencies and exposed to film for various periods of time at room temperature.

Example of Nondenaturing Western Blotting Application

This nondenaturing Western blotting methodology can be applied to examine the relationship between intermediates along the productive folding pathway and those species associating along the aggregation pathway. As described previously, metastable folding intermediates and multimeric aggregation intermediates of P22 tailspike polypeptide chains have been isolated by nondenaturing gel electrophoresis.[5] To characterize the structural epitopes of these intermediates, the nondenaturing Western blot procedure was used to determine whether these productive and aggregating conformers display native-like structural epitopes.[30]

As described earlier, monoclonal antibodies (MAbs) against tailspike polypeptide chains were originally isolated from hybridomas derived from mice immunized with native tailspike. Seven of the derived antibodies (designated "anti-native") were found to recognize native tailspike protein by the ELISA screening assay.[27] An additional four antibodies (designated as "anti-intermediate") were found to recognize nonnative intermediates and thermally denatured protein.

To characterize the structural epitopes of aggregation intermediates, reactivity of the aggregates with a set of MAbs was screened by nondenaturing Western blotting. These antibodies were tested for reactivity with native tailspike protein and refolding tailspike polypeptide chains. The refolding conditions were chosen, as discussed earlier, to generate multimeric aggregation intermediates and a minor fraction of protrimer folding intermediate and native protein. Samples of purified native protein and refolding tailspike were loaded in triplicate onto a nondenaturing gel (9% acrylamide) and electrophoresed for 4 hr in the cold.

The silver-stained gel (Fig. 4A) showed a distribution of folding and aggregation intermediates. Because the sample was taken immediately after dilution, a minimal fraction of tailspike had refolded into the native trimer.

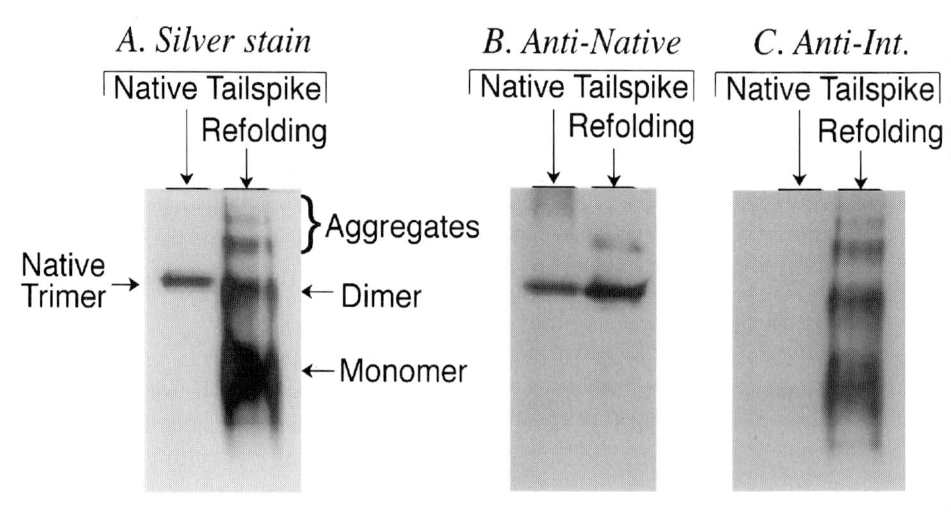

FIG. 4. Screening of MAbs for reactivity against native tailspike and *in vitro* refolding tailspike. Refolding was performed at 100 μg/ml, 0.8 *M* urea, in 40 m*M* phosphate buffer, pH 7.6, at 20°, and aliquots of sample were taken 0.5 min after dilution. Purified native protein (0.3 μg) and refolding and aggregating tailspike (2 μg) were loaded onto a 9% acrylamide nondenaturing gel and electrophoresed for 4 hr at 4°. Bands were visualized by silver staining (A) and immunolabeling using MAbs against native tailspike (B) and against nonnative tailspike chains (C). Adapted from M. A. Speed, T. Morshead, D. I. C. Wang, and J. King, *Prot. Sci.* **6,** 99 (1997). Reprinted with permission of Cambridge University Press.

The majority of polypeptide chains were in the monomer, dimer, and trimer aggregate states, with some higher order aggregates seen faintly at the top of the gel.

The reactivity of anti-native tailspike MAbs (anti-N) with the folding and aggregation intermediates and purified native protein is shown in Fig. 4B. Anti-native tailspike MAbs recognized the purified native tailspike and the fraction of tailspike chains that refolded into the native conformation. These antibodies did not bind to the ladder of aggregation intermediates. In addition, the protrimer folding intermediate, migrating with an electrophoretic mobility similar to the trimer aggregate, was detected by anti-N MAbs. This result indicated that the protrimer intermediate displayed the native-like epitope recognized by the anti-N MAbs.

The reactivity of anti-intermediate tailspike MAbs is shown in Fig. 4C. Anti-intermediate MAbs reacted with the aggregates and not with the native species. The distribution of multimeric intermediates was similar to that of silver-stained bands. The higher order multimers at the top of the gel appeared to display antigenic reactivity similar to the early multimeric intermediate precursors. Because there was no apparent cross-reactivity of the anti-intermediate MAbs with the native tailspike protein, the multimeric aggregation intermediates displayed only epitopes that were nonnative. This suggests that an early folding intermediate without native-like epitopes was susceptible to aggregation rather than the protrimer folding intermediate.

Additional Applications

The nondenaturing Western blotting technique can be utilized for additional applications in characterizing folding and aggregation reactions. Using a set of monoclonal antibodies against native and nonnative conformers, one can characterize the structural epitopes of folding and aggregation intermediates to determine the nature of the critical intermediate at the junction between the productive folding and aggregation pathways. Nondenaturing Western blotting can also be used to compare the structural epitopes displayed by intermediates generated during *in vitro* refolding and aggregation versus *in vivo* folding and inclusion body formation. Mildly denaturing conditions can be used in order to determine whether antibodies could recognize partially unfolded or even fully denatured protein. Probing the conformation of the folding and aggregation intermediates can lead to a greater understanding of the structural basis of aggregation and the possibility of controlling the kinetic competition between productive folding and aggregation.

Acknowledgments

John Konz kindly provided the photograph shown in Fig. 3A. This research was supported by the National Institutes of Health (GM17, 980 to JK) and by the National Science Foundation's Engineering Research Center Initiative (8803014).

[23] Deposition of Soluble Amyloid-β onto Amyloid Templates: With Application for the Identification of Amyloid Fibril Extension Inhibitors

By William P. Esler, Evelyn R. Stimson, Patrick W. Mantyh, and John E. Maggio

Introduction

Several human diseases are now recognized to be caused by the formation of insoluble amyloid from naturally occurring peptides and proteins. Amyloid-forming monomers are not intrinsically pathological; rather it is their conformation and assembly state that are most important to their cytotoxicity. Because the chemical information for forming these amyloids apparently lies in the primary structure of the monomers, there has been significant interest in using *in vitro* systems to study amyloid formation and growth. The amyloid-forming monomer in Alzheimer's disease, a ≅40-mer peptide known as Aβ, has been widely employed in *in vitro* systems to study the processes involved in amyloid assembly and growth *in vivo*.

In vitro, the assembly process from soluble Aβ monomers to fibrillar Aβ assemblies similar by several criteria to senile plaques of Alzheimer's disease is a complex one involving multiple phases and intermediates. The two fundamental parts of the process are assembly of monomeric Aβ into a template and growth of the template by deposition of additional Aβ (Fig. 1a). The former process, called aggregation, and the latter process, called deposition, are fundamentally distinct *in vitro,* and there is good reason to believe they are also distinct *in vivo.*[1]

Mechanistic differences between these processes[2–5] require the use of

[1] J. E. Maggio and P. W. Mantyh, *Brain Pathol.* **6,** 147 (1996).
[2] J. T. Jarrett and P. T. Lansbury Jr., *Biochemistry* **31,** 12345 (1992).
[3] W. P. Esler, E. R. Stimson, J. R. Ghilardi, H. V. Vinters, J. P. Lee, P. W. Mantyh, and J. E. Maggio, *Biochemistry* **35,** 749 (1996).
[4] H. Naiki and K. Nakakuki, *Lab. Invest.* **74,** 374 (1996).
[5] A. Lomakin, D. S. Chung, G. B. Benedek, D. A. Kirschner, and D. B. Teplow, *Proc. Natl. Acad. Sci. U.S.A.* **93,** 1125 (1996).

0076-6879/99 $30.00

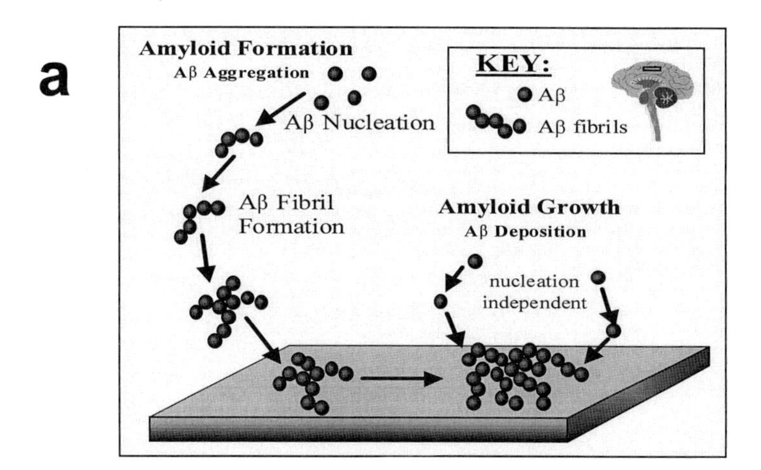

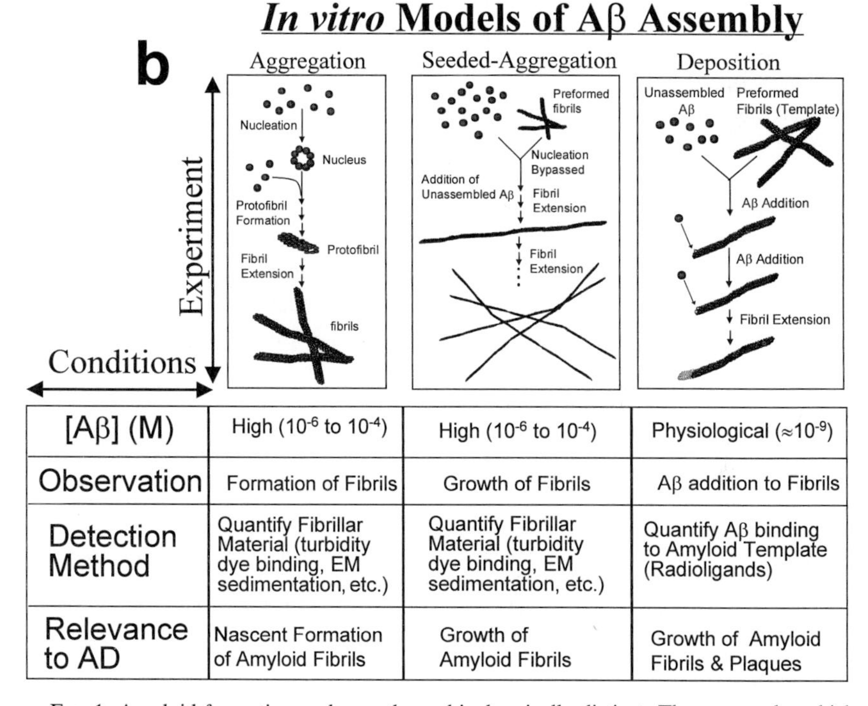

	Aggregation	Seeded-Aggregation	Deposition
[Aβ] (M)	High (10^{-6} to 10^{-4})	High (10^{-6} to 10^{-4})	Physiological ($\approx 10^{-9}$)
Observation	Formation of Fibrils	Growth of Fibrils	Aβ addition to Fibrils
Detection Method	Quantify Fibrillar Material (turbidity dye binding, EM sedimentation, etc.)	Quantify Fibrillar Material (turbidity dye binding, EM sedimentation, etc.)	Quantify Aβ binding to Amyloid Template (Radioligands)
Relevance to AD	Nascent Formation of Amyloid Fibrils	Growth of Amyloid Fibrils	Growth of Amyloid Fibrils & Plaques

FIG. 1. Amyloid formation and growth are biochemically distinct. The process by which fibrillar $A\beta$ amyloid deposits form and grow in Alzheimer's disease is complex, involving distinct chemical steps (a). The first and rate-limiting step in the assembly of solution phase $A\beta$ is the nucleation-dependent formation of a nidus or "seed." Once present, the assembled $A\beta$ template recruits additional $A\beta$ to deposit and promote amyloid growth by a nucleation-independent mechanism. The processes of nascent amyloid formation and amyloid growth may be modeled readily *in vitro* (b) as described in the text.

different methods to examine the growth or propagation of amyloid than are used to study its nascent formation (Fig. 1). Aβ deposition, the process of amyloid growth by the association of individual soluble Aβ molecules with a preexisting amyloid template,[6,7] can be modeled readily *in vitro* under physiologically relevant conditions and importantly at physiological (n*M*) Aβ concentrations. Such assays may elucidate key molecular interactions[3,8–11] in amyloidosis and may be exploitable in the design of antiamyloid therapeutics.[1,10] High-throughput Aβ deposition assays[10] may provide a means for the identification and optimization of such compounds. The methods described here have been developed for the examination of Aβ deposition, but are adaptable for the study of other amyloids.

Comparison of Aβ Amyloid Formation and Growth Assays

Because nascent formation (Fig. 1b, left-hand side) of Aβ amyloid-like fibrils *in vitro* is a higher order kinetic process rate-limited by nucleation,[2] the earliest stages of Aβ amyloid formation may, in general, be studied only at supersaturating Aβ concentration. Such aggregation assays suffer from the limitation that they must be performed at Aβ concentrations (μ*M* to m*M*) many orders of magnitude above physiological (n*M*). Recent observations that the assembly of exogenous and endogenous Aβ into higher molecular weight oligomers at physiological Aβ concentrations in conditioned cell media[12] may allow for the investigation of Aβ nucleation under more physiological conditions. Assays used to examine the *de novo*

[6] P. W. Mantyh, E. R. Stimson, J. R. Ghilardi, C. J. Allen, C. E. Dahl, D. C. Whitcomb, S. R. Vigna, H. V. Vinters, M. E. Labenski, and J. E. Maggio, *Bull. Clin. Neurosci.* **56,** 73 (1991).

[7] J. E. Maggio, E. R. Stimson, J. R. Ghilardi, C. J. Allen, C. E. Dahl, D. C. Whitcomb, S. R. Vigna, H. V. Vinters, M. E. Labenski, and P. W. Mantyh, *Proc. Natl. Acad. Sci. U.S.A.* **89,** 5462 (1992).

[8] J. P. Lee, E. R. Stimson, J. R. Ghilardi, P. W. Mantyh, Y.-A. Lu, A. M. Felix, W. Llanos, A. Behbin, M. Cummings, M. Van Criekinge, W. Timms, and J. E. Maggio, *Biochemistry* **34,** 5191 (1995).

[9] W. P. Esler, E. R. Stimson, J. R. Ghilardi, Y.-A. Lu, A. M. Felix, H. V. Vinters, P. W. Mantyh, J. P. Lee, and J. E. Maggio, *Biochemistry* **35,** 13914 (1996).

[10] W. P. Esler, E. R. Stimson, J. R. Ghilardi, A. M. Felix, Y.-A. Lu, H. V. Vinters, P. W. Mantyh, and J. E. Maggio, *Nature Biotechnol.* **15,** 258 (1997).

[11] W. P. Esler, E. R. Stimson, J. R. Ghilardi, J. Fishman, H. V. Vinters, P. W. Mantyh, and J. E. Maggio, *Biopolymers* **49,** 505 (1999).

[12] M. B. Podlisny, D. M. Walsh, P. Amarante, B. L. Ostaszewski, E. R. Stimson, J. E. Maggio, D. B. Teplow, and D. J. Selkoe, *Biochemistry* **37,** 3602 (1998).

formation of amyloid[2,13] and adaptations of these assays for screening[14] are discussed elsewhere in this volume.

Later steps in the amyloid pathway (Fig. 1b, middle and right-hand side) may also be examined readily by monitoring the addition of soluble Aβ to a preexisting amyloid template. These experiments fall into two categories. In the first (Fig. 1b, middle), preformed amyloid fibrils are added to the Aβ aggregation assay described earlier to bypass the rate-limiting nucleation step.[2] In such seeded aggregation experiments, the amount of preassembled peptide is necessarily small relative to the amount of unassembled peptide initially present. Such experiments are typically performed at high or saturating Aβ concentrations to promote rapid assembly and to facilitate detection.

In the second type of Aβ amyloid growth assay (Fig. 1b, right-hand side), the addition of Aβ at physiological concentrations onto preexisting amyloid template is measured.[3,6,7] In contrast to the Aβ aggregation assays described previously,[2] the growth of preexisting amyloid by Aβ "deposition" onto a preexisting amyloid template (Fig. 1b, right-hand side) is quite favorable at physiological (≈n*M*) Aβ concentrations *in vitro*.[3,6,7] However, as would be expected for a slowly progressing disease, *in vitro* growth of Alzheimer's disease amyloid at physiological Aβ concentrations is a slow process, with the amount of Aβ added to the template over hours or days representing a very small fraction (on a molar basis) of the material initially present. In such experiments, it is not feasible to detect differences in the total quantity of amyloid present. Instead, the amount of Aβ peptide monomer added onto the template is quantified using radioisotopes[7,15] or other high-sensitivity methods to distinguish between the peptide initially present as template and the peptide added.

Overview of Aβ Deposition Assays

Because Aβ deposition experiments are performed by monitoring the deposition of a labeled Aβ peptide from solution onto a preexisting template and quantifying the amount of the peptide bound (Fig. 2), key considerations include the source of amyloid template, the means to track the depositing peptide, and the method to separate the bound or deposited tracer from the free tracer remaining in solution. Natural amyloid from

[13] D. A. Kirschner, H. Inoue, L. K. Duffy, A. Sinclair, M. Lind, and D. J. Selkoe, *Proc. Natl. Acad. Sci. U.S.A.* **84,** 6953 (1987).

[14] H. LeVine III, *Prot. Sci.* **2,** 404 (1993).

[15] A. L. Biere, B. Ostaszewski, E. R. Stimson, B. T. Hyman, J. E. Maggio, and D. J. Selkoe, *J. Biol. Chem.* **271,** 32916 (1996).

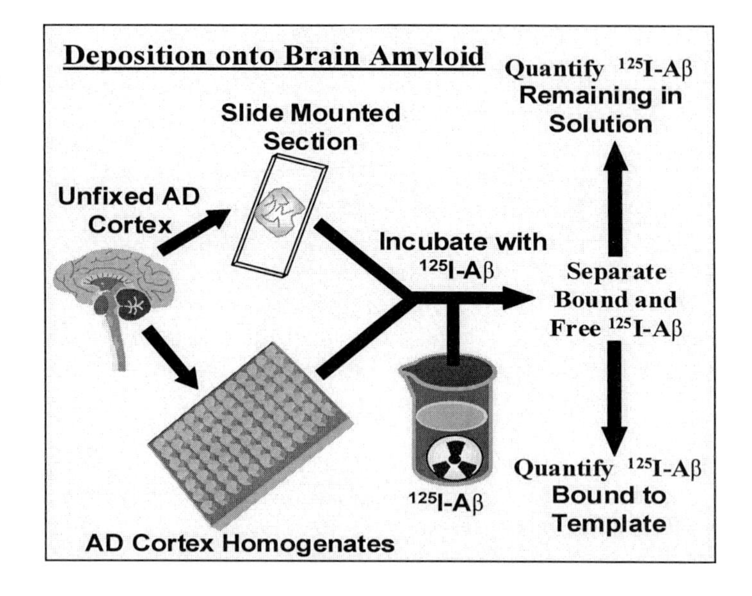

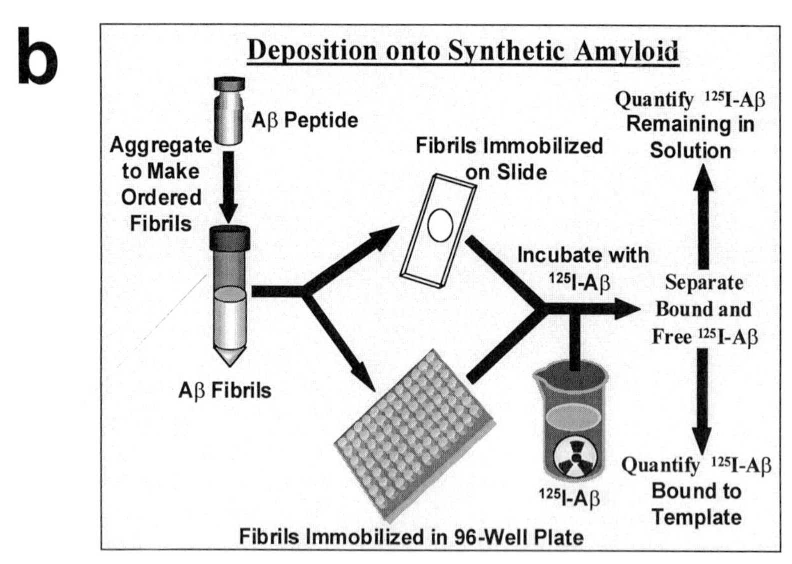

FIG. 2. Assays to monitor $A\beta$ deposition. Amyloid templates used in $A\beta$ deposition assays can be authentic brain amyloid in preparations of AD cortex or amyloid fibrils prepared *in vitro* from synthetic $A\beta$. For $A\beta$ deposition assays using authentic brain amyloid (a), unfixed AD cortex tissue is prepared as sections or homogenates. The tissue preparations are then incubated with a solution of unassembled labeled (e.g., ^{125}I)$A\beta$ at a physiological ($\approx$nM) concentration. After the desired incubation interval the amount of labeled $A\beta$ deposited onto the amyloid in the tissue or the change in solution $A\beta$ is quantified. For deposition assays using synthetic templates (b), $A\beta$ fibrils are formed *in vitro* and immobilized onto glass microscope slides or the wells of multiwell plates. The template is then incubated with labeled $A\beta$ and the amount of $A\beta$ deposited quantified as described earlier.

tissue specimens in the form of slide-mounted sections or amyloid-enriched tissue homogenates can be used as template in deposition assays (Fig. 2a).[3,6,7] Alternatively, templates formed *in vitro* by aggregating synthetic, recombinant, or purified Aβ (or other amyloidogenic protein) into fibrils (synthetic amyloid) may also be used (Fig. 2b).[10] Advantages and limitations of each template (discussed later) must be considered when designing a study and in selecting a method to label and track the depositing peptide.

Template Selection

In vitro Aβ deposition onto brain amyloid (Fig. 3) and synthetic Aβ fibrils (synthaloid) (Fig. 4) share several biochemical properties, including time dependence, kinetic order, structure activity relations, and pH dependence, demonstrating that the processes of deposition onto each template are similar. The amyloid plaques in AD tissue preparations (Fig. 2a) are heterogeneous in nature,[16,17] containing other components in addition to the predominate Aβ. Although these factors contribute to the relevance of the system to the disease state, they also contribute to its complexity. The density of the various amyloid subtypes is also variable between patients and individual specimens from a single patient, making quantitative comparisons difficult. Deposition onto AD cortex preparations in the form of either slide-mounted sections (which preserve morphology but prevent true replicates) or homogenates (which allow replicates but lack morphology) is labor-intensive and not readily adaptable for high throughput. Further, these assays require relatively large amounts of well-characterized postmortem human tissue (and the risks in handling such specimens).

Deposition onto synthetic templates prepared *in vitro* (Fig. 2b) allows examination of Aβ : template interactions and Aβ association in the absence of these potentially complicating factors. Production of synthetic Aβ amyloid templates is highly reproducible, allowing for controlled comparison between batches. Although this method may lack some of the relevance of the tissue-based assays, it affords a means to systematically evaluate the effect of alterations in both the depositing peptide and the template. The synthaloid system is also suitable for high-throughput experimentation and robotics.

Preparation of Deposition Templates

Tissue-based Aβ deposition assays (Fig. 2a) utilize human postmortem tissue preparations as slide-mounted sections for radioligand autoradiogra-

[16] T. Dyrks, E. Dyrks, T. Hartmann, C. Masters, and K. Beyreuther, *J. Biol. Chem.* **267,** 18210 (1995).

[17] S. M. Dudek and G. U. Johnson, *Brain Res.* **651,** 129 (1994).

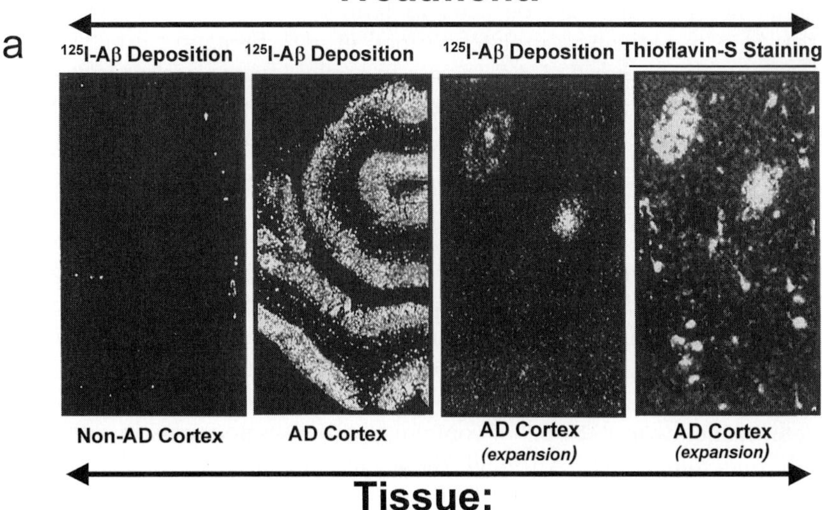

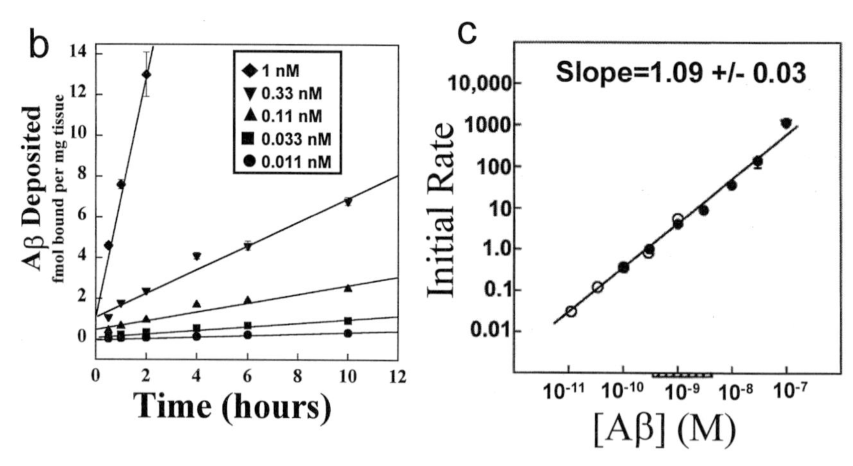

FIG. 3. *In vitro* deposition of Aβ onto amyloid in AD tissue preparations. (a) Dark-field autoradiograms of Aβ deposition using brain sections. Whereas labeled Aβ shows focal "binding" onto AD cortex, little or no radiolabeled Aβ binding is observed with similar preparations of age-matched non-AD cortex. Higher magnification reveals that the sites of Aβ deposition correlate with the location of amyloid plaques detected by tinctorial markers for amyloid. Aβ deposition in brain tissue homogenates follows linear time dependence over the times examined (b). Log–log plots (c) of deposition rate [determined from linear regression of time course data in (b)] vs soluble Aβ concentration indicate that Aβ deposition is first order in solution phase Aβ concentration. The relative deposition "activity" of Aβ analogs can also be determined in this manner.

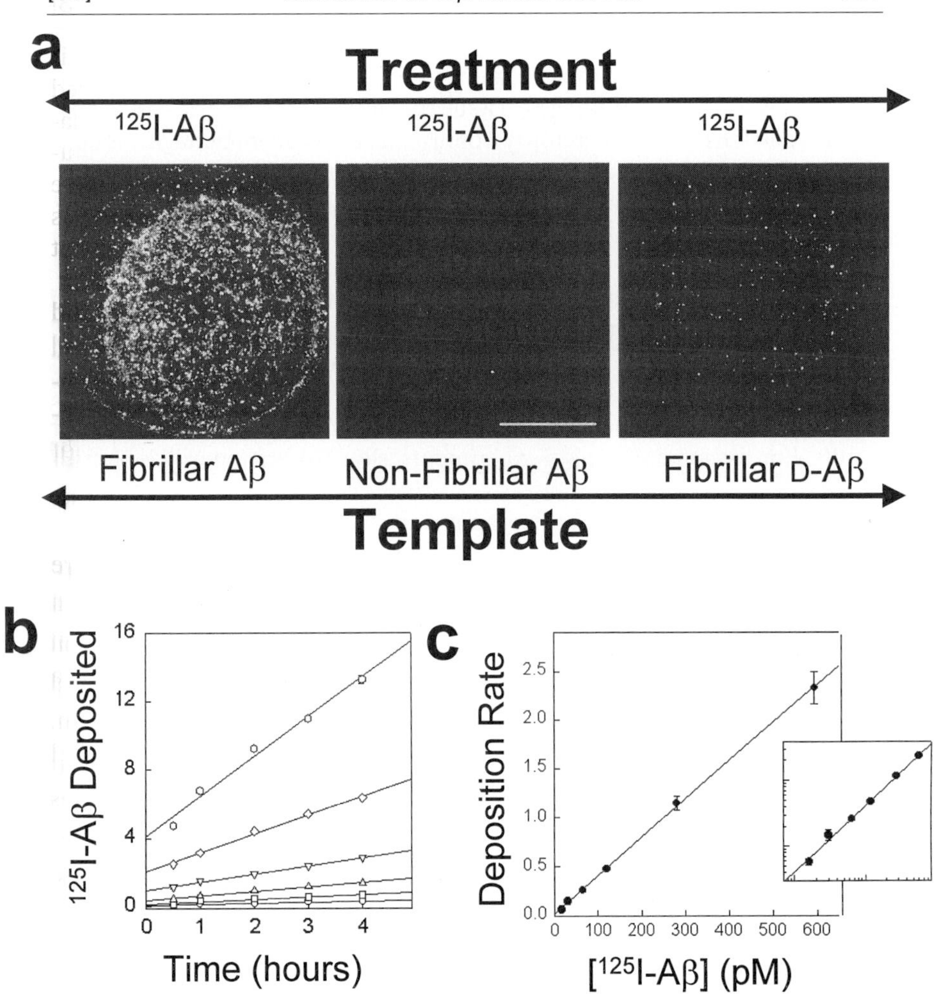

FIG. 4. *In vitro* deposition of Aβ onto synthetic amyloid preparations. (a) Dark-field autoradiograms of [^{125}I]Aβ deposition using synthaloid. Whereas labeled Aβ shows focal deposition onto a template of immobilized fibrillar Aβ, little or no radiolabeled Aβ binding is observed on a similar template prepared from unassembled Aβ or fibrils composed of all D-Aβ. Bar: 4 mm. Aβ deposition onto synthaloid follows linear time dependence over the times examined (b). Log–log plots of deposition rate vs Aβ concentration indicate that Aβ deposition is first order in solution phase Aβ concentration (c, inset). Note the similarity in Aβ deposition onto brain amyloid (Fig. 3) and synthaloid (Fig. 4).

phy and as tissue homogenates for quantitative kinetic experiments. It is important that any tissue used for deposition experiments be characterized carefully by a trained pathologist. Tissue that has met the criteria of Khachaturian[18] and CERAD[19] for autopsy-confirmed AD is recommended. Nonhuman brain tissue, particularly some transgenic models,[20,21] is also suitable for use in Aβ deposition assays if a significant amount of Aβ amyloid is present. Binding of labeled Aβ and Aβ assemblies to amyloid-naive rat tissue preparations[22] is not relevant to amyloid growth.

In addition to areas of AD brain that contain large numbers of amyloid deposits (frontal, occipital, and parietal lobes of the cerebral cortex, and hippocampus), it is also important to use amyloid-free tissue such as non-AD cortex as a control. Although it is possible to use fixed tissue in deposition assays, unfixed snap-frozen material is preferred and yields far superior results. For tissue homogenate experiments, slow frozen material is also acceptable.

Tissue sections for autoradiography experiments (protocol 1 later) are prepared by sectioning frozen AD or non-AD tissue specimens using a cryostat microtome. Snap-frozen tissue blocks about 1–2 cm in cross section and containing areas of both gray and white matter are embedded in a cryostat media (e.g., Tissue-Tek) and are sectioned at about 10 to 25 μm. The sections are thaw-mounted on gelatin-subbed microscope slides and stored with desiccant at $-20°$. Tissue homogenates for deposition studies (protocol 2 later) are prepared using a modification[3] of a standard membrane receptor preparation.[23] For comparison of different tissue samples, values should be normalized to protein content in the membrane homogenate.

Synthetic templates for Aβ deposition assays (protocol 3 later) are prepared from pure (>95%) and well-characterized synthetic starting material. As formation of well-ordered fibrils of the peptide, rather than amorphous insoluble material, is critical for both the relevance of the experiment and the quality of the template, the method by which the aggregates are prepared is critical. The importance of clean, aggregate-free starting mate-

[18] Z. S. Khachaturian, *Arch. Neurol.* **42,** 1097 (1985).

[19] M. Gearing, S. S. Mirra, J. C. Hedreen, S. M. Sumi, L. A. Hansen, and A. Heyman, *Neurology* **45,** 461 (1995).

[20] D. Games, D. Adams, R. Alessrini, R. Barbour, P. Bathelette, C. Blackwell, A. Carr, J. Clemens, T. Donaldson, F. Gillepsie, T. Guido, S. Hagopian, K. Johnson-Wood, K. Khan, M. Lee, P. Leibowitz, I. Lieberburg, S. Little, E. Masliah, L. McConlogue, M. Montoya-Zavala, L. Mucke, L. Paganini, E. Penniman, M. Power, D. Schenk, P. Suebert, B. Snyder, F. Soriano, H. Tan, J. Vitale, S. Wadsworth, B. Wolozin, and J. Zhao, *Nature* **373,** 523 (1995).

[21] F. M. LaFerla, B. T. Tinkle, C. J. Bieberich, C. C. Hudenschild, and G. Jay, *Nature Genet.* **9,** 21 (1995).

[22] T. A. Good and R. M. Murphy, *Biochem. Biophys. Res. Commun.* **207,** 209 (1995).

[23] H. P. Too and M. R. Hanley, *Biochem. J.* **252,** 545 (1988).

rial cannot be overemphasized. Whereas other suppliers may produce acceptable material, we have found that Aβ(1–40) from the Biopolymers Laboratory at the Center for Neurologic Diseases, Harvard Medical School (Boston, MA) and from Quality Controlled Biochemicals (QCB, Hopkinton, MA) yields reproducibly good template preparations, presumably a result of the lack of preassembled oligomers in the starting material. As Aβ aggregates formed at different pH have different morphological properties and abilities to seed Aβ aggregation at physiological pH,[24] we recommend preparing synthetic amyloid templates at or around physiological pH. To minimize locally high Aβ concentrations that would result in incomplete dissolution, the buffer must be added rapidly. The resulting solution should be clear with no signs of incomplete dissolution or flocculation. Following initial mixing, the solution is incubated at room temperature with vigorous agitation. We prefer the largest magnetic stir bar that the bottom of the tube can accommodate spun at the highest sustainable speed. Although fibrils will form in the absence of agitation, we find that vigorous mixing during incubation results in the formation of many short fibrils rather than relatively few longer fibrils formed in the absence of agitation. As the Aβ deposition rate is almost certainly proportional to the number of individual fibrils rather than their length, aggregates prepared with vigorous agitation have considerably greater activity per milligram. Vigorous vortexing and vigorous stirring during incubation both produce acceptable results, but we find the latter yields preparations with somewhat greater activity. Following a 12- to 24-hr incubation, >95% of the Aβ(1–40) initially present typically has aggregated into a sedimentable form. Aggregates produced by the method appear as mats of short Aβ fibrils under electron microscopy. Aggregates typically have a mean diameter of 12 μm[25] that compares favorably with the diameter of plaque cores (5–30 μm) purified from AD brain.[26] Importantly, synthaloid aggregates injected into rat brain[25] produced cellular changes consistent with those observed around senile plaques in AD tissue.

As Aβ aggregates may be amorphous or well-ordered amyloid fibrils, it is appropriate to characterize preparations that are to be used in deposition assays to ensure fibrillar morphology. Transmission electron microscopy (EM)[13,27] provides the best method to determine if the aggregates in a given preparation are appropriate. Aβ amyloid fibrils have a characteristic

[24] S. J. Wood, L. MacKenzie, B. Maleeff, W. R. Hurle, and R. Wetzel, *J. Biol. Chem.* **271,** 4086 (1996).

[25] D. T. Weldon, S. D. Rogers, J. R. Ghilardi, M. P. Finke, J. P. Cleary, E. O'hare, W. P. Esler, J. E. Maggio, and P. W. Mantyh, *J. Neurosci.* **18,** 2161 (1998).

[26] D. J. Selkoe and C. R. Abraham, *Methods in Enzymol.* **134,** 388 (1986).

[27] D. M. Walsh, A. Lomakin, G. B. Benedek, M. M. Condron, and D. B. Teplow, *J. Biol. Chem.* **272,** 22364 (1997).

fibril diameter of 7 to 9 nm and fibril lengths of several hundred nanometer.[13] Tinctorial properties, such as Congo red birefringence, may also be examined. It is critical to examine the ability of the aggregates to serve as a template for deposition of the monomer from solution. This is accomplished easily by using a simplified system where immobilization of the aggregates (see later) will not play a role, e.g., centrifugation to separate bound from free. For typical fibrillar Aβ template preparations, about 10,000 disintegrations per minute (dpm) of [^{125}I] Aβ from a 100 p*M* solution will deposit onto 1 or 2 μg of aggregates in a few hours.

The significant advantage of the synthaloid assay for high-throughput experimentation is that centrifugation is not required to separate bound and free material. To achieve separation of the Aβ peptide bound to the template during the deposition assay from that remaining in solution, the templates are typically immobilized onto a surface (e.g., wells in a multiwell plate) using a polymer. For Aβ aggregates, we find that using 0.1% type B gelatin (Fisher or Baker) works well; other polymers, such as heparin and laminin, also produce satisfactory results. In developing an immobilization protocol, it is critical to examine the effect of the polymer on nonspecific ligand binding onto the plate and the ability of the polymer to immobilize the template stably. For the Aβ aggregates produced as described previously approximately half of the material added in the immobilization step is removed in the first few minutes of washing whereas the remainder stays stably immobilized for at least 2 days of incubation in aqueous buffers at room temperature.

To immobilize Aβ aggregates, we dilute the 10^{-4} *M* (total concentration) aggregated Aβ prepared as described into a warm (about 50°) 0.1% (1 mg/ml) gelatin solution to yield a final concentration ranging from 0.1 to 100 μg Aβ per ml (10 to 20 μg/ml typical). The Aβ aggregate suspension is then mixed thoroughly and is dispensed into the wells of a multiwell plate (e.g., 100 μl/well for a 96-well plate). As the Aβ aggregates are in suspension, it is critical to mix thoroughly (e.g., vortex) frequently during this process to ensure an even distribution of aggregates in each well. The plastic composition of the plate can have significant effects on the signal/noise ratio of binding assays (see later). It is also important to include control wells (e.g., polymer only, polymer and unaggregated peptide, or aggregated reverse sequence peptide) in each experiment. Another consideration is the means of quantitation, e.g., if wells must be counted in single tubes, rather than in an intact multiwell plate, a flexible polyvinyl chloride plate (Dynex Technologies, Chantilly, VA) is appropriate, with separation of wells accomplished easily with a pet nail clipper. After dispensing template, the plates are then heated in a warm (45–57°) incubator until dry (4–24 hr, depending on incubator and load). The same suspension can be used to immobilize fibrils onto gelatin-subbed glass microscope slides for use in autoradiography experiments. Dry plates can be stored in sealed plastic bags for at least 6 months at room temperature without loss of activity.

Selection of Method to Quantify Deposition

In deposition assays performed at physiological concentrations, the method used to quantify the amount of peptide bound to the template must be sensitive enough to accurately detect a relatively small addition of material relative to the amount of material already present in the template. To accurately quantify the amount of Aβ deposited under physiologically relevant conditions and physiological Aβ concentrations (the amount of peptide added onto a microgram of template in a few hours is on the order of a few fmols), we prefer radioisotopes to track the peptide. Whereas Aβ tracers using fluorophores, biotin, and other methods[28,29] have been prepared and adequately track Aβ at higher concentrations, the need to follow the peptide at physiological (n*M*) concentrations and below has encouraged the use of radioisotopes for Aβ deposition studies. Considerations of detection, synthesis, half-life, and purification of Aβ, as for the majority of peptide radioligands, have led to use of ^{125}I as the isotope of choice.

Preparation of Radiolabeled Aβ Tracer

Considerations for Radiolabeling

Radioiodine is incorporated easily into peptides with tyrosine residues by oxidative methods; in the case of Aβ, into Tyr.[10] Several studies[12,15,27,30,31] have established that[^{125}I]Aβ behaves indistinguishably from native Aβ in a variety of experiments, including those described here. General methods of peptide radioiodination have been described previously,[32,33] and the

[28] R. Prior, D. D'Urso, R. Frank, I. Prikulis, S. Cleven, R. Ihl, and G. Pavlakovic, *Am. J. Pathol.* **148,** 1749 (1996).

[29] W. Garzon-Rodriguez, M. Sepulveda-Becerra, S. Milton, and C. G. Glabe, *J. Biol. Chem.* **272,** 1037 (1997).

[30] W. P. Esler, E. R. Stimson, J. M. Jennings, J. R. Ghilardi, P. W. Mantyh, and J. E. Maggio, *J. Neurochem.* **66,** 723 (1996).

[31] B. P. Tseng, W. P. Esler, E. R. Stimson, J. R. Ghilardi, H. V. Vinters, P. W. Mantyh, and J. E. Maggio, *Biochemistry,* in press (1999).

[32] H. P. Too *et al., Meth. Neurosci.* **6,** 232 (1991).

[33] J. E. Maggio and P. W. Mantyh, *in* "Receptor Localization" (M. Ariano, ed.), p. 17. Wiley, New York, 1998.

specific application to amyloid peptides is highlighted here. As the process of radioiodination oxidizes the methionine residue at position 35 of Aβ to methionine sulfoxide, the peptide is returned to its native (thioether) form by subsequent chemical reduction as described later.[34] In order to prepare a tracer of high specific activity, the separation of radioiodinated from unlabeled Aβ by reversed-phase high-performance liquid chromatography (HPLC) is necessary.

In aqueous solution at physiological concentrations, Aβ does not show any significant aggregation over days,[1,31] but the radiochemical synthesis of labeled peptide is done at significantly higher concentrations, and care must be taken to avoid aggregation during the synthesis and purification of tracers. It is important to begin with a pure sample of Aβ free of preexisting "seeds" that could accelerate aggregation and to minimize handling throughout the procedure. The tendency of Aβ to aggregate, stick to surfaces, and change conformation as a function of solvent makes it a particularly challenging peptide to label effectively. Brief descriptions of the procedures described later have been published.[6,7,15] At the time of this writing, we have not found commercial preparations of [^{125}I]Aβ to be reproducibly reliable in deposition assays, but these products may improve in the future.

Considerations for Sample and Apparatus Preparation

All procedures potentially involving volatile radioiodine must be performed by experienced personnel in an approved iodination hood equipped with a charcoal filter in an approved facility. Considerations of low-reaction volumes, low Aβ solubility, storage, and convenience suggest that dry samples of Aβ are most appropriate for iodination; the use of frozen aliquots may risk aggregation during freeze-thaw and conformational changes induced by solvent changeover. In consideration of purification of labeled peptide after radioiodination, it is necessary to choose an HPLC column and solvent system that can adequately separate the components of the reaction (including unlabeled peptide, mono- and di-iodinated peptide, and their sulfoxides), as validated with components generated from nonradioactive iodine, ^{127}I. Typically a shallow gradient of organic solvent is required, as the separation of some forms of the peptide may be less than 1% acetonitrile on a typical TFA/H_2O/CH_3CN system.

[34] R. A. Houghten and C. H. Li, *Anal. Biochem.* **98,** 36 (1979).

Considerations for Labeling and Purification Procedures

Because Aβ can adsorb to surfaces, all glassware and plasticware should be checked for adsorption. Polypropylene is usually much better than polystyrene. For similar reasons, solid-phase oxidative iodination reagents seem less effective than soluble chloramine-T for labeling Aβ. Excess oxidizing agent or reaction time causes oxidative damage and decreases yield.

After iodination, it is necessary to separate unincorporated ^{125}I from the mixture because HI is volatile and increases the risk of radioisotope exposure by inhalation. Removing peptides from unincorporated ^{125}I is conveniently done by use of a reversed-phase cartridge (C_{18} Sep-Pak or equivalent), activated before loading by wetting with water-miscible organic solvent (e.g., alcohol; see later) and then washing with aqueous phase (e.g., 10 m*M* TFA; see later). Time is a variable that increases amyloid accumulation not only in the brain but also in reaction vessels. It is important to minimize the time reaction mixtures stand when the system is thermodynamically unstable.

Procedure for Radiolabeling

Typically, 5 nmol of dry monomeric peptide in a 12 × 75-mm polypropylene tube is dissolved completely in 50 μl of 0.5 *M* phosphate buffer, pH 7.5 (0.5 *M* borate buffer at pH 8.0 works equally well). The solution is mixed gently with 1 mCi of Na ^{125}I (10 μl of Amersham IMS-30 or other relatively concentrated carrier-free product). Chloramine-T (10 μg freshly dissolved in 10 μl of distilled water) is added and mixed to distribute the oxidant evenly throughout the mixture. The reaction is allowed to proceed for 20 to 90 sec (time is adjusted to accommodate the individual peptide analog; peptides vary in activity and propensity to oxidative damage). The reaction is quenched with 100 μg of sodium metabisulfite ($Na_2S_2O_5$) in 20 μl of distilled water and is then acidified and diluted with aqueous TFA solution (500 μl of 0.010 *M* and 25 μl of 1 *M*).

Bovine serum albumin [BSA 25 μl of a 2% (w/v) solution] is then added for efficient transfer to the reversed-phase cartridge for the separation of free iodine, salt, and so on from the peptide. The cartridge is then eluted in a stepwise gradient consisting of 0.5 ml each of the 10, 10, 20, 40, 50, 80, 90, 95, and 100% alcohol mixtures. [These are the percent of a 1:1 (v/v) mixture of absolute methanol and ethanol with the 10 m*M* aqueous TFA solution.] During elution of the cartridge, low flow rates (0.5 ml/min) are used. Each fraction is collected separately; the peptides elute in fractions 6, 7, and 8, whereas free iodine elutes in the first fractions. At this point,

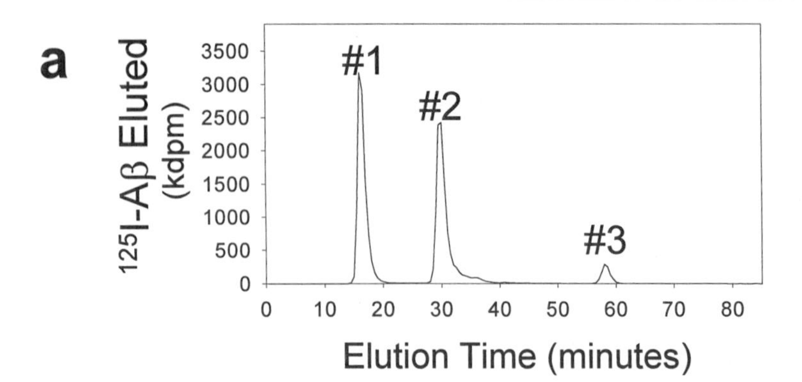
a
125I-Aβ Eluted (kdpm)
3500
3000
2500
2000
1500
1000
500
0
#1
#2
#3
0 10 20 30 40 50 60 70 80
Elution Time (minutes)

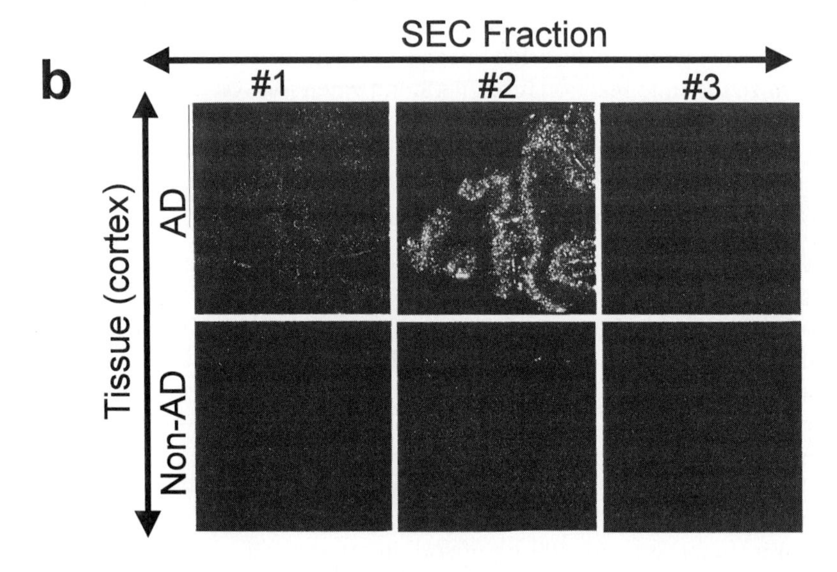
b
SEC Fraction
#1
#2
#3
Tissue (cortex)
AD
Non-AD

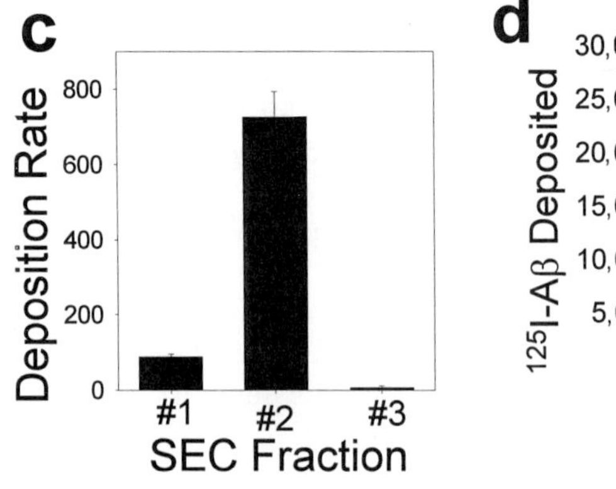
c
Deposition Rate
800
600
400
200
0
#1
#2
#3
SEC Fraction

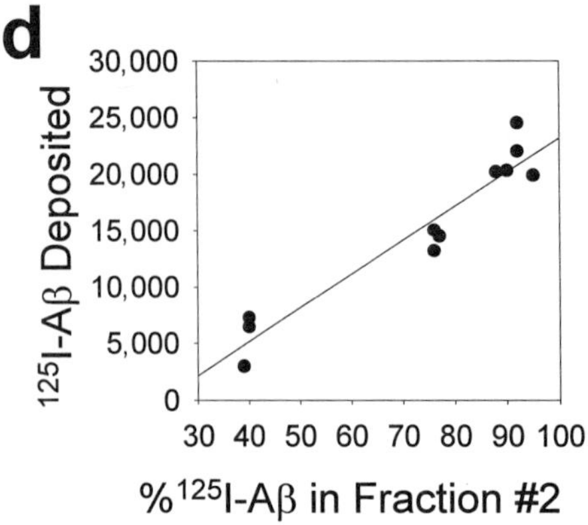
d
125I-Aβ Deposited
30,000
25,000
20,000
15,000
10,000
5,000
0
30 40 50 60 70 80 90 100
%125I-Aβ in Fraction #2

all fractions, reaction vessel, and the cartridge may be counted in a gamma counter to determine the distribution of isotope. Peptide fractions are collected and the solution is azeotroped (to remove the alcohol) under a gentle flow of nitrogen (this must be done in an iodination hood; volatile ^{125}I may still be present). When the alcohol has been removed, neat 2-mercaptoethanol is added to a final concentration of 20% for reduction (see Houghten and Li[34] for relative rates, optimal temperatures, concentrations, pH, peptide stability, etc.). Reduction is necessary only for Aβ peptides or analogs containing methionine.

The tightly sealed tube containing thiol and ^{125}I is placed in a 90° heating block in the radioiodination hood. The integrity of the vessel at this step, under positive pressure, is critical; a popped tube could scatter dangerous quantities of radioiodine. After the reduction has proceeded for 90 min (it probably will not be complete at this point), the hot radioactive solution is drawn into a 1-ml syringe containing 10 mM TFA and is injected into the HPLC, and the chromatographic run is initiated. [Vydac C_8 and C_{18} columns with acetonitrile at room temperature work well with many Aβ peptides; however, for Aβ(1–42) or Aβ(1–43), a Zorbax 300 column and isopropanol at an elevated temperature work better.] The order of elution for most peptides using most columns and solvents is the iodinated fragments from oxidative damage, cold fragments, oxidized cold peptide, oxidized iodinated peptide, cold reduced peptide, and iodinated reduced peptide. Peptides with modified side chains may be scattered throughout the chromatogram as minor impurities. Fractions are collected, and 2-mercaptoethanol (0.5 to 1% final concentration) is added to the fraction(s) with the desired product to prevent the air oxidation of methionine.

The purified tracer, now approximately 2000 Ci/mmol, is stored in its elution buffer at −20° until used. For best results, the tracer should be used within a month, but the working half-life is highly dependent on the type of experiment in which it is applied (Fig. 5). For example, immunoassay measurements are more sensitive to decay than deposition experiments.

FIG. 5. Characterization of [^{125}I]Aβ by size-exclusion chromatography. (a) SEC of a poorly prepared commercial batch of ligand showing unassembled (peak 2) as well as assembled or degraded material [peak 1 (void volume) and peak 3]. Only the unassembled material (peak 2) is active in deposition onto tissue-derived (b) or synthetic (c) amyloid. The activity of several [^{125}I]Aβ preparations correlates well (r^2 = 0.91) with their content of unassembled (peak 2) peptide (d).

Quality Control for Radiolabeled Aβ

If prepared, purified, or stored improperly, $[^{125}I]A\beta$ may be oxidized, assembled into high molecular weight species, or degraded. It is therefore advisable to test each batch of $[^{125}I]A\beta$ synthesized or obtained from commercial sources to determine if the material is suitable for quantitative experiments. Two relatively quick and simple experiments that are sensitive to the quality of radiolabeled Aβ are size-exclusion chromatography (SEC)[27,31] and radioimmunoassay (RIA).[27,30] SEC (Fig. 5) is useful to determine if assembled or degraded Aβ is present in a stock of tracer. Properly prepared$[^{125}I]A\beta$ elutes from SEC as a single low molecular weight peak.[27,31] Poorly prepared Aβ tracer from a commercial source (Fig. 5a) or an extensively aged (>8 weeks) tracer,[31] however, displays multiple SEC peaks that correspond to assembled or degraded material (peaks 1 and 3) as well as unassembled Aβ peptide (peak 2). Although such multicomponent tracers may appear "active" in deposition assays, the only real active component is the unassembled material (Figs. 5b and 5c). As the deposition rate is proportional to the fraction of the material that elutes as the unassembled peptide (Fig. 5d), it is critical that radiolabeled Aβ used in quantitative experiments contains little or no assembled or degraded peptide. Anti-Aβ RIA[30] is quite sensitive to the presence of assembled or degraded material in a tracer preparation. When a properly prepared tracer is fresh, anti-Aβ titer curves[30,35] show maximum antibody binding near 100% and minimum binding close to zero, whereas poor or aged preparations show much less specific binding to antibody.

Performing Deposition Assays

In developing a deposition assay, several factors must be considered. The most important of these is a method to quantify the amount of depositing protein bound to the template and the specificity of this association. The most straightforward method to quantify deposition is to monitor the amount of radiolabeled protein added onto the template with time. To quantify the amount of the radiolabeled protein deposited readily and accurately, a convenient method to separate bound and free tracer must be developed. Four ways to achieve this for deposition assays using radioligands are filtration, centrifugation (sedimentation), immobilization, and proximity-based assays. As is true for all quantitative radioligand-binding

[35] A. Tamaoka, R. N. Kalaria, I. Lieberburg, and D. J. Selkoe, *Proc. Natl. Acad. Sci. U.S.A.* **89,** 1345 (1992).

assays, the means used to separate bound and free radioligand must be quick relative to the dissociation rate. In general, immobilization-based and filtration-based assays allow shorter wash times relative to centrifugation assays. A proximity-based assay (such as scintillation proximity) eliminates the need for washing as only the bound material is in close enough proximity to the scintillant to produce a signal. We have used filtration-based[7] and centrifugation-based[3] strategies for tissue homogenate experiments and template immobilization for tissue section autoradiography[6] and synthetic amyloid-based experiments (both microplate and autoradiography).[10]

Specificity is also important in deposition assays, so at least one negative control should be included in every experiment. Amyloid-free tissue preparations should be included as control templates for tissue-based experiments whereas aggregates prepared from all D-Aβ or reverse sequence Aβ should be used as a control template for synthetic amyloid-based assays. It is equally important to include inactive depositing peptides as tracers. For Aβ deposition assays, background binding to tissue in the absence of template should be less than 5%. If a lower signal/noise ratio is observed, the quality of the depositing peptide tracer or the template (see earlier discussion) should be examined. Increasing the amount of template may improve the signal/noise ratio of an assay, but it is often the case that high background binding in the absence of template is responsible for the poor signal to noise ratio. To minimize nonspecific Aβ binding to plastic, a blocking protein (e.g., BSA) should be present in all deposition experiments. In some cases, it may be necessary to pretreat surfaces with buffer containing BSA prior to the addition of the Aβ. We find that polyvinyl chloride (PVC) assay plates are excellent for minimizing nonspecific losses of Aβ, although other plastics can be acceptable.

Tissue Amyloid Deposition Experiments

Homogenates of tissue with a significant amyloid burden can be used for rapid and reproducible quantitative measurements. For [^{125}I]Aβ deposition experiments using brain homogenates, the method described[3] (protocol 2) is recommended. Deposition rates (Fig. 3c) can be determined from the slopes of lines fitted to points of deposition versus time or single point assays can be used for screening.

Although homogenate-based experiments are useful for quantitative measurements of deposition rate, tissue section autoradiography experiments (protocol 1) are useful for examining the sites of radioligand binding. To conserve the amount of radiolabeled peptide, the sections are placed horizontally on a rack and a small volume (1 ml) of the radioligand solution is placed directly on top of the section. To prevent unwanted spreading of the radioligand solution, the edges and area of the glass slide surrounding the section should be carefully wiped dry with a tissue. Care should be taken to avoid touching the section and to avoid desiccation. To minimize evaporation of the radioligand solution during incubation, the rack holding the sections should be placed (prior to incubation) in a humid environment. Following incubation the slides are washed and then dried overnight in a sealed slide box containing desiccant. The slides can then be placed against X-ray film and exposed in a light-tight cassette. As amyloid deposits are typically small ($\approx$20 to 40 μm) and focal, the use of intensifying screens is undesirable.

Autoradiograms can be produced from the film negative using conventional darkroom techniques or imaging (e.g., with a CCD camera attached to a dissecting microscope). Alternatively, films can be scanned using a photographic negative scanner (e.g., Polaroid Sprintscan or Hewlett Packard Photosmart). For quantitative experiments, the density of radioligand bound to tissue amyloid in sections can be determined quantitatively using microdensitometry.[3,9] Following the deposition experiment, the tissue sections can be stained using standard histological techniques. Film autoradiograms (Fig. 3a) lack the resolution of micrographs. Higher resolution autoradiograms (silver halide emulsions rather than film) are used[3,9] for the optimal comparison of deposition sites with histological staining and to confirm radioligand deposition onto individual amyloid plaques (Fig. 3a)

Synthetic Amyloid Deposition Experiments

Synthetic amyloid-based assays (Fig. 2b) are a straightforward system for the study of template-mediated amyloid assembly. The use of immobilized templates allows rapid washing and thus allows the study of earlier or later events than can be studied conveniently with tissue-based systems and also provides a model system for high-throughput experiments. In general, synthetic amyloid-based assays are performed by incubating the labeled amyloidogenic monomer with a preformed template. Following incubation, the bound and free ligand are readily separable by washing, without centrifugation or filtration. Because tissue preparations and biological fluids are not involved, the protease inhibitors used in the tissue-based assays are not necessary.

For Aβ deposition assays using synthetic amyloid, the template is first rinsed (preincubated) with buffer containing blocking protein (50 m*M* Tris-HCl, 0.1% BSA, pH 7.5) to remove any fibrils not immobilized securely. Later the preincubation buffer is removed and radiolabeled Aβ (10 to 900 p*M*) is added to each well in the presence or absence of test compounds. Following incubation (typically 0.5–4 hr), the Aβ solution is removed and each well is washed with buffer, and the amount of [^{125}I]Aβ deposited is quantified by γ counting. The amount of [^{125}I]Aβ bound to identically prepared wells without template is less, usually much less, than 5% of [^{125}I]Aβ bound to Aβ aggregate/polymer-treated wells and can be subtracted as background. Synthetic amyloid templates can also be used in autoradiography experiments (protocol 4).

Identifying Inhibitors of Amyloid Fibril Extension

Current anti-Aβ aggregation screens may be useful in identifying inhibitors of the earlier stages of Aβ amyloid formation. However, because there appears to be no reliable correlation between the ability of a compound to inhibit nascent amyloid formation and its ability to inhibit amyloid growth,[10] compounds identified in aggregation assays may be of little use in halting further Aβ deposition once an amyloid template is present. Because AD patients typically have an abundance of brain amyloid, which appears capable of supporting further Aβ deposition by an aggregation-independent mechanism, at the time of diagnosis, compounds that inhibit the initial stages of amyloid formation would not, *a priori,* be expected to have any effect on the growth of existing amyloid deposits. Thus, Aβ nucleation inhibitors are likely to be more useful prophylactically than therapeutically. Conversely, because the degree of dementia may be correlated more closely with the density of mature plaques rather than with total amyloid burden, inhibiting Aβ deposition postdiagnosis may prove useful therapeutically in slowing the progression of AD.

Protocols for Aβ Deposition Experiments

Protocol 1. Deposition Autoradiography Using AD Tissue Sections

Materials

AD cortex sections: Slide mounted sections of unfixed snap-frozen AD cortex

Non-AD cortex sections: Slide mounted sections of unfixed snap-frozen non-AD cortex (control)

Deposition buffer: 50 m*M* Tris-HCl, pH 7.5, at room temperature containing 0.1% (w/v) BSA, 0.004% bacitracin, 0.0002% chymostatin, 0.0002% leupeptin

Wash buffer: 50 m*M* Tris-HCl, pH 7.5, at room temperature

Radiolabeled Aβ: [^{125}I]Aβ (prepared as described in text or obtained from commercial sources)

Method

1. Place sections horizontally on a rack over wet paper towels or a tray containing distilled water. Cover each section with about 1 ml of the deposition buffer and incubate at room temperature for 30 min.
2. Dilute the stock radiolabeled Aβ into deposition buffer to yield the desired peptide concentration (30–1000 p*M*, typically 100 p*M*).
3. Drain the buffer from the sections and dry the edges of the slide with a tissue. Place about 1 ml of the diluted radiolabeled Aβ solution onto the section.
4. After the desired incubation time, pour or aspirate the radiolabeled Aβ solution from the section and wash (3×3 min) by submersing in wash buffer. After washing, dip the section twice into a beaker of distilled water. Dry the slides and place in a slide box containing desiccant to dry overnight.
5. Place the slides section-side down against a piece of X-ray film (Kodak, Rochester, NY X-Omat AR or substitute) in an autoradiography cassette. After the desired exposure (typically 3 to 10 days), develop and fix the film according to the manufacturer's instructions. If the sections are to be used for histology after the deposition experiment, expose the film at −20°. For higher resolution images, emulsion autoradiography[7] should be used.

Protocol 2. Deposition onto Brain Homogenate

Preparation of Homogenates

Materials

Cortex tissue: 2–20 of autopsy-confirmed AD cortex enriched by dissection for gray matter

Homogenization buffer: 50 m*M* Tris–HCl pH 7.5, at 4° containing 10%, (w/v) sucrose, 0.01% bacitracin, 0.002% soybean trypsin inhibitor, 0.002% chick egg trypsin inhibitor, 1 m*M* benzamidine

Resuspension buffer: 50 m*M* Tris–HCl, pH 7.5, at 4°

Method

1. Thaw AD cortex in 5 volumes (w/v) of homogenization buffer. Homogenize tissue with Polytron (5 × 10 sec on setting 6) and centrifuge homogenate at 10,000*g* for 20 min at 4°.
2. Discard supernatant (S1) and resuspend pellet (P1) in 10 volumes (initial w/v) using Polytron (5 × 10 sec on setting 6). Centrifuge at 40,000*g* for 15 min.
3. Discard supernatant (S2) and resuspend pellet (P2) in 10 volumes (initial w/v) using Polytron (5 × 10 sec on setting 6). Centrifuge at 40,000*g* for 15 min.
4. Discard supernatant (S3) and resuspend pellet (P3) in 1 volume (initial w/v). Aliquot (1 ml or 0.5 g equivalent weight of tissue) to microcentrifuge tubes and centrifuge at 15,000*g* for 30 min at 4°. Discard supernatants (S4) and freeze pellets on dry ice. Store at −20° until use (<4 months).

Deposition Assay

Materials

AD tissue homogenates:	Prepared as described earlier
Deposition buffer:	50 m*M* Tris-HCl, pH 7.5, at room temperature containing 0.1 % (w/v) BSA, 0.004% bacitracin, 0.0002% chymostatin, 0.0002% leupeptin
Wash buffer:	50 m*M* Tris-HCl, pH 7.5, at room temperature
Radiolabeled Aβ:	[^{125}I]Aβ (prepared as described in text or obtained from commercial sources)

Method

1. Add deposition buffer to frozen homogenates (0.5 g equivalent weight of tissue) in the microcentrifuge tube (prepared earlier) to bring the total volume up to 1 ml. Resuspend the pellet and incubate at room temperature for 30 min. If more than one microcentrifuge tube of tissue is used in the experiment, pool the homogenates in a larger container and mix thoroughly.
2. Dilute the stock radiolabeled Aβ into deposition buffer to yield the desired peptide concentration (30–1000 p*M*, typically 100 p*M* final).
3. Aliquot tracer solution to a 96-well plate (150 μl per well). At time zero, add 20 μl of well-mixed tissue homogenate suspension to each well and mix thoroughly. For control wells, 20 μl of buffer or prepara-

tions of non-AD cortex or cerebellum should be substituted for the AD cortex homogenate.

4. After the desired incubation time (typically 3 hr), the plates are centrifuged (600*g* × 6 min) and the supernatant is discarded. The tissue pellets are washed twice by resuspension/centrifugation in wash buffer.
5. The amount of radioactivity bound to the tissue in the wells is quantified by gamma counting.

Protocol 3. Deposition onto Synthetic Amyloid (Synthaloid) Plates

Preparation of Synthetic Amyloid Plates

Materials

Aβ peptide:	5 mg of synthetic Aβ(1–40) (HPLC purified, >95%, lyophilized), enough to prepare about thirty 96-well plates
Aggregation buffer:	10 m*M* Na_2HPO_4/NaH_2PO_4, 100 m*M* NaCl, pH 7.5, 0.02% NaN_3
Immobilization polymer:	Gelatin (type B) powder

Method

1. Prepare aggregates: Transfer the peptide to a 15-ml round-bottom polypropylene tube. Add 12 ml of aggregation buffer rapidly and mix thoroughly. It is critical that the resulting solution be clear with no signs of incomplete dissolution or flocculation. Transfer one-half of the Aβ solution to a second round-bottom tube. Incubate the solution for about 24 hr at room temperature under vigorous agitation (magnetic stir plate). After incubation, the particulate should be readily visible in the tubes. Pool the material and store at −20° until use. Characterize the aggregates as described in the text.
2. Prepare plates: Heat two 50-ml conical polypropylene tubes each containing 40 ml of distilled water in a 57° incubator. To each tube add 42 mg of gelatin. To one of the tubes, add 2 ml of the aggregate suspension prepared as described earlier. Vortex the aggregate stock suspension thoroughly before diluting. To the second tube containing gelatin, add 2 ml of aggregation buffer without Aβ. Aliquot the aggregate polymer solution (100 μl per well) into 6 of the 8 wells per row in 96-well assay plates. Agitate the tube containing the suspension often while aliquoting to prevent aggregates from settling. In the other 2 wells per row, aliquot 100 μl of the polymer solution

containing no Aβ aggregates; these wells will serve as negative controls and for the determination of background binding. Dry the plates at 45–57° until the liquid has evaporated (4–12 hr). Store the plates in a dry environment until use.

Deposition Assay Protocol

Materials

Synthetic amyloid plate:	Prepared as described earlier or obtained from commercial sources (QCB, Hopkinton, MA)
Deposition buffer:	50 m*M* Tris-HCl, pH 7.5, at room temperature containing 0.1% (w/v) BSA
Wash buffer:	50 m*M* Tris-HCl, pH 7.5, at room temperature
Radiolabeled Aβ:	[^{125}I]Aβ (prepared as described or obtained from commercial sources)

Method

1. Add 200 μl of deposition buffer to each well in the synthetic amyloid plate to wash off unattached Aβ aggregates.
2. Dilute the stock radiolabeled Aβ into deposition buffer to yield the desired peptide concentration (30–1000 p*M*, typically 100 p*M*).
3. After about 30 min, remove the buffer from the wells and discard. At time zero, add 150 μl of the diluted tracer solution prepared in step 2 to each of the wells.
4. After the desired incubation time (typically 3 hr), remove the tracer solution from the wells and discard. Rinse the wells with the wash buffer (3 × 2 min, 200 μl per well). Quantify the amount of radiolabeled Aβ deposited by gamma counting the washed empty wells.

Protocol 4. Deposition Autoradiography Using Synthetic Amyloid Slides

Preparation of Synthetic Amyloid Slides

Materials

Aβ aggregates:	Prepared as described earlier (protocol 3)
Immobilization polymer:	Gelatin (type B) powder

Method

1. Heat two 15-ml conical polypropylene tubes containing 5 ml of distilled water to 50°. To each tube, add 5 mg of gelatin and swirl to dissolve.
2. To one tube, add 125 μl of the aggregate suspension prepared earlier. Vortex the aggregate stock suspension well before diluting.
3. Place a 50-μl drop of the aggregate gelatin suspension onto a gelatin-subbed slide. About 3 separate drops will fit on each slide. Agitate the tube containing the suspension often to avoid settling.
4. To the second tube containing gelatin, add 125 μl of aggregation buffer (protocol 3). Using this solution, prepare polymer-treated slides that contain no Aβ aggregates. These slides will serve as negative controls and for determining background binding. Dry the slides in a 50° incubator until all of the liquid has evaporated (1–2 hr). Store the slides in a dry environment until use.

Deposition Assay Protocol

Materials

Synthetic amyloid slides: Prepared as described earlier or obtained from commercial sources (QCB), Hopkinton, MA)
Deposition buffer: 50 m*M* Tris-HCl, pH 7.5, at room temperature containing 0.1% (w/v) BSA
Wash buffer: 50 m*M* Tris-HCl, pH 7.5, at room temperature
Radiolabeled Aβ: $[^{125}I]$Aβ (prepared as described in text or obtained from commercial sources)

Method. Steps 1 through 5 as in protocol 1.

Note: For all these protocols, compounds to be tested for effects on Aβ deposition rate (candidate accelerants or inhibitors) may be added to preincubations, incubations, or both.[10] The assays described here are quite tolerant of organic solvents, e.g., all perform well in 10% (v/v) dimethyl sulfoxide.

Acknowledgments

We thank D. Selkoe for the gift of anti-Aβ antiserum and H. Vinters for the gift of human brain tissue. Work in the authors' laboratories has been supported by the National Institutes of Health, the American Health Assistance Foundation, and the Veterans' Administration.

[24] Membrane Filter Assay for Detection of Amyloid-like Polyglutamine-Containing Protein Aggregates

By ERICH E. WANKER, EBERHARD SCHERZINGER, VOLKER HEISER, ANNIE SITTLER, HOLGER EICKHOFF, and HANS LEHRACH

Introduction

The accumulation of polyglutamine-containing protein aggregates in neuronal intranuclear inclusions (NIIs) has been demonstrated for several progressive neurodegenerative diseases such as Huntington's disease (HD),[1,2] dentatorubral pallidoluysian atrophy (DRPLA),[2,3] and spinocerebellar ataxia (SCA) types 1,[4,5] 3,[6] and 7.[7] Furthermore, it has been shown *in vitro* that the proteolytic cleavage of fusion proteins of glutathione *S*-transferase (GST) and the polyglutamine-containing huntingtin peptide coded for by the first exon of the HD gene[8] leads to the formation of insoluble high molecular weight protein aggregates with a fibrillar or ribbon-like morphology[9] reminiscent of β-amyloid fibrils in Alzheimer's disease and scrapie prion rods.[10,11]

Insoluble, ordered protein aggregates (amyloids) are commonly ana-

[1] M. DiFiglia, E. Sapp, K. O. Chase, S. W. Davies, G. P. Bates, J. P. Vonsattel, and N. Aronin, *Science* **277,** 1990 (1997).

[2] M. W. Becher, J. A. Kotzuk, A. H. Sharp, S. W. Davies, G. P. Bates, D. L. Price, and C. A. Ross, *Neurobiol. Dis.* **4,** 387 (1998).

[3] S. Igarashi, R. Koide, T. Shimohata, M. Yamada, Y. Hayashi, H. Takano, H. Date, M. Oayke, T. Sato, A. Sato, S. Egawa, T. Ikeuchi, H. Tanaka, R. Nakano, K. Tanaka, I. Hozumi, T. Inuzuka, H. Takahashi, and S. Tsuji, *Nature Genet.* **18,** 11 (1998).

[4] P. J. Skinner, B. T. Koshy, C. J. Cummings, I. A. Klement, K. Helin, A. Servadio, H. Y. Zoghbi, and H. T. Orr, *Nature* **389,** 971 (1997).

[5] A. Matilla, B. T. Koshy, C. J. Cummings, T. Isobe, H. T. Orr, and H. Y. Zoghbi, *Nature* **389,** 974 (1997).

[6] H. L. Paulson, M. K. Perez, Y. Trottier, J. Q. Trojanowski, S. H. Subramony, S. S. Das, P. Vig, J.-L. Mandel, K. H. Fischbeck, and R. N. Pittman, *Neuron* **19,** 333 (1997).

[7] M. Holmberg, C. Duyckaerts, A. Durr, G. Cancel, I. Gourfinkel-An, P. Damier, B. Faucheux, Y. Trottier, E. C. Hirsch, Y. Agid, and A. Brice, *Hum. Mol. Genet.* **7,** 913 (1998).

[8] HDCRG, *Cell* **72,** 971 (1993).

[9] E. Scherzinger, R. Lurz, M. Turmaine, L. Mangiarini, B. Hollenbach, R. Hasenbank, G. P. Bates, S. W. Davies, H. Lehrach, and E. E. Wanker, *Cell* **90,** 549 (1997).

[10] C. B. Caputo, P. E. Fraser, I. E. Sobel, and D. A. Krischner, *Arch. Biochem. Biophys.* **292,** 199 (1992).

[11] S. B. Prusiner, M. P. Mc Kinley, K. A. Bowman, D. C. Bolton, P. E. Bendheim, D. F. Groth, and G. G. Glenner, *Cell* **35,** 349 (1983).

METHODS IN ENZYMOLOGY, VOL. 309
0076-6879/99 $30.00

lyzed by electron microscopy (EM),[9,12] sedimentation,[13] fluorescence staining,[14] turbidity,[15] or quasi-elastic light scattering (QLS).[16,17] All these existing methods have complementary strengths and weaknesses. EM, for instance, is useful for the identification of amyloid fibrils and determining their morphology, but this technique is not suitable for the quantification of amyloids because of the clumping of fibrils.[9] Sedimentation has been used to separate insoluble amyloid fibrils from soluble monomeric protein. However, this method does not distinguish ordered amyloid fibrils from insoluble amorphous aggregates.[13,18] The thioflavin fluorescence assay has been widely used to examine the formation of Aβ amyloids *in vitro,*[14,19] but a general application of this assay for the detection of other amyloid-like aggregates such as huntingtin fibrils has not been demonstrated. This is because not all ordered protein aggregates bind thioflavin. Both turbidity[15] and QLS[16,17] measurements have been used successfully for the quantification of amyloid fibrils formed *in vitro*. However, they are not specific for amyloids. Also, relatively large amounts of recombinant proteins or peptides are needed for the quantification of fibrillar structures with these methods.

For the detection and quantification of small amounts of polyglutamine-containing protein aggregates we have developed a rapid and sensitive filter retardation assay, which should be suitable for high-throughput screenings of drugs that prevent aggregate formation. This assay is based on the finding that the polyglutamine-containing protein aggregates are insoluble in sodium dodecyl sulfate (SDS) and are retained on a cellulose acetate filter, whereas the monomeric forms of the HD exon 1 protein do not bind to this filter membrane. The captured aggregates are then detected by simple immunoblot analysis using specific antibodies. This article describes in detail the use of the filter retardation assay for the identification and quantification of huntingtin protein aggregates formed *in vitro* and *in vivo* in a transient expression system using COS cells.

[12] P. T. Lansbury Jr., *Biochemistry* **31,** 6865 (1992).

[13] B. Soreghan, J. Kosmoski, and C. Glake, *J. Biol. Chem.* **269,** 28551 (1994).

[14] H. Naiki, K. Higuchi, M. Hosokawa, and T. Takeda, *Anal. Biochem.* **177,** 244 (1989).

[15] J. T. Jarrett, E. P. Berger, and P. T. Lansbury, Jr., *Biochemistry* **32,** 4693 (1993).

[16] A. Lomakin, D. S. Chung, G. B. Benedek, D. A. Kirschner, and D. B. Teplow, *Proc. Natl. Acad. Sci. U.S.A.* **93,** 1125 (1996).

[17] Y. Georgalis, E. B. Starikov, B. Hollenbach, R. Lurz, E. Scherzinger, W. Saenger, H. Lehrach, and E. E. Wanker, *Proc. Natl. Acad. Sci. U.S.A.* **95,** 6118 (1998).

[18] J. Maggio, W. P. Esler, E. R. Stimson, J. M. Jennings, J. R. Ghilardi, and P. W. Mantyh, *Science* **268,** 1920 (1995).

[19] H. I. LeVine, *Prot. Sci.* **2,** 404 (1993).

Construction of Plasmids

Standard protocols for DNA manipulations are followed.[20] *IT-15* cDNA sequences[8] encoding the N-terminal portion of huntingtin, including the CAG repeats, are amplified by polymerase chain reactions (PCR) using the oligonucleotides ES25 (5′-TGGGATCCGCATGGCGACCCTGG AAAAGCTGATGAAGG-3′) and ES27 (3′-CTCCTCGAGCGGCGG TGGCGGCTGTTGCTGCTGCTGCTG-5′) as primers and the plasmids pCAG20 and pCAG51 as template.[9] Conditions for PCR are as described.[21] The resulting cDNA fragments are gel purified, digested with *Bam*HI and *Xho*I, and inserted into the *Bam*HI–*Xho*I site of the expression vector pGEX-5X-1 (Pharmacia, Piscataway, NJ), yielding pCAG20ΔP and pCAG51ΔP, respectively. The plasmids pCAG20ΔP-Bio and pCAG51ΔP-Bio are generated by subcloning the PCR fragments obtained from the plasmids pCAG20 and pCAG51 into pGEX-5X-1-Bio. pGEX-5X-1-Bio is created by ligation of the oligonucleotides BIO1 (5′-CGCTCGAGGG TATCTTCGAGGCCC AGAAGATCGAGTGGCGATCACCATGAG C-3′) and BIO2 (5′-GGCCGCTCATGGTGATCGCCACTCGATCTTCT GGGCCTCGAAGATACCCTCGAGCG-3′), after annealing and digestion with *Xho*I, into the *Xho*I–*Not*I site of pGEX-5X-1. Plasmids with the *IT-15* cDNA inserts are sequenced to confirm that no errors have been introduced by PCR. The construction of plasmids pTL1-CAG20, pTL1-CAG51, and pTL1-CAG93 for the expression of huntingtin exon 1 proteins containing 20, 51, and 93 glutamines in mammalian cells has been described.[22]

Structure of GST-HD Fusion Proteins

The amino acid sequence of the GST-HD fusion proteins encoded by the *Escherichia coli* expression plasmids pCAG20ΔP, pCAG51ΔP, pCAG20ΔP-Bio, and pCAG51ΔP-Bio is shown in Fig. 1. The plasmids pCAG20ΔP and pCAG51ΔP encode fusion proteins of GST and the N-terminal portion of huntingtin containing 20 (GST-HD20ΔP) and 51 (-HD51ΔP) polyglutamines, respectively. In these proteins the proline-rich region located immediately downstream of the glutamine repeat is deleted.[9]

[20] J. Sambrook, E. F. Fritsch, and T. Maniatis, *in* "Molecular Cloning: A Laboratory Manual," 2nd Ed. Cold Spring Harbor Laboratory Press, Plainview, NY, 1989.

[21] L. Mangiarini, K. Sathasivam, M. Seller, B. Cozens, A. Harper, C. Hetherington, M. Lawton, Y. Trottier, H. Lehrach, S. W. Davies, and G. P. Bates, *Cell* **87,** 493 (1996).

[22] A. Sittler, S. Wälter, N. Wedemeyer, R. Hasenbank, E. Scherzinger, G. P. Bates, H. Lehrach, and E. E. Wanker, *Mol. Cell* **2,** 427 (1998).

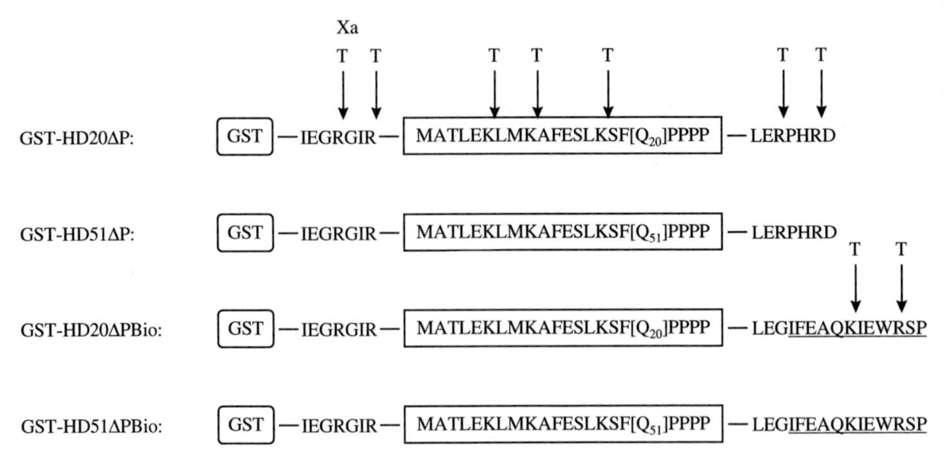

FIG. 1. Structure of GST-HD fusion proteins. The amino acids sequence corresponding to the N-terminal portion of huntingtin is boxed and the amino acids corresponding to the biotinylation site are underlined. Arrows labeled (Xa) and (T) indicate cleavage sites for factor Xa and trypsin, respectively.

The fusion proteins GST-HD20ΔPBio and -HD51ΔPBio are identical to GST-HD20ΔP and -HD51ΔP, except for the presence of a biotinylation site[23] at their C termini.

Strains and Media

Escherichia coli DH10B (BRL) is used for plasmid construction, and *E. coli* SCS1 (Stratagene) is used for the expression of GST-HD fusion proteins. Transformation of *E. coli* with plasmids and ligation mixtures is performed by electroporation using a Bio-Rad Gene Pulser (Richmond, CA). Transformed cells are spread on LB plates supplemented with appropriate antibiotics.[20] For expression of GST fusion proteins, cells are grown in liquid TY medium (5 g NaCl, 5 g yeast extract, and 10 g tryptone per liter) buffered with 20 m*M* MOPS/KOH (pH 7.9) and supplemented with glucose (0.2%), thiamine (20 μg/ml), and ampicillin (100 μg/ml). COS-1 cells are grown in Dulbecco's modified Eagle's medium (GIBCO-BRL) supplemented with 5% (w/v) fetal calf serum (FCS) containing penicillin (5 U/ml) and strepomycin (5 μg/ml); transfection is performed as described.[24]

[23] P. J. Schatz, *Biotechnology* **11,** 1138 (1993).

[24] A. Sittler, D. Devys, C. Weber, and J.-L. Mandel, *Hum. Mol. Genet.* **5,** 95 (1996).

Purification of GST Fusion Proteins

The procedure for the purification of GST fusion proteins is an adaption of the protocol of Smith and Johnson.[25] Unless indicated otherwise, all steps are performed at 0–4°.

1. Inoculate 100 ml TY medium with a single colony containing the expression plasmid of interest and incubate the culture at 37° overnight with shaking.
2. Inoculate 1.5 liter TY medium with the overnight culture and grow at 37° until an OD_{600} of 0.6 is reached.
3. Add isopropyl-β-D-thiogalactopyranoside (IPTG) to a final concentration of 1 m*M* and continue to grow the culture at 37° for 3.5 hr with vigorous shaking.
4. Chill the culture on ice, and harvest the cells by centrifugation at 4000*g* for 20 min.
5. Wash cells with buffer A [50 m*M* sodium phosphate (pH 8), 150 m*M* NaCl, and 1 m*M* EDTA] and, if necessary, store the cell pellet at −70°.
6. Resuspend cells in 25 ml buffer A, add phenylmethylsulfonyl fluoride (PMSF) and lysozyme (Boehringer Mannheim) to 1 m*M* and 0.5 mg/ml, respectively, and incubate on ice for 45 min.
7. Lyse cells by sonication (2 × 45 sec, 1 min cooling, 200–300 W), and add Triton X-100 to a final concentration of 0.1% (v/v).
8. Centrifuge the lysate at 30,000*g* for 30 min and collect the supernatant.
9. Add 5 ml of a 1 : 1 slurry of GST-agarose (Sigma, St. Louis, MO), equilibrated previously in buffer A, and stirr the mixture for 30 min.
10. Pour the slurry into a 1.6-cm-diameter column, wash once with 40 ml buffer A containing 1 m*M* (PMSF) and 0.1% Triton X-100, and wash twice with 40 ml buffer A containing 1 m*M* PMSF.
11. Elute the protein with 5 × 2 ml buffer A containing 15 m*M* reduced glutathione (Sigma). Analyze aliquots of the fractions by SDS–PAGE and combine the fractions containing the purified GST fusion protein.
12. Dialyze the pooled fractions overnight against buffer B [20 m*M* Tris–HCl (pH 8), 150 m*M* NaCl, 0.1 m*M* EDTA, and 5% (v/v) glycerol], aliquot, freeze in liquid nitrogen, and store at −70°.

Typical yields are 10–20 mg for GST-HD20ΔP and -HD51ΔP and 5–10 mg for GST-HD20ΔPBio and-HD51ΔPBio per liter of bacterial culture. The protein concentration is determined using the Coomassie protein assay reagent from Pierce with bovine serum albumin (BSA) as a standard.

[25] D. B. Smith and K. S. Johnson, *Gene* **67,** 31 (1988).

Proteolytic Cleavage of GST Fusion Proteins

The GST–huntingtin fusion proteins (2 μg) are digested with bovine factor Xa (New England Biolabs) or with modified trypsin (Boehringer Mannheim, sequencing grade) at an enzyme/substrate ratio of 1 : 10 (w/w) and 1 : 20 (w/w), respectively. The reaction is carried out in 20 μl of 20 m*M* Tris–HCl (pH 8), 150 m*M* NaCl, and 2 m*M* $CaCl_2$. Incubations with factor Xa are performed at 25° for 16 hr. Tryptic digestions are at 37° for 3 to 16 hr. Digestions are terminated by the addition of 20 μl 4% (w/v) SDS and 100 m*M* dithiothreitol (DTT), followed by heating at 98° for 5 min.

Isolation of Amyloid-like Protein Aggregates from Transfected COS-1 Cells

COS-1 cells transfected with the mammalian expression plasmids are harvested 48 hr after transfection. The cells are washed in ice-cold phosphate-buffered saline (PBS), scraped, and pelleted by centrifugation (2000*g*, 10 min, 4°). Cells are lysed on ice for 30 min in 500 μl lysis buffer [50 m*M* Tris–HCl (pH 8.8), 100 m*M* NaCl, 5 m*M* $MgCl_2$, 0.5% (w/v) Nonidet P-40 (NP-40), 1 m*M* EDTA] containing the protease inhibitors PMSF (2 m*M*), leupeptin (10 μl/ml), pepstatin (10 μg/ml), aprotinin (1 μg/ml), and antipain (50 μg/ml). Insoluble material is removed by centrifugation for 5 min at 14,000 rpm in a microfuge at 4°. Pellets containing the insoluble material are resuspended in 100 μl DNase buffer [20 m*M* Tris–HCl (pH 8.0), 15 m*M* $MgCl_2$], and DNase I (Boehringer Mannheim) is added to a final concentration of 0.5 mg/ml followed by incubation at 37° for 1 hr. After DNase treatment the protein concentration is determined by the dot metric assay (Geno Technology) using BSA as a standard. Incubations are terminated by adjusting the mixtures to 20 m*M* EDTA, 2% (w/v) SDS, and 50 m*M* DTT, followed by heating at 98° for 5 min.

Dot-Blot Filter Retardation Assay

The filter assay used to detect polyglutamine-containing huntingtin protein aggregates has been described.[9] Denatured and reduced protein samples are prepared as described above, and aliquots corresponding to 50–250 ng fusion protein (GST-HD20ΔP and GST-HD51ΔP) or 5–30 μg extract protein (pellet fraction) of COS-1 cells are diluted into 200 μl 2% SDS and filtered on a BRL dot-blot filtration unit through a cellulose acetate membrane (Schleicher and Schuell, Keene, NH, 0.2-μm pore size) that has been preequilibrated with 2% SDS. Filters are washed twice with 200 μl 0.1% SDS and are then blocked in TBS (100 mM Tris–HCl, pH 7.4, 150 mM NaCl) containing 3% nonfat dried milk, followed by incubation with the anti-HD1 antibody (1:1000).[9] The filters are washed several times in TBS and are then incubated with a secondary anti-rabbit antibody conjugated to horseradish peroxidase (Sigma, 1:5000) followed by ECL (enhanced chemiluminescence, Amersham) detection. The developed blots are exposed for various times to Kodak (Rochester, NY) X-OMAT film or to a Lumi-Imager (Boehringer Mannheim) to enable quantification of the immunoblots.

For detection and quantification of polyglutamine-containing aggregates generated from the protease-treated fusion proteins GST-HD20ΔPBio and -HD51ΔPBio, the biotin/streptavidin–AP detection system is used. Following filtration, the cellulose acetate membranes are incubated with 1% (w/v) BSA in TBS for 1 hr at room temperature with gentle agitation on a reciprocal shaker. Membranes are then incubated for 30 min with streptavidin–alkaline phosphatase (Promega, Madison, WI) at a 1:1000 dilution in TBS containing 1% BSA, washed three times in TBS containing 0.1% (v/v) Tween 20 and three times in TBS, and finally incubated for 3 min with either the fluorescent alkaline phosphatase substrate AttoPhos or the chloro-substituted 1,2-dioxetane chemiluminescence substrate CDP-Star (Boehringer Mannheim) in 100 mM Tris–HCl, pH 9.0, 100 mM NaCl, and 1 mM $MgCl_2$. Fluorescent and chemiluminescent signals are imaged and quantified with the Boehringer Lumi-Imager F1 system and LumiAnalyst software (Boehringer Mannheim).

Results

As shown previously,[9] removal of the GST tag from the HD exon 1 protein containing 51 glutamines (GST-HD51) by site-specific proteolytic cleavage results in the formation of high molecular weight protein aggregates, seen as characteristic fibrils or filaments on electron microscopic examination. Such ordered fibrillar structures were not detected after proteolysis of fusion proteins containing only 20 (GST-HD20) or 30 (GST-HD30) glutamines, although light-scattering measurements[17] revealed that some form of aggregation also occurred with these normal repeat-length proteins. In the present study, truncated GST-HD exon 1 fusion proteins with or without a C-terminal biotinylation tag[23] were used. These fusion proteins contain either 20 or 51 glutamines but lack most of the proline-rich region located downstream of the glutamine repeat.[9] Potential factor Xa and trypsin cleavage sites within the GST-HD fusion proteins are shown in Fig. 1. The proteins GST-HD20ΔP and -HD51ΔP were expressed in *E. coli* and affinity-purified under native conditions. They were then digested overnight with trypsin or factor Xa protease to promote the formation of polyglutamine-containing huntingtin aggregates. Figure 2A shows an immunoblot of a cellulose acetate membrane to which the native GST-HD20ΔP and -HD51ΔP proteins and their factor Xa and trypsin cleavage products have been applied. As expected from our previous studies using fusions of GST and the full-length HD exon 1 protein,[9] only the cleavage products of GST-HD51ΔP were retained by the filter and were detected by the huntingtin-specific antibody HD1, indicating the formation of high molecular weight protein HD51ΔP aggregates from this fusion protein. Scanning electron microscopy of the material retained on the surface of the membrane revealed bunches of long fibrils or filaments (Fig. 2B), which were not detected after filtration of the uncleaved GST-HD51ΔP preparation or the protease-treated GST-HD20ΔP preparation. These results indicate that an elongated polyglutamine sequence but not the proline-rich region in the HD exon 1 protein is necessary for the formation of high molecular weight protein aggregates *in vitro*.

To examine whether polyglutamine-containing aggregates are also formed *in vivo,* HD exon 1 proteins with 20, 51, or 93 glutamines (without a GST tag) were expressed in COS-1 cells. Whole cell lysates were prepared, and after centrifugation, the insoluble material was collected and treated with DNase I to achieve maximal resuspension. The resulting protein mixture was then boiled in SDS and analyzed using the dot-blot filter retardation assay. Figure 2C shows that insoluble protein aggregates are being formed in transfected COS cells expressing the HD exon 1 protein with 51 and 93 glutamines but not in COS cells expressing the normal exon 1 allele with 20 glutamines or in nontransfected control cells. Thus, as observed *in vitro* with purified GST fusion proteins, the formation of high molecular weight aggregates *in vivo* occurs in a repeat length-dependent way and requires a polyglutamine repeat in the pathological range. In addition, like the *in vitro* aggregates, HD exon 1 aggregates formed *in vivo* are resistant to boiling in 2% (w/v) SDS as well as to 8 *M* urea.

To monitor the *in vitro* formation of polyglutamine-containing aggregates without the need for a specific antibody, a modified filter retardation assay was developed. In this assay, streptavidin-conjugated alkaline phosphatase (AP) is used to detect the insoluble protein aggregates retained on the cellulose acetate filter membrane. Streptavidin binds specifically to the biotinylation tag[23] that has been added C-terminal to the polyglutamine tract in the fusion proteins GST-HD20ΔPBio and -HD51ΔPBio (Fig. 1).

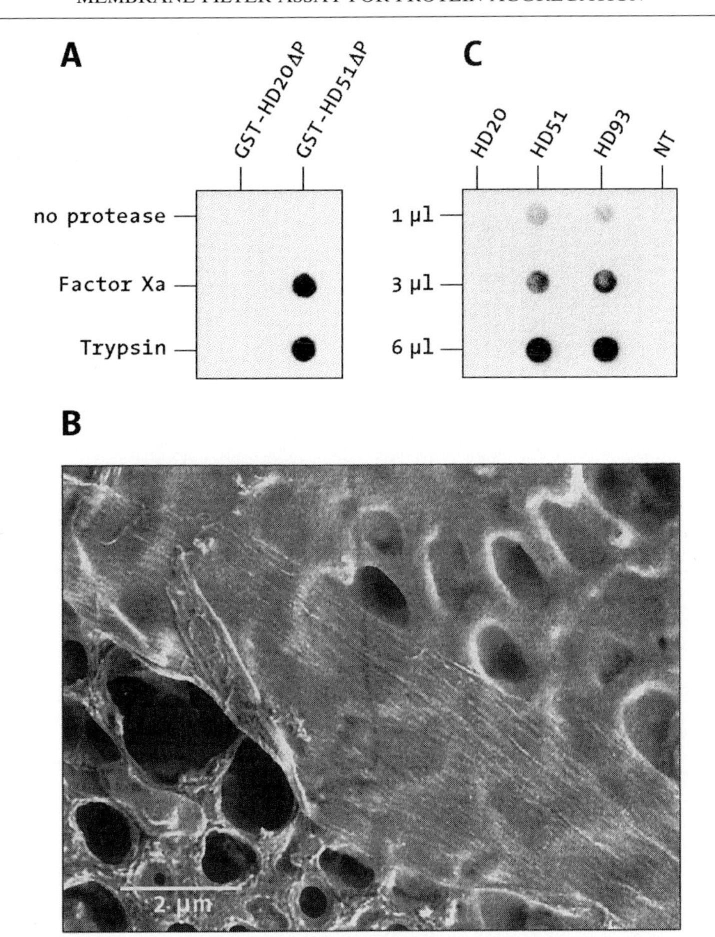

FIG. 2. Detection of polyglutamine-containing protein aggregates formed *in vitro* and in transfected COS-1 cells using the dot-blot filter retardation assay. (A) Purified GST-HD20ΔP and -HD51ΔP fusion proteins (250 ng) and their factor Xa and trypsin cleavage products were applied to the filter as indicated. The aggregated proteins retained by the cellulose acetate membrane were detected by incubation with the anti-HD1 antibody. (B) Scanning electron micrograph of aggregated GST-HD51ΔP trypsin cleavage products retained on the surface of the cellulose acetate membrane (Heinrich Lündsdorf, GBF Braunschweig, Germany). (C) Dot-blot filter retardation assay performed on the insoluble fraction isolated from transfected and nontransfected COS-1 cells. COS-1 cells were transfected transiently with the plasmids pTL1-CAG20, -CAG51, and -CAG93 encoding huntingtin exon 1 proteins with 20 (HD20), 51 (HD51), and 93 (HD93) glutamines, respectively. The pellet fractions obtained after centrifugation of whole cell lysates were subjected to DNase I digestion, boiled in 2% SDS, and portions of 1, 3, and 6 μl were filtered through a cellulose acetate membrane. The aggregated huntingtin protein retained on the membrane was detected with the anti-HD1 antibody. NT, nontransfected cells.

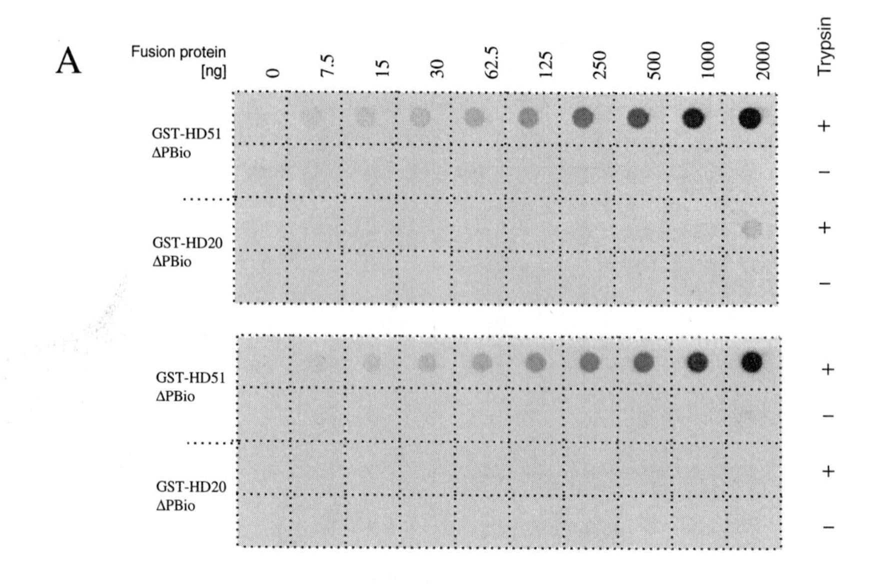
A
Fusion protein [ng]
0
7.5
15
30
62.5
125
250
500
1000
2000
Trypsin
GST-HD51 ΔPBio
GST-HD20 ΔPBio
GST-HD51 ΔPBio
GST-HD20 ΔPBio
+
−
+
−
+
−
+
−

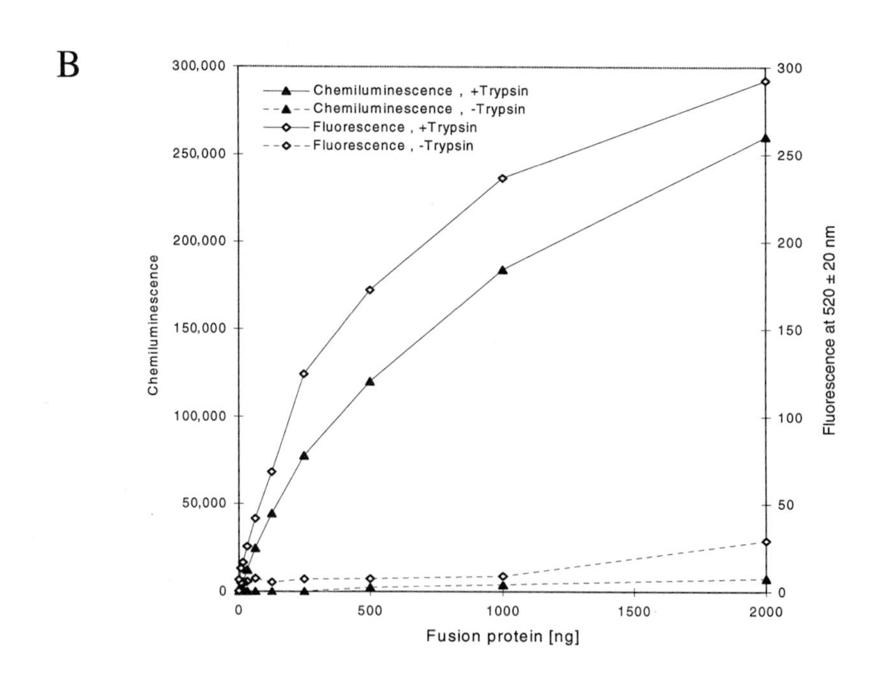
B
Chemiluminescence , +Trypsin
Chemiluminescence , -Trypsin
Fluorescence , +Trypsin
Fluorescence , -Trypsin
Chemiluminescence
0
50,000
100,000
150,000
200,000
250,000
300,000
Fluorescence at 520 ± 20 nm
0
50
100
150
200
250
300
0
500
1000
1500
2000
Fusion protein [ng]

Figure 3A shows that the modified aggregation assay gives results comparable to those obtained with nonbiotinylated fusion proteins in that insoluble aggregates are produced from the trypsin-treated GST-HD51ΔPBio protein but not from the uncleaved GST-HD51ΔPBio protein or the corresponding 20 repeat samples. Using either fluorescent (AttoPhos) or chemiluminescent (CDP-Star) substrates for alkaline phosphatase, it is possible to capture and quantify the filter assay results with the Boehringer Lumi-Imager F1 system. With both AP substrates, aggregates formed from as little as 5–10 ng of input GST-HD51ΔPBio protein were readily detected on the cellulose acetate membrane, and signal intensities increased linearly up to 250 ng of fusion protein applied to the filter (Fig. 3B).

Conclusions

Our results demonstrate that the cellulose acetate filter retardation assay can be a useful tool for the identification, structural characterization, and quantification of SDS-insoluble polyglutamine-containing protein aggregates formed *in vitro* and *in vivo*. In addition to the histochemical identification of amyloids, this assay may be useful in detecting insoluble protein aggregates in all types of human and animal amyloidoses, including the polyglutamine diseases, and also in screening compound libraries for potential aggregation inhibitors. Currently, attempts to develop a microtiter plate-based high-throughput filter retardation assay to identify chemical compounds that slow down the rate of formation of polyglutamine-containing fibrils *in vitro* are in progress. The amyloid-binding agents arising from this screen then will be tested in a HD cell culture model system and in the HD animal model[21,26] for their therapeutic potential.

[26] S. W. Davies, M. Turmaine, B. A. Cozens, M. DiFiglia, A. H. Sharp, C. A. Ross, E. Scherzinger, E. E. Wanker, L. Mangiarini, and G. P. Bates, *Cell* **90,** 537 (1997).

FIG. 3. Detection and quantification of aggregates formed *in vitro* from biotinylated GST-HD exon 1 fusion proteins. Various amounts of the fusion proteins GST-HD51ΔPBio and -HD20ΔPBio were filtered through a cellulose acetate membrane after a 3-hr incubation at 37° in the presence or absence of trypsin as indicated. (A) Images of the retained protein aggregates, detected with streptavidin–AP conjugate using either a fluorescent (top) or a chemiluminescent AP substrate (bottom). (B) Quantification of signal intensities obtained for the GST-HD51ΔPBio dots seen in A. Fluorescence and chemiluminescence values are arbitrary units generated by the Lumi-Imager F1 and LumiAnalyst software (Boehringer Mannheim).

Acknowledgments

We thank Rudi Lurz and Heinrich Lünsdorf for preparing the scanning electron micrographs. The work described in this article has been supported by grants from the HDSA, Deutsche Forschungsgemeinschaft (Wa1151/1), and the European Union (Project: Eurohunt, BMH4-CT96-0244).

[25] Analysis of Fibril Elongation Using Surface Plasmon Resonance Biosensors

By DAVID G. MYSZKA, STEPHEN J. WOOD, and ANJA LEONA BIERE

Introduction

Commercially available surface plasmon resonance (SPR) biosensors, such as BIACORE, have revolutionized the characterization of macromolecular interactions. These instruments can be used to monitor binding events in real time without labeling, making them convenient for studying a wide variety of biological systems. Biosensors are routinely used to provide insight into the reaction kinetics and thermodynamics for saturable bimolecular reactions such as ligand–receptor or antigen–antibody interactions.[1] However, biosensors have the ability to provide detailed information on polymerization and aggregation reactions as well. There is a growing interest in kinetic analysis of fiber formation, mainly due to the increasing body of evidence implicating the involvement of peptide or protein fibers in human diseases, with one prominent example being Alzheimer's disease.[2] One of the pathological criteria for Alzheimer's disease are neuritic plaques, which predominantly consist of fibrillar amyloid β-peptide (Aβ), a 40–42 amino acid peptide.[3] Unfortunately, in the case of Aβ, kinetic data regarding fibrillogenesis are still very limited. This is partly due to the lack of adequate techniques and partly to the difficulties in handling peptides that aggregate readily. Using Aβ as a model system, we illustrate how biosensors have the potential to provide detailed information on aggregation and polymerization processes. Because this represents a new application for biosensors, this article focuses on the experimental methods that can be used to monitor fibril elongation.

[1] D. G. Myszka, *Curr. Opin. Biotech.* **8,** 50 (1997).
[2] D. J. Selkoe, *Science* **275,** 630 (1997).
[3] D. B. Teplow, *Amyloid* **5,** 121 (1998).

0076-6879/99 $30.00

Surface Plasmon Resonance Biosensors

SPR biosensor technology and its applications continue to evolve at a rapid pace. There are now a number of companies that produce biosensors, including Affinity Sensors (Cambridge, UK; www.affinity-sensors.com), Biacore AB (Uppsala, Sweden; www.biacore.com), BioTuL Bio Instruments GmbH (Munich, Germany; www.biotul.com), and XanTec (Münster, Germany; xantec.com). Information on how the different technologies work can be found at the manufactures' web sites. BIACORE instruments continue to be the most commonly used, therefore, this article will describe the experimental methods in terms of this system. BIACORE 2000 and the newly released 3000 version of the technology offer four flow cells in which reactions can be monitored simultaneously. Typically the surface for one flow cell is used for internal referencing, making it possible to correct for refractive index changes, nonspecific binding, and instrument drift.[4] This improves the quality of data dramatically, allowing the direct detection of small molecular mass compounds (<500 Da)[5] and the ability to describe binding responses with simple interaction models.[6,7]

Advantages

SPR biosensors offer a number of advantages for studying fibril elongation. The lack of labeling requirements allows one to study unmodified peptides directly. Because binding reactions are monitored in real time, it is possible to collect kinetic data for rapid binding events on the seconds time scale. BIACORE is ideally suited to study elongation events because contact times can be varied easily in order to collect data for both short-term reversible and long-term polymerization reactions. The high sensitivity available with BIACORE allows the detection of small molecular weight peptides to large fibrils as well as the decay of fibrils directly. It is also possible to test how other proteins or compounds affect fibril stability. Compared to other interaction technologies, biosensor experiments typically consume small amounts of material (1–50 μg) and binding studies can be performed under a wide variety of conditions. BIACORE sensor chips can be reused depending on the stability of the immobilized sample. A carefully designed assay will use the same chip for more than 100 binding experiments. Immobilized sample chips can be conveniently removed from the instrument and stored in a refrigerator for later use.

[4] T. A. Morton and D. G. Myszka, *Methods Enzymol.* **295,** 268 (1998).

[5] P. O. Markgren, M. Hämäläinen, and U. H. Danielson, *Anal. Biochem.* **265,** 340 (1998).

[6] L. D. Roden and D. G. Myszka, *Biochem. Biophys. Res. Commun.* **225,** 1073 (1996).

[7] J. K. Stuart, D. G. Myszka, L. Joss, R. S. Mitchell, S. M. McDonald, Z. Xie, S. Takayama, J. C. Reed, and K. R. Ely, *J. Biol. Chem.* **273,** 22506 (1998).

Limitations

SPR biosensors have their own limitations and caveats. It is important to establish that the binding responses are due to the interaction of interest and that the binding reaction rates are unaffected by experimental artifacts. Some of these artifacts include nonspecific binding, bulk refractive index changes, avidity effects, mass transport limitations, surface heterogeneity, instrument drift, and injection noise.[1] A detailed discussion on improving the experimental design and implementing advanced data analysis methods can be found in Morton and Myszka.[4]

SPR Biosensor Assay

In an SPR experiment, the target molecule or "ligand" is immobilized onto the sensor chip surface, and the potential binding partner, referred to as the "analyte," flows through a narrow channel in contact with the surface. As the analyte binds to the immobilized ligand the increase in mass changes the refractive index of the solvent near the sensor surface. The SPR detector monitors the change in refractive index in real time and displays the response in what are referred to as sensorgrams.[8]

Figure 1A shows a schematic of the flow cell and SPR detector used in BIACORE. To study Aβ–fibril interactions, fibrils of the peptide are attached to the sensor surface, which is actually the ceiling of the flow cell. The surface area for these microfluidic cells is on the order of 1 mm^2 and cell volume is approximately 20 nl. To monitor a binding interaction, Aβ is passed through the flow cell in solution, and a sensogram of the time-dependent change in refractive index is recorded.

Figure 1B highlights the essential features of a sensorgram for a typical experiment. While running buffer is passed over the flow cell, there should be no change in the response, as there is no mass change at the surface. During the association phase, reactant is injected at a constant concentration. If binding occurs, the accumulation of mass changes the refractive index near the surface and an increase in the response is observed over time. For simple bimolecular reactions, binding reactions may reach equilibrium, resulting in no further change in response over time. This occurs because the same number of complexes are formed on the surface as are broken down. During the dissociation phase, running buffer replaces the sample plug and the decay rate of the complex is monitored. Because binding data

[8] U. Jönsson, L. Fägerstam, B. Ivarsson, B. Johnsson, R. Karlsson, K. Lundh, S. Löfås, B. Persson, H. Roos, I. Rönnberg, R. Sjölander, E. Stenberg, R. Ståhlberg, C. Urbaniczky, H. Östlin, and M. Malmqvist, *Bio. Tech.* **11,** 620 (1991).

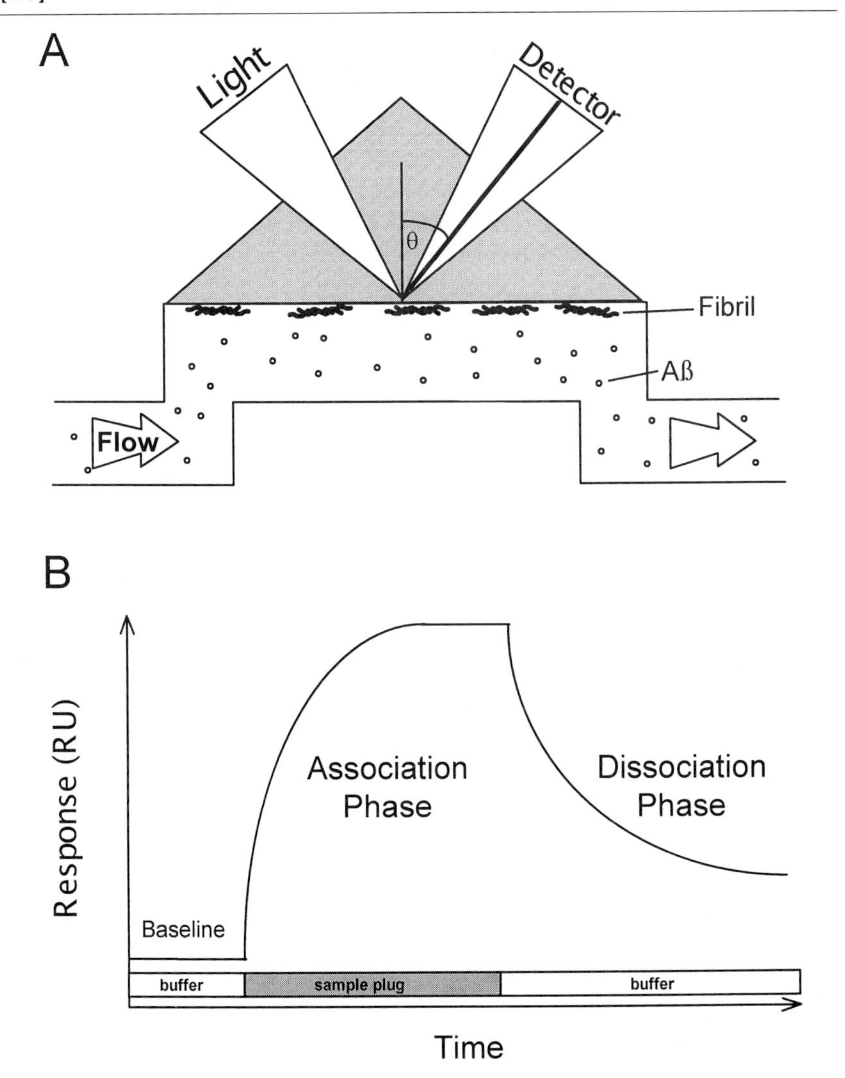

FIG. 1. Schematic of SPR biosensor assay. (A) Salient features of the BIACORE flow cell and detection system. For the present study, fibrils were immobilized on the sensor surface and Aβ was passed across the surface in solution. (B) A typical sensorgram collected for a saturable binding reaction. Reactant is injected at a constant concentration during the association phase and the sensor surface is washed with running buffer during the dissociation phase.

are collected in real time, it is possible to extract detailed kinetic information about a reaction.

Experimental Design

Successful application of biosensor technology hinges on designing the experiment properly. Although this article focuses on the analysis of Aβ–fibril interactions, where possible we will provide information that may be applicable to similar systems. Because no two biological systems are alike, it is impossible to provide protocols that will work in each case. As with any high-resolution biophysical technique, however, it is important to start with pure reagents that are well characterized.

Aβ–Fibril Preparation

Commercial lots of Aβ vary with respect to purity and degree to which preformed aggregates are initially present on solubilization. An ultrapure (>98% purity) Aβ(1–40) catalog number 03-138 from Quality Controlled Biochemicals (Hopkinton, MA; www.qcb.com) was used in the present study. Fibrillar aggregates are prepared as follows: 1 mg of lyophilized peptide is solubilized to 0.86 mg/ml in sterile water for 5 min at room temperature with gentle swirling. Following solubilization (as monitored by visual inspection) the material is diluted to 0.43 mg/ml using 2× Dulbecco's phosphate-buffered saline (PBS), pH 7.4 (GIBCO-BRL, Gaithersburg, MD). This mixture is then stirred vigorously for 24 hr at room temperature. Following this incubation, the aggregates are stored frozen at −20° and are stable up to 8 months.

To test Aβ binding, a fresh batch of lyophilized peptide is solubilized at 10 μ*M* in BIACORE running buffer (10 m*M* HEPES, 150 m*M* NaCl, 3.4 m*M* EDTA, and 0.005% Tween 20, pH 7.4), kept on ice, and centrifuged immediately before using.

Background Binding

The standard CM5 biosensor chips from Biacore Inc. contain a carboxymethyl dextran layer that coats a gold surface required for the plasmon resonance response.[9] The dextran layer is important for minimizing nonspecific interactions of proteins with the biosensor surface as well as providing a flexible anchor that is conducive for measuring protein interactions. One of the limitations with the carboxymethyl dextran surface is that it is highly negatively charged. This can promote ionic interactions with some proteins

[9] B. Johnsson, S. Löfås, G. Lindquist, Å. Edström, R.-M. Müller Hillgren, and A. Hansson, *J. Mol. Recog.* **8,** 125 (1995).

leading to a high level of background binding. Before beginning any biosensor experiments it is essential to test both reagents over an underivatized surface to ensure that there is not a high level of background binding. Given the aggregation potential of the Aβ system, it is imperative to show that neither Aβ nor fibril complexes interact with the sensor surface. For these studies, each protein preparation is injected over an unmodified surface at the highest concentration that will be used in future binding studies (0.43 mg/ml). Importantly, neither Aβ nor the fibril preparation bound to a blank sensor surface, demonstrating that nonspecific binding would not be a problem for the Aβ system. If a high level of background binding is observed, sensor chips with different surface chemistries may minimize this effect. For example, the pioneer chip B1 available from BIACORE, which has 10-fold fewer carboxyl groups on the dextran compared to the CM5 chip, is useful for minimizing nonspecific binding of basic proteins.

Fibril Immobilization

A significant advantage of biosensors is their ability to monitor the immobilization process itself. Amine-coupling chemistry is often employed because it is convenient to use and easy to control.[9] The best way to configure the biosensor assay to collect data on fibril elongation is to immobilize the fibrils and monitor the binding of Aβ from solution. This mimics the *in vivo* mechanism of elongation where soluble Aβ is thought to bind to fibrils in the growing plaque.[10] To immobilize the fibrils the carboxymethyl dextran surface is first activated with a 7-min injection of a mixture of 0.4 *M* *N*-ethyl-*N*′-(3-dimethylaminopropyl)carbodiimide and 0.1 M *N*-hydroxysuccinimide in water. Next, a fibril preparation is diluted 10-fold (0.043 mg/ml) into 10 m*M* acetate, pH 4.0, and injected for 1-min intervals until the desired level of immobilization is achieved. The low ionic strength and low pH of the acetate buffer help preconcentrate protein into the dextran layer, which promotes chemical cross-linking. In the final coupling step the remaining activated groups are blocked with a 7-min injection of 1 *M* ethanolamine, pH 8.0. The difference between the response before coupling and after blocking denotes the total amount of fibril preparation immobilized. We used this immobilization procedure to couple approximately 1000 RU of fibril onto the surface of flow cell 2.

As mentioned previously, some versions of BIACORE instruments have multiple flow cells that can be monitored simultaneously. It is generally very informative to couple different amounts of reactants on the different

[10] J. E. Maggio, E. R. Stimson, J. R. Ghilardi, C. J. Allen, C. E. Dahl, D. C. Whitcomb, S. R. Vigna, H. V. Vinters, M. E. Labenski, and P. W. Mantyh, *Proc. Natl. Acad. Sci. U.S.A.* **89,** 5462 (1992).

surfaces within the same sensor chip to test if the binding reactions are dependent on surface density. Because BIACORE 2000 has four flow cells, we repeated the coupling procedure on flow cell 3 with a longer contact time of 5 min during the fibril injections. This resulted in coupling three times the amount of fibril onto flow cell 3 (~3000 RU).

Reference Surfaces

Reference surfaces are essential for improving the quality of biosensor data. By passing the reactant over an identically treated surface, it is possible to correct for refractive index changes, nonspecific binding, injection noise, and instrument drift. In order to create similar chemical environments on the reference surface, it is important to carry out the same immobilization procedure without adding ligand. If the molecular mass of the reactant is very low (<1000 Da), it is useful to immobilize the same amount of a noninteracting protein. This helps normalize the excluded volume between the reaction and the reference surface.[5] For Aβ experiments, two reference surfaces were created by activating and deactivating the surfaces in flow cells 1 and 4 with the same immobilization conditions used for fibril coupling.

Association Time

In BIACORE it is easy to control the association time by changing the injection volume and/or the flow rate. In normal injection modes, BIACORE 2000 can inject samples of 5–750 μl at flow rates from 1 to 100 μl/min. This allows association phase times from as short as 3 sec to longer than 1 hr. If required, much longer association phases can be used by loading samples directly through the running buffer syringe.[11] To test for initial reversible binding reactions, it is important to use a short association phase. We limited the association phase to 12 sec by injecting a 20-μl sample of Aβ at a flow rate of 100 μl/min. With the instrument operating in series mode, the same Aβ sample was passed over all four flow cells on the sensor chip. At the end of the injection, the sensor surfaces were washed with running buffer for 60 min to collect dissociation phase data. Figure 2A shows two examples of the binding responses obtained for Aβ during the 12-sec association phase and Fig. 2B shows the dissociation phase out to 30 min.

Replicating Binding Experiments

An essential and often overlooked aspect of SPR biosensor analysis is the demonstration that the observed binding responses are reproducible.

[11] D. G. Myszka, M. D. Jonsen, and B. J. Graves, *Anal. Biochem.* **265,** 326 (1998).

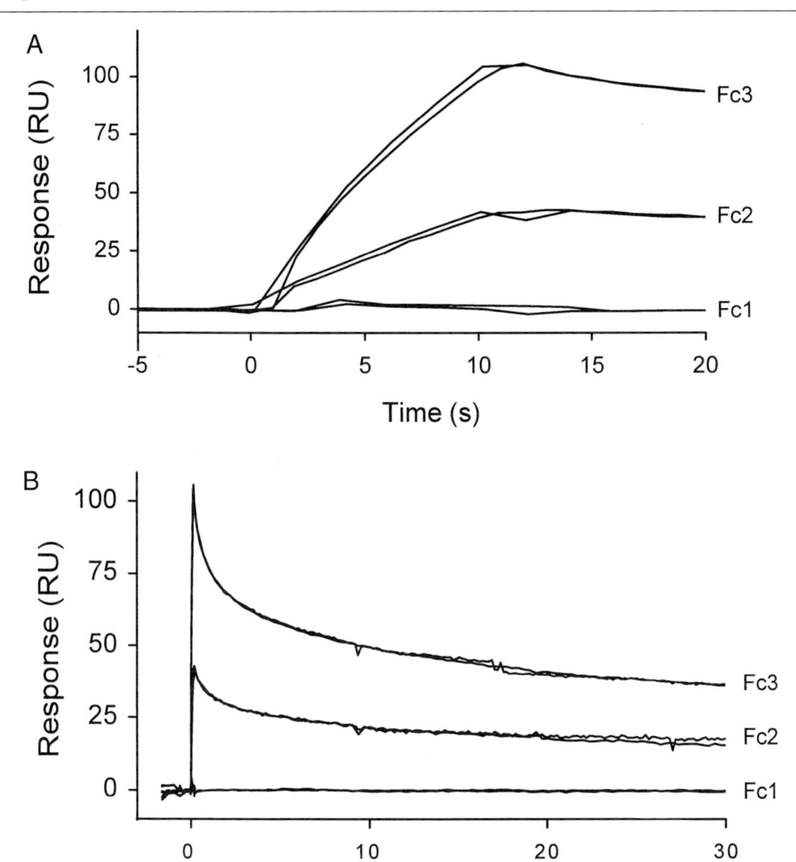

FIG. 2. Replicate responses for Aβ binding to fibril and control surfaces. (A) Association phase from 0 to 12 sec Aβ contact time. (B) Dissociation phase out to 30 min. A 20-μl sample of freshly dissolved Aβ (10 μM) was injected through all four biosensor flow cells at a flow rate of 100 μl/min. The dissociation phase was monitored for 60 min after which time the Aβ injection was repeated. Flow cell 2 (Fc2) and flow cell 3 (Fc3) contained ~1000 and ~3000 RU of immobilized fibril, respectively. Flow cell 1 (Fc1) and 4 were left blank. The response from flow cell 4 was subtracted from flow cells 1–3 to correct for refractive index changes and instrument drift. The overlay plots were created by setting the injection time and baseline to zero at the start of each injection. The running buffer was 10 m*M* HEPES, 150 m*M* NaCl, 3.4 m*M* EDTA, and 0.005% Tween 20, pH 7.4, and the instrument was at 20°.

In BIACORE this is easy to do by simply repeating the sample injection. First it is important to ensure that all of the previously formed complexes have broken down and the surface is regenerated for another binding cycle. If complexes still exist after a reasonable amount of dissociation time, a surface can usually be regenerated with a mild acid or alkaline wash step. Data shown in Fig. 2B indicate that a portion of the response attributed to the Aβ injection did not return to baseline. Normally we would want to remove this bound material with a regeneration step. However, given the fact that the fibril is a polymer of Aβ, no attempt was made to remove the remaining material out of a concern that the regeneration step may disrupt the fibril itself. Therefore, after 1 hr of washing with running buffer, the Aβ injection was simply repeated with the same concentration of peptide. To create the overlay plots shown in Fig. 2, the baseline responses and injection times were set to zero at the start of each Aβ injection.

Assessing Binding Responses

The shapes of the binding responses provide information about the binding reaction. In the case of the Aβ–fibril interaction, a larger binding response was observed from the sensor surface in flow cell 3 than in flow cell 2. This is consistent with the fact that more of the fibril preparation was immobilized onto flow cell 3 and demonstrates that the binding responses are dependent on the fibril surface concentration. No binding response was observed on the sensor surface without fibrils (flow cell 1). It is also important to note that the baseline response from this control surface is stable and does not drift. The binding responses observed from the fibril surfaces were also very reproducible as shown by repeating the Aβ injections as described earlier. This high reproducibility suggests that the Aβ that remains bound to the fibril after the first injection does not change the binding capacity for subsequent injections.

The kinetics of the Aβ–fibril interaction are interesting in that the dissociation phase is clearly multiphasic. One fraction of the bound peptide dissociates rapidly whereas other portions dissociate in one or two slower kinetic phases. The dissociation phase for a simple bimolecular reaction would normally appear as a single exponential decay. Therefore, the shapes of the binding responses indicate the Aβ–fibril-binding mechanism is complex.

Model Testing

An advantage of using biosensors to characterize a binding mechanism is that it is easy to vary a wide variety of reaction parameters. It is possible to vary ligand surface density, analyte concentration, association phase time, flow rate, and which reactant is immobilized on the sensor surface. This information can provide important clues as to the binding mechanism involved in complex formation.

Fibril Elongation Model

The current model of Aβ fibrillogenesis evolved from a multitude of studies that envision fiber elongation to occur via end growth and to include several steps. The diagram shown in Fig. 3 illustrates the main steps in the elongation reaction. An Aβ monomer (or potentially dimer) needs to recognize its respective binding site at the fiber tip, bind to it, and regenerate a new binding site, which can then be recognized by another monomer. John Maggio coined the term "dock-and-lock" mechanism to describe the binding and the template regeneration step, respectively.[10] The initial binding ("docking") step between Aβ (A) and the fibril tip (B) is readily reversible. The second step referred to as "locking" constitutes an isomerization of the bound peptide to form a more stable complex and the generation of a new binding site (B'). When the next monomer binds, it would inhibit the dissociation of the previously bound peptide, which we refer to as "blocking." Whereas the model shown in Fig. 3 may be a simplification of the polymerization process, it does suggest a number of experiments that can be performed on the biosensor to test its general validity.

Monitoring Fibril Decay

The chemical surface and temperature control system used in BIACORE 2000 and 3000 instruments are very stable, which produces a stable

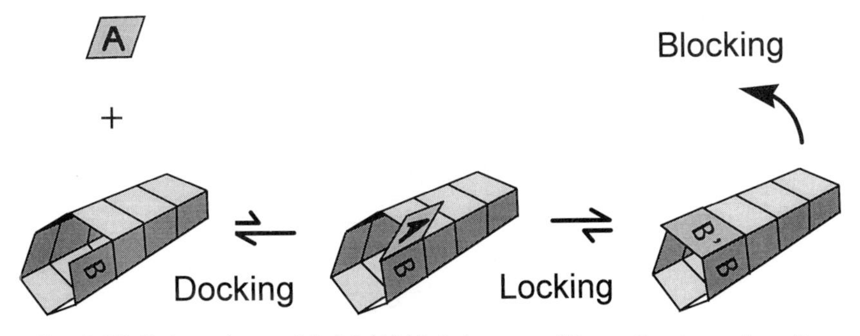

FIG. 3. Fibril elongation model. Aβ (A) binds in a reversible reaction to a unique site on the end of the fibril tip (B) referred to as "docking." Once bound, the conformation of the peptide rearranges, leading to a more stable complex referred to as "locking" and regenerating the recognition site on the end of the fibril (B'). The continued binding of Aβ monomers to the fibril during polymerization leads to a "blocking" of the release of earlier bound peptides.

baseline response. In combination with internal reference surfaces, as described earlier, it is possible to monitor the binding as well as the dissociation of low molecular weight peptides from larger complexes. We noticed early on in our Aβ experiments that fibrils immobilized onto the sensor surface exhibited a slow linear background decay as compared to a reference surface. Figure 4 shows responses obtained from a freshly prepared fibril surface before and after injection of Aβ. In this example, the Aβ injections are actually repeated three times and are overlaid in Fig. 4, again demonstrating the high level of reproducibility of the assay. The slow background decay of the fibril surface is evident compared to a reference line. After the Aβ injection, about half of the complexes dissociate with a short half-life ($t_{1/2}$) of approximately 3 min, whereas the remaining material assumes a slower decay rate similar to the background fibril decay. The observation that some of the bound Aβ does not completely dissociate, even after 60 min, suggests that this material has been incorporated into the growing fibril. The fact that the slow dissociation rate ($t_{1/2} > 45$ min) matches the background decay of the fibril suggests that this decay could represent depolymerization of the fibril. The Aβ that dissociates rapidly from the surface may represent initially docked material that is not "locked" into the mature fibril conformation.

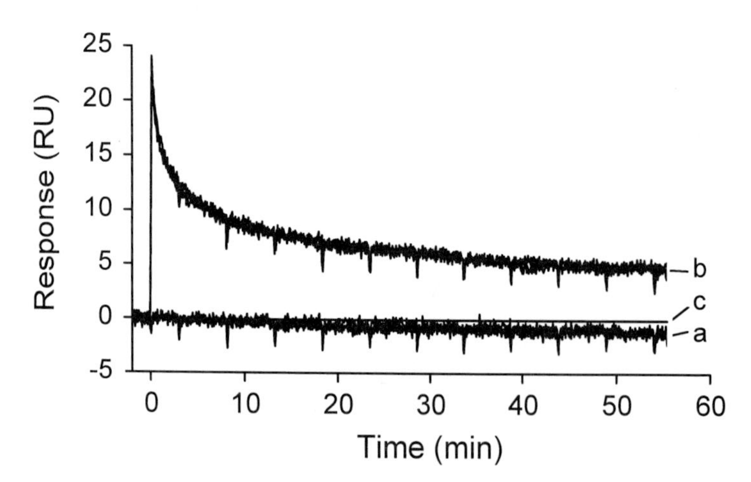

FIG. 4. Monitoring background fibril decay. Decay of a freshly prepared fibril surface was monitored for 1 hr (a). The response shown by (b) represents Aβ (10 μM) that was injected over the same fibril surface for 6 sec and the dissociation phase was monitored for 1 hr. The solid line (c) represents a reference baseline. The background decay and Aβ injections were each replicated three times and overlaid in the plot.

Varying Association Phase Time

As mentioned earlier, one parameter that can be varied easily in BIACORE is the association phase time. This can provide information about the interaction that is useful in resolving the binding mechanism. For example, in Fig. 5A, the same concentration of Aβ was injected over a fibril surface for different lengths of contact time. As the association phase was

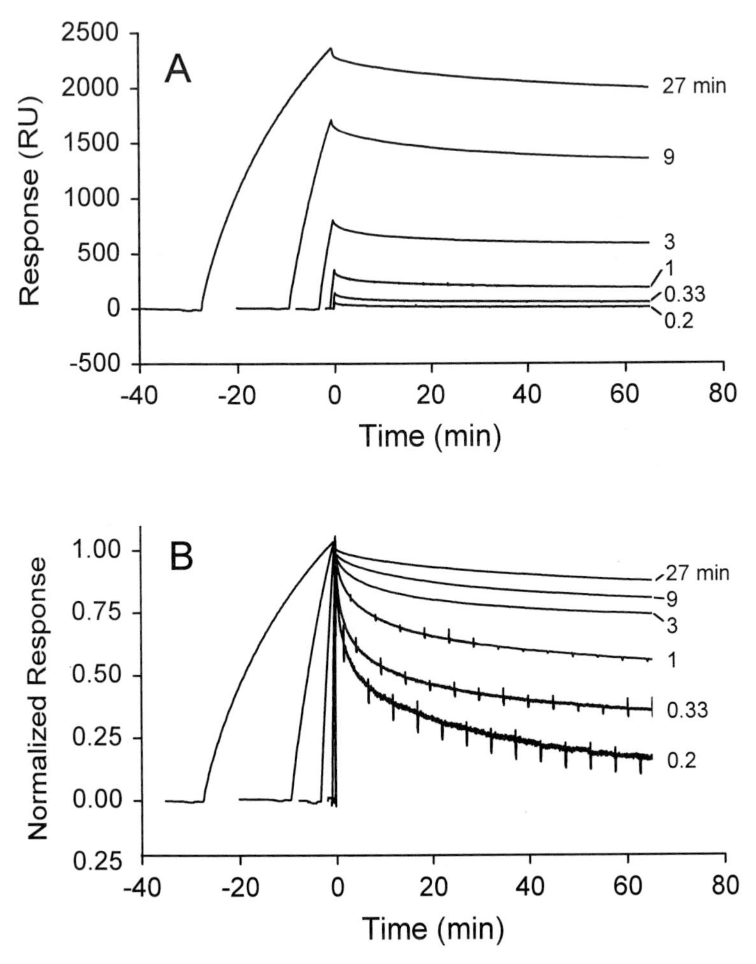

FIG. 5. Varying the Aβ association phase time. (A) A 10 μM sample of Aβ was injected over a fibril surface for 0.2, 0.33, 1, 3, 9, and 27 min as labeled at the right. The responses were overlaid by zeroing the response on the *y* axis prior to injection and zeroing the time on the *x* axis at the start of the dissociation phase. (B) Normalized dissociation phase. Data were normalized by setting the response at the start of the dissociation phase to 1.

increased from 12 sec to 27 min, the amount of peptide captured onto the surface increases dramatically. It is interesting to note that even with a prolonged association phase, the binding reactions do not reach saturation or equilibrium. The binding responses for a simple bimolecular reaction would normally plateau once equilibrium was reached. Failure to plateau during the Aβ injections suggests that Aβ may be undergoing a polymerization process. This is consistent with the mechanism where Aβ generates a new binding site once incorporated into the fibril, and so the surface reactions never reach equilibrium.

Normalizing Dissociation Phase

One way of directly comparing the binding responses for the different association phase experiments described earlier is to normalize the binding responses. This is done by setting the binding response at the start of dissociation phase to 1 for each experiment. For a simple bimolecular interaction the dissociation rate should be independent of analyte concentration and contact time. The normalized responses shown in Fig. 5B indicate that at very short contact times the majority of the bound peptide dissociates rapidly. As the contact times are increased, a greater proportion of the response takes on a slower dissociation rate. The decrease in dissociation rate with increasing contact time suggests that more of the peptide is integrated into the fibril the longer Aβ is exposed to the fibril surface.

Varying Association Time and Aβ Concentrations

The goal of this experiment is to again test how the apparent dissociation rate changes with contact time, but this time starting with the same amount of Aβ captured for each experiment. Because the amount of Aβ bound by the same fibril surface is dependent on both the association time and the Aβ concentration, both parameters can be varied with the goal of obtaining a constant amount of Aβ captured at the start of the dissociation phase. Using a freshly prepared fibril surface, a high concentration of Aβ (10 μM) was injected for a short contact time of 3 sec (5 μl at 100 μl/min = 0.05 min). These conditions resulted in capturing 20 RU of Aβ at the start of the dissociation phase. The dissociation phase was then monitored for 1 hr. In the next injection the Aβ concentration was reduced threefold and injected for longer period of time (20 sec). We repeated this process until a data set was collected with an association phase, for the 20 RU target, of 20 min. Figure 6 shows an overlay of the responses after baseline correction at the start of the injection and alignment on the time axis at the start of the dissociation phase. In comparing these response curves we see that the longer the association time, a greater proportion of

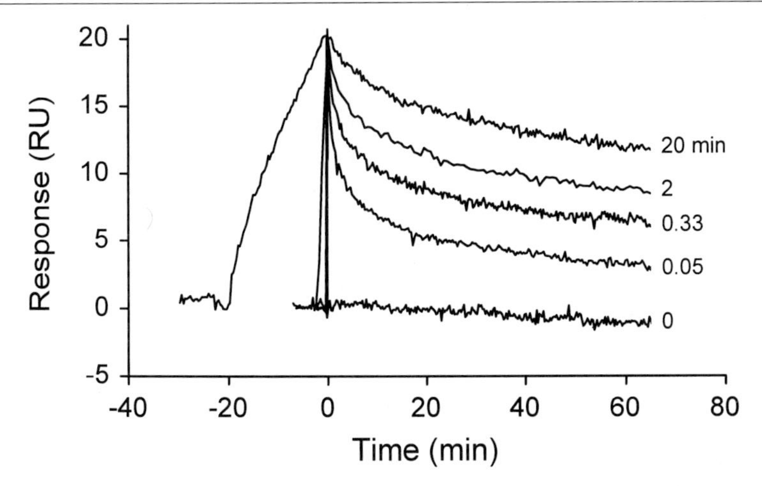

FIG. 6. Varying contact time and Aβ concentration. The concentration and contact time of Aβ were varied with the intent on keeping the amount of peptide bound at the end of the association phase constant at 20 RU. Data were first collected from the 1000 RU fibril surface without injection peptide (0). Different dilutions of Aβ were injected from 0.05 to 20 min. The response before each injection was zeroed on the y axis and at the start of the dissociation phase on the x axis.

the bound Aβ assumes a slower dissociation rate. This is consistent with this material becoming incorporated into the growing fiber. It is interesting to note how the responses from the different contact times assume a similar slow background decay rate as seen for the fibril surface alone.

Determining the exact amount of contact time for a given Aβ concentration was done partly through trial and error. Because the dilute Aβ concentrations required longer contact times, it was possible to stop the association phase manually as soon as the desired level of binding was achieved. Recently released versions of the BIACORE control software allow one to dial in the precise amount of response that is desired for a given injection. These automated injection systems will make it much easier to perform the experiments described in this section.

Flat Surfaces

Up to this point all of the biosensor data were collected using sensor surfaces that contain a carboxymethyl dextran matrix. Whereas the dextran matrix works well for this system, it may affect long-term fibril elongation and it may not be suitable for other aggregation systems. Also, the presence of dextran makes it difficult to perform a detailed structural analysis of the surface. Techniques such as atomic force microscopy have been used to image fibrils[12,13] but are difficult to employ on dextran surfaces. BIACORE offers a flat carboxymethylated sensor surface called C1, which does not contain a dextran layer. Amine coupling procedures can be used to immobilize reactants onto this flat surface. Before using the C1 chip we established that neither Aβ nor the fibril preparation bound nonspecifically to the flat surface. We immobilized a fibril sample onto these surfaces using the same amine coupling procedures described earlier for the CM5 dextran chip. We observed the same background decay rate for the fibril and a similar binding response for Aβ from the flat surface as on the dextran chip. The flat surface opens up the possibility of removing the sensor chip from the instrument and performing AFM to determine the structure of the surface bound material.

Varying Experimental Conditions

An advantage of biosensors is the ability to monitor binding reactions under a wide variety of experimental conditions. For instance, with BIACORE 2000 it is possible to measure the temperature dependence of the reaction rate constants and equilibrium constants from 4 to 40°.[14] Solvent conditions may be varied across a wide range to test the effects of pH and ionic strength on reactions. It is also possible to monitor how other buffer components such as metal ions and lipids affect the assembly process. Together this information can provide important insights into an aggregation mechanism.

Data Analysis

Sophisticated nonlinear global fitting routines have been developed to interpret the kinetics for both simple and complex reactions recorded on BIACORE.[4,15] We demonstrated previously how globally fitting biosensor data provide a more stringent test for the reaction models and return better estimates for the parameter values.[16] The ability to globally fit data with a simple 1:1 interaction model validates the sensor technology as a tool for monitoring binding kinetics.[6,7] The ability to resolve complex reactions

[12] J. D. Harper, C. M. Lieber, and P. T. Lansbury, *Chem. Biol.* **4,** 119 (1997).

[13] L. Narhi, S. J. Wood, S. Steavenson, Y. Jiang, G. M. Wu, D.Anafi, S. A. Kaufman, F. Martin, K. Sitney, P. Denis, J. C. Louis, J. Wypych, A. L. Biere, and M. Citron, *J. Biol. Chem.* **274,** 9843 (1999).

[14] D. G. Myszka, *Methods Enzymol.* **315,** in press (2000).

[15] D. G. Myszka and T. A. Morton, *TIBS* **23,** 149 (1998).

[16] T. A. Morton, D. G. Myszka, and I. M. Chaiken, *Anal. Biochem.* **227,** 176 (1995).

involving mass transport,[17,18] multiple surface-binding sites,[19] and multiple analyte complexes[20] demonstrates the utility of using global analysis to characterize more elaborate binding mechanisms. The software program we use, CLAMP, is available to the public at the web site of Huntsman Cancer Institute (www.hci.utah.edu/groups/biacore). Many of the data analysis functions available in the program have been incorporated into BIAevaluation software from BIACORE.

We are currently developing the appropriate models required to resolve the reaction kinetics associated with polymerization reactions. Unlike simple bimolecular reactions, polymerization reactions are indefinite associations. Therefore, reaction models must account for the regeneration of binding sites. By globally fitting data collected under different conditions it should be possible to characterize reactions that undergo linked equilibrium such as the one proposed for the Aβ–fibril interaction. However, in analyzing long-term polymerization reactions, one may also need to take into consideration the nonlinear response of the SPR detector. Molecules that bind further from the sensor surface give a smaller response. This is probably not a concern for short-term reactions, but could complicate the analysis of binding events that extend the polymer away from the sensor surface. More experimental work is required to assess how significant this effect will be on the observed binding response.

Conclusions

The experimental methods and results described in this article demonstrate that SPR sensors can be used to gain important insights into Aβ–fibril elongation. The ability to monitor the association phase for both short and long periods of time makes it possible to collect data on different binding events. Current biosensor data are consistent with a reversible "docking" reaction of Aβ onto the fibril followed by a "locking" step that results in the formation of a new fibril site. Repeating the binding cycle results in "blocking" of the previously bound peptides and fibril extension. Future work will be aimed at resolving the kinetic rate constants associated with each of these steps. The small amounts of material required for biosensor

[17] D. G. Myszka, T. A. Morton, M. L. Doyle, and I. M. Chaiken, *Biophys. Chem.* **64,** 127 (1997).
[18] D. G. Myszka, H. Xiaoyi, M. Dembo, T. A. Morton, and B. Goldstein, *Biophys. J.* **75,** 583 (1998).
[19] D. G. Myszka, P. G. Arulanantham, T. Sana, Z. Wu, T. A. Morton, and T. L. Ciardelli, *Prot. Sci.* **5,** 2468 (1996).
[20] L. A. Joss, T. A. Morton, M. L. Doyle, and D. G. Myszka, *Anal. Biochem.* **261,** 203 (1998).

experiments make it an ideal technology to study how other proteins or small molecules affect fibril assembly. The convenience of the biosensor assay will make it a valuable tool to shed new light on Aβ fibrillogenesis as well as other polymerization and aggregation systems.

[26] Methods for Studying Protein Adsorption

By VLADIMIR HLADY, JOS BUIJS, and HERBERT P. JENNISSEN

Introduction

Proteins are interfacially active molecules; a statement that is demonstrated easily by the spontaneous accumulation of proteins at interfaces.[1-4] Why do proteins show the propensity to adsorb to interfaces and why do they adsorb so tenaciously? For some proteins, the tendency to adsorb is due to the nature of side chains present on the surface of the protein. Protein is an amphoteric polyelectrolyte.[5] Its amino acids have different characteristics: some are apolar and like to be buried inside the protein globule, whereas others are polar and charged and are often found on the outside protein surface. A strong, long-ranged electrostatic attraction between a charged adsorbent and oppositely charged amino acid side chains will lead to a significant free energy change favoring the adsorption process. In other cases, the interfacial activity of the protein may be driven by its marginal structural stability.[6] The compactness of the native structure of the protein is due to the optimal amount of apolar amino acid residues. The stability of such a structure depends on the combination of hydrophobic interactions between the hydrophobic side chains, hydrogen bonds between the neighboring side chains and along the polypeptide chains, and the Coulomb interactions between charged residues and van der Waals interactions. An adsorbent surface can "compete" for the same interactions and minimize the total free energy of the system by unfolding the protein

[1] J. D. Andrade, *in* "Surface and Interfacial Aspects of Biomedical Polymers. 2. Protein Adsorption" (J. D. Andrade, ed.), p.1. Plenum Press, New York, 1985.

[2] J. D. Andrade and V. Hlady, *Adv. Polym. Sci.* **79,** 1 (1986).

[3] M. Malmsten, ed., "Biopolymer at Interfaces." Dekker, New York, 1998.

[4] J. L. Brash and P. W. Wojciechowski, eds., "Interfacial Phenomena and Bioproducts." Dekker, New York, 1996.

[5] C. Branden and J. Tooze, "Introduction to Protein Structure." Garland Publishing, New York, 1991.

[6] C. A. Haynes and W. Norde, *J. Colloid Interface Sci.* **169,** 313 (1995).

0076-6879/99 $30.00

structure: the adsorption process may result in a surface-induced protein denaturation.[7,8] Elements of the secondary structure of the protein (α helix and β sheet) together with the supersecondary motifs form a compact globular domain. Some proteins are built from more than one domain. In a multidomain protein, it is possible that one domain will dominate the adsorption property of the whole macromolecule at a particular type of interface. For example, acid-pretreated antibodies bind with their constant fragments to a hydrophobic surface.[9]

In order to completely characterize and predict protein adsorption, one would like to have a quantitative description of adsorption. This description is typically obtained by measuring the adsorption isotherm, adsorption and desorption kinetics, conformation of adsorbed proteins, number and character of protein segments in contact with the surface, and other physical parameters related to the adsorbed protein layer, such as layer thickness and refractive index. This article describes a selected set of techniques and protocols that will provide answers about the mechanism of protein adsorption onto and desorption from surfaces. The reader is referred to the specialized monographs[1–4] and a review[10] on protein adsorption for a more comprehensive coverage of various aspects of protein–surface interactions.

Description of Protein Adsorption

A. Adsorption Isotherms

The adsorption isotherm is a function that relates the measured adsorbed amount of a protein (per unit area), Γ_p, to the solution concentration of protein, c_p. Typically, Γ_p increases sharply at low solution concentrations of protein and levels off at higher protein concentrations approaching a limiting Γ_p value. The existence of a Γ_p adsorption "plateau" has been interpreted as a sign that the adsorbing surface is "saturated" with protein molecules; any further increase in the solution protein concentration typically does not affect Γ_p. The amount of adsorbed protein at the "plateau" of the adsorption isotherm is often close to the amount that can fit into a closed-packed monolayer; hence, the notion of a saturating monolayer coverage is often applied to protein adsorption.

[7] V. Hlady and J. Buijs, *in* "Biopolymer at Interfaces" (M. Malmsten, ed.), p. 181. Dekker, New York, 1998.
[8] J. Buijs and V. Hlady, *J. Colloid Interface Sci.* **190,** 171 (1997).
[9] I.-N. Chang and J. N. Herron, *Langmuir* **11,** 2083 (1995).
[10] J. J. Ramsden, *Q. Rev. Biophys.* **27,** 41 (1993).

Under ideal conditions, the shape of the adsorption isotherm can provide information about the affinity between protein and surface. The evaluation of the affinity, however, does require a model for the protein–surface interactions; a model from which the adsorption isotherm function, $\Gamma_p(c_p)$, can be derived and compared with the experimental results. Most experimentally measured protein adsorption isotherms may not be the true equilibrium isotherms.[1,2,11] It may take a relatively long time for a protein adsorbing to a surface to reach true equilibrium. Other processes, such as protein conformational change, may run concurrently with the adsorption process and affect the adsorbed amount. Hence, the information contained in the adsorption isotherm may not refer to identical molecules. The existence of several protein conformers in solution, each with slightly different adsorptivity, may result in an isotherm that will reflect the competition between conformers for a limited adsorption surface area.

1. Adsorption to Two-Dimensional Lattice of Binding Sites. It has been shown that a protein molecule adsorbing to a solid surface with immobilized alkyl residues will bind multivalently to several residues at the same time.[12,13] The contact points with alkyl residues can be considered as individual binding sites situated in either a homogeneous or a heterogeneous two-dimensional lattice.[14] In principle, therefore, the amount of adsorbed protein under standard conditions also depends on the surface concentration of binding sites on the solid phase. On the basis of multivalence, which implies a specific geometry of sites, protein adsorption on such tailor-made lattices has been compared to a molecular recognition process.[15]

Terminologically, just as a coenzyme is viewed as being a small entity, i.e., a ligand in binding to a macromolecular enzyme, a protein can be similarly viewed as being a small entity, i.e., a ligand in comparison to the large macroscopic solid surface it is being adsorbed to. For simplicity, therefore, protein molecules in protein adsorption studies (adsorbates) have been defined as ligands.[12,16]

To account for the two fundamental parameters governing protein adsorption on natural and artificial surfaces, one may distinguish two categories of protein adsorption isotherms: (a) a lattice site binding function and

[11] H. P. Jennissen, *in* "Surface and Interfacial Aspects of Biomedical Polymers." (J. D. Andrade, ed.), Vol. 2, p. 295. Plenum Press, New York, 1985.

[12] H. P. Jennissen, *Biochemistry* **15,** 5683 (1976).

[13] H. P. Jennissen, *Hoppe-Seyler's Z. Physiol. Chem.* **359,** 1201 (1976).

[14] H. P. Jennissen, *J. Chromatogr.* **159,** 71 (1979).

[15] H. P. Jennissen, *Makromol. Chem., Macromol. Symp.* **17,** 111 (1988).

[16] H. P. Jennissen and G. Botzet, *Int. J. Biol. Macromol.* **1,** 171 (1979).

(b) a bulk ligand binding function.[15,17,18] Equations describing these binding functions have been derived previously and are similar power functions as is the Hill equation.[17,18]

The lattice site binding function governs the binding of an immobilized residue, constituting a surface lattice site, to a complementary site (patch, pocket) on the protein adsorbed from the bulk solution. The more lattice sites that interact simultaneously with a protein molecule, the higher the affinity of binding will be.[19] Protein adsorption, therefore, increases as a function of the surface concentration of lattice sites (at a constant equilibrium concentration of free bulk protein) according to Eq.(1)[12,13,17]:

$$\theta_s/(1 - \theta_s) = K_s(\Gamma^s{}_{Res})^{n_s} \tag{1}$$

where θ_s is the fractional saturation of binding units with protein on the adsorbent as a function of the surface concentration of lattice sites in a binding unit. $\Gamma^s{}_{Res}$ is the surface concentration of lattice sites (i.e., immobilized residues), K_S is the lattice site adsorption constant, and n_s is the lattice site adsorption coefficient. For the case $n_s = 1$, Eq. (1) reduces to a rectangular hyperbola. Saturation of binding occurs when the complementary binding sites of the protein in a binding unit on the adsorbent surface cannot accommodate further lattice sites.[17] The corresponding half-saturation constant $K_{S,0.5}$ relates to K_S according to $K_{S,\,0.5} = (K_s)^{1/n_s}$.[12,16]

The bulk ligand (i.e., protein) binding function, which corresponds to a classical protein adsorption isotherm, can be described by Eq. (2) [13,16,17]:

$$\theta_B/(1 - \theta_B) = K_B c_p{}^{n_B} \tag{2}$$

where θ_B is the fractional saturation of binding units on the surface at a constant lattice site concentration with the bulk protein as independent variable, C_p is the bulk protein concentration at equilibrium, K_B is the bulk-ligand adsorption constant, and n_B is the bulk-ligand adsorption coefficient. For the case $n_B = 1$, Eq.(2) also reduces to a rectangular hyperbola. Saturation occurs when the solid phase surface cannot accommodate additional protein molecules. The half-saturation constant $K_{B,0.5}$ in this case relates to K_B according to $K_{B,0.5} = (K_B)^{1/n_B}$.[12,15,16]

B. Adsorption Kinetics

The basic parameter of any adsorption kinetics is the number of protein molecules adsorbed per unit area in any increment of time, $d\Gamma_p/dt$. It is

[17] H. P. Jennissen, *J. Chromatogr.* **215,** 73 (1981).
[18] H. P. Jennissen and G. Botzet, *J. Mol. Recogn.* **6,** 117 (1993).
[19] H. P. Jennissen, *Adv. Enzyme Regul.* **19,** 377 (1981).

generally accepted that the adsorption/desorption process comprises the following subprocesses: transport toward the interface, attachment at the interface, detachment from the interface, and transport away from the interface.[2]

Each of these steps can, in principle, determine the overall rate of the adsorption process. Hence, one needs to design the adsorption kinetics experiment in which only one of the four steps will dominate. For example, transport of protein from solution toward the interface will usually occur by diffusion or by a combination of convective–diffusive processes. At low protein concentration, the rate of transport may be slower than the rate of actual attachment of protein to the surface. As a result, the measured adsorption kinetics may only contain information about the transport and not about the actual attachment kinetics.

C. Single Protein vs Multiprotein Adsorption

It is not unlikely that more than one type of protein is present in solution. This situation is common for all body fluids. In a multiprotein adsorption, differences between the interactions of each protein with adsorbing surface, as well as lateral protein–protein interactions, will determine the outcome of the adsorption. The competitive nature of the protein–surface interactions often result in transient dominance of one protein over others.[20] The solution protein exchange with a surface protein may be direct or through an adsorption site made available transiently by desorption of an adsorbed protein. Models for such protein–protein exchange usually involve a notion of monolayer coverage based on the assumption that a finite surface capacity exists for each competing species.

D. Adsorbent Surfaces and the Role of Water

Protein–surface interactions are influenced by the physical state of both the adsorbent and the solution environment. The important factors, including surface energy, its polar and nonpolar contributions, surface charge, and surface roughness, all have to be considered in defining the role of the solid–solution interface. The use of well-defined surfaces with a known surface density of protein-binding sites helps in the interpretation of protein interfacial characteristics. In all studies, the surface for protein adsorption will have to be characterized thoroughly, or even engineered specifically to fit the aim of the study. Adsorbents with immobilized protein-binding

[20] P. A. Cuypers, G. M. Willems, H. C. Hemker, and W. T. Hermens, *Makromol. Chem. Macromol. Symp.* **17,** 155 (1988).

residues are particularly suitable for fundamental protein adsorption studies. In contrast to model surfaces, "real" surfaces, even when they are not characterized rigorously, are more interesting from a practical point of view.

Most surfaces acquire electric charge when exposed to ionic solution. Although a charged protein is expected to prefer adsorption onto an oppositely charged surface, the osmotic pressure of counterions, the desolvation of the charged groups, and burying of the charges into a low dielectric medium may, in fact, oppose adsorption. Considered alone, electrostatic interactions may not fully account for the attachment of proteins to a charged interface.

Similar to the role that they play in maintaining the compactness of protein structure, hydrophobic interactions between an adsorbent surface and protein are of utmost importance. Proteins with a significant fraction of solution-exposed hydrophobic residues will tend to adsorb strongly to hydrophobic adsorbents. In some cases, dehydration of the hydrophobic surface of an adsorbent alone may tip the adsorption-free energy balance in favor of strong adsorption.

III. Review of Methods for Studying Protein Adsorption

A. Solution Depletion Technique

Solution depletion is one of the simplest methods to study protein adsorption. One measures a concentration change in bulk solution prior to and after adsorption, Δc_p. In the solution depletion technique, any protein concentration change is attributed to the adsorbed layer, i.e., $\Gamma_p = \Delta c_p V/A_{tot}$, where V is the total volume of protein solution and A_{tot} is the total area available for adsorption. Choices of protein concentration measuring technique are numerous: UV absorption, fluorescence, colorimetric methods, and many other, including the use of radioisotope-labeled or fluorescently labeled proteins. The solution depletion method requires a high surface area material such as beaded and particulate adsorbents (see Section IV,B,1).

Chromatographic adsorption media are generally in a particulate or beaded form. For example, 4% porous agarose beads (Sepharose 4B) have a surface area of ca. 8 m^2/ml of packed gel available for protein adsorption.[17] For ground quartz glass powder of a mean particle diameter of ca. 15 μm, a specific surface area of ca. 7 m^2/g has been determined by the BET-isotherm method with N_2 (H. P. Jennissen and A. Remberg, 1993, unpublished). Both types of adsorbents can be modified further chemically for protein adsorption studies.

In principle, protein adsorption on particulate or beaded materials can be measured either by zonal chromatography or by batch methods. Both methods lead to comparable results.[17] Such chromatographic methods, with the exception of complex competitive analytical chromatographic systems of the zonal and the continuous type,[21] are generally only applicable to low-affinity adsorption systems.[17] These systems will therefore not be treated here. Because protein adsorption to natural and artificial surfaces is generally of high affinity, simple chromatographic methods developed for the measurement of such interactions will be treated. One should, however, keep in mind that the column in such high-affinity chromatographic systems is washed with buffer after the protein adsorption step so that the free equilibrium bulk concentration of the protein is not clearly defined. This effect generally has little influence on the adsorbed amount of protein so that valuable information on the adsorptive behavior of proteins and surfaces is nevertheless obtained.

Batch methods, however, are especially suited for measuring both high- and low-affinity protein–surface interactions under equilibrium conditions. Because batch methods involve stirring with the aim of suspending the particles in the stirred protein solution, they are best suited for beaded low-density materials (e.g., gel beads). If the particle size is too large or the density is too high (glass, inorganic material, or metal powders), very high stirring rates may have to be employed with the danger of protein denaturation ("egg-beater effect"). If the surface of the particles is too irregular (e.g., with sharp edges) the "egg-beater effect" will be increased. The methods reported in Section IV,B,1 have been tested extensively with agarose beads (Sepharose 4B) and proteins varying in size from calmodulin (17 kDa) to phosphorylase kinase (1.2 MDa). Under the conditions described, no significant denaturation of these proteins was detected.

Batch methods are very well suited for the determination of adsorption and desorption isotherms, including adsorption hysteresis,[16,22] and also for monitoring sorption kinetics[23] of proteins on particulate or beaded materials. Such studies are generally hampered by complicated sampling procedures usually involving centrifugation steps to obtain a bead-free supernatant. In addition to the long time intervals needed for centrifugation, this separation method may also influence the equilibrium by concentrating effects beads finally localized in the pellet. A novel method, described in Section IV,B,1, was developed to circumvent centrifugation.[12]

[21] I. M. Chaiken, *J. Chromatogr.* **376,** 11 (1986).

[22] V. Hlady and H. Furedi-Milhofer, *J. Colloid Interface Sci.* **69,** 460 (1979).

[23] H. P. Jennissen, *J. Colloid Interface Sci.* **111,** 570 (1986).

B. Optical Techniques

In the category of optical techniques, the most commonly used ones are ellipsometry,[24] variable angle reflectometry,[25] and surface plasmon resonance.[26] In ellipsometry, polarized light is reflected obliquely from the interface under investigation. Optical properties of the reflecting surface can be determined from changes in the phase and amplitude of the reflected polarized light. When an adsorbed protein layer is present at the interface, these changes can be related to the thickness and the refractive index of the layer. The method of ellipsometry has been used extensively for *in situ,* real time single protein adsorption studies.[27–29] The method of variable angle reflectometry[25] is somewhat similar to ellipsometry. The method involves measurements of reflection coefficients of the polarized beam at angles of incidence in the vicinity of the Brewster angle. The technique of surface plasmon resonance (SPR) is described elsewhere in this volume.[29a]

C. Spectroscopic Techniques

Whereas optical techniques rely mostly on the interactions between light and an adsorbed protein layer, spectroscopic techniques for protein adsorption studies depend on the interaction of photons with some parts of the adsorbed protein molecules. The magnitude of the spectroscopic signal can often be related to the amount of adsorbed proteins, whereas at the same time spectral information can yield information on the conformation of adsorbed proteins. Fluorescence spectroscopy, infrared absorption (IR), Raman scattering, and circular dichroism (CD) are examples from this category of techniques. Surface-sensitive spectroscopic techniques require that the photons are (mostly) interacting with adsorbed protein molecules and much less so with bulk solution dissolved proteins. The use of surface-sensitive spectroscopic techniques in protein adsorption has been reviewed.[7]

Fluorescence emission spectroscopy offers a great specificity and sensitivity in protein studies. It can utilize both intrinsic protein fluorescence

[24] H. G. Thompson, "A User's Guide to Ellipsometry." Academic Press, San Diego, 1993.

[25] P. Schaaf, P. Dejardin, and A. Schmitt, *Langmuir* **3,** 3 (1987).

[26] J. Davies, ed., "Surface Analytical Techniques for Probing Biomaterial Processes." CRC Press, Boca Raton, FL, 1996.

[27] P. A. Cuypers, Ph.D. Thesis, Rijksuniversiteit Limburg, Limburg, 1976.

[28] P. A. Cuypers, G. M. Willems, H. C. Hemker, and W. T. Hermens, *Ann. N.Y. Acad. Sci.* **516,** 224 (1987).

[29] M. Malmsten, *J. Colloid Interface Sci.* **172,** 106 (1995).

[29a] D. G. Myszka, S. J. Wood, and A. L. Biere, *Methods in Enzymology* **309** [25] (1999) (this volume).

and a variety of extrinsic fluorophores attached to protein molecules.[30,31] Many variants of fluorescence spectroscopy used to study proteins in solution can be applied to protein molecules adsorbed at interfaces. The most common way to achieve a surface sensitivity is to excite adsorbed protein fluorescence with an evanescent surface wave generated by the total internal reflection of the excitation beam.[32,33] The technique of total, internal reflection fluorescence spectroscopy (TIRF), described in detail in Section IV,C, is often used to study kinetics of protein adsorption.[33–35]

In IR spectroscopy the spectral region between 1100 and 1700 cm^{-1} can provide information on global properties of the polypeptide conformation. In fact, it was one of the earliest methods employed to study the secondary structure of proteins.[36] The so-called amide I band, located in the wavenumber region approximately between 1600 and 1700 cm^{-1} (i.e., at wavelengths around 6 μm), is the most frequently studied parameter in protein IR studies. This characteristic band primarily represents the C=O stretching vibrations of this chemical group in the protein backbone. The frequency of this vibration depends on the nature of the hydrogen bonding in which the C=O group participates, which makes it highly sensitive to the secondary structure adopted by the polypeptide chain, e.g., α helices, β sheets, turns, and random coil structures, and thus provides "fingerprints" for protein secondary structure elements.[37] In addition to "classical" ATR-IR spectroscopy, Fourier transform infrared reflection absorption spectroscopy (FT-IRAS) and Fourier transform-attenuated total reflection infrared spectroscopy (ATR-FTIR) are used commonly in protein adsorption studies. ATR-FTIR examples include studies of the adsorbed amount of plasma protein onto surfaces of different commercial polymers[38–40] and studies of

[30] J. R. Lakowicz, "Principles of Fluorescence Spectroscopy." Plenum Press, New York, 1983.

[31] A. P. Demchenko, *in* "Topics in Fluorescence Spectroscopy" (J. R. Lakowicz, ed.), Vol. 3, p. 65. Plenum Press, New York, 1992.

[32] D. Axelrod, T. P. Burghardt, and N. L. Thompson, *Annu. Rev. Biophys. Bioeng.* **13,** 247 (1984).

[33] V. Hlady, R. A. Van Wagenen, and J. D. Andrade, *in* "Surface and Interfacial Aspects of Biomedical Polymers" (J. D. Andrade, ed.), Vol. 2, p. 81. Plenum Press, New York, 1985.

[34] N. L. Thompson, T. P. Burghardt, and D. Axelrod, *Biophys. J.* **33,** 435 (1981).

[35] B. K. Lok, Y.-L. Cheng, and C. R. Robertson, *J. Colloid Interface Sci.* **91,** 104 (1983).

[36] A. Elliott and E. J. Ambrose, *Nature* **165,** 921 (1950).

[37] M. Levitt and J. Greer, *J. Mol. Biol.* **114,** 181 (1977).

[38] R. I. Leiniger, D. J. Fink, R. M. Gendreau, T. M. Hutson, and R. J. Jakobsen, *Trans. Am. Artif. Intern. Organs* **29,** 152 (1983).

[39] K. K. Chittur, D. J. Fink, R. I. Leininger, and T. B. Hutson, *J. Colloid Interface Sci.* **111,** 419 (1986).

[40] D. J. Fink, T. B. Hutson, K. K. Chittur, and R. M. Gendreau, *Anal. Biochem.* **165,** 147 (1987).

secondary structure alterations in adsorbed proteins after being subjected to different sorbents[41–44] and protein aggregation.[45,46]

Autoradiography and Microscopic Techniques

The use of radioisotope-labeled proteins in protein adsorption studies is very common. The detection of adsorbed protein utilizes radioisotope counting or autoradiography. The latter method can provide spatially resolved information about adsorbed proteins on a micron-sized scale.

Almost all of the techniques described earlier will average over many thousands, if not millions, of adsorbed protein molecules. Hence, the local distribution of adsorbed protein molecules is unknown. Spatial information about adsorbed species requires some form of high-resolution microscopy. The scanning force microscopy technique can be applied to dynamically image protein adsorption or record the topography of an adsorbed protein layer.[47] Advances in near-field scanning optical microscopy (NSOM) and the capability of single molecule detection, including monitoring spectroscopic properties of a single adsorbed protein molecule,[48] have promising potential for protein adsorption studies.

IV. Protocols for Protein Adsorption Studies

The following sections contain a set of protocols used by the authors in studying protein adsorption. The reader is referred to the original papers for additional information not included due to space limitations. Some of the following protocols are applied to a particular protein; they can be applied to other proteins with minimum modifications.

A. Protocols for Protein Labeling and Surface Modifications

The radioisotope of iodine (^{125}I) is often selected as a probe for the quantitative determination of the amount of protein adsorbed to the various

[41] J. S. Jeon, R. P. Sperline, and S. Raghavan, *Appl. Spectrosc.* **46,** 1644 (1992).
[42] J. Buijs, W. Norde, and J. W. T. Lichtenbelt, *Langmuir* **12,** 1605 (1996).
[43] S.-S. Cheng, K. K. Chittur, C. N. Sukenik, L. A. Culp, and K. Lewandowska, *J. Colloid Interface Sci.* **162,** 135 (1994).
[44] M. Müller, C. Werner, K. Grundke, K. J. Eichhorn, and H. J. Jacobash, *Mikrochim. Acta* **14,** 671 (1997).
[45] A. Ball and R. A. L. Jones, *Langmuir* **11,** 3542 (1995).
[46] H. H. Bauer, M. Muller, J. Goette, H. P. Merkle, and U. P. Fringeli, *Biochemistry* **33,** 12276 (1994).
[47] C.-H. Ho, D. W. Britt, and V. Hlady, *J. Mol. Recogn.* **9,** 444 (1996).
[48] J. J. Macklin, J. K. Trautman, T. D. Harris, and L. E. Brus, *Science* **272,** 255 (1996).

types of surfaces for several reasons: elemental iodine is chemically reactive; it can be bonded covalently to protein; an iodine atom is relatively small, so it is expected to have minimal influence on protein properties; radiation from ^{125}I is detected easily; and handling of ^{125}I can be made reasonably safe.

^{125}I emits strong γ-radiation with the energy of 35 keV during its conversion to ^{125}Te.[49] The half-life of the radioisotope (59.6 days) gives a good balance between the efficiency of radioactivity detection and safety.

1. Protein Labeling with ^{125}I Using Iodine Monochloride. There are several chemical routes to label protein with ^{125}I. Typically, an oxidant incorporates ^{125}I into a protein molecule by replacing the hydrogen at the *o*-position of the tyrosine benzene ring with ^{125}I. Commonly used oxidants are iodine monochloride, ICl, and chloramine-T. The ICl protocol for labeling fibrinogen using the modified procedure from McFarlane[50] is given later. This reaction is considered to be much more gentle compared to the chloramine-T method; the ICl method has been shown to minimize damage to the fibrinogen molecule.[51]

A 2-ml solution of 5 mg/ml fibrinogen (Calbiochem, La Jolla, CA) in pH 9.0 glycine buffer, freshly made 25 μl of 1 m*M* ICl in pH 8.5 glycine buffer, and 1 mCi of $Na^{125}I$ (carrier free, 97%, Amersham Life Science, Cleveland, OH) are mixed together. The iodination reaction takes place at 4° for 1 hr. The reaction mixture is eluted through a PD-10 minicolumn (Pharmacia Piscataway, NJ) with 0.01 *M* phosphate-buffered saline (PBS, pH 7.4), and a 2-ml fraction is collected after the first eluted 2.5 ml. The protein concentration is determined by measuring the UV absorption at 280 nm using the extinction coefficient of 513,400 cm^{-1} M^{-1}.[52]

The degree of fibrinogen labeling is computed from the protein and protein-bound ^{125}I concentrations. Total ^{125}I in protein solution is determined by measuring solution radioactivity and comparing it with the ^{125}I standard (i.e., $Na^{125}I$ solution). The amount of free ^{125}I present in ^{125}I-labeled fibrinogen solution is determined using protein precipitation in trichloroacetic acid (TCA, Sigma, St. Louis, MO). The control sample is made of a mixture of 45 μl of bovine serum albumin (BSA) solution (20 mg/ml), 50 μl of PBS, and 5 μl of ^{125}I-labeled fibrinogen solution. The test sample is prepared from 45 μl of BSA solution (20 mg/ml), 50 μl of TCA (20% solution), and 5 μl of ^{125}I-labeled fibrinogen solution. Both samples are centrifuged at 15,000 rpm in a minicentrifuge (Model 235A, Fisher) for 10 min. Five microliters of both the control sample and the supernatant of

[49] D. M. Tollefen, J. R. Feagler, and P. W. J. Majerus, *J. Biol. Chem.* **246,** 2646 (1971).

[50] A. S. McFarlane, *J. Clin. Invest.* **42,** 346 (1963).

[51] N. Ardaillou and M. J. Larrieu, *Thrombosis Res.* **5,** 327 (1974).

[52] A. Haeberli, "Human Protein Data." VCH Verlags, Weinheim, 1992.

the precipitated test sample are counted for γ-radiation using the γ-counter (Model 170M, Beckman Instruments). The ratio of radioactivity in the supernatant of the precipitated sample to that in the control sample should be less than 0.05, i.e., less than 5% of total ^{125}I is free in the solution.

2. Protein Labeling with ^{125}I Using Chloramine-T. If a higher degree of protein labeling is required, the chloramine-T method can be used. The following protocol is shown applied to the labeling of bovine serum albumin.[49] An aliquot of $Na^{125}I$ solution amounting to a 0.3 mCi radioactivity (a few microliters at most) is added to a 0.5-ml solution of 1.5 mg/ml BSA (fraction V, ICN, Costa Mesa, CA). Subsequently, 50 μl of freshly prepared 4 mg/ml chloramine-T (Kodak, Rochester, NY) solution made in deionized water is added and allowed to react under gentle mixing for 1 min. Immediately thereafter, 50 μl of a 4.8-mg/ml sodium metabisulfite ($Na_2S_2O_3$, Mallinckrodt, Phillipsburg, NJ) solution is added to the mixture to stop the oxidation reaction. Free ^{125}I is removed as described in Section IV,A,1 or by passing the mixture through two Sephadex G-25 minicolumns (ca. 6 cm long, 1 cm in diameter) packed with a coarse grade resin of Sephadex G-25 (Pharmacia) under a centrifugation force of 40*g*. Each of these columns will accommodate about half the sample volume (0.3 ml). Free ^{125}I in the protein fraction can be determined using the TCA precipitation protocol (see Section IV,A,1).

3. Protein Labeling with Fluorescein Isothiocyanate. Many protein molecules contain one or several tyrosine, tryptophan, or phenylalanine residues that are intrinsically fluorescent in UV.[31] The applications of protein intrinsic fluorescence in the adsorption experiments are technically limited by the short wavelengths of excitation and emission and by the low quantum yields.[53] Because of that limitation, labeling of proteins with extrinsic fluorescent probes is a common practice.[54]

In the protocol that follows, fluorescein isothiocyanate (FITC) is used as a model fluorescent label for human serum albumin (HSA). Isothiocyanate moiety reacts selectively with the amine groups of protein, although it is also known to react reversibly with thiols and the phenol groups of tyrosine.[54] The FITC reaction with proteins generally gives a satisfactory yield and is quite a simple procedure. The dye has a high absorbance and emits strong fluorescence in aqueous solution. The fluorescein absorption band matches the wavelength of the commonly used Ar^+ ion on laser line at 488 nm. FITC is readily water soluble, and the isothiocyanate group is reasonably stable in most buffer solutions for proteins.

[53] V. Hlady, D. R. Reinecke, and J. D. Andrade, *J. Colloid Interface Sci.* **111,** 555 (1986).

[54] R. P. Haugland, "Handbook of Fluorescent Probes and Research Chemicals." 5th ed. Molecular Probes, Eugene, OR, 1994.

The labeling of human serum albumin (ICN) with FITC (Isomer I, Aldrich, Milwaukee, WI) follows the method of Coons *et al.*[55] Sixty milligrams of HSA is dissolved in 3 ml of carbonate–bicarbonate buffer (CBB, 0.1 *M*, pH 9.2). Of the freshly made FITC solution (1 mg/ml) in CBB, 0.6 ml is added to the HSA solution and the reaction is allowed to take place at room temperature for 3 hr. The protein–FITC mixture is then loaded on a PD-10 minicolumn (Pharmacia). The column is eluted with PBS (0.01 *M*, pH 7.4). The first colored fraction (protein-dye conjugate) is collected and its absorption at 280 (A_{280}) and 494 (A_{494}) nm is measured by a UV–visible spectrophotometer. The molar concentrations of FITC, c_{FITC}, and HSA, c_{HSA}, are determined as

$$c_{\mathrm{FITC}} = A_{494}/\varepsilon_{\mathrm{FITC}} \tag{3}$$

and

$$c_{\mathrm{HSA}} = [A_{280} - 0.308A_{494}]/\varepsilon_{\mathrm{HSA}} \tag{4}$$

where $\varepsilon_{\mathrm{FITC}}$ and $\varepsilon_{\mathrm{HSA}}$ are the extinction coefficients of FITC (76,000 cm^{-1} M^{-1})[54] and HSA (35,279 cm^{-1} M^{-1}),[52] respectively. A constant, 0.308, is the ratio of FITC absorption at 280 nm to that at 494 nm. The degree of labeling ($c_{\mathrm{FITC}}/c_{\mathrm{HSA}}$) is typically around unity.

A similar protocol can be used to label other proteins. The degree of labeling can be adjusted by changing the protein/dye molar ratio in the reaction mixture. The final result, however, will vary from protein to protein because it will depend on the number of reactive amine groups at pH 9.2 and their accessibility.

4. Chemical Modification of Silica Surfaces. In any surface modification it is absolutely critical to start the chemical reaction with a clean surface. The following surface modification protocols are shown applied to flat fused silica plates or glass beads. Identical protocols can be applied to oxidized silicon wafers. The fused silica plates (ESCO Products, Oak Ridge, NJ) or glass bead (diameter 5–50 μm, Duke Scientific, Palo Alto, CA) surfaces are first cleaned by submerging them in piranha solution (70% H_2SO_4 : 30% H_2O_2) for 15 min and subsequently rinsed thoroughly with deionized water. After drying in a vacuum desiccator, the surfaces are further cleaned with a radio-frequency generated oxygen plasma (200 mTorr, 50 W) (PLASMOD, Tegal Corp., Richmond, CA) for 2 min. An alternative cleaning procedure consists of immersing the silica plates (or beads) in hot chromosulfuric acid at 80° for 30 min, rinsing the surfaces thoroughly with deionized water and drying them in a vacuum desiccator. The cleanliness of flat silica surfaces

[55] A. M. Coons, H. J. Crech, R. N. Jones, and E. J. Berliner, *J. Immunol.* **45,** 159 (1942).

can be verified by measuring the water contact angle either by a sessile drop or by the Wilhelmy plate technique.[56] The clean silica surface should be fully wettable.

a. APS SILANIZATION. In the surface modification with 3-aminopropyltriethoxy silane (APS, Aldrich), the silica surfaces are reacted with a fresh ethanol solution of APS [5%(v/v) APS, 5% (v/v) deionized water, and 90%(v/v) absolute ethanol] for 30 min at room temperature, followed by rinsing thoroughly with deionized water and with absolute ethanol two times and by subsequent curing in the vacuum oven, which has been flushed with nitrogen three times at 80° overnight.

b. DDS SILANIZATION. In the surface modification with dimethyldichlorosilane (DDS, United Chemical Technologies, Bristol, PA), the silica surfaces are reacted with a freshly prepared dry toluene solution of DDS [10% (v/v) DDS and 90% (v/v) toluene] for 30 min at room temperature. The slides are then rinsed consecutively with absolute ethanol three times, with deionized water five times, and with ethanol again followed by curing in the vacuum oven (see earlier discussion). In both cases the silanized silica slides are stored at room temperature and used within 72 hr.

c. GPS SILANIZATION. In the surface modification with 3-glycidoxypropyltrimethoxysilane (GPS, United Chemical Technologies), the surfaces are placed in an Erlenmeyer wide-mouthed reaction vessel to which 1% (v/v) GPS and 0.2% (v/v) triethylamine in dry toluene is added. The vessel needs to be equipped with a $CaCl_2$ drying tube and is suspended in a heated water bath, where it is heated to 70° for 8 hr, stirred overnight at room temperature, and heated another 8 hr before washing with toluene and acetone. The GPS-modified silica surfaces are then dried in a vacuum oven.

d. LIGAND IMMOBILIZATION TO MODIFIED SILICA SURFACES. It is sometimes of interest to measure binding of protein from solution onto a surface that carries a particular ligand. Both APS- and GPS-modified silica can further be used to immobilize various ligands. In the case of APS-modified silica, the terminal surface group is an amine moiety that reacts with the isothiocyanate group, *N*-hydroxysuccinimide ester, and many other groups.[57] The APS-modified surfaces can also be "activated" with a cross-linker such as glutaraldehyde. The GPS-modified surface contains terminal epoxide groups that react with primary amines or sulfhydryls or it can be activated further with a cross-linker such as 1,1′-carbonyldiimidazole (CDI).

[56] J. D. Andrade, L. M. Smith, and D. E. Gregonis, *in* "Surface and Interfacial Aspects of Biomedical Polymers" (J. D. Andrade, ed.), Vol. 1, p. 249. Plenum Press, New York, 1985.

[57] G. T. Hermanson, "Bioconjugate Techniques." Academic Press, San Diego, 1996.

The reader is referred to specialized monographs that contain a large number of protocols for ligand and protein immobilization to surfaces.[57–59]

B. Protocols for Protein Adsorption Experiments

1. Adsorption Experiments Using High Surface Area Adsorbents. The following four protocols show the use of high surface area adsorbents. In each protocol the protein concentration in solution needs to be determined after adsorption took place. Protein concentration can be measured directly, according to the method of Lowry *et al.*,[60] by measuring the UV absorbance or fluorescence of the protein, using ^{125}I or FITC labels, or, in the case of adsorption of enzymes, by measuring the enzymatic activity of the protein.

a. SATURATING SAMPLE-LOAD COLUMN METHOD. This method can be employed for all sorts of gel particles (low density) and also for glass, inorganic material, or metal powders. The saturation sample-load method should be employed if one wants to compare the *absolute capacity* of different gels or powders at a defined bulk protein concentration. Methods employed for the measurement of adsorption of the enzyme phosphorylase *b* to butyl-Sepharose 4B particles at 5°, described briefly later, can be found in more detail elsewhere.[12,16,23] Before use, the protein, e.g., phosphorylase *b*, is dialyzed extensively against the adsorption buffer (buffer A containing 10 m*M* tris(hydroxymethyl)aminomethane/maleate, 5 m*M* dithioerythreitol, 1.1 *M* ammonium sulfate, 20% sucrose, pH 7.0).[16] Sucrose is included in buffer A to minimize nonspecific adsorption to the agarose gel backbone. Before adsorption the substituted beaded or particulate adsorbent is equilibrated with adsorption buffer. Low-volume Plexiglas columns (1 cm i.d. × 12 cm height) containing 1–3 ml packed gel or larger volume columns (with a large cross section for fast flow) with the dimensions 2 cm i.d. × 15 cm height filled with 10–20 ml packed gel can be employed. A purified protein solution (phosphorylase *b*) or a crude extract is applied by pump or gravity to the columns until no more enzyme is adsorbed, i.e., until the protein concentration in the run-through is identical to that in the applied sample. The columns are then washed with a 10- to 50-fold buffer volume until significant protein can no longer be detected in the run-through. The elution of protein is either accomplished by specific affinity agents, salts,[61] or, for protein balance studies, by denaturants, e.g., urea, SDS applied to the

[58] G. T. Hermanson, A. K. Mallia, and P. K. Smith, "Immobilized Affinity Ligand Techniques." Academic Press, San Diego, 1992.

[59] S. S. Wong, "Chemistry of Protein Conjugations and Cross-Linking." CRC Press, Boca Raton, FL, 1993.

[60] O. H. Lowry, N. J. Rosebrough, A. L. Farr, and R. J. Randall, *J. Biol. Chem.* **193,** 265 (1951).

[61] H. P. Jennissen and L. M. G. Heilmeyer, Jr., *Biochemistry* **14,** 754 (1975).

column in buffer. The adsorbed amount of protein is calculated from difference measurements of the amount applied and the amount in the runthrough or by elution of the adsorbed protein by a strong denaturing detergent mixture such as 0.1 *M* NaOH, 1% SDS and subsequent determination of the protein amount in this eluent.

b. LIMITED SAMPLE-LOAD COLUMN METHOD. This method can also be employed for all sorts of gel particles and for glass, inorganic material, or metal powders. The limited sample-load method is a good screening procedure for a series of adsorbents or protein samples. In this method a defined amount of protein (e.g., 1 mg) in a defined sample volume is applied to each column under identical conditions. The applied nonsaturating amount is dimensioned so as to be 100% adsorbed (i.e., no protein in runthrough) on an adsorbent displaying the expected maximal affinity and capacity. The method allows a comparison of relative adsorption capacities and affinities of adsorbents on a quantitative chromatographic basis. This method, employed for quantitative evaluation of the adsorption of calmodulin, fibrinogen, and peptides to beaded agarose adsorbents of varying hydrophobicity, is exemplified next.

In the case of calmodulin quantitative adsorption, chromatography is performed on a column (0.9 × 12 cm) containing 2 ml of packed gel of various alkyl agaroses. The gel is washed and equilibrated with 20 volumes of buffer B (20 m*M* Tris–HCl, 1 m*M* $CaCl_2$, pH 7.0). One milligram of purified calmodulin is applied to a sample volume of 1 ml (in buffer B) and 1-ml fractions are collected. The column is then washed with 9 ml buffer B and then with 9 ml buffer C (= buffer B + 0.3 *M* NaCl). The adsorbed calmodulin is eluted with buffer D [20 m*M* Tris–HCl, 0.3 *M* NaCl, 10 m*M* ethylene glycol bis(β-aminoethyl ether) *N,N,N',N'*-tetraacetic acid, i.e., EGTA, pH 7.0]. The EGTA chelates the calmodulin-bound Ca^{2+}, thereby changing the conformation of calmodulin to a more hydrophilic species leading to elution. For quantifying tightly bound, i.e., "irreversibly bound," calmodulin on highly hydrophobic columns elution is continued with 7.5 *M* urea and finally with 1% SDS added to the adsorption buffer.[62]

Quantitative hydrophobic adsorption chromatography of fibrinogen is performed on the same column type containing 2 ml of packed gel. The gel is washed and equilibrated with 20 volumes buffer E (50 m*M* Tris–HCl, 150 m*M* NaCl, 1 m*M* EGTA, pH 7.4). As sample 1 mg purified fibrinogen is applied in a volume of 1 ml (in buffer E) and 1.5-ml fractions are collected. The column is then washed with 15 ml buffer E and eluted with 7.5 *M* urea and at high hydrophobicity of the gel with 1% SDS.[62]

Quantitative hydrophobic adsorption chromatography of a tripeptide,

[62] H. P. Jennissen and A. Demirolgou, *J. Chromatogr.* **597,** 93 (1992).

Trp–Trp–Trp (Paesel & Lorei, Frankfurt, 98.5% purity) is performed on a column (0.9 × 12 cm) containing 2 ml of packed gel. The gel is washed and equilibrated with 20 volumes buffer F (1 m*M* sodium β-glycerophosphate, pH 7.0). As sample 1 mg Trp-tripeptide is applied in a volume of 1 ml (in buffer F) and 1.5-ml fractions are collected. The column is then washed with 15 ml buffer F and eluted with 1% SDS.[62] Calmodulin, fibrinogen, and the tripeptide in the obtained fractions are measured directly according to the method of Lowry *et al.*[60] employing BSA or the Trp-tripeptide as standard.

In the examples of quantitative analytical chromatography given earlier, the experiments can be performed at room temperature, good flow is usually achieved by gravity, and only a fresh, nonregenerated gel should be employed for each experiment.[62]

c. BATCH DEPLETION METHOD: ADSORPTION. In the batch adsorption method, the nonadsorbed protein is typically separated from adsorbed protein by a centrifugation process.[22] A faster sampling method was developed to circumvent the slow centrifugation step.[12] The decisive components that make this method a simple procedure are easy to make: (a) stainless-steel sampling tips for disposable syringes and (b) thermostatted beakers of volumes varying from 20 to 500 ml, which allow magnetic stirring of the contents. The top end of the sampling tip (diameter 7 mm[15]) is constructed similar to a column adapter and is covered by a stainless-steel grid (screen) with a pore size of ca. 10–20 μm soldered to the steel adapter so as to exclude the gel beads or particulate adsorbent material during sampling. The bottom end of the sampling tip corresponds to that part of a hypodermic needle (cannula) that fits onto a syringe. Gel beads/particle-free buffer can now be aspirated directly form the particle suspension being stirred in the thermostatted beaker. The aspirated sample volume can be varied from 100 to 3000 μl, but should be held small compared to the total volume of the stirred gel suspension so that an influence on the bead concentration remains negligible. The very short sampling time of 3–5 sec allows a sampling time resolution of ca. 10 sec[23] in kinetic studies with two persons sampling. Often the measurement of such high-resolution kinetics is not necessary. However, the monitoring of low-resolution kinetics is essential for establishing the existence of equilibrium in the case of adsorption and desorption isotherms.[16]

In the batch method, the procedure used for determining the protein concentration is of prime importance for the sensitivity of the method and the range of the isotherm. In the case of phosphorylase *b*, three methods were employed. Protein determinations according to Lowry[12,60] can be performed reliably in volumes of ca. 500 μl down to ca. 5–10 μg/ml. The catalytic activity of phosphorylase *b* allows the determination of the enzyme to concentrations of 2–3 μg/ml in very small sample volumes down to 50 μl.[16] A radioactive labeling of protein allows protein determinations in the range of 200–500 ng/ml.[12,16] Somewhat lower sensitivity is expected for proteins labeled with FITC.

In typical adsorption isotherms on the beaded alkyl agarose adsorbent (alkyl-Sepharose 4B) the free concentration of phosphorylase *b* at apparent equilibrium[16] is usually between 0.05 and 0.5 mg/ml. The adsorbed amount of protein is calculated from the difference between the initial and the final protein concentration, i.e., from Δc_p. A parallel experiment with control gel (unsubstituted Sepharose 4B) yields the amount of nonspecifically adsorbed protein, which is subtracted. Volume changes due to sampling have to be considered in these calculations. The amount of adsorbed protein is expressed in milligrams per milliliter of packed adsorbent or per meter squared surface area.[12,17] The calculated amount of enzyme corresponds to the amount that can be released from the gel in the presence of 0.1 *M* NaOH, 1% SDS.[16] A new incubation mixture is usually employed for each measurement, i.e., for each data point on an adsorption isotherm or kinetics plot.

Before measuring the actual adsorption isotherm, the adsorption kinetics should be determined in the protein concentration range that the isotherm is expected to cover. This is important for guaranteeing that equilibrium is being obtained for all isotherm points. The adsorption time curve is determined by adding 0.5–1.5 ml of equilibrated packed Sepharose measured in a thermostatted, graduated column[61] to 20 ml of buffer containing phosphorylase *b* in an initial concentration of approximately 0.01–1.5 mg/ml. The adsorption isotherm can be calculated directly from such adsorption time curves. Once the latter is known, the isotherms can be determined by only measuring the time end points of adsorption at equilibrium. The incubation mixture of 20 ml is stirred in a thermostatted Plexiglas beaker (2.5 cm i.d. × 9 cm, stirring bar 1.5 cm) until equilibrium is reached. A homogeneous incubation mixture is usually obtained when the speed of the stirring bar exceeds 150 rpm. For a good mixing, speeds of 500–700 rpm are generally recommended. Alternatively, incubation beakers of 60 ml volume (beaker size 3.5 cm i.d × 9 cm, stirring bar 3 cm, stirring velocity 450 rev/min) may be employed. The capacity of the adsorbent for the protein ligand in these systems is independent of the stirring rate. Examination of the substituted Sepharose under a phase microscope (Leitz) demonstrated that stirring at 700 rpm for 2 hr (20-ml beaker) did not lead to fragmentation of the beads. Samples of 0.2–0.5 ml gel-free buffer are obtained by suction with a disposable 2-ml plastic syringe through a stainless-steel sampling tip (pore diameter of steel grid 20 μm, see earlier) that excludes the Sepharose 4B spheres (bead diameter ca. 40–190 μm, Pharmacia). After each sampling procedure, the steel grid is washed by pressing 2–4 ml of H_2O, 0.5 *M* NaCl, H_2O, and, finally, acetone, respectively, through the grid with the syringe. The grid is then removed and dried in a stream of pressurized air and is ready for reuse. The washing and drying procedure of the sampling tip takes about 1 min. A fresh disposable syringe is employed for each sample.

d. BATCH REPLETION METHOD: DESORPTION. When protein desorption is measured by dilution experiments, as is described in this section, then the protein concentration in the bulk solution increases in time (after addition of loaded gel) through dissociation from the solid surface until a new equilibrium is reached (repletion). Because the amount of protein released from such a surface is usually very small,[16] sensitive methods of detection have to be applied in order to measure the new equilibrium concentration of protein. In general, only an isotope-labeling procedure (e.g., ^{3}H, ^{125}I) allows reliable determinations of the protein concentration in desorption experiments. Should such a labeling procedure be chosen, then it is essential to make any dilution of the specific radioactivity of the labeled protein only with cold-labeled protein and not with unlabeled protein.[16] This procedure avoids the generation of two different protein populations, i.e., heterogeneity, in the adsorption–desorption mixture.

Before measuring the actual desorption isotherm, the desorption kinetics should be determined in the selected protein concentration range to ensure attainment of equilibrium. In measuring desorption by dilution, the adsorbent gel is first loaded with protein in an adsorption experiment (60–90 min). At apparent equilibrium, the gel loaded with protein is isolated from the bulk buffer solution (time, 10–20 min) by first removing the buffer carefully with the sampling tip and then applying the gel slurry for further concentration to a small thermostatted, graduated column that is allowed to run "dry" by gravity. In this way, the liquid phase is effectively removed from the gel (= loaded gel), which can then be diluted into a beaker for desorption measurements. As shown previously, the concentration of gel beads in this way has no measurable effect on the attained adsorption equilibrium.[16] The protein is in the adsorbed state on the gel prior to dilution for a maximal time of 70–100 min. For the following dilutions of the gel volume by factors of 1:10 to 1:100, the freshly loaded gel (0.5–2.0 ml packed gel) is blown from the small column into a fresh protein-free buffer solution stirred in a thermostatted beaker of 2.5 cm i.d. (20 ml) or 3.5 cm i.d. (60 ml) as described previously.[16] For high dilutions (e.g., 1:500 and 1:1000), a thermostatted 500-ml beaker (10 cm i.d. $\times$ 14 cm, stirring bar 8 cm, 200 rpm) is employed. An equally efficient mixing of the agarose buffer suspension is obtained at 700 (beaker 2.0 cm i.d.), 450 (beaker 3.5

cm i.d.), and 200 (beaker 10 cm i.d.) rpm. Under these conditions the desorption rate is independent of the stirring velocity.

After dilution of the protein-loaded gel into the buffer solution the desorption mixture is incubated for 90–120 min until a new apparent equilibrium is reached (time independence). Samples of 1–3 ml are taken at regular intervals during this time using the steel grid method (sampling tip) for protein concentration determination assay such as radioactivity counting. As control unsubstituted Sepharose 4B is incubated for 90 min prior to dilution with a concentration of free protein, so chosen for adsorption, that a time-independent concentration of free protein, comparable to that obtained with the substituted gel at desorption equilibrium, is obtained.[16] The amount of adsorbed protein is then calculated as described earlier.

2. *Adsorption Experiments Using Flat Surface Adsorbents.* The following protocols describe the use of flat, low surface area, fused silica in protein adsorption experiments. The advantage of flat silica is that its surface can be modified chemically (see Section IV,A,4) to carry a specific chemical group, immobilized ligand, or another protein. Together with similarly modified, smooth silicon wafers, these surfaces can be used in ellipsometry and atomic force microscope protein adsorption studies. Transparent flat silica surfaces are also needed for evanescent wave fluorescence spectroscopy (see Section IV,C). Due to the small surface area of flat surfaces, the solution depletion methods cannot be used at all and the detection of adsorbed protein has to be carried out directly on the flat adsorbent.

a. Radiolabeled Protein Adsorption onto a Flat Silica Surface. Fused silica plates small enough to fit into the counting chamber of a γ-counter ($2.5 \times 0.9 \times 0.1$ cm, area 5.2 cm^2) are cut with a diamond saw (Buehler) and their sides are polished. The plates are cleaned and modified chemically as desired. The adsorption chambers are prepared from a glass tube (1.0 cm i.d.). The length of the chamber is chosen so that it can accommodate three small plates. One end of the glass tube is drawn to allow 3/16-inch i.d. plastic tubing to fit tightly: the tubing leads to a waste container. The other end is fitted with a rubber stopper pierced with a 1 : 1/2-inch 18-gange syringe needle. When joined with a three-way stopcock, the addition of protein and buffer solutions to the adsorption chamber is easy.

Three silica plates are placed in each chamber and the chamber is prewetted with buffer solution. The buffer is drained and 10 ml of ^{125}I-labeled protein solution is injected slowly, completely filling the chamber. Different concentrations of protein are used in each chamber: the ^{125}I-labeled protein concentration is set as needed for an adsorption isotherm determination. Adsorption is allowed to proceed for 30 min after which each chamber is flushed with 6 volumes of buffer and drained. Each plate is removed carefully, placed in a capped scintillation vial, and counted for 1 min in a γ-counter (Model 170M, Beckman Instruments). One-hundred microliter aliquots of each ^{125}I-labeled protein solution are counted separately. Counts obtained from the ^{125}I-labeled protein solution are used to construct a calibration curve (i.e., radioactivity counts vs protein amount) from which the amount of adsorbed protein is calculated.[53] The protocol can also be applied to study adsorption of one ^{125}I-labeled protein from a multiprotein solution mixture.

b. AUTORADIOGRAPHY AND PROTEIN ADSORPTION FROM FLOWING SOLUTION. In some applications one is interested in protein adsorption taking place from a flowing rather than from a quiescent solution. When protein adsorption takes place in a narrow and long flow chamber the adsorbed amount will not be the same at the beginning and at the end of the chamber even if the adsorbing surface is identical everywhere in the chamber. This effect is due to the depletion of protein along the flow direction. The interplay between the flow and the adsorption is even more complex in the case of adsorption from the multiprotein solution. In such cases one needs to record the protein-adsorbed amount at each surface position along the flow direction. One way of achieving this spatial resolution is to use ^{125}I-labeled protein and autoradiography.[63] A simple protein adsorption experiment that employs a dual rectangular channel flow cell and adsorbed protein detection using autoradiography is described. A schematic of the autoradiography flow cell can be found in Lin *et al.*[64] The cell support is made of a poly(methyl methylacrylate) (PMMA) block (dimensions: $2.54 \times 2.54 \times 7.62$ cm) with dual ports. A surface-modified silica plate (dimensions: $0.1 \times 2.54 \times 7.62$ cm) serves as the other interior surface. A gasket of silicone sheet (0.05 cm thick) is placed between the PMMA block and the silica plate. The silicone rubber gasket defines the dimensions of the flow channel: $0.05 \times 0.5 \times 6$ cm. The dual-flow channel cell is assembled using two aluminum plates and six screws to provide a watertight sealing. Prior to the protein adsorption experiment, both the channels are filled with buffer solution.

In the actual experiment the protein adsorption is started by flowing the protein solution through the first channel of the flow cell with a given flow rate. In the present example, a 0.4-mg/ml solution of ^{125}I-labeled low-density lipoprotein, [^{125}I]LDL, is used and the flow is 0.84 ml/min. After a desired period of time, the flow is switched to the buffer solution to

[63] E. Hahn, *Am. Lab.,* July, 64 (1983).
[64] Y.-S. Lin, V. Hlady, and J. Janatova, *Biomaterials* **13,** 61 (1992).

remove the unbound proteins from the flow cell and allow for desorption. At the end of the adsorption–desorption cycle, 3 ml of 0.6% glutaraldehyde in buffer solution is introduced into the flow channel to fix the adsorbed protein molecules for 5 min. The flow channel is then emptied by introducing the air. Subsequently, another adsorption–desorption experiment is carried out in the second flow channel using the same surface. The flow cell is disassembled and the surface with fixed adsorbed protein is subjected to autoradiography. A DDS-treated silica plate with a set of known amounts of [^{125}I]LDL is prepared separately to serve as an autoradiography calibration standard for the quantification of adsorbed protein. The known amounts of [^{125}I]LDL are applied as small droplets to the DDS-treated plate and dried completely.

The autoradiographs are obtained using 5 × 7-inch photosensitive film (X-OMAT AR, Kodak, Rochester, NY) in a light-tight cassette. Both plates are covered with Scotch tape to avoid the contamination of film. The two plates are placed in a polyethylene bag and are brought in contact with an autoradiographic film. The film is exposed to the samples at low temperature (−70°) for 21 days. No intensifying screen is used as it degrades the spatial resolution of the autoradiographs. After a desired exposure time (typically 2–3 weeks), the exposed film is processed in an automated developer system.

The optical density of the developed film can be recorded by any good quality imaging densitometer. The densitometer is first used to record the light source and then the autoradiographs of adsorbed protein and calibration plates. An example of a digitized autoradiography image of [^{125}I]LDL adsorbed along a C_{18} silica gradient surface[65] (upper half, Fig. 1) and of a plate with protein standards (lower half, Fig. 1) is shown in Fig. 1.

The integrated optical density, IOD, of each protein spot on the calibration plate is computed as a sum of background-subtracted optical density, OD_{cal}:

$$OD_{cal} = \log(\mathrm{Int}/\mathrm{Int}_o)_{spot} - \log(\mathrm{Int}/\mathrm{Int}_o)_{back} \tag{5}$$

[65] The so-called "gradient" surface was originally designed as an experimental tool to investigate the effect of protein-binding residue surface density in a single protein adsorption experiment.[66] In Fig. 1, [^{125}I]LDL adsorption onto an octadecyldimethylsilyl (C_{18})-silica gradient surface is shown: approximately one-half of the silica plate is fully covered with C_{18} residues and the other half of the silica plate is left unmodified with a short gradient of C_{18} chain surface density in between the two halves.

[66] H. Elwing, S. Welin, A. Askendahl, U. Nilsson, and I. Lundström, *J. Colloid Interface Sci.* **119,** 203 (1987).

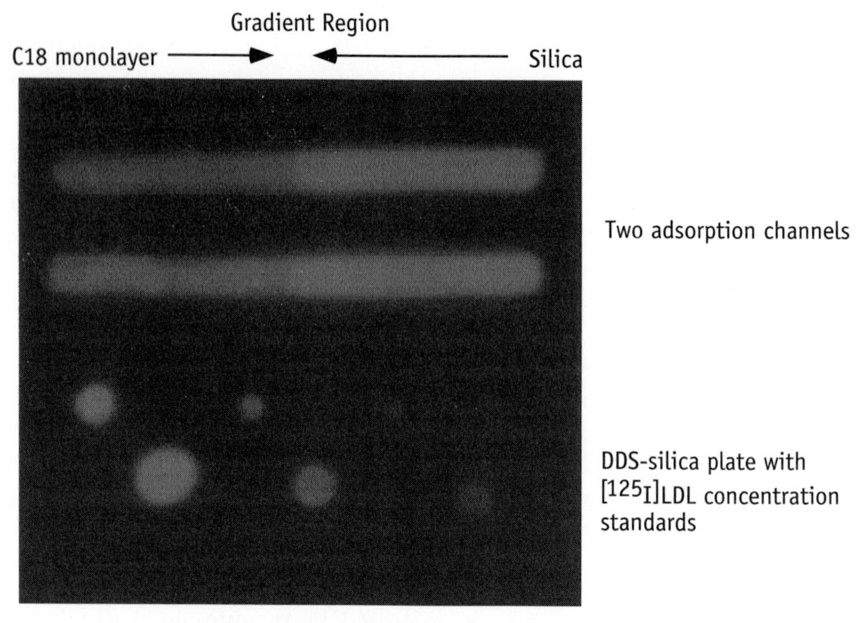

FIG. 1. Autoradiograph of [^{125}I]LDL adsorbed along the C_{18} silica gradient surface in two adsorption channels (top) and a DDS silica plate with [^{125}I]LDL standards (bottom).

measured at each individual "pixel" of the spot using the Beer–Lambert equation, where Int is the intensity of the light transmitted through the autoradiograph to the individual pixel of the CCD camera in the area of the protein spot, Int_o is the intensity of the light source reaching the camera pixel without the autoradiograph, and the subscripts "spot" and "back" represent the contribution from the protein spot area and background, respectively. Using this method, IOD of each protein standard is determined first. From the IOD of the standards, a calibrating relationship between the amount of protein (in mass per area unit) and IOD per pixel (in counts) is obtained. Using this calibration curve, the adsorbed amount of protein found in each channel is then calculated as a function of the gradient position. The adsorbed amount can also be presented as a two-dimensional map. Similar experiments, each performed with a different protein concentration, can be used to construct an adsorption isotherm for any given position on the adsorbent surface.

C. Protocols for Using Total Internal Reflection Fluorescence Spectroscopy

1. Intrinsic Protein Fluorescence TIRF Experiments. Among the techniques that provide information about the conformational state of protein molecules, fluorescence spectroscopy can be applied to both solution and adsorbed protein molecules.[7] A variant of fluorescence spectroscopy for interfacial studies, the total internal reflection fluorescence (TIRF) technique, is based on the phenomenon of total internal reflection: when an incident light beam impinges on the planar interface at an angle (Θ_i) greater than the critical angle (Θ_c), the beam is totally reflected. In the case when the interface is the one between the solid (refractive index, n_1) and the solution (refractive index, n_2), the totally reflected light creates an evanescent surface electromagnetic wave in the solution, thereby selectively exciting molecules that are close to the interface. The intensity (I) of the evanescent wave is a function of the distance (z) from the interface into the solution:

$$I(z) = I_o \exp(-2z/d_p) \tag{6}$$

where I_o is the intensity of the electromagnetic wave right at the interface and d_p is the characteristic decaying distance, or so-called penetration depth of the evanescent wave, and it is related to the optical properties of the two phases:

$$d_p = (\lambda_o/4\pi)\ [n_2{}^2 \sin^2\Theta_i\ n_1{}^2)^{-1/2}] \tag{7}$$

where λ_o is the wavelength of the incident light in vacuum.

Although most of the protein adsorption TIRF studies are performed using fluorescently labeled proteins, it is also possible to utilize the intrinsic fluorescence of the tryptophan residues in UV.[8,33] Advantages of intrinsic protein fluorescence spectroscopy are that the tryptophanyl fluorescence emission is sensitive to changes of protein conformation and that no labeling is required. The schematic of the UV TIRF apparatus is given in Fig. 2. The excitation light source is a 300-W high-pressure Xe lamp (R300-3, ILC Technology). The wavelength of excitation is selected by a monochromator (0.1 m, f/4.2, H10 1200 UV, ISA Inc.). The light is pseudo-collimated by a lens [L1, focal length (f.l.) 8 cm] and passed through a polarizer. The beam is typically polarized vertically with respect to the incident plane. The focused beam (L2, f.l. 10 cm) is directed normal to the face of the 70°-cut dovetail-fused silica prism where it totally reflects at the interface between the sorbent surface and the solution. The adsorbent surface consists of a quartz slide, which is optically coupled to the prism using glycerol. The quartz surface exposed to the protein solution can be modified chemically. A silicon rubber gasket is used to separate the slide from a black-anodized aluminum support creating a flow-through space, which is filled with solutions. Solutions are injected into the cell using syringes and the flow rate is controlled by a syringe pump. Fluorescence emitted from the interface is collected through the prism, collimated by a lens (L3, f.l. 3 cm), focused by another lens (L4, f.l. 9 cm) onto the slit of the emission monochromator

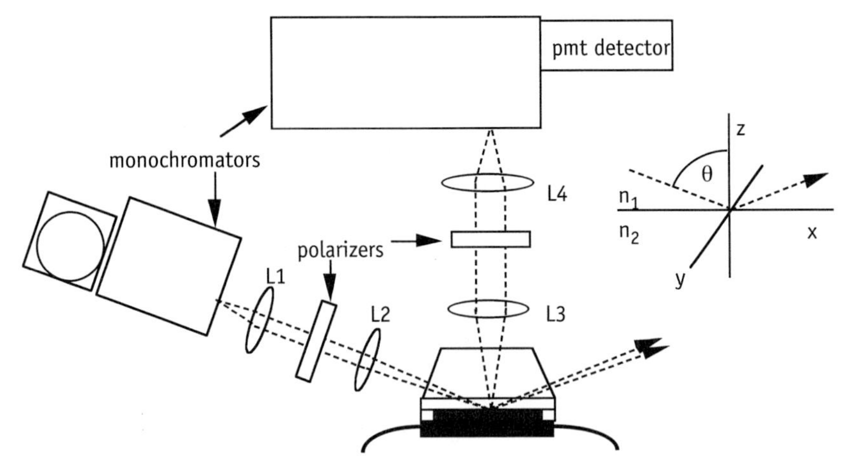

FIG. 2. Schematic drawing of the UV TIRF setup. (Inset) Coordinate system. The solid/liquid interface is within the *xy* plane. The excitation beam is polarized normal to the plane of incidence (*xz* plane) and is incident at the solid/liquid interface at an angle, Θ, measured from the normal to the interface. Redrawn after J. Buijs and V. Hlady, *J. Colloid Interface Sci.* **190,** 171 (1997).

(0.64 m, f/5.2, HR-640, ISA Inc., Metuchen, NJ), and subsequently detected by a cooled photomultiplier tube (R928, Hamamatsu, Bridgewater, NJ). The photomultiplier signal is sent into a photon counter (SR 400, Stanford Research, Sunnyvale, CA) controlled by a PC computer for fluorescence emission measurements and data acquisition. Final alignment of the TIRF setup is performed by optimizing the fluorescent signal emitted from a cell filled with a solution that generates a high fluorescence. In addition to optimizing the fluorescent signal, accurate measurements require a low background light intensity. The background intensity can be lowered by shielding the pathway from the cell to the emission monochromator, by placing a cutoff filter in the excitation pathway, which will reduce the light intensity at emission wavelengths, and by performing experiments in the dark.

2. Calibration of TIRF Using Fluorescence Standards. To quantify the adsorbed amount of proteins, the fluorescence intensity has to be calibrated. One method to perform such a calibration is by using standard fluorescence solutions.[8,53] In general, the fluorescence intensity emitted by molecules excited by the evanescent wave is proportional to the probability of absorption, characterized by its extinction coefficient, ε, the fluorescence quantum yield, ϕ, the concentration of molecules, c, multiplied by the intensity of the evanescent wave, I. Both the concentration and the intensity are a function of the distance from the interface into the cell, z. The observed fluorescence signal, S, is written as

$$S = f\varepsilon\phi \int_0^\infty c(z)I(z)dz \tag{8}$$

where f is an instrumental factor, which accounts for efficiency of fluorescence detection. The instrumental factor, f, is calibrated using a set of standard fluorescence solutions.[53] In the UV range the TIRF set-up is typically calibrated by recording the fluorescence intensities of a number of tryptophan solutions with different concentrations. A suitable tryptophan solution is, for example, prepared using L-5-hydroxytryptophan (TRP). The fluorescence intensity of standard solutions emanates from adsorbed and nonadsorbed molecules in the evanescent field and from the fluorescence excited by scattered light. By increasing the fluorophore concentration, a point is reached for which the scattered light is entirely absorbed by solution molecules. After that point is reached, the fluorescence from nonadsorbed fluorophores present in the volume excited by the evanescent surface wave increases linearly with concentration. This linear increase, which reflects the increment in the measured fluorescence signal per increment in concentration (in units of absorption) of TRP solutions, is directly used to convert the fluorescence signal from adsorbed proteins into surface concentrations. The conversion is performed by normalizing, i.e., dividing, the fluorescence signal from an adsorbed protein layer to the fluorescence signals obtained from standard TRP solutions. The fluorescence of TRP emanates from molecules in solution, which are excited by the evanescent wave, and is derived by integration of the observed fluorescence signal, S. Because the size of an adsorbed protein is approximately two orders of magnitude less than the penetration depth of the evanescent wave, the protein concentration profile is considered as a quantity equal to the surface concentration, Γ_p, positioned at $z = 0$. After rewriting the normalized fluorescence intensity the surface concentration is expressed as

$$\Gamma_p = S_p \cdot \frac{d}{2\varepsilon_p} \cdot \frac{1}{\Delta S_t / \Delta(c_t \varepsilon_t)} \cdot \frac{\varphi_t}{\varphi_p} \tag{9}$$

where the subscripts p and t refer to quantities of the protein and tryptophan standards, respectively. The remaining unknown parameters are the fluorescence quantum yields. Their ratio, φ_t/φ_p, is established from the ratio of the fluorescence intensity increments per unit of absorption for the respective bulk solutions.[8] Note that the quantum yield of fluorophores is sensitive to its environment, and therefore, the fluorescence intensity in bulk solution can differ from that in the adsorbed state.

3. Protein Adsorption Kinetics Experiments Using TIRF. Before a protein adsorption measurement is started, the sensitivity of the TIRF setup is calibrated using three tryptophan solutions with increasing concentrations, thus providing the value for $\Delta S_t/\Delta(c_t\varepsilon_t)$ in Eq. (9). After the cell is rinsed with water and the last tryptophan solution is removed, the cell is filled with a buffer solution and the background signal intensity at the relevant emission wavelengths are recorded. Protein solutions flow through the cell at a desired flow rate. The adsorption kinetics are followed by monitoring the fluorescence signal at the emission wavelengths for a desired period of time. In the case of intrinsic protein fluorescence the emission wavelength is typically recorded at 340 nm. The fluorescence signal during the desorption part of the experiment is followed for a desired period of time while a buffer solution is flowing through the cell. The adsorption–desorption cycle is then repeated with another protein concentration using either the same or a pristine adsorbent surface.

4. Fluorescence Emission Spectra and Quenching Experiments. In addition to providing information about Γ_p, the fluorescence emission from adsorbed protein can also be interpreted with respect to the conformation of adsorbed molecules by examining the fluorescence emission spectra and studying the solvent accessibility of the fluorophore in the protein by fluorescence quenching experiments. With the protein molecules still adsorbed on the surface, the fluorescence emission intensity is recorded as a function of the emission wavelength. The wavelength of maximum fluorescence intensity depends on the energy dissipation between excitation and emission of the fluorophore. Fluorescence emission spectra of the tryptophanyl residue in proteins are very sensitive to the polarity of its local surrounding. Usually, a more polar local environment results in a larger energy dissipation of the excited state and thus in a red-shifted emission maximum wavelength. A tryptophanyl residue surrounded by water has an emission maximum at 348 nm, whereas the tryptophanyl fluorescence maximum inside protein structures varies from 308 nm for azurin to 352 nm for glucagon.[30] The shift in the emission maximum wavelength is sometimes difficult to interpret because a change in the local environment of the fluorophore can alter the duration between excitation and emission as well as the amount of energy dissipation per time unit. Nevertheless, fluorescence emission differences between solution and adsorbed protein can be used as an indication for structural changes in the local surrounding of the fluorescent reporter group.[8]

Quenching of fluorescence occurs when a quencher collides with the excited fluorophore group, thereby adsorbing the energy of the excited state. As quenching occurs on molecular contact between the quencher and the fluorescent group, this process yields information on the accessibility of fluorophores for the quencher and, hence, about the permeability of protein molecule. A large amount of organic or inorganic molecules can act as quenchers.[30] One should take into account that the action of charged quenchers is also affected by electrostatic interactions. Collisional quenching of fluorescence is described by the Stern–Volmer equation:

$$F_o/F = 1 + K[Q] \qquad (10)$$

where F_o and F are the fluorescence intensities in the absence and presence of quencher, respectively, Q is the concentration of the quencher, and K the Stern–Volmer quenching constant. By plotting the ratio F_o/F against the concentration of the quencher, a linear relation should be obtained for collisional quenching, and the Stern–Volmer quenching constant, K, indicates how accessible the fluorophore is. The quenching experiment is performed by injecting buffer solutions with increasing concentrations of the quencher while measuring the fluorescence intensity at the maximum emission wavelength. The adsorbed protein quenching is compared with the quenching of the same protein in bulk solution or at some other standard conditions. As desorption of proteins from the surface might occur, the fluorescence intensity of the sample without the presence of a quencher in the buffer solution is measured between the quenched fluorescence intensity measurements.

[27] Monitoring Protein Assembly Using Quasielastic Light Scattering Spectroscopy

By ALEKSEY LOMAKIN, GEORGE B. BENEDEK, and DAVID B. TEPLOW

Introduction

The process of protein assembly is vital for the survival of all organisms, from those as complex as humans to those as simple as viruses. Common macromolecular assembly reactions include actin and tubulin polymerization, collagen and myosin fibrillization, and formation of bacterial flagella and viral capsids. Ordered protein assembly is also required to produce multisubunit structures with enzymatic, transport, or other activities, such as bacterial aspartate transcarbamoylase, ribosomes, proteasomes, ion channels, receptor complexes, and transcription complexes.

In addition to their role in the normal physiology of living organisms, protein assembly reactions have been found to be associated with, or causative of, an increasing number of human diseases. Examples include Alzheimer's disease,[1] prion diseases,[2] and a large variety of systemic amy-

[1] D. J. Selkoe, *J. Neuropathol. Exp. Neurol.* **53,** 438 (1994).
[2] A. L. Horwich and J. S. Weissman, *Cell* **89,** 499 (1997).

0076-6879/99 $30.00

loidoses.[3] In each case, proteins that exist normally in a soluble, disaggregated state assemble into ordered polymers, which then form insoluble deposits. Because the proteins involved participate in normal physiologic processes, therapeutic strategies targeting their production may be unfeasible. Therefore, efforts have been directed at inhibiting and/or reversing the aberrant assembly and deposition of these proteins. To achieve this goal, the key factors controlling the assembly process must be elucidated. This requires both model systems in which reproducible and experimentally manipulable protein assembly occurs and the ability to monitor the process with high sensitivity and resolution. This article discusses the use of quasielastic light scattering spectroscopy (QLS)[4] for monitoring protein assembly reactions.

QLS is an optical method for the determination of diffusion coefficients of particles undergoing Brownian motion in solution.[5,6] Knowledge of diffusion coefficients allows one to determine many features of the molecules under study, including their size, shape, and flexibility. Temporal changes in these parameters provide important information about the kinetics and structural transitions that occur during protein assembly. The QLS method is rapid, sensitive, noninvasive, and quantitative, which makes it useful for thermodynamic studies. Although QLS spectrometers may be constructed relatively easily and are also available commercially, obtaining accurate and meaningful information from a QLS experiment requires thorough knowledge of the theoretical and practical aspects of the technique. This point cannot be overemphasized.

"Quick Start" Guide for the Reader

The first three sections of this article address the theoretical underpinnings of the QLS method and include general discussions of experimental techniques of light scattering, of quasielastic light scattering from macromolecules in solution, and of methods of analyzing QLS data. Next are presented sections on unique features of the QLS monitoring of fibril growth and important practical considerations in the execution of the QLS experiment. Finally, a number of real-world examples of QLS studies of amyloid β-protein (Aβ) fibril formation are presented. Those readers already con-

[3] P. Westermark, *Am. J. Pathol.* **152,** 1125 (1998).

[4] The method of QLS is also referred to as "dynamic light scattering" and "photon correlation spectroscopy." These, and several other terms, are interchangeable.

[5] R. Pecora, "Dynamic Light Scattering: Applications of Photon Correlation Spectroscopy." Plenum Press, New York, 1985.

[6] K. S. Schmitz, "An Introduction to Dynamic Light Scattering by Macromolecules." Academic Press, Boston, 1990.

versant in the theory of QLS or who may wish to simply understand the types of data obtainable using the method are encouraged to focus first on the last section of the article and then, if desired, to examine the penultimate two sections on unique and practical aspects of QLS studies of fibrillogenesis. For those readers new to the QLS method or interested in applying the method in their own laboratories, starting at the beginning is highly recommended.

General Principles

Light Scattering

The propagation of light may be viewed as a continuous rescattering of the incident electromagnetic wave from every point of the illuminated medium. The amplitude of each secondary wave is proportional to the polarizability at the point from which this wave originates. If the medium is uniform, rescattered waves all have the same amplitude and interfere destructively in all directions except in the direction of the incident beam. If, however, at some location the index of refraction differs from its average value, the wave rescattered at this location is not compensated for and some light can be observed in directions other than the direction of incidence, i.e., light scattering occurs. The scattering of light may thus be viewed as a result of microscopic heterogeneities within the illuminated volume. Macromolecules and supramolecular assemblies are examples of such heterogeneities.

Light-Scattering Techniques

Light scattering is a versatile approach for noninvasive, fast, and accurate study of scattering media. The simplest light scattering technique is turbidimetry. Turbidimetry measures attenuation in the intensity of light as it passes through a scattering medium. This attenuation is characterized quantitatively by the extinction coefficient, which is determined by the concentration, molecular weight, and size of the scatterers. Turbidity is a good indicator of phenomena such as aggregation, agglutination or phase separation, in which large scatterers alter the macroscopic properties of the medium. It is widely used to detect and monitor formation and subsequent aggregation of Aβ fibrils.

Static light scattering probes concentration, molecular weight, size, shape, orientation, and interactions among scattering particles by measuring the average intensity and polarization of the scattered light. Static light scattering measurements done at different scattering angles provide information on the molecular weight, size, and shape of the scattering particles. Measurements of the intensity of light scattering as a function of concentration yield the second virial coefficient, which is the key characteristic of the strength of attractive or repulsive interactions between solute particles.

Quasielastic light scattering probes the relatively slow fluctuations in concentration, shape, orientation, and other particle characteristics by measuring the correlation function of the scattered light intensity. Fast vibrations of small chemical groups, which lead to significant changes in the frequency of the scattered light, are the domain of Raman spectroscopy. These latter two methods, which probe the dynamics of the scatterers, are intrinsically more complicated than static light scattering, as they involve measurements of spectral characteristics or related correlation properties of the scattered light.

The spectrum of the scattered light is affected by the motion of the scattered particles. Fast motion leads to large changes, which can be measured using classical interferometers. The slow motion associated with the diffusion of macromolecules in solution leads to tiny changes in the frequency of the scattered light. Only with the advent of lasers has it become practical to observe these small spectral changes indirectly by measuring the correlation function of the scattered light intensity, i.e., by QLS.

Light Scattering from Macromolecules in Solution

In solutions there are two equivalent ways to describe light scattering. One way is to consider the solution as a homogeneous medium and ascribe light scattering to the spatial fluctuations in the concentration of a solute. An alternative way is to consider each individual solute particle as a heterogeneity and therefore as a source of light scattering. The first approach is more appropriate for solutions of small molecules in which the average distance between scatterer centers is small compared to the wavelength of light. The second approach is more appropriate for solutions of large macromolecules and colloids, when the average distance between particle centers is comparable to the wavelength of light. In the case of amyloid fibrils, the size of the solute particles themselves becomes comparable to the wavelength of light. In this case, the description of the effects of orientational motion and deformation of the solute particles is much more straightforward when these particles are treated as individual scatterers.

Using these considerations, we will describe light scattering in a solution of macromolecules as a result of the interference of electromagnetic waves scattered by individual solute particles with a refractive index not matching that of the solvent. Thus, the first question to address is the intensity of the light scattered by a single particle and its dependence on the mass and the shape of the particle. We first consider an aggregate composed of m monomers. Let the amplitude of the electromagnetic wave scattered by an individual monomer be E_0 (at the point of observation). If the size of the aggregate is small compared to the wavelength of light, λ, all waves scattered by individual monomers interfere constructively and the resulting wave has an amplitude $E = mE_0$. The intensity of a light wave is proportional to its amplitude squared. Thus the intensity of the light scattered by the aggregate is proportional to the aggregation number squared, $I = m^2 I_0$, where I_0 is the intensity of scattering by a monomer. The quadratic dependency of scattering intensity on the size of the scatterer is the basis for optical determination of the molecular weight of macromolecules, for various turbidimetry techniques, and for understanding many natural phenomena, from critical opacification to cataractogenesis in the eye lens.[7]

If an aggregate particle is not small compared to λ, the interference of the electromagnetic waves scattered by the constituent monomers is not all constructive and the phases of these waves must be taken into account. If the phase of a wave scattered at the origin is used as a reference, the phase of a wave scattered at a point with radius vector $\mathbf{r}$ is $\mathbf{q} \cdot \mathbf{r}$ (Fig. 1). The vector $\mathbf{q}$ is called the "scattering vector" and is a fundamental characteristic of any scattering process. Its length is $q \equiv |\mathbf{q}| = 4\pi n/\lambda \cdot \sin \theta/2$, where n is the refractive index of the medium, λ is the wavelength of light, and θ is the scattering angle. Partial cancellation of waves scattered by different parts of the large aggregate reduces the intensity of light scattering by a factor of $|\alpha|^2$, where α is an averaged value of the phase factors $\exp(i\mathbf{q} \cdot \mathbf{r})$ for all monomers. The factor α should be averaged over all possible orientations of the particle. The result of this averaging yields the structure factor, $S(q)$. Expressions for the structure factors for particles of various shapes can be found elsewhere.[8]

Method of Quasielastic Light Scattering Spectroscopy

Motion of Particles in Solution Causes Temporal Fluctuations in Scattered Light Intensity

We are now in a position to analyze the light scattered from a collection of N solute molecules. For the moment, we will assume that the solution is sufficiently dilute so that the scattering particles move independently. At the observation point we again have a sum of waves scattered by individual particles, but this time each particle could be at any random location within

[7] G. B. Benedek, *Appl. Opt.* **10,** 459 (1971).
[8] H. C. van de Hulst, "Light Scattering by Small Particles." Dover, New York, 1981.

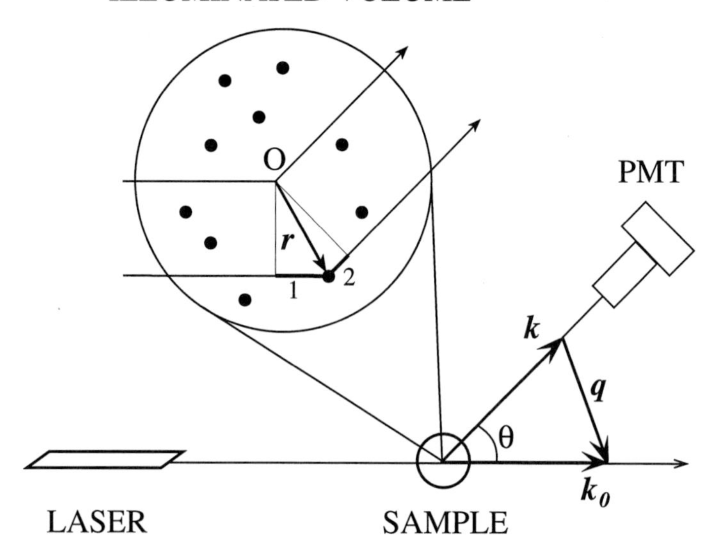

FIG. 1. The scattering vector **q**. The path traveled by a wave scattered at the point with radius vector **r** differs from the path passing through the reference point O by two segments, 1 and 2, with lengths l_1 and l_2, respectively. The phase difference is $\Delta\phi = k(l_1 + l_2)$, where $k \equiv |\mathbf{k}| = |\mathbf{k}_0| = 2\pi n/\lambda$ is the absolute value of the wave vector **k** (or $\mathbf{k}_0$). The segment l_1 is a projection of **r** on the wave vector of the incident beam $\mathbf{k}_0$, i.e., $l_1 = \mathbf{r} \cdot \mathbf{k}_0/k$. Similarly, $l_2 = -\mathbf{r} \cdot \mathbf{k}/k$, and thus $\Delta\phi = \mathbf{r} \cdot (\mathbf{k}_0 - \mathbf{k}) = \mathbf{rq}$. Vector $\mathbf{q} = \mathbf{k}_0 - \mathbf{k}$ is called the scattering vector.

the scattering volume (the intersection of the illuminated volume and the volume from which the scattered light is collected). The size of the scattering volume is much bigger than q^{-1} (with the exception of nearly forward scattering, where $q \sim \theta \approx 0$) and the phases of the waves scattered by different particles vary dramatically. As a result, the average amplitude of the scattered wave is proportional to $N^{1/2}$. Thus the average intensity of the scattered light is simply N times the intensity scattered by an individual particle, as expected. The local intensity, however, fluctuates from one point to another around its average value. The spatial pattern of these fluctuations in light intensity, called an interference pattern or "speckles," is determined by the positions of the scattering particles. As the scattering particles move, the interference pattern changes in time, resulting in temporal fluctuations in the intensity of light detected at the observation point. The essence of the QLS technique is to measure the temporal correlations in the fluctuations in the scattered light intensity and to reconstruct the physical characteristics of the scatterers from these data.

Coherence Area

There is a characteristic size for speckles in the interference pattern. If the intensity of the scattered light is above average at a certain point it will also be above the average within an area around this point where phases of the scattered waves do not change significantly. This area is called the coherence area. Within different coherence areas, fluctuations in the intensity of light collected are statistically independent. Therefore, increasing the size of the light-collecting aperture in a QLS spectrometer beyond the size of a coherence area does not lead to improvement of the signal-to-noise ratio because the temporal fluctuations in the intensity are averaged out. For a monochromatic source, the scattered light is coherent within a solid angle of the order of λ^2/A, where A is the cross-sectional area of the scattered volume perpendicular to the direction of the scattering. Because the coherence angle is fairly small, powerful (10–500 mW) and well-focused laser illumination and photon-counting techniques are used in QLS.

Correlation Function

The photodetector signal in QLS is, in fact, random noise. Information is contained only in the spectrum or correlation function of this random signal (the spectrum is a Fourier transform of the correlation function). The correlation function of the signal $i(t)$, which in the particular case of QLS is the photocurrent, is defined as

$$G^{(2)}(\tau) = \langle i(t)i(t + \tau)\rangle \tag{1}$$

The notation $G^{(2)}(\tau)$ is introduced to distinguish the correlation function of the photocurrent from the correlation function of the electromagnetic field $G^{(1)}(\tau)$, which is the Fourier transform of the light spectrum:

$$G^{(1)}(\tau) = \langle E(t)E^*(t + \tau)\rangle \tag{2}$$

In Eqs. (1) and (2), angular brackets denote an average over time t. This time averaging, an inherent feature of the QLS method, is necessary to extract information from the random fluctuations in the intensity of the scattered light.

For very large delay times τ, the photocurrents at moment t and $t + \tau$ are completely uncorrelated and $G^{(2)}(\infty)$ is simply the square of the mean current $\bar{i}^2$. At $\tau = 0$, $G^{(2)}(0)$ is obviously the mean of the current squared $\overline{i^2}$. Since for any $i(t)$, $\overline{i^2} \geq \bar{i}^2$, the initial value of the correlation function is always larger than the value at a sufficiently long delay time. The characteristic time within which the correlation function approaches its final value is called correlation time. For example, in the most practically important case of a correlation function that decays according to an exponential law $\exp(-\tau/\tau_c)$, the correlation time is the parameter τ_c.

In the majority of practical applications of QLS, the scattered light is a sum of waves scattered by many independent particles and therefore displays Gaussian statistics. This being the case, there is a relation between the intensity correlation function $G^{(2)}(\tau)$ and the field correlation function $G^{(1)}(\tau)$:

$$G^{(2)}(\tau) = I_0^2(1 + \gamma|g^{(1)}(\tau)|^2) \quad (3)$$

Here $g^{(1)}(\tau) \equiv G^{(1)}(\tau)/G^{(1)}(0)$ is the normalized field correlation function, I_0 is the average intensity of the detected light, and γ is the efficiency factor. For perfectly coherent incident light and for scattered light collected within one coherence area, the efficiency factor is 1. If light is collected from an area J times larger than the coherence area, fluctuations in light intensity are averaged out and the efficiency factor is of the order of $1/J \ll 1$. Low efficiency makes the quality of measurements extremely vulnerable to fluctuations in the average intensity caused by the presence of large dust particles in the sample or by instability of the laser intensity. It is thus highly desirable to collect the scattered light from within one coherence area. Unfortunately, this is not always possible because the coherence area is small and the intensity of light might be too low. If fewer than one photon is registered per correlation time τ_c, the statistical accuracy of the correlation function deteriorates rapidly. The optimal size of the light-collecting aperture thus varies depending on the conditions of the experiment.

Determination of Correlation Function

In modern instruments the correlation function is determined digitally. The number of photons registered by the photodetector within each of a number of short consecutive intervals is stored in the correlator memory. Each count in a given interval (termed the "sample time" and denoted Δt) represents the instantaneous value of the photocurrent $i(t)$. The series of K counts held in the correlator memory is termed the "digitized copy" of the signal. According to Eq. (1), to obtain the correlation function $G^{(2)}(\tau)$ at $\tau = n\Delta t(n = 1 \ldots K)$, the average product of counts separated by n sample times should be determined. The number n is referred to as a channel number. Up to K channels, in principle, can be measured simultaneously, but usually a smaller subset of M equidistant or logarithmically spaced channels is used. Clearly, the shortest delay time at which the correlation function is measured by the procedure described earlier is Δt (channel 1). The longest delay time cannot exceed the duration of the digitized copy, $K\Delta t$. Thus, it is important that the correlation time τ_c fit into the interval $\Delta t \ll \tau_c \ll K\Delta t$. This condition determines the choice of the sample time for the particular measurement.

To increase the statistical accuracy with which the correlation function is determined, it is essential to maximize the number of count pairs whose products are averaged within the measurement time. If the correlation function is being measured in M channels simultaneously, ideally M products should be processed for each new count, i.e., during sample time Δt. The instrument capable of doing this is said to be working in the "real-time regime." The real-time regime means that the information contained in the signal is processed without loss. The minimal sample time Δt and the number of channels M capable of being processed in real time are the most critical characteristics of a correlator. Modern commercial correlators, with several hundred channels for different delay times, can work in real time at sample times as low as a fraction of a microsecond. This capability is sufficient to measure diffusion coefficients of the smallest macromolecules.

QLS System

A QLS system consists of four elements: the sample, the optical equipment, the correlator, and the data analysis software (Fig. 2). To obtain satisfactory results, all four elements must meet certain criteria. The correlator should work close to the real-time regime for sample times significantly shorter than the correlation times of the molecules under investigation. The optical setup should collect at least one photon per correlation time per coherence area. Because the hardware characteristics of most commercial instruments are fixed, the just-described requirements determine whether a particular particle system can be studied by QLS at all. Among the more well-known suppliers of research grade spectrometers are Brookhaven Instruments (Holtsville, NY), Malvern Instruments (Malvern, UK), ALV (Langen, Germany), and Otsuka Electronics (Shiga, Japan). Simple, basic instruments are manufactured by Precision Detectors (Franklin, MA) and Protein Solutions (Charlottesville, VA). If instrument performance is suitable for a particular study, the quality and reliability of the results will be determined by the other two elements of the QLS system, namely the sample and the data analysis software. The rest of this article focuses on these two elements of the QLS experiment and provides practical examples taken from studies of Aβ fibrillogenesis.

Brownian Motion

Temporal fluctuations in the intensity of the scattered light are caused by the Brownian motion of the scattering particles. The smaller these

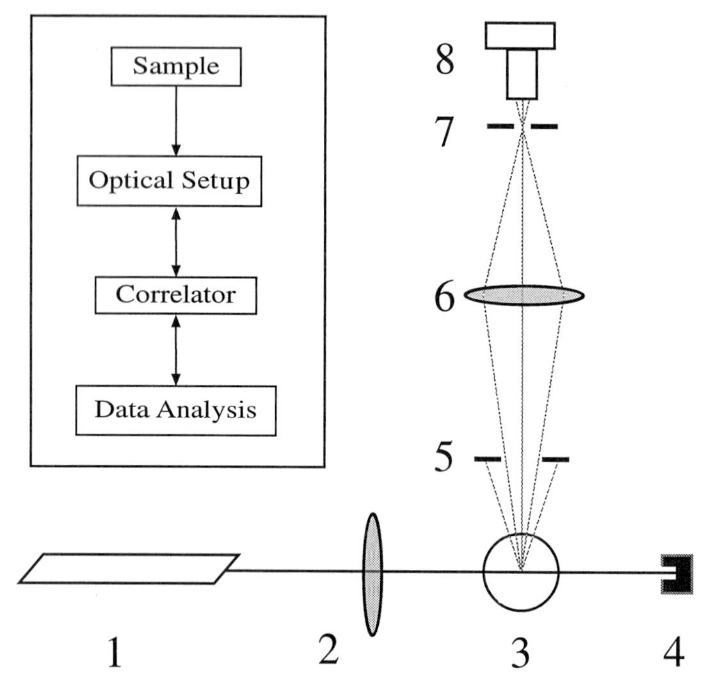

FIG. 2. A QLS instrument. The four key components of a QLS apparatus are presented in block diagram form within the box. Adjacent is a typical scheme for the optical arrangement of the spectrometer, which includes (1) laser; (2) focusing lens (to direct light to center of sample cuvette); (3) sample holder; (4) beam stop; (5) pinhole controlling the solid angle inside which the scattered light is collected; (6) focusing lens (to focus image of the beam within sample on pinhole 7); (7) pinhole, which determines the volume from which the scattered light is registered by photodetector; and (8) photodetector.

particles are, the faster they move. Although each particle moves randomly, in a unit time more particles leave regions of high concentration than leave regions of low concentration. This results in a net flux of particles along the concentration gradient. Brownian motion is thus responsible for the diffusion of the solute and is characterized quantitatively by the diffusion coefficient, D. The laws of diffusive motion stipulate that over time δt the displacement Δx of a Brownian particle in a given direction is characterized by the relationship $\overline{\Delta x^2} = 2D\delta t$.

Determination of Diffusion Coefficient D

As explained earlier, temporal fluctuations in scattered light intensity are caused by the relative motions of particles in solution. Two spherical waves scattered by a pair of individual particles have, at the observation point, a phase difference of $\mathbf{q} \cdot \mathbf{r}$, where $\mathbf{r}$ is the (vector) distance between particles. As the scattering particles move over distance $\Delta x \approx q^{-1}$ along the vector $\mathbf{q}$, phases for all pairs of particles change significantly and the intensity of the scattered light becomes completely independent of its initial value. The correlation time, τ_c, is thus the time required for a Brownian particle to move a distance q^{-1} along the vector $\mathbf{q}$. As stated earlier $\overline{\Delta x^2} = 2D\delta t$, thus for $\Delta x \approx q^{-1}$, $\tau_c \sim 1/Dq^2$. Rigorous mathematical analysis of the process of light scattering by Brownian particles leads to the following expression for the correlation function of the scattered light:

$$|g^{(1)}(\tau)| = \exp(-Dq^2\tau) \tag{4}$$

Determination of Sizes of Particles in Solution

According to Eqs. (3) and (4), measurement of the intensity correlation function allows evaluation of the diffusion coefficients of the scattering particles. The diffusion coefficient in an infinitely dilute solution is determined by particle geometry. For spherical particles, the relation between the radius R and its diffusion coefficient D is given by the Stokes–Einstein equation:

$$D = \frac{k_B T}{6\pi\eta R} \tag{5}$$

Here k_B is the Boltzmann constant, T is the absolute temperature, and η is solution viscosity. For nonspherical particles it is customary to introduce the apparent hydrodynamic radius R_h^{app}, defined as:

$$R_h^{app} = \frac{k_B T}{6\pi\eta D^{app}} \tag{6}$$

where D^{app} is the diffusion coefficient measured in the QLS experiment.

For nonspherical particles, it is important to note that the diffusion coefficient is actually a tensor—the rate of particle diffusion in a certain direction depends on the particle orientation relative to this direction. Small particles, as they diffuse over a distance q^{-1}, change their orientation many times. QLS measures the average diffusion coefficient for these particles. Particles of a size comparable to, or larger than, q^{-1} essentially preserve their orientation as they travel a distance smaller than their size. For these particles, the single exponential expression of Eq. (4) for the field correlation function is not strictly applicable. This fact is particularly important in QLS applications designed for studying long fibrils and will be discussed in more detail later.

For particles small compared to q^{-1}, the hydrodynamic radius is calculated numerically, and in some cases analytically, for a variety of particles shapes. In studies of fibrils, the important analytical formula is that for the prolate ellipsoid, where a is long axis length and p is the ratio of lengths of the short axis to the long axis:

$$R_{\mathrm{h}} = \frac{a}{2}\sqrt{1-p^2} \Bigg/ \ln \frac{1+\sqrt{1-p^2}}{p} \tag{7}$$

The numerical expression for the hydrodynamic radius of a long cylinder can be found in de la Torre and Bloomfield.[9] Effects of flexibility are discussed in the monograph by Schmitz.[6] For thin, rigid objects of length L short compared to q^{-1}, the following asymptotic form of Eq. (7) is useful, where d is the particle diameter:

$$R_{\mathrm{h}}^{\mathrm{app}} = \frac{L}{2\ln(2L/d)} \tag{8}$$

All of the formulas connecting the diffusion coefficient or hydrodynamic radius to particle geometry are strictly applicable only for infinitely dilute solutions. At finite concentrations, two additional factors affect the diffusion of particles significantly: viscosity and interparticle interactions. Viscosity generally increases with the concentration of macromolecular solute. According to Eq. (5), this leads to a lower diffusion coefficient and therefore to an increase in the apparent hydrodynamic radius. Interactions between particles can act in either direction. If the effective interaction is repulsive, which is usually the case for soluble molecules (otherwise they would not be soluble), local fluctuations in concentration tend to dissipate faster, meaning higher apparent diffusion coefficients and lower apparent hydrodynamic radii. If the interaction is attractive, fluctuations in concentration dissipate slower and the apparent diffusion coefficients are lower. Thus, depending on whether the effect of repulsion between particles is strong enough to overcome the effect of increased viscosity, both increasing and decreasing types of concentration dependence of the hydrodynamic radius are observed.[10] In this context, it should be noted that the interaction between large particles (as compared to q^{-1}) generally leads to a nonexponential correlation function that does not take the form of Eq. (4) and therefore cannot be described completely by a single parameter D^{app}.

[9] J. G. de la Torre and V. A. Bloomfield, *Quart. Rev. Biophys.* **14,** 81 (1981).
[10] M. Muschol and F. Rosenberger, *J. Chem. Phys.* **103,** 10424 (1995).

Data Analysis

Polydispersity and Mathematical Analysis of QLS Data

Polydisersity can be an inherent property of the sample, for instance when polymer solutions or protein aggregation are studied, or it can be a consequence of impurities or deterioration of the sample. In the first case, the polydispersity itself is often an object of interest, whereas in the second case it is an obstacle. In both instances, polydispersity complicates data analysis significantly.

For polydisperse solutions, Eq. (4) for the normalized field correlation function must be replaced with

$$|g^{(1)}(\tau)| = \frac{1}{I_0} \sum_i I_i \exp(-D_i q^2 \tau) \tag{9}$$

In this expression, D_i is the diffusion coefficient of particles of the ith kind and I_i is the intensity of light scattered by all of these particles. $I_i = N_i I_{0,i}$, where N_i is the number of particles of ith kind in the scattering volume and $I_{0,i}$ is the intensity of the light scattered by each such particle. For a continuous distribution of scattering particle size, Eq. (9) is generalized as

$$|g^{(1)}(\tau)| = \frac{1}{I_0} \int I(D) \exp(-Dq^2 \tau) dD \tag{10}$$

Here $I(D)dD \equiv N(D)I_0(D)dD$ is the intensity of light scattered by particles having their diffusion coefficient in interval $[D, D + dD]$, $N(D)dD$ is the number of these particles in the scattering volume, and $I_0(D)$ is the intensity of light scattered by each of them. The goal of the mathematical analysis of QLS data is to reconstruct as precisely as possible the distribution function $I(D)$ [or $N(D)$] from the experimentally measured function $G^{(2)}_{\exp}(\tau)$.

It should be noted that polydispersity is not the only source of nonsingle exponential correlation functions of scattered light. Even in perfectly monodisperse solutions, interparticle interactions, orientation dynamics of asymmetric particles, and conformational dynamics or deformations of flexible particles will lead to a much more complicated correlation function than described by Eq. (4). These effects are usually insignificant for scattering by particles small compared to the length of the inverse scattering vector q^{-1}, but become important, and often overwhelming, for larger particles. In those cases, QLS probes not the pure diffusive Brownian motion of the scatterers, but also other types of dynamic fluctuation in the solution. Fortunately, the relaxation times of these other types of fluctuations rarely depend on the scattering vector as Dq^2, which is characteristic for the diffusion process. Thus, in principle, measurement of the correlation function at several different angles of scattering, and therefore at several different q, allows polydispersity to be distinguished from multimodal relaxation of a nondiffusive nature.

Deconvolution of Correlation Function, an "Ill-Posed" Problem

The values of $G^{(2)}_{\text{exp}}(\tau)$ contain statistical errors. We have described previously the features of the QLS instrument that are essential for minimizing these errors. It is equally important to minimize the distorting effect that experimental errors in $G^{(2)}_{\text{exp}}(\tau)$ have on the reconstructed distribution function $I(D)$. The distribution $I(D)$ is a nonnegative function. *A priori* then, a nonnegative function $I(D)$ should be sought that produces, via Eqs. (3) and (10), the function $G^{(2)}_{\text{theor}}(\tau)$, which is the best fit to experimental data. Unfortunately, this simplistic approach does not work. The underlying reason is that the corresponding mathematical minimization problem is "ill-posed,"[11] meaning that dramatically different distributions $I(D)$ lead to nearly identical correlation functions of the scattered light and therefore are equally acceptable fits to experimental data. For example, addition of a fast oscillating component to the distribution function $I(D)$ does not change $G^{(2)}_{\text{theor}}(\tau)$ considerably, as the contributions from closely spaced positive and negative spikes in the particle distribution cancel each other. Three approaches for dealing with this ill-posed problem are discussed.

Direct Fit Method

The simplest approach is the direct fit method. Here the functional form of $I(D)$ is assumed *a priori* (single modal, bimodal, Gaussian, etc.). The parameters of the assumed function that lead to the best fit of $G^{(2)}_{\text{theor}}(\tau)$ to $G^{(2)}_{\text{exp}}(\tau)$ are then determined. This method is only as good as the original guess of the functional form of $I(D)$. Moreover, using the method can be misleading because it may confirm nearly any *a priori* assumption made. It is also important to note that the more parameters there are in the assumed functional form of $I(D)$, the better experimental data can be fit but the less meaningful the values of the fitting parameters become. In practice, typical QLS data allow reliable determination of about three independent parameters of the size distribution of the scattering particles.

Method of Cumulants

The second approach is not to attempt to reconstruct the shape of the scattering particle distribution but instead to focus on so-called "stable"

[11] A. N. Tikhonov and V. Y. Arsenin, "Solution of Ill-Posed Problems." Halsted Press, Washington, 1977.

characteristics of the distribution, i.e., characteristics that are insensitive to possible fast oscillations. In particular, these stable characteristics are moments of the distribution or closely related quantities called cumulants.[12] The first cumulant (moment) of the distribution $I(D)$, the average diffusion coefficient $\overline{D}$, can be determined from the initial slope of the field correlation function. Indeed, using Eq. (10), it is straightforward to show that

$$-\frac{d}{d\tau}\ln|g^{(1)}(\tau)|_{\tau\to 0} = \frac{1}{I_0}\int I(D)Dq^2dD \equiv \overline{D}q^2 \tag{11}$$

The second cumulant (moment) of the distribution can be obtained from the curvature (second derivative) of the initial part of the correlation function. As in the direct fit method, the accuracy of the real QLS experiment allows determination of at most three moments of the distribution $I(D)$. The first moment, $\overline{D}$, can be determined with better than $\pm 1\%$ accuracy. The second moment, the width of the distribution, can be determined with an accuracy of ± 5–10%. The third moment, which characterizes the asymmetry of the distribution, usually can be estimated with an accuracy of only about $\pm 100\%$.

Regularization

The regularization approach combines the best features of both of the previous methods. The advantage of the cumulant method is that it is completely free from bias introduced by *a priori* assumptions about the shape of $I(D)$, assumptions that are at the heart of the direct fit method. However, reliable *a priori* information on the shape of the distribution function, in addition to experimental data, improves the quality of results obtained by the QLS method significantly. The regularization method assumes that the distribution $I(D)$ is a smooth function and seeks a nonnegative distribution producing the best fit to experimental data. As discussed earlier, the ill-posed nature of the deconvolution problem means that distributions differing by the presence or absence of a fast oscillating function produce very similar correlation functions. The regularization requirement that the distribution should be sufficiently smooth eliminates this ambiguity, allowing unique solutions to the minimization problem. There are several methods that utilize this approach for reconstructing the scattering particle distribution function from QLS data. All of these methods impose the condition of smoothness on the distribution $I(D)$ but differ in the specific mathematical approaches used for this purpose. The most popular program,

[12] D. E. Koppel, *J. Chem. Phys.* **57,** 4814 (1972).

originally developed by Provencher,[13] is called CONTIN. We have developed and used our own algorithm.[14]

All regularization algorithms produce similar results and incorporate the use of a parameter that determines how smooth the distribution has to be. The choice of this parameter is one of the most difficult and important parts of the regularization method. If the smoothing is too strong, the distribution will be very stable but will lack details. If the smoothing is too weak, false spikes can appear in the distribution. The "rule of thumb" is that the smoothing parameter should be just sufficient to provide stable, reproducible results in repetitive measurements of the same correlation function. Two facts are helpful for choosing the appropriate smoothing parameter. First, the lower the statistical errors of the measurements, the smaller the smoothing parameter can be without loss of stability. This will yield finer resolution in the reconstructed distribution $I(D)$. Second, narrow distributions generally require much less smoothing and can be reconstructed much better than wide distributions. This is because oscillations in narrow distributions are effectively suppressed by nonnegativity conditions.

The moments of the distribution reconstructed by the regularization procedure coincide closely with those obtained by other methods. However, the regularization procedure, in addition, gives unbiased (apart from smoothing) information on the shape of the distribution. This shape cannot be extracted through use of the direct fit method or from cumulant analysis. In a typical QLS experiment, regularization analysis can resolve a bimodal distribution with two narrow peaks of equal intensity if the diffusion coefficients corresponding to these peaks differ by more than a factor of ~2.5.

Unique Features of QLS Monitoring of Fibril Growth

Effect of Fibril Length

The study of protein fibrils using QLS requires very careful interpretation of data. As fibrils grow in length, three important factors come into play. First, when the length of the fibril becomes comparable to q^{-1}, the time of diffusion of the fibril over distance Δx in the direction of vector $\mathbf{q}$, such that $q\Delta x \approx 1$, becomes dependent on the orientation of the fibril relative to this direction. As a result, even a monodisperse distribution of such fibrils will produce a multiexponential correlation function. In addition, because the intensity of the scattering by each fibril also depends on its

[13] S. W. Provencher, *Comput. Phys. Commun.* **27,** 213 (1982).

[14] T. G. Braginskaya, P. D. Dobitchin, M. A. Ivanova, V. V. Klyubin, A. V. Lomakin, V. A. Noskin, G. E. Shmelev, and S. P. Tolpina, *Phys. Scripta* **28,** 73 (1983).

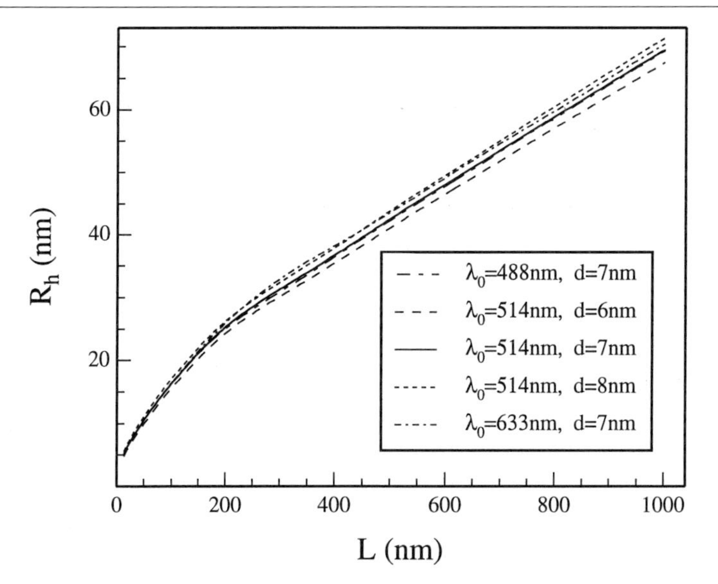

FIG. 3. Calibration curve relating R_h of a rigid rod to the rod length L. Data are shown for rod diameters of 6, 7, and 8 nm and for scattering vectors corresponding to 90° scattering in aqueous solutions with incident light of wavelength 488, 514, or 633 nm. Adapted from A. Lomakin, D. B. Teplow, D. A. Kirschner, and G. B. Benedek, *Proc. Natl. Acad. Sci. U.S.A.* **94,** 7942 (1997). Copyright (1997) National Academy of Sciences, U.S.A.

orientation relative to **q**, the orientational diffusion of the fibril makes a significant contribution to the correlation function. Nevertheless, in most instances, all these effects can be subsumed into the definition of the apparent hydrodynamic radius, R_h^{app}, of the fibril. In this definition, the inverse average relaxation time of the multiexponential correlation function given by the left side of Eq. (11) is used to calculate the apparent diffusion coefficient, D^{app}, which is then used in Eq. (6) to calculate R_h^{app}. Figure 3 presents numerically calculated curves of $R_h^{app}(L)$ for several different q. Using data in Fig. 3, fibril length can be determined from the value of R_h^{app}. It should be noted that when L remains relatively small, $R_h^{app}(L)$ is insensitive to variations in wavelength and fibril diameter.

Effect of Fibril Flexibility

When the length of the fibril becomes comparable to the persistence length,[15] the effects of fibril flexibility must be taken into account. Flexibility adds even more complexity to the anisotropic diffusion and orientational

[15] O. Kratky and G. Parod, *Rec. Trav. Chim.* **68,** 1106 (1949).

dynamics of a fibril. Effects of flexibility on the correlation function have been discussed in detail by Maeda and Fujume.[16,17] For the same length, a flexible fibril will have a smaller R_h^{app} than a rigid one. Fibrils that are much longer than the persistence length l_p form coils with apparent hydrodynamic radii of $(3\sqrt{\pi}/8)R_g$, where $R_g = \sqrt{2Ll_p}$ is the radius of gyration of the coil and L is the contour length of the fibril. This is a much weaker dependence on $R_h^{app}(L)$ than that for rigid rods [Eq. (8)]. If, in addition, the radius of gyration is large compared to q^{-1}, the correlation function of the scattered light is then determined by the diffusion of the segment of the coil of size q^{-1} and becomes independent of the total fibril length.

Effect of Fibril–Fibril Interaction

The third, and potentially most serious, factor affecting the interpretation of QLS data on growing fibrils is fibril–fibril interaction. Even in the absence of attraction between fibrils, fibril–fibril interactions become significant in what is known as the "semidilute" concentration regime. This regime occurs when fibrils grow sufficiently long so that average fibril length becomes comparable to or longer than the average distance between fibrils. Let us consider fibrils of diameter d and average fibril length L occupying volume fraction ϕ of the solution. The volume of an individual fibril is $\nu \sim d^2L$ and the number N of fibrils in a unit volume is $N = \phi/\nu$. The average distance between fibrils is $N^{-1/3}$. The semidilute regime starts when $N^{-1/3} \sim L$, and, therefore, $L^3 = Ld^2/\phi$. From these estimates, it is clear that fibril–fibril interactions become important when fibrils reach the length L^*, which is related to fibril diameter by the following equation:

$$L^* = d/\sqrt{\phi} \tag{12}$$

Regardless of the concentration, sufficiently long rigid fibrils always enter the semidilute regime. In a semidilute solution, fibrils cannot move perpendicular to their long axes because of caging by other fibrils. They only diffuse along their axes.[18] The dynamic behavior of semidilute solutions of rods has been discussed by Zero and Pecora[19] and reviewed thoughtfully by Russo.[20] Qualitatively, fibril–fibril interaction in a semidilute solution

[16] T. Maeda and S. Fujume, *Macromolecules* **17,** 1157 (1984).

[17] T. Maeda and S. Fujume, *Macromolecules* **17,** 2381 (1984).

[18] M. Doi and S. F. Edwards, *J. Chem. Soc. Faraday II* **74,** 569 (1978).

[19] K. Zero and R. Pecora, *in* "Dynamic Light Scattering: Applications of Photon Correlation Spectroscopy" (R. Pecora, ed.), p. 59. Plenum Press, New York, 1985.

[20] P. S. Russo, *in* "Dynamic Light Scattering" (W. Brown, ed.), p. 512. Claredon Press, Oxford, 1993.

slows down the diffusion of fibrils dramatically and, if unaccounted for, leads to gross overestimation of fibril length.

Practical Aspects of QLS Experiments

Temperature Control

The diffusion coefficient D depends on the temperature T both explicitly and through the solvent viscosity η. In aqueous solutions, this dependency is strong. Indeed, near room temperature, the factor T/η in the Stokes–Einstein relation between D and R [Eq. (5)] changes ~3% per 1°. To obtain consistent results, it is thus essential to stabilize the temperature of the sample precisely. It is also important to avoid local heating of the sample by the focused laser beam, which if unaccounted for can cause systematic underestimation of the size of the scattering particles. In addition, local heating can cause convection and thermal lens effects, which can affect the quality of the QLS measurements adversely. The degree of local heating depends on light absorption by the sample solution. In most cases, if the intensity of the laser is adjusted so that several photocounts per coherence area per correlation time are produced, overheating of the scattering volume is rarely a problem.

Laser Modes and Stability

In QLS experiments, the laser must operate in the single mode regime, i.e., it should generate a single transverse mode (called TEM_{00}). Different transverse modes have different cross-sectional intensity profiles and very close frequencies. The mode TEM_{00} produces a beam with a nearly Gaussian intensity profile. If the laser were to generate several transverse modes simultaneously, not only would the beam have an irregular intensity profile, but more importantly, optical "beating" between these modes would be registered by the photodetector. This would result in significant distortion of the correlation function.

Laser stability is another requirement for successful QLS experimentation. Fluctuations of laser intensity can be caused by mechanical vibration, thermal expansion, and fluctuations in power supply performance. Depending on their origin, intensity fluctuations of the incident beam can occur as oscillations, large abrupt changes due to "mode hops," or slow drifting of intensity. To account for the fluctuations in laser intensity, the factor I_0^2 in the expression for the intensity correlation function $G^2(\tau)$ in Eq. (3) should be replaced by the correlation function of the laser intensity, $\langle I_0(0) \cdot I_0(t) \rangle = I_0^2 + \langle \delta I(0) \cdot \delta I(\tau) \rangle$. Here $\delta I(t)$ is the deviation of the laser intensity at moment t from its average value I_0. From examination of Eq. (3), we see that with a low efficiency factor γ, even small fluctuations in laser intensity can have significant effects on the determination of $|g^1(\tau)|$. It is therefore important to use a well-stabilized laser and to maximize the efficiency of measurements by collecting light from a single coherence area. If mode hops happen during the measurements, $\delta I \sim I_0$. In this case, the resulting data should be discarded.

Sample Purity

Samples monitored by QLS must be optically pure. This concept is quite different from the concept of chemical purity. In the latter case, one or a few large inert particles would not be expected to affect the phenomenon under study. However, in the case of QLS, the fact that the intensity of light scattered by a particle is proportional to the square of the particle mass (for particles small compared to the wavelength of the incident light) means that even a small weight fraction of large scatterers can completely dominate scattering from a population of relatively small particles. A common problem associated with monitoring protein assembly processes is the presence of very large particles, such as dust particles and protein aggregates. These particles occasionally drift inside the scattering volume, increasing the total scattering intensity several times. This is as detrimental to the QLS experiment as are mode hops and can make QLS measurements unusable.

Three approaches are commonly used to deal with large, biologically irrelevant particles: *pre facto* particle removal, careful control of the illuminated sample volume, and *post facto* data manipulation. The easiest way to remove large impurities from solution is by filtration. Standard 0.22-μm filters generally are too porous to be of use. We have found that 20-nm Anatop filters are satisfactory. In cases where the particles of interest are too big to pass through the 20-nm filter, their scattering intensity is usually great enough so that dust and other large particles do not significantly corrupt the resulting data set. It should be noted that dust in the air has a strong tendency to adsorb electrostatically to charged groups on the surfaces of the empty cuvette. These adsorbed particles can be suspended in the sample solution during the process of expulsion of the sample from the filter. If the cuvette is not to be filled completely, it is desirable to introduce the sample into the bottom of the cuvette to avoid washing dust off the cuvette walls. This is typically done by first sealing the cuvette with Parafilm, attaching a narrow gauge needle to the filter bottom, piercing the Parafilm seal with the needle, and positioning the needle tip at the cuvette bottom.

Centrifugation is another effective way to remove large impurities from the solution, provided that the sample is spun in the same sealed cuvette in which the QLS measurements will be done. Transferring the sample into another cuvette after centrifugation defeats the purpose of the procedure. Typical airborne dust can be pelleted in 30 min at 5000*g*. However, there are always very "flaky" dust particles that will not sediment by this procedure.

Successful measurement of the hydrodynamic radius of low molecular weight molecules is absolutely dependent on the optical purity of the sample. In studies of the amyloid β-protein, we have used the intrinsic filtering potential of high-performance liquid chromatography (HPLC) column packings, and a continuous flow procedure for washing the QLS cuvette, to produce optically pure Aβ samples. We first prepared cuvettes by heating the top 20 mm of standard 6×50-mm glass test tubes in the flame of a small torch. The tops of each tube are then pulled to form narrow capillaries. A similar procedure is performed on disposable glass micropipettes in order to form a junction between the HPLC detector and the cuvette. In this case, the untreated end of the micropipette is inserted into the HPLC detector outflow line while the pulled tip of the micropipette is inserted into the cuvette so that its tip is at the very bottom. In this way, "filtered" buffer is constantly washing the interior of the cuvette from the bottom out through the narrow capillary top. When the peak of interest has passed the detector and filled the cuvette, the micropipette is removed from the cuvette and the end of the cuvette is immediately fire sealed. This procedure, although somewhat cumbersome, provides excellent dust-free samples. We strongly recommend it for any QLS study of peptides and small proteins.

The second approach for dealing with large particles involves careful focusing of the laser beam in order to minimize the scattering volume. Dust particles or aggregates pass through a small scattering volume less frequently and in fewer numbers than through larger volumes. In addition, high brightness of the beam improves the signal-to-noise ratio, allowing collection of the light from fewer coherence areas. This increases the value of the important efficiency factor γ in Eq. (3).

Despite diligent efforts to produce optically pure samples and a well-focused beam, large particles often pass through the illuminated volume. As a result, significant fluctuations in the intensity of the scattered light are produced. These fluctuation are of a different nature than the standard fluctuations in the interference pattern and are difficult to account for. Some correlators allow suppression of data acquisition during spikes of intensity. These algorithms are called "software dust filters" and involve establishing arbitrary cutoff levels for the intensity of the scattered light. In addition, large particles can be identified by data analysis software, which reconstructs particle size distributions. Thus, in principle, regularization algorithms can alleviate the effect of dust on the reconstructed distribution of particle sizes. However, these *post facto* approaches have limited usefulness. It is thus most advantageous to minimize the effects of irrelevant large particles by removing them *pre facto*.

Assembly of Amyloid β-Protein (Aβ)

Aβ Fibrillogenesis

This section illustrates how QLS techniques may be applied to investigate protein assembly using examples from studies of the kinetics of Aβ fibrillogenesis. Aβ plays a seminal role in the pathogenesis of Alzheimer's disease, during which this peptide accumulates as amyloid deposits in many regions of the brain.[21] Fibrillogenesis of Aβ is associated with neuronal damage, thus efforts directed at inhibiting or reversing this process could have therapeutic value. An elucidation of the morphologic and kinetic stages of fibrillogenesis would permit thoughtful targeting and design of therapeutic agents.[22] As a step toward understanding the molecular mechanism of Aβ fibrillogenesis, we have used QLS to monitor quantitatively the temporal evolution of the fibril length distribution in solutions of synthetic Aβ *in vitro*.

It should be emphasized that the ability to monitor quantitatively the fibril length distribution allows determination of the rate constants for fibril nucleation and elongation, k_n and k_e, respectively.[23,24] Knowledge of these quantities is sufficient to model the fibrillogenesis process in a given system. Importantly, examination of the effects of perturbations of the system on these kinetic parameters can provide data about key elements controlling the fibrillogenesis process. The following examples demonstrate how the QLS approach has been used to determine rate constants for Aβ fibrillogenesis,[23,24] to examine the type of macromolecular assemblies formed during Aβ polymerization,[25] to determine the effects of primary structure

[21] D. J. Selkoe, *Science* **275,** 630 (1997).

[22] D. B. Teplow, *Amyloid Int. J. Exp. Clin. Invest.* **5,** 121 (1998).

[23] A. Lomakin, D. S. Chung, G. B. Benedek, D. A. Kirschner, and D. B. Teplow, *Proc. Natl. Acad. Sci. U.S.A.* **93,** 1125 (1996).

[24] A. Lomakin, D. B. Teplow, D. A. Kirschner, and G. B. Benedek, *Proc. Natl. Acad. Sci. U.S.A.* **94,** 7942 (1997).

[25] D. M. Walsh, A. Lomakin, G. B. Benedek, M. M. Condron, and D. B. Teplow, *J. Biol. Chem.* **272,** 22364 (1997).

changes on fibrillogenesis kinetics,[26] and to elucidate thermodynamic properties of the fibrillogenesis reaction.[27]

Quantitative Monitoring of Earliest Stages of Fibrillogenesis

Aβ fibrils have diameters $d \approx$ 6–10 nm and have linear densities $\rho \approx$ 1.6 nm^{-1}.[23,28] At a total Aβ concentration C = 0.1 mM, a solution of Aβ fibrils occupies a volume fraction $\phi = CN_A\pi d^2/4\rho \approx 2 \times 10^{-3}$, where N_A is Avogadro's number and d is fibril diameter (in this case 8 nm). According to Eq. (12), the transition to the semidilute regime in this solution starts when fibrils are approximately 150 nm long. This length is close to the estimated persistence length of Aβ fibrils[29] and to the typical value of ~100 nm for q^{-1}. These values mean that the quantitative analysis of fibril assembly by QLS is most appropriate when fibril length does not exceed 100–150 nm. The QLS method is thus ideal for examining the critical early stages of fibril assembly, i.e., the nucleation and elongation of nascent fibrils. Among the methods used to monitor Aβ fibrillogenesis, which include Congo red and thioflavin T binding, turbidimetry, sedimentation, filtration, atomic force microscopy, and electron microscopy, only QLS can provide directly an absolute, quantitative measure of Aβ particle size during fibril nucleation and elongation. At later stages, when solutions contain longer fibrils, structural information can still be obtained, but it is more qualitative in nature.

Measurement of Fibril Elongation Rates

By relating R_h^{app} to L, QLS can be used to determine temporal changes in fibril length, i.e., the elongation rate dL/dt. This is an extremely powerful capability because it enables the mechanistic study of fibril elongation. Determination of the concentration dependence of dL/dt is one method for studying the mechanism of polymer elongation reactions. The elongation rate and the rate of monomer addition at the fibril tip, dN/dt, are related by the equation $dN/dt = \rho dL/dt$, where ρ is the linear density of monomers

[26] D. B. Teplow, A. Lomakin, G. B. Benedek, D. A. Kirschner, and D. M. Walsh, *in* "Alzheimer's Disease: Biology, Diagnosis and Therapeutics" (K. Iqbal, B. Winblad, T. Nishimura, M. Takeda, and H. M. Wisniewski, eds.), p. 311. Wiley, Chichester, England, 1997.

[27] Y. Kusumoto, A. Lomakin, D. B. Teplow, and G. B. Benedek, *Proc. Natl. Acad. Sci. U.S.A.* **95,** 12277 (1998).

[28] C. L. Shen, M. C. Fitzgerald, and R. M. Murphy, *Biophys. J.* **67,** 1238 (1994).

[29] C. L. Shen and R. M. Murphy, *Biophys. J.* **69,** 640 (1995).

within the fibril. To determine dL/dt, the slope of the curve $L(t)$ is measured when the most reliable connection between R_h^{app} and L exists, i.e., during the initial nucleation and elongation of fibrils when $L \leq 150$ nm. To do so, $L(t)$ is reconstructed from the time dependency of the apparent R_h, which in turn is derived from $D^{app}(t)$ according to Eq. (6). This process is illustrated in the following example, in which temporal changes in the distribution of apparent diffusion coefficients D^{app} were measured during the fibrillogenesis of Aβ(1–40), one of the two major forms of the peptide found in amyloid deposits *in vivo*.[22] This approach resulted in the discovery of a novel mechanism for Aβ fibril nucleation, namely nucleation within Aβ micelles.[23]

The experiment is performed by dissolving HPLC-purified Aβ(1–40)

[NH_2-DAEFRHDSGYEVHHQKLVFFAED
VGSNKGAIIGLMVGGVV-*COOH*]

at concentrations ranging from 25 μM to 1.7 mM in 0.1 N HCl. For each concentration, ~200 μl of sample is placed in a 5-mm-diameter glass test tube and then centrifuged at 5000g for 30 min to pellet dust particles and large aggregates. The tube is then placed into the QLS spectrometer, and the intensity and correlation function of the scattered light are measured periodically over the next 20–50 hr. Each measurement lasts 5–30 min, depending on the intensity of the signal. During the monitoring period, the sample remains at ~22°. Figure 4A shows distributions of D^{app} during fibrillogenesis of Aβ(1–40) at a total concentration $C_0 = 0.25$ mM. Initially, the distributions are unimodal and narrow (data not shown), reflecting the presence of a relatively homogeneous population of scattering particles with large diffusion coefficients. Within 1–2 hr, an asymmetry develops in the distribution, characteristic of the presence of larger particles (Fig. 4A, 1.7 hr). As time passes, a bimodal distribution develops in which the population of larger scatterers displays a decreasing apparent diffusion coefficient. Finally, the more rapidly diffusing particles disappear, at which point the distribution of large particles becomes fixed (data not shown).

Using Eq. (6), $R_h^{app}(t)$ can be derived for each concentration C_0. Applying this process to data of Fig. 4A yields Fig. 4B. Here we see that the rapidly diffusing scatterers have an R_h = 5–7 nm,whereas the R_h distribution of the larger particles emerges from that of the smaller and displays a temporally increasing mean value and intensity. By averaging across the entire distribution, one obtains the mean hydrodynamic radius, $\overline{R_h}$. The time dependence of $\overline{R_h}$ is shown in Fig. 5. The initial $\overline{R_h}$ for each C_0, determined by extrapolation of the linear portion of each growth curve to $t = 0$, is 4 nm. This value is concentration independent. However, for $C_0 > 0.1$ mM, especially at higher concentrations, an apparent lag phase

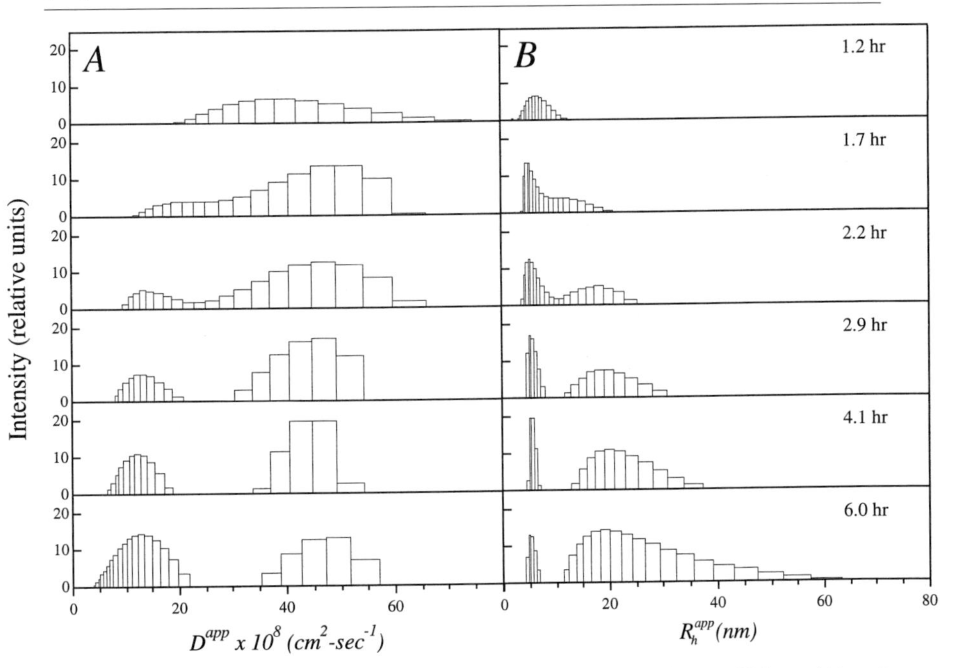

FIG. 4. Temporal evolution of the distribution of apparent diffusion coefficients (A) and apparent hydrodynamic radii (B) during fibrillogenesis of Aβ(1–40).

is observed during which time $\overline{R_h}$ remains constant at ~7 nm. Using the numerical relationship between R_h^{app} and L shown in Fig. 3, curves of $L(t)$ are constructed for each concentration of Aβ examined. Determination of the initial slopes of these curves yielded values for dL/dt (Fig. 6) and enabled assessment of the concentration dependence of the initial elongation rate. Interestingly, the kinetics differ depending on whether C_0 is above or below ~0.1 mM. We refer to this concentration as the critical concentration c^*. Below this concentration, initial rates of fibril elongation are proportional to C_0, but above this concentration, the fibrillogenesis kinetics is concentration independent. In fact, over a full decade of concentrations above c^* (0.17–1.7 mM), experimental measurements of $\overline{R_h}(t)$ yielded essentially the same results (Fig. 5). The concentration 0.1 mM thus constitutes a boundary above and below which very different fibrillogenesis behavior occurs. The final fibril length, L_f, also shows a different concentration dependence above and below c^* (Fig. 6). For $C_0 < c^*$, L_f is related inversely to concentration. However, for $C_0 > c^*$, L_f is independent of concentration and is significantly lower than the final average L_f values observed in experiments done at $C_0 < c^*$.

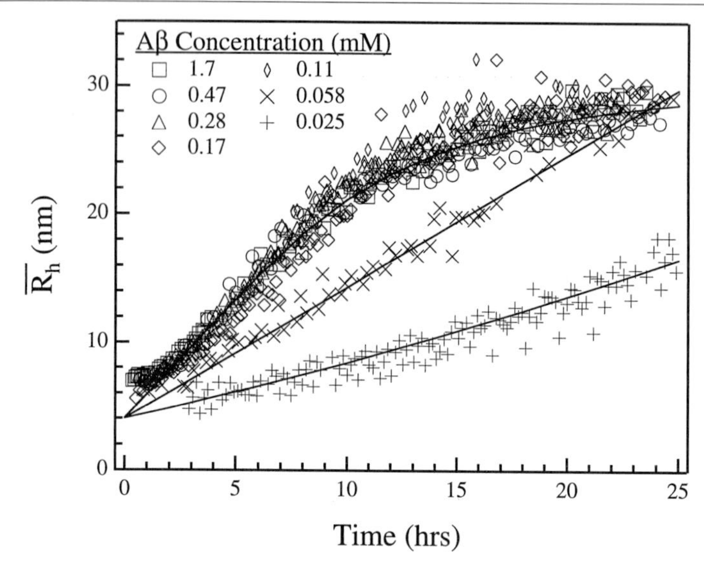

FIG. 5. Concentration dependence of Aβ fibrillogenesis. Solid lines have been drawn through data to assist the eye. Adapted from A. Lomakin, D. S. Chung, G. B. Benedek, D. A. Kirschner, and D. B. Teplow, *Proc. Natl. Acad. Sci. U.S.A.* **93,** 1125 (1996). Copyright (1996) National Academy of Sciences, U.S.A.

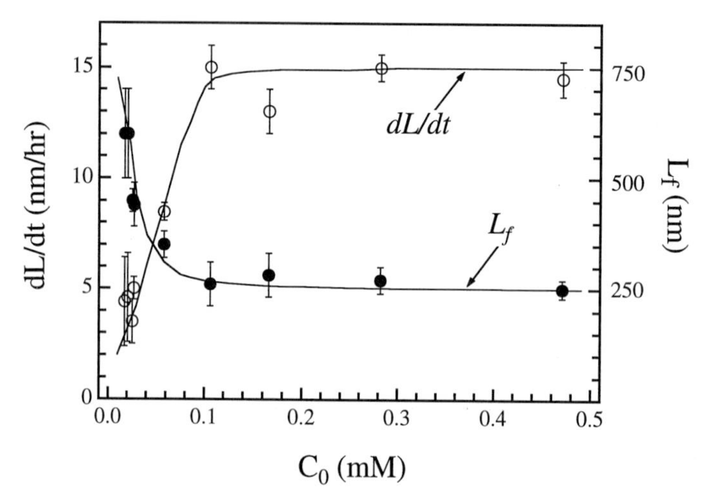

FIG. 6. Concentration dependence of fibril elongation rate (dL/dt) and final fibril length (L_f). Adapted from A. Lomakin, D. S. Chung, G. B. Benedek, D. A. Kirschner, and D. B. Teplow, *Proc. Natl. Acad. Sci. U.S.A.* **93,** 1125 (1996). Copyright (1996) National Academy of Sciences, U.S.A.

These observations formed the basis for development of a model of Aβ fibril nucleation and elongation in which micellization occurs in the concentration domain $C_0 > c^*$.[23] Micelles are the 7-nm particles seen immediately on peptide dissolution and are nidi for Aβ fibril nucleation. Nascent nuclei, the 4-nm particles seen in all experiments, are then extended by Aβ monomer addition at their ends. A full mathematical formulation of this model has been published.[24] Importantly, this formulation provides values for the key kinetic rate constants of the model, nucleation rate k_n and elongation rate k_e. It also enables determination of the time-dependent changes in monomer and micelle concentration and in the concentration and size distribution of fibrils as fibrillogenesis proceeds. The ability to monitor quantitatively the earliest nucleation and elongation events occurring during Aβ fibrillogenesis is a unique and particularly powerful characteristic of QLS.

Simultaneous Monitoring of Intensity and Hydrodynamic Radius Discriminates between Fibrillogenesis and Amorphous Aggregation

As shown earlier, the intensity of the light scattered by a single small particle composed of m monomers is $I = m^2 I_0$. In a monodisperse system of n scatterers, the total intensity of the scattered light would be simply $I = nm^2 I_0$. In a closed system, the total number of monomers $M = nm$ remains constant and thus aggregation is characterized by a proportionality between scatterer size and scattered light intensity, i.e., $I = MmI_0$. In polydisperse systems, m should be replaced with the weight-averaged aggregation number of the scatterers.

The hydrodynamic radius of an aggregate depends on its geometry. For a dense spherical particle, $R_h \sim m^{1/3}$. For thin rigid rods, their length is proportional to monomer number, i.e., $L \sim m$, and Eq. (8) shows that R_h increases nearly proportionally to the size of aggregate. Therefore, for linear elongation, normalized curves of the temporal change in R_h and I will be similar, whereas amorphous and other types of nonlinear aggregation produce a disproportionate increase in I relative to R_h. This is illustrated in Fig. 7, where the temporal changes in $\overline{R_h}$ and intensity observed during fibrillogenesis of wild-type Aβ(1–40) and of a mutant Aβ molecule associated with amyloidosis in a Dutch kindred are plotted.[30] Here, the wild-type molecule displays a proportionality between R_h and I as fibrils grow. In contrast, during the period from 5 to 30 hr, the "Dutch" peptide displays a fourfold increase in I whereas $\overline{R_h}$ increases <50%. This is consistent with fibril–fibril bundling of the Dutch peptide.

[30] E. Levy, M. D. Carman, I. J. Fernandez-Madrid, M. D. Power, I,. Lieberburg, S. G. van Duinen, G. T. A. M. Bots, W. Luyendijk, and B. Frangione, *Science* **248,** 1124 (1990).

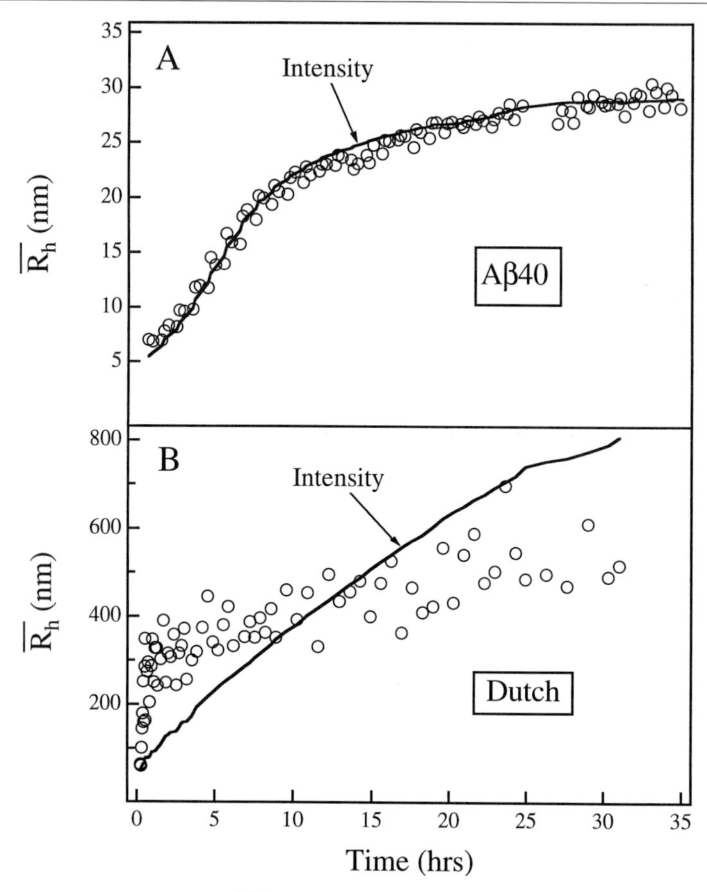

FIG. 7. Temporal changes in $\overline{R_h}$ and scaled intensity observed during fibrillogenesis of wild-type Aβ (A) and Dutch Aβ (B). Adapted from D. B. Teplow, *in* "Molecular Biology of Alzheimer's Disease: Genes and Mechanisms Involved in Amyloid Generation" (C. Haass, ed.), p. 163. Harwood Academic, Amsterdam, 1998.

The just-described considerations relating I and $\overline{R_h}$ are valid only while aggregates remain small compared to q^{-1}. As soon as aggregate size exceeds this value, no further increase of intensity will be observed. In addition, the relationship between $\overline{R_h}$ and intensity is not that straightforward in a polydisperse system. With these caveats in mind, it is nevertheless valuable to recognize that out of all possible aggregates of a given mass, linear rods have the smallest intensity of scattering and the largest $\overline{R_h}$. Thus, linear fibrillization produces the least possible increase of intensity of scattering as a function of $\overline{R_h}$.

Effects of Primary Structure Changes on Nucleation and Elongation Rates

The Dutch mutation mentioned earlier causes a $Glu^{22} \rightarrow$ Gln amino acid substitution within the $A\beta$ sequence, which leads to severe amyloid deposition within cerebral blood vessels, resulting in fatal strokes. In an effort to correlate this neuropathology with the biophysical behavior of the mutant peptide, the kinetics of fibrillogenesis of the Dutch peptide was studied using QLS.[26] The peptide was dissolved at a concentration of 0.1 m*M* in 0.1 *N* HCl at room temperature and the temporal change in $\overline{R_h}$ and in the average intensity of the scattered light were measured. Relative to wild-type $A\beta$, a 200-fold increase in initial elongation rate was observed (Fig. 8). Furthermore, an 8-fold increase in final fibril length was seen. In the concentration domain $C_0 > c^*$, the final fibril size is given by the equation $L_f = (k_e c^*/k_n)^{1/2}/\rho$.[23] The ratio of nucleation rates therefore can be estimated as $k_n^{\text{Dutch}}/k_n^{\text{wild type}} \approx 3$. This demonstrates that the $Glu^{22} \rightarrow$ Gln mutation accelerates both nucleation and elongation of fibrils. The ability to quantify differences in nucleation and elongation rates provides the means to identify key structures, such as Glu^{22} in $A\beta$, controlling the kinetics of protein assembly reactions.

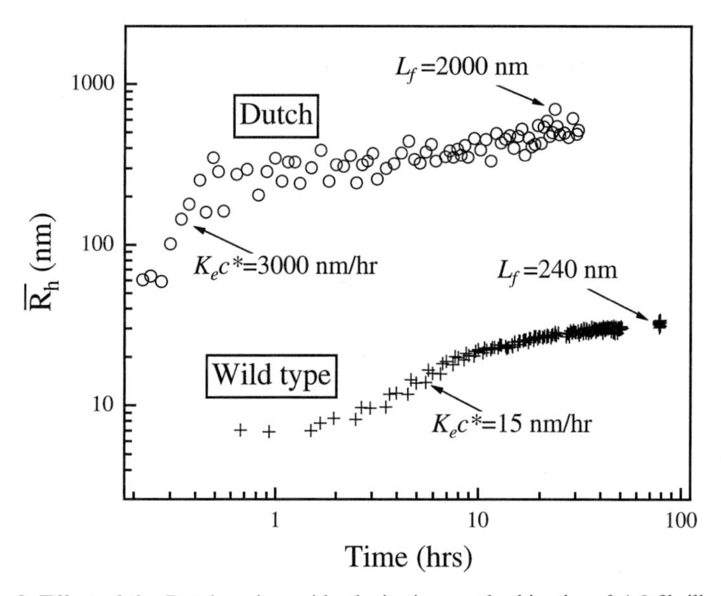

FIG. 8. Effect of the Dutch amino acid substitution on the kinetics of $A\beta$ fibrillogenesis (note the log scales). Adapted from D. B. Teplow, *in* "Molecular Biology of Alzheimer's Disease: Genes and Mechanisms Involved in Amyloid Generation" (C. Haass, ed.), p. 163. Harwood Academic, Amsterdam, 1998.

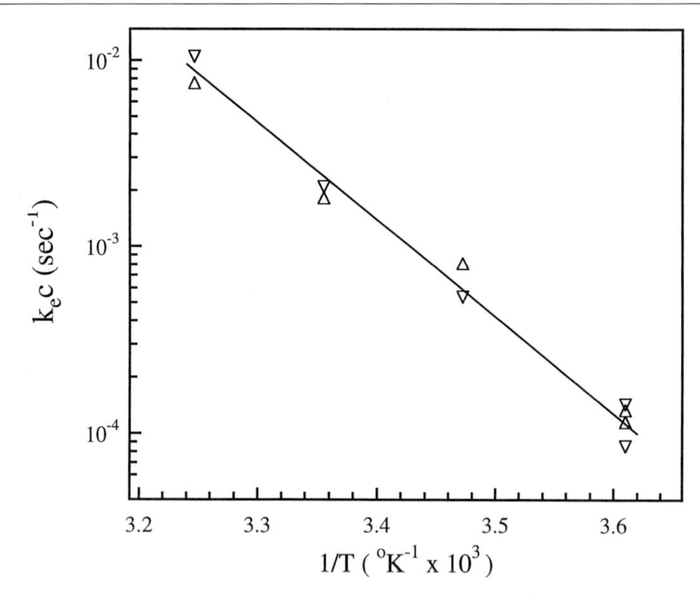

FIG. 9. Arrhenius plot of initial fibril elongation rate versus inverse temperature. Two sets of samples (▽, △) were prepared and their elongation rates were determined in 0.1 *N* HCl at four different temperatures. Resulting data were fit well by a straight line ($r = -0.99$) whose slope yielded an activation energy $E_A = 22.8$ kcal/mol. Adapted from Y. Kusumoto, A. Lomakin, D. B. Teplow, and G. B. Benedek, *Proc. Natl. Acad. Sci. U.S.A.* **95,** 12277 (1998). Copyright (1998) National Academy of Sciences, U.S.A.

Thermodynamics of Aβ Fibrillogenesis

The QLS approach allows one to address fundamental questions about the mechanism of the fibril elongation reaction. Previous studies have shown that the temporal increase in thioflavin T binding that occurs during Aβ fibrillogenesis depends on temperature.[31] Because of their low resolution, dye-binding approaches do not allow determination of whether the temperature dependence exhibited is due to the fibril elongation reaction, fibril–fibril association, or other processes. We therefore used QLS to study the temperature dependence of Aβ fibrillogenesis quantitatively. We determined the fibril elongation rate k_e in a series of experiments in which Aβ(1–40) was allowed to fibrillize at temperatures ranging from 4 to 40°. In each case, we could show that the function $k_e(T)$ obeyed the Arrhenius relationship $k_e = A \exp(-E_A/kT)$, where A is a preexponential constant and E_A is the activation energy for the reaction (Fig. 9).[27] Using this relationship, and examining the monomer addition reaction in the context of transi-

[31] H. Naiki and K. Nakakuki, *Lab. Invest.* **74,** 374 (1996).

tion-state theory, one can deduce the thermodynamic characteristics of the activation process. We found that the activation energy E_A was ~23 kcal/mol and that the activation entropy ΔS was ~53 cal/mol-deg. These numbers show that the fibril elongation reaction requires a significant input of energy and involves substantial conformational changes, e.g., unfolding of Aβ at the fibril tip or increased disorder of associated solvent molecules. The activation process thus may provide targets for therapies aimed at preventing the formation of Aβ conformers involved in fibril elongation.

Summary

This article discussed the principles and practice of QLS with respect to protein assembly reactions. Particles undergoing Brownian motion in solution produce fluctuations in scattered light intensity. We have described how the temporal correlation function of these fluctuations can be measured and how mathematical analysis of the correlation function provides information about the distribution of diffusion coefficients of the particles. We have explained that deconvolution of the correlation function is an "ill-posed" problem and therefore that careful attention must be paid to the assumptions incorporated into data analysis procedures. We have shown how the Stokes–Einstein relationship can be used to convert distributions of diffusion coefficients into distributions of particle size. In the case of fibrillar polymers, this process allows direct determination of fibril length, enabling nucleation and elongation rates to be calculated. Finally, we have used examples from studies of Aβ fibrillogenesis to illustrate the power these quantitative capabilities provide for understanding the molecular mechanisms of the fibrillogenesis reaction and for guiding the development of fibrillogenesis inhibitors.

Acknowledments

We gratefully acknowledge the critical comments provided by Drs. Dominic Walsh, Youcef Fezoui, Jayanti Pande, Neer Asherie, and George Thurston. This work was supported by Grant 1P01AG14366 from the National Institutes of Health, through the generosity of the Foundation for Neurologic Diseases, and by Amgen/MIT and Amgen/Brigham and Women's Hospital research collaboration agreements.

[28] Flow Cytometric Characterization of Amyloid Fibrils

By JONATHAN WALL and ALAN SOLOMON

Flow Cytometric Analysis of Aggregates: The Search for Fibrils

Classical approaches to monitoring aggregative phenomena involve measurements of turbidity and light scattering,[1–2a] as well as fluorimetric/spectroscopic techniques, where the availability of suitable probes permits.[3–5] Studies of biological polymerizations have been, and remain, largely dependent on these technologies, which are employed extensively in the study of fibrillogenesis. A major hurdle in the *in vitro* analysis of fibrillogenesis is the ability to discern and quantitate fibrils with respect to amorphous aggregates when both are present in the same solution. This article describes a method that utilizes both light scattering and amyloidophilic fluorescent probes to facilitate the rapid screening of fibril suspensions using flow cytometry. The fluorescent probe thioflavin T (ThT) has been widely employed for the clinical detection of amyloid as well as for monitoring *in vitro* fibrillogenesis (see, e.g., Wall *et al.*[5a] and LeVine,[5b] this volume and references therein). Although not generally considered a fluorescent probe, Congo red is used in concert with ThT to label fibrils for flow cytometric analysis. The fluorescent capabilities of Congo red have been exploited previously to measure the growth of fungal hyphae by epifluorescence microscopy with impressive results.[6]

The technique enables analyses of scattering and multiple fluorescence intensities to be performed independently and simultaneously and on individual particles in suspension. Samples for analysis can come from any time point within a fibrillogenesis experiment, allowing reaction kinetics to be monitored. This methodology may be used to confirm the presence and

[1] C.-L. Shen, G. Scott, F. Merchant, and G. M. Murphy, *Biophys. J.* **65,** 2383 (1993).
[2] S. J. Tomski and R. M. Murphy, *Arch. Biochem. Biophys.* **294,** 630 (1992).
[2a] A. Lomakin, G. B. Benedek, and D. B. Teplow, *Methods Enzymol.* **309** [27] 1999 (this volume).
[3] E. M. Castano, F. Prelli, T. Wisniewski, A. Golabek, R. A. Kumar, C. Soto, and B. Fragione, *Biochem J.* **306,** 599 (1995).
[4] H. Naiki, K. Higuchi, M. Hosokawa, and T. Takeda, *Anal. Biochem.* **177,** 244 (1989).
[5] J. M. Buzan and C. Frieden, *Proc. Natl. Acad. Sci. U.S.A.* **93,** 91 (1996).
[5a] J. Wall, C. L. Murphy, and A. Solomon, *Methods Enzymol.* **309** [14] 1999 (this volume).
[5b] H. LeVine III, *Methods Enzymol.* **309** [18] 1999 (this volume).
[6] H. Matsuoko, H.-C. Yang, T. Homma, Y. Nemoto, S. Yamada, O. Sumita, K. Takatori, and H. Kurata, *Appl. Microbiol. Biotechnol.* **43,** 102 (1995).

METHODS IN ENZYMOLOGY, VOL. 309
0076-6879/99 $30.00

relative abundance of fibrils in suspension as well as providing certain morphological and clinical data.

Method

Measurements are made using a Becton-Dickinson FACScan flow cytometer (San Jose, CA) with a 15-mW, air-cooled, argon ion laser providing a 20 × 64-μm elliptical excitation beam at 488 nm, with standard 530 nm (fluorescein), 585 nm (phycoerythrin/propidium iodide), and >650 nm (rhodamine) bandpass filters. The cytometer is computer controlled using a Becton-Dickinson FACStation; data analysis is performed using the CellQuest software package, version 1.2 (Becton-Dickinson). While any available flow cytometer would be suitable for this type of analysis, the following experimental parameters relate solely to the FACScan and should be extrapolated for use on other instruments. Gain settings used for the analysis of synthetic fibrils are shown in Table I; for comparison, the optimum settings for routine immunophenotyping are also presented. The major difference in instrument settings is the forward angle light-scatter (FSC) detector voltage, which is an order of magnitude greater than that used for cells, indicating the low FSC signal of fibrils. All other values are comparable.

For the analysis of fibrils, a threshold value is set at channel 52 using the FSC data to remove signals arising from dust and minute particulate debris. This is lower than the value of 124 used routinely for immunophenotyping to include the smaller fibrillar aggregates. Compensation for spectral overlap of the three fluorescence signals is as follows: FL1 = 1.2% of FL2; FL2 = 19.8% of FL1; FL2 = 0.0% of FL3; and FL3 = 25.1% of FL2.

TABLE I
FACScan Settings for Amyloid Fibril Detection[a]

Gain option	FSC (lin)	SSC (lin)	FL1 (log)	FL2 (log)	FL3 (log)
		Settings for fibril detection			
Amplifier gain	6.64	1.00	1.00	1.00	1.00
Detector (V)	10	308	674	476	524
		Settings for immunophenotyping			
Amplifier gain	2.31	1.00	1.00	1.27	1.00
Detector (V)	1	333	664	473	550

[a] Comparison of gain settings used for the detection of synthetic fibrils during immunophenotyping using the Becton-Dickinson FACScan. lin and log refer to linear and logarithmic modes of data collection. FSC, foward scatter (<1° to 10° angle); SSC, side scatter (90° angle); FL1, FL2, and FL3 refer to fluorescence signals from the three photomultipliers corresponding to the 530-, 585-, and >650-nm bandpass filters, respectively.

Between 0.5 and 1 ml of a fibril suspension at <0.3 mg/ml in aqueous buffer is filtered (6-μm membrane) and placed in the flow cytometer; 10,000 data points are acquired for subsequent analysis. The flow cytometric examination of fibrils proceeds as in Scheme I.

Measurements of FSC, SSC, and FL1, FL2, and FL3 (for definitions, see Table I) are made during each acquisition. Synthetic and extracted amyloid fibrils exhibit a linear relationship on a plot of FSC vs SSC (Fig. 1A, color insert). This relationship alone, however, is insufficient to proclaim the presence of fibrils, as for example, a suspension of *Escherichia coli* bacteria will also yield a similar pattern. In contrast, amorphous aggregates exhibiting neither tinctorial nor electron microscopic properties of fibrils, in this instance, heat-aggregated Bence Jones protein (BJp), appear stochastically distributed (Fig. 1B, color insert).

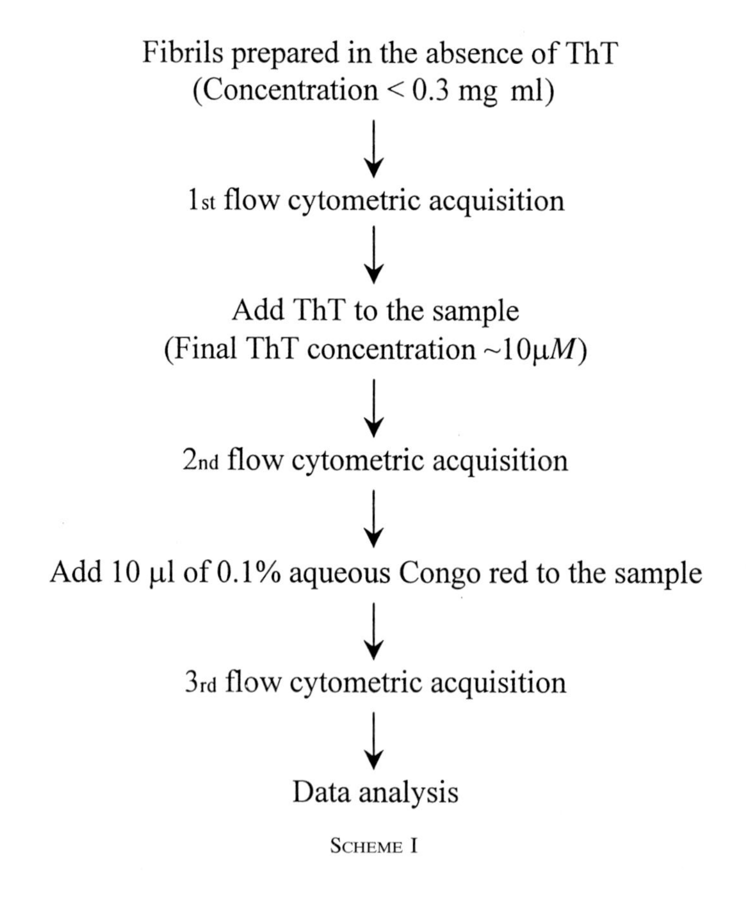

SCHEME I

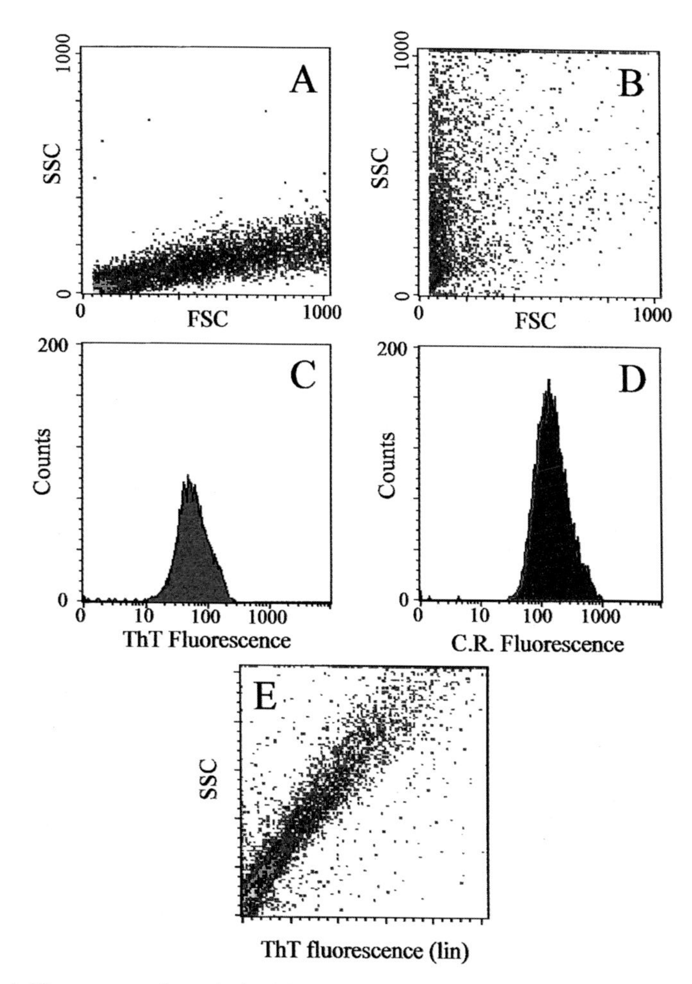

FIG. 1. Flow cytometric analysis of fibrils. (A) Linear dependence found on a plot of FSC vs SSC characteristic of fibrillar aggregates. Sample consists of synthetic fibrils composed of IgLC V_L peptide in phosphate-buffered saline (PBS) at 50 μg/ml. (B) Stochastic distribution of FSC vs SSC characteristic of amorphous aggregates obtained using a suspension of heat-aggregated IgLC V_L peptide in PBS at 50 μg/ml. (C) ThT fluorescence in FL1 associated with fibrils. Same sample as A in the presence of 10 μM ThT. (D) Congo red (C.R.) fluorescence in FL3 associated with fibrils. Same sample as A in the presence of 0.001% aqueous Congo red. (E) Linear dependence of ThT fluorescence intensity (FL1—linear data) on SSC. Same sample as A. All data were collected using the Becton-Dickinson FACScan Flow Cytometer using the instrument settings shown in Table I. In (A), (B), and (E) plots are displayed on a logarithmic density scale; levels are calculated as a percentage of the maximum event number and have logarithmic spacing between them. Log density scale for (A), (B), and (E): yellow, 50% peak height; purple, 25% peak height; orange, 12% peak height; blue, 60% peak height; pink, 3% peak height; green, 1% peak height. Peak height is defined as the maximum number of events.

The fluorescence intensity of ThT (FL1), collected during the second acquisition, is then plotted versus counts in histogram form (Fig. 1C, color insert). The presence of fibrils is indicated by a population distribution of particles with a high mean fluorescence intensity (i.e., generally >10). The ThT^+ material represents fibrils. Fluorescence data collected during the first acquisition, in the absence of either dye, can be used to establish values of sample autofluorescence. A more impressive fluorescence signal arises from the Congo red captured in FL3 during the third acquisition (displayed in histogram form in Fig. 1D, color insert). The addition of Congo red often diminishes ThT fluorescence intensity in this acquisition due to its more efficient absorption of the excitation wavelength as well as the emitted ThT fluorescence wavelength.

Criteria, therefore, for the detection of fibrils by flow cytometry are a linear dependence of FSC with SSC; positive FL1 in the presence of ThT; and positive FL3 in the presence of Congo red. Synthetic fibrils composed of V_L proteins and $A\beta(25–35)$ meet all three criteria, as do extracted human light chain (AL) amyloid fibrils. In contrast, a suspension of *E. coli* lacks signals in both FL1 and FL3, but displays a linear SSC vs FSC plot, and heat-precipitated BJp meets none of these requirements. Mixtures of amorphous and fibrillar material in suspension can be examined to determine the percentage of fibril particles. In this case, the population of particles meeting the fibril criteria is expressed as a percentage of total particles analyzed. Mixed samples will be bimodal when viewed as FL1 or FL3 histograms, fibrils exhibiting ThT and CR positivity, and amorphous material that appears negative for both dyes.

The relationship between SSC and FL1/FL3 is intriguing, as all three are indirect measurements of the number concentration of polymerized monomer. Berne demonstrated theoretically that, with certain caveats, the intensity of scattered light at any angle greater than forward is proportional to the number of isotropic units forming a rod-like polymer.[7] This model has been tested and applied successfully in the investigation of neurotubule polymerization.[8] Similarly, the use of ThT in quantifying amyloid fibrils is dependent on the assumption that the concentration of polymerized monomer is directly proportional to the fluorescence intensity. By extension, the same can be assumed for Congo red. Data obtained from an examination of fibril suspensions, plotted as FL1 or FL3 (collected in linear mode) vs SSC, demonstrate that fluorescence intensity and SSC are related directly and linearly (Fig. 1E, color insert). Thus, the flow cytometric examination of fibrils can provide valuable information relating to the morphology of

[7] B. J. Berne, *J. Mol. Biol.* **89,** 755 (1974).

[8] F. Gaskin, C. R. Cantor, and M. L. Shelanski, *J. Mol. Biol.* **89,** 737 (1974).

fibril populations as well as determining their presence and relative abundance rapidly in a solution of mixed aggregates.

Semiquantitative Analysis of AL Amyloid Burden by Flow Cytometry

The clinical diagnosis of amyloidoses is achieved by the histologic examination of suspect tissues using alkaline–Congo red staining. Positive diagnoses, which exhibit the characteristic blue-green birefringence under polarizing light microscopy, are usually confirmed by electron microscopy. Although satisfactory for discerning positive or negative samples, neither technique is capable of yielding a quantitative measure of amyloid burden. This information may be clinically important in determining the progression of the disease as well as providing information on the efficacy of treatment in reducing tissue amyloid. Patients with suspected immunoglobulin light chain (AL) amyloidosis routinely undergo a fat biopsy or fine needle aspirate to confirm the presence of amyloid in the fat tissue. This technique has rapidly become the first choice of physicians, as it is an expeditious and relatively noninvasive procedure, with respect to rectal or visceral biopsies. Flow cytometric evaluation of samples has been applied to provide a semiquantitative measurement of amyloid burden, based on the number of "positive" particles per unit weight of tissue.

Method

A 30-mg sample (wet weight) of abdominal fat is obtained by biopsy and processed according to a microextraction method similar to those described elsewhere in this volume.[9] Briefly, the sample is suspended in 1 ml of high-performance liquid chromatography-pure water and vortexed for approximately 1–2 min. This is followed by sonication on ice using a 5-mm-diameter sonication probe (Tekmar Sonic Disruptor, Cincinnati, OH). Four 20-sec pulses at 3 W are delivered, allowing a 1-min cooling period between each. The resulting milky suspension is then partitioned by centrifugation at 17,000*g* for 30 min at room temperature. This yields a slightly opaque supernatant covered with a dense layer of fat cells. A pellet may be present, dependent on the amount of amyloid in the sample. The supernatant is removed carefully, together with any pelleted material using a Pasteur pipette, avoiding contamination by the upper fatty layer.

Three control samples should be prepared: two negative controls consisting of (1) an amyloid-free abdominal fat sample of equal wet weight

[9] B. Kaplan, R. Hrncic, C. L. Murphy, G. Gallo, D. T. Weiss, and A. Solomon, *Methods Enzymol.* **309** [5] 1999 (this volume).

prepared in parallel with the test sample and (2) a preparation of nonfibrillar aggregates; a 1-mg/ml suspension of heat-denatured Bence Jones protein or IgLC V_L is used routinely. The positive control is a preparation of synthetic fibrils, composed of IgLC V_L,[5a] confirmed as fibrillar by spectroscopic, microscopic, or cytometric analyses.

Samples are analyzed as described in Scheme I. The negative fat sample is used first to determine the position of the negative gates for FL1 and FL3. As before, a sample of 10,000 particles is observed. Negative gates are positioned on histograms of FL1 and FL3 vs counts such that >97% of the particles are left of the gate (see Fig. 2). Any particles with higher fluorescence intensity than the cutoff gate will be deemed positive. The positive control is then analyzed to validate the positioning of the gate, i.e., >97% of the particles should be to the right of the determined cutoff value. Correct demarcation of the negative cutoff is then confirmed using the amorphous-aggregate control, again >97% of the particles should reside on the left of the gate. Having established the negative cutoff, experimental samples can be analyzed in the same way as the controls. Data should be examined to provide an absolute value of the number of positive particles in each sample so that this value can be expressed as a percentage of the total

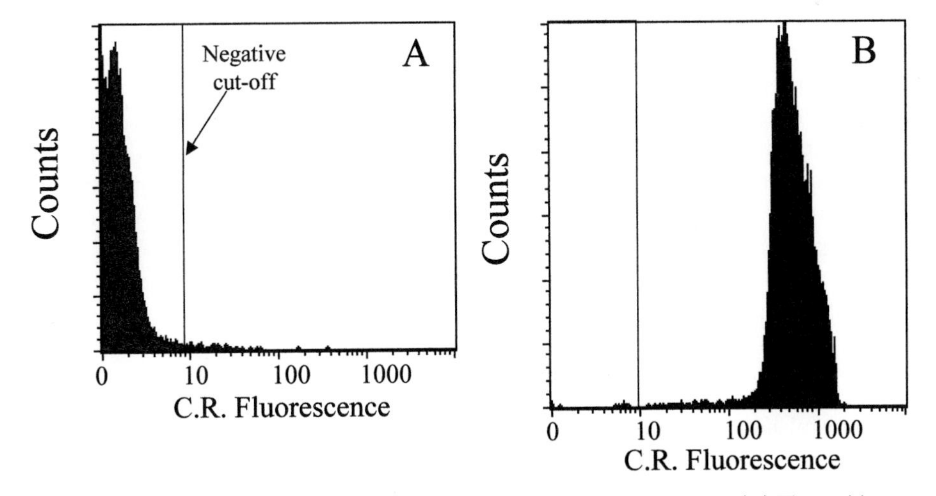

FIG. 2. Positioning of the negative cutoff gate when determining the ABI. (A) The position of the gate is determined using a negative control sample, i.e., a preparation of nonfibrillar aggregates prepared by heat denaturation. The protein sample is in PBS at 0.1 mg/ml and 0.001% aqueous Congo red. A similar result is obtained using an amyloid-free, abdominal fat sample. (B) To confirm the positioning of the gate, a positive control sample is used. This consists of a preparation of synthetic IgLC V_L fibrils in PBS at 50 μg/ml. The majority (>97%) of the particles fall within the "amyloid-positive" region. The negative cutoff gate is positioned as in A.

number of particles interrogated. Application of the following equation will yield a quantitative parameter, the amyloid-burden index (ABI):

$$\mathrm{ABI} = \frac{(x + y)}{2}$$

where x is the percentage of positive particles in FL1 and y is the percentage of positive particles in FL3. Clearly, if every particle is positive, as should be the case for a synthetic fibril preparation, an ABI value of 100 will be obtained. Conversely, the negative fat sample and amorphous aggregates will yield an ABI of zero.

Data obtained to date using this technique are in excellent agreement with clinical findings and the Congo red staging system developed and utilized in our program. In order to maintain consistency with clinical diagnostic techniques, an arbitrary negative cutoff value for the ABI has been adopted, i.e., ABI > 12 is regarded as positive. This value will be modified as the patient data base grows and a more accurate representation is determined. The ABI cutoff value is used in order to reconcile differences between flow cytometric results and those from other clinical procedures.

Caution must be exercised in applying the ABI, as it has proven inadequate in two cases of confirmed amyloid deposition in the fat. Both were characterized by "rare" amyloid deposits associated intimately with an abnormally extensive collagenous matrix. In these instances, the collagenous material, which retains the amyloid particles during the microextraction procedure, remains in the upper fatty layer. Because the ABI is quantitative and related to the weight of biopsied material, serial analyses from the same patient will provide information on the amount and rate of deposition, which may relate to overall disease progression. In a similar vein, ABI may be used in the *in vivo* testing of antiamyloid therapeutics.

Acknowledgments

We thank Valerie Brestel and Rich McCoig for assistance in the preparation of this manuscript, Richard Andrews for flow cytometry analyses, and Rudi Hrncic for assistance in the clinical application of flow.

[29] Screening for Pharmacologic Inhibitors of Amyloid Fibril Formation

By HARRY LEVINE III and JEFFREY D. SCHOLTEN

Background

The deposition of amyloid fibrils is responsible for the pathology of a number of diseases. Their resistance to biological clearance is a hallmark of amyloidosis; however, systemic amyloidoses such as AL (immunoglobulin light chain) and AA (serum amyloid A protein) have been shown to reduce or resolve on removal of the source of precursor protein or on treatment with a pharmacologic agent that blocks fibril formation.[1] Thus, interfering with amyloid fibril formation by a small organic molecule pharmacologic therapy could be a treatment option. A variety of such ligands have been described,[2] although they lack suitable profiles as therapeutics. For Alzheimer's disease (AD), whose amyloidosis affects the largest number of people, peptides[3] and monoclonal antibodies[4] have also been proposed as potential therapeutic agents, although it is unclear how blood–brain barrier penetration and cellular penetration will be achieved with these large, polar molecules.

Amyloid fibrils composed of the Aβ peptide (1–40) and (1–42) are deposited in plaque structures and within blood vessel walls concentrated in the associative brain areas of AD patients. Besides being diagnostic of the disease along with neurofibrillary tangles (NFTs) and genetic linkage with the precursor protein in familial forms of AD, other evidence has accumulated for the role of various forms of these peptides in eliciting the neuronal degeneration in affected brain regions, leading to the clinically observed dementia.[5] While it may not be the singular cause of all cases of AD, the Aβ peptide can be at the very least a permissive or predisposing factor in the disease. Hence, pharmacologic agents with the ability to cross the blood–brain barrier that block Aβ fibril formation and are well tolerated

[1] F. Tagliavini, R. A. McArthur, B. Canciani, G. Giaccone, M. Porro, M. Bugiani, P. M.-J. Lievens, O. Bugiani, E. Peri, P. Dall'Ara, M. Rocchi, G. Poli, G. Forloni, T. Bandiera, M. Varasi, A. Suarato, P. Cassutti, M. A. Cervini, J. Lansen, M. Salmona, and C. Post, *Science* **276,** 1119 (1997).

[2] T. Bandiera, J. Mansen, C. Post, and M. Varasi, *Curr. Med. Chem.* **4,** 159 (1997).

[3] L. O. Tjernberg, J. Naslund, F. Lindqvist, J. Johansson, A. R. Karlstrom, J. Thyberg, L. Terenius, and C. Nordstedt, *J. Biol. Chem.* **271,** 8545 (1996).

[4] B. Solomon, R. Kopple, E. Hanan, and T. Katzav, *Proc. Natl. Acad. Sci. U.S.A.* **93,** 452 (1996).

[5] D. J. Selkoe, *J. Biol. Chem.* **271,** 18295 (1996).

0076-6879/99 $30.00

on chronic administration are being pursued as potential disease-modifying therapeutics for AD.[6]

There is now considerable evidence that amyloid fibril formation constitutes a protein misfolding process[7] in which intermolecular β-sheet interactions become stabilized abnormally, either through amino acid changes introduced by mutations, under altered environmental conditions, or by interactions with other molecules known as "pathological chaperons." Interestingly, the physical structural analysis of fibers (amyloid fibrils do not form crystals) by X-ray diffraction and by high-resolution electron and atomic force microscopy indicates that a common structural format is assumed by all amyloid fibrils studied, regardless of primary amino acid sequence.[8] Such conservation of structure may reflect similar underlying organizational principles, implying that certain inhibitors of amyloidogenesis in one system may work for a variety of amyloid fibrils. Besides broadening the therapeutic potential and providing therapy for amyloidoses that only would have been beneficiaries of orphan drug development, general applicability would facilitate testing compounds for amyloid diseases for which animal models are inadequate or not available.

Rationale

Considering amyloid fibril formation as a protein-folding problem focuses the targeting of inhibitors either on the destabilization of the interactions that ultimately result in building the amyloid fibril or on a stabilization of the monomer, dimer, or small oligomers. If atomic level resolution structures of amyloid fibrils were available, drug design would be much simplified. Because fibril structure lacks sufficient supramolecular repeat order for detailed X-ray crystallographic analysis, more traditional pharmacological compound screening methods are used, followed by medicinal chemical lead optimization of inhibitor structure based on activity against amyloid fibril formation. This procedure requires selection of an assay format that is representative of the fibril-forming reaction and is compatible with the number and types of compounds to be assayed. This is not always an easy task, especially if the relationship of the *in vitro* reaction mechanism to the *in vivo* situation is unclear. In the case of the Aβ peptide, a definitive reaction mechanism remains to be established even *in vitro,* and there is the possibility that several pathways to fibril formation may coexist.

[6] H. LeVine III, *ID Res. Alert.* **1,** 1 (1996).
[7] R. Wetzel, *Cell* **86,** 699 (1996).
[8] M. Sunde and C. Blake, *Adv. Prot. Chem.* **50,** 123 (1997).

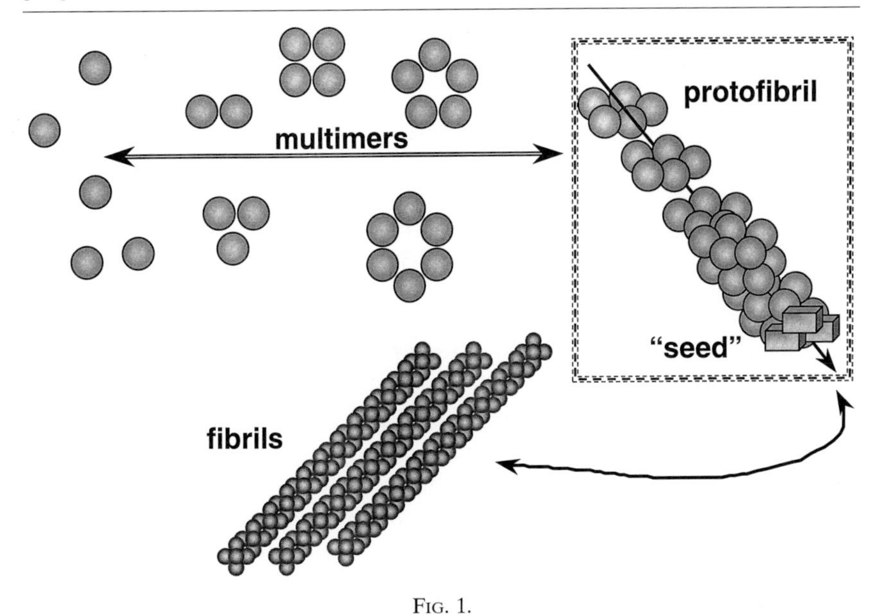

Fig. 1.

One plausible reaction scheme for *in vitro* Aβ amyloid fibril formation that is supported by recent atomic force microscopy observations[9] is presented in Fig. 1. Even without the complications introduced by the biological milieu in which the peptide finds itself after cleavage from the precursor protein (cell surfaces, membrane compartments, binding proteins, and other molecules such as heparan sulfate proteoglycans, peptide degradation, mixtures and variable concentrations of Aβ40 and Aβ42, pH, ionic, and dielectric environment), it is obvious that the reaction is significantly more complicated than a typical enzymatic reaction. Fibril formation is not necessarily a serial process, and several reactions could be proceeding in parallel, each of which could have different biological consequences. The reaction usually chosen for targeting is the fibril formation reaction shown boxed, normally performed at high, micromolar concentrations with synthetic Aβ(1–40) or (1–42). This choice reflects the practicality of rapid (hours), reproducible assays whose end product (amyloid fibrils) can be verified by physical methods[10] and by cellular assays.[11] Depending on the initial state of the

[9] J. D. Harper, S. S. Wong, C. M. Lieber, and P. T. Lansbury, Jr., *Chem. Biol.* **4,** 119 (1997).
[10] J. D. Harper and P. T. Lansbury, Jr., *Annu. Rev. Biochem.* **66,** 385 (1997).
[11] L. L. Iverson, R. J. Mortishire-Smith, S. J. Pollack, and M. S. Shearman, *Biochem. J.* **311,** 1 (1995).

peptide and environmental conditions, different reaction steps become rate limiting and the structural dependence of inhibitor activity for that rate-limiting step may be different. The worse case scenario would be to have multiple pathways contributing similar kinetics, which could lead to a nonsensical structure–activity relationship (SAR). Assays in which other parts of the reaction scheme such as low oligomeric species[12] and accretion of nanomolar concentrations of monomer/low oligomer species onto preformed fibrils or plaques (plaque growth) have been developed,[13] but effective inhibitors of these reactions have not been published.

Properly established screening assay formats provide qualitative data about a large number of compounds but often leave something to be desired in the form of false positives and may not give accurate IC_{50} values. In screening large libraries of compounds, the compound concentration is adjusted to obtain a convenient hit rate, usually 0.1% or less. The advantage of large compound libraries with related structures is that not all positive compounds need be identified in the initial screen, just enough to weed out the false positives and to provide a base from which to begin an SAR profile. The number of compounds to be tested can be limited by making educated guesses about critical topological arrangements by hand or by computerized searching with topological or structure similarity programs.

Positive compounds or "hits," compounds inhibiting amyloid fibril formation by the chosen cutoff criteria (say >50% inhibition at 30 μM compound), are rechecked in the original screening assay and are then validated by titration in another assay that may be more labor-intensive but is designed to be quantitative and to avoid possible artifacts from the screening assay. Several different verification assays may be required at this stage to address both of these concerns. After a series of compound classes are defined to be active, SAR analysis and compound structural optimization can begin.

Assay for β(1–40) Fibril Formation

Radiolabeled β(1–40), most commonly ^{125}I-labeled tyrosine-10, is employed in the screening assays largely because of the ability to quantify the amount of fibril formation. A caveat to the use of radioiodinated peptides is that the iodine adds a large hydrophobic substituent to the aromatic ring,

[12] W. Garzon-Rodriguez, M. Sepulveda-Becerra, S. Milton, and C. G. Glabe, *J. Biol Chem.* **272,** 21037 (1997).

[13] W. P. Esler, E. R. Stimson, J. R. Ghilardi, A. M. Felix, Y.-A. Lu, H. V. Vinters, P. J. Mantyh, and J. E. Maggio, *Nature Biotechnol.* **15,** 258 (1997).

which could cause the peptide not to reflect the chemical properties of the unmodified molecule accurately. In the presence of excess unlabeled peptide, however, it appears that ^{125}I-labeled β(1–40) accurately tracks the nonradioactive peptide in the assays described here.

Extrinsic probes for amyloid fibrils such as thioflavin T (ThT) or Congo red are susceptible to optical artifacts and interaction with compound binding sites that are difficult to interpret when large numbers of compounds are being evaluated. Such assays are useful at the next level of testing, which will be discussed in a later section. Some newer fluorescence methods that may be able to avoid these artifacts with probes built into the β(1–40) peptide.

Two types of screening assays are described here, one based on spontaneous fibril formation[14,15] and the other based on seeded fibril growth.[16] Fluorometric or radiometric detection can be used. The kinds of inhibitory compounds detected can be different with the different assay formats. There are several schools of thought on selecting the optimal therapeutic target Aβ species. Stabilizing the monomeric peptide in a nonamyloidogenic conformation is one way to inhibit amyloid fibril formation. The self-seeded assay described in this article is likely to be less sensitive to inhibitors that bind stoichiometrically with the Aβ monomer than is the spontaneous fibril formation assay. Targeting a sparsely populated intermediate that was rate limiting in fibril formation would presumably minimize the CSF drug levels required for efficacy, assuming that the monomeric Aβ species was not itself toxic to cells. Other assay formats that target different parts of the fibril-forming pathway, such as fibril extension[13] and dimerization,[12] are described elsewhere in this volume. β(1–40) has been selected as the test peptide as methods exist for radioiodination and purification,[17,18] [^{125}I] Tyr-10 β(1–40) (>1000 Ci/mmol), and [2,3-^{3}H]Ala-2 β(1–40) (48 Ci/mmol) are available commercially from Amersham. Although available commercially from several sources in varying purity, the β(1–42) peptide is hard to purify, and fibril formation is extremely hard to control well enough for large-scale assays. The corresponding radiolabeled β(1–42) forms are not yet available commercially.

[14] S. J. Hays, H. LeVine III, and J. D. Scholten, International Patent WO 9716191 (1997).
[15] S. J. Hays, H. LeVine III, and J. D. Scholten, International Patent WO 9716194 (1997).
[16] H. Naiki and K. Nakakuki, *Lab. Invest.* **74,** 374 (1996).
[17] J. E. Maggio, E. R. Stimson, J. R. Ghilardi, C. J. Allen, C. E. Dahl, D. C. Whitcomb, S. R. Vigna, H. V. Vinters, M. W. Labenski, and P. W. Mantyh, *Proc. Natl. Acad. Sci. U.S.A.* **89,** 5462 (1992).
[18] H. LeVine, III, *Neurobiol. Aging.* **16,** 755 (1995).

Materials and Methods

Synthetic β(1–40) peptide, TFA salt (BACHEM, Torrance, CA) is stored in the lyophilized form at −80°. For the spontaneous fibril formation assay the peptide is dissolved at 10 mg/ml in 1,1,1,3,3,3-hexafluoro-2-propanol (HFIP) (Aldrich, Milwaukee, WI) at 37° for 10 min. Peptide batches that do not dissolve under these conditions are already uselessly fibrillized and are either returned to the manufacturer or depolymerization can be attempted after removal of the solvent by the addition of glass-distilled 80% (w/v) formic acid. For the seeded fibril assay, the peptide is dissolved in ice-cold high-performance liquid chromatography (HPLC)-grade water at 2.2 mg/ml (pH 1.6) and frozen immediately in 50- to 100-μl aliquots at either −20° or −80°. Upon storage at either temperature the peptide will seed spontaneously over the period of weeks to several months, forming small amounts of an uncharacterized species that passes through a 0.2-μm filter. Fresh unopened lyophilized peptide or peptide disaggregated by treatments involving neat trifluoroacetic acid,[19] concentrated formic acid,[20] and/or HFIP[21,22] exhibits a prolonged lag phase when assayed at pH values >pH 6 before fibril formation is apparent. Different batches of peptide from the same supplier and material from various suppliers will display altered kinetics of fibril formation. Comparable irreproducibility is observed for all methods following fibril formation and is generally ascribed either to the conformational ambiguity of Aβ peptides that leads to its amyloidogenicity or to the presence of variable contamination by an incompletely deblocked peptide. Any batch of Aβ peptide should be characterized under standard conditions before experimental use. Matrix-assisted laser dissociated mass spectroscopy (MALDI-MS) and capillary electrophoresis are particularly useful in detecting impurities, although the amount of impurity required for interference will depend on the identity of the contaminant.

Spontaneous Fibril Formation Assay

The conditions for amyloid fibril formation from depolymerized β(1–40) are derived from those described previously.[14,15] [^{125}I]β(1–40) (15–50,000 cpm/assay) is dried down from the acetonitrile/TFA storage solvent and dissolved in HFIP containing 2.5 mg/ml unlabeled β(1–40), and the mixture is diluted to 0.5 mg/ml peptide with HPLC-grade water. Ten microliters of this peptide solution is added to 25 μl of 25 m*M* sodium phosphate

[19] S. C. Jao, K. Ma, J. Talafous, R. Orlando, and M. G. Zagorski, *Amyloid Int. J. Exp. Clin. Invest.* **4,** 240 (1997).

[20] A. E. Roher, K. C. Palmer, V. Chau, and M. J. Ball, *J. Cell Biol.* **107,** 2703 (1988).

[21] S. J. Wood, W. Chan, and R. Wetzel, *Chem. Biol.* **3,** 949 (1996).

[22] C. J. Barrow, A. Yasuda, P. T. Kenny, and M. G. Zagorski, *J. Mol. Biol.* **225,** 1075 (1992).

buffer, pH 6.0, containing a compound of interest [<5% dimethyl sulfoxide (DMSO)] in each well of a polypropylene (Costar, Acton, MA) 96-well plate. The residual HFIP (6.75%) does not appear to effect the fibril-forming reaction. After sealing the plate with adhesive-backed aluminum film (Beckman, Fullerton, CA) to prevent evaporation, the plate is incubated for 2–6 hr at room temperature. The assay conditions are designed to allow less than 60% of the added peptide to form fibrils, thus allowing detection of precipitation or adsorption of peptide caused by compounds. Two hundred microliters of 25 m*M* sodium phosphate buffer, pH 6.0, is added to stop the reaction and the β(1–40) fibrils are collected by vacuum filtration through 0.2-μm pore size GVWP filters in 96-well plates (Millipore, Bedford, MA) after washing the filters with a further 50 μl of buffer. The radioactivity of the filters can be determined either by impregnating the dry filters with a melted solid scintillant such as Multilex (Wallac, Gaithersburg, MD) and quantifying with a Wallac Betamax 96-well scintillation counter or by punching out the filters into tubes and using gamma scintillation counting. The latter counting procedure avoids color quench corrections, which are necessary with the scintillant method.

Seeded Fibril Assay

The conditions for amyloid fibril formation from a preformed endogenous seed are those described by Naiki and Nakakuki[16] and the extent of fibril formation can be determined either fluorometrically with ThT or radiometrically with ^{125}I-labeled β(1–40) by filtration. To measure the effect of an added component on the rate of Aβ(1–40) fibril formation, 15 μl of an ice-cold reaction mix composed of 35 μM Aβ(1–40) peptide in 50 m*M* sodium phosphate, pH 7.5, 150 m*M* NaCl, and 0.02% NaN_3 is pipetted into the wells of a 96-well plate on ice containing the additive and mixed with compound. In a radiometric filtration assay, −15,000 cpm/well of [^{125}I]Y10 β(1–40) (~600 Ci/mmol) is included in the reaction mix. The plate is sealed with an aluminum film (Beckman) and the reaction is initiated by bringing the plate to 37°, incubating unstirred. Aβ(1–40) fibril formation is highly temperature dependent, with the rate increasing dramatically between 25° and 37°.[16] A more rapid rate of reaction can be obtained by shaking the samples gently. Up to 2% DMSO can be tolerated in the assay before fibril formation ceases. No effect on the reaction rate is seen at 0.1% DMSO. The reaction time will depend on the state of the peptide used, temperature, and stirring and must be determined empirically. Typically, a detectable reaction is obtained between 3 and 6 hr under the indicated conditions and reaches equilibrium overnight.

Fluorescence-detected fibril formation is terminated after removal of the sealing film by the addition of 250 μl/well of 5 μM ThT in 50 m*M*

TABLE I
ASSAY METHODS FOR MEASURING AMYLOID FIBRIL FORMATION

Heterogeneous[a]	Homogeneous[b]
Filtration	Dye binding
Sedimentation	ThT
Accretion (plaque growth)	CR
Oligomerization	FRET or polarization
	Accretion
	Fibril formation
	Oligomerization
	Light scattering

[a] Heterogeneous assays require separation of fibrils from solution.

[b] Homogeneous assays do not require separation of fibrils. If no extrinsic probe is added (i.e., light scattering and FRET or polarization/lifetime), measurements can be made at several different times on the same reaction. FRET, fluorescence resonance energy transfer; CR, Congo red; and ThT, thioflavin T.

glycine–NaOH, pH 8.5, and the amyloid-specific fluorescence is read within 1–2 min in a fluorescence plate reader such as the PerSeptive Biosystems Cytofluor 2350 (Perkin-Elmer Biosystems, Foster City, CA) or equivalent instrument (filters: excitation 440/20 nm; emission 485/20 nm for ThT) in round-bottom polystyrene 96-well microtiter plates (Corning, Corning, NY). The reaction is allowed to proceed to less than 30% to equilibrium in order to be able to detect precipitation or adsorption of peptide caused by compounds. The radiometric reaction is terminated after removal of the sealing film by the addition of 200 μl/well of ice-cold assay buffer, and the radiolabeled fibrils are collected by vacuum filtration, washed with an additional 200 μl/well of cold assay buffer, and the retained radioactivity determined as for the spontaneous fibril formation assay.

Potential Artifacts

There is no perfect assay system, particularly for β amyloid. Each method has its own blind spots and instances that require careful interpretation. Some advantages and disadvantages of a number of popular methods for measuring amyloid fibril formation are listed in Table I. Filtration methods require that the filaments be of sufficient size to be retained (here 0.2 μm). False-positive readings result from compounds that produce fibril fragmentation or nucleate large numbers of short fibrils, reducing the pep-

TABLE II
CHARACTERISTICS OF AMYLOID FIBRIL FORMATION INHIBITOR ASSAYS[a]

Method	Advantages	Disadvantages
Filtration (HET)	Rapid, high capacity, inexpensive, simple equipment	Requires labeling of peptide, false positives with fragmentation, cannot distinguish ppt from fibril, adsorption to filter, <0.2-μm fibrils lost, need high peptide concentrations
Sedimentation (HET)	Low absorption, inexpensive, simple equipment	False positives with fragmentation, cannot distinguish ppt from fibril, small fibrils lost, low capacity, high peptide concentrations needed
Plaque growth (HET)	Rapid, high capacity, inexpensive, simple equipment, low and high concentrations of peptide useable	Depends on quality of label, metal ion-catalyzed, different reaction mechanism
Turbidity (HOM)	Rapid, high capacity, absorbance plate reader, no labels or probes required	Expensive, large volume high peptide concentration, fibril size $\propto$ wavelength agitation required, all ppts measured, color interference (optical artifacts)
Congo red (HOM)	Rapid, high capacity, inexpensive, absorbance plate reader	Optical artifacts, extrinsic probe interference, high peptide concentrations needed, detects unknown species
Thioflavin T (HOM)	Rapid, high capacity, inexpensive, fluorescence plate reader	See Congo Red; >octamer detected, not all fibrils detected (e.g., amylin 20–29)
FRET (HOM)	Rapid, high capacity, fluorescence plate reader, low or high concentrations of peptide	Requires bulky chemical modification, optical artifacts, intrinsic probe interference, interpretation, fairly expensive equipment
Fluorescence polarization or lifetimes (HOM)	See FRET, avoid optical artifacts	See FRET

[a] Heterogeneous (HET) assays require separation of fibrils from solution. Homogeneous (HOM) assays do not require separation of fibrils. If no extrinsic probe is added (i.e., light scattering and FRET or polarization/lifetime), measurements can be made at several different times on the same reaction. ppt, precipitate.

tide concentration below that needed to sustain fibril growth. False negatives or inconclusive data result if the compounds induce precipitation of peptide or adsorb to the filter and bind peptide. For this reason, secondary assay systems are needed to verify hits and to evaluate the mechanism of action of compounds. A number of these are listed in Table II and are described elsewhere in this volume. Classical arbiters of fibril structure are Pd/C rotary shadowed or negative staining electron microscopy or atomic force microscopy, techniques that can resolve fibril structure. Optical artifacts in the ThT-based assay arise from color absorbance of test compounds at the excitation and/or emission wavelengths or from light scattering due to the precipitation of compound/peptide complexes or compound insolubility. These errors can be reduced by dilution to reduce precipitation, but little can be done to prevent them with plate readers whose light path includes the whole sample volume. Another potential artifact is competition of the test compound for the ThT-binding site on the Aβ fibrils. Fortunately, this can be controlled for in part by studying displacement of ThT from preformed Aβ fibrils by the test compound.

Clearly, finding candidate molecules that interfere with Aβ fibril formation is only the first step in defining a therapeutic candidate. Much structural optimization of the compounds will be required to obtain the desired efficacy, stability, safety, and novelty for commercial development. Further testing of interesting compounds to ascertain their toxological and pharmacokinetic characteristics in cellular and animal models of amyloidosis is required. Several potential systems for this analysis are described in this volume. Only after a viable clinical candidate compound has emerged will it be possible to test the hypothesis suggested from clinical responses of several systemic amyloidoses that intervening in β-amyloid fibril deposition or accelerating its turnover will have a salutary effect on the progression of Alzheimer's disease.

[30] Design and Testing of Inhibitors of Fibril Formation

By MARK A. FINDEIS and SUSAN M. MOLINEAUX

Introduction

Amyloid β-peptide (Aβ) is an approximately 40-residue proteolytic fragment of amyloid precursor protein (APP). Genetic, neuropathologic, and transgenic modeling studies implicate the expression and accumulation of Aβ as a necessary but not necessarily, by itself, sufficient step in the pathogenesis of AD.[1,2] Under normal conditions, monomeric Aβ appears

[1] D. J. Selkoe, *Science* **275,** 630 (1997).

[2] Alzheimer's Disease and Related Disorders Association, Inc. "Alzheimer's Disease: Statistics" fact sheet (1996) and references therein.

METHODS IN ENZYMOLOGY, VOL. 309 0076-6879/99 $30.00

to be a nonpathogenic molecule produced during metabolism of APP.[3–6] However, as Aβ undergoes the process that eventually produces amyloid fibrils, it becomes extremely toxic to neuronal cells.[7–10] The mechanism by which toxicity occurs is not well understood, but it clearly involves perturbation of Ca^{2+} homeostasis, oxidative injury, and apoptosis. Neurotoxicity has also been demonstrated through microglial activation mediated by Aβ fibrils.[11–15] Soluble amyloid-derived diffusible ligands (ADDLs) have been described as a form of Aβ that may account for direct neuronal toxicity.[16] Regardless of the exact structure of the toxic form of Aβ and the mechanism by which toxic effects are exerted, suppression or prevention of the transition of Aβ from monomeric to toxic oligomeric and polymeric species has thus emerged as a goal in the development of a therapy for AD.[1,2,17–19] This article discusses our strategy for designing and testing inhibitors of the polymerization of Aβ to identify leads for further development into potential therapeutics for AD.

Several groups have described efforts to define the structural basis for the fibrillogenic nature of Aβ and approaches to developing inhibitors of

[3] J. Hardy, *Trends Neurosci.* **4,** 154 (1997).

[4] N. Y. Barnes, L. Li, K. Yoshikawa, L. M. Schwartz, R. W. Oppenheim, and C. E. Milligan, *J. Neurosci.* **18,** 5869 (1998).

[5] C. Haass, M. G. Schlossmacher, A. Y. Hung, C. Vigo-Pelfrey, A. Mellon, B. L. Ostaszewski, I. Lieberburg, E. H. Koo, D. Schenk, D. B. Teplow, and D. J. Selkoe, *Nature* **359,** 322 (1992).

[6] M. Shoji, T. E. Golde, J. Ghiso, T. T. Cheung, S. Estus, L. M. Shaffer, X.-D. Cai, D. M. McKay, R. Tintner, B. Frangione, and S. G. Younkin, *Science* **258,** 126 (1992).

[7] M. S. Shearman, C. I. Ragan, and L. L. Iversen, *Proc. Natl. Acad. Sci. U.S.A.* **91,** 1470 (1994).

[8] K. Ueda, Y. Fukui, and H. Kageyama, *Brain Res.* **639,** 240 (1994).

[9] A. Lorenzo and B. A. Yankner, *Proc. Natl. Acad. Sci. U.S.A.* **91,** 12243 (1994).

[10] C. J. Pike, D. Burdick, A. J. Walencewicz, C. G. Glabe, and C. W. Cotman, *J. Neurosci.* **13,** 1676 (1993).

[11] M. P. Mattson and R. E. Rydel, *Nature* **382,** 674 (1996).

[12] S. D. Yan, X. Chen, J. Fu, M. Chen, H. Zhu, A. Roher, T. Slattery, L. Zhao, M. Nagashima, J. Morser, A. Migheli, P. Nawroth, D. Stern, and A. M. Schmidt, *Nature* **382,** 685 (1996).

[13] J. El Khoury, S. E. Hickman, C. A. Thomas, L. Cao, S. C. Silverstein, and J. D. Loike, *Nature* **382,** 716 (1996).

[14] J. A. London, D. Biegel, and J. S. Pachter, *Proc. Natl. Acad. Sci. U.S.A.* **93,** 4147 (1996).

[15] L. Meda, M. A. Cassatella, G. I. Szendrei, L. Otvos, P. Baron, M. Villalba, D. Ferrari, and F. Rossi, *Nature* **374,** 647 (1995).

[16] M. P. Lambert, A. K. Barlow, B. A. Chromy, C. Edwards R. Freed, M. Liosatos, T. E. Morgan, I. Rozovsky, B. Trommer, K. L. Viola, P. Wals, C. Zhang, C. E. Finch, G. A. Krafft, and W. L. Klein, *Proc. Natl. Acad. Sci. U.S.A.* **95,** 6448 (1998).

[17] S. R. Diehl, *Nature Med.* **1,** 120 (1995).

[18] D. J. Selkoe, *Nature* **375,** 734 (1995).

[19] D. B. Schenk, R. E. Rydel, P. May, S. Little, J. Panetta, I. Lieberburg, and S. Sinha, *J. Med. Chem.* **38,** 4141 (1995).

amyloid formation. The phenylalanyl-phenylalanyl dipeptide at position 19–20 has been identified previously as pivotal for the aggregation of Aβ and the deposition of monomeric Aβ onto plaque.[20,21] Substitutions of this sequence have been reported to result in inhibitors of aggregation.[22] Oxidation of Met-35 to the sulfoxide has been reported to increase aggregation.[23] Hydrophobic clustering in the carboxyl-terminal domain of Aβ has been suggested to be energetically important in fibrillogenesis.[24]

While larger peptides (various analogs of Aβ, modified forms of Aβ,[23] and other proteins such as apolipoprotein E, α_2-macroglobulin, and transthyretin) can have inhibitory effects on fibrillogenesis, such materials are less appealing as leads for drug development than much smaller molecules. The avid binding of dyes such as Congo red[25,26] and thioflavin T[27,28] to amyloid and the inhibition by 4′-deoxy-4-iododoxorubicin (IDOX) of prion protein amyloid formation,[29] Immunoglobulin G (IgG) light chain amyloidosis,[30] and insulin fibrillogenesis[31] suggest the possibility of preparing therapeutics based on these compounds. However, IDOX has acute cytotoxicity that precludes its use in chronic therapy. More recently, several groups, including ours, have begun to work with small fragments of the Aβ sequence, which have been reported to inhibit Aβ polymerization.[32,33]

[20] W. P. Esler, E. R. Stimson, J. R. Ghilardi, Y.-A. Lu, A. M. Felix, H. V. Vinters, P. W. Mantyh, J. P. Lee, and J. E. Maggio, *Biochemistry,* **35,** 13914 (1996).

[21] T. Tomiyama, S. Asano, Y. Furiya, T. Shirasawa, N. Endo, and H. Mori, *J. Biol. Chem.* **269,** 10205 (1994).

[22] C. Hilbich, B. Kisters-Woike, J. Reed, C. L. Masters, and K. Beyreuther, *J. Mol. Biol.* **228,** 460 (1991).

[23] S. W. Snyder, U. S. Ladror, W. S. Wade, G. T. Wang, L. W. Barrett, E. D. Matayoshi, H. J. Huffaker, G. A. Krafft, and T. F. Holzman, *Biophys. J.* **67,** 1216 (1994).

[24] P. H. Weinreb, J. T. Jarrett, and P. T. Lansbury, Jr., *J. Am. Chem. Soc.* **116,** 10835 (1994).

[25] S. J. Wood, R. Wetzel, J. D. Martin, and M. R. Hurle, *Biochemistry* **34,** 724 (1995).

[26] W. E. Klunk, J. W. Pettegrew, and D. J. Abraham, *J. Histochem. Cytochem.* **37,** 1273 (1989).

[27] H. LeVine, III, *Prot. Sci.* **2,** 404 (1993).

[28] H. LeVine, III, *Amyloid Int. J. Exp. Clin. Invest.* **2,** 1 (1995).

[29] F. Tagliavini, R. A. McArthur, B. Canciani, G. Giaccome, M. Porro, M. Bugiani, P. M.-J. Lievens, O.Bugiani, E. Peri, P. Dall'Ara, M. Rocchi, G. Poli, G. Forloni, T. Bandiera, M. Varasi, A. Suarato, P. Cassutti, M. A. Cervini, J. Lansen, M. Salmona, and C. Post, *Science* **276,** 1119 (1997).

[30] L. Gianni, V. Bellotti, A. M. Gianni, and G. Merlini, *Blood* **86,** 855 (1995).

[31] G. Merlini, E. Ascari, N. Amboldi, V. Bellotti, E. Arbustini, V. Perfetti, M. Ferrari, I. Zorzoli, M. G. Marinone, P. Garini, M. Diegoli, D. Trizio, and D. Ballinari, *Proc. Natl. Acad. Sci. U.S.A.* **92,** 2959 (1995).

[32] C. Soto, M. S. Kindy, M. Baumann, and B. Frangione, *Biochem. Biophys. Res. Commun.* **226,** 672 (1996).

[33] L. O. Tjernberg, J. Näslund, F. Lindqvist, J. Johansson, A. R. Karlström, J. Thyberg, L. Terenius, and C. Nordstedt, *J. Biol. Chem.* **271,** 8545 (1996).

This article describes elements of how we have approached the design and testing of such compounds.

Methods

Strategy: Design and Assay

Amyloid fibrils are characterized by an extensive repeating structure that is hypothesized to be primarily an antiparallel cross β-pleated sheet. Various models have been proposed to account for the folding of amyloidogenic proteins and peptides into cross β-fibrillar forms, but high-resolution data are still lacking. In the absence of such structural data, "rational drug design" is neither practical nor necessarily desirable. However, because Aβ has such a high affinity for itself, and the nature of these interactions has begun to be elucidated,[22,23,25,34–38] Aβ itself represents a starting point for developing leads for inhibition of its polymerization. The goal of this strategy was to combine the high-affinity, self-recognition of Aβ with a structural perturbation that interferes with the extended ordered structure. First, we examined the effect of longer Aβ-derived peptides on polymerization with the expectation that altered packing effects might disturb fibrillogenesis and result in inhibition. Although such compounds can act as inhibitors of polymerization, many of the Aβ analogs and subsequences described in the literature can form fibrils themselves and some of these are toxic as well. The second strategy was to modify the peptides to introduce a nonpeptidic pharmacophore that introduces a steric or conformational effect that augments peptide–Aβ interactions to effect antifibrillogenesis and, additionally, imparts desirable drug-like properties such as improved blood–brain barrier permeability and biostability. This strategy is diagrammed schematically in Fig. 1.

Amyloid formation has been characterized in a variety of ways. Amyloid fibril formation can be monitored by electron microscopy and fibril diffraction,[35] dye-binding birefringence and fluorescence shifting,[27,28] light scattering[39,40] atomic force microscopy,[41] and cellular toxicity.[7] For evaluating a

[34] C. J. Pike, M. J. Overman, and C. W. Cotman, *J. Biol. Chem.* **270,** 23895 (1995).

[35] P. E. Fraser, D. R. McLachlan, W. K. Surewicz, C. A. Mizzen, A. D. Snow, J. T. Nguyen, and D. A. Kirschner, *J. Mol. Biol.* **244,** 64 (1994).

[36] C. J. Barrow, A. Yasuda, P. T. M. Kenny, and M. G. Zagorski, *J. Mol. Biol.* **225,** 1075 (1992).

[37] C. J. Barrow and M. G. Zagorski, *Science* **253,** 179 (1991).

[38] C. Hilbich, B. Kisters-Woike, J. Reed, C. L. Masters, and K. Beyreuther, *J. Mol. Biol.* **218,** 149 (1991).

[39] J. T. Jarrett and P. T. Lansbury, Jr., *Cell* **73,** 1055 (1993).

[40] J. T. Jarrett, E. P. Berger, and P. T. Lansbury, Jr., *Biochemistry,* **32,** 4693 (1993).

[41] J. D. Harper, S. S. Wong, C. M. Lieber, and P. T. Lansbury, Jr., *Chem. Biol.* **4,** 119 (1997).

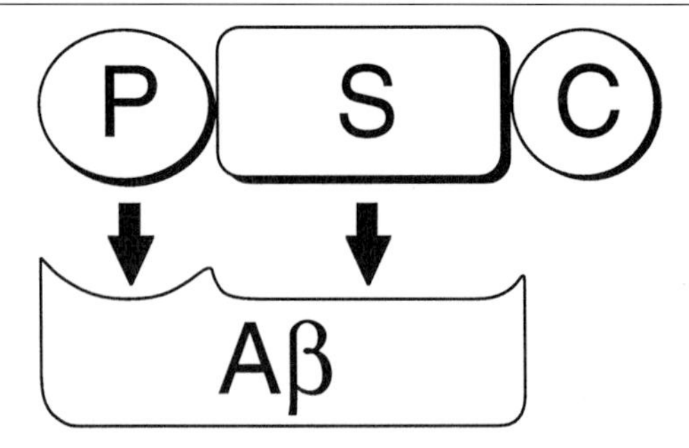

FIG. 1. Schematic depiction of a design strategy for developing inhibitors of amyloid β-peptide polymerization. Inhibitors are designed to have three components: P, S, and C. These components represent organic groups or pharmacophores ("P") adding binding energy and improved drug-like properties to a specificity element ("S") based on the sequence of Aβ, and one or more chemical modifications ("C") to improve biostability or other enhanced pharmacokinetic properties (e.g., C-terminal amidation to suppress exoproteolysis). An appropriate P–S–C combination is expected to result in high-affinity binding to Aβ and an effective biostability/bioavailability profile to afford robust leads for therapeutic development.

drug lead, we wanted a well-defined and reasonably short time course assay that would allow discrimination among potential amyloid inhibitors of varying potencies. For this purpose we chose to use *nucleation* and *extension* assays adapted from those described by Jarrett and Lansbury[39] and Jarrett *et al.*[40]

The general features of a nucleation assay are outlined in Fig. 2. The extent of formation of amyloid is observed over time. The amyloidogenic peptide or protein is dissolved in an appropriate buffer under conditions where it is initially in an unpolymerized state. Under controlled conditions of agitation, a period of time is observed during which no polymerization is detected (e.g., by light scattering or dye binding) but prefibrillar or soluble but highly ordered forms of amyloid are formed. Once a sufficient amount of this prefibrillar material has accumulated in the sample, fibril formation is rapid, continuing until the unpolymerized peptide is consumed, whereupon the reaction plateaus. Such a nucleated polymerization reaction profile has three major characteristics: (1) the time period during which nucleation occurs but fibril formation is not observed, (2) the rate of the fibrillogenesis reaction once fibrillogenesis has started, and (3) the level of fibril formation at the end of the polymerization process once it has plateaued.

In practice we use the first and third characteristics to evaluate and compare inhibitors (Fig. 2). A compound that interferes with nucleation will delay the onset of fibril formation. Inhibitory potency can then be expressed as the ratio of the time to nucleation in the presence of the

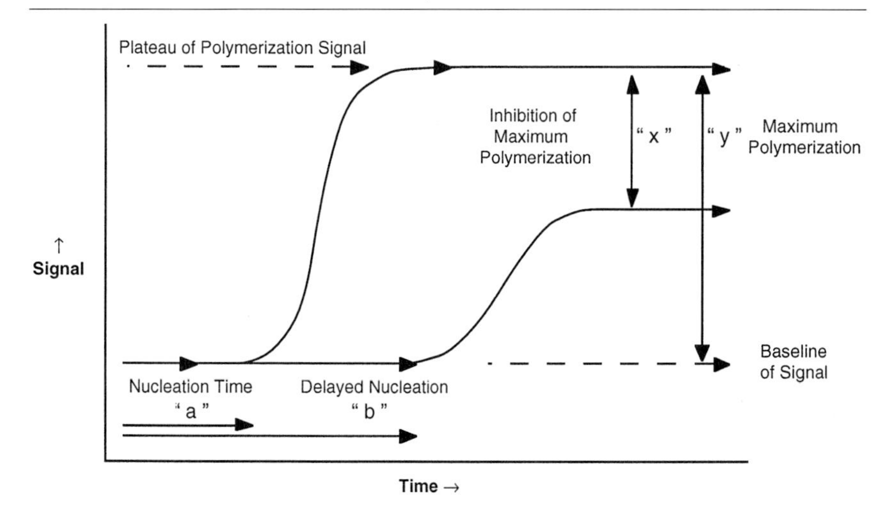

FIG. 2. Scheme of nucleation assay. Over the time course of a polymerization reaction, polymerization is monitored by the measurement of turbity in an ultraviolet spectrometer or by fluorescent dye binding in a fluorimeter. Data are summarized as lag times (time of delayed nucleation divided by time of control nucleation: b/a) and percentage inhibition (extent of lowering of polymerization plateau divided by the control level of polymerization: x/y).

inhibitor over the time to nucleation in the absence of test compound. An inhibitor or antagonist of polymerization will thus have a ratio or *lag* greater than one. An inhibitor can also reduce the amount of polymerization observed at the end of the assay. Any reduction in final polymerization level can be expressed as the *percentage inhibition* of polymerization. We routinely use both lag and percentage inhibition to characterize compounds tested as inhibitors.

The features of an extension assay are outlined in Fig. 3. In this assay, the preformed amyloid is combined with monomeric Aβ to seed the monomer so that it is incorporated into fibrils immediately. In comparison with the nucleation assay, there is no lag time in the extension assay. This assay is a first-order kinetic model for the addition of Aβ onto the ends of existing fibrils.[42] In this assay the primary effect of an inhibitor is to slow the rate of fibrillogenesis. Further characterization is possible at the level of fibril morphology, e.g., by electron microscopy or atomic force microscopy, to observe differences not reflected in the kinetic result of the assay. It may be possible that inhibitors can display alternative effects through binding differentially to various forms of Aβ in the assay. An inhibitor that bound preferentially to nuclei in this assay might delay polymerization (introduce

[42] H. Naiki and K. Nakakuki, *Lab. Invest.* **74,** 374 (1996).

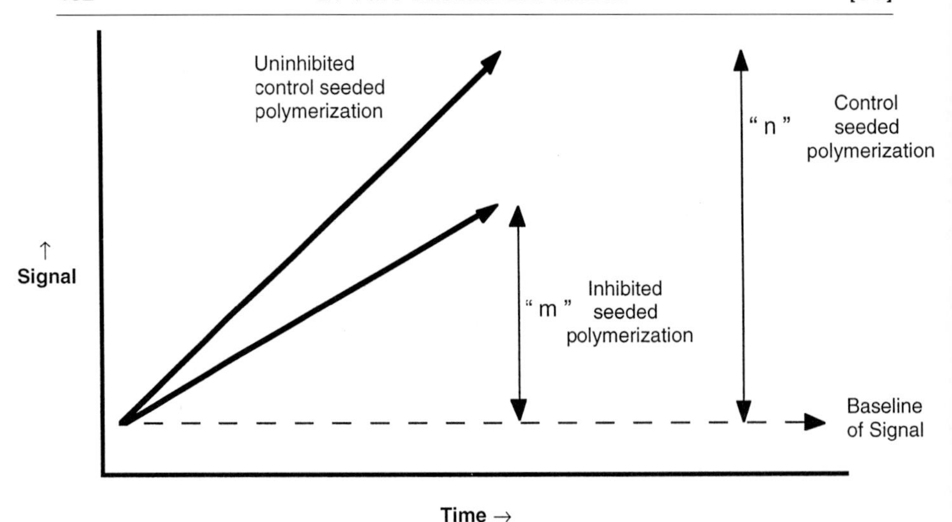

FIG. 3. Scheme of extension assay. Seeding of monomeric Aβ with prepolymerized Aβ initiates polymerization immediately (zero lag time in comparison with the nucleation assay). An inhibitor that suppresses the addition of monomer to seed and the further nucleation of monomer will reduce the initial rate of the polymerization reaction. Inhibition can be expressed as the ratio of the slopes (rates) of the control and inhibited polymerization reactions (m/n).

a lag period) but still allow a normal rate of polymerization once initiated. With the peptide-based compounds that we have studied, we have not seen this pattern in extension assay data; compounds exhibit the same rank order of potency in the nucleation assay and in the extension assay.

β-Amyloid Sequence Analysis for Secondary Structure

Secondary structural modeling of Aβ is useful in evaluating the structural propensities of Aβ and the relative importance of different residues for their contributions to different potential local structures. Experimentally, however, the folded structure of unpolymerized (monomeric?) Aβ is quite variable. Under different conditions of solvent and pH, Aβ can be predominantly in helical, random coil, or extended conformation. Polymerized Aβ or fragments of Aβ, as characterized by fibril diffraction[35] and solid-state nuclear magnetic resonance techniques,[43] retains substantial cross β-sheet structure.

Given this variability in experimental data, secondary structure prediction is noteworthy for what it cannot tell us: a well-defined structure of

[43] P. T. Lansbury, Jr., P. R. Costa, J. M. Griffiths, E. J. Simon, M. Auger, K. J. Halverson, D. A. Kocisko, Z. S. Hendsch, T. T. Ashburn, R. G. S. Spencer, B. Tidor, and R. G. Griffin, *Nature Struct. Biol.* **2,** 990 (1995).

Aβ under well-defined conditions. Results from predictive algorithms can, however, be useful in several ways. Relative structural propensities (e.g., among analyses of alanine-scan Aβ analogs) can be useful to identify potentially important subsequences of Aβ that might serve as targets or starting points for inhibitor design. Comparisons are potentially informative in evaluating Aβ structure or amyloidogenicity or in providing ideas about the interaction of Aβ or portions of its sequence with itself.

One predictive method we have used is the GOR method.[44] An example of the output from this program is listed in Table I. We have run this program using the published BASIC code modified to run under GWBASIC on a PC-type computer in automatic mode as described by the authors using standard BASIC and DOS conventions to save output as text files stored on disk. The program was modified further to explicitly note when calculated structural propensities for a given position are equal for any two of the four structural states evaluated (helix, extended, turn, coil). Using a word processing program such as Microsoft Word it is straightforward to generate computer text files containing large arrays of sequences to evaluate (an alanine or proline scan, deletion scans, etc.). This program provides output that includes both the calculated favored state and the numerical values of propensities for the other states. Inspection of this latter data allows comparison of the extent to which a structural state is favored.

Structural Models of Aβ

"Rational drug design" in the usual sense cannot currently be undertaken with Aβ. The absence of definitive structural data on the fibrillar and nonfibrillar forms of Aβ means that any model of the Aβ structure is probably flawed to some extent. Even so, reasonable assumptions can be introduced to establish a hypothesis about how inhibitors work. We have assumed that the primary structural feature of Aβ fibrils is β strand. It is unknown at this time whether monomeric units of Aβ within fibrils exist as small β sheets in a continuous β-sheet helix as reported for transthyretin[45] or in a different manner based on antiparallel β-helical protofibrils as proposed in an alternative model.[46]

If one assumes the targeted binding site on Aβ has β structure, then inhibitors based on Aβ-derived peptide sequences can be expected to have a propensity for β structure and the ability to form high-affinity complexes with Aβ. Further, if non-β structure is important for binding to a non-β

[44] B. Robson and J. Garnier, "Introduction to Proteins and Protein Engineering," Chapter 9 and Appendix III. Elsevier, New York, 1986.

[45] C. Blake and L. Serpell, *Structure* **4,** 989 (1996).

[46] N. D. Lazo and D. T. Downing, *Biochemistry* **37,** 1731 (1998).

TABLE I

OUTPUT FROM THE GOR ALGORITHM[a] FOR SECONDARY STRUCTURAL PREDICTION APPLIED TO THE SEQUENCE OF PRIMATE Aβ(1–40)[b]

SEQUENCE= DAEFRHDSGYEVHHQKLVFFAEDVGSNKGAIIGLMVGGVV

PREDICTION H=HELIX E=EXTENDED T=TURN C=COIL
HY=HYDROPHOBICITY Q=CHG Lexp2=(CHAR'C LENGTH)exp2

PREDICTION	H	E	T	C	HY	Q	L72
1 D->H:	150	−89	−74	−74	−1.2	−1	241
2 A->H:	165	−43	−115	−115	−1	0	135
3 E->H:	158	−15	−117	−117	−.7	−1	144
4 F->H:	144	−19	−53	−53	2.8	0	308
5 R->H:	99	−76	11	11	.3	1	136
6 H->TC:	70	−165	87	87	1.1	.3	308
7 D->C:	28	−259	176	186	−1.2	−1	241
8 S->T:	41	−267	151	141	−1.2	0	347
9 G->TC:	79	−222	127	127	0	0	36.8
10 Y->H:	110	−120	69	69	2.1	0	308
11 E->H:	133	−55	−37	−37	−.7	−1	144
12 V->H:	152	−42	−60	−60	1.4	0	411
13 H->H:	165	−45	−48	−48	1.1	.3	308
14 H->H:	135	−55	−43	−43	1.1	.3	308
15 Q->H:	96	−53	−51	−51	−.1	0	120
16 K->H:	103	−28	−55	−55	−.9	1	136
17 L->H:	107	18	−56	−56	2	0	136
18 V->H:	144	68	−65	−65	1.4	0	411
19 F->E:	109	131	−103	−103	2.8	0	308
20 F->HE:	151	151	−153	−163	2.8	0	308
21 A->H:	160	97	−145	−145	−1	0	135
22 E->H:	158	−15	−52	−52	− .7	−1	144
23 D->H:	128	−124	51	51	−1.2	−1	241
24 V->H:	107	−197	65	65	1.4	0	411
25 G->TC:	52	−242	117	117	0	0	36.8
26 S->TC:	49	−222	121	121	−1.2	0	347
27 N->TC:	34	−176	87	87	−.7	0	89.5
28 K->TC:	48	−98	50	50	−.9	1	136
29 G->TC:	−1	−22	17	17	0	0	36.8
30 A->E:	0	67	−35	−35	−1	0	135
31 I->E:	−24	162	−121	−121	4	0	403
32 I->E:	−31	242	−166	−166	4	0	403
33 G->E:	−71	278	−128	−128	0	0	36.8
34 L->E:	−58	208	−81	−81	2	0	136
35 M->E:	−77	163	−38	−38	1.8	0	136
36 V->E:	−98	118	−15	−25	1.4	0	411
37 G->E:	−131	83	32	12	0	0	36.8
38 G->E:	−111	58	12	−13	0	0	36.8
39 V->E:	−66	108	−80	−110	1.4	0	411
40 V->E:	−41	128	−125	−160	1.4	0	411

[a] From B. Robson and J. Garnier, "Introduction to Proteins and Protein Engineering," Chapter 9 and Appendix III. Elsevier, New York, 1986.

[b] Predicted conformation for a specified position is indicted by H (helical), E (Extended), T (turn), or C (random coil). Subsequent columns list the calculated propensities for the different conformations along with charge and the characteristic square length of the residue at that position.

structured element of Aβ, then the corresponding subsequence peptide can be expected to exist in the parent conformation partially or have access to that conformation through a low-energy pathway. Implicit in this low-resolution "modeling" is the recognition that the precise nature of the toxic form of Aβ that should be targeted by an inhibitor may be multicomponent. Also, within the model we assume that an inhibitor that binds to one component to suppress the formation of toxic forms of Aβ will also bind to other species. It is hoped that resolution of the uncertainties inherent in these assumptions will be resolved with further studies.

Synthesis of Peptide-Based Inhibitors

Much of our work, and that of other groups, has been based on the study of Aβ-derived peptides and their modifications. Because the synthesis of Aβ, particularly the 1–42 sequence, has been problematic historically, a few comments are warranted on the synthesis of related compounds. We and many other laboratories have found that standard fluorenylmethoxycarbonyl-based (FMOC) solid-phase peptide synthesis, particularly using contemporary *O*-benzotriazole-*N*,*N*,*N*′,*N*′-tetramethyluronium-hexafluorophosphate (HBTU) and related coupling protocols, is reliable for the synthesis of peptide sequences taken from or modified from the sequence of Aβ. Precise protocols should be selected based on the particular apparatus, whether automated or manual, that an investigator has available. Following deprotection and cleavage from resin, we purify crude synthesis products by HPLC (RP) using C_{18} reversed-phase columns and acetonitrile–water–trifluoroacetic acid gradients. Precise conditions for preparative chromatography are typically selected based on an analytical run with a 10–85% acetonitrile gradient.

Amino-terminally modified compounds are routinely synthesized by several routes. We have made a wide variety of compounds using amino-terminal modification with carboxylic acids. Depending on the structure of the modifying reagent, one or more of the coupling protocols used in standard peptide synthesis will provide the desired product. Some reagents of interest are available commercially in activated form (e.g., acid chlorides or activated esters) and can be coupled to free amine peptide resins directly in the presence of an appropriate base such as *N*-methylmorpholine or diisopropylethylamine.

Quality of Aβ

The robustness of one's assays is highly dependent on the quality of Aβ used.[47] Several reliable commercial sources of Aβ(1–40) and Aβ(1–42)

[47] B. A. Yankner, *Neurobiol. Aging* **13,** 615 (1992).

now exist. Several years ago, commercial Aβ was at times characterized by its toxicity, indicating that it was already polymerized to some degree. Even now, Aβ that started out well synthesized and purified may "go bad" and care must be exercised in how it is handled. RP-HPLC analysis will reveal the presence of any overt impurities. Discerning between "good" and "bad" lots of Aβ, however, often can be achieved with HPLC only with the use of shallow gradients and qualitative comparison of the chromatographic profiles. Highest quality lots of Aβ will have a single peak with limited tailing whereas lesser quality lots will have an observably broader rear shoulder that may even display a rippled pattern due to the presence of impurities such as deletion peptides.

In practice, the ultimate test of a lot of Aβ is if it performs well in an assay in which the performance of high-quality Aβ is well characterized. This kind of observation requires ongoing vigilance to know your assay well and to be able to recognize the signs of reduced performance. Examples of decreasing Aβ quality can be inferred, for instance, from decreasing lag times in a nucleation assay (see later), incomplete solubility, and increasing background toxicity of "monomeric" Aβ.

When preparing solutions of Aβ, one must account for the amount of net peptide in the sample being used. Typical variability in the amount of salt, type(s) of counterion, and residual moisture affords Aβ preparations that are 70–80% peptide by weight.

Nucleation Assay

This assay is adapted from published procedures.[40] Aβ(1–40) is dissolved in dimethyl sulfoxide (DMSO) as a 40× stock (2.0 or 0.2 mM) prior to dilution into buffer (150 mM NaCl, 10 mM sodium phosphate, pH 7.4). The Aβ peptide stock solution in DMSO is diluted into the buffer to a final concentration of 50 or 5 μM in all wells. A master stock of each sample is prepared and added in 200-μl aliquots in triplicate in 96-well titer plates (Ultra Low Cluster, Costar). Inhibitors are added as concentrated DMSO solutions so that the final concentration of DMSO is 5% in all the wells. Plates are rotary shaken at 27° using a titer plate shaker (Lab-Line, Model 4625) at a rate of 700 rpm as determined with a tachometer. The progress of Aβ polymerization is monitored by measuring turbidity[40] as the apparent UV absorbance (for 50 μM Aβ mixtures) in a Bio-Rad (Richmond, CA) Model 450 microplate reader equipped with a 405-nm filter. Alternatively, polymerization is monitored by measuring the fluorescence of a 160-μl aliquot of sample at each time point. Samples are diluted into 40 μl of a 5× buffer (250 μM sodium phosphate, pH 6.5, 50 μM thioflavin T) in a separate 96-well plate,[27,28] and the samples are read in a CytoFluor II fluorescence spectrophotometer (Perseptive Biosystems, Framingham, MA) ($\lambda_{ex} = 450$ nm, $\lambda_{em} = 482$ nm). Fluorescence is expressed in arbitrary units. After graphing of the time course of controls and test mixtures, results are summarized as lag and percentage inhibition values (see earlier discussion).

Extension Assay

In this assay, a 100 μM solution of $A\beta$(1–40) monomeric peptide is incubated with a 5 μM solution of preassembled fibrils. Fibril growth is monitored by measuring the change in the fluorescence signal of thioflavin T binding over time. A 5 mM $A\beta$(1–40) solution in DMSO is prepared and diluted with 2× phosphate-buffered saline (PBS: 274 mM NaCl, 5.4 mM KCl, 16 mM Na_2HPO_4, 2.8 mM KH_2PO_4, pH 7.2) to yield a 2× stock of 200 μM $A\beta$(1–40) monomer solution. A stock solution of $A\beta$(1–40) fibrils is prepared by incubating a 200 μM solution of peptide in 1× PBS/4% DMSO for 8 days at 37°. The resulting suspension of polymerized $A\beta$ is cooled to 4° and sheared using a sonicator probe (Branson Ultrasonics, Model 250, 1/8 inch tapered tip). Samples of the sheared $A\beta$ suspension (2 μl) are removed for fluorescence measurement with thioflavin T, and sonication is stopped when the fluorescence signal has reached a plateau. Electron microscopic observation indicates that the sonicated fibrils have an average size of 25 nm.

Test compounds are prepared as 1.5 mM stocks in DMSO and diluted to 2× stocks in PBS/4% DMSO. The extension assay is performed in triplicate in 96-well plates (polystyrene, Costar). For each sample, 75 μl of a 200 μM $A\beta$ monomer solution in 2× PBS/4% DMSO, 3.75 μl of a 200 μM solution of sonicated $A\beta$ fibrils, and 60 μl of a 2× stock of test compound in 1× PBS/4% DMSO are added for a final volume of 150 μl. The final concentration of $A\beta$ is 100 μM, and the test compound is in the range of 3–100 μM. The assay mixture is incubated at 37° for 6 hr. Aliquots (10 μl) are removed at 1-hr intervals and added to 200 μl of a solution containing 10 μM thioflavin T and 50 μM sodium phosphate, pH 6.5. Fluorescence is measured in a CytoFluor II fluorescence spectrophotometer (Perseptive Biosystems) ($\lambda_{ex} = 450$ nm, $\lambda_{em} = 482$ nm).[27]

Evaluation of Results

Active compounds that inhibit the polymerization of $A\beta$ at a mole ratio of one to one relative to $A\beta$ can be observed in both assays. The higher concentration of $A\beta$ in the extension assay makes it difficult to assay less soluble compounds at a high concentration relative to the $A\beta$. Highly potent compounds, however, will be inhibitory at submolar equivalents in either assay. For compounds that are highly potent, assaying at lower concentrations of inhibitor allows comparison and differentiation between compounds.

Further characterization of the effects of inhibitors is possible through the comparison of control and inhibited polymerization mixtures using techniques not discussed here. Representative techniques of potential utility can include fibril diffraction, electron microscopy, and atomic force microscopy to characterize fibril abundance and morphology and various cellular toxicity assays to correlate the inhibition of polymerization with the suppression of toxicity.

Section III

Aggregate and Precursor Protein Structure

A. Aggregate Morphology
Articles 31 through 33

B. Molecular Level Aggregate Structure
Articles 34 through 38

C. Characterization of Precursor Protein Structure
Articles 39 and 40

[31] Electron Microscopy of Prefibrillar Structures and Amyloid Fibrils

By ELLEN HOLM NIELSEN, MADS NYBO, and SVEN-ERIK SVEHAG

Introduction

Several techniques, such as X-ray crystallography, light scattering, fluorescence spectrometry, size exclusion chromatography, atomic force microscopy, and transmission electron microscopy, have been employed in studies of structural intermediates of fibril formation and fibrillar assembly of amyloid proteins.

Electron microscopy, with a resolution of approximately 2 nm, offers a useful technique for the ultrastructural characterization of preprotofilaments, protofilaments, and mature fibrils formed during *in vitro* fibrillogenesis. For contrast enhancement of specimens, negative staining[1] is applied. This article outlines the methodology used in our laboratory for electron microscopic examination of negatively stained prefibrillar structures and amyloid fibrils.

Peptide Preparation

The $A\beta$-peptide used influences the kinetics of the fibril formation. We have used $A\beta 1-42$ (Bachem, Bubendorf, Switzerland), which forms fibrils within a few hours of incubation at 37°. In order to decelerate the fibril formation enabling us to investigate early intermediates and prefibrillar structures, we have performed the *in vitro* studies at low concentrations of $A\beta 1-42$ (170 μg/ml), as the kinetics of fibril formation is highly concentration dependent. Because Tris buffers, even at low concentrations, may cause uncontrolled fibril formation, they should be avoided.

The peptide (1 mg/ml) is dissolved in double-distilled water without additives to achieve as native conditions as possible,[2] and all preincubation steps are carried out on ice to avoid uncontrollable fibril formation. Immediately after dissolving the peptide, the preparation is centrifuged at 30,000*g* and 4° for 1 hr to eliminate possibly preformed $A\beta$ fibrils. The $A\beta 1-42$ protein concentration in the supernatant is about 170 μg/ml and is determined by the use of a bicinchoninic acid protein kit (Pierce, Rockford,

[1] C. E. Hall, *J. Biophys. Biochem. Cytol.* **1,** 1 (1955).
[2] D. B. Teplow, *Amyloid Int. J. Exp. Clin. Invest.* **5,** 121 (1998).

METHODS IN ENZYMOLOGY, VOL. 309
0076-6879/99 $30.00

IL)[3] with bovine serum albumin (BSA) as a standard. The supernatant is aliquoted and kept at −60° until experiments are performed.

Fibril Formation

The Aβ1–42 supernatant is incubated for 0–24 hr at 37° with slow agitation, and at different times samples are collected and immediately prepared for electron microscopy to avoid uncontrolled fibril formation.

Preparation of Grids

Preparation of the grids is of vital importance for obtaining electron micrographs of sufficient resolution. In our laboratory we use copper grids, 400 mesh, coated with a supporting holey Triafol plastic film.[4] The holes are covered by a thin carbon film for application of the sample.

Cleaning Grids

The procedure starts with a thorough cleaning of a glass slide with trichloroethylene, drying it with gauze, and dusting it with lens paper. The slide is dipped for 2–3 min in a 0.1% solution of Triton X-100 (Merck, Darmstadt, Germany) in double-distilled water and allowed to dry in a vertical position.

Coating with Triafol Film

The glass slide is placed in a −20° freezer for 1–2 min and, when removed, condensed water in the form of minute droplets will precipitate on the surface of the glass. The size of the water droplets will depend on the temperature and humidity of the room, with the higher humidity, the larger the droplets. If the air is very dry it may be necessary to keep the glass slide in a stream of moist air for a few minutes.

As soon as the slides have been removed from the freezer, a solution of 0.4% Triafol-foil (Merck, Darmstadt, Germany) in ethyl acetate is dropped on one side of the glass and allowed to cover its entire surface in a thin layer. The surplus of the Triafol solution is allowed to drain. Water drops on the glass produce the perforations in the Triafol film, and the size of the drops determines the size of the holes in the film. When the Triafol film is dry and after cutting the edges of the film along the two short sides

[3] P. K. Smith, R. I. Krohn, G. T. Hermanson, A. K. Mallia, F. H. Gartner, M. D. Provenzano, E. K. Fukimoto, N. M. Goeke, B. J. Olson, and D. C. Klenk, *Anal. Biochem.* **150,** 76 (1985).

[4] A. Fukami and K. Adachi, *J. Electr. Microsc.* (*Tokyo*) **14,** 112 (1965).

of the glass, the film is easily floated off onto a surface of distilled water in a dish by immersing the slide with the film side facing up at about a 30° angle to the surface of the water. The copper grids, cleaned in 1,2-dichloroethane, are placed upside down on the floating film, and the grids are covered by a filter paper strip. By slowly turning the paper strip over, all the grids are pulled out of the water now laying on the filter paper and covered by the Triafol film.

When the holes in the Triafol film are not perforated completely, the grids are ionized by glow discharge in atmospheric air for 10–30 sec in a vacuum of $2 \times 10^{-1\,\mathrm{mbar}}$ (BAL-TEC MED 020, Liechtenstein), which makes the film thinner and the last, thin diaphragm left in the holes to disappear.

Coating with Carbon

In order to reinforce the Triafol film, a 3- to 4-nm-thick layer of carbon is deposited onto the film with the BAL-TEC MED 020 system by evaporating a double, 1-cm-long, spun carbon thread in a vacuum of $1.3 \times 10^{-5\,\mathrm{mbar}}$ (BAL-TEC, Liechtenstein). The holes in the plastic film are not covered by a carbon film by this procedure.

Coating with Ultrathin Carbon Film

The holes in the Triafol film now have to be covered with a very thin carbon film and only these carbon-covered perforations are used for microscopy. A small piece of mica is cleaved into sheets and a sheet is placed in the BAL-TEC evaporator with the freshly cleaved, clean side up. By evaporating a single, 1-cm-long, spun carbon thread in a vacuum of $1.3 \times 10^{-5\,\mathrm{mbar}}$, the mica sheet will be covered by a thin (1.5–2 nm) carbon film.

A dish is filled with double-distilled water and the mica with the thin carbon film, of which the edges have been cut, is then immersed into the water at a 30–40° angle, causing the film to float off onto the surface of the water. The Triafol film-covered grids are placed on a metal net a few millimeters over the bottom of the dish. The water is slowly withdrawn and as the level of the water is lowered, the carbon film is gently pushed around with a forceps until it settles on the surface of the grids. The grids are picked up with a finely tipped forceps and allowed to dry on a piece of filter paper. Only the carbon-coated holes in the Triafol film are used for examination of specimens by electron microscopy.

Staining

Freshly prepared carbon films are hydrophobic, resulting in the poor adhesion of most protein molecules. The films can be made hydrophilic by ionizing with glow discharge in atmospheric air, but glow-discharged films have to be used within 30 min to 1 hr. This can be impractical when several grids have to be stained. As a wetting agent we prefer to use the antibiotic bacitracin (Fluka, Buchs, Switzerland), which is added to the protein sample at a concentration of 50 μg/ml. The stock solution of bacitracin (500 μg/ml in double-distilled water) is stable for 2–3 weeks at 4°.[5]

During staining, the grids are held in a stand by a fine watchmakers tweezers. A 7-μl drop of the Aβ1–42 supernatant (see peptide preparation) is left on the grid for approximately 2 min. The specimen drop is gently flushed away with 15–20 drops of freshly prepared aqueous (double-distilled water) 2% (w/v) uranyl acetate, pH 4.4, with a strip of filter paper as a drain. The last drop is drained off, leaving only a thin film on the grid. When the grid is air-dried, it is ready for microscopy.[6] We have found that uranyl acetate is superior to other negative stains such as uranyl formate and phosphotungstic acid due to its higher penetrability and minimal grain size. Note that uranyl salts are both toxic and radioactive and should be handled with care. The stained grids can be stored at room temperature without adverse effects.

Electron Microscopy

It is important not to use a higher illumination intensity than required, as the electron beam will readily deteriorate the specimen and increase the granularity of the uranyl salt. The samples are photographed at ×50,000 magnitude and the film plates (Kodak Electron Microscope film 4489, (Rochester, NY) are given a ×2.3 secondary magnification. Electron micrographs of prefibrillar structures and Aβ1–42 fibrils are shown in Figs. 1 and 2.

Critical Steps

Preparation steps requiring the most experience are preparation of the holey Triafol film, the carbon film supporting the specimen, and staining with uranyl acetate.

When the holes in the Triafol film are too small, they may have a thin diaphragm left; furthermore, the areas that can be used for microscopy are reduced. If the perforations are too large, the plastic film will disrupt easily under the electron beam. The thickness of the carbon film covering the perforations in the Triafol film is also critical. If the film is too thick, it will

[5] D. W. Gregory and B. J. S. Pirie, *J. Microsc.* **99,** 251 (1973).
[6] J. Tranum-Jensen, *Methods Enzymol.* **165,** 357 (1988).

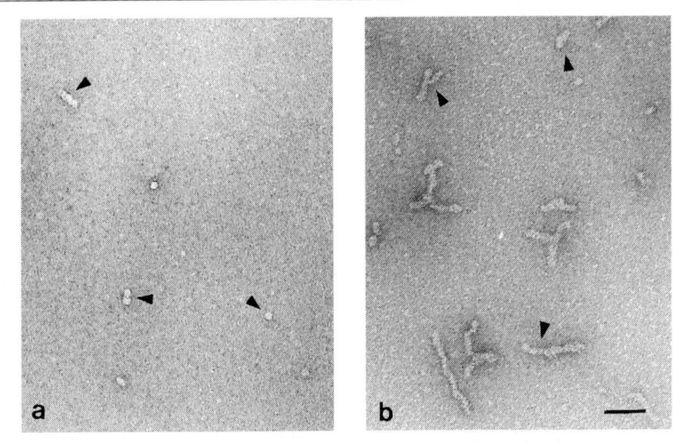

FIG. 1. Electron micrographs of amyloid intermediates in Aβ1–42 fibrillogenesis. (a) Monomer, dimer, and trimer of globular structures seen after 0–0.5 hr of incubation. (b) The globular subunits appear to fuse, forming protofilaments of varying shapes after 0.5–2 hr of incubation. Bar: 50 nm.

reduce the image contrast and resolution of the micrographs. A film that is too thin will be difficult to handle and will often break during its handling or under the electron beam. When specimens are stained with uranyl acetate it is important to have the correct thickness of the uranyl layer, which forms a cast of the embedded structures; too thick a layer will cover fine

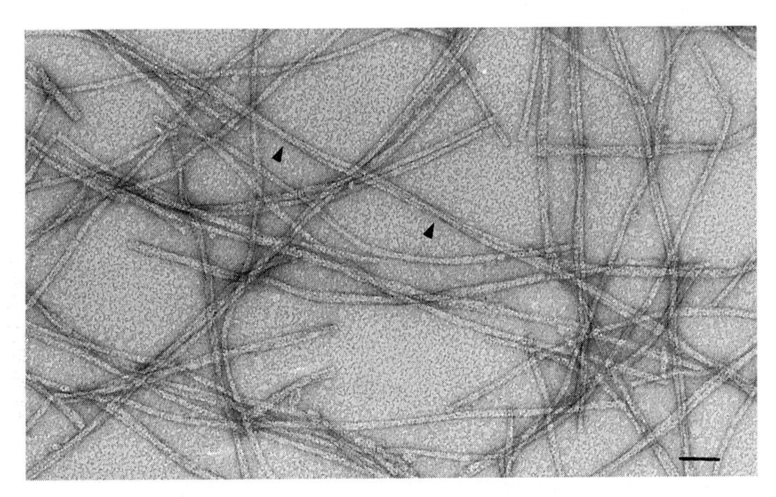

FIG. 2. Electron micrograph of mature Aβ1–42 fibrils after 24 hr of incubation. Arrows indicate intertwined fibrils with a periodicity of about 220 nm. Bar: 50 nm.

structures whereas too thin a layer provides insufficient image contrast for photography. It is also important to achieve a rather even distribution of both the sample and the uranyl salt on the grid.

Acknowledgments

We thank Ms. Kirsten Junker and Inger Margrethe Rasmussen for careful, skilled preparation of Aβ preparations and specimens for electron microscopy. This study was supported by the Danish Medical Research Council, The Danish Association against Rheumatism, and Lily Benthine Lund's Foundation.

[32] *In Situ* Electron Microscopy of Amyloid Deposits in Tissues

By SADAYUKI INOUE and ROBERT KISILEVSKY

Introduction

Amyloidosis is a widespread multiform disorder, and amyloids are proteinaceous usually extracellular tissue deposits characterized by positive Congo red staining, red-green birefringence of Congo red-stained tissues under polarized light, and the presence of cross β molecular conformation as detected by X-ray diffraction. Ultrastructurally they are composed mainly of assemblies of fibrils (Fig. 1) approximately 10 nm in diameter.[1-3] Fibrils of various types of amyloid have been observed in the electron microscope both *in situ* in the tissue[4-7] as well as after isolation.[1,8-10] Although all amyloids have the same structural and tinctorial properties, they are quite

[1] A. S. Cohen, *Int. Rev. Exp. Pathol.* **4,** 159 (1965).
[2] A. S. Cohen and E. Calkins, *Nature* **183,** 1202 (1959).
[3] G. G. Glenner and D. L. Page, *Int. Rev. Exp. Pathol.* **15,** 1 (1976).
[4] F. B. Gueft and J. J. Ghidoni, *Am. J. Pathol.* **43,** 837 (1963).
[5] T. Miyakawa, S. Katsuragi, and R. Kuramoto, *Virch. Arch. Cell Pathol.* **56,** 21 (1988).
[6] T. Miyakawa, S. Katsuragi, and K. Watanabe, *in* "Amyloid and Amyloidosis: Proceedings of 5th International Symposium on Amyloidosis, Hakone, Japan" (T. Isobe, S. Araki, F. Uchino, S. Kato, and E. Tsubura, eds.), p. 573. Plenum Press, New York, 1990.
[7] R. D. Terry, N. K. Gonatas, and M. Weiss, *Am. J. Pathol.* **44,** 269 (1964).
[8] A. S. Cohen, T. Shirahama, and M. Skinner, *in* "Electron Microscopy of Proteins" (J. R. Harris, ed.), p. 165. Academic Press, London, 1982.
[9] T. Shirahama and A. S. Cohen, *J. Cell Biol.* **33,** 679 (1967).
[10] G. D. Sorensen and E. Finke, *in* "Amyloidosis: Proceedings of Symposium of Amyloidosis, University of Groningen, The Netherlands" (E. Mandema, L. Luinen, J. H. Scholten, and A. S. Cohen, eds.), p. 184. Excerpta Medica Foundation, Amsterdam, 1968.

0076-6879/99 $30.00

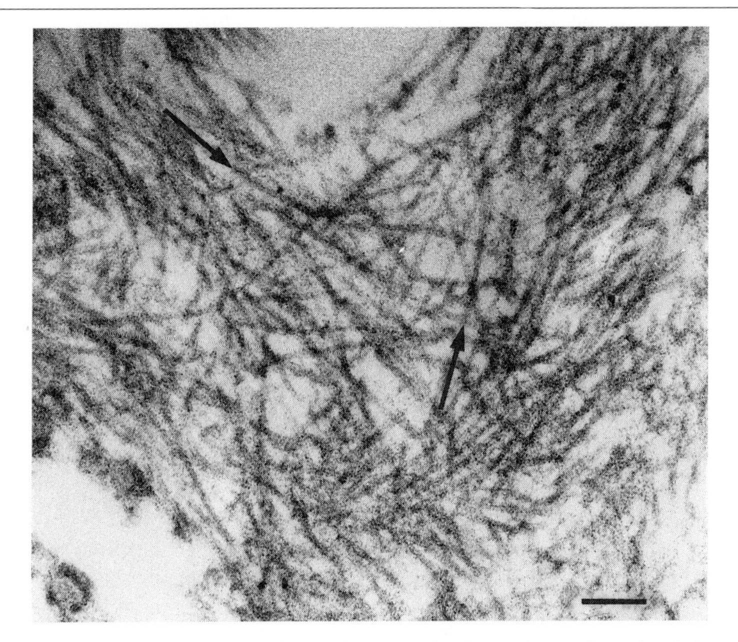

FIG. 1. Aggregates of AA fibrils (arrows) in deposit of experimental murine AA amyloid. They are straight or gently curved entities approximately 10 nm in width. Thin sectioned material. Bar: 100 nm.

diverse chemically, with each form of protein being associated with a specific disease or pathological process. More detailed examinations of the ultrastructure and composition of the fibrils *in situ* have been done in our laboratories by means of high-resolution electron microscopy and ultrastructural immunohistochemical labeling.[11,12] Because amyloid deposits are composed mainly of "amyloid fibrils," high-resolution ultrastructural characterization of these fibrils *in situ* is the main subject of this article.

Previous Ultrastructural Studies of Amyloid Fibrils

The ultrastructure of amyloid fibrils has been studied with isolated preparations[1,8–10] or *in situ*.[4–7] Isolated fibrils visualized mainly by negative staining were composed of a set of two or more slowly twisting, parallel subunits variously referred to as filaments or protofibrils. Fibrils observed *in situ* in thin sectioned tissues were reported to be tubular structures with surface cross-banding patterns indicative of a helical structure. Miyakawa

[11] S. Inoue and R. Kisilevsky, *Lab. Invest.* **74,** 670 (1996).
[12] S. Inoue, M. Kuroiwa, K. Ohashi, M. Hara, and R. Kisilevsky, *Kidney Int.* **52,** 1543 (1997).

and collaborators found Aβ amyloid fibrils in both plaques[6,13] and vessels[5] to be hollow rods in which filaments were arranged as tightly coiled helices, each turn of which contained five globular subunits 3–5 nm in width.

Systematic High-Resolution Observation of Amyloid Fibrils *in Situ*

The ultrastructure of amyloid fibrils was examined in detail in our laboratories with "high-resolution" (definition, see Ref. 14) electron microscopy; i.e., a 100-kV electron microscope with a resolution in the range of 2–5 nm to the limit of the instrument's capability was used in these observations. To search for the possible presence of substructures within individual amyloid fibrils *in situ,* a "systematic microdissection" of the fibril was done by examining sections selected from variously oriented fibrils. Of particular interest were those that were cut parallel to the axis of the fibril and at different levels of depth as one went from the surface to the interior of the fibril. The anatomical relationship, or order, of substructures observed at various levels of depth was determined by examining selected sections cut slightly obliquely against the axis of the fibril so that a gradual transition from one substructure to the next could be confirmed. The relationship between electron microscope images and the thickness of the fibril or sections is discussed later under "Systematic Microdissection of Amyloid Fibrils *in Situ.*" Thus, it was possible to "systematically" identify substructures of the fibril from its surface progressively into its interior.

With this method, *in situ* fibrils of experimental murine AA amyloid[11] and hemodialysis-associated β_2-microglobulin amyloid[12] were examined and compared with previous reports.[4–7] The approximately 12-nm-wide fibril of experimental AA amyloid was found to be composed of a core consisting of tightly coiled 3-nm-wide ribbon-like "double-tracked" structures enclosing assemblies of 3.5-nm-wide pentagonal frames at the center. These 3-nm-wide double tracks were identified previously as a form of chondroitin sulfate proteoglycan (CSPG).[15] Similarly, using electron microscopic images of purified amyloid P component (AP) and immunoperoxidase staining, the 3.5-nm-wide pentagonal frames were shown to be the subunits of the AP.[16] The core was wrapped in a surface layer composed of random assemblies of another type of double-tracked structure with a

[13] T. Miyakawa, S. Katsuragi, K. Watanabe, A. Shimoji, and Y. Ikeuchi, *Acta Neuropathol.* **70,** 202 (1986).

[14] J. M. Cowley, *in* "Principles and Techniques of Electron Microscopy" (M. A. Hayat, ed.), Vol. 6, p. 40. Van Nostrand Reinhold, New York, 1976.

[15] S. Inoue, *Cell Tissue Res.* **279,** 291 (1995a).

[16] S. Inoue, *Cell Tissue Res.* **263,** 431 (1991).

width of 4.5–5 nm that was distinct from 3-nm-wide CSPG double tracks. These wider double tracks were also characterized previously, in the manner described earlier, as a form of heparan sulfate proteoglycan (HSPG).[17] Along the outermost surface of the fibril, loose assemblies of 1- to 2-nm-wide flexible filaments were associated (Figs. 2A and 2B), and these filaments were found with immunogold labeling (Fig. 2C) to be composed of AA protein.[11] A similar organization has been demonstrated for Aβ[18] and transthyretin (TTR) amyloid in familial amyloidotic polyneuropathy.[19] The fibrils of β_2-microglobulin amyloid were intensely curved in contrast to AA, Aβ, and ATTR where the amyloid fibrils were straight or only gently curved, but their ultrastructural organization resembled that of AA amyloid fibrils.[12] They were composed of a core and a surface layer onto which loose assemblies of 1-nm-wide flexible filaments were externally associated. The core was identical to that of AA amyloid fibrils and was composed of CSPG and AP. The flexible filaments at the surface of the fibril were composed again of a protein specific to this type of amyloid, i.e., β_2-microglobulin. The only difference in the structure between these two amyloid fibrils was the nature of the surface layer proteoglycan, which in the case of β_2-microglobulin fibrils was composed of random assemblies of 3-nm-wide CSPG double tracks instead of 4.5- to 5-nm-wide HSPG double tracks as seen in AA deposits (Fig. 3). The difference in glycosaminoglycans in terms of their binding properties[20] may result in the less rigid, frequently curved structure of β_2-microglobulin amyloid fibrils.

We have also examined *in situ* AA protein filaments present at the outermost surface of experimental AA amyloid fibrils after cryofixation[21] followed by freeze substitution.[22] These latter two methods are known to significantly improve the quality of tissue preservation and allow examination of tissues in a state very close to their original living state (reviewed by Chan and Inoue[23]). Our preliminary results indicate that in living conditions these peripheral 1-nm-wide flexible filaments may, in fact, be organized in the form of 3-nm-wide tight helices (3-nm-wide "helical rods"), which in turn appear to be arranged parallel to the axis of the fibril and to one another with a uniform center-to-center distance of 5 nm.

[17] S. Inoue, D. Grant, and C. P. Leblond, *J. Histochem. Cytochem.* **37,** 597 (1989).
[18] S. Inoue, M. Kuroiwa, and R. Kisilevsky, *Brain Res. Rev.* **29** 218 (1999).
[19] S. Inoue, M. Kuroiwa, M. J. Saraiva, A. Guimarães, and R. Kisilevsky, *J. Struct. Biol.* **124,** 1 (1998).
[20] B. Casu, M. Petitou, M. Provasoli, and P. Sinaÿ, *Trends Biochem. Sci.* **13,** 221 (1988).
[21] H. Y. Elder, *Tech. Immunocytochem.* **4,** 1 (1989).
[22] D. M. R. Harvey, *J. Microsc.* **127,** 209 (1981).
[23] F. L. Chan and S. Inoue, *Microsc. Res. Techn.* **28,** 48 (1994).

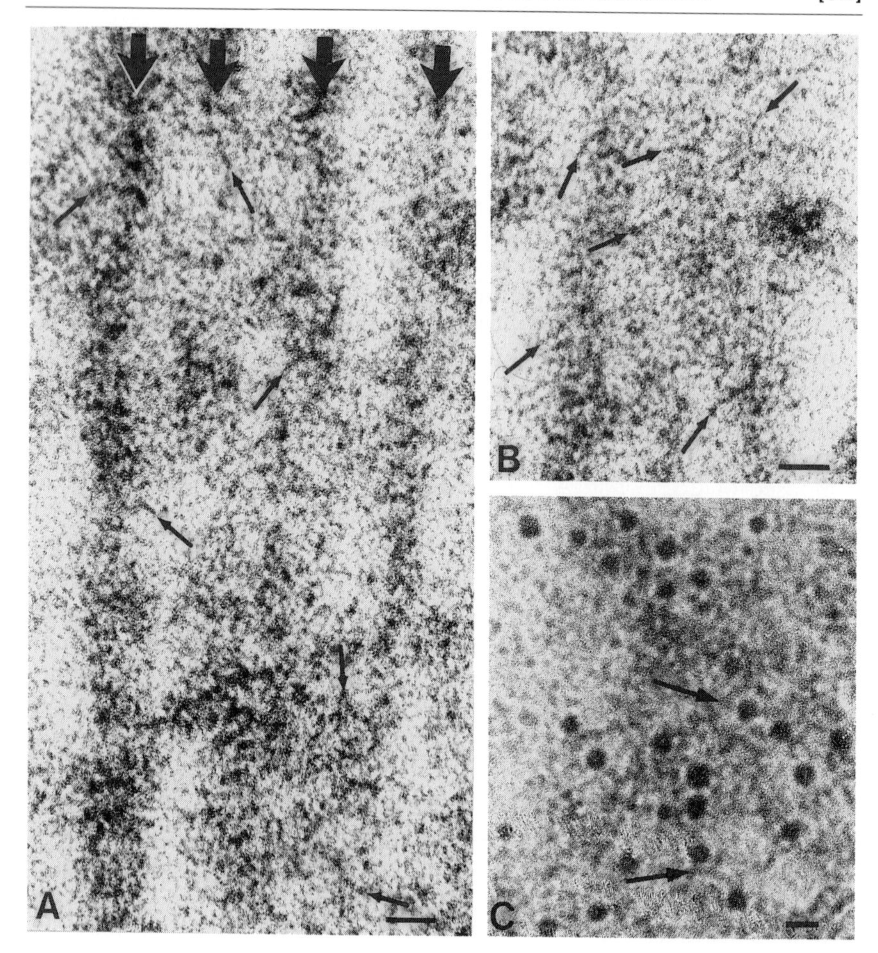

FIG. 2. Filaments composed of AA amyloid protein on the exterior surface of the fibril (see "Systematic High-Resolution Observation of Amyloid Fibrils *in Situ*" for methodology). (A) Among four parallel fibrils (bold arrows), the one at the far left was sectioned close to its surface, whereas the others were cut slightly above their surface. All of them are associated with numerous flexible filaments approximately 1 nm in thickness. (B) Section cut at a level some distance away from the surface of fibrils. Filaments 1 nm in thickness are present (arrows). (C) Immunogold labeling for AA protein. Flexible 1-nm-wide filaments are associated preferentially with 5-nm gold particles (arrows). Bars: 10 nm.

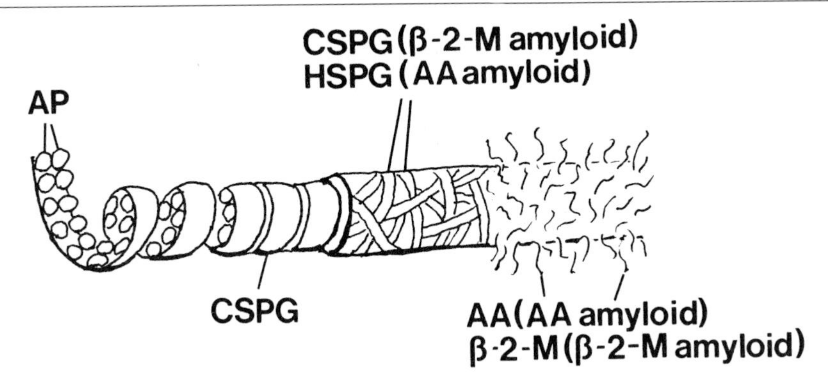

FIG. 3. Schematic drawing of an *in situ* amyloid fibril. The core of the fibril is composed of amyloid P component (AP) and chondroitin sulfate proteoglycan (CSPG). The surface layer is made up of heparan sulfate proteoglycan (HSPG) in experimental AA amyloid fibrils or CSPG in the case of hemodialysis-associated β_2-microglobulin (β_2-M) amyloid. The outer surface of the fibril is associated with 1-nm-wide flexible filaments composed of a protein specific to each type of amyloid, AA protein and β_2-M, respectively.

Similarity of Amyloid Fibrils *in Situ* to Connective Tissue Microfibrils

The size and general morphology of microfibrils[15,24] closely resemble those of fibrils of experimental murine AA amyloid[11] and hemodialysis-associated β_2-microglobulin amyloid.[12] The core of these two types of amyloid is composed of helically wound 3-nm-wide ribbon-like "double tracks" identified as CSPG enclosing a central assembly of AP, which is identical to the core of microfibrils. The core is wrapped in a layer of 4.5- to 5-nm-wide ribbon-like double tracks identified as HSPG in the case of experimental AA amyloid, as well as microfibrils, and of 3-nm-wide CSPG double tracks in the case of β_2-microglobulin amyloid. In the case of microfibrils, the layer of HSPG is thicker and fibrillin is also associated periodically along the length of the fibril.[24] The main difference between amyloid fibrils and connective tissue microfibrils is that the outermost surface of amyloid fibrils is associated with loose assemblies of 1-nm-wide flexible filaments composed of a protein specific to each type of amyloid, while the surface of the microfibril is associated with fibrillin. Similar ultrastructure and composition have been observed in the fibril of other types of amyloid, including Aβ amyloid of Alzheimer's disease,[18] AL amyloid,[25] and FAP amyloid.[19]

[24] S. Inoue, *Cell Tissue Res.* **279,** 303 (1995b).

[25] S. Inoue, M. Kuroiwa, and R. Kisilevsky, submitted for publication.

Comparison of *in Situ* and Isolated AA Amyloid Fibrils

There is a significant difference in the ultrastructural organization of amyloid fibrils observed *in situ* in thin sections and after their isolation. Fibrils observed *in situ* were reported to be tubular[4–7] or, in more detail, microfibril-like structures,[11,12] whereas isolated fibrils were composed of a set of two or more slowly twisting parallel protofibrils.[1,8–10] In an attempt to explain this disagreement, fibrils of experimental murine AA amyloid were isolated with the distilled water washing method and their ultrastructure was compared with that of fibrils observed *in situ.*[26] As described earlier, fibrils observed *in situ* were made up of a microfibril-like core 8–9 nm in width and a surface layer of HSPG to which 1-nm-wide flexible AA protein filaments were externally associated. During isolation with the standard distilled water washing procedure, the HSPG layer and AA protein filaments detached from the core of the fibril and became dispersed into the water. The remaining cores were loosened and generally lost their fibrillar appearance, whereas the detached 1-nm-wide filaments in distilled water became more conspicuous. These filaments coiled themselves into 3-nm-wide tight helices (helical rods). The helical rods then assembled into the characteristic slowly twisting sets of two parallel protofibrils that resembled previously reported "isolated amyloid fibrils."

Methods

Induction of Experimental Murine AA Amyloid

Splenic AA amyloid is induced in 6- to 8-week-old female CD_1 mice (weight 25–30 g, Charles Rivers, Montreal, Quebec, Canada) with a method described previously.[27] The induction protocol consists of 200 μg of amyloid enhancing factor (AEF)[28] (as amyloid fibril protein) given intravenously in a lateral tail vein injection. It is followed by an inflammatory stimulus by either (a) a subcutaneous injection in the back of 0.5 ml of a 2% aqueous solution of silver nitrate or (b) 25 μg of *Salmonella minnesota* lipopolysaccharide (Sigma, St. Louis, MO) in phosphate-buffered saline (PBS) given

[26] S. Inoue, M. Kuroiwa, R. Tan, and R. Kisilevsky, *Amyloid Int. J. Exp. Clin. Invest.* **5,** 99 (1998).

[27] L. Brissette, I. Young, S. Narindrasorasak, R. Kisilevsky, and R. Deeley, *J. Biol. Chem.* **264,** 19327 (1989).

[28] M. L. Baltz, D. Caspi, C. R. K. Hind, A. Feinstein, and M. B. Pepys, *in* "Amyloidosis" (G. G. Glenner, E. F. Osserman, E. P. Benditt, E. Calkins, A. S. Cohen, and D. Zucker-Franklin, eds.), p. 115. Plenum Press, New York, 1986.

intraperitoneally every other day for 4 weeks. Animals are killed 6 days after induction in the case of (a) or 4 weeks in the case of (b) by CO_2 narcosis, and the spleen tissue is isolated after *in situ* whole body perfusion with 0.9% saline (4°).

Preparation of Amyloid Tissues for Thin Section Electron Microscopy

Standard Method (Applied to Experimental Murine AA Amyloid)[11]

1. Freshly isolated spleens of mice with experimental AA amyloid are cut into small blocks (approximately 1 mm^3) and are immersed in a fixative containing 3% glutaraldehyde in 0.1 *M* sodium cacodylate buffer, pH 7.4, for 3 hr at room temperature.
2. The blocks are washed with cacodylate buffer and postfixed with 1% (w/v) osmium tetroxide in the same buffer for 1.5 hr at 4°.
3. After rinsing with distilled water the blocks are stained *en bloc* with 2% (w/v) uranyl acetate for 1 hr at room temperature, dehydrated in a graded series of acetone, and embedded in Epon.
4. Thin sections 40–60 nm in thickness are counterstained with uranyl acetate (5 min) followed by lead citrate (2 min) before examination in the electron microscope.

Tissues with Other Types of Amyloid

The standard method as described earlier or in modified forms has been used in our laboratories in processing other types of amyloid.

1. (a) Tissue of familial amyloidotic polyneuropathy (TTR Met-30) obtained through sural nerve biopsies[19] and (b) prefixed (with 10% buffered formalin) portions of spleen and liver obtained through autopsy from a patient with AL amyloid[25] are prepared with the standard method described previously.
2. Tissue of hemodialysis-associated β_2-microglobulin amyloid[12] obtained through autopsy is prepared with the standard method except that a fixative containing 2.5% (v/v) glutaraldehyde in 0.1 *M* phosphate buffer, pH 7.2, is used as the primary fixative for 2 hr at room temperature.
3. Pieces of the frontoparietal cortex of the brains of patients with Alzheimer's disease[18] obtained through autopsies and prefixed with 10% formalin are washed with 0.1 *M* sodium cacodylate buffer, pH 7.4, for 3 hr. They are cut into small blocks (approximately 1 mm^3). The blocks are postfixed with 1% osmium tetroxide in the same

buffer for 1.5 hr at 4° and processed further for embedding as described earlier.

High-Resolution Examination of Amyloid Fibrils *in Situ*

High-resolution electron microscopy is not a difficult method to use if one has access to a 100-kV electron microscope and if one closely follows several important steps in the maintenance and use of the microscope. The importance of these steps in high-resolution work is apparent through common sense, but some of them may often be neglected. The following major ones are listed.

Proper Maintenance and Use of Electron Microscope and Photographic Enlarger

Electron Microscope

1. The use of a high-performance 100-kV electron microscope is essential for high-resolution observations.
2. Perform routine maintenance work, particularly alignment of the column and correction of astigmatism, by following instructions from the manufacturer.
3. Calibrate magnification, particularly in the high magnification range (×40,000–×60,000) by means of a standard such as catalase crystals.
4. Select a smaller objective lens aperture, to the extent at which images are barely bright enough for adequate focusing, for higher resolution and contrast of the image.

Electron Microscopy

1. Once a grid square with favorable areas for photography is found, stabilize the entire square of the section and its immediate vicinity that is attached to grid bars. This is done by a uniform preexposure scanning with the electron beam as described in Fig. 4. It is quite effective in preventing microdrift of the image due to beam-induced movement of the specimen during photography. In addition to good focusing of the image, this procedure is critically important in obtaining satisfactory high-resolution micrographs.
2. To keep radiation damage of specimens to a minimum, maintain the duration of exposure as well as intensity of the electron beam as low as possible.
3. Focus images with great care. Good focus is critical for high-resolution electron microscopy.

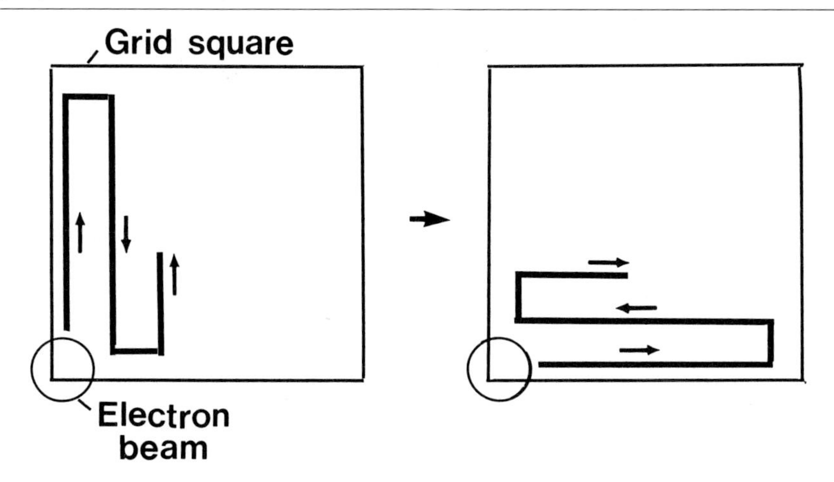

FIG. 4. Schematic drawing showing a method of stabilizing thin sections with uniform preexposure to an electron beam. A grid square is systematically scanned back and forth with a narrowed beam (circle) in one direction and then in the perpendicular direction.

4. Try to limit the photographic exposure time to a minimum (approximately 1 sec) to minimize the effect of possible minute drift of the image.
5. After washing, expose negatives to a jet of running water before drying. This removes extremely fine particles of rust often present in running water that can adhere to the photographic emulsion during washing. Because of their extremely small size, they are usually harmless in conventional electron microscopy with low or moderate magnifications, but can easily ruin high-resolution micrographs.

Photographic Enlargement

1. When these steps are followed properly, the resultant accurately focused negatives of astigmatism-free and drift-free images can easily withstand further additional photographic enlargement of ×10 to ×15. Therefore, it is possible to ultrastructurally examine amyloid fibrils at magnifications of ×600,000 to ×800,000 without major difficulty.
2. In preparing final prints of electron micrographs, accurate focusing in an enlarger is again of critical importance. Photographic enlargement should be regarded as the final important step of electron microscopy. Careless focusing during photographic enlargement can easily spoil high-resolution images painstakingly recorded in negatives.

3. Determination of the degree of enlargement is done by simultaneously enlarging a scale on a transparent base with 1-mm divisions and it is usually determined in three digits.

"Systematic Microdissection" of Amyloid Fibrils, in Situ

The relationship between section thickness and observed electron microscope images is as follows. Although little is known[29] about the actual mechanisms of section staining, it is apparent that the adsorption of heavy metals begins at the surface of the section, which is in direct contact with staining solutions. The speed of penetration of heavy metals through the thickness of the section depends on a combination of various factors, including the condition of fixation and embedding of the specimen, the nature of the staining solutions, and the duration of staining.[29] For example, although some authors[29a] reported that the stain penetrated completely through the thickness of the sections, under the specific conditions used in our laboratories for amyloid fibrils, the stain did not seem to penetrate very deeply. Therefore, despite the fact that section thickness (40–60 nm) exceeds the diameter of amyloid fibrils by far, the limited depth of the section actually stained is small enough to eliminate structural overlap when observed in the electron microscope.

1. Take pictures of amyloid deposits at magnifications in the range of ×30,000 to ×50,000.
2. Enlarge negatives approximately ×2.5 to prepare prints of a total magnification of ×75,000 to ×125,000.
3. With a hand magnifier, select areas where the fibril is sectioned parallel to its axis. In such areas, various distinct types of substructures of the fibril may be seen depending on the depth at which it is sectioned.
4. Similarly, select areas in which the fibril is sectioned, this time, slightly obliquely against its axis. In such areas, a gradual transition of a specific substructure of the fibril to another may be seen.
5. Photographically enlarge all of the selected areas ×15 to show detailed ultrastructure. Thus, the "systematic" ultrastructural visualization of substructures of the amyloid fibril may be established beginning from the surface to the interior in stepwise fashion.

[29] M. A. Hayat, "Principles and Techniques of Electron Microscopy: Biological Applications," Vol. 1. Van Nostrand Rheinhold, New York, 1970.

[29a] A. Peters, P. L. Hinds, and J. E. Vaughn, *J. Ultrastruct. Res.* **36,** 37 (1971).

Immunohistochemical Labeling of Constituents of Amyloid Fibrils

Immunolabeling for AA Protein

1. Mice with experimental AA amyloidosis are perfused, under anesthesia, with cold 4% (w/v) formaldehyde in 0.1 *M* PBS, and spleens are isolated and cut into small blocks (approximately 1 mm^3).
2. Tissue blocks are immersed in the same fixative for 30 min at 4° and are then transferred into the same fixative with the pH raised to 9.0[30] and kept overnight at 4°.
3. The blocks are washed, dehydrated in a series of methanol, and embedded in Lowicryl K4M at −35°.
4. Thin sections are mounted on nickel grids and floated on the surface of a drop of 1% bovine serum albumin (BSA) in PBS and then over a drop of anti-AA antiserum (dilution 1:100) (prepared by the method of Kisilevsky *et al.*[31]) for 1 hr at room temperature. Control sections are exposed to nonimmunized rabbit serum.
5. The grids are washed by passing them, successively, through PBS (three times) and a set of drops of 1% BSA in PBS.
6. The grids are floated on the surface of drops of 0.05 *M* Tris–HCl, pH 7.6, and then of goat antirabbit immunoglobulins coupled to 5-nm gold particles (dilution 1:15) (Janssen Pharmaceutica, Antwerp, Belgium).
7. The grids are washed by floating them on the surface of drops of Tris buffer and then by exposure to a gentle stream of the same buffer followed by distilled water.
8. The sections are counterstained with uranyl acetate for 4 min followed by lead citrate for 15 sec before examination in the electron microscope.

Protein of Other Types of Amyloid and Nonamyloid Proteins

1. In step 4 just described, rabbit antihuman β_2-microglobulin antiserum (dilution 1:100; DAKO Corporation, Carpinteria, CA) is used in the case of hemodialysis-associated β_2-microglobulin amyloid[12]; rabbit antihuman transthyretin (prealbumin) antiserum (prediluted; DAKO) for familial amyloidotic polyneuropathy (FAP) amyloid[19];

[30] W. D. Eldred, C. Zucker, H. J. Karten, and S. Yazulla, *J. Histochem. Cytochem.* **31,** 285 (1983).

[31] R. Kisilevsky, S. Narindrasorasak, C. Tape, R. Tan, and L. Boudreau, *Amyloid Int. J. Exp. Clin. Invest.* **1,** 174 (1994).

and monoclonal mouse antihuman Aβ antiserum (dilution 1:100; DAKO) for the β amyloid of Alzheimer's disease.[18] In the last case, antimouse immunoglobulins coupled to 5-nm gold particles (dilution 1:15; Janssen) are used in step 6 described earlier.

2. For the labeling for HSPG, rabbit anti-HSPG antiserum (dilution 1:100; prepared with HSPG purified from mouse EHS tumor as antigen. This antiserum reacts with human tissues[32]) is used. Monoclonal antimouse CSPG antiserum (dilution 1:100; Sigma, St. Louis, MO) is used for the labeling for CSPG.

Amyloid P Component

This component in the subunit form is too small (3.5 nm) for immunogold labeling. However, it can be identified with immunoperoxidase staining, and the detailed procedure that involves the use of anti-AP antiserum raised in rabbit[33] has been described previously.[16]

Pretreatments

Thin sections of glutaraldehyde- and osmium tetroxide-fixed, Epon-embedded tissues can also be used for immunohistochemical labeling after pretreatment with sodium metaperiodate ($NaIO_4$).[34] Thin sections on nickel grids are exposed to a saturated aqueous solution of $NaIO_4$ for 1 hr followed by washing with distilled water.

Occasionally, unmasking of antigenic sites is necessary in certain types of amyloid as in the case of immunolabeling for Aβ. This is done by pretreatment of thin sections with 10% hydrogen peroxide for 15 min followed by washing with distilled water.

Isolation of Amyloid Fibrils and Preparation for High-Resolution Observations

Isolation of Experimental AA Amyloid Fibrils

The fibrils are isolated with the distilled water extraction technique as described by Pras *et al.*[35]

[32] J. R. Hassell, P. Gehron-Robey, H. J. Barrach, J. Wilczek, S. I. Rennard, and G. R. Martin, *Proc. Natl. Acad. Sci. U.S.A.* **77,** 4494 (1980).

[33] M. Skinner, A. S. Cohen, T. Shirahama, and E. S. Cathcart, *J. Lab. Clin. Med.* **84,** 604 (1974).

[34] M. Bendayan and M. Zollinger, *J. Histochem. Cytochem.* **31,** 101 (1983).

[35] M. Pras, M. Schubert, D. Zucker-Franklin, A. Rimon, and E. C. Franklin, *J. Clin. Invest.* **47,** 924 (1968).

1. Spleen tissue is homogenized and washed extensively in 0.15 *M* saline at 4° with centrifugations between washes at 10,000 rpm with a Beckman Model J2-21 centrifuge with a JA-17 rotor.
2. Saline washes continue until the OD_{280} of the supernatant reaches less than 0.1.
3. The tissue is then washed in distilled water with centrifugations between washes at 18,000 rpm in a Beckman L8-70 centrifuge and a Ti SO.2 rotor. The first supernatant is discarded while the second and third supernatants are pooled for the fibrils.

Preparation for High-Resolution Observations

The isolated material is pelleted by centrifugation and is prepared for both embedding and negative staining.

Embedding and Thin Sectioning

1. The isolated specimen in distilled water is centrifuged at 70,000 rpm (100,000*g*) for 1 hr at 4°.
2. The pellet formed is fixed with glutaraldehyde and osmium tetroxide and is processed further with the standard method described earlier.

Negative Staining

1. A small drop of isolated specimen in distilled water is placed on a grid with Formvar supporting film. The amount of fluid is adjusted by blotting with pieces of filter paper and then it is allowed to dry.
2. The specimen on supporting film is stained negatively by placing a small drop of 1% potassium phosphotungstate, pH 7.0, on it, and the amount is adjusted with pieces of filter paper and then allowed to dry.

[33] Analysis of Amyloid-β Assemblies Using Tapping Mode Atomic Force Microscopy Under Ambient Conditions

By TOMAS T. DING and JAMES D. HARPER

Scope

The purpose of this article is to draw attention to atomic force microscopy[1] (AFM) as an extremely valuable tool for the study of amyloid assembly processes. AFM has only recently been used by researchers in the amyloid field, although it is an attractive imaging technique for a number of reasons. Nanometer-level resolution, relatively simple specimen preparation, and new milder imaging modes make AFM an excellent alternative to negative stain electron microscopy (EM) for imaging sensitive biological specimens. Several studies have demonstrated the capability of AFM to resolve numerous distinct β amyloid protein (Aβ) assemblies (including multiple fibril morphologies).[2–8] These studies revealed that *in vitro* Aβ fibril formation follows a more complex pathway than initially suspected, involving multiple intermediate and product assemblies of Aβ. Elucidating the relationship of these species is required to understand the *in vitro* fibril formation process and may provide insight into the Alzheimer's disease (AD) process as well. This article presents a brief introduction of the use of AFM to study Aβ fibrillogenesis, with particular focus on describing methods for imaging in air of Aβ-amyloid physisorbed to mica. Although

[1] G. Binnig, C. F. Quate, and C. Gerber, *Phys. Rev. Lett.* **56,** 930 (1986).

[2] J. D. Harper, S. S. Wong, C. M. Lieber, and P. T. Lansbury, Jr., *Chem. Biol.* **4,** 119 (1997).

[3] J. D. Harper, C. M. Lieber, and P. T. Lansbury, Jr., *Chem. Biol.* **4,** 951 (1997).

[4] M. P. Lambert, A. K. Barlow, B. A. Chromy, C. Edwards, R. Freed, M. Liosatos, T. E. Morgan, I. Rozovsky, B. Trommer, K. L. Viola, P. Wals, C. Zhang, C. E. Finch, G. A. Krafft, and W. L. Klein, *Proc. Natl. Acad. Sci. U.S.A.* **95,** 6448 (1998).

[5] T. Oda, P. Wals, H. H. Osterburg, S. A. Johnson, G. M. Pasinetti, T. E. Morgan, I. Rozovsky, W. B. Stine, S. W. Snyder, T. F. Holzman, G. A. Krafft, and C. A. Finch, *Exp. Neurol.* **136,** 22 (1995).

[6] A. E. Roher, M. O. Chaney, Y. M. Kuo, S. D. Webster, W. B. Stine, L. J. Haverkamp, A. S. Woods, R. J. Cotter, J. M. Tuohy, G. A. Krafft, B. S. Bonnell, and M. R. Emmerling, *J. Biol. Chem.* **271,** 20631 (1996).

[7] A. P. Shivji, M. C. Davies, C. J. Roberts, S. J. B. Tendler, and M. J. Wilkinson, *Prot. Pept. Lett.* **3,** 407 (1996).

[8] W. B. Stine, Jr., S. W. Snyder, U. S. Ladror, W. S. Wade, M. F. Miller, T. J. Perun, T. F. Holzman, and G. A. Krafft, *J. Prot. Chem.* **15,** 193 (1996).

0076-6879/99 $30.00

the methods described in this article were optimized for investigation of Aβ fibril formation, they should also be useful for the study of other protein assembly processes with relatively minor modifications to solution and specimen preparation procedures.

Introduction

Fibrillar amyloid deposits in the brain containing Aβ peptide, a 39–43 amino acid peptide produced by proteolysis of a transmembrane protein of unknown function (the amyloid precursor protein), are a key pathological hallmark of AD.[9] The 40 and 42 amino acid variants (Aβ40 and Aβ42) are the predominant variants found in AD amyloid fibrils and differ only by the presence of an additional two hydrophobic residues at the C terminus of Aβ42. The current inability to follow the details of the amyloid formation process with time in a single individual makes it impossible to reconstruct the *in vivo* pathway to Aβ amyloid fibrils based on neuropathological data. Therefore, this laboratory and others[10] study Aβ amyloid formation *in vitro* on the assumption that the *in vitro* pathway may be analogous to the situation in AD. This reductionist approach to understanding AD is justified in part by the ability of synthetic Aβ to form fibrils under aqueous conditions *in vitro* (see Refs. 10 and 11 and subsequent publications) that are indistinguishable from those which form a defining pathological feature in the brains of AD patients.[12–14] *In vitro* studies of fibril formation led to the hypothesis that amyloid formation is a nucleation-dependent polymerization process[15,16] and suggested that the increased production of Aβ42, which is more prone to spontaneous nucleation, could provide a plausible explanation for early onset familial forms of AD.[17,18] Observations from subsequent examinations of the effects of early onset disease-associated mutations

[9] D. J. Selkoe, *J. Biol. Chem.* **271,** 18295 (1996).

[10] J. D. Harper and P. T. Lansbury, Jr., *Annu. Rev. Biochem.* **66,** 385 (1997).

[11] C. Hilbich, B. Kisters-Woike, J. Reed, C. L. Masters, and K. Beyreuther, *J. Mol. Biol.* **218,** 149 (1991).

[12] D. Burdick, B. Soreghan, M. Kwon, J. Kosmoski, M. Knauer, A. Henschen, J. Yates, C. Cotman, and C. Glabe, *J. Biol. Chem.* **267,** 546 (1992).

[13] P. A. Merz, H. M. Wisniewski, R. A. Somerville, S. A. Bobin, C. L. Masters, and K. Iqbal, *Acta Neuropathol.* (*Berl.*) **60,** 113 (1983).

[14] H. K. Narang, *J. Neuropathol. Exp. Neurol.* **39,** 621 (1980).

[15] A. E. Roher, K. C. Palmer, E. C. Yurewicz, M. J. Ball, and B. D. Greenberg, *J. Neurochem.* **61,** 1916 (1993).

[16] P. T. Lansbury, Jr., *Neuron* **19,** 1151 (1997).

[17] P. T. Lansbury, Jr., *Acc. Chem. Res.* **29,** 317 (1996).

[18] J. T. Jarrett, E. P. Berger, and P. T. Lansbury, Jr., *Biochemistry* **32,** 4693 (1993).

(which have all been shown to increase the amount or ratio of the longer variants compared to the shorter variants[19]) support this hypothesis. A more complete understanding of fibril formation in the *in vitro* model may provide additional insights into the mechanism of Aβ amyloid plaque formation in AD.

In particular, better understanding of the molecular events leading to the formation of the first fibril nucleus or seed and the subsequent rapid growth phase is needed. Understanding this phase of fibril growth could provide clues about critical steps in Aβ assembly that occur in individuals with AD prior to the formation of amyloid plaques and the onset of symptoms. For this reason, investigators have been interested in analytical methods suitable for the detection and morphological characterization of small Aβ fibril assembly intermediates ("Aβ oligomers") that may exist prior to nucleation. Detection of small oligomeric Aβ species has been reported using a variety of techniques, including analytical ultracentrifugation,[20] size-exclusion chromatography (SEC),[21] quasi-elastic light scattering (QLS),[22] electron microscopy,[21] and atomic force microscopy.[2–8] Analytical ultracentrifugation, SEC, and QLS provide information about the size and the general shape (i.e., spherical or rod like) of the aggregates, but do not supply detailed information about aggregate morphology.

Negative stain electron microscopy, turbidometry, sedimentation assays, Congo red binding, and thioflavin T fluorescence have long been the standard tools among amyloid researchers for following amyloid aggregation. Negative stain EM is a useful technique for detecting and imaging of fibrillar amyloid aggregates. However, the staining procedures required to visualize protein assemblies can produce variable results for amyloid samples. Negative stain electron micrographs often do not resolve fine details in surface morphology that distinguish aggregate types and can also provide insufficient contrast for detecting very small species reliably. The remaining methods mentioned are useful in monitoring the rate or extent of Aβ aggregation *in vitro* and/or the detection of amyloid in biological specimens, but in their present form these methods may not equally detect or distinguish the different Aβ aggregate morphologies.[23]

[19] D. J. Selkoe, *Science* **275,** 630 (1997).

[20] S. W. Snyder, U. S. Ladror, W. S. Wade, G. T. Wang, L. W. Barrett, E. D. Matayoshi, H. J. Huffaker, G. A. Krafft, and T. F. Holzman, *Biophys. J.* **67,** 1216 (1994).

[21] D. M. Walsh, A. Lomakin, G. B. Benedek, M. M. Condron, and D. B. Teplow, *J. Biol. Chem.* **272,** 22364 (1997).

[22] A. Lomakin, D. S. Chung, G. B. Benedek, D. A. Kirschner, and D. B. Teplow, *Proc. Natl. Acad. Sci. U.S.A.* **93,** 1125 (1996).

[23] S. J. Wood, B. Maleeff, T. Hart, and R. Wetzel, *J. Mol. Biol.* **256,** 870 (1996).

Atomic Force Microscopy

Successful AFM analysis of biological specimens (including amyloid assemblies) involves the integration of three basic steps: sample preparation, scanning the sample, and interpretation of topographical data. Each of these steps will be discussed briefly in the following sections, and details of procedures used successfully for routine AFM analysis of Aβ assemblies will be given.

Sample Preparation

Preparation of Seed-Free Aggregation Solutions

To allow the greatest opportunity to observe prenucleation assembly events, it is important to remove preexisting seeds from solutions. Several approaches to generating seed-free Aβ solutions have been used, including presolubilization in organic solvents [e.g., 1,1,1,3,3,3-hexafluoro-2-propanol (HFIP) or dimethyl sulfoxide (DMSO)] and size-exclusion chromatography.

In this laboratory we have found that dilution of concentrated DMSO stocks of Aβ peptide into aqueous buffer (10 m*M* phosphate, 100 m*M* NaCl, pH 7.4) provides consistent results and recovery of peptide. Lyophilized Aβ is dissolved at 10 to 15 mg/ml in DMSO, sonicated for 5–10 min in a water bath sonication unit, and filtered using a microspin membrane filtration unit (0.2 μm or smaller pore size). It is important to include the filtration step during DMSO stock preparation and to use DMSO stocks promptly, as Aβ42 fibril formation in aged DMSO stock solutions has been observed (Ding and Harper, 1997, unpublished observation). The resulting DMSO stock solutions are typically between 1.5 and 2.5 m*M* and allow convenient preparation of aqueous solutions with Aβ concentrations up to 250 μM with DMSO constituting less than 10% of the final solution by volume. The best imaging specimens are produced from aggregating solutions that are left undisturbed following initial dilution and mixing except for gentle mixing immediately prior to the removal of aliquots for specimen preparation. Constant stirring or vortexing promotes the formation of large Aβ aggregates (bundles of protofibrils and fibrils), which make specimen preparation less reproducible and interfere with subsequent AFM analysis.

Seed-free solutions have also been prepared by SEC, which fractionates protein and protein aggregates on the basis of molecular weight under nondenaturating conditions (using, e.g., a Sephadex-type HPLC column).[21] There are at least two advantages with size-exclusion chromatography over DMSO stock solutions: organic solvents can be avoided and low molecular weight intermediates can be detected and removed from the solution.[21] The SEC fractions should be used immediately after collection.

Specimen Preparation

Methods of specimen preparation from aqueous samples for AFM analysis all have the same goal: to attach the species of interest (e.g., the protein fibrils) firmly to a support. For optimal resolution, the support must be flat, i.e., have a roughness several orders of magnitudes lower than the height of the adsorbed specimen, and the particles on the surface should be well dispersed. Ideally, the specimen should be adsorbed onto a suitable substrate directly from a physiological buffer without further modifications that may alter its structure or impair its function. The yield of surface-adsorbed biomaterial depends on multiple factors, including the chemical properties of the substrate, the pH and ionic strength of the adsorption buffer, the concentration of macromolecule, and the incubation time.[24,25]

The method commonly used is to adsorb the biomaterial directly from solution onto the desired substrate. The two most popular substrates used are mica and highly ordered pyrolytic graphite (HOPG). Mica is a relatively inexpensive material that is easily cleaved (by carefully peeling off the top layer with a piece of transparent tape) to produce a clean hydrophilic surface with large (several hundred μm^2) regions that are atomically smooth. HOPG has a higher hydrophobicity as compared to mica, and it is cleaved in the same way to produce clean surfaces. One disadvantage of HOPG is that the atomically smooth areas are typically smaller than those on mica. The boundaries between these smooth regions can produce fibril-like or "DNA-like" artifacts that can complicate the interpretation of images obtained on this substrate.

To prepare amyloid specimens for AFM imaging, we incubate 3- to 5-μl aliquots of Aβ40 or Aβ42 aggregation solutions on the surface of freshly cleaved mica, usually for 30 to 60 sec. Following incubation, the excess unbound protein and buffer salts are rinsed away with filtered water. The rinsing is done gently by applying the water with a micropipettor (typically two 50-μl rinses are sufficient) to the edge of the mica support and allowing it to run across the surface as the substrate is tilted with its edge touching an absorbant material. The remaining water on the surface can be allowed to dry slowly in a covered container or removed with a gentle stream of filtered compressed air or nitrogen. Once the mica surface is dry the specimen is ready for immediate imaging under ambient conditions, and further rinsing with water should be avoided as it can often completely remove the adsorbed protein.

[24] D. J. Müller and A. Engel, *Biophys. J.* **73,** 1633 (1997).

[25] D. J. Müller, M. Amrein, and A. Engel, *J. Struct. Biol.* **119,** 172 (1997).

All aggregate species may not be equally well adsorbed, therefore it is not possible to quantitate the relative amounts of species present in solution based on their abundance on the mica substrate, nor is it possible to rule out the existence of additional species that might be adsorbed very poorly. There is little reason to suspect, however, that the ability of a certain species to adsorb will change with time. Thus, the relative amounts of a species observed at each time point during an experiment should provide reliable qualitative information about the changes in abundance of a particular morphology during the course of aggregation.

Scanning the Sample

The design and operation of a typical AFM for imaging biological specimens have been described in detail elsewhere.[26] Briefly, an AFM creates images of surfaces by measuring the position of a silicon probe tip mounted on the end of a sensitive cantilever as a sample is moved underneath. On atomically flat surfaces the AFM is capable of atomic resolution. However, with biological samples, resolution is limited to a few nanometers due to tip convolution effects.[27] The AFM instrument can be operated in several modes, with the most common modes being the contact mode (also called the constant force or height mode) and the tapping mode (also called the intermittent contact mode).[26] In the contact mode (the first AFM imaging mode developed), the tip remains in constant contact with the surface throughout a scan. The contact mode works well for hard noncompressible surfaces, which resist damage due to lateral shear forces generated by the sweeping motion of the tip. However, compressible composite samples, such as biomolecules adsorbed to substrates for imaging, do not always withstand these forces and can be deformed or dislocated by the tip in this mode of operation.

In the tapping mode, the cantilever is oscillated at its resonant frequency (60–400 kHz depending on the cantilever) and positioned above the sample so that it contacts the surface briefly at the bottom of its swing. The time the tip contacts the surface is very short when compared to the time required to trace across the surface. As a result, shear forces are almost completely eliminated, making this mode of imaging better suited for imaging relatively soft biomolecules.[28]

[26] C. Bustamante and D. Keller, *Phys. Today* **48,** 32 (1995).

[27] T. Thundat, X. Zheng, S. Sharp, D. Allison, R. Warmack, D. Joy, and T. Ferrell, *Scan. Microsc.* **6,** 903 (1992).

[28] R. J. Colton, D. R. Baselt, Y. F. Dufrêne, J.-B. D. Green, and G. U. Lee, *Curr. Opin. Chem. Biol.* **1,** 370 (1997).

Optimization of Tapping Mode Scan Parameters

A force is generated between the sample and the tip due to the cantilever oscillation when operating in the tapping mode. This force depends mainly on two parameters: the amplitude of free oscillation (A_0) and the set point amplitude (A_{sp}). The force that the cantilever is generating on the sample is related to the ratio A_{sp}/A_0. As the tip is brought close to the sample surface, the vibrational characteristics (e.g., resonance frequency and amplitude) of the cantilever change due to tip–sample interactions and the contamination or hydration layer on the tip and sample. The observed amplitude of oscillation is maintained at A_{sp} by adjusting the vertical position of the sample.

Tapping too hard, i.e., low A_{sp} and/or high A_0, can reduce image resolution and overall quality by damaging the sharp tip or deforming the protein. Tapping too lightly, however, is also problematic because adhesive forces between the sample and the cantilever tip (e.g., from capillary interactions or tip contamination) might cause damping of the cantilever oscillation. Because amplitude damping is recorded as higher in topography, the result could be erratic or anomalous height data.[29] Therefore, the best parameters for imaging represent a compromise between these two situations: tapping hard enough to avoid anomalous cantilever oscillation damping, but lightly enough that damage to the tip and deformation of the sample is minimized.

A convenient way to find the suitable scan parameters for imaging Aβ specimens (as well as other similar biological specimens) in the tapping mode is to engage the cantilever and adjust the set point amplitude so that any increase in the set point results in the tip pulling off the surface. This should provide imaging with the minimum tapping force for the chosen initial drive amplitude and cantilever oscillation amplitude. Tapping forces sometimes can be minimized further and image quality improved by incremental reductions of the cantilever oscillation amplitude used for scanning. This is achieved by first reducing the drive amplitude slightly while scanning (this should cause the tip to disengage from the surface and features in the image to disappear) and then decreasing the set point slowly until the image sharpens again. This process can be repeated until the minimum drive amplitude/set point combination is found that allows stable imaging without introducing image artifacts from adhesive interactions. The gain parameters control the properties of the feedback loop and are normally adjusted slightly below the point at which oscillations or other feedback-induced artifacts occur.[30]

[29] R. Brandsch, G. Bar, and M.-H. Whangbo, *Langmuir* **13,** 6349 (1997).
[30] H. G. Hansma and J. H. Hoh, *Annu. Rev. Biophys. Biomol. Struct.* **23,** 115 (1994).

Acquisition of Topographical Data

Once suitable scan parameters have been selected, the position of the sample under the tip is controlled by a piezoelectric ceramic-based scanner that can reposition the sample under the tip in three dimensions in angstrom scale increments. Keeping track of the movements required to raster the sample underneath the tip while maintaining a constant cantilever deflection (in the contact mode) or oscillation amplitude (in the tapping mode) at each position in the region results in a digitized topographic map of the surface. Visual representations of these data can then be generated by assigning each point in the area a color or brightness based on its relative height. For images in this article, the smooth surface of the substrate will appear dark gray with the brightness of the adsorbed features increasing as a function of their height.

Interpretation of Images

The resulting topographical data for the area can also be used to obtain three-dimensional measurements of adsorbed features with a high degree of precision, which can be displayed as images. However, it is important to consider several limitations that arise as a result of the physical nature of the technique when discussing dimensions measured by AFM as well as by any other physical technique. In general, the height of an object above the surface can be determined, but the lateral dimension in the plane of the image (the width) is difficult to measure accurately due to the shape and the finite size of the tip.[26] The overestimation of width is related to the height and true width of the features and the shape and diameter of the silicon tip, which is difficult to determine.[26]

Because of the difficulties associated with comparisons of width, we use measurements of the height rather than the width as estimates of the diameters of Aβ aggregate species. However, it is also important to note that there are a number of reasons why the height of features measured by AFM in air may misrepresent the actual diameters of features as they exist in solution: (1) The force applied by the cantilever may compress soft biological features. (2) Dehydration of the protein following the removal of buffer prior to imaging could reduce measured dimensions. (3) Height measurements may be reduced if features deform during the adsorption process to maximize the surface area of contact to the substrate (similar to the deformation that results when a spherical water droplet adheres to a surface). (4) The cross section of filaments and other species may not be circular or square.[28] Differences in cantilevers, tips, instrument parameters, and/or sample preparation conditions can result in variability in the measured dimensions of the same species in different scans. Therefore, comparisons of height between morphologies should be made, when possible, from single images clearly resolving both species. This is done to ensure that variability in measurements among scans does not cause the relative dimensions of morphologies to be misrepresented. When this is not possible the diameter of a distinct morphology present with each species should be used as a reference to allow comparisons of their relative heights to be made with more confidence. For example, protofibrils (for definitions of structures, see later) are generally not observed in the same specimen with type 2 fibrils, but standardization of their diameters relative to type 1 fibrils commonly present alongside each species can be used to accurately compare their heights.

Classification of Distinct Morphologies Based on AFM

When Aβ40 and Aβ42 are dissolved in aqueous solution and allowed to aggregate, the first-formed species are not the amyloid fibrils. This section describes a representative time course for the aggregation of a 45 μM solution of Aβ40 at room temperature in aqueous buffer (10 mM phosphate, 100 mM NaCl, pH 7.4) and shows clear images and section analysis data of the morphologies described. The overall assembly process was much more rapid in the case of the 42 amino acid variant Aβ42 when compared to the truncated variant Aβ40. For comparison, AFM reveals the presence of fibrils in unstirred, aqueous (pH 7.4) solutions of Aβ42 (20 μM) within 5–8 days of initial dilution whereas similar solutions of Aβ40 (45 μM) typically have no detectable fibrils during the first 2–3 weeks of incubation.[16]

In this experiment the adsorbed material from an aliquot removed after 4 days consisted mostly of small species with uniform heights of about 4 nm (Fig. 1). A significant population of the ~4-nm height features at 4 days appeared to be roughly spherical, whereas many others were clearly elongated,[2,21] with lengths between 40 and 60 nm being the most common. We refer to this population of ~4-nm height features as protofibrils to reflect their probable role as fibril assembly intermediates.[2,3]

While retaining a constant height, the length of typical protofibrils increased over 3 weeks and commonly reached lengths of 100–200 nm. In general, protofibrils appear to be a flexible species as evidenced by the random curvatures they exhibit following adsorption to the mica surface. The average diameter of protofibrils measured from a single representative image is 4.4 $\pm$ 0.4 nm. Similar average height measurements of protofibrils from a range of different samples (based on $\geq$50 individual measurements in each case) range from 3.1 $\pm$ 0.2 to 4.4 $\pm$ 0.6 nm. This ~30% variability in average height measurements from different images is typical even when

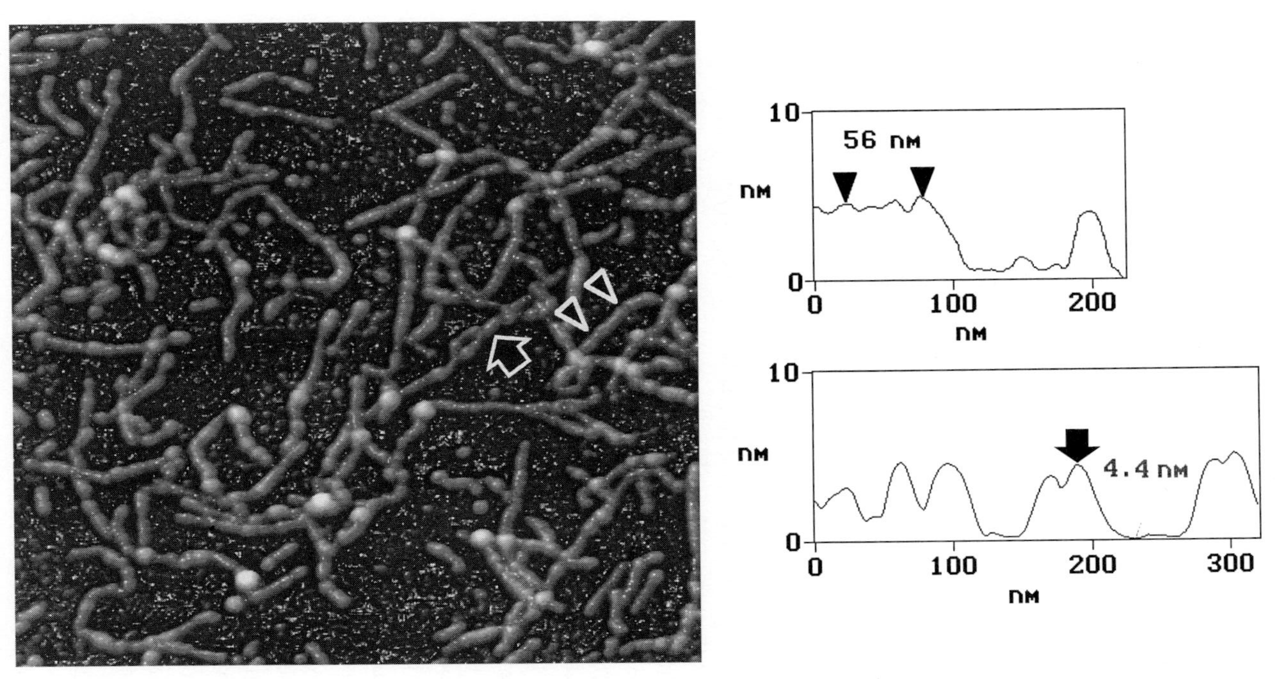

FIG. 1. Representative image and sections of Aβ40 protofibrils. As discussed in the text, the brightness of features in the image increases proportionally with increases in relative height above the surface. Arrows and arrowheads in the 1-μm^2 image correspond to the similarly marked features in the sections shown at the right. (Top right) An axial section of a protofibril showing 56 nm total distance between arrowheads indicating three periods in the axial substructure of the protofibril. (Bottom right) A cross section of protofibril indicating the 4.4-nm measured diameter of this representative feature.

the same type of probe and similar set points are used for imaging (which should result in similar tapping forces) and illustrates the importance of comparing morphologies from within a single scan whenever possible to ensure the greatest accuracy. In well-resolved scans, some protofibrils show small increases in diameter at regular intervals along their axis. Analyzing longitudinal sections (along the fibril axis) of protofibrils from a typical scan shows that these peaks have an average spacing of 20 ± 4.7 nm (Fig. 1).[2]

In frequently sampled solutions after about 4 weeks of incubation, the first long Aβ40 species resembling prototypical amyloid fibrils appeared (designated type 1 fibrils since they were the earliest type of fibrils to be observed). A well-resolved image of type 1 fibrils is shown in Fig. 2. A modulation in diameter with an average period of 43 nm characterizes type 1 fibrils and is most likely the result of a left-handed helical twist of two or more filament subunits.[3] Apparent helicity in AFM images of filaments could also result from tip-induced artifacts. However, constant left-handed helicity for all fibrils in a single scan regardless of orientation with regard to the scan direction rules out this possibility for type 1 fibrils. The winding of filament subunits to form these fibrils is also suggested by the 8- to 10-nm diameter of the type 1 fibril (Fig. 2), and by the observation of partially wound and branched type 1 fibrils.[3]

After 5 weeks, the protofibrils had disappeared completely and a second fibril type (type 2 fibril) appeared that had a smaller diameter than the type 1 fibril and was characterized by a segmented rather than a beaded appearance (Fig. 3).[2] The type 2 fibril commonly appears along with type 1 fibrils at time points following the disappearance of protofibrils. Although they lack the regular height increases that characterize both protofibrils and type 1 fibrils, they do show less regular discontinuities along their axis (Fig. 3). The interval between these dislocations commonly measured between 100 and 200 nm, with the intervals along a single fibril typically varying over a smaller range (± 20 nm).

Although type 1 and type 2 fibrils are the predominant fibril morphologies, fibrillar species that do not fit into these categories are also observed in some specimens (not shown). Generally, the diameters of these fibrils are similar to or slightly larger (diameter $\sim$10–12 nm) than the two fibril types described earlier. These may represent fibrils formed from different numbers of filaments (as suggested by previous observations of multiple higher order assemblies of filaments in calcitonin[31] and amylin filaments[32])

[31] H. H. Bauer, U. Aebi, M. Haner, R. Hermann, M. Muller, and H. P. Merkle, *J. Struct. Biol.* **115,** 1 (1995).

[32] C. S. Goldsbury, G. J. Cooper, K. N. Goldie, S. A. Muller, E. L. Saafi, W. T. Gruijters, M. P. Misur, A. Engel, U. Aebi, and J. Kistler, *J. Struct. Biol.* **119,** 17 (1997).

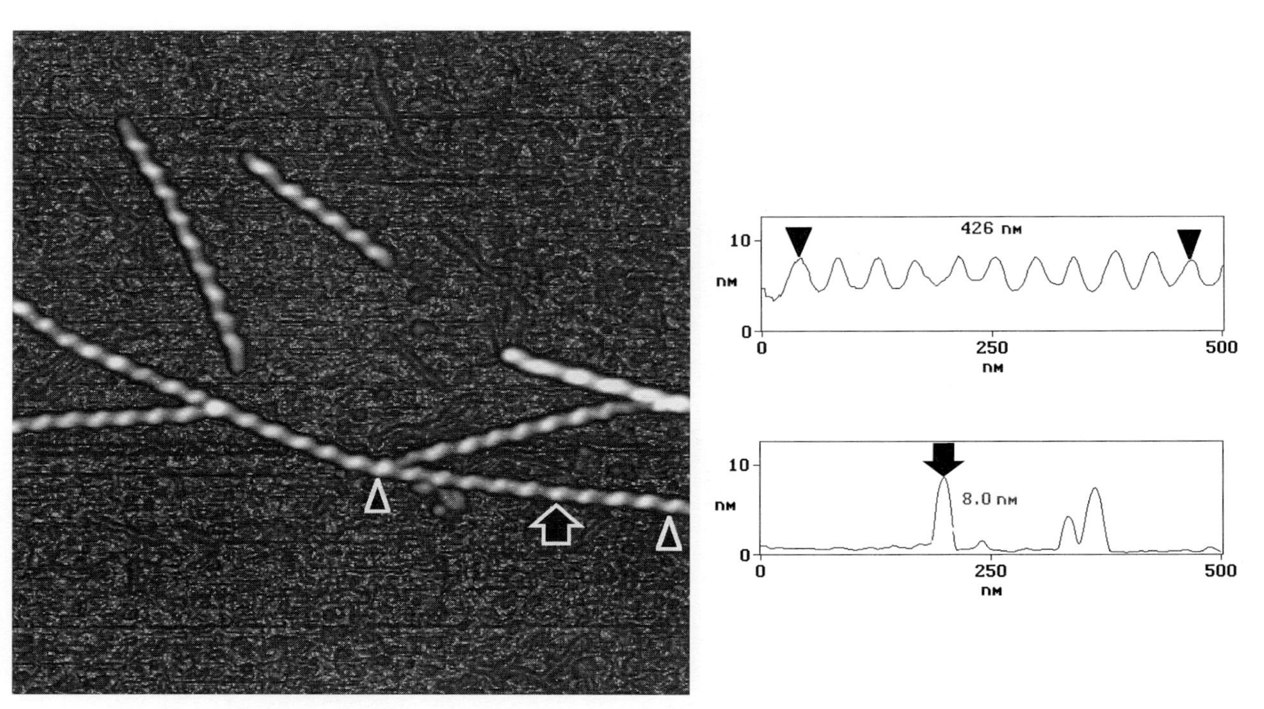

FIG. 2. Representative image and sections of Aβ40 type 1 fibrils. Arrows and arrowheads in the 1-μm^2 image correspond to the similarly marked features in the sections shown at the right. (Top right) An axial section of a type 1 fibril showing 426 nm total distance between arrowheads indicating 10 periods in the axial substructure. (Bottom right) A cross section of type 1 fibril indicating the 8.0-nm measured diameter of this representative feature.

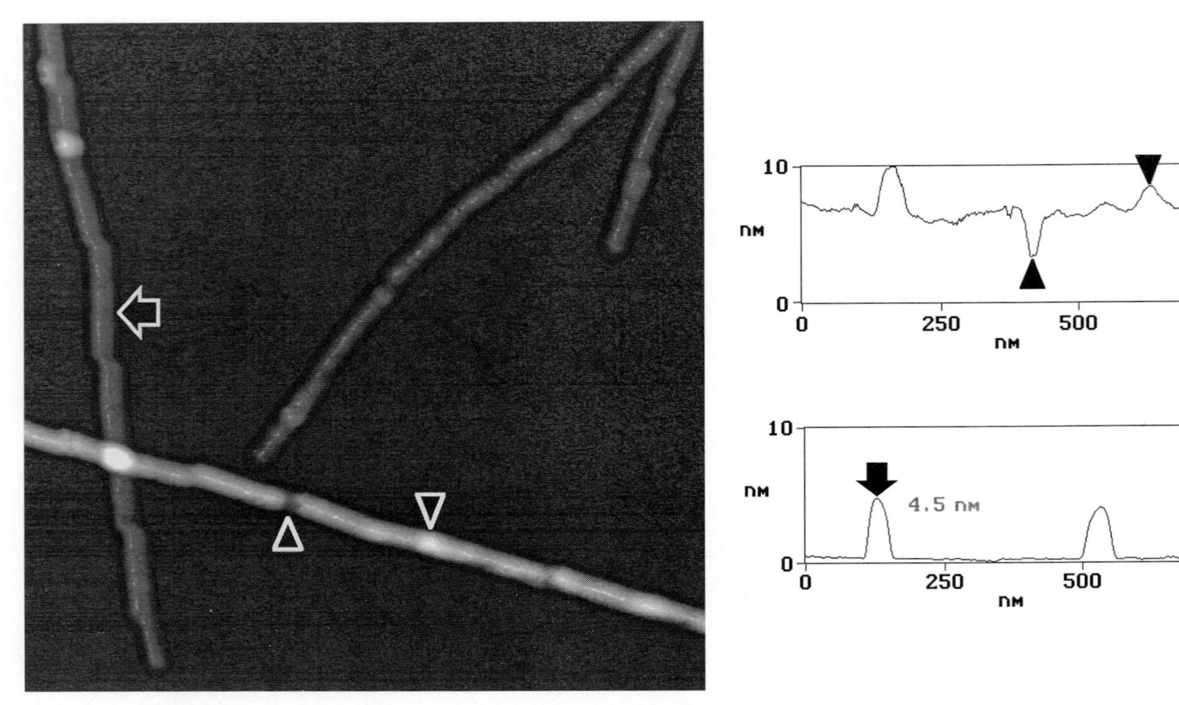

FIG. 3. Representative image and sections of Aβ40 type 2 fibrils. Arrows and arrowheads in the 1-μm^2 image correspond to the similarly marked features in the sections shown at the right. (Top right) An axial section of a type 2 fibril showing both the decrease in diameter that is typical at the characteristic dislocations (left arrowhead) and an increase in diameter that is an occasional feature in some type 2 fibrils (right arrowhead). (Bottom right) A cross section of type 2 fibril indicating the 4.5-nm measured diameter of this representative feature.

or alternate lateral association modes of similar numbers of filaments (similar to the polymorphism of flagella[33]).

Significant variability in the time course for fibril formation among similar samples is common. Protofibrils have been observed in completely undisturbed samples, up to 8 weeks after dilution, and incubation at room temperature. These samples were identical (in Aβ40 concentration and peptide stock source) to the solution described earlier, which was frequently disturbed to evenly suspend aggregates prior to imaging for time course studies, yet they contained only protofibrils at a time well after the time course sample contained only fibrils. This observation is consistent with previous observations using turbidity[34,35] and Congo red binding,[23] which show slower progression toward fibrils observed in unstirred samples compared to similar samples that have been stirred or shaken.

Alzheimer's disease senile plaque amyloid fibrils (presumably Aβ fibrils) with an 8- to 10-nm diameter and two filament wound morphology (30–50 nm periodicity) have been observed in previous EM studies of tissue sections[13] and purified plaque material.[12,14] These reports support the hypothesis that the type 1 fibril morphology is an important product of Aβ fibril formation *in vivo* and that *in vitro* systems that produce this morphology are likely to be useful models for the AD pathogenic process.

Fibril morphologies reminiscent of type 1 and type 2 fibrils have been observed by other researchers studying *in vitro* Aβ fibril formation as well. Stine *et al.*[8] analyzed aged samples of Aβ40 (10 to 500 μM aqueous solutions) by EM and AFM. They observed "periodic" and "smooth" morphologies. The periodic morphologies had a thickness of 8–12 nm and axial periodicity of 25 nm, whereas the smooth morphologies were featureless with a height of 4–6 nm (imaged in contact mode under isopropanol). Seilheimer *et al.*[36] also noticed the occurrence of different types of fibrils using electron microscopy, which they call "immature" and "mature" fibrils, when they compared electron micrographs of aggregating Aβ samples. The mature fibrils are thinner than the immature fibrils and with a segmented periodicity of 150 nm.

Small oligomeric Aβ species (Fig. 4) have been identified by AFM[2–8] as well as other methods.[20,22,37] Lambert *et al.*[4] identified small Aβ morphol-

[33] K. Hasegawa, I. Yamashita, and K. Namba, *Biophys. J.* **74,** 569 (1998).

[34] S. J. Wood, W. Chan, and R. Wetzel, *Chem. Biol.* **3,** 949 (1996).

[35] K. C. Evans, E. P. Berger, C. G. Cho, K. H. Weisgraber, and P. T. Lansbury, Jr., *Proc. Natl. Acad. Sci. U.S.A.* **92,** 763 (1995).

[36] B. Seilheimer, B. Bohrmann, L. Bondolfi, F. Muller, D. Stuber, and H. Dobeli, *J. Struct. Biol.* **119,** 59 (1997).

[37] M. B. Podlishny, B. L. Ostaszewski, S. L. Squazzo, E. H. Koo, R. E. Rydell, D. B. Teplow, and D. J. Selkoe, *J. Biol. Chem.* **270,** 9564 (1995).

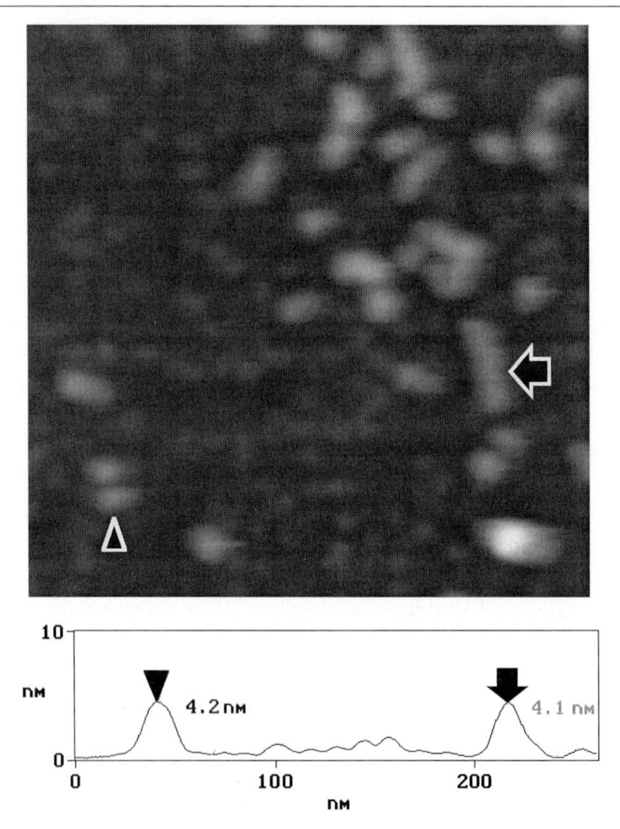

FIG. 4. Representative image and sections of Aβ40 globular particles. (Top) A 250-nm image of Aβ40 (16 μM) after 2 days of incubation at room temperature (left) with sections along the indicated lines (right). (Bottom) A 500-nm image of Aβ40 (45 μM) after 4 days of incubation at room temperature (left) with a section along the indicated line (right) showing an example of a large globular aggregate. Arrows and arrowheads in the image correspond to the similarly marked features in the sections shown at the right where the heights are indicated.

ogies that he referred to as "Aβ derived diffusible ligands" (ADDLs). ADDLs are neurotoxic, 5- to 6-nm globular aggregates (measured by AFM), prepared by aggregation of Aβ42 in the presence of clusterin or at low temperatures.[4] Small globular and elongated species have also been observed by AFM when imaging samples under 2-propanol[5,6,8] or samples that have been coated with platinum carbon[7]. The smallest Aβ40 particle detected directly by QLS (at pH 1.0) had a hydrodynamic radius of 7 nm.[22] When Walsh *et al.*[21] examined the included peak from their size exclusion separation of Aβ40 aggregates (at pH 7.4) by QLS, they were able to detect particles with an even smaller hydrodynamic radius (1.8 nm), interpreted to be consistent with extended monomeric or globular dimeric Aβ40.

Conclusion

AFM analysis of the topography of species adsorbed from solutions of Aβ aggregating under aqueous conditions *in vitro* represents the first generation of AFM experiments aimed at studying amyloid formation. These initial experiments have provided new information that has significantly increased our understanding of the molecular events involved in amyloid formation *in vitro* by the Aβ peptide. Progress toward improving image resolution and utilizing dynamic imaging capabilities in an aqueous environment suggests that this already powerful tool will become even more powerful for studying the amyloid formation process in the very near future. Work using carbon nanotube AFM tips[38] has significantly improved topographical resolution of biomolecular assemblies[39,40] and has opened new avenues for nanometer scale biological and chemical discrimination of biomolecular surfaces by measuring the strength of binding interactions using covalently functionalized nanotube tips.[41,42] Goldsbury *et al.*[43] have developed conditions for imaging assemblies of another amyloid forming protein, amylin, under aqueous conditions that allowed the generation of time-lapse images of individual growing amylin fibrils and protofibrils. These developments promise to open new and extremely productive avenues for studying the molecular events that occur during amyloid fibril formation.

[38] H. Dai, J. H. Hafner, A. G. Rinzler, D. T. Colbert, and R. E. Smalley, *Nature* **384,** 147 (1996).
[39] S. S. Wong, J. D. Harper, P. T. Lansbury, Jr., and C. M. Lieber, *J. Am. Chem. Soc.* **120,** 603 (1998).
[40] S. S. Wong, A. T. Woolley, T. W. Odom, J. L. Huang, P. Kim, D. V. Vezenov, and C. M. Lieber, *Appl. Phys. Lett.* **73,** 3465 (1998).
[41] S. S. Wong, E. Joselevich, A. T. Woolley, C. L. Cheung, and C. M. Lieber, *Nature* **394,** 52 (1998).
[42] R. Merkel, P. Nassoy, A. Leung, K. Ritchie, and E. Evans, *Nature* **397,** 50 (1999).
[43] C. Goldsbury, J. Kistler, U. Aebi, and T. Arvinte, *J. Mol. Biol.* **285,** 33 (1999).

[34] X-Ray Fiber Diffraction of Amyloid Fibrils

By LOUISE C. SERPELL, PAUL E. FRASER, and MARGARET SUNDE

Introduction

Amyloid is an ordered structure generated by the polymerization of amyloidogenic proteins. It is a high molecular weight, insoluble material and therefore the atomic structure cannot be investigated by conventional X-ray crystallography or nuclear magnetic resonance (NMR). However, information about its overall fibrillar structure can be obtained by X-ray fiber diffraction, particularly if high-resolution data can be collected.

The earliest reported use of this technique, investigating serum amyloid A and light chain amyloid, reported meridional reflections at 4.68 Å and equatorial reflections at 9.8 Å.[1-3] These diffraction patterns (Fig. 1) are consistent with fibrils composed of polypeptide chains extended in the so-called cross-β conformation, a structure that had earlier been identified as a possible conformation for polypeptide chains on the grounds of model building by Pauling and Corey[4] and which was described for insect silk (*Crysopa*) by Geddes and co-workers.[5] The meridional reflection indicates a regular structural repeat of 4.68 Å along the fibril axis, and the equatorial reflection indicates a structural spacing of 9.8 Å perpendicular to the fibril axis. A β-sheet structure (or more precisely a β ribbon), organized so that the sheet (or ribbon) axis is parallel to the fibril axis, with its constituent β strands perpendicular to the fibril axis, fulfills these spacing requirements (Fig. 2). The structural repeat of 4.68 Å along the fiber axis corresponds to the spacing of adjacent β strands and the 10- to 12-Å spacing perpendicular to the fiber axis corresponds to the face-to-face separation of the β sheets. This latter can only occur if the amyloid fibrils, or protofilaments, are composed of two or more β sheets. Subsequent analyses by X-ray diffraction on a variety of amyloids,[6-11] and also by solid-state NMR tech-

[1] E. D. Eanes and G. G. Glenner, *J. Histochem. Cytochem.* **16,** 673 (1968).
[2] L. Bonar, A. S. Cohen, and M. Skinner, *Proc. Soc. Exp. Biol. Med.* **131,** 1373 (1967).
[3] G. G. Glenner, E. D. Eanes, and D. L. Page, *J. Histochem. Cytochem.* **20,** 821 (1972).
[4] L. Pauling and R. Corey, *Proc. Natl. Acad. Sci. U.S.A.* **37,** 729 (1951).
[5] A. J. Geddes, K. D. Parker, E. D. T. Atkins, and E. A. J. Beighton, *J. Mol. Biol.* **32,** 343 (1968).
[6] M. J. Burke and M. A. Rougvie, *Biochemistry* **11,** 2435 (1972).
[7] D. A. Kirschner, C. Abraham, and D. A. Selkoe, *Proc. Natl. Acad. Sci. U.S.A.* **83,** 503 (1986).
[8] W. Turnell, R. Sarra, J. O. Baum, D. Caspi, M. L. Baltz, and M. B. Pepys, *Mol. Biol. Med.* **3,** 409 (1986)
[9] P. Gilchrist and J. Bradshaw, *Biochim. Biophys. Acta* **1182,** 111 (1993).

0076-6879/99 $30.00

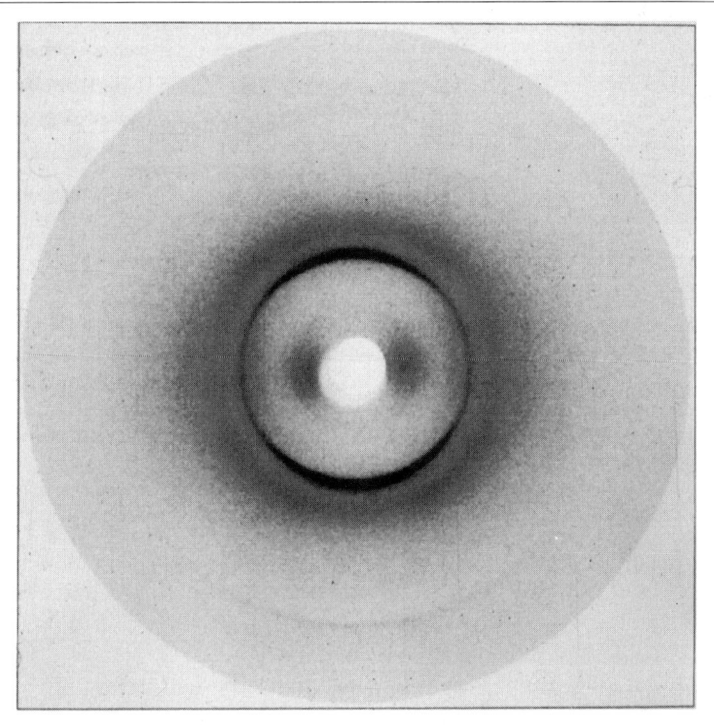

FIG. 1. Characteristic cross-β X-ray fiber diffraction pattern. The fibril sample is composed of a 10 residue peptide with the sequence of the A strand of transthyretin, which forms amyloid fibrils spontaneously when dissolved in water. The fibrils were aligned in a magnetic field to improve alignment. The diffraction pattern shows the dominant reflections at 4.7–4.8 Å on the meridian and approximately 10 Å on the equator.

niques,[12] have confirmed that the protein chains in all amyloid fibrils have a predominantly cross-β structure. Fiber diffraction data indicate that the β sheets in amyloid probably contain a mixture of parallel and antiparallel hydrogen-bonded β strands. The strong, sharp 4.7-Å reflection on the meridian indicates that the β strands must be perpendicular to the fiber axis. A purely antiparallel arrangement of strands would give rise to a smallest repeat spacing of 9.6 Å. The presence of only the intense 4.7-Å meridional

[10] J. T. Nguyen, H. Inouye, M. A. Baldwin, R. Fletterick, F. E. Cohen, S. B. Prusiner, and D. A. Kirschner, *J. Mol. Biol.* **252,** 412 (1995).

[11] C. C. F. Blake and L. C. Serpell, *Structure* **4,** 989 (1996).

[12] P. T. Lansbury, P. R. Costa, J. M. Griffiths, E. J. Simon, M. Auger, K. J. Halverson, D. A. Kocisko, Z. S. Hendsch, T. T. Ashburn, R. G. S. Spencer, B. Tidor, and R. G. Griffin, *Nature Struct. Biol.* **2,** 990 (1995).

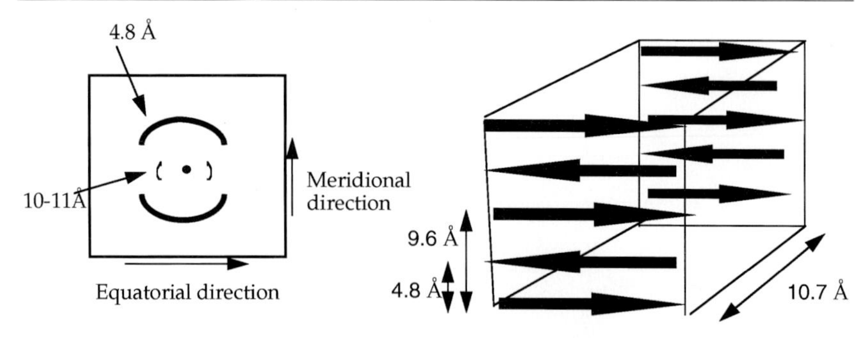

FIG. 2. Underlying structure that gives rise to the cross-β reflections. On the left, a schematic representation of the characteristic 4.8-Å meridional and approximately 10-Å equatorial reflection, which are the two main features of the cross-β pattern from amyloid. The diagrammatic representation of the β-sheet structure, shown on the right, indicates the molecular spacings that give rise to these reflections. The interstrand spacing in the direction of the fibril axis (4.8 Å) gives rise to the meridional reflection, and the intersheet spacing (10–11 Å) is perpendicular to this and produces the equatorial diffraction.

reflection could be due to a predominance of parallel β strands or to the fact that the 2_1 symmetry of an antiparallel β sheet gives rise to the systematic absence of this reflection.

Although evidence shows that a residual α-helical structure is present in some amyloid fibrils,[13,14] fiber diffraction evidence does not support the idea that α helices play a significant part in the ordered structure of the fibrils. If the fibrils contained a large amount of an ordered, repeating α-helical structure, this would give rise to reflections at 5.4 and 1.5 Å, corresponding to the helical pitch and rise per residue, and these reflections are not observed in any of the amyloid fiber diffraction patterns that have been examined.

The reflections at approximately 4.7 Å on the meridian and 10 Å on the equator are seen in all amyloid fiber diffraction patterns and submit to the same interpretation, with some slight variation allowed in the precise spacings of the two reflections, to reflect the structural characteristics of different amyloids. For example, calcitonin fibrils grown *in vitro*[9] and insulin fibrils produced by heating and cooling the protein under acidic conditions both give fiber diffraction images exhibiting strong, sharp meridional reflections at 4.76 Å and more diffuse equatorial reflections at approximately 10 Å.[6] Purified amyloid cores isolated from senile plaques associated with

[13] J. D. Termine, E. D. Eanes, D. Ein, and G. G. Glenner, *Biopolymers* **11,** 1103 (1972).

[14] K.-M. Pan, M. A. Baldwin, J. T. Nguyen, M. Gasset, A. Serban, D. Groth, I. Melhorn, Z. Huang, R. J. Fletterick, F. E. Cohen, and S. B. Prusiner, *Proc. Natl. Acad. Sci. U.S.A.* **90,** 10962 (1993).

Alzheimer's disease show reflections at 4.76 Å and approximately 10.6 Å[7] and prion rods exhibit a prominent interstrand spacing at 4.72 Å and an equatorial spacing of 8.82 Å.[10] High-resolution fiber diffraction studies of transthyretin amyloid have mainly been performed on fibrils composed of the Val30Met variant TTR.[11,15,16] Many studies of synthetic fibrils formed from peptides corresponding to fragments of the Aβ peptide[17–20] and the prion protein[10,21–23] have been reported. These have used fiber diffraction to characterize the nature of fibrils formed from various fragments[17–20,24–28] and to determine the effect of pH, solution ions, and residue charge on fibril structure.[19,26,28] All of these patterns are consistent with a cross-β structure, and the commonalities observed in the meriodional diffraction patterns over the medium- and high-angle regions can only occur if all of these amyloid fibrils have well-defined and closely similar molecular structures. Fiber diffraction data indicate that different amyloid fibrils actually share a common molecular skeleton, the "protofilament" core structure, which is a continuous β-sheet helix.[11,29] A model of the core structure of the protofilament was constructed using the fiber diffraction pattern given by Val30Met transthyretin amyloid fibrils.[11] This model consists of four β sheets, hydrogen bonded in the direction of the fiber axis, with the β strands running perpendicular and the sheets twisting around a central

[15] C. Terry, A. M. Damas, P. Oliveira, M. J. M. Saraiva, I. L. Alves, P. P. Costa, P. M. Matias, Y. Sakaki, and C. C. F. Blake, *EMBO J.* **12,** 735 (1993).

[16] A. Damas, M. P. Sebastião, F. S. Domingues, P. P. Costa, and M. J. Saraiva, *Amyloid Int. J. Exp. Clin. Invest.* **2,** 173 (1995).

[17] D. A. Kirschner, H. Inouye, L. Duffy, A. Sinclair, M. Lind, and D. A. Selkoe, *Proc. Natl. Acad. Sci. U.S.A.* **84,** 6953 (1987).

[18] P. Gorevic, E. Castano, R. Sarma, and B. Frangione, *Biochem. Biophys. Res. Commun.* **147,** 854 (1987).

[19] P. E. Fraser, J. T. Nguyen, W. K. Surewicz, and D. A. Kirschner, *Biophys. J.* **60,** 1190 (1991).

[20] H. Inouye, P. E. Fraser, and D. A. Kirschner, *Biophys. J.* **64,** 502 (1993).

[21] J. H. Come, P. E. Fraser, and P. T. Lansbury, *Proc. Natl. Acad. Sci. U.S.A.* **90,** 5959 (1993).

[22] F. Tagliavini, F. Prelli, L. Verga, G. Giaccone, R. Jarma, P. Gorevic, B. Ghetti, F. Passerini, E. Ghibaudi, G. Forloni, M. Salmona, O. Bugiani, and B. Frangione *Proc. Natl. Acad. Sci. U.S.A.* **90,** 9678 (1993).

[23] H. Inouye and D. A. Kirschner *J. Mol. Biol.* **268,** 375 (1997).

[24] K. Halverson, P. E. Fraser, D. A. Kirschner, and P. T. Lansbury, *Biochemistry* **29,** 2639 (1990).

[25] C. B. Caputo, P. E. Fraser, I. E. Sobel, and D. A. Kirschner, *Arch. Biochem. Biophys.* **292,** 199 (1992).

[26] P. E. Fraser, J. T. Nguyen, D. T. Chin, and D. A. Kirschner, *J. Neurochem.* **59,** 1531 (1992).

[27] P. E. Fraser, J. T. Nguyen, H. Inouye, W. K. Surewicz, D. J. Selkoe, M. B. Podlisny, and D. A. Kirschner, *Biochemistry* **31,** 10716 (1992b).

[28] P. E. Fraser, D. R. McLachlan, W. K. Surewicz, C. A. Mizzen, A. D. Snow, J. T. Nguyen, and D. A. Kirschner, *J. Mol. Biol.* **244,** 64 (1994).

[29] M. Sunde, L. C. Serpell, M. Bartlam, M. B. Pepys, P. E. Fraser, and C. C. F. Blake, *J. Mol. Biol.* **273,** 729 (1997).

axis such there is a 15° twist between one strand and the adjacent one (see Ref. 11). This underlying cross-β substructure appears to be the building block of all different amyloid fibrils.[23]

Several reports of fiber diffraction studies of *ex vivo* amyloid samples have noted the presence of lipid or other contaminants in samples that give rise to non-cross-β reflections. These reflections are usually characterized as being of different appearance from the "amyloid" reflections (lipid-derived reflections are usually sharper than amyloid-derived reflections) and as being variable in intensity between one fiber sample and another. It is also possible to reduce the intensity of the foreign reflections by repeated washing of the amyloid fibrils.[2,16] Synthetic amyloid fibrils formed *in vitro* from pure protein components do not exhibit any anomalous or foreign reflections.[30,31] The presence of lipid in samples most commonly results in characteristic reflections at 4.13[2,16] and 2.59 Å.[11] Fiber diffraction has also been used to study ligand and fibril interactions. Sulfate binding to Alzheimer Aβ peptide-derived fibrils gives rise to an intense 65-Å meridional reflection that presumably arises from the periodic deposition of sulfate along the long axis of the fibrils.[26] This approach may give insight into the nature of binding of other ions and interacting molecules such as Congo red.

Sample Preparation

From ex Vivo Tissue Material

Amyloid fibrils are usually isolated and purified according to the methods of Pras *et al.*[32] and Nelson *et al.*[33] Lipid can be removed using the method of Damas and colleagues,[16] and the salt concentration in fibril samples should be minimized.

From Synthetic Peptides

Peptides homologous to various regions of the Aβ peptide,[19,34,35] islet-associated polypeptide IAPP,[36–38] transthyretin,[30,39] and other proteins have

[30] J. A. Jarvis, D. J. Craik, and M. C. J. Wilce, *Biochem. Biophys. Res. Commun.* **192,** 991 (1993).

[31] L. C. Serpell, D.Phil. Thesis, University of Oxford (1995).

[32] M. Pras, M. Schubert, D. Zucker-Franklin, A. Rimon, and E. Franklin, *J. Clin. Invest.* **47,** 924 (1968).

[33] S. Nelson, M. Lyon, J. Gallager, E. Johnson, and M. B. Pepys, *Biochem. J.* **275,** 67 (1991).

[34] C. Hilbich, B. Kisters-Woike, J. Reed, C. Masters, and K. Beyreuther, *J. Mol. Biol.* **218,** 149 (1991).

[35] A. Lomakin, D. S. Chung, G. B. Benedek, D. A. Kirschner, and D. B. Teplow, *Proc. Natl. Acad. Sci. U.S.A.* **93,** 1125 (1996).

[36] S. B. P. Charge, E. J. P. De Koning, and A. Clark, *Biochemistry* **34,** 14588 (1995).

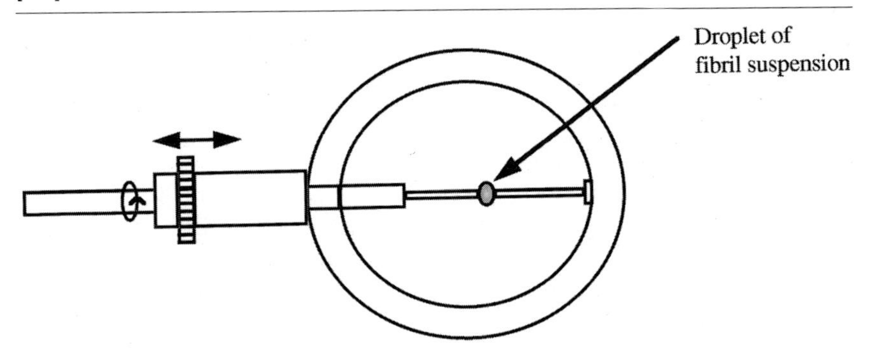

FIG. 3. Stretch frame apparatus used to prepare fibril samples. The droplet of fibril suspension is placed between the ends of two glass capillaries, one of which is attached to a screw. The capillaries can therefore be drawn apart slowly and in small increments to encourage gradual alignment of fibrils as the droplet dries.

been shown to form amyloid-like fibrils *in vitro*. The peptides are generally synthesized using conventional solid-phase synthesis methods and purified by reversed-phase high-performance liquid chromatography (HPLC). Depending on the nature of the amyloidogenic sequence, the peptides may be dissolved in water or in a range of organic solvents. The nature of the solvent can be manipulated to control the rate and extent of fibril formation. Reducing the rate of fibril formation, or imposing a need for slow concentration of the sample, may facilitate the alignment of fibrils, thereby improving the subsequent quality and information content of the diffraction pattern. The use of buffer solutions should be minimized, as salt crystals give strong diffraction spots that can overwhelm the amyloid fiber diffraction image.

Alignment of Fibrils

Stretch Frame Method. Glass capillary tubes of 1 mm diameter (Clark Electromedical Instruments, Reading, UK) are cut into lengths of 2–3 cm and prepared by dipping the tube ends into melted beeswax to form a plug in the mouth of the tube of about 1 mm in length. The capillaries are placed at opposite ends of the stretch frame (see Fig. 3) and secured with plasticine. A suspension of fibrils extracted from tissue or formed *in vitro* in a buffer of low salt concentration or in water is prepared at a concentration of 5–20 mg/ml. An aliquot (10–20 μl) of the amyloid fibril suspension is placed

[37] T. T. Ashburn and P. T. Lansbury, *J. Am. Chem. Soc.* **115,** 11012 (1993).
[38] P. Westermark, U. Engstrom, K. Johnson, G. Westermark, and C. Betsholtz, *Proc. Natl. Acad. Sci. U.S.A.* **87** 5036 (1990).
[39] A. Gustavsson, U. Engstrom, and P. Westermark, *Biochim. Biophys. Acta* **175,** 3 (1991).

between the two capillaries. The wax serves to maintain the surface tension on the droplet. The droplet is allowed to dry at room temperature, and during drying the distance between the ends of the capillaries is increased slowly, by small increments, to stretch out the fibrils and encourage alignment.

Magnetic Alignment of Synthetic Amyloid Fibrils. Quartz or glass capillary tubes (0.7 or 1 mm diameter, W. Müller, D-1000 Berlin 27, Germany) are treated with Sigmacote (Sigma, St. Louis, MO) or any general siliconizing reagent. Initially, capillaries are washed with concentrated HCl, water, and methanol and allowed to dry (60°), rinsed thoroughly with undiluted siliconizing reagent, and dried in an oven at 80–100°.

A solution of amyloid forming peptide in distilled water (10–20 mg/ml) is drawn up into a siliconized glass capillary tube to give a length of solution of 2–4 mm. The capillary tube is sealed at one end using wax and placed into a 2-T magnet (Charles Supper Co.) so that the bottom of the fibril solution lies between the two poles of the magnet (see Fig. 4). The solution is allowed to dry gradually under ambient conditions of humidity and temperature. Under some solvent conditions, e.g., with hexafluoro-2-propanol present, the sample may be maintained in a humidity chamber to ensure that some hydration is retained. When it has dried to a disk the capillary tube is sealed with wax. The high degree of orientation introduced in these samples often gives rise to birefringence (Fig. 5).

Fiber Diffraction Data Collection

Fiber diffraction images from most amyloid fibril samples can be collected from any in-house source of X rays equipped with a suitable detector such as film or a MAR research image plate. However, the information content of X-ray diffraction images of amyloid fibrils is strongly dependent on relative fibril orientation, and the amount of diffracting material posi-

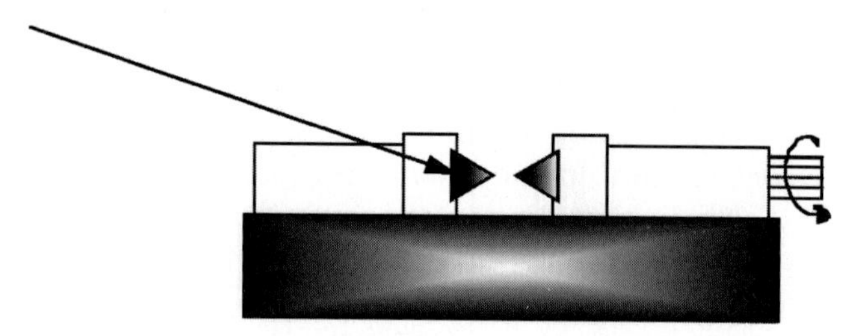

Fig. 4. Magnetic alignment of fibril samples. A 2-T magnet is used to align fibrils and synthetic peptides as they dry in the field. The poles of the magnet can be moved in and out to allow placement of the capillary tube. From Charles Supper Co., Inc.

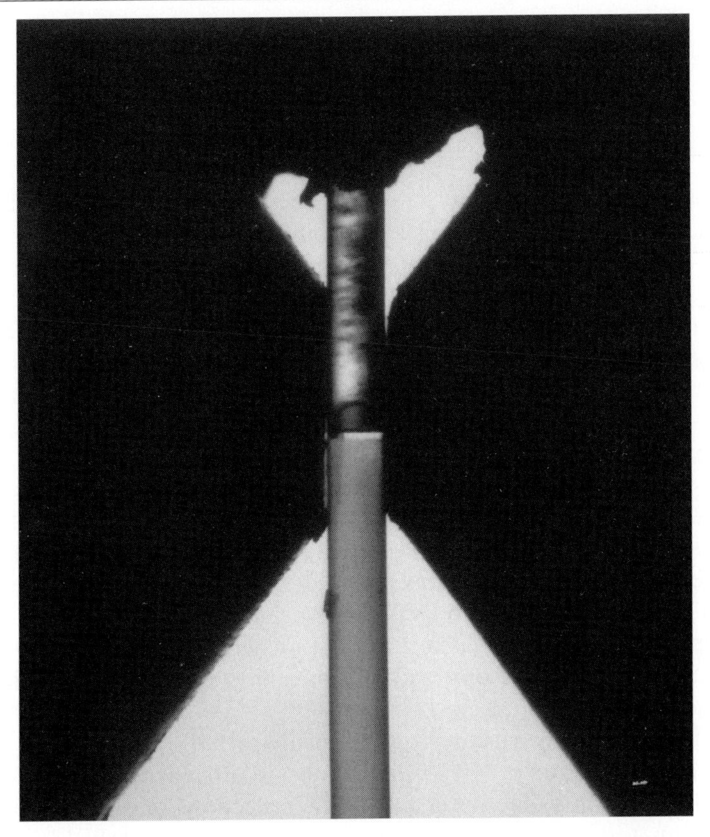

FIG. 5. Birefringence observed in a sample of highly orientated fibrils. A high concentration solution of a synthetic peptide is placed in a siliconized glass capillary maintained in the presence of a magnetic field and it is slowly allowed to dry as the amyloid fibrils form. As the fibrils align, a strong birefringence is observed in the specimen under cross-polarized light.

tioned within the X-ray beam and fibril samples from *ex vivo* sources tend to be poorly aligned and may only exhibit the strong 4.7-Å reflection and diffuse 8- to 12-Å reflections with in-house sources. High-resolution data from these relatively poorly aligned samples will only be obtained from the high brilliance, coherent beam lines available at synchrotron sources such as Daresbury, United Kingdom (http://www.dl.ac.uk), the ESRF in Grenoble, France (http://www.esrf.fr), and APS/Biocat at Illinois Tech. Air scatter may also pose a problem with in-house sources of X rays, which are usually generated from Cu Kα rotating anodes and have a wavelength of 1.5 Å. The use of X-ray beams with shorter wavelengths at synchrotrons removes this problem, which is particularly advantageous for the collection of low-angle data. Alternatively, a helium chamber may be fitted between the specimen and the detector to remove air-scatter problems.

Data Interpretation

Amyloid samples are usually exposed to the X-ray beam with the long axis of the fibrils more or less perpendicular to the direction of the beam. The X-ray reflections are then distinguished by their direction with reference to the fiber axis and their distance from the center of the pattern: meridional reflections are defined as those lying parallel to the fiber axis, whereas equatorial reflections are those positioned at right angles to the fiber axis. If the long axes of the individual fibrils within a bundle of fibrils, or the bundles themselves, deviate from the mean fiber axis, the corresponding reflections will be drawn out into arcs whose angular dispersion is related to the relative dispersion within the fibrillar bundles. In the most extreme unaligned case, where fibrils lie at all possible angles, the reflections are drawn out into complete rings, giving no indication as to whether they are meridional or equatorial. Although it is advantageous to obtain diffraction data from well-aligned samples, all that is required in the first instance is for the meridional and equatorial reflections to be separated, having arcs as opposed to rings. This should be possible for most amyloid fibril specimens.

Data can be viewed using programs such as IPDISP,[40] Mosflm,[41] CCP13 program FIX,[42] Profida,[43] and others. For partially aligned diffraction patterns the first step is measurement of the reflections on the meridian and on the equator. The presence of reflections at approximately 4.7 and 10 Å that are characteristic of the cross-β pattern is obviously essential. The presence of a 2.4-Å reflection on the meridian is common when the specimen diffracts well, as this is the harmonic of the 4.7- to 4.8-Å reflection. Measurement of other reflections on the meridian will allow indexing of these reflections to a unit cell, which is basically the smallest common factor of all of the reflections. In the case of TTR amyloid,[11] 115.5 Å was found to be the repeating unit. Following this we found that many other amyloid fibrils diffraction patterns show the same repeating unit.[29] Interpretation of equatorial reflections is more problematic and requires more in-depth study. When amyloid samples produce X-ray diffraction patterns with a

[40] CCP4-Collaborative Computational Project. Number 4. *Acta Cryst. D50,* 760 (1994). CCP4 http:/gserv.dl.ac.uk/CCP/CCP4.

[41] Mosflm, contact A. G. W. Leslie, MRC Laboratory of Molecular Biology, Cambridge, UK.

[42] CCP13 http://wserv.dl.ac.uk/SRS/CCP13.

[43] M. Lorenz and K. C. Holmes, *J. Appl. Crystallogr.* **26,** 82 (1993).

range of well-defined equatorial reflections, it is possible to analyze the substructure of the fibrils further to gain insight into the arrangements of protofilaments. Burge[44,45] has shown that a short-range order in bundles of filaments gives rise to a series of "non-Bragg" reflections whose spacings are dependent on the center-to-center separation, a, of the filaments in the bundle. For filaments in contact, the value of a corresponds to the diameter of the filament. The center-to-center separation of fibrils may be influenced by sample preparation; dehydration of samples may distort the native protofilament packing observed in cross sections of embedded fibrils and give rise to different lateral organization of protofilaments. Use of the technique on the Val30Met transthyretin amyloid fibrils from familial amyloidotic polyneuropathy[11] gives a diameter of 64 Å for the protofilament as compared to approximately 60 Å from electron microscopy.[46]

When a highly aligned fiber diffraction specimen reveals a well-oriented pattern, the CCP13 program suite[42] can be applied for indexing, profile fitting, and integration of the diffraction spots. For example, magnetically aligned Aβ peptides revealed extremely well-oriented fiber diffraction patterns with Bragg spacing along layer lines.[20] Inouye and Kirschner[23] have analyzed these patterns in detail, along with fiber diffraction patterns obtained from prion-related fibrils. They have identified so-called "beta (β) crystallites," which constitute the protofilaments, have determined unit cells for these crystallites, and have studied the close packing of the protofilaments in the fibrils.[23,47] Given these very well-oriented patterns, it is possible to estimate particle size by measuring the breadth of reflections. This has been achieved for PrP synthetic fibers[23] from which a measurement of the width of an equatorial reflection at 9.04 Å of 0.02274 $Å^{-1}$ corresponds to a coherence length of 44 Å (see Refs. 23 and 29).

Concluding Remarks

Use of fiber diffraction has shown similarities in the core structure of amyloid. Further improvements in the alignment of fibers, using magnetic fields, could lead to more detailed models of fibril protofilament. Information to be gained by use of this technique is limited by the rotational averaging that occurs and which has the effect that data can only be interpreted to the level of the ordered backbone structure. Details of side chain

[44] R. E. Burge, *Acta Crystallogr.* **12,** 285 (1959).
[45] R. E. Burge, *J. Mol. Biol.* **7,** 213 (1963).
[46] L. C. Serpell, M. Sunde, P. E. Fraser, P. K. Luther, E. Morris, O. Sandgren, E. Lundgren, and C. C. F. Blake, *J. Mol. Biol.* **254,** 113 (1995).
[47] S. B. Malinchik, H. Inouye, K. E. Szumowski, and D. A. Kirschner, *Biophys. J.* **74,** 537 (1998).

packing can only be obtained from specimens that exhibit the preferred orientation of fibrils, with ordered three-dimensional packing of the constituent molecules, and therefore give rise to single crystal-like diffraction patterns with Bragg reflections. These relatively ordered specimens are the exception and are unlikely in disease-state amyloid samples. However, the combination of molecular spacing data obtained from X-ray fiber diffraction patterns, with phase information from electron microscopic analysis, now promises to yield important insights into the structure of amyloid.

Acknowledgments

The authors thank Colin Blake for essential involvement with the fiber diffraction studies of amyloid in Oxford. We are grateful to Colin Nave (SRS, Daresbury Laboratory), Trevor Greenhough (SRS, Daresbury Laboratory and Keele University), and Trevor Forsyth (Keele University) for advice on sample preparation and data collection; Karl Harlos (Oxford) and Bjarne Rasmussen (ESRF, Grenoble) for help with data collection; and Richard Denny (Daresbury Laboratory and Imperial College), John Squire (Imperial College), and Mark Bartlam (LMB, Oxford) for advice on data analysis. LCS acknowledges the Medical Research Council (UK) for support. PEF was supported by the Alzheimer's Society of Ontario and the Ontario Mental Health Foundation. MS was supported by the Medical Research Council (UK), the E.P.A. Cephalosporin Junior Research Fellowship from Lady Margaret Hall (Oxford), and The Oxford Centre for Molecular Sciences, which is funded by the BBSRC, EPSRC, and MRC.

[35] Solid-State Nuclear Magnetic Resonance of Protein Deposits

By DAVID WEMMER

The central problem in applying physical methods to the study of protein deposits is the intermediate degree of order. The molecules aggregate into assemblies sufficiently large that the overall tumbling correlation time becomes very long, making normal "solution" nuclear magnetic resonance (NMR) line widths very large. However, the long-range order is not sufficient to give high-resolution diffraction data, although fiber diffraction data have been very useful in understanding the basic features of amyloid aggregates. Spectroscopic methods such as Fourier transform infrared spectroscopy (FTIR) can be used, but cannot generally give accurate structural information, especially on a site-resolved basis. One of the few methods that can still be applied, and can yield detailed local structural information,

0076-6879/99 $30.00

is solid-state NMR. There are a variety of experiments that have been applied in peptide systems, producing useful information; however, these techniques are demanding both of the spectrometers and sample preparation methods. This article describes the major approaches, with examples of work on protein and peptide aggregates. The challenges of working on larger systems are also discussed.

In solution NMR spectra there are just two parameters to consider for each spin: the chemical shift and spin–spin (often just called "J") couplings. The very rapid tumbling motion of molecules (tumbling correlation times $\tau_C \approx 10$ nsec at 15 kDa, with τ_C scaling linearly with molecular weight) leaves just the isotropic average of these interactions and reasonably sharp resonances. However, in solid samples, in which molecular motion is generally very limited, the full anisotropic nature of couplings must be considered. These provide the desired structural information but also force complexity in designing experiments to separate the various interactions. As a first step in understanding the experiments needed to analyze peptide structures, each of the relevant interactions is discussed in the following section. Experiments to separate the interactions and the structural interpretation of the measured parameters will then be presented with some examples from studies of protein deposits.

Nuclear Magnetic Resonance Interactions

For the present discussion a basic understanding of NMR is assumed: the phenomena of nuclear spin and its quantization, the splitting of energy levels on interaction with a magnetic field, and the spectroscopic observation of transitions between levels either through a direct absorption experiment or through use of short radio frequency (r.f.) pulses and a Fourier transform. There are many texts that can provide a good description of these basic principles.[1] Although many nuclei can be observed by NMR, discussion is restricted to those that are commonly present in proteins, and are most useful for solid-state NMR, namely ^{1}H, ^{13}C, and ^{15}N. Although spectrometers operating at many frequencies are used for solid-state NMR, there are usually compromises that favor moderate fields, e.g., 400 MHz frequency for ^{1}H, which corresponds to 100 MHz for ^{13}C and 40 MHz for ^{15}N, although a trend toward higher fields has become clear over recent years.

[1] T. C. Farrar and E. D. Becker, "Pulse and Fourier Transform NMR." Academic Press, New York, 1971.

Chemical Shielding

Chemical shielding occurs because electrons in bonds surrounding the observed nuclei react to the applied magnetic field, reducing the field actually present at the nucleus. The magnitude of the reaction from the electrons depends on their density around the nucleus and also on constraints from the orbitals that they occupy. These features are intrinsically quite anisotropic, i.e., the chemical shift depends strongly on the relative orientation of a molecule and the direction of the applied magnetic field. Formally, this is described as a second rank tensor interaction. To fully describe the chemical shielding, three principal values of the coupling tensor are required, generally called σ_{11}, σ_{22}, and σ_{33}. Any orientation of the field can be specified with three Euler angles, and then the chemical shift for that orientation can be calculated.[2] In a solid sample for which all the molecules are aligned in one direction (e.g., a single crystal), a single resonance will be seen at the chemical shift corresponding to that orientation. However, if the molecules are oriented randomly, as in a powder or amorphous solid, then all angles will be present. The observed line will be broad, covering the full range from σ_{11} to σ_{33}. The line shape is determined by the probability of finding molecules at each chemical shift, the result being called a powder pattern (Fig. 1). The chemical shift observed in solution is the average of the principal values of the shielding tensor, $\sigma_i = 1/3(\sigma_{11} + \sigma_{22} + \sigma_{33})$. The width of the powder pattern for a spin in a single asymmetric environment can be as large as the range of isotropic shifts for all spins of that type. For example, the carbonyl carbon anisotropy ($\sigma_{33} - \sigma_{11}$) is about 150 ppm, where the range of typical isotropic shifts is about 180 ppm.

Line Shapes and Magic Angle Spinning

Observing powder patterns in solid samples is feasible. However, if many different chemical environments are present for the observed nucleus, then overlap of the powder patterns becomes a problem, and it can be difficult to resolve and correctly associate the different tensor values with one another. In addition, having the signal spread over a wide range of shifts lowers sensitivity relative to a solution experiment in which all spins have the same frequency. To regain the simplicity of solution spectrum, or nearly so, magic angle sample spinning can be applied. In considering the orientation dependence of the chemical shielding, it was realized long ago that mechanical rotation of the sample led to a reduction in width of the powder pattern, the scaling going as $1/2(3\cos^2\theta - 1)$, where θ is the angle

[2] M. Mehring, "Principles of High-Resolution NMR in Solids," 2nd ed. Springer, Berlin, 1983.

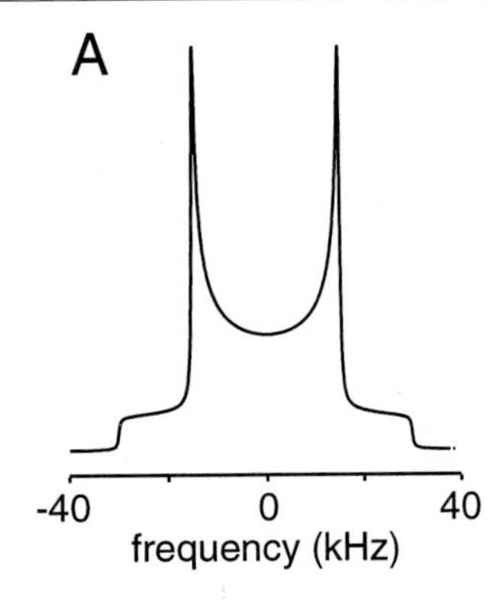

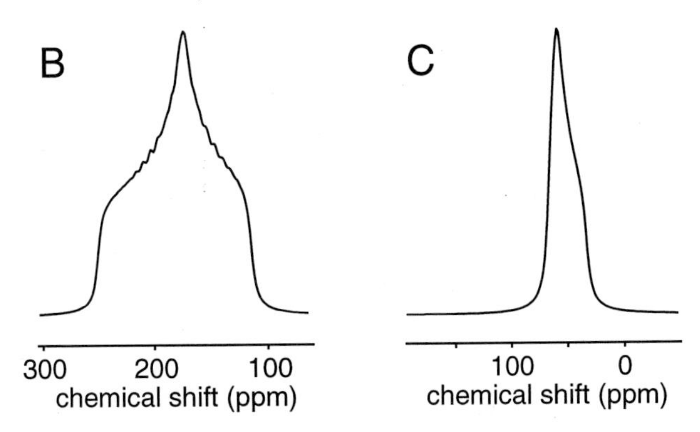

FIG. 1. Simulated powder spectra for (A) a dipolar coupling of 30 kHz, the approximate value for a directly bonded ^{13}C–^{1}H coupling; (B) chemical shift anisotropy of a carbonyl carbon (σ_{11} = 243 ppm, σ_{22} = 184 ppm, σ_{33} = 107 ppm); and (C) chemical shift anisotropy of an α carbon (σ_{11} = 65 ppm, σ_{22} = 57 ppm, σ_{33} = 32 ppm). Each simulated pattern was broadened by 500 Hz.

between the rotation axis and the magnetic field. Setting θ to 54.7° (the magic angle) leads to a width of zero, and observation of a peak at the average chemical shift σ_i just as in solution. The rotation leads to a modulation of the chemical shift, producing side bands on the isotropic peak, spaced at the rotation frequency. The intensity of the side bands follows the intensity of the powder pattern at the spot where the side band occurs. For slow spinning rates, many side bands occur within the powder pattern width, but for high rates all but the central band can be eliminated. With modern probes, spinning rates of over 20 kHz (1.2 million rpm) can be achieved fairly routinely and is enough to eliminate all side bands except in the broadest carbon spectra for which a couple of side bands will remain. As described further in a following section, the tensor values can be recovered from the spinning spectra if desired. The spectra with magic angle spinning (MAS) recover both sensitivity and resolution lost in powder spectra.

Dipolar Couplings

In solution, only isotropic couplings transmitted through bonds need to be considered. These couplings are not very strong, direct $^1H–^{13}C$ couplings being ≈140 Hz, $^1H–^{15}N$ ≈ 90 Hz, and typical three bond $^1H–^1H$ couplings ≤10 Hz. However, in solids one must also consider direct dipole–dipole couplings, which are generally much larger. The magnetic field generated by each magnetic dipole adds to the applied field at the sites of nearby spins. This dipolar field, like chemical shielding, is orientation dependent and is also dependent on the distance between the spins. The magnitude of the coupling between two spins I and S goes as $D_{IS} \propto \gamma_I\gamma_S(1 - 3\cos^2\theta)/r_{IS}^3$, where γ_I and γ_S are the gyromagnetic ratios of the spins I and S, θ is the angle between the internuclear vector and the magnetic field, and r_{IS} is the distance between spins I and S. With isotropic tumbling of molecules in solution, this coupling averages exactly to zero and hence does not affect the solution spectrum directly. However, dipolar couplings in solids can be very large, reaching about 40 kHz for two protons. If just two spins are coupled, then the orientation dependence gives rise to a powder pattern, similar to that for chemical shifts, although with axial symmetry (equivalent to $\sigma_{11} = \sigma_{22}$), and superimposed on the spectrum reflected about the center (Fig. 1). The symmetry axis corresponds to the internuclear vector, which is the bond direction if the atoms are bonded to one another (e.g., a $^1H–^{13}C$ or $^1H–^{15}N$ pair).

A factor that makes homonuclear dipolar couplings quite different from chemical shifts comes from the spin operator part of the Hamiltonian describing the coupling. The chemical shift involves just the z axis component of the spin, just the spin operator I_z appearing in the Hamiltonian. However, for a dipolar coupling, the spin operator part involves both spin I and S as $I_zS_z - 1/4(I_+S_- + I_-S_+)$. The latter term is often referred to as a "flip-flop" term because it can induce a transition from a state where I is up and S is down to one with I being down and S being up. When many spins are present, each coupled to many others, this term leads to homogeneous broadening, a situation in which any particular spin does not really respond at an individual frequency, but rather responds over a range of frequencies. The line shape resulting in this case is usually a Gaussian shape, with a width of approximately the largest dipolar coupling present. Thus, although the dipolar coupling has structural information of interest, the distance entering in the $1/r^3$ dependence, this is usually lost through the general broadening from all of the other spins present.

Heteronuclear Decoupling

The flip-flop terms are only energy conserving and hence only have an effect for spin pairs in which spins are of the same type, e.g., ^1H–^1H or ^{13}C–^{13}C. However, when the proton in a ^1H–^{13}C pair is coupled to other protons, the spectrum of the ^{13}C will be broadened by the flip flops of the protons. Hence ^{13}C spectra in the presence of many protons are also broad Gaussian peaks. Because different nuclear types have rather different frequencies, it is possible to use radio frequency pulses to do "heteronuclear spin decoupling." Irradiation of the protons, for example, while observing ^{13}C can remove all effects of the ^1H–^{13}C couplings. The decoupling r.f. field must be strong, causing rotation of the protons at 50 kHz or more (preferably $\geq$100 kHz to really remove all effects). Under proton decoupling, ^{13}C or ^{15}N then gives chemical shielding powder pattern line shapes as discussed earlier. In principle, because dipolar couplings have the same angular dependence as chemical shielding, they too can be removed by magic angle spinning. However, the spinning rate must be near the value of the strongest couplings before narrowing of the peaks is seen, unlike the chemical shielding which gives side bands even for slow spinning rates. The difference is that the dipolar couplings are homogeneous. Because ^1H–^1H couplings can be as much as 40 kHz (faster than even the best spinners can achieve), proton decoupling must be used with MAS to obtain sharp spectra for ^{13}C or ^{15}N. If there are ^{13}C–^{13}C, ^{15}N–^{15}N, or ^{13}C–^{15}N couplings, these are much weaker (due to lower γs and longer distances) and will be removed by MAS. If information from them is desired, then they must be "reintroduced," using methods discussed in later sections.

Homonuclear Decoupling

Homonuclear decoupling (removing ^1H–^1H couplings) using pulses can also be achieved, although it is more difficult than heteronuclear decoupling. Because sufficiently rapid physical rotation of the sample cannot be accomplished for homonuclear decoupling, the spins are rotated in "spin space" through the use of r.f. pulses. The basic idea is that if the spins spend an equal amount of time pointed along the x, y, and z axes, then on the average it is equivalent to their pointing along a vector in the (1,1,1) direction, which is at the magic angle with respect to the magnetic field along z. The transfer of spins between axes must occur at a rate faster than the magnitude of the dipolar coupling to cause efficient averaging, but this can be accomplished with strong r.f. pulses (pulse powers up to kW) with short delays between them. Phase cycling is important to remove pulse artifacts and errors, leading to sequences such as MREV8[2] for practical use. These sequences can be combined with MAS to remove the shielding anisotropy. For protons, even the best pulse sequences leave residual broadening of several tenths of a ppm. Because chemical shift ranges of protons are smaller than ^{13}C or ^{15}N in ppm, and protons are abundant, this leads to low effective resolution. When measurements of ^{1}H–^{13}C or ^{1}H–^{15}N couplings are desired, then homonuclear ^{1}H–^{1}H decoupling must be applied while observing ^{13}C or ^{15}N, removing the effects of the proton flip-flop terms. The heteronuclear ^{1}H couplings are somewhat reduced by the pulse sequence, but by a predictable amount. Under such conditions a coupling to a single nearby proton can be measured accurately, even if there are a moderate number of more distant ^{1}H couplings.

Extracting Structural Parameters: Chemical Shifts

Chemists interpret chemical shifts to give information about molecular structure. The value of the shift for any nucleus is sensitive to the bonding arrangement near it, fortunately including the values of dihedral angles. Quite some time ago it was realized that the shift of the H_α of an amino acid was downfield relative to the random coil value if that amino acid was in a β sheet, but was upfield if the residue was in a helix. When uniform isotope labeling became common for proteins, solution studies showed that shifts of the C_α and C_β were also sensitive to local conformation.[3] By combining the direction and magnitude shifts for α and β carbons to make an "index," it is possible to very reliably map regions of secondary structure using just the shifts.[4] Although the principles of this were established using proteins in solution, using isotropic shifts from solids is equally valid. Peptides from the prion protein have been shown to have different conformations when dried from different solvents using this approach (Fig. 2) Amino acids that take on more than one conformation can also be identified. The measurements are simple, just requiring a MAS ^{13}C spectrum and measurement of the peak centers. Identification (assignment) of peaks is difficult at natural abundance (Fig. 3), but with peptides enriched at single

[3] D. S. Wishart and B. D. Sykes, *Methods Enzymol.* **239,** 363 (1994).
[4] D. S. Wishart and B. D. Sykes, *J. Biomol. NMR* **4,** 171 (1994).

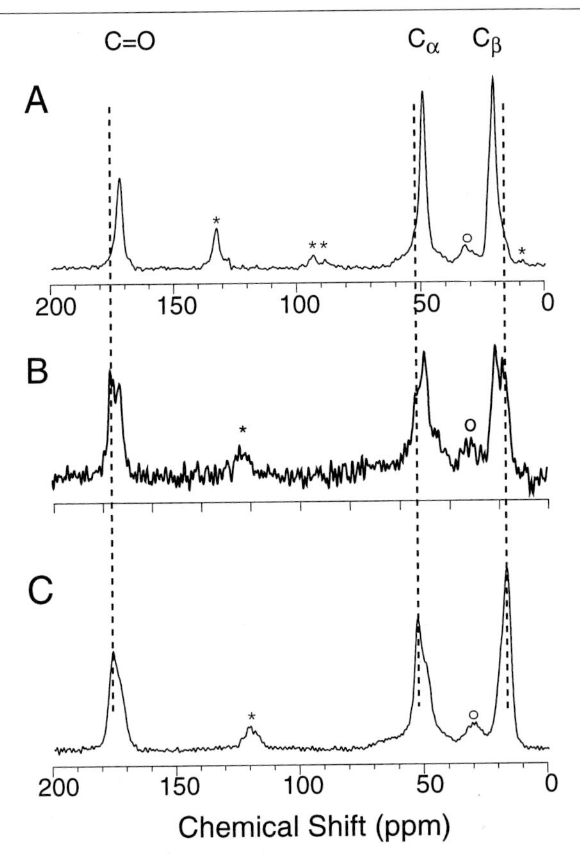

FIG. 2. Magic angle spinning ^{13}C spectra of the prion H1 peptide (numbering from the intact protein: M_{109} KHMAGAAAAGAVV_{122}), labeled at Ala-115 with equal amounts of label in the C=O, C_α and C_β, giving resonances as labeled. For (A) the peptide was lyophilized from 50% acetonitrile/50% water (a β-sheet promoting solvent); in (B) the sample was lyophilized from hexafluoro-2-propanol (a helix promoting solvent), then exposed to air at 30% relative humidity for 30 min.; and in (C) the sample was lyophilized from hexafluoro-2-propanol and kept dry. All spectra were taken with cross-polarization (2 msec transfer from protons), with a 1.5-sec delay between scans. The samples were spun at about 3 kHz. Peaks with an asterisk are spinning side bands, whereas those with a circle are natural abundance signals. Vertical dashed lines indicate the positions of the β-sheet form.

sites, sensitivity is high (measurements can be done with a few milligrams of peptide) and assignment is obvious.

While the isotropic chemical shift can categorize residues into helix or sheet, this single parameter cannot further define the conformation. However, if all three tensor elements are determined for $C_\alpha(\sigma_{11}, \sigma_{22}$, and

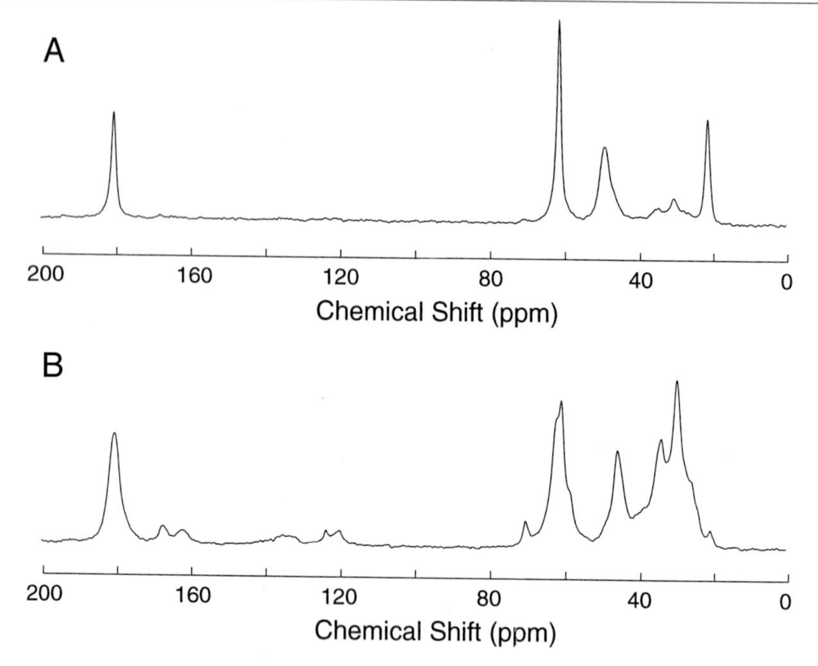

FIG. 3. ^{13}C MAS spectra are shown for the leucine zipper portion (a 33 amino acid peptide) of the GCN4 protein. ^{13}C labels were included for C9 C=O, L19 C_α, A24 C_β, and G31 C_α. Spectrum (A) is a fully labeled peptide, the peaks from labeled sites are prominent, and natural abundance peaks are visible only near 30 ppm. In (B) the labeled peptide has been diluted to 1% in natural abundance peptide; the difficulty in selecting out labeled peaks under these conditions is clear. Spectrum (A) is the result of 2048 scans, whereas (B) represents 53,248 scans. Both samples were spinning at about 10 kHz.

σ_{33}), the backbone angles ϕ and ψ can be determined with reasonable accuracy. With a single label, and a simple molecule so that natural abundance signals do not interfere, the tensor elements can be measured from a simple nonspinning powder spectrum. However, these conditions are often not met, but there are several measurement techniques that allow recovery of the tensor elements even while spinning. The simplest is to measure the intensities of the spinning side bands under fairly slow spinning conditions and fitting them with equations relating them to the tensor elements, as shown by Herzfeld and Berger.[5] The needed spectra are just one dimensional and can be collected with good sensitivity. Overlap of side bands with different peaks or side bands of them can be a problem, but if several different spinning rates are used and data are fitted together, fairly

[5] J. Herzfeld and A. E. Berger, *J. Chem. Phys.* **63,** 6021 (1980).

accurate tensor values can be obtained. Another approach is to use switched angle spinning. In this method the anisotropic chemical shifts evolve during a period in which the spinning angle is adjusted away from the magic angle value.[6] For the detection period the angle is jumped back to the magic angle for highest resolution, giving a two-dimensional (2D) spectrum. The tensor elements are then obtained by fitting the line shape in the ω_2 dimension. It is also possible to use 180° pulses to interfere with the averaging during the indirect time period, then detect without them to get high resolution during detection.[7] A newer version of the sequence gives improved analysis.[8] An alternative method, VACSY, uses the rotor angle as one Fourier variable to correlate isotropic and anisotropic shifts,[9] again in a 2D spectrum.

Unlike the use of an empirical index for interpretation, to get ϕ and ψ from the tensor elements requires extensive calculations. Laws and coworkers[10] have shown that *ab initio*-calculated chemical shieldings can be quite accurate. To interpret tensor elements, a set of calculations is done, with ϕ and ψ incremented over their full ranges, calculating the tensor elements at each dihedral angle pair. From these shielding calculations, probability surfaces can be derived; only the regions of ϕ and ψ which give correct matches to the experimental values (within measurement accuracy) are accepted. The regions of possible solutions can be narrowed further by restricting solutions only to the "allowed" regions of the Ramachandran map for the appropriate residue. This method has been experimentally calibrated[11] by measuring tensor elements for small peptides whose conformation has been determined crystallographically, finding the accuracy to be ca. $\pm 10°$, although this may improve with better calibration. Calculations have also indicated[12] that the shielding tensor values for some amino acids are also sensitive to the side chain angle χ_1, although experimental data are not yet available for calibration.

In principle, further structural information might be available through determination of the orientation of the shielding tensor in the molecular

[6] A. Bax, N. M. Szeverenyi, and G. E. Maciel, *J. Magn. Res.* **55,** 494 (1983).

[7] A. Bax, N. M. Szeverenyi, and G. E. Maciel, *J. Magn. Res.* **51,** 400 (1983).

[8] R. Tycko, G. Dabbagh, and P. A. Mirau, *J. Magn. Res.* **85,** 265 (1989).

[9] L. Frydman, G. C. Chingas, Y. K. Lee, P. J. Grandinetti, M. A. Eastman, G. A. Barrall, and A. Pines, *J. Chem. Phys.* **97,** 4800 (1992).

[10] D. D. Laws, H. Le, A. C. de Dios, R. H. Havlin, and E. Oldfield, *J. Am. Chem. Soc.* **117,** 9542 (1995).

[11] J. Heller, D. D. Laws, M. Tomaselli, D. S. King, D. E. Wemmer, A. Pines, R. H. Havlin, and E. Oldfield, *J. Am. Chem. Soc.* **119,** 7827 (1997).

[12] R. H. Havlin, H. Le, D. D. Laws, A. C. de Dios, and E. Oldfield, *J. Am. Chem. Soc.* **119,** 11951 (1997).

frame. To determine the tensor orientation in the molecular frame, an experiment to correlate the shielding with an interaction with known orientation must be done. The most straightforward second interaction is a dipolar coupling of a bonded pair, with its principal direction along the bond axis. A REDOR experiment (described later for determining distances) can be used with a dephasing period[13] to give side band intensities that depend on the relative orientation of dipolar and shift tensors. The accuracy is estimated to be on the order of $\pm 5°$.

Experiments to Determine Dihedral Angles

A number of experiments have been described to use the intrinsically anistropic nature of interactions in solids to determine relative orientations. In the past, these used nonrotating samples, and hence suffer from relatively low resolution and sensitivity. However, several new variations have evolved that exploit magic angle spinning, using r.f. pulses to recover the anisotropic interactions. Feng and co-workers[14] developed a method for an H–C–C–H fragment (requiring double labeling of the carbons). As usual in solid-state experiments, magnetization is transferred from the protons to ^{13}C by cross-polarization. Their method then uses a special pulse sequence on the carbons (called C7, discussed further later) to create double quantum coherence in the spinning sample. This coherence then evolves for a period in which the protons are decoupled from one another (using an MREV8 sequence). This evolution creates a DQ side band pattern, which depends on relative values of the C–H dipolar couplings. The side band patterns are calculated, and experiments are matched with simulated spectra as a function of angle. It is estimated that the angular accuracy is about $\pm 20°$ when the protons are at the maximum separation and $\pm 10°$ when they are at the minimal separation. The method was tested with small organic compounds, but has not yet been applied to peptides or proteins. Due to the requirement for the H–C–C–H atom types this would be most applicable to side chain angle determinations.

An analogous experiment has been done with ^{1}H–^{15}N–^{13}C$_{\alpha}$–^{1}H groups in amino acids,[15] determining the backbone angle ϕ. In this case, magnetization on the ^{13}C from cross-polarization is allowed to evolve under the dipolar coupling to the ^{15}N, using 180° pulses to interfere with the averaging

[13] J. M. Goetz and J. Schaefer, *J. Magn. Res.* **129,** 222 (1997).

[14] X. Feng, Y. K. Lee, D. Sandström, M. Edén, H. Maisel, A. Sebald, and M. H. Levitt, *Chem. Phys. Lett.* **257,** 314 (1996).

[15] M. Hong, J. D. Gross, and R. G. Griffin, *J. Phys. Chem. B* **101,** 5869 (1997).

due to spinning, creating an antiphase state, which is then converted into a combination of zero- and double-quantum coherences. These evolve under the proton couplings, again with proton homonuclear decoupling (MREV8) applied, yielding a dipolar side band pattern. To simulate these patterns, which by comparison with experimental data gives the angle of interest, a measure of the direct C–H and N–H dipolar couplings is needed. These are determined independently with direct dipolar-shift correlation experiments (DIPSHIFT).[16] The accuracy of bond angles determined with this approach is estimated to be about $\pm10°$ when the C–H and N–H bonds are near antiparallel and $\pm20°$ when they are near parallel. The requirement that spinning be slow enough to give a good dipolar sideband pattern, but fast enough to give good narrowing can be problematic. A variation in this experiment, which effectively amplifies the magnitude of the dipolar coupling, has been described.[17] The enhanced coupling increases the accuracy by at least twofold.

Another qualitatively similar experiment, executed with either of two closely related pulse sequences,[18,19] can be used to determine the ψ angle in peptides. Because ψ is defined on one end by a carbonyl, the proton dipolar couplings are replaced instead by those to ^{15}N, i.e., $^{15}N-^{13}C_\alpha-^{13}C=O-^{15}N$. The fundamental measurement still arises from the effects of the dipolar couplings on a double-quantum coherence. The experiment begins with cross-polarization from protons to ^{13}C, followed by evolution of the ^{13}C under a MELODRAMA-4.5 or C7 sequence, which creates ^{13}C double-quantum coherence. While this coherence evolves, pulses are used on the ^{15}N to restore the coupling to the ^{13}C and also to remove effects of ^{15}N chemical shift anisotropy and isotropic shift. The signal returns to ^{13}C to be detected with proton decoupling. Simulations are used to calculate the average evolution of the ^{13}C coherence under the nitrogen dipolar couplings. It is found that the DQ dephasing is quite sensitive to the angle ψ when the two C–N groups are near the trans geometry, $\psi \approx 160°$, the β-sheet region (as seen with C–H groups as well), although it is much less sensitive for angles less than 120°. Fortunately, distances, measured with methods described later, are more sensitive to ψ in the low angle range.

[16] M. G. Munowitz, R. G. Griffin, G. Bodenhausen, and T. H. Huang, *J. Am. Chem. Soc.* **103,** 2529 (1981).

[17] M. Hong, J. D. Gross, C. M. Rienstra, R. G. Griffin, K. K. Kumashiro, and K. Schmidt-Rohr, *J. Magn. Res.* **129,** 85 (1997).

[18] P. R. Costa, J. D. Gross, M. Hong, and R. G. Griffin, *Chem. Phys. Lett.* **280,** 95 (1997).

[19] X. Feng, M. Edén, A. Brinkman, H. Luthman, L. Eriksson, A. Gräslund, O. N. Antzutkin, and M. H. Levitt, *J. Am. Chem. Soc.* **119,** 12006 (1997).

Experiments to Determine Distances

The dipolar coupling between two spins has structural information in the form of the internuclear distance, but this valuable information can rarely be extracted from line shapes for nonspinning samples. Due to resolution considerations, measurements in peptides have involved ^{13}C and/or ^{15}N enriched at specific sites. Line shapes in static samples then arise from a combination of chemical shielding anisotropy of the observed spin and the dipolar coupling. In principle, the dipolar coupling can be observed as a splitting of the powder pattern features, but large widths of powder patterns lead to relatively low sensitivity, and natural abundance signals from all nonenriched sites in the peptide may obscure some features. However, a variety of clever pulse sequences have now been developed that recover the dipolar coupling information for samples spinning at the magic angle. The fundamental issue is that magic angle spinning, which gives sharp lines for observation, also eliminates dipolar coupling. Therefore, some special mechanism for recoupling must be added during part of the sequence.

The first and simplest of the recoupling methods is termed rotational resonance, often designated R^2. The R^2 phenomenon was first observed when the rate of spinning of a sample happened to match the difference in chemical shift between two resonances. The resonances (which were sharp, individual lines for other spinning rates) broadened with a complex, split line shape. The two spin types were physically close in the compound being studied, and it was realized that the splitting of the line shape arose from a partial reintroduction of the dipolar coupling. An oversimplified way of thinking about this is that the central line of each spin has the same resonance frequency as one of the side bands of the resonance from the other spin (indeed the effect remains if any side band for one resonance matches the central line of the other, although the magnitude of the effect declines as the order of the side band gets higher). Because some of the spins have the same effective frequency, the "flip-flop" term in the dipolar coupling is energy conserving and hence manifests itself. It is possible to fit the splitting to estimate the distance between the pair of spins. However, the magnitude of the splitting is rather small, and hence the splitting is only apparent for rather short distances. Raleigh and co-workers[20] realized that the flip-flop terms could still have an effect through the transfer of polarization, behavior rather similar to the nuclear Overhauser effect (NOE) used in solution NMR. Their approach is implemented by applying a 180° pulse selectively to one of the resonances, followed by a mixing

[20] D. P. Raleigh, M. H. Levitt, and R. G. Griffin, *Chem. Phys. Lett.* **146,** 71 (1988).

delay with the rotational resonance condition of $\omega_r = n\Delta\omega$ met (ω_r is the spinning rate, $\Delta\omega$ is the difference in chemical shift, and n is an integer), and then a 90° pulse and observation of the resonances with decoupling applied. During the mixing period, the flip-flop term swaps magnetization from the inverted spins to those not inverted, and vice versa. As the mixing time is varied this leads to an oscillating decay of magnetization, with the oscillations weaker as the distance increases. The magnetization exchange curve can be fitted to give the value of the dipolar coupling and hence the distance (Fig. 4), although an estimate of the intrinsic dephasing rate is needed. For pairs of ^{13}C spins, distances of a little over 6 Å have been determined. The accuracy probably depends somewhat on circumstances

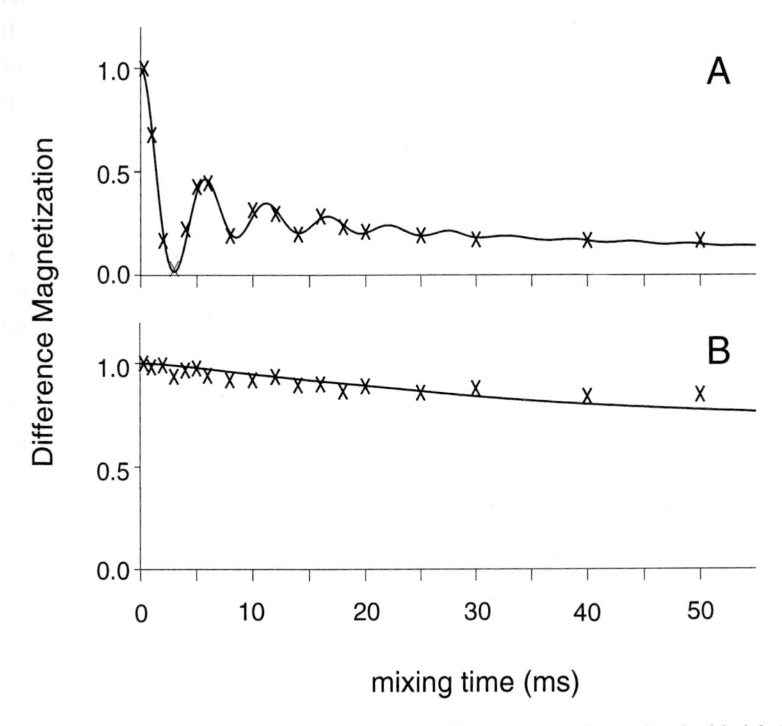

FIG. 4. Rotational resonance magnetization transfer curves are shown for doubly labeled H1 peptides (numbering from the intact protein). (A) C=O of Gly-114 to C_α of Ala-115: fitted parameters were $r = 2.37$ Å, dephasing time was 9.5 msec, and the inhomogeneous line width was 150 Hz. The crosses are experimental points, the solid line is the best fit. A total of 512 averages were done for each point. (B) C=O of Ala-115 to C_α of Ala-117: fitted $r = 5.78$ Å with dephasing time and line width taken from the experiment shown in (A) Seven hundred and four scans were averaged for each point, crosses are the experimental data, and the line is the best fit.

(some issues are discussed later) but is probably in the range of ±0.5–1 Å for amorphous materials such as amyloid. For ^{15}N pairs the maximum distance would be considerably shorter, as the lower gyromagnetic ratio makes dipolar couplings smaller.

Despite the apparent simplicity of the rotational resonance approach, there are quite a number of potential problems that can affect the accuracy of results. First, at least one of the spins involved must have a reasonably large shift anisotropy in order to give reasonable side band intensity at the resonance of the second spin. The width of shielding tensors increases (in Hz) with increasing field strength, making it beneficial to work at high field. There must also be a reasonable separation in chemical shift between the spins observed, as a reasonable spinning rate must be achieved. The rate of magnetization exchange is affected by the side band intensities, and hence by the relative anisotropies of the two shielding tensors involved, including both orientation and magnitude. Because the orientations, at least, are usually not known, this introduces an uncertainty into the analysis, although of modest magnitude in typical cases. The analysis is complicated in cases for which there is inhomogeneous broadening of the MAS lines[21] (different spins in the sample having somewhat different chemical shifts due to conformation, environment, etc.). The inhomogeneous nature of the lines means that parts of the line do not meet the resonance condition well, slowing magnetization exchange. It has also been found to be important to have very high proton decoupling field strengths during the experiment.[22] The spinning rate must be precisely set (the resonance condition to be matched to a few Hertz) and accurately maintained over the full time of the experiment. The overall stability of the spectrometer is important as intensity differences are being measured, with a full set of mixing times often taking over a day to collect.

A variation on this method, termed rotational resonance tickling,[23] removes the effects of distributions in chemical shift and residual broadening from incomplete heteronuclear decoupling. In this version of the experiment the spinning frequency is set close to but not on rotational resonance (e.g., 250 Hz away). This would normally suppress magnetization transfer, but a radiofrequency field is added, introducing a modulation that can add to that from sample rotation. When the sum of the modulation frequencies matches the chemical shift offset, magnetization exchange occurs. The experiment is done by sweeping the r.f. strength to vary the effective modula-

[21] J. Heller, R. Larsen, M. Ernst, A. C. Kolbert, M. Baldwin, S. B. Prusiner, D. E. Wemmer, and A. Pines, *Chem. Phys. Lett.* **251,** 223 (1996).

[22] O. Peersen, M. Groesbeek, S. Aimoto, and S. O. Smith, *J. Am. Chem. Soc.* **117,** 7228 (1995).

[23] P. R. Costa, B. Sun, and R. G. Griffin, *J. Am. Chem. Soc.* **119,** 10821 (1997).

tion frequency. By altering the rate of the sweep, a varying amount of time is spent "in resonance." Internuclear distances that are short can mix over a larger range of frequencies, and hence give a larger effect even at high sweep rates. In addition, the sweep rate at which magnetization exchange approaches the long mixing time value in a normal rotational resonance experiment is also characteristic of the distance between spins. Sweeping through the rotational resonance makes this approach less sensitive to inhomogeneous broadening. It is also less sensitive to effects of incomplete decoupling, which limits the distance range in other "r.f. driven mixing" experiments described later. The accuracy of the rotational resonance tickling-derived distances is estimated to be ±0.1 Å.

The rotational resonance approach is powerful, but is restricted to measurement of one homonuclear distance at a time. A variety of different pulse sequences have now been described that use r.f. pulses to remove the effects of chemical shifts during mixing, reactivating the dipolar coupling, and hence achieving simultaneous mixing for many spins at once. Accomplishing the desired averaging is complex and requires coordination of the sample rotation and pulses. There are two basic variations of these experiments, one type transfers population information (like the rotational resonance experiment discussed earlier) and the other leads to dephasing of the detected signal. In either case the measured effect is directly related to the strength of the dipolar coupling and hence can be analyzed to give a distance.

The initial work on dephasing using heteronuclear dipolar couplings while magic angle spinning was termed REDOR[24] (rotational echo double resonance). It was shown that a combination of 180° pulses on both ^{15}N and ^{13}C spins (Fig. 5) in doubly labeled amino acids allowed dephasing through their dipolar coupling. The magnitude of the effect is measured as a dephasing (loss of signal) on the ^{13}C. By removing the ^{15}N pulses there is a perfect control experiment; all other variables except the dipolar coupling remain unchanged. Difference signals between experiments with and without the ^{15}N pulses, with a varied number of rotor revolutions for dipolar coupling to have an effect, are interpreted using a universal curve depending only on the known rotation rate and the dipolar coupling. The limit in distance is determined by the T_2 value of the observed spin (dephasing by other mechanisms). Natural abundance signals can contribute a small amount to dephasing, which may affect measurements of longer distances. For ^{15}N–^{13}C pairs the range is fairly short (perhaps to 5 Å), with the low gyromagnetic ration of ^{15}N giving rise to weak couplings. The approach is better for other pairs, such as ^{31}P–^{13}C (but phosphorous usually does not

[24] T. Gullion and J. Schaefer, *J. Magn. Res.* **81,** 196 (1989).

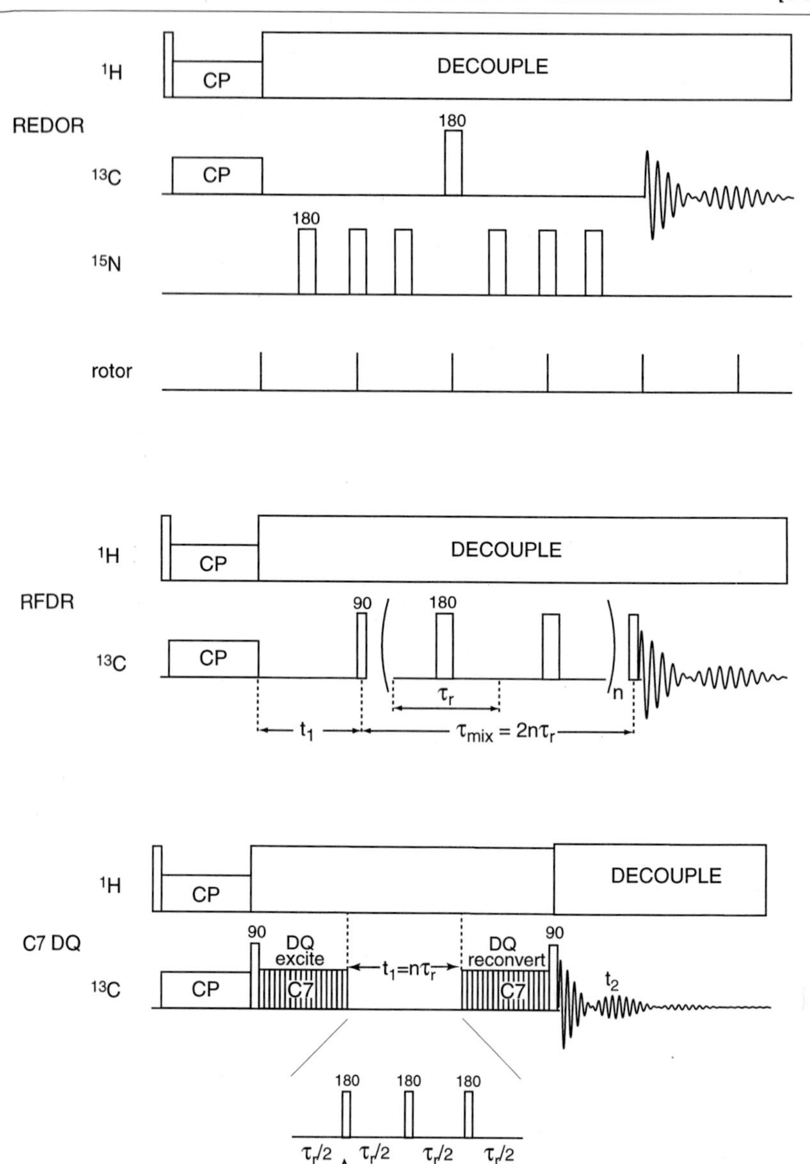

FIG. 5. Pulse sequences are shown for REDOR, RFDR, and C7DQ experiments. CP represents cross-polarization from ^{1}H to enhance the ^{13}C signals.

occur in peptides) or ^{19}F–^{13}C (fluorinated amino acids being substituted for normal ones). With ^{19}F–^{13}C, a distance of 8 Å in a peptide was determined.[25]

A variant termed TEDOR[26] (transferred-echo double resonance NMR) first cross-polarizes from protons to the ^{15}N, then uses 180° pulse trains, first on ^{13}C then ^{15}N with a pair of 90° pulses between. This allows coherence transfer via the dipolar coupling through an INEPT mechanism, selecting only for those ^{13}C spins near ^{15}N as the coherence transfer function depends on the coupling strength. A Fourier transform of the signal amplitude as a function of the dipolar evolution time gives a dipolar spectrum. The greatest strength of this approach seems to be its exceptional selectivity; ^{13}C spins away from the ^{15}N are eliminated completely.

For homonuclear spin pairs, a sequence (called DRAMA, dipolar recovery at the magic angle) uses a repeated cycle of 90° and 180° pulses synchronized with the rotation to eliminate the averaging of the dipolar coupling while retaining the resolution of a MAS spectrum. In this two-dimensional technique the evolution due to dipolar coupling occurs during the first time period, containing the pulses, which is incremented by increasing the number of pulse cycles, followed by the second period during which the MAS spectrum is acquired. Transformation of the first time period gives a powder dipolar spectrum for each line. The design of the sequence makes it work properly only when the chemical shift difference between the coupled spins is small. Another sequence (SEDRA, simple excitation for the dephasing of rotational-echo amplitudes) uses only 180° pulses and eliminates this problem, but does not work if the chemical shift difference is zero. Subsequent versions, with modified pulse cycles, have reduced the sensitivity to CSA and resonance offset.[27] Fairly short 180° pulses are required to avoid dephasing during the pulses. Further elaborations of this general idea have led to "windowless" sequences that allow dipolar dephasing, while compensating well for CSA, resonance offsets, and r.f. field inhomogeneity. DRAWS (dipolar recoupling with a windowless sequence) uses pulse cycles that coincide with the rotor period.[28] This sequence gives observable dephasing up to about 5.5-Å distances for ^{13}C–^{13}C pairs, as with other methods longer for higher gyromagnetic nuclei. An interesting variation has been described that uses averaging over multiple rotor periods (although the pulses are still synchronized with rotor motion) but has a

[25] S. M. Holl, G. R. Marshall, D. D. Beusen, K. Koclolek, A. S. Redlinski, M. T. Leplawy, R. A. McKay, S. Vega, and J. Schaefer, *J. Am. Chem. Soc.* **114,** 48309 (1992).

[26] A. W. Hing, S. Vega, and J. Schaefer, *J. Magn. Res.* **96,** 205 (1992).

[27] R. Tycko and S. O. Smith, *J. Chem. Phys.* **98,** 932 (1993).

[28] D. M. Gregory, D. J. Mitchell, J. A. Stringer, S. Kiihne, J. C. Shiels, J. Callahan, M. A. Mehta, and G. P. Drobny, *Chem. Phys. Lett.* **246,** 654 (1995).

sevenfold symmetric phase cycle, termed C7[29] (Fig. 5). This scheme has little sensitivity to chemical shift or r.f. imperfections and yields a high efficiency of excitation of double-quantum coherence (which is dependent on the magnitude of the dipolar coupling).

Another homonuclear experiment was suggested to drive a rotational resonance-like magnetization transfer with a series of phase-cycled 180° pulses (Fig. 5), removing chemical shifts but not dipolar couplings. This experiment was termed RFDR[30] (radio frequency-driven recoupling) and was demonstrated to allow simultaneous magnetization transfer for multiple pairs of spins. The theoretical basis for these experiments has been described.[31] The resulting spectrum is very much like a solution NOESY data set in which cross-peaks connect pairs of resonances that are allowed to undergo magnetization transfer. It has clear value for easy interpretation, but may be less quantitative than other methods for distance determination.

Another experiment that transfers population has been demonstrated,[32] but in this case the correlations are between spinning side bands of the CSA patterns. The sample was a tripeptide labeled at two specific carbonyls, diluted in normal peptide. A population distribution on ^{13}C was created by a cross-polarization from protons, followed by the t_1 delay and a 90° pulse. After a mixing time the populations were probed with another 90° pulse, quite analogous to the NOESY sequence. The sample was spun at the magic angle at a moderate rate, and during the mixing time proton decoupling was turned off. Due to ^1H couplings the ^{13}C resonances were then broad during the mixing period, and magnetization exchange could occur. Only the two carbonyls within the same peptide could exchange, and their relative orientations are defined by the molecular conformation. The relative orientation controls the amounts of exchange occurring between different spinning side bands in the spectrum. The overall distribution of exchange cross-peaks was compared with those predicted as a function of ϕ and ψ. The best fit was very near the crystallographically determined value for this peptide. It is somewhat difficult to estimate the potential accuracy of this approach in more complex systems, with likely lower S/N ratios for experimental data.

A new constant time variant of the RFDR experiment has been described.[33] In this version, RFDR sequences are applied to allow evolution

[29] Y. K. Lee, N. D. Kurur, M. Helmle, O. G. Johannessen, N. C. Nielsen, and M. H. Levitt, *Chem. Phys. Lett.* **242,** 304 (1995).

[30] A. E. Bennett, J. H. Ok, R. G. Griffin, and S. Vega, *J. Chem. Phys.* **96,** 8624 (1992).

[31] D. K. Sodickson, M. H. Levitt, S. Vega, and R. G. Griffin, *J. Chem. Phys.* **98,** 6742 (1993).

[32] D. P. Weliky and R. Tycko, *J. Am. Chem. Soc.* **118,** 8487 (1996).

[33] A. E. Bennett, D. P. Weliky, and R. Tycko, *J. Am. Chem. Soc.* **120,** 4897 (1998).

under a homonuclear dipolar coupling, separated by pair 90° pulses to convert to double-quantum coherence and back (a DQ filter). A set of experiments is done with different lengths for each of the RFDR periods, but maintaining the overall time of evolution from creation of the coherence to the detection. This means that any loss of coherence in the experiment, whatever the source, is constant. By varying the distribution of time spent in the different RFDR periods, however, one can measure the time dependence of development of a double-quantum signal or the rate at which dephasing occurs. These are again related to local structural parameters, and when combined with the magnetization exchange experiment described earlier is fairly restrictive on values of ϕ and ψ that are consistent with all observations.

Requirements for Measurements

Modern NMR spectrometers have great flexibility; however, there are a number of specific requirements for solid-state measurements that are not met in common spectrometers used for solution measurements. The first is very strong r.f. fields. The pulse lengths during many experiments must be kept short. In addition, deoupling fields must be very strong[22] to completely remove effects of the heteronuclear couplings. Obtaining strong fields requires high power amplifiers, reaching kilowatt levels. The probes must be specially designed to handle such high power, as well as to carry out the rapid magic angle spinning that is now routinely done. Spinning speeds in the single digit kilohertz range are very routine, with the highest rates getting to and above 20 kHz now. Smaller samples can be spun faster, and although the sample volumes are then also smaller, the sensitivity remains quite good because the r.f. coil characteristics can be optimized. Isotopically labeled samples are often diluted in unlabeled material, with the total sample being around 50 mg to obtain a sufficient signal-to-noise ratio in a reasonable time, although considerably less is required for some experiments. The rotation rate requirements depend very much on the particular samples and experiments being used. It is important that the overall stability of the systems (field, r.f. strengths, temperatures, spinning rate) be very good.

Generation of Labeled Samples

To have both sensitivity and selectivity in the experiments described earlier it is necessary to have enrichment of ^{13}C and/or ^{15}N at specific sites in the molecules under study. For small peptides and proteins this can be accomplished by direct chemical synthesis, using appropriate monomers enriched at the desired sites. In experienced laboratories, peptides of up to about 100 residues can be made. With some minor sequence requirements, peptides can also be chemically ligated to give yet larger molecules.[34] There are commercial sources for many labeled amino acids and resources (e.g., the Stable Isotope Resource at Los Alamos[35]) for obtaining others. Labels in larger proteins can be inserted by expressing the protein in bacteria and feeding the labeled amino acid (sometimes requiring an auxotrophic strain).[36] This labels all copies of that particular amino acid in the protein. Some further selectivity can be achieved by double labeling, using [^{13}C]carboxyl label in one residue type and α-[^{15}N]amino in another. If these two amino acids occur sequentially in the protein, then the peptide bonds formed between them will be double labeled, and pulse sequences can be used to select only signals from those sites. Incorporation of labels at single sites in larger proteins can be done with *in vitro* translation.[37] Initially this could only be done on a rather small scale, but improvements[38] suggest that (for some proteins at least) this method could be used to generate tens of milligrams of protein. Labeling two sites is more difficult because two different codon/suppresser tRNA combinations are then needed; however, such methodology is also being developed. Uniform labeling of large proteins has become routine for solution NMR by feeding bacteria-labeled sugar and ammonia. However, resolution becomes a major issue for solid-state NMR work. A modeling study[39] suggested that using three-dimensional correlation experiments of several types would be sufficient for the assignment of residues for peptides up to around 30 amino acids in length.

Applications of Solid-State NMR to Protein Aggregates

As can be seen from the dates of most of the papers cited in this article, the methodology in solid-state NMR for addressing questions of peptide and protein conformation has developed fairly recently and is continuing. This, together with technical issues of isotope labeling and solid-state NMR hardware, has limited the number of applications, but the field is poised to begin making a substantial contribution to understanding aggregating systems.

[34] T. W. Muir, P. E. Dawson, and S. B. Kent, *Methods Enzymol.* **289,** 266 (1997).
[35] A resource supported by the NCRR Division of NIH, see http://cst.lanl.gov/sir
[36] L. P. McIntosh and F. W. Dahlquist, *Quart. Rev. Biophys.* **23,** 1 (1990).
[37] J. A. Ellman, B. F. Volkman, D. Mendel, P. G. Schultz, and D. E. Wemmer, *J. Am. Chem. Soc.* **114,** 959 (1992).
[38] T. Kigawa, Y. Muto, and S. Yokoyama, *J. Biomol. NMR* **6,** 129 (1995).
[39] R. Tycko, *J. Biomol. NMR* **8,** 239 (1996).

The most studied aggregate derives from fragments of the Aβ peptide involved in Alzheimer's disease. It was shown that much of the behavior of the natural 42 amino acid peptide, in terms of aggregation and amyloid formation, is recapitulated in much shorter fragments. Spencer *et al.*[40] have undertaken extensive solid-state NMR work on residues 34–42, fragment (β34–42): H_2N-LMVGGVVIA-COOH. A set of rotational resonance measurements was done using a combination of C_α and carbonyl labels. Most of the pairs had distances that were consistent with an extended β-sheet conformation, as anticipated. However, a distance within the Gly–Gly pair seemed to be too short, and a *cis*-peptide bond there was suggested.[40] However, further measurements with improved accuracy and a better understanding of intermolecular effects later led to reconsideration of this. More extensive R^2 measurements and a determination of the relative orientation of the shift tensors for the carbonyls led to the final conclusion[41] that the bond was in fact *trans*. An important addition to the intrapeptide distances, other label pairs were used to address intermolecular contacts. The intra- and intermolecular effects are distinguished by doing dilutions of labeled peptides in unlabeled ones. The intramolecular exchange of magnetization is unaffected, but of course intermolecular exchange is suppressed as the amount of unlabeled peptide increases. A set of intermolecular short distances (which could not be determined accurately, but must be fairly short) were combined with minimum restraints (from labels that did not exchange magnetization in an intermolecular way) to define the sheet structure as antiparallel, and set the register of the strands relative to one another, assuming normal β-sheet hydrogen bonding. Two possible packing arrangements were found[42]: one in which peptides are related by a twofold rotation and the other in which the relation is by a twofold screw axis. The characteristics of the sheet formed are consistent with many other data about the dimensions and spacings in amyloid.

The conformation of a peptide fragment of the prion protein has also been analyzed with solid-state NMR.[43] This 14 amino acid fragment was shown to have conformational lability in solution (converting between helix

[40] R. G. S. Spencer, K. J. Halverson, M. Auger, A. E. McDermott, R. G. Griffin, and P. T. Lansbury, Jr., *Biochemistry* **30,** 10382 (1991).

[41] P. R. Costa, D. A. Kocisko, B. Q. Sun, P. T. Lansbury, Jr., and R. G. Griffin, *J. Am. Chem. Soc.* **119,** 10487.

[42] P. T. Lansbury, Jr., P. R. Costa, J. M. Griffiths, E. J. Simon, M. Auger, K. J. Halverson, D. A. Kocisko, Z. S. Hendsch, T. T. Ashburn, R. G. S. Spencer, B. Tidor, and R. G. Griffin, *Nature Struct. Biol.* **2,** 990 (1995).

[43] J. Heller, A. C. Kolbert, R. Larsen, M. Ernst, T. Bekker, M. Baldwin, S. B. Prusiner, A. Pines, and D. E. Wemmer, *Prot. Sci.* **5,** 1655 (1996).

and extended conformations in different solvent conditions[44]), and it was thought that this might relate to the conformational transition of PrP between the native and the infectious form. It was shown that the peptide could be trapped as helix or extended forms in the solids and that hydration of the helix form promoted its conversion to an extended form (Fig. 2). Distances determined with rotational resonance indicated distances intermediate between helix and fully extended sheet; intermolecular interactions have not yet been investigated. An analysis of chemical shift tensors showed that the extended form ϕ, ψ values were in ranges expected for β-sheet conformations.[45] A glycine near the center of the sequence showed clear conformational heterogeneity (population of two different states at least) in the extended form.

Although not specifically involved in protein aggregates, a number of larger systems have been studied with solid-state NMR methods.[46] Some of these have been membrane associated,[47] whereas others have been too large for solution NMR but amenable to solid-state NMR for the determination of bound ligand or substrate geometry[48] or to determine the relative geometry of domains.[49] Such studies demonstrate that single labeled sites can be detected in large proteins (100 kDa and above) with sufficient sensitivity to do experiments such as REDOR. Most of these studies have been applied in systems where a good deal of structural information is available, and local information about the rearrangement of structural units, or relative positions of different bound molecules is needed. Structural information available at the start of the study helps design the experiments, i.e., where to put isotope labels so that distances can be measured. A challenging aspect of working on systems with less known at the start is design of a labeling/measurement strategy that will ultimately give a global structure that will help in understanding the aggregation phenomenon. One approach is to first do local structure, identifying secondary structure elements from distances, angles, or a combination thereof. Once these are identified, distances between elements may be measured with distance determination experiments. Those with the longest range are likely to be most efficient because they are more likely to give a positive result (an actual distance rather than a lower bound). Care must be used to distinguish inter- and intramolecular effects, as both are required for a full understand-

[44] H. Zhang, K. Kaneko, J. T. Nguyen, T. L. Livshits, M. A. Baldwin, F. E. Cohen, T. L. James, and S. B. Prusiner, *J. Mol. Biol.* **250,** 514 (1995).

[45] J. Heller, Ph.D. thesis, University of California Berkeley, 1997.

[46] L. M. McDowell and J. Schaefer, *Curr. Opin. Struct. Biol.* **6,** 624 (1996).

[47] S. J. Opella, *Nature Struct. Biol. Suppl.,* 845 (1997).

[48] A. M. Christensen and J. Schaefer, *Biochemistry* **32,** 2868 (1993).

[49] C. A. Klug, K. Tasaki, N. Tjandra, C. Ho, and J. Schaefer, *Biochemistry* **36,** 9405 (1997).

ing of aggregation behavior. In addition, it will be necessary to develop and use experiments to correlate the positions of side chains, which usually control tertiary structure and packing. It is still in the early days of solid-state NMR analysis of aggregates, but experiments are being developed at a rapid rate. One can expect to see more results on protein aggregates in the near future.

[36] Fourier Transform Infrared Spectroscopy in Analysis of Protein Deposits

By SANGITA SESHADRI, RITU KHURANA, and ANTHONY L. FINK

Introduction

The intrinsic insoluble nature of protein deposits (as in amorphous aggregates, inclusion bodies, amyloid fibrils, etc.) places severe restrictions on the availability of methods for ascertaining the structure of the material. Fourier transform infrared (FTIR) spectroscopy is well suited for determining structural features of proteins, both in solution and as deposits. Proteins in the form of solutions, thin films (hydrated or dry), solids (including lyophilized or spray-dried powders), or suspensions of precipitates in various solvents can be used for FTIR analysis. The most commonly used methods for FTIR analysis of protein deposits are KBr pellets (in which the dry solid is dispersed in a KBr disk) and diffuse reflectance for dry samples, thin films [using transmission or attenuated total reflectance (ATR) modes] for samples in solution or suspension, and attenuated total reflectance mode for solid, liquid, or suspended samples. For dry solid samples there are several sampling techniques, including powder, mull, alkali halide pellet or disk, and film. Powdered samples can be difficult to analyze due to the high incidence of scattered light, which results in a loss of energy transmitted to the detector. To reduce scattering, samples should be ground to a powder of 5-μm particle size or less, smaller than the wavelength of the radiation.

Infrared spectroscopy is a form of vibrational spectroscopy; for proteins, the major bands of interest are the amide I, II, and III bands, which absorb in the 1600–1700, 1500–1600, and 1200–1350 cm^{-1} regions, respectively. FTIR spectra are customarily shown in units of wave numbers. The wave number is defined as the reciprocal of the wavelength, $\nu = 1/\lambda$, with units of cm^{-1}, and is thus proportional to the frequency of the radiation. Most

0076-6879/99 $30.00

commercial instruments measure the mid-IR region, within which proteins absorb in the 4000- to 400-cm^{-1} range. The amide I band arises predominantly (80%) from the C=O stretching mode of the amide functional group; the remaining contribution (20%) arises from C–N stretching. The amide II band arises from N–H bending and C–N stretching vibrations. The amide III absorption is attributed to C–N stretching vibrations coupled to N–H in plane bending vibrations, with weak contributions from C–C and C=O stretching.[1]

As each of the secondary structural motifs in proteins is associated with a characteristic hydrogen bonding pattern between the amide C=O and N–H groups, each type of secondary structure gives rise to characteristic amide I and II absorption bands.[2] It is important to note, however, that the resulting frequencies are also very sensitive to the local environment, mostly due to transition dipole coupling and variations in φ and ψ angles.[1] A number of investigations suggest that one of the hallmarks of aggregated proteins is a substantial increase in the amount of β structure (Fig. 1); this is true for both amorphous and fibrillar deposits. In many cases, but not all, the new β structure is found at low frequencies in the IR spectrum, e.g., 1625 cm^{-1} and lower. The lower frequencies probably represent the pairing of intermolecular (antiparallel) β strands, which have shorter H bonds than many β sheets found in globular proteins, due to fewer constraints on the peptide backbone. In the case of all-helical native proteins, aggregation leads to dramatic changes in IR spectra (Fig. 1), whereas in the case of native all-β proteins the effects are more subtle but, as shown in Fig. 2, often result in a substantial shift in the position of the major β component.

Problem of H_2O Absorption

Protein deposits usually originate from H_2O solutions, rather than D_2O, especially for *in vivo* samples, and thus the amide bonds will involve NH rather than ND bonds. The major problem in examining aqueous samples of a protein is the very strongly absorbing H–O–H bending vibrations of H_2O near 1640 cm^{-1}, the same region in which the amide I band absorbs. In many cases of protein deposits this will not be a problem due to the relatively low concentration of water present in the sample.

In order to detect the signal from the protein in the presence of the potentially far-larger signal from the water, excellent signal/noise resolution

[1] S. Krimm and J. Bandekar, *Adv. Prot. Chem.* **38,** 181 (1986).
[2] M. Jackson and H. H. Mantsch, *Crit. Rev. Biochem. Mol. Biol* **30,** 95 (1995).

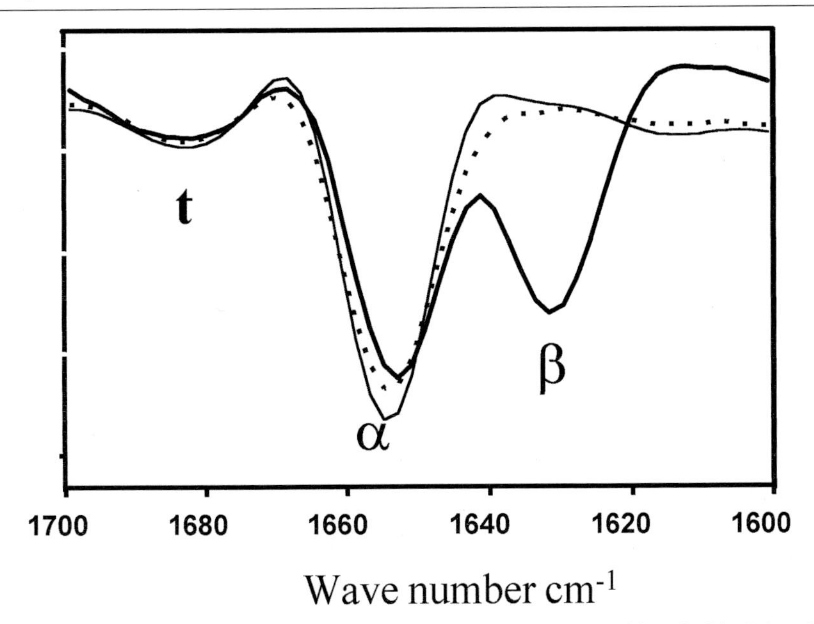

FIG. 1. Second-derivative FTIR spectrum of the amide I region of interleukin-2 (an all-helical protein) in native (thin-solid and dotted lines) and inclusion body (thick solid line) forms. The dotted line is from a precipitate of the native protein salted out by ammonium sulfate, and the thin solid line is for the native state in a hydrated thin film.

is required. It is best that the combined protein–water absorbance at 1640 cm^{-1} not exceed a value of 1 absorbance unit in order to conserve the linear relationship between absorbance and concentration, especially if quantitative information is desired. For transmission FTIR (i.e., solution samples) this necessitates the use of a short path length on the order of 4–10 μm and fairly high protein concentrations, generally on the order of 1–50 mg/ml. Frequently the problem is overcome by using D_2O in place of H_2O. The D–O–D bending vibration is shifted to 1220 cm^{-1}, eliminating its absorbance contribution in the amide I region. However, it is critical to fully exchange all H by D: partial deuteration leads to amide I band components with frequencies in between those of fully H or fully D substitution. In many cases of protein deposits, e.g., inclusion bodies or *in vivo* amyloid fibrils, it is difficult to prepare a fully exchanged sample in D_2O. Because the more buried amide hydrogen atoms will be less accessible to the solvent, and hence H/D exchange, the discrimination between solvent-inaccessible and solvent-accessible protons to exchange with D_2O may

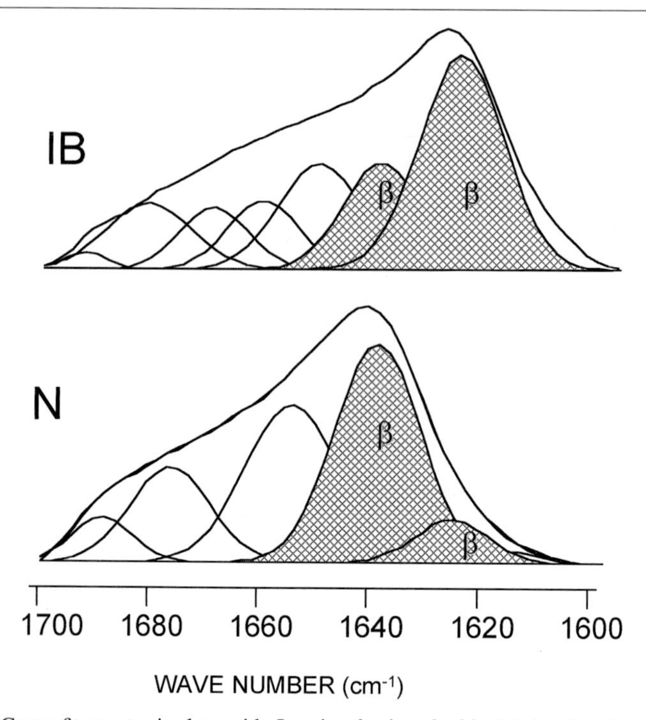

FIG. 2. Curve-fit spectra in the amide I region for interleukin-1β showing the component bands. The top spectrum is for the protein in the form of inclusion bodies, and the lower spectrum is for the native state. Both spectra were acquired in the ATR mode as thin films.

permit useful structural information to be obtained in certain cases, e.g., fibrillar deposits.[3]

KBr Pellets and Related Methods for Dry Solid Samples

To minimize scattering, samples can be dispersed in a medium of similar refractive index such as in a mineral oil mull or alkali halide[4] pellet. Preparation of particles fine enough for inclusion in mulls and pellets requires vigorous grinding in order to achieve small and homogeneous particles. Because preparation of both pellets and mulls requires some degree of mechanical processing, the conformation of the protein may be affected. Because proteins are only marginally stable at room temperature, the poten-

[3] B. W. Caughey, A. Dong, K. S. Bhat, D. Ernst, S. F. Hayes, and W. S. Caughey, *Biochemistry* **30,** 7672 (1991).

[4] H. K. Chan, B. Ongpipattanakul, and J. Au-Yeung, *Pharm. Res.* **13,** 238 (1996).

tial for conformational change due to the sample processing should be considered when analyzing protein secondary structure.

In preparing the pellet, pressures on the order of 5–10 kbar are required to compress the alkali halide and sample to form a transparent disk.[5] Such pressures are potentially high enough to denature proteins. It is known that pressure can induce conformational changes of proteins in solution.[6] Adverse effects on the bioactivity of some proteins have been reported in pressure-induced compaction of the solid state,[7] and there have been reports confirming that this may happen in the formation of KBr pellets for FTIR, e.g., loss of enzyme activity and an increase in high molecular weight aggregates.[4]

Differences between two forms of solid-state samples, the pellet and film, were noted in infrared spectra of both myoglobin and bovine serum albumin.[8] However, some polypeptides in the lyophilized form[9] have been successfully analyzed by this approach, and we have found that spectra of lyophilized lysozyme and a fibrillar peptide[10] in KBr pellets are very similar to those obtained using both thin films and solid samples with ATR. Because *some* proteins may undergo conformational changes when compressed during sample preparation using KBr pellets, it is important in using this method to check if the sample preparation results in conformational change, which may or may not be easily accomplished. If the sample is affected by high pressure, then techniques that do not involve the use of high pressures such as diffuse reflectance and ATR are possible alternatives.

Methodology for KBr Pellets

A small amount of KBr is transferred into a beaker and dried in a 90° oven overnight. Immediately after removal from the oven, about 0.5 g KBr is added to 0.5–0.75 mg of protein or peptide sample and mixed well using a mortar and pestle. Approximately 250 mg of KBr and sample mixture is placed in the chamber used to apply pressure and kept under vacuum for 5 min. High pressure, 5–10 kbar, is then applied using a Carver press to form the translucent, thin pellet. The pellet is placed in a transmission holder and the spectrum is recorded. The background spectrum is collected

[5] F. S. Parker, "Applications of Infrared Raman, and Resonance Raman Spectroscopy in Biochemistry." Plenum Press, New York, 1983.

[6] R. J. Siilva and G. Wever, *Annu. Rev. Phys. Chem.* **44,** 89 (1993).

[7] M. J. Groves and C. D. Teng, *in* "Stability of Protein Pharmaceutical," Part A. Plenum Press, New York, 1992.

[8] K. Kaiden, T. Matsui, and S. Tanaka, *Appl. Spectrosc.* **41,** 180 (1987).

[9] A. C. Dong, S. J. Prestrelski, S. D. Allison, and J. F. Carpenter, *J. Pharm. Sci.* **84,** 415 (1995).

[10] N. D. Lazo and D. T. Downing, *Biochem. Biophys. Res. Commun.* **235,** 675 (1997).

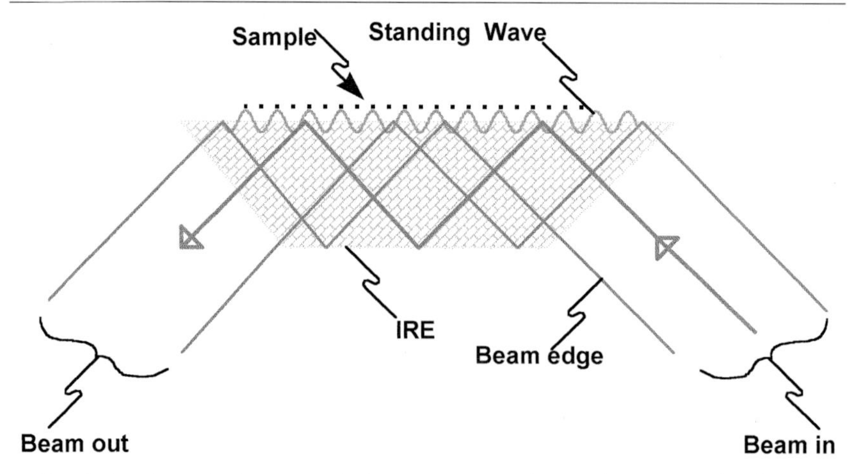

FIG. 3. A schematic of a trapezoidally shaped internal reflection element (IRE) and the basis of attenuated total reflectance (ATR) FTIR. The number of internal reflections is determined by the size and geometry of the IRE.

with a KBr pellet in which the protein was omitted, or the empty pellet holder.

Attenuated Total Reflectance FTIR

In ATR–FTIR the sample is placed in contact with the surface of material having a high refractive index, known as the internal reflection element (IRE). The IRE is usually made of germanium or zinc selenide. When infrared radiation penetrates the IRE at an angle above the critical angle of incidence, total internal reflection of this incident radiation produces an evanescent (standing) wave at the boundary between the IRE and the sample (Fig. 3). This reflected beam penetrates beyond the crystal and can be absorbed by materials in contact with its surface. The depth of penetration varies, depending on the material of the IRE and the wavelength. Because the IR beam is not transmitted through the sample, spectra are unaffected by turbidity. This makes ATR–FTIR an excellent choice for characterization of the structure of aggregated proteins. Several different ATR cell designs are available; we favor out-of-compartment, horizontal, trapezoidal-shaped IREs, which may be used in either a trough or flow-cell configuration. ATR may be used with samples in solution,[11] as a

[11] K. A. Oberg and A. L. Fink, *Anal. Biochem.* **256,** 92 (1998).

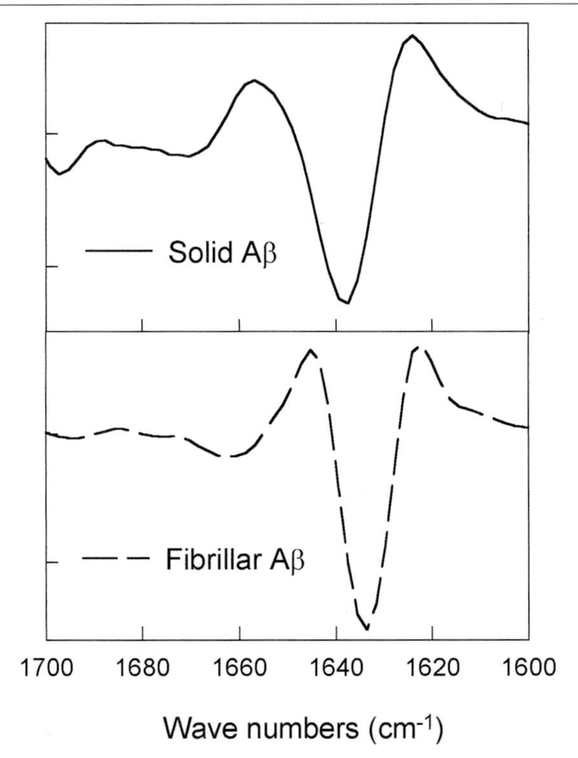

FIG. 4. Second-derivative spectra of the Alzheimer's Aβ40 peptide in the solid state following chemical synthesis and RP-HPLC purification (top) and after fibril formation (bottom). Identical spectra for the solid sample were obtained using KBr disks or ATR.

thin-film (starting with either solution or suspension), as a suspension, or in the dry solid state (Fig. 4).

Experimental Details

ATR has been used to analyze spectra of soluble and insoluble proteins and peptides (including inclusion bodies and fibrils) and for solid (powdered) proteins/peptides.[9,12,13] The method has also been used to monitor the kinetics of aggregate formation.[13] In ATR experiments the slit can be opened to its maximum to maximize light throughput. For many instruments, sensitivity can be increased by using a liquid nitrogen-cooled MCT

[12] K. Oberg, B. A. Chrunyk, R. Wetzel, and A. L. Fink, *Biochemistry* **33,** 2628 (1994).

[13] H. H. Bauer, M. Muller, J. Goette, H. P. Merkle, and U. P. Fringeli, *Biochemistry* **33,** 12276 (1994).

detector. Depending on the details of the specific apparatus used, the system may need to be purged with dry nitrogen or dry air to minimize contributions from water vapor in the air. Spectra are typically recorded at 2 or 4 cm^{-1} resolution by coadding 1000–10,000 interferograms. Background spectra are collected using a clean IRE. Depending on the length of time the experiment takes and the instrument and environmental conditions, it may be necessary to acquire additional background spectra, as well as water vapor spectra. The latter are obtained most readily by reducing the dry gas purge and collecting a spectrum using the clean IRE. Details of the solution ATR method have been published previously.[11]

Thin Film ATR

Probably the best form in which to study deposited protein samples is as a film deposited from solution or suspension. There are few or no interfering bands from either solvents or dispersion media, and the sample preparation does not require extremes of mechanical processing. However, care must be taken to avoid precipitation of salts as large crystals or aggregates if at high concentrations. Also, films of proteins can be examined under polarized radiation in order to learn more about the orientation of infrared-active functional groups, which in turn can provide insight into molecular orientation.

Thin-film ATR has been used to determine the secondary structure of inclusion bodies, folding aggregates, and amyloid fibrils. For soluble and insoluble proteins, 50 to 100 μl of 0.5 to 1 mg/ml solution or suspension, respectively, is placed on the IRE and the sample is dried down using nitrogen gas or dry air or a vacuum. The thin films have been shown to maintain protein molecules in a hydrated state and to conserve their three-dimensional structure.[14,15] The limited penetration depth in thin-film ATR is a substantial advantage in that it minimizes the contribution of any liquid water present. ATR samples containing materials of a high refractive index may result in significant shifts in the observed band positions, making comparisons between samples difficult. Hygroscopic materials tend to stick to the surface of the IRE and give a higher signal. ZnSe IREs are preferable to germanium for studying solid and thin-film samples due to their lower refractive index and hence higher signal-to-noise ratio.

Inclusion bodies, and several other kinds of cellular protein deposits, can typically be prepared as follows: frozen cell paste is suspended in water, lysed mechanically or with lysozyme, and centrifuged at 1000*g* for 30 min.

[14] E. Goormaghtigh, V. Cabiaux, and J. M. Ruysschaert, *Eur. J. Biochem.* **193,** 409 (1990).

[15] J. Safar, P. P. Roller, G. C. Ruben, D. C. Gajdusek, and C. J. Gibbs, Jr., *Biopolymers* **33,** 1461 (1993).

The pellet is resuspended in cell suspension buffer, homogenized, and centrifuged at 3000–5000g for 2 hr. This pellet may need to be resuspended in cell suspension buffer with 0.4 M urea or 0.1–0.2% nonionic detergent (e.g., Triton X-100 or Tween), homogenized, and centrifuged at 1000g for 30 min for further purification. The resulting pellet containing the inclusion bodies or insoluble protein is washed several times with water and can be stored at −20°. For ATR analysis the pellet may be suspended in water to form a thin film as discussed earlier.

ATR for Solid (Powdered) Proteins

Depending on the sample (probably the size of the particles is most critical), 1 to 5 mg of powder is simply spread on a large horizontal IRE and the spectrum collected. In some cases much less material may be needed to achieve an adequate signal-to-noise ratio. This method is well suited for the study of lyophilized proteins and peptides.

Diffuse Reflectance

The diffuse reflectance (DRIFT) method is most suited for dry samples of protein, in which either the neat protein powder or the protein diluted with dry KBr or KCl is placed directly in the sample cup. In diffuse reflectance the incident radiation, which is scattered in all directions by repeated reflections and refractions at the surface of particles, is collected over a very large scattering area. The diffusely reflected radiation contains a specular reflection component, a multiply refracted component, multiply reflected components, and other components that undergo a series of refractions and reflections. Although the specular component comes from the very first layer of the sample, the path length of the rest of the individual IR rays varies because each ray of incident radiation is scattered differently by particles in the sample powder. It has been reported that the radiation reaches about 3 mm deep from the surface. Therefore, it should be kept in mind that the diffuse reflection spectrum always contains both absorption and reflected spectral components. Diffuse reflectance has been used to monitor the aggregation of *S*-carboxymethylkeratin on thermal denaturation in the dry powder state.[16]

Methodology for Diffuse Reflectance

Diffuse reflectance FTIR protein spectra are typically gathered by filling a sample cup with a mixture of 400 mg of dry KBr with 0.5 mg of solid

[16] J. Koga, K. Kawaguchi, E. Nishio, N. Ikuta, I. Abe, and T. Hirashima, *J. Appl. Polym. Sci.* **37,** 2131 1989.

protein/protein formulation: microscale accessories are also available, requiring much smaller amounts of sample. The background spectra can be obtained using the empty cell or dry KBr powder. In some cases it may be desirable to use the dry sample without the addition (dilution) of KBr.

Transmission Mode: Thin Films

Thin films made from drying an aqueous suspension of aggregated protein (e.g., amyloid fibrils[17]) onto thin silver chloride disks can be used in transmission mode. As with ATR thin films, drying is accomplished either with the aid of a vacuum or by passage of dry air or nitrogen.

Transmission FTIR has also been used to study the interaction and stabilization of lyophilized proteins.[18] Dehydration-induced changes in the conformation seen by FTIR may not be supported by other techniques.[19] An increase in the degree of aggregation has been observed as a function of storage time for proteins dried in the absence of stabilizers. This aspect of protein stabilization and its application in the formulation of protein drugs is discussed in detail elsewhere in this volume.[20]

Protein Gels

Most proteins form gels on thermal denaturation at high protein concentrations, and protein gels have also been observed under other conditions for certain proteins, again at high concentrations. Such samples may be analyzed using FTIR in the transmission mode. For example, heat-gelled samples can be prepared *in situ* in a standard thermostated transmission cell.[21]

Methodology for Transmission Studies

In conventional transmission infrared spectroscopy the IR beam passes directly through the solution; due to minimal effects of IR light scattering, samples in the form of thin films or suspensions can be analyzed by transmission IR spectroscopy. Transmission mode FTIR has been used to study suspensions of aggregated proteins.[22] For conventional transmission FTIR,

[17] J. D. Termine, E. D. Eanes, D. Ein, and G. G. Glenner, *Biopolymers* **11,** 1103 (1972).
[18] S. J. Prestrelski, N. Tedeschi, T. Arakawa, and J. F. Carpenter, *Biophys. J.* **65,** 661 (1993).
[19] J. A. Rupley and G. Careri, *Adv. Prot. Chem.* **41,** 37 (1991).
[20] J. F. Carpenter, *Methods Enzymology* **309** [16] 1999 (this volume).
[21] A. H. Clark, D. H. P. Saunderson, and A. Suggett, *Int. J. Peptide Prot. Res.* **17,** 353 (1981).
[22] S. D. Allison, A. Dong, and J. F. Carpenter, *Biophys. J.* **71,** 2022 (1996).

protein concentrations in the 2- to 60-mg/ml (usually >30 mg/ml for samples in H_2O) range are required. For protein solutions in H_2O the path length must be $\leq$10 μm, and a 6-μm spacer is frequently used. In contrast, in D_2O, longer path lengths, typically 25–50 μm, can be used. The amount of sample required is fairly small, about 10–50 μl. For studies in D_2O, it is desirable to have complete H/D exchange; for most proteins, this entails exchanging under conditions of marginal stability, such as heating the sample at 50° for 1 hr.

Kinetics of Aggregation

The aggregation process may be followed in real time using either transmission cells or ATR. For example, aggregation of the peptide hormone calcitonin was monitored using ATR–FTIR,[13] and the aggregation of chymotrypsinogen on thermal denaturation was monitored using the transmission mode with a standard thermostated transmission cell.[23] Caution must be used in interpreting the results from ATR–FTIR experiments when starting with protein *solutions* due to the strong physisorption of most proteins to the IRE: this often leads to conformational changes in the protein.[11]

Infrared Microscopy

FTIR microscopy can be used to study the conformation of proteins in tissue samples, e.g., it has been used to observe the IR spectroscopic features of amyloid plaques *in situ.*[24] The FTIR microscope consists of a beam condenser, which focuses the IR radiation onto a small area, thereby allowing investigation of small samples with high optical throughput. With this technique, undesirable tissue areas can be masked and specific areas of interest in the tissue can be investigated. In addition, the use of a movable stage allows tissue to be mapped spectroscopically. The signal-to-noise ratio of the spectrum acquired with the desirable small apertures is very poor, as only a very small amount of IR radiation from the source passes through the aperture and the tissue to reach the detector. To improve spectral quality, the intensity of the infrared source must be increased; however, this results in the generation of heat, which may damage the tissue. This problem can be overcome using synchrotron infrared microspectroscopy.[24] The intrinsic brilliance of the synchrotron light source is 100–1000 times

[23] A. A. Ismail, H. H. Mantsch, and P. T. T. Wong, *Biochim. Biophys. Acta* **1121,** 183 (1992).

[24] L. P. Choo, D. L. Wetzel, W. C. Halliday, M. Jackson, S. M. LeVine, and H. H. Mantsch, *Biophys. J.* **71,** 1672 (1996).

greater than a traditional thermal emission source. Furthermore, the source is highly collimated (thus improving spatial resolution) and nonthermal, thereby eliminating the risk of tissue damage from excess heat production, although increasing the chance of radiation damage.

Data Analysis

Water and Buffer Subtraction

As noted, the presence of H_2O in the sample can pose the problem of a large contribution of liquid water absorption (1640 cm^{-1}) to the IR spectrum, as well as smaller contributions from water vapor in the air. In addition, most samples will also have contributions from buffer components. These contributions must be removed prior to analysis of the protein spectrum. This is usually accomplished by subtracting the spectrum of the buffer treated in the same manner as the sample (e.g., a thin film). The two criteria used for subtracting the liquid water contribution from protein spectra are (1) the region between 2000 to 1800 cm^{-1} is flat and (2) the amide I and amide II bands are well separated. This method does have its shortcomings and it is not always possible to get an absolutely flat spectrum between 2000 to 1800 cm^{-1}, but for most practical purposes it does work fairly efficiently. The water vapor contribution is subtracted from the spectrum after the liquid water subtraction.

Isotope-Edited Spectra

In some cases it may be possible to specifically and selectively label the sample with isotopes, e.g., ^{13}C in the carbonyl group. Because IR spectroscopy measures vibrational modes, which are sensitive to the mass of the atoms, it is possible, in principle, to study specific bonds due to the change in frequency associated with replacing a given atom by its isotope. This is most feasible in small peptides and has been used in the study of a fragment of the Aβ peptide that forms amyloid fibrils.[25]

Data Analysis

The feasibility of quantitative high-resolution IR spectra of proteins has been made possible by technical developments that overcome inherent limitations such as low sensitivity, spectral interference of surrounding environment such as H_2O, and overlapping component peaks in the amide bonds, characteristic of condensed-phase samples. Even a small protein

[25] T. T. Ashburn, M. Auger, and P. T. Lansbury, *J. Am. Chem. Soc.* **114,** 790 (1991).

will have many different IR-absorbing components; the overlap of these signals leads to broad spectra, which can, however, be analyzed to extract the underlying waveforms.

Two general methods have been used to analyze IR spectra of proteins, especially their secondary structure. One type of analysis, the frequency-based approach, involves deconvolution or resolution enhancement and assignment of the component bands. The alternative approach is based on factor analysis, using pattern recognition methods. The most commonly used software packages are available commercially and are based on Galactic Corp.'s GRAMS.

Each of the conformational states of the polypeptide chain has a distinct signature in the amide I region of the infrared spectrum. Because each of these conformational entities contributes to the vibrational spectrum, the observed amide I band contours are complex composites; they consist of many overlapping component bands that represent different structural elements, such as α helices, parallel or antiparallel β sheets, turns, and unordered or irregular structures.[26] The fundamental difficulty encountered in the analysis of such amide I envelopes arises from the fact that the widths of these component bands are usually greater than the separation between the maxima of adjacent peaks.

Deconvolution Methods

These methods, often referred to as resolution enhancement methods, decrease the widths of infrared bands, allowing for increased separation and better identification of the overlapping components. The two most commonly used methods are Fourier self-deconvolution (FSD) and second-derivative spectroscopy.[27]

Secondary structure assignment requires knowledge of the frequencies where the individual secondary structural components of a protein will absorb. Such assignments have been made using model proteins and polypeptides in which the structure has been determined by alternate methods such as X-ray crystallography, as well as from theoretical calculations. Secondary structure assignments of component bands have been made most frequently using the amide I region. Deconvolution of other spectral areas (e.g., amide III) may yield information on side chain interaction (tertiary structure) as well, but the necessary model systems and assignments are not yet well developed.

[26] W. K. Surewicz and H. H. Mantsch, "Spectroscopic Methods Determining Protein Structure in Solution." VCG Publishers, New York, 1996.

[27] H. H. Mantsch, D. J. Moffat, and H. L. Casal, *J. Mol. Struct.* **173,** 285 (1988).

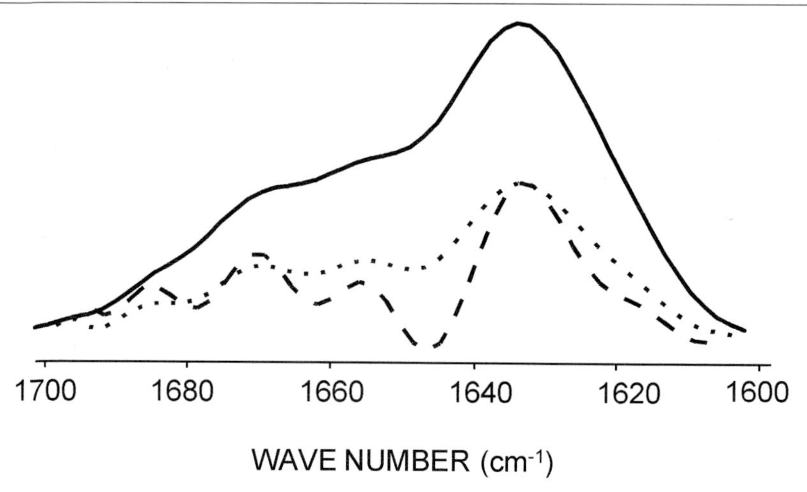

FIG. 5. Resolution enhancement/deconvolution of the amide I region of the IR spectrum. The solid line is the spectrum after the subtraction of water (liquid and vapor) and buffer components. The dotted line is the Fourier self-deconvoluted spectrum, and the dashed line represents the inverted second derivative.

Peak Assignment

The preferred method to deconvolute the amide I region is as follows: first both FSD and second derivative spectra are run to determine the positions of the components (Fig. 5). A robust analysis can only be assured if both methods give identical peak locations. Curve fitting can be done using the band positions derived from the FSD and second derivative spectra to fit the raw spectrum to a combination of several Gaussian–Lorentzian peaks. In robust analyses the band positions vary by less than ± 2 cm^{-1} from the deconvolution values. In a typical curve-fitting procedure the heights of individual peaks are allowed to vary, and the widths are constrained, e.g., to a maximum of 25 cm^{-1}, during the fit. The area under each peak is then used to compute the percentage of the individual component contributing to the amide I region. Quantitative analysis of band components can also be done by curve fitting to the FSD-deconvoluted spectrum. It has been suggested that quantitative information may also be obtained from curve fitting the second-derivative spectra,[28] although this is controversial: certainly the second-derivative spectra allow for qualitative estimates of component areas; however, errors may arise in curve fitting to baseline-corrected second-derivative spectra. In all curve-fitting proce-

[28] A. Dong, P. Huang, and W. S. Caughey, *Biochemistry* **29,** 3303 (1990).

TABLE I
CHARACTERISTIC FREQUENCIES (WAVE NUMBERS) FOR PROTEIN SECONDARY STRUCTURE ASSIGNMENTS OF AMIDE REGION IN H_2O

Wave number (cm^{-1})	Assignment
1615–1643	Extended chain/β sheet[a]
1647–1654	Disordered
1650–1660	α Helix
1651–1663	Loops
1663–1695	Turns
1692–1697	β Sheet

[a] There are also potential significant side chain band contributions in the 1600- to 1620-cm^{-1} region.

dures to the amide I band it is assumed that the extinction coefficients are identical for each type of secondary structure and consequently that the component peak area is directly proportional to the fraction of the secondary structure accounting for that component. However, this may not be a valid assumption.

Secondary structure assignments in amide I are shown in Table I. It should be noted that there are exceptions to the assignments shown in Table I, reflecting the effects of local conformational effects in the protein on the vibrational frequencies, as well as in the attribution of observed component bands. Differences in the length and direction of hydrogen bonds lead to variations in the strength of the hydrogen bonds for different secondary structures, and hence the vibrational frequencies of the amide C=O groups, resulting in characteristic amide I frequencies.[2] Stronger hydrogen bonds lead to lower amide I absorption frequencies. In addition, the local environment about the carbonyl can significantly affect the frequency of the CO stretch, due to transition dipole coupling.[28] Typically, proteins containing predominantly β-sheet structure show peaks between 1623–1643 cm^{-1}, whereas those which are predominantly α-helical show amide I peaks between 1648–1658 cm^{-1}. The less intense absorption peaks between 1675–1695 cm^{-1} may be assigned to high frequency β-sheet components arising from transient dipole coupling[29] and turns.

Given the prevalence of the β structure in aggregated protein, it would be desirable to differentiate between parallel and antiparallel β sheets. Calculations show that the parallel pleated β sheet and the antiparallel β sheet have different frequencies (1642 and 1630 cm^{-1}), which correspond

[29] W. K. Moore and S. Krimm, *Proc. Natl. Acad. Sci. U.S.A.* **72,** 4933 (1975).

closely to the observed frequencies for poly(Ala) in these conformations.[1] The sensitivity of the carbonyl stretch to transition dipole coupling effects and its local environment, however, mean that significant deviations from these theoretical values are observed. In fact, it is clear from examination of spectra of proteins whose structure is well established from X-ray crystallography and nuclear magnetic resonance that both types of β sheets may be observed at frequencies throughout the 1623- to 1643-cm^{-1} range. Thus, in the absence of additional corroborative evidence, e.g., from the corresponding Raman spectrum, it is likely to be erroneous to assume that strongly absorbing components in the vicinity of 1640 cm^{-1} will reflect the parallel β sheet and that those in the vicinity of 1625 cm^{-1} will reflect the antiparallel β sheet. Figure 6 shows the curve-fit spectra of three representative proteins in their native states, interleukin-2 (IL-2), a four-helix bundle protein, haptoglobin, an example of an $\alpha + \beta$ protein, and the variable domain (V_L) of an immunoglobulin, an all-β protein. These spectra illustrate the typical features of proteins that are predominantly of well-defined secondary structure, as well as some of the difficulties encountered in secondary structure assignment.

One of the biggest problems with the secondary structure analysis of proteins using amide I FTIR is that the α helical and the unordered structures show bands fairly close together, especially in H_2O. This sometimes causes complications and errors in the analysis. In contrast, β-sheet/extended chain bands are well separated from those of other types of secondary structure. As noted, aggregated proteins have large β-structure contributions and thus are less affected by such problems.

Factor Analysis

Factor analysis is a mathematical method that is used increasingly to analyze multicomponent spectra. In particular, it may be used to determine how many pure components are present in a series of mixture spectra, what the pure spectral components are, and how much of each pure component is present. Factor analysis as a method to analyze infrared amide bands in order to extract quantitative information on protein secondary structure was introduced independently by Lee *et al.*[30] and Dousseau and Pezolet.[31] In the procedure described by Lee *et al.*,[30] a calibration set is first generated from IR spectra of proteins for which the quantitative information regarding the individual secondary structural components can be deduced independently from X-ray crystallography. Factor analysis is then used to generate eigenspectra, which may be combined linearly to reconstruct the original

[30] D. C. Lee, P. I. Haris, D. Chapman, and R. C. Mitchell, *Biochemistry* **29,** 9185 (1990).
[31] F. Dousseau and M. Pezolet, *Biochemistry* **29,** 8771 (1990).

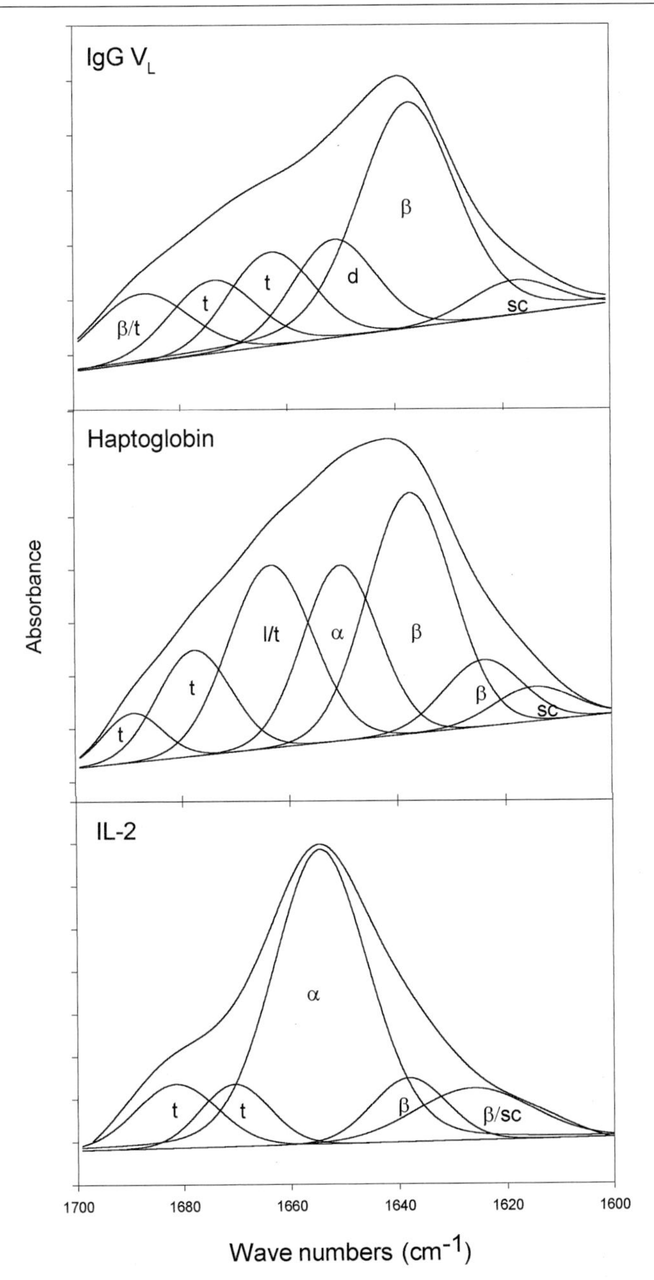

FIG. 6. Representative spectra of proteins in their native states, interleukin-2 (IL-2), a four-helix bundle protein, haptoglobin, an $\alpha + \beta$ protein, and the variable domain (V_L) of an immunoglobulin, and all-β protein. sc, side chain contributions; l, loop; t, turn.

spectra. Multiple linear regression analysis is subsequently used to identify the eigenspectra that correlate with the variations in the properties of interest (i.e., % α helix, β sheet, and turns). In the analysis of the spectrum of an unknown protein, the factor loadings required to reproduce this spectrum are substituted in the regression equation for each property to determine the secondary structure content. This yielded best results when the spectral range analyzed was limited to the amide I region (1600–1700 cm^{-1}) and spectra measured in H_2O.

Dousseau and Pezolet[31] performed detailed analysis of IR spectra of proteins using both classical least-squares and partial least-squares methods. Correlation between the secondary structure content calculated by the classical least-squares method and those estimated from X-ray data were relatively poor. The corresondence between X-ray data and infrared estimates of the secondary structure was improved considerably when the analysis based on the classical least-squares was replaced by that employing the partial least-squares (or factor analysis) method. Best results were obtained when both amide I and II regions of the spectra were used to generate the calibration set and when it was assumed that the protein secondary structure of proteins is composed of only four types of structure: ordered and disordered helices, β sheet, and undefined conformation. Attempts to introduce turns as a distinct type of secondary structure resulted in a loss of accuracy. Furthermore, no distinction could be made between parallel and antiparallel β sheets.

Singular value decomposition methods may also be used for the analysis of IR spectra using an appropriate basis set. It is worth noting, in the context of factor analysis, that unless the basis set has a good representation of spectra from known samples of similar structure to that in the unknown, significant errors may result. This is particularly likely in the case of protein deposits consisting of partially folded conformations and basis sets of native proteins.

[37] Stable Isotope-Labeled Peptides in Study of Protein Aggregation

By MICHAEL A. BALDWIN

This article was inspired by studies carried out over several years on synthetic peptides aimed at elucidating aspects of the behavior of the prion protein (PrP) that causes fatal neurodegenerative diseases, often referred to

0076-6879/99 $30.00

as prion diseases or the transmissible spongiform encephalopathies. These diseases are unique in that they can arise in sporadic, familial, and infectious forms, and the infectious agent is a modified form of a normal cellular protein. Such studies were necessitated by the difficulty of isolating and purifying sufficient PrP from natural sources and difficulties encountered in developing an effective expression and purification system for recombinant PrP to determine three-dimensional structures. Furthermore, it was anticipated that dissection of the entire sequence into shorter fragments might allow a detailed analysis of the structural characteristics of regions of PrP most likely to be involved in its different functions, including its pathogenic behavior.

PrP occurs as two physically and biologically distinct isoforms: the normal cellular form, PrP^{C}, and the protease-resistant infectious isoform, PrP^{Sc}, which accumulates in amyloid plaques. An N-terminally truncated version of PrP^{Sc} (PrP 27–30) was first isolated from scrapie-infected hamster brain, N-terminally sequenced and cloned.[1–5] Subsequently, both isoforms of the full-length protein were purified, and although they are dramatically different in their physical properties,[6] they were found to have the same N-terminal sequence.[7–10] Later studies showed that both isoforms have identical amino acid sequences and chemical posttranslational modifications.[11] It has been established that PrP^{c} interacts with PrP^{Sc} to give more PrP^{Sc}, thus the formation of PrP^{Sc} must involve a conformational change. PrP^{C} is protease sensitive and largely α helical, whereas PrP^{Sc}, when truncated by limited proteolysis (PrP27–30), shows typical amyloid properties,

[1] S. B. Prusiner, D. C. Bolton, D. F. Groth, K. A. Bowman, S. P. Cochran, and M. P. McKinley, *Biochemistry* **21,** 6942 (1982).

[2] D. C. Bolton, M. P. McKinley, and S. B. Prusiner, *Science* **218,** 1309 (1982).

[3] S. B. Prusiner, D. F. Groth, D. C. Bolton, S. B. Kent, and L. E. Hood, *Cell* **38,** 127 (1984).

[4] B. Oesch, D. Westaway, M. Wälchli, M. P. McKinley, S. B. H. Kent, R. Aebersold, R. A. Barry, P. Tempst, D. B. Teplow, L. E. Hood, S. B. Prusiner, and C. Weissmann, *Cell* **40,** 735 (1985).

[5] B. Chesebro, R. Race, K. Wehrly, J. Nishio, M. Bloom, D. Lechner, S. Bergstrom, K. Robbins, L. Mayer, J. M. Keith, C. Garon, and A. Haase, *Nature* **315,** 331 (1985).

[6] R. K. Meyer, M. P. McKinley, K. A. Bowman, M. B. Braunfeld, R. A. Barry, and S. B. Prusiner, *Proc. Natl. Acad. Sci. U.S.A.* **83,** 2310 (1986).

[7] J. Hope, L. J. D. Morton, C. F. Farquhar, G. Multhaup, K. Beyreuther, and R. H. Kimberlin, *EMBO J.* **5,** 2591 (1986).

[8] K. Basler, B. Oesch, M. Scott, D. Westaway, M. Wälchli, D. F. Groth, M. P. McKinley, S. B. Prusiner, and C. Weissmann, *Cell* **46,** 417 (1986).

[9] D. C. Bolton, P. E. Bendheim, A. D. Marmorstein, and A. Potempska, *Arch. Biochem. Biophys.* **258,** 579 (1987).

[10] E. Turk, D. B. Teplow, L. E. Hood, and S. B. Prusiner, *Eur. J. Biochem.* **176,** 21 (1988).

[11] N. Stahl, M. A. Baldwin, D. B. Teplow, L. Hood, B. W. Gibson, A. L. Burlingame, and S. B. Prusiner, *Biochemistry* **32,** 1991 (1993).

forming fibrils that stain with Congo red dye to give green/gold birefringence.[12,13] Fourier transform infrared (FTIR) spectroscopy and circular dichroism (CD) confirmed that PrP^{Sc} is rich in β sheet.[14–18]

Peptides as Structural Models

As structural models for the behavior of proteins, synthetic peptides have advantages and disadvantages. With careful attention to detail, peptides of up to 50–70 amino acids can be prepared by stepwise solid-phase peptide synthesis (SPPS) and isolated in high purity within a few days, much quicker than the time time taken to create a recombinant protein from scratch. It is easy to introduce mutations and even modified or unnatural amino acids.[19] However, sequences of 100 amino acids or more are very difficult to synthesize, whereas length is not an issue for recombinant methods. One way to circumvent this problem is to use so-called orthogonal synthesis, making sections of a protein by SPPS and then chemically ligating these peptides to construct a larger entity. It is also possible to ligate a synthetic peptide containing unique elements such as isotopic labels to a larger recombinant fragment.

The physical properties and secondary structure of synthetic peptides may mirror those of the same region in the protein, but this is not always true. Thus a number of peptides derived from an N-terminal region of PrP have been shown by different workers to adopt β-sheet structure and to form amyloid fibrils, e.g., 109–122 (referred to as H1),[20] 106–126,[21] and 117–133.[22] The behavior of this region is believed to be the same in PrP^{Sc}.

[12] S. B. Prusiner, M. P. McKinley, K. A. Bowman, D. C. Bolton, P. E. Bendheim, D. F. Groth, and G. G. Glenner, *Cell* **35,** 349 (1983).

[13] M. P. McKinley, R. K. Meyer, L. Kenaga, F. Rahbar, R. Cotter, A. Serban, and S. B. Prusiner, *J. Virol.* **65,** 1340 (1991).

[14] B. W. Caughey, A. Dong, K. S. Bhat, D. Ernst, S. F. Hayes, and W. S. Caughey, *Biochemistry* **30,** 7672 (1991).

[15] M. Gasset, M. A. Baldwin, R. J. Fletterick, and S. B. Prusiner, *Proc. Natl. Acad. Sci. U.S.A.* **90,** 1 (1993).

[16] K.-M. Pan, M. Baldwin, J. Nguyen, M. Gasset, A. Serban, D. Groth, I. Mehlhorn, Z. Huang, R. J. Fletterick, F. E. Cohen, and S. B. Prusiner, *Proc. Natl. Acad. Sci. U.S.A.* **90,** 10962 (1993).

[17] J. Safar, P. P. Roller, D. C. Gajdusek, and C. J. Gibbs, Jr., *J. Biol. Chem.* **268,** 20276 (1993).

[18] J. Safar, P. P. Roller, D. C. Gajdusek, and C. J. Gibbs, Jr., *Prot. Sci.* **2,** 2206 (1993).

[19] G. B. Fields, *Methods Enzymol.* **289** (1997).

[20] M. Gasset, M. A. Baldwin, D. Lloyd, J.-M. Gabriel, D. M. Holtzman, F. Cohen, R. Fletterick, and S. B. Prusiner, *Proc. Natl. Acad. Sci. U.S.A.* **89,** 10940 (1992).

[21] G. Forloni, N. Angeretti, R. Chiesa, E. Monzani, M. Salmona, O. Bugiani, and F. Tagliavini, *Nature* **362,** 543 (1993).

[22] J. H. Come, P. E. Fraser, and P. T. Lansbury, Jr., *Proc. Natl. Acad. Sci. U.S.A.* **90,** 5959 (1993).

However, peptide 104–122 is unstructured both in solution and as a dried film and will not form fibrils unless it is seeded with H1.[23] Thus the choice of the optimum regions to be synthesized is not always obvious and may only be discovered by trial and error, requiring the synthesis of many overlapping peptides.

Structural information derived from peptides using various spectroscopic methods, including FTIR and solid-state nuclear magnetic resonance (NMR), may be enhanced by the introduction of isotope labels. Protocols are well established for the uniform isotopic labeling of recombinant proteins by expression in isotopically enriched media. Thus ^{13}C- and/or ^{15}N-labeled proteins are produced routinely for structural analysis by NMR. It is also possible to label every amino acid of one kind in a recombinant protein, such as all methionines, and a specific site may be labeled by *in vitro* translation. However, the ability to easily introduce one or more stable isotopes at *selected positions* in a synthetic peptide is unique to chemical methods.

Peptide Synthesis

The preparation of peptides by SPPS as pioneered by Merrifield[24] involves the stepwise addition of chemically activated amino acids to the N terminus of the growing peptide attached to a solid resin. This field has been covered in detail in another volume in this series and will be summarized only briefly here.[19] Reactive side chains are protected from unwanted reactions by chemical groups that are resistant to reagents used during the synthesis. The N terminus is also protected with a chemical group that is removed after addition to the peptide to allow reaction with the next amino acid. Merrifield[24] employed the *tert*-butyloxycarbonyl group (*t*-BOC) for N-terminal protection, which is removed at each cycle with TFA. Finally the peptide is cleaved from the resin and the side chain protecting groups are removed in a single step with anhydrous hydrofluoric acid. More recent methods employ the 9-fluorenylmethoxycarbonyl (FMOC) protecting group. Because this is base labile there is no need to expose the peptide to acid conditions at every cycle and alternative side chain protecting groups can be used that are cleaved under less harsh conditions. This eliminates the need for highly toxic HF gas, which requires careful handling in a specialized apparatus. Similarly, new classes of highly active acylating reagents have been developed that reduce the time required to couple each amino acid from approximately 1 hr to 20 min.

[23] J. Nguyen, M. A. Baldwin, F. E. Cohen, and S. B. Prusiner, *Biochemistry* **34,** 4186 (1995).
[24] R. B. Merrifield, *J. Am. Chem. Soc.* **85,** 2149 (1963).

Although peptide synthesis can be carried out manually, the repetitive steps involved lend themselves to automation, and commercial synthesizers are available from several manufacturers. At the "black box" level these require the operator only to introduce amino acid cartridges in the desired order and to ensure that the reagent bottles are full. FMOC has replaced *t*-BOC as the method of choice for most applications; in practice there may be little choice as most commercial synthesizers today are designed only for FMOC chemistry. Despite this, many researchers engaged in the synthesis of long peptides of >70 residues still believe *t*-BOC to be superior, consequently they go to great lengths to maintain their older synthesizers in good working order, particularly the venerable Applied Biosystems Model 430 (PE Biosystems, Foster City, CA).

Amino acid coupling reactions are less than 100% efficient, resulting in deletion peptides, sometimes lacking a single amino acid. With increasing length of the peptide chain, the chromatographic properties between the target peptide and such impurities become more similar, making separation difficult. Capping any unreacted amino groups after the addition of each amino acid to prevent further reaction, e.g., with acetic anhydride, produces families of much shorter impurity peptides that are removed more easily.

Isotope Labels

Many but not all amino acids can be purchased with specific atoms replaced by one or more of the heavy stable isotopes ^{2}H, ^{13}C, or ^{15}N. Established commercial suppliers include Cambridge Isotope Laboratories (Andover, MA) and Isotec (Miamisberg, OH), and there is an NIH resource for stable isotopes at Los Alamos, supported by the National Center for Research Resouces. The most common labeled amino acids can be purchased with either *t*-BOC or FMOC N-terminal protection. They are expensive, ranging from approximately $400 to several thousand dollars per gram, consequently it is important to plan a strategy that places the labels in the most effective positions. The cost depends greatly on the position of the label in the amino acid, e.g., a recently quoted price for N-*t*-BOC-[^{13}C]alanine labeled at C-1 was ~$500 per gram whereas the C-2 and C-3 analogs cost ~$750 per 0.25 g. The cost in dollars/gram for *t*-BOC and FMOC protected compounds is approximately the same, but the higher molecular mass of the FMOC group makes the *t*-BOC compounds less expensive on a molar basis, e.g., the M_r of unlabeled *t*-BOC-alanine is 189 compared with 311 for the FMOC analog. The quantity required depends on the scale of the synthesis, which in turn depends on the amount of peptide required. For structural studies we usually employ a 0.25 m*M* scale, which for a 10 residue peptide of M_r 1000 gives a theoretical yield of 250 mg. Losses in synthesis and purification might limit the true yield to 100 mg, which is sufficient for all biophysical characterization. Longer peptides give higher theoretical yields, but in practice they give lower yields on a percentage basis, thus an expectation of about 100 mg of purified product remains reasonable. The high yields of SPPS are achieved by using reagents in large excess, thus each amino acid may be applied in a fourfold excess, i.e., 1 m*M* or 0.25 g for one of the common protected amino acids having an M_r of about 250. In practice we may apply the labeled amino acid in only a twofold excess, at some risk of reducing the overall yield of peptide. Under these conditions, 1 g of labeled amino acid is sufficient for about eight reaction cycles.

Purification and Analysis

Peptides are cleaved from the resins and purified by standard reversed-phase high-performance liquid chromatography (HPLC) methods. In general we use semipreparative C_4 columns of 10 mm diameter, capable of separating up to 50 mg peptide in a single loading. However, chromatographic resolution is compromised under these conditions and a further separation of each of the eluted components may be carried out on a 4.6-mm analytical column, collecting multiple fractions across each peak. We normally use standard water/acetonitrile/TFA buffers, but 1,1,1,3,3,3-hexafluoro-2-propanol (HFIP) has been employed for very hydrophobic amyloid peptides.

There are a number of approaches to confirming the identity of synthetic peptides. Although an HPLC trace may show a single peak, coelution of deletion or truncation peptides having identical retention times is common. We routinely analyze a small aliquot of each fraction from the analytical HPLC by either electrospray mass spectrometry (ESIMS) or laser-assisted matrix desorption/ionization mass spectrometry (MALDIMS).[25] For ESIMS, a few microliters of each HPLC fraction can be sprayed directly into the ion source, although TFA used in the HPLC reduces the sensitivity of detection. For MALDIMS, 1 μl of each fraction is dried with 1 μl of a standard matrix solution, such as α-cyano-4-hydroxycinnamic acid, available predissolved in high purity from Hewlett Packard. Both of these techniques reveal coeluting impurities and are capable of accurately measuring peptide M_r values, generally to within 0.1 Da or better. This is sufficient to identify fractions containing peptides of the correct molecular composition, although this information does not confirm the correct sequence of the amino acids. Careful attention to detail in loading the correct sequence of

[25] J. R. Chapman, ed., "Protein and Peptide Analysis by Mass Spectrometry." Humana Press, Totowa, NJ, 1996.

amino acids into the peptide synthesizer, or entering the correct sequence into the computer that controls the selection of amino acids, makes it extremely unlikely that a peptide of the correct mass will have the wrong sequence. However, peptide sequences can be confirmed by either mass spectrometric or Edman sequencing. Although most laboratories are not equipped for such measurements, they can be contracted out.

An increased M_r will confirm the presence of heavy isotopes in a peptide, especially if an unlabeled control is available for comparison. Edman sequencing cannot identify the location of the labels as it is unable to distinguish between normal isotopes and heavy isotopes, but this is achieved readily by tandem mass spectrometry (MS/MS) with collision-induced dissociation (CID). The molecular ion is selected in one mass analyzer (MS1) and is fragmented by collisions with an inert gas, then the fragments are separated in a second mass analyzer (MS2). An example of the use of this is given in Fig. 1 for the PrP peptide H1 with the sequence MKHMA GAAAAGAVV-NH_2. This corresponds to PrP residues 109–122, but for the purpose of this discussion these residues will be identified as 1–14. This was synthesized for solid-state NMR without labels and with two different combinations of two ^{13}C labels in alanine-7 and -8. The M_r for the unlabeled peptide was measured as 1282.7 (calculated 1282.66), whereas both doubly labeled peptides were found to be 2 Da higher, as anticipated for the replacement of two ^{12}C's by ^{13}C's. Figure 1 shows a portion of MS/MS spectra for the three versions: (A) unlabeled, (B) having heavy isotopes at the carbonyl of Ala-7 and the α carbon of Ala-8, and (C) having the labels reversed, i.e., at the α carbon of Ala-7 and the carbonyl of Ala-8. In this particular peptide the lysine residue close to the N terminus favors retention of charge by the N-terminal fragments, giving so-called **a** and **b** ions.[26] Thus cleavage of an amide bond between residue 7 and 8 gives $\mathbf{b}_7$, whereas cleavage between the α carbon and the carbonyl carbon in residue 7 gives $\mathbf{a}_7$, which is lacking the final C=O and is therefore 28 Da lower in mass than $\mathbf{b}_7$ in an unlabeled peptide. Mass differences between successive **a** and **b** ions allow the positions of the heavy isotopes to be determined precisely. Because $\mathbf{b}_6$ is at m/z 656 for all three species, including the unlabeled control, we can conclude that no ^{13}C labels are present in the first six residues. However, because $\mathbf{a}_7$ for compound C is 1 Da higher than the equivalent ion for A or B, the α carbon of the seventh residue in C is ^{13}C. The mass difference of 28 Da between $\mathbf{a}_7$ and $\mathbf{b}_7$ for peptides A and C corresponds to $^{12}C{=}O$, but the difference of 29 Da for peptide B is due to $^{13}C{=}O$. Similar logic allows the positions of the other labels to be identified unambiguously.

[26] K. Biemann, *Methods Enzymol.* **193,** 886 (1990).

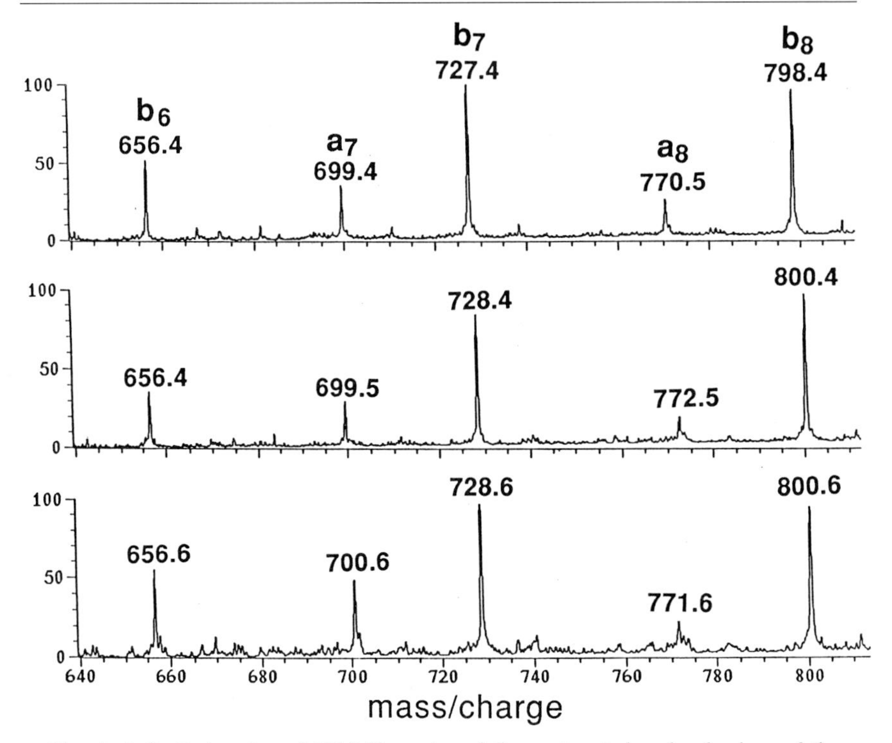

FIG. 1. A limited region of MS/MS spectra of the protonated molecular ions of three analogs of H1: (top) without isotopic labels, (middle) with ^{13}C atoms at the carbonyl of Ala-7 and the α carbon of Ala-8, and (bottom) with ^{13}C atoms at the α carbon of Ala-7 and the carbonyl of Ala-8. The marked peaks represent N-terminal fragment ions formed by cleavages between an α carbon and a carbonyl, giving the **a** ion series, and of an amide bond between a carbonyl and a nitrogen, giving the **b** ion series, respectively.

Spectra shown in Fig. 1 were obtained with a high-resolution magnetic sector tandem mass spectrometer employing liquid secondary ionization (LSIMS). The availability of such instruments is limited, but similar experiments can be conducted at lower resolution using the more widely available triple quadrupole mass spectrometer with electrospray ionization (ESI) or a MALDI time-of-flight mass spectrometer with a reflectron or ion mirror that allows postsource decay spectra to be monitored. By ESI, most peptides give multiply charged ions and the interpretation of CID spectra may be more complicated. In general, if a doubly charged ion can be selected for fragmentation, most of the fragments will be singly charged. In this instance, it was fortuitous that the lysine at the N terminus favored **a** and **b** ions. A basic residue close to the C terminus would favor charge retention by the

C-terminal fragments, giving predominantly **x** and **y** ions, which are equally effective for determining the sequence. In many cases, both N- and C-terminal fragment ions are observed.

For spectroscopic studies, HPLC fractions identified by mass spectrometry as containing the target peptides without other peptide impurities should be lyophilized, resuspended in dilute HCl, and lyophilized a second time to remove residual TFA, as this interferes with the FTIR amide I band used for secondary structural analysis. For most subsequent experiments, stock solutions of known concentration can be prepared by dissolving a known weight of the dry purified peptide in water. Short hydrophobic peptides such as H1 frequently have very limited solubility in water, in which case organic solvents mixed with water can be employed, e.g., acetonitrile or HFIP. It may be necessary to prepare solutions fresh for fibrillogenic peptides rather than use stock solutions as these may dissolve initially but then come out of solution as amyloid fibrils. If the degree of solubilization is in doubt, the solution should be filtered and the concentration determined by amino acid analysis. A number of methods have been described for this, including the following due to Tarr.[27] For this, aliquots are dried in glass tubes and hydrolyzed overnight with 6 *N* HCl and phenol *in vacuo* at 110°. Residual HCl is removed on a vacuum centrifuge and the hydrolyzate is dissolved in water. Samples are derivatized with *o*-phthalaldehyde (Pierce, Rockford, IL) and separated by HPLC (C_{18} column: solvent A, 46 m*M* sodium acetate/4 m*M* acetic acid; solvent B, methanol) using a nonlinear gradient designed to optimize the separation of all amino acids. Peaks are detected by fluorescence and are quantitated relative to 25- and 125-pmol amino acid standards. If fibril formation is suspected, this may be confirmed by electron microscopy.

Preparation of Proteins by Peptide Ligation

Many workers have achieved the successful ligation of peptides to form proteins by either chemical or enzymatic methods and, in many instances, have demonstrated that the synthetic product has normal biological action. We have ligated two peptides of 52 and 53 amino acids, respectively, to form a truncated PrP of 105 amino acids (Ball *et al.,* in preparation). We are also able to make this entire species by SPPS, but ligation allows a cassette system to be adopted. For example, we wish to probe the effects of mutations in the H1 region, such as the incorporation of a number of known pathogenic mutations. The H1 sequence is in the N-terminal fragment, so the preparation of a stock of the invariant C-terminal peptide

[27] G. E. Tarr, *in* "Methods of Protein Microcharacterization" (J. E. Shively, ed.), p. 154. Humana Press, Clifton, NJ, 1986.

allows this to be ligated to any number of variants of the N terminus. Thus to make a new molecule requires the synthesis of a 52-mer rather than a 105-mer. For isotope incorporation this is particularly advantageous as the yield of purified peptide by SPPS decreases significantly for species of ~100 residues, with a consequent loss of the expensive labeled amino acids.

Unfortunately, there is little consensus as to the ideal approach to peptide ligation and many strategies employ reagents that are not available commercially. This situation may improve quite rapidly if the commercial benefits of protein syntheis compared with recombinant expression become more widely appreciated. One succesful approach requires ligation at a cysteine residue, which therefore is dependent on the target protein having an appropriately located cysteine(s).[28,29] This was the case with the PrP analog referred to earlier, a single cysteine coming exactly at the halfway point of the molecule. The peptides to be ligated are selected such that the cysteine residue is at the N terminus of the C-terminal peptide, which is synthesized by conventional SPPS, cleaved, and purified. The first residue of the N-terminal peptide is coupled to a special resin through a thioester linkage such that cleavage from the resin leaves a thioester at the C terminus. In the condensation between the two peptides, nucleophilic attack by the cysteine thiol links both peptides through a new thio ester bond, which then rearranges, with the amino group displacing the thiol to revert back to cysteine. The ligation is straightforward and occurs in high yield. Chromatographic separation and purification are much easier compared with the analogous product made by SPPS as there are no impurities of comparable size.

FTIR Experiments with Isotopically Labeled Peptides

Because FTIR is covered in a separate article in this volume,[30] only issues specific to the analysis of isotopically labeled peptides will be considered here. Infrared spectroscopy monitors the vibrations of molecular bonds. If an atom within a vibrating bond is replaced by a heavier isotope, this increases the moment of inertia, thereby reducing the frequency of vibration.[31] The natural frequency of the infrared carbonyl bond vibration that contributes 90% of the amide I absorption in FTIR is ~1645–1650 cm^{-1} for unstructured regions of a peptide or protein.[32] If the amide

[28] P. E. Dawson, T. W. Muir, I. Clark-Lewis, and S. B. H. Kent, *Science* **266,** 776 (1994).
[29] J. P. Tam, Y.-A. Liu, C.-F. Liu, and J. Shao, *Proc. Natl. Acad. Sci. U.S.A.* **92,** 12485 (1995).
[30] S. Seshadri, R. Khurana, and A. L. Fink, *Methods Enzymol.* **309** [36] 1999 (this volume).
[31] S. Krimm and J. Bandekar, *in* "Advances in Protein Chemistry" (C. B. Anfinsen, J. T. Edsall, and F. M. Richards, ed.), Vol. 38, p. 181. Academic Press, Orlando, FL, 1986.
[32] D. M. Byler and H. Susi, *Biopolymers* **25,** 469 (1986).

I band were due solely to the carbonyl vibration, the relationship in Eq. (1) would predict that replacing the normal ^{12}C by ^{13}C would cause a reduction of 37 cm^{-1} in the frequency ν.[33] Thus we would anticipate a shift to ~1610 cm^{-1}.

$$\nu_{C=O} = (1/2\pi)(k/\mu)^{1/2} \quad (1)$$

where k is the force constant for the CO stretch and μ is the reduced mass, $M_C M_O / M_C + M_O$.

In the absence of heavy isotopes, dipole–dipole interactions in intermolecular antiparallel β sheets cause a splitting of the FTIR absorption into a weak high frequency band (~1690 cm^{-1}) and a strong low frequency band (~1625 cm^{-1}).[32] The low frequency band is unique to the β sheet, thus FTIR is a powerful method for detecting the presence of this secondary structural feature. However, these interactions can be disrupted by a single ^{13}C in place of a ^{12}C at a carbonyl carbon. If the label is embedded in the β sheet, the normal $^{12}C{=}O$ low frequency peak is shifted to higher frequencies and a new, anomalously intense, lower frequency peak appears due to the $^{13}C{=}O$ vibrations. The greater the interactions between isotopic labels in adjacent strands, the lower the frequency of the $^{13}C{=}O$ peak. This approach, which is described frequently as isotope-edited FTIR, was employed to identify regions of the Alzheimer's Aβ peptide involved in the β sheet[33] and to study the structural basis of pancreatic amyloid formation.[34] We have used this methodology to modulate the spectrum of H1, labeling at each of the carbonyl carbons in turn between Met-112 and Val-122 (Fig. 2) (Baldwin *et al.,* unpublished data).

Unlabeled peptide H1 as a thin film deposited from HFIP gave an FTIR spectrum typical of α helix with the amide I band maximum at ~1660 cm^{-1}. However, exposure to water caused an instantaneous shift to a β-sheet spectrum with a strong signal at ~1628 cm^{-1} and a weaker peak at ~1700 cm^{-1} (Fig. 2, bottom trace). This behavior was consistent with other experiments that showed that H1 could form amyloid fibrils that stain with Congo red dye.[20] H1 peptides specifically labeled at the carbonyl carbon of Met-112 or Ala-113 gave almost identical spectra with very little evidence of any contribution from the heavy isotope, except for a weak shoulder in the peak at 1613 cm^{-1}, due to the vibrations of isolated $^{13}C{=}O$ bonds, which at this point in the peptide were not incorporated in the β sheet. However, labels at Gly-114 through Val-121 all had dramatic effects, shifting the ^{12}C absorption to higher frequencies by as much as 7 cm^{-1} and giving

[33] K. J. Halverson, I. Sucholeiki, T. T. Ashburn, and P. T. Lansbury, Jr., *J. Am. Chem. Soc.* **113,** 6701 (1991).

[34] T. T. Ashburn, M. Auger, and P. T. Lansbury, Jr., *J. Am. Chem. Soc.* **114,** 790 (1992).

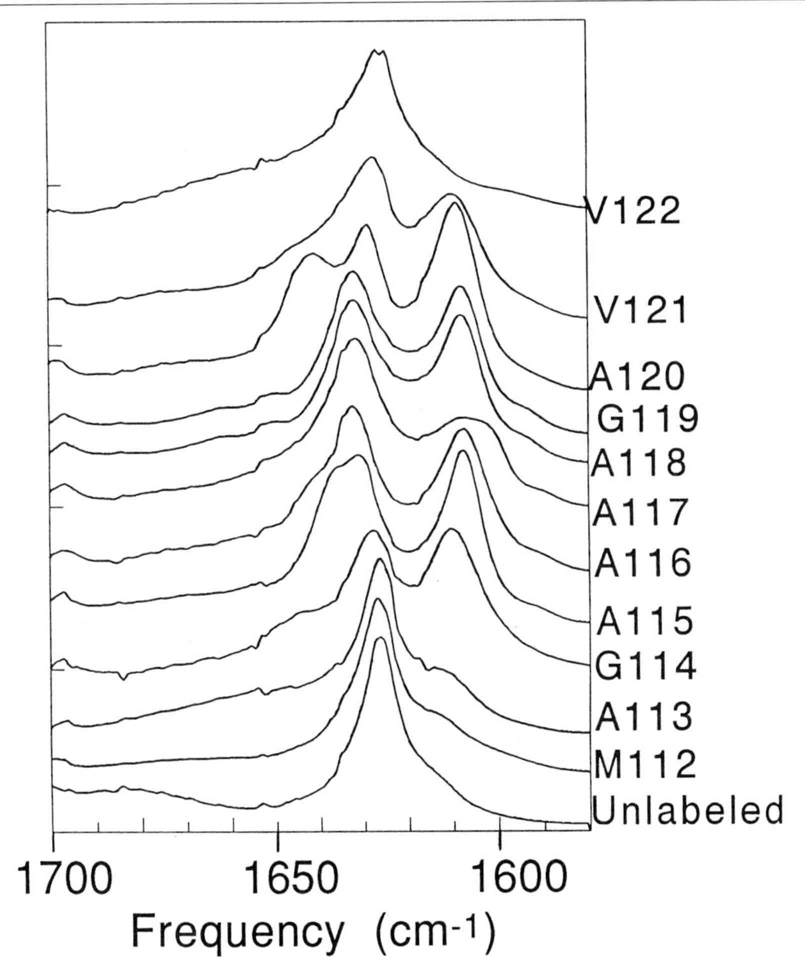

FIG. 2. Amide I bands from FTIR spectra of thin films of H1 peptides deposited from aqueous solution. Except for the lowest trace, each spectrum was obtained from a peptide having a ^{13}C label at the carbonyl carbon of a single amino acid throughout the region Met-111 to Val-122.

new ^{13}C peaks, mostly at 1608 cm^{-1}, in some instances more intense than the ^{12}C peaks. Some apparently anomalous results such as the ^{13}C doublet at 1608 and 1602 cm^{-1} for Ala-117, the ^{12}C doublets at 1632 and 1636 for Ala-115, and 1629 and 1642 for Ala-120 were reproducible and must reflect irregularities in the structure. There were no major discontinuities in the sheet that would be caused by a hairpin turn, suggesting that these H1 sheets are formed by the intermolecular association of peptides, each containing a single strand. A peptide extended at the C terminus to residue 125 with three additional residues GGL was prepared with $^{13}C{=}O$ at Val-122. This gave a similar spectrum to the Val-121 H1 peptide with a $^{12}C{=}O$ shift and a new low frequency $^{13}C{=}O$ peak. Thus the termination of H1 at residue 122 had artificially constrained the formation of the β sheet.

Interactions between Peptides

A unique feature of prion diseases is the ability of the pathogenic protein PrP^{Sc} to induce a structural conversion of PrP^{C} to create more PrP^{Sc}. Having established that peptide H1 had a strong tendency to convert from α helix to β sheet, it was of interest to probe whether this peptide could induce a similar transition in other unstructured or helical peptides. Using both CD and FTIR, structural changes in mixtures of peptides were monitored over periods of several days.[23] A longer peptide 104H1 (PKTNMKHMA GAAAAGAVV) was unstructured in 30% AcN and its CD spectrum showed no change over 24 hr. When a 10:1 104H1:H1 mixture (molar ratio) was incubated under the same conditions for the same time period, it changed from unstructured to β sheet. However, this effect was very sensitive to the primary amino acid sequence. Mixing 104H1 having the Syrian hamster amino acid sequence (SHa104H1) with mouse H1 (MoH1), which differed from the hamster analog SHaH1 at only 2 of the 14 amino acids, the rate of conversion was reduced greatly, even though MoH1 alone was as strongly structured as SHaH1.

These experiments were unable to distinguish whether the fibrils formed by seeding 104H1 with H1 were a homogeneous mixture of H1 and 104H1 or whether these formed fibrils independently. However, we were able to determine this using isotope editing. When a ^{13}C-labeled H1 was mixed with unlabeled H1, the dilution of the label in the intermolecular sheet caused a shift in the low frequency $^{13}C{=}O$ vibration from 1604 to 1610 nm, and an identical shift was observed when labeled H1 was mixed with unlabeled 104H1.[23] Thus we can conclude that these different species mix intimately at a molecular level as they form fibrils. In contrast, a species-dependent phenomenon was observed when ^{13}C-labeled SHaH1 was mixed with the unlabeled Mo peptide and monitored by FTIR. Although the ^{13}C signal was reduced in intensity, its position remained constant.[23] Thus the mouse and hamster peptides resulted in the formation of separate β sheets in which the isotope label in the SHaH1 sheet was not diluted by the ^{12}C of the mouse peptide. Consequently, the IR spectrum was the sum of the independent contributions from each sheet. These effects are a physical manifestation of the phenomenon that causes species barriers, i.e., the relatively low efficiency of transmission of prion disease between different species compared with transmission within species. For example, scrapie prions from mice are rarely transmissible to hamsters, and vice versa.[35]

Solid-State Nuclear Magnetic Resonance

Several peptides were subjected to analysis by NMR, both in solution and in the solid state; the solid-state experiments are also referred to elsewhere in this volume.[36] High-resolution ^{13}C NMR spectra of peptides can be obtained by using a combination of cross-polarization and magic angle spinning (CPMAS) (see Fig. 3). The ^{13}C NMR resonances of the backbone carbons in polypeptides depend on the conformation, be it helix, sheet, or coil. The conformation-dependent chemical shift variations for C-1, Cα, and Cβ of typically 5–8 ppm can be used to characterize the secondary structure, provided that (i) the ^{13}C label is more than two residues away from the end of a chain and (ii) neither neighboring residue is a proline.[37] Changes in the chemical shifts obtained from an equimolar mixture of three ^{13}C-containing H1 peptides, each labeled at one of the three carbons of Ala-115, are shown in Ref. 36. Experimental data were compared with literature values for [^{13}C]alanine in peptides known to have α-helix and β-sheet structures. The chemical shifts in H1 were found to be dependent on whether the peptides were dissolved and lyophilized from 50% acetonitrile, which gave values indicative of the β sheet, or from pure HFIP and protected from moisture, which gave significantly different values corresponding to α helix. Samples that had been lyophilized from HFIP and then exposed to water vapor demonstrated movements in the peak positions appropriate to an α-helix to β-sheet transition.[38]

The ^{13}C-containing peptides described in the section on FTIR were also studied by solid-state NMR.[38] When the peptides were lyophilized from either aqueous media or acetonitrile, the chemical shifts were consistent with the β sheet observed by FTIR (Fig. 2). However, when they were lyophilized from pure HFIP, spectra displayed unresolved doublets at positions characteristic of both α-helical and β-sheet conformations. Nevertheless, it was possible to conclude that 60 $\pm$ 5% of the peptide molecules were converted to the α helix. These experiments confirmed the N terminus of the β sheet in H1 to be either Ala-113, which demonstrated partial β-sheet

[35] M. Scott, D. Foster, C. Mirenda, D. Serban, F. Coufal, M. Wälchli, M. Torchia, D. Groth, G. Carlson, S. J. DeArmond, D. Westaway, and S. B. Prusiner, *Cell* **59,** 847 (1989).

[36] D. E. Wemmer, *Methods Enzymol.* **309** [35] 1999 (this volume).

[37] H. Saito, *Magn. Reson. Chem.* **24,** 835 (1986).

[38] J. Heller, A. C. Kolbert, R. Larsen, M. Ernst, T. Bekker, M. A. Baldwin, S. B. Prusiner, A. Pines, and D. E. Wemmer, *Prot. Sci.* **5,** 1655 (1996).

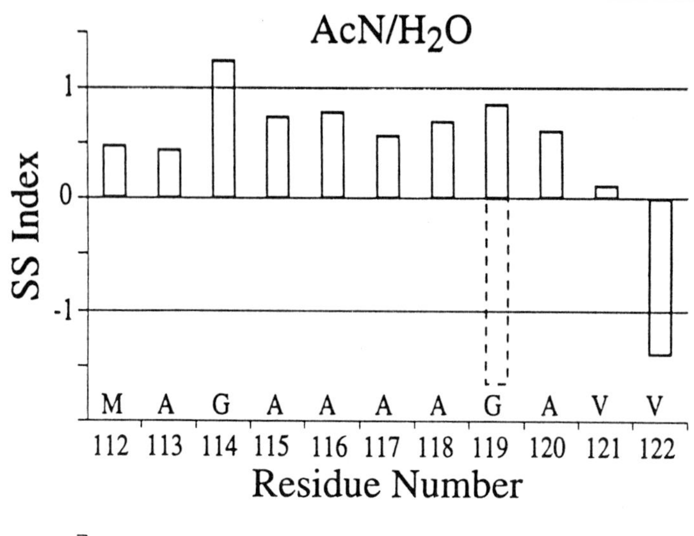

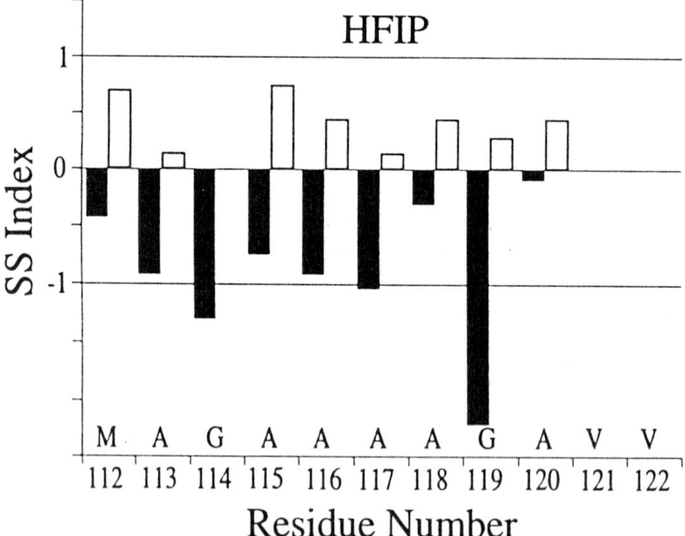

FIG. 3. CPMAS NMR secondary structure (SS) indices for the labeled carbonyl carbons of the same ^{13}C–H1 peptides illustrated in Fig. 2 (reproduced with permission from Ref. 38). Peptides were lyophilized from acetonitrile/water (top) or HFIP (bottom). SS indices of +1 (−1) represent perfect agreement with chemical shifts of residues of that type in β-sheet (α-helix) conformations.

character, or Gly-114. The chemical shifts also confirmed the C terminus of the sheet to be at Val-121; the observation from FTIR that Val-122 was also involved in the β sheet if the peptide C terminus was extended was not confirmed by CPMAS NMR. Both FTIR and NMR indicated an unusual feature near the C terminus of the β sheet, with FTIR showing a doublet for the ^{12}C vibration of Ala-120 at 1629 and 1642 cm^{-1} and NMR giving two distinct conformations for Gly-119.

Distance measurements in doubly or multiply labeled peptides can also be carried out by solid-state NMR using techniques such as rotational resonance. Because the theoretical basis of these experiments and some applications to amyloid peptides are discussed by Wemmer,[36] this topic will not be considered here, except to note that other physical methods capable of giving high-resolution structures of peptides and proteins such as solution NMR and X-ray crystal diffraction are either impossible or very difficult to apply to amyloids. Therefore, any method that can yield distance restraints in amyloid peptides is potentially valuable. Solid-state NMR of specifically labeled peptides appears to be particularly promising.

Acknowledgments

I thank Stan Prusiner for support and encouragement. Haydn Ball made important contributions to the synthetic program. I am grateful for peptide syntheses carried out by Tatiana Bekker, David Lloyd, and Sherman Jew and for advice and assistance from Steve Kent, David King, James Tam, and Jim Wells. Financial support was provided by NIH Grant AG10770. Mass spectrometry was carried out in the UCSF Mass Spectrometry Facility (Director A. L. Burlingame), supported by the Biomedical Research Technology Program of the National Center for Research Resources, NIH NCRR BRTP RR01614.

[38] Mapping Protein Conformations in Fibril Structures Using Monoclonal Antibodies

By Erik Lundgren, Hakan Persson, Karin Andersson, Anders Olofsson, Ingrid Dacklin, and Gundars Goldsteins

Introduction

Sixteen proteins are known to aggregate and form amyloid fibrils. These have very similar biophysical and biochemical properties, independent of which proteins form the building blocks. Fibrils have an ordered structure containing β strands arranged perpendicular to the fibril direction.[1] This

[1] G. G. Glenner, *N. Engl. J. Med.* **302,** 1283 (1980).

0076-6879/99 $30.00

obviously implies conformational changes, but their character or the mechanisms involved to achieve them are only partly known.

The three-dimensional structure is known for many amyloid-forming proteins, mostly from X-ray diffraction studies. Much less is known about the structure of the involved proteins in fibrils, mainly because available techniques for high-resolution studies are not sufficiently informative for fibrillar structures. The methods described here are designed to develop monoclonal antibodies specific for structural changes associated with amyloid formation.

There is no obvious correlation between the native structure of amyloidogenic proteins and β structure-rich amyloid fibrils. Two necessary steps can be envisaged as a general mechanism for the formation of amyloid, independent of the protein involved. The first step is a change in conformation leading to β structures, even in proteins with a dominance of α helices, as exemplified by prion proteins and lysozyme. The second step is self-aggregation with the formation of fibrils. This sequence of events necessarily means that the structural changes in the first stage open up new surfaces, which are important for the aggregation step.

The plasma protein transthyretin (TTR) is a useful model for studies of amyloid formation. It is a homotetramer, with a subunit made up of two β sheets each containing four β strands.[2] TTR forms amyloid as a result of point mutations. The highest frequency of mutations is seen in a region close to the mouth of a central channel in the molecule, called the edge region.[3] It is known that TTR dissociates into partly denatured monomers on amyloid formation.[4,5] A core structure is assumed to be preserved when TTR is trapped in amyloid.

A key question is therefore which denaturation events occur and what factors govern their expression. One approach taken has been to resolve crystal structures of TTR mutants, which are clinically important for the development of amyloid. However, they show in principle a wild-type fold or minor changes without disclosing features, which explain self-aggregation or packaging.

Another strategy presented here is to generate monoclonal antibodies, with specificity for epitopes not present in the wild-type conformation, but in mutants with the amyloidogenic fold. In other words, to generate monoclonals, which should decorate amyloid fibrils but not the wild-type

[2] C. C. F. Blake, M. J. Geisow, S. J. Oatley, B. Rérat, and C. Rérat, *J. Mol. Biol.* **121,** 339 (1978).

[3] L. C. Serpell, G. Goldsteins, I. Dacklin, E. Lundgren, and C. C. F. Blake, *Amyloid Int. J. Exp. Clin. Invest.* **3,** 75 (1996).

[4] W. Colon and J. W. Kelly, *Biochemistry* **31,** 8654 (1992).

[5] S. L. McCutchen, Z. Lai, G. J. Miroy, J. W. Kelly, and W. Colon, *Biochemistry* **34,** 13527 (1995).

conformation of a given protein, or which should detect mutant proteins, which have an amyloidogenic fold, either during the partly denaturing conditions, which triggers an amyloidogenic fold, or carry such a fold spontaneously. Some mutants have been constructed that spontaneously form amyloid *in vitro* in physiological pH.[6] They have been used to generate monoclonals with specificity for TTR species with an amyloidogenic fold. They also show specific binding to *ex vivo* amyloid, but not wild-type TTR. An alternative would be to screen for monoclonals with specificity for epitopes that disappear on amyloid formation, which is not dealt with in this article.

Experimental Strategy

Partial unfolding of the intermediate stage can be assumed to result in the exposure of cryptic epitopes not present in the native fold. The generation of monoclonal antibodies with specificity for such cryptic epitopes will be useful for elucidation of the structural changes. Cryptic epitopes may be exposed on the surface of amyloid or reside in domains necessary for self-aggregation. In the latter case the antibodies should be expected to block amyloidogenesis. The flow chart in Fig. 1 shows the experimental strategy for the demonstration of such epitopes by the generation of monoclonal antibodies, using TTR as an example.

The strategy is based on two principal concepts: (1) generation of proteins with an amyloidogenic fold and (2) using these proteins for screening of antibodies with specificity for the amyloidogenic but not the wild-type fold. The key factor is the screening conditions. The antibodies we have generated so far react with amyloid fibrils. Another possibility would be to screen for antibodies, which block *in vitro* amyloid generation, provided that amyloid can be formed in conditions where antibodies bind.

Comments

1. Mutants of TTR have been constructed and expressed in *Escherichia coli* using standard procedures. The amyloidogenic fold is used here as an operational definition as the structure of mutants that spontaneously forms amyloid in physiological pH or the structure obtained after denaturing treatments such as acid treatment. Several double or triple mutants are unstable, as they start to aggregate on storage, seen as undefined smears on polyacrylamide gel electrophoresis in

[6] G. Goldsteins, K. Andersson, A. Olofsson, I. Dacklin, Å. Edvinsson, V. Baranov, O. Sandgren, C. Thylén, S. Hammarström, and E. Lundgren, *Biochemistry* **36,** 5346 (1997).

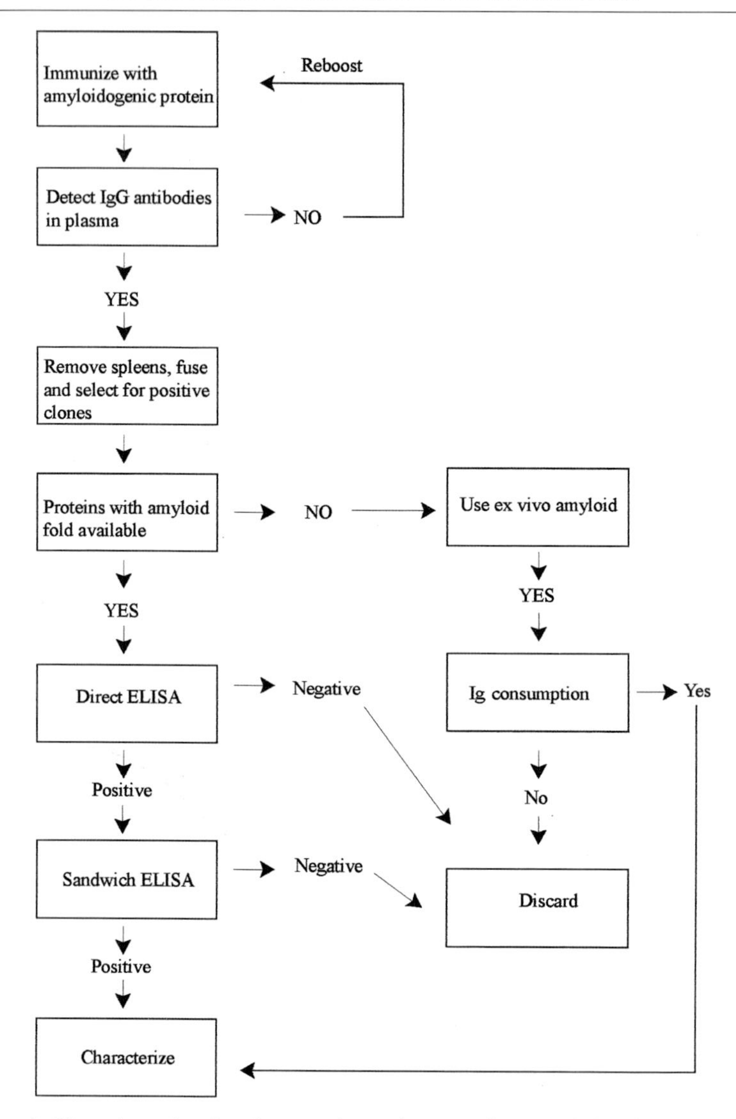

FIG. 1. Flow chart showing the experimental approach as used for the generation of monoclonal antibodies detecting cryptic epitopes on double or triple mutants of TTR.

native, nondenaturing conditions. Finally, the aggregates transform into amyloid.

2. A series of monoclonals has been generated that only detects mutants with an amyloidogenic fold. Thus, TTR V30M or the highly amyloido-

genic mutant TTR L55P has low reactivity with such antibodies, which might reflect the inability of these mutants to form amyloid *in vitro* in physiological pH. However, in low pH they do react when amyloid is formed.

3. To map the epitopes detected by these antibodies, a simple and reliable method has been used.[7] It is based on the cloning of small DNA fragments and fusion of them to a part of a phage protein encoded by the T7 gene *10*. One should remember that the method only gives a minimal epitope, comprising less than 5–10 amino acids.

ELISA Techniques

Standard enzyme-linked immunosorbent assays (ELISA) are used for the approach. They are modified for different purposes in the following protocols and are either direct ELISAs, where the antigen is coated directly to the plates, or indirect or sandwich ELISAs, where the antigen is first captured on polyclonal antibodies.

Direct ELISA

1. Coat the protein in a concentration of 5 or 10 μg/ml on a microtiter plate (Nunc, Immunosorp, Denmark) using 0.1 ml/well. A suitable coating buffer is phosphate-buffered saline (PBS), pH 7.4, or 50 m*M* sodium bicarbonate, pH 9.0. By incubation on a rotary shaker a higher coating efficiency is achieved. Incubate overnight at 4° and then wash with 0.05% Tween 20 in PBS.
2. Block unoccupied sites with 5% skimmed milk or 0.5% bovine serum albumin (BSA), diluted in PBS, followed by washing three times with 0.05% Tween 20 in PBS. All further washing steps are done in the same way.
3. Add 100 μl of each undiluted hybridoma supernatant, dilutions of serum, or monoclonal antibodies, incubate for 1 hr, and wash three times.
4. Add 100 μl antimouse immunoglobulin G (IgG) horseradish peroxidase (HRP)-conjugated antibodies and incubate for 1 hr. Titrate a suitable concentration with a control antibody.
5. Develop the plate using 5 m*M* 2,2′-azinobis (3-ethylbenzthiazoline-6-sulfonic acid) (ABTS) in 50 m*M* phosphate–citrate buffer, pH 5.2.

[7] K. K. Stanley and J. Herz, *EMBO J.* **6,** 1951 (1987).

Sandwich-Type ELISA

1. Coat polyclonal antibodies ideally from rabbit or sheep directed against the protein of interest to the microtiter plate using the method just described.
2. Add the protein of interest in a suitable concentration in 100 μl after blocking of the plate followed by subsequent washing as described earlier and incubate for 1 hr. We found that in the case of TTR, 10 μg/ml works well in 5% BSA in TBS. It is important that the protein of interest be purified in a gentle way, avoiding denaturing conditions. If possible, preservation of structural integrity or function should be verified. Continue the procedure as described earlier.

Immunization of Mice for Production of Monoclonal Antibodies

At least five mice should be immunized. It is important to test the animals for the generation of antibodies of the IgG type before sacrifice and fusion. The reason is that several amyloid-forming proteins are conserved and are not immunogenic. In the case of TTR it is not uncommon to find antibodies of the IgM isotype, which probably represent so-called natural low specificity antibodies, which are not suitable for all applications. The following protocol has been found convenient for TTR.

Immunization

Inject 100 μg TTR together with Freund's complete adjuvant subcutaneously. Boost after 3 weeks intravenously with 50 μg TTR in Freund's incomplete adjuvant. Test mice sera for specific antibodies after another week. Repeat every 3 weeks until IgG reactivity is detected. We found that it was necessary to use two or three boosters before reactivity was found and the animals could be sacrificed and the spleens taken out for fusion.

To test for the presence of mouse IgG with specificity for the antigen to be studied, one might follow the protocol for the direct ELISA. In the case of TTR, a concentration of 10 μg/ml was found suitable in 0.5% BSA in Tris-buffered saline, pH 7.4 (TBS). Add the serum to be tested in dilutions (10^{-1}–10^{-6}) and incubate for 1 hr at 37°. Several commercial antimouse IgG isotype-specific antibodies exist that can be detected with HRP-conjugated antibodies.

Screening of Hybridoma Supernatants

The strategy to select hybridoma cell lines producing amyloid-specific monoclonal antibodies is to identify positive clones reacting with the epitope of the protein of interest with an amyloid fold and exclude specificities for epitopes exposed on the same protein with wild-type conformation. A simple and rapid method is to directly coat the antigen to a microtiter plate, using the direct ELISA protocol as the first step. Attachment of the protein to the plastic surface of the plate will cause partial denaturation and exposure of previously buried epitopes. Note that epitopes of interest for amyloid formation on mutant proteins might be destroyed in the direct ELISA. Therefore, screening should always include both ELISA methods. By using the wild-type protein, monoclonals directed against an introduced mutation are avoided.

Samples are scored positive when giving an absorbance exceeding average background with three standard deviations.

The second step in the screening process is to exclude monoclonal antibodies that react with the wild-type conformation of the protein. Discard antibodies reacting with proteins with nonamyloidogenic conformation, e.g., wild-type TTR. The sandwich ELISA protocol is a convenient method where the native structure of the antigen is preserved on immobilizing.

Clones that have lost their reactivity toward the wild type are putative candidates for being selectively reactive toward a certain fold or a cryptic epitope of the protein.

Alternative Methods for Screening

When proteins lack an amyloidogenic fold, the screening is more complicated. Low pH is used for the generation of amyloid *in vitro* for several amyloid-forming proteins. Acid conditions obviously do not represent a very favorable screening condition for hybridoma supernatants, as antibodies will not bind to its cognate antigen in low pH nor can, e.g., *ex vivo* amyloid be attached easily to plastic surfaces. However, a simple protocol can be used if an *ex vivo*- or *in vivo*-generated amyloid is available, which works well and can be adapted for screening conditions of hybridoma supernatants.

Mix the hybridoma supernatant with the *ex vivo*- or *in vivo*-generated stable amyloid. Centrifuge the fibrils for 20 min at 20,000*g* and 37° and discard the pellets. Ideally the amount of added monoclonal antibodies should be equal to the amount of putative binding sites on the added amyloid fibrils, although this is not known during screening. The supernatant should be tested for loss of IgG using the same direct ELISA protocol as mentioned earlier for the detection of anti-TTR IgG antibodies in mouse serum after immunization. The criteria for positive scoring are less stringent in this case. We found that samples where the signal is less than two standard deviations below the average background level could be analyzed further.

Epitope Mapping of Cryptic Antibodies

A simple system for rapid epitope mapping was devised by Stanley and Herz.[7] The approach is to generate a fragmented cDNA library of the antigen linked to a fusion protein. Peptide fragments are usually less stable than native proteins, which could be overcome by the creation of a fusion protein. The following protocol is based on the commercially available kit NovaTope system (Novagen Inc., Madison, WI) where the fragment library is fused to the N-terminal 260 amino acids of the phage T7 gene *10*. The library is screened by standard colony lift methods, and immunodetection is performed using the antibodies to be tested. Plasmid DNA is prepared from rescreened positive clones and sequenced.

DNase I Digestion

DNA can be either the whole plasmid carrying the cDNA of interest or just the purified cDNA fragment. DNA should be pure and fairly concentrated, not less than 2 μg/μl is recommended. Make dilutions of DNase I in 1× DNase I buffer (0.05 *M* Tris–HCl, pH 7.5, 0.05 mg/ml BSA plus 10 m*M* $MnCl_2$). Avoid Mg^{2+} ions as it will only nick DNA. To find a DNase I concentration giving fragments of 50–100 bp, make a series of dilutions from 1.5×10^{-3} to 4.44×10^{-4} U/μl to digest 10 μg DNA in a total volume of 10 μl 1× DNase I buffer. Incubate at room temperature and stop the reactions after exactly 10 min by the addition of 2 μl 6× stop buffer (100 m*M* EDTA, 30% glycerol, tracking dyes). Analyze fragment size by running 1 μl from each dilution by agarose gel electrophoresis (2%) next to polymerase chain reaction markers from 50 to 2000 bp.

Fractionation of DNA Fragments

Any procedure giving good recovery could be used, although it should be noted that the recovery of small fragments (<200 bp) is better from agarose gels by electroelution or by using low melting point agarose. The following procedure has been successful for 50- to 100-bp fragments. Pool DNase I digest reactions containing DNA fragments around 50–100 bp and run on a 1.5% low melting agarose gel at 4° adjacent to DNA size markers. Excise bands of the desired size, transfer to 2-ml microcentrifuge tubes, add 6 volumes of buffer (20 m*M* Tris–HCl, pH 8.0, 1 m*M* EDTA, pH 8.0), and melt the agarose at 65° for 5–10 min. Cool down to room temperature and extract carefully with phenol, phenol–chloroform–isoamyl alcohol (25:24:1) and chloroform–isoamyl alcohol. Precipitate the DNA and resuspend in 30 μl TE buffer (10 m*M* Tris–HCl, 1 m*M* EDTA, pH 8.0). Determine the DNA concentration by reading the A_{260} of a 100×

dilution in water. The minimum concentration should be 0.1 μg/μl. If the average size of the fragments is 100 bp, 6 μg corresponds to about 100 pmol DNA and 200 pmol of ends.

Cloning of Fusion Genes

The fragmented target cDNA should be blunted and deoxyadenylate (dA)-tailed before cloning into the pTOPE T-vector. This type of vector with a T overhang is advantageous compared to blunt end cloning, as it minimizes the risk for tandem insertions. The ends of 1-μg fragmented cDNA are blunted in buffer [0.05 *M* Tris–HCl, pH 8.0, 5 m*M* $MgCl_2$, 0.1 mg/ml BSA, 5 m*M* dithiothreitol (DTT)], 0.1 m*M* each of dATP, dTTP, dCTP, and dGTP, and 1–2 units T4 DNA polymerase in a total volume of 25 μl at 11° for 20 min. Inactivate the enzyme at 75° for 10 min. For dA-tailing the entire blunt reaction is adjusted to 1× dA tailing buffer (10 m*M* Tris–HCl, pH 9.0, 50 m*M* KCl, 0.01% gelatin, 0.1% Triton X-100), 1.25 U *Tth* DNA polymerase in 85 μl. Incubate at 70° for 15 min and extract with 1 volume chloroform–isoamyl alcohol. The DNA concentration in the aqueous phase will be about 11.8 ng/μl. The ratio between the T-overhang vector and the fragment insert is crucial, as too much fragment will decrease ligation efficiency. It is recommended that 0.2 pmol of fragments (≈6 ng for 100 bp size) be ligated to 0.03 pmol (50 ng) pTOPE T-vector in 20 m*M* Tris–HCl, pH 7.6, 5 m*M* $MgCl_2$, 5 m*M* DTT, 0.5 m*M* ATP, and 2–3 Weiss units T4 DNA ligase in a total volume of 10 μl. Incubate at 16° overnight. Transform supercompetent Nova Blue(DE3) cells (20 μl) with 1 μl ligation mix in 1.5-ml tubes by incubation on ice, followed by heat shock to 42°, phenotypic expression in 80 μl SOC medium for 1 hr at 37°, and an appropriate amount plated on LA plates (1000 to 2000 clones/plate) containing 50 μg/ml carbenecillin and 15 μg/ml tetracycline.

Screening

Note that this method only detects continuous epitopes and all antibodies may not be suitable for screening. It is advantageous to use fragments of 50–100 bp, which will give overlapping epitope sequences of 15–30 amino acids. Reduce the amount of clones to be screened by using cDNA cleaved from its vector instead of the whole plasmid. Identify positive clones after ligation and transformation into competent cells using immunodetection with the monoclonal antibodies of interest. Chill the plates at 4° for 30 min and overlay with nitrocellulose membranes (e.g., Hybond C super) for 1 min. Mark orientation of the filters carefully. Lyse the bacteria for 15 min by chloroform vapor and later denature in 20 m*M* Tris–HCl, pH 7.9, 6 *M* urea, and 0.5 *M* NaCl for 15 min at room temperature. Block the filters for unspecific antibody binding by 5% skimmed milk in TBST (20 m*M* Tris–HCl, pH 7.6, 137 m*M* NaCl, 0.05% Tween 20) for 1 hr at room temperature. Perform 5× 5-min washes in TBST. Incubate with primary antibody at a suitable dilution (1:200 to 1:10,000 is appropriate for antisera, ascites fluids, or purified antibodies) in 5% milk/TBST for 1 hr at room temperature. Wash as described earlier. As the secondary antibody, sheepαmouse Ig-HRP (DAKO) diluted 1/3000 is recommended. Incubate in 5% milk/TBST for 1 hr at room temperature. After washing, the signal is detected by enhanced chemiluminescence (ECL). Enrich positive clones by replating and rescreening. The DNA should be prepared and sequenced by standard methods.

Control specificity of the cloned fragments by immunoblotting total cell lysates of isolated clones. Inoculate the bacteria in LB medium plus 50 μg/ml carbenecillin and 15 μg/ml tetracycline and grow for 2 hr. Harvest cells and prepare total cell lysates by mixing with 2× sample buffer, separate on SDS–PAGE, and immunoblot with the test antibodies. Strip the filter and immunoblot with an anti-T7 gene *10* antibody to confirm the integrity of the fusion protein. Figure 2 shows the pattern obtained with two monoclonal antibodies (MAb) raised against mutants of TTR with an amyloid fold. From the immunoreactivities and the sequences obtained, minimal epitopes comprising positions 55–61 (MAb47) and 39–44 (MAb15), respectively, could be defined in the example shown. It should be stressed that the method defines minimal epitopes and that the actual epitopes are probably larger.

Characterization of Monoclonal Antibodies

Several experiments could be designed to characterize the obtained antibodies. Not unexpectedly, we found that some of them are convenient for immunohistochemical purposes. If methods are available for the generation of amyloid *in vitro,* they could be tested for the ability to block amyloid formation. One important possibility is to use them for obtaining structural information. Thus, with the strategy used for screening, they should detect cryptic epitopes not exposed on the wild-type conformation.

Demonstration of Cryptic Epitopes

Demonstration that the expression of cryptic epitopes is restricted to the partly denatured potentially amyloidogenic fold could be done by the ELISA protocols described earlier.

Thus, test in parallel the reactivity of interesting monoclonal antibodies by both the direct and the sandwich ELISA using dilutions of the antigen. In the case of TTR, 0.01–10 μg/ml has been found to be suitable. Naturally,

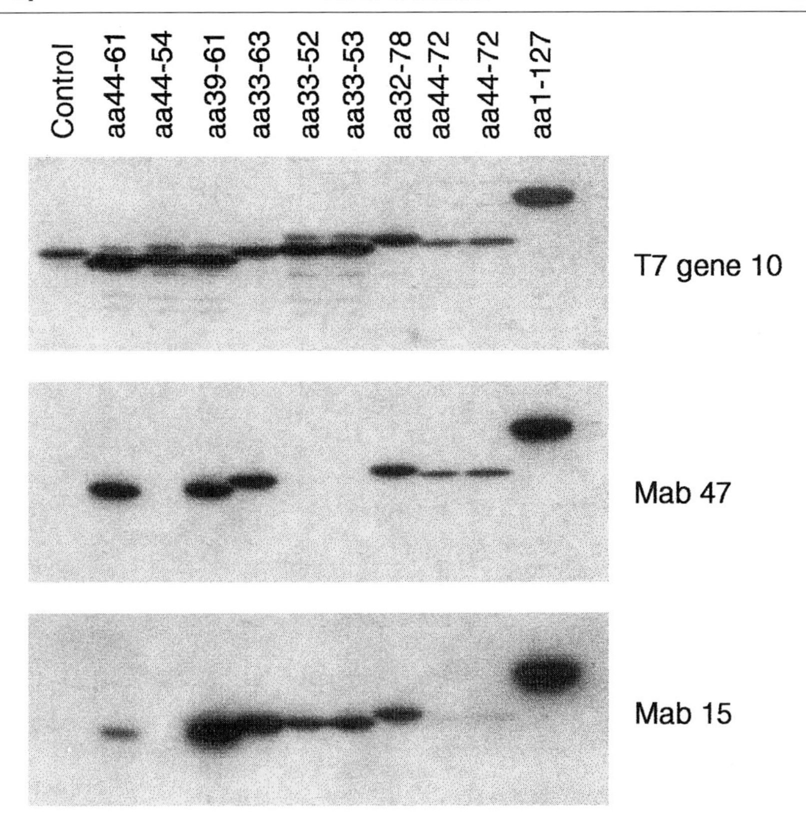

FIG. 2. Epitope mapping of monoclonal antibodies with specificity for cryptic epitopes of amyloidogenic mutants of TTR. An immunoblot of fusion proteins derived from a fragmented cDNA library and the phage T7 gene *10* is shown. (Top) Amino acid positions of TTR cDNA fragments blotted with an anti-T7 gene *10* antibody. (Middle and bottom) Blotted with two monoclonal antibodies isolated by screening for reactivity with cryptic epitopes of amyloidogenic TTR. Control depicts the religated plasmid without insert.

epitopes detectable in the direct ELISA but not in the indirect ELISA represent a conformational change. It is up to the investigator to show that these cryptic epitopes are associated with a change in folding, which predisposes for amyloid formation. We have been able to show that cryptic epitopes expressed on certain TTR mutants are markers for an amyloidogenic conformation. Note that the reactivity in both ELISA systems might be due to neo-epitopes caused by mutations introduced or by the presence of N-terminal methionine in *E. coli*-produced proteins. Figure 3 shows an example where some mutants of TTR have been used to characterize reactivity with two monoclonal antibodies.

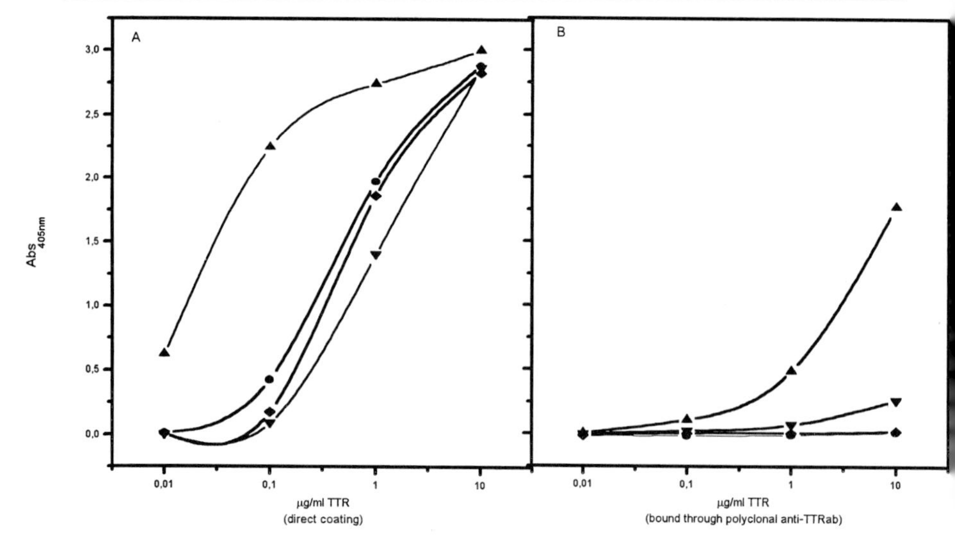

Fig. 3. Immunoreactivity of wild type and some mutants of TTR. (A) Direct ELISA and (B) sandwich ELISA: (●) TTR wild type; (▲) mutant TTR G53S, E54D, L55S; (▼) mutant TTR del53-55; and (◆) mutant TTR V30M. Only the two mutants[6] (TTR G53S, E54D, L55S and TTR del53-55) that spontaneously form amyloid in physiological pH react in the indirect ELISA, whereas several mutants, including the clinically important TTR V30M and TTR L55P, do not react. In contrast, both wild type and all mutant TTR react in the direct ELISA.

Immunoprecipitation

An alternative way to assay for immunoreactivity in nondenaturating conditions is to use immunoprecipitation, as shown in the following protocol.

Aliquot 25 μl of sheep antimouse IgG DynaBeads M450 (Dynal AS, Norway) in suspension in 1.5-ml test tubes and wash three times for 5 min with TBST. Add 100 μl of monoclonal antibody to be tested (0.02 mg/ml in TBST) to each tube and incubate for 30 min. After three washes with TBS, add 25 μl TTR mutant to be tested (0.05 mg/ml in TBS) and 75 μl TBS. Incubate tubes on a rocking table for 30 min at room temperature. Repeat the washing step to remove unbound TTR.

Add 25 μl of SDS–PAGE sample buffer to each tube, vortex briefly, and apply 10 μl on a 12% SDS–PAGE gel. Immunoblot on a PVDF membrane and develop employing polyclonal rabbit anti-TTR antibodies and HRP-labeled goat antirabbit secondary antibodies with ECL detection.

Figure 4 shows an experiment according to the protocol, where either wild-type TTR or a highly amyloidogenic mutant was immunoprecipitated

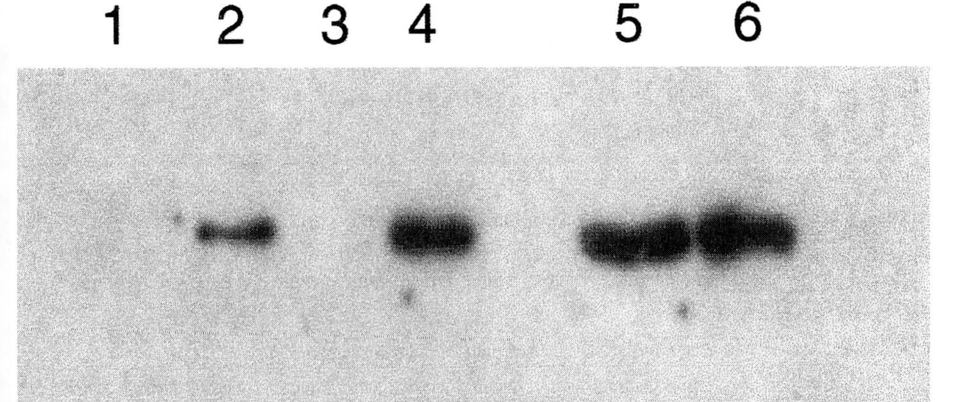

FIG. 4. Immunoprecipitation of wild-type TTR (lanes 1, 3, and 5) or the highly amyloidogenic mutant TTR G53S, E54D, L55S[6] (lanes 2, 4, and 6). Monoclonal antibody 1 (lanes 1 and 2), monoclonal antibody 2 (lanes 3 and 4), and rabbit anti-TTR polyclonal antibodies (lane 5 and 6) were used, and immunoblot was performed with the rabbit antibody. Only mutant proteins precipitated with the two monoclonal antibodies, whereas the positive control precipitated both TTR variants.

with a monoclonal antibody. Only the mutant protein was precipitated with the monoclonal antibody.

Binding to ex Vivo-Extracted Amyloid Fibils

Binding of monoclonals to amyloid can be done as described in the method described earlier for the screening of hybridoma supernatants. An alternative way is to measure the ability of the unlabeled antibody to compete with the iodinated antibody for the consumption by *ex vivo*-extracted amyloid of labeled antibody. Use a gentle method to iodinate the monoclonal antibodies to be tested. We have used vitreous amyloid from patients with the TTR V30M mutation as antigen, but other sources are probably just as good. Disperse amyloid through microtip sonication for 3×5 sec on level 3 (Heat System, Ultrasonicator XL) and suspend in TBS. Dilute the radiolabeled antibody to give a molar excess of TTR amyloid compared to the iodinated antibody and then mix with unlabeled monoclonal in dilutions ranging from 100 to 0 μg/tube. The total volume should be 200 μl/tube. Add 100-μl aliquots of the dispersed vitreous TTR amyloid and incubate for 1 hr, followed by centrifugation for 20 min at 12,000g. Measure radioactivity in 100 μl of the supernatant in a γ counter. Use a monoclonal antibody of the same isotype with irrelevant specificity as a negative control.

Precautions

For studies on binding of monoclonals to amyloid it is important that the *in vivo*-isolated amyloid has been purified in a gentle way. In case of TTR, pure amyloid could be isolated from the vitreous body of the eye. Brief sonication to disperse the fibrils has not been shown to affect the end result and is recommended. If *ex vivo*-isolated amyloid is limiting for the initial screening, the first approach to be tested is to use pure protein (*in vivo* isolated or recombinant) that has been treated or mutated so that it expresses an amyloid fold. It is important that the formed amyloid fibrils are stable in physiological conditions. We have found that a common reason for failures to obtain monoclonals with specificity for amyloid or amyloid precursor peptides is that the production of amyloid *in vitro* suffers from low yield or variable results. The admixture of unspecific aggregates with amyloid should be tested by electron microscopy (negative uranyl-acetate staining) for judgment of the homogeneity of the amyloid preparation.

Possible Applications

Mapping of the epitopes on the polypeptide will disclose new structures of TTR, which are exposed as a result of partial denaturing, and give clues for the structural changes in a pathway leading from the native conformation to the structural changes associated with self-aggregation of the molecule. In addition to their role for elucidating structural changes, the antibodies could be tested for the ability to block amyloid formation in the assay conditions used.

This article exemplifies the approach with antibodies that detects cryptic epitopes on highly amyloidogenic mutants of TTR, but not on wild-type protein, or on some clinically important mutants such as TTR V30M or TTR L55P unless they are partially denatured. Such antibodies are useful for studies of factors or conditions that initiate the denaturing events leading to generation of an amyloidogenic fold.

Antibodies have been described that react with amyloid generated *in vitro* from the β peptide of Alzheimer's disease.[8] Such antibodies have been shown to be able to dissolve amyloid plaques and have been proposed to be useful for therapeutic purposes. The monoclonals found so far directed against cryptic epitopes on TTR with an amyloidogenic fold do not have this property nor do they block amyloid formation. The reason is that the mutants generated so far generate a mixture of amyloid and unspecific

[8] B. Solomon, R. Koppel, D. Frankel, and E. Hanan-Aharon, *Proc. Natl. Acad. Sci. U.S.A.* **94,** 4109 (1997).

aggregates as judged by electron microscopy. To screen for monoclonal antibodies blocking amyloid formation requires mutants of amyloid-forming proteins, which generate amyloid in conditions, where antibodies would bind, i.e., at physiological pH, and which give amyloid with a high yield. The generation of blocking antibodies or fragments of them offers the intriguing potential for therapeutic use or for the development of such reagents.

Fv fragments of the found antibodies could be generated for further studies of structural requirements for amyloid formation by cocrystallization with TTR or TTR mutants.

[39] Analysis of Protein Structure by Solution Optical Spectroscopy

By WILFREDO COLÓN

Introduction

Experimental observations in recent years that the mechanism by which a protein is converted to amyloid is related intimately to conformational changes have caused an increased interest in the various methodologies for analyzing protein conformation and structure.[1–3] Thus, research associated with amyloid diseases has truly become an interdisciplinary field, involving scientists from areas as diverse as clinical medicine and biophysical chemistry. The increasing volume of literature on biophysical studies of amyloidogenic peptides and proteins has made it necessary for amyloid researchers to have a basic understanding of, and perhaps apply, some of the common spectroscopic methods available to gain structural information on amyloidogenic proteins. This article provides a practical guide to amyloid researchers on the most commonly used spectroscopic techniques to study protein structure: ultraviolet visible (UV/Vis) fluorescence, and circular dichroism (CD) spectroscopy.

Protein Structure

Because of their diverse amino acid sequence, which results in a unique three-dimensional (3D) structure that is essential for function, the structural

[1] J. W. Kelly, *Curr. Opin. Struct. Biol.* **6,** 11 (1996).
[2] R. Wetzel, *Cell* **86,** 699 (1996).
[3] G. Taubes, *Science* **271,** 1493 (1996).

0076-6879/99 $30.00

complexity of proteins is unparalleled in nature. Four levels of organization describe the structure of proteins. The primary (1°) structure of a protein refers to the amino acid sequence, which by its unique composition determines the higher levels of structure. The secondary (2°) structure is the local repetitive 3D structure of the peptide backbone. Due to the rigid and planar peptide unit (O=C–N–) and the allowed ϕ and ψ torsion angles of the N–C_α and C–C_α backbone bonds, respectively, two types of secondary structure, the α helix and β sheet, predominate. The tertiary (3°) structure refers to the global 3D structure of the protein that is formed by the packing of the various elements of 2° structure. Finally, the quaternary (4°) structure refers to the association of two or more protein molecules to form an oligomeric structure. As the mechanism by which the 1° structure of a protein determines its 3° structure is not known, there has been an enormous amount of work aimed at understanding the mechanism of protein folding. In recent years, interest in protein folding has surged because of the increasing number of diseases in which protein misfolding and aggregation play an important pathological role.[1–5]

Apart from the determination of 3D structure by X-ray and nuclear magnetic resonance (NMR) methods, most experimental approaches to studying protein structure involve optical spectroscopy to monitor the changes in 2° and 3° structure as the solvent conditions are changed. Structural information is usually obtained by studying the effect of temperature, pH, ligands, and solvent composition on stability and conformation. Temperature denaturation is the method of choice in assessing the thermodynamic effect of amino acid substitution.[6,7] However, temperature denaturation often does not fully unfold proteins[8] and is limited for conventional kinetic studies of protein folding as large changes in temperature cannot be accomplished in the millisecond time scale of stopped-flow experiments. Thus, the most widely used method to study protein conformational changes involves chemical denaturation by urea or guanidine hydrochloride (GuHCl).[9]

The choice of technique for determining the overall structure or conformation of a protein in solution will depend on several factors, including quantity of protein available, protein concentration requirements, access to instrumentation, available expertise, and structural information required.

[4] P. J. Thomas, B. Qu, and P. L. Pedersen, *TIBS* **20,** 456 (1995).

[5] A. L. Horwich and J. S. Weissman, *Cell* **89,** 499 (1997).

[6] J. A. Schellman, *Annu. Rev. Biophys. Biophys. Chem.* **16,** 115 (1987).

[7] C. N. Pace, *TIBTECH* **8,** 93 (1990).

[8] C. R. Matthews, *Annu. Rev. Biochem.* **62,** 653 (1993).

[9] C. N. Pace, B. A. Shirley, and J. A. Thomson, *in* "Protein Structure: A Practical Approach" (T. E. Creighton, ed.), p. 311. Oxford Univ. Press, Oxford, 1990.

Table I shows a list of the most common spectroscopic methods for probing protein structure. The various methods differ significantly in their resolution, instrument price, and expertise required. UV/Vis spectroscopy is the least expensive and simplest of the methods, but also provides the least resolution. In contrast, NMR provides atomic resolution, but requires expensive equipment and specialized training. Very often several low-resolution techniques are combined to obtain complementary structural information. Of the techniques mentioned in Table I, this article focuses on UV/Vis, CD, and fluorescence spectroscopy because they are the fastest and easiest to perform and because the instruments are relatively inexpensive and readily available. Emphasis will be given to experimental aspects, and theoretical discussions will be kept to a minimum. Special applications of infrared (IR) and NMR in the solid state are discussed in this volume, and several reviews on Raman spectroscopy are available elsewhere.[10–12]

UV/Vis Spectroscopy

When a molecule absorbs light, electrons are excited (promoted) from their lowest energy ground state to a higher energy excited state. The group within a molecule responsible for light absorption is called a chromophore. The wavelength at which a chromophore absorbs light depends on the energy difference between the ground state and higher energy levels, as energy (E) and wavelength (λ) are inversely related by the following equation:

$$E = hc/\lambda = h\nu \tag{1}$$

where h is Planck's constant, c is the speed of light, and ν is the frequency. Hence, absorption will occur at a wavelength whose energy matches the ΔE between electronic levels. Absorbance (A) is linearly related to concentration as indicated by the Beer–Lambert law:

$$A = \varepsilon lc \tag{2}$$

where ε is the molar absorption coefficient (in M^{-1} cm^{-1}), l is the light path length (in cm), and c is the molar concentration. Absorbance data can also

[10] D. C. DeNagel, D. J.-Y. Ho, and J. F. R. Kuck, *in* "Biological Applications of Raman Spectroscopy" (T. G. Spiro, ed.), p. 47. Wiley, New York, 1987.

[11] S. Krimm and N.-T. Yu, *in* "Biological Applications of Raman Spectroscopy" (T. G. Spiro, ed.), p. 1. Wiley, New York, 1987.

[12] L. G. Tensmeyer and E. W. Kauffman II, *in* "Spectroscopic Methods for Determining Protein Structure in Solution" (H. A. Havel, ed.), p. 69. VCH Publishers, New York, 1996.

TABLE I

SPECTROSCOPIC METHODS FOR DETERMINING PROTEIN STRUCTURE

Spectroscopic method	Structure	Concentration range (m*M*)	Advantages	Disadvantages
UV absorption	3°	0.01–0.1	Equipment is inexpensive and easy to use	Very low resolution
CD	2° (far UV), 3° (near UV)	0.01–0.1 0.05–0.5	Equipment is easy to use; preferred method for determining 2° structure content of proteins; best method for comparing effect of solvent composition on 2° and 3° structure	Low resolution, aromatic residues can interfere in far-UV region; equipment is relatively expensive
Fluorescence	3°	0.0001–0.01	High sensitivity allows study of low protein concentrations; relatively inexpensive and easy to use; structure compactness and dynamics information	Only monitors aromatic residues and must be at least semiburied
IR absorption	2°	0.5–2.0	Equipment is relatively inexpensive and easy to use; can provide 2° structure information on solid samples	Water interference; requires high protein concentration
Raman	2°, 3°	>0.5	Minimal interference by water; can provide 2° structure information on solid samples	Slow; equipment not as generally available
NMR	2°, 3°	1–10	Highest resolution (atomic); provides information on protein dynamics	Expensive equipment; difficult to use; requires large amounts of protein; limited to small proteins MW < 40,000

TABLE II

ABSORBANCE AND FLUORESCENCE PROPERTIES OF AROMATIC AMINO ACIDS IN PROTEINS[a]

Amino acid	ε (M^{-1} cm^{-1})	λ_{max} (nm)	Emission λ_{max} (nm)	Quantum yield (Φ)
Trp	5600	280	348	0.20
Tyr	1400	275	303	0.14
Phe	200	257	282	0.04

[a] From C. R. Cantor and P. R. Schimmel, "Biophysical Chemistry, Part II." Freeman, San Francisco, 1980.

be reported as percent transmittance (T) as A and T are related by the following equation:

$$A = -\log(T) \tag{3}$$

where T equals I/I_0, the ratio of the intensities of the transmitted (I) and incident (I_0) light.

Proteins absorb light in the ultraviolet region of the spectrum, mainly because of the peptide group and the aromatic residues tryptophan (Trp), tyrosine (Tyr), and phenylalanine (Phe). The peptide group absorbs strongly below 230 nm due to $n \rightarrow \pi^*$ and $\pi \rightarrow \pi^*$ electronic transitions, whereas aromatic residues absorb in the near-UV region between 250 and 290 nm due to $\pi \rightarrow \pi^*$ electronic transitions. The absorption maxima (λ_{max}) and absorption coefficients for the aromatic amino acids are shown in Table II.

Because the absorption coefficients for Trp and Tyr are much higher than that of Phe, the contribution of the latter is trivial in a protein containing tryptophans or tyrosines. Therefore, to determine structural information about proteins in solution using absorption, one usually relies on the intrinsic Trp and Tyr residues. It is important to realize that not all Tyr or Trp residues in a protein are necessarily going to be good conformational probes, as they may be exposed to the solvent under all experimental conditions and therefore be silent to conformational changes.

Interpreting Absorption Spectra

The absorption maximum of a protein in the near-UV region depends on the environment of the aromatic amino acids. In particular, Trp and Tyr are sensitive to their neighboring residues and to the polarity of the environment. When placed in a nonpolar environment, Trp, Tyr, and Phe will have a higher λ_{max} and ε than in a polar environment.[13,14] Therefore,

[13] S. Yanari and F. A. Bovey, *J. Biol. Chem.* **235,** 2818 (1960).

[14] J. E. Bailey, G. H. Beaven, D. A. Chignell, and W. B. Gratzer, *Eur J. Biochem.* **7,** 5 (1968).

unless these residues are already exposed in the native protein, protein unfolding (and perhaps small conformational changes) will expose them to the solvent, resulting in a blue shift (shorter wavelength) in the absorption spectrum. The blue shift is a result of the increased energy difference between the ground state and the excited state of the chromophore caused by the increased polarity of its environment. Of course, if they are exposed to the solvent in the native state they will be very sensitive to changes in solvent polarity, but will most likely be insensitive to protein conformational changes.

The spectrum of Tyr is particularly sensitive to changes in pH because of its titratable OH group, which has a pK_a of about 10 when free in solution. When the OH group of Tyr is deprotonated, there is an increase in λ_{max} and ε.[15,16] By monitoring the spectral changes of a Tyr-containing protein when the pH is raised, it is possible to infer its solvent accessibility within the protein. For instance, if the spectrum of a protein containing one Tyr does not change at a pH where a change is observed for the free amino acid, the Tyr is most likely buried in the nonpolar interior of the protein, unless its OH group is involved in a hydrogen bond, in which case a shift in the pK_a of Tyr may be observed. In contrast, if the spectrum change as a function of pH is similar to that of the free amino acid, this indicates that the Tyr residue is exposed to the solvent. Neighboring residues can also influence the pK_a of Tyr significantly. A very different pK_a than that of free Tyr suggests that the Tyr residue may be in a very polar environment, perhaps surrounded by lysines, arginines, or carboxyl groups.[16] When multiple Tyr residues are present in a protein, the number that are exposed to the solvent can be determined by comparing the spectral changes with those of an equimolar solution of free Tyr.

Difference Spectroscopy

As the name implies, difference spectroscopy (DS) involves measuring the difference between the absorption spectra of a protein (or nucleic acid) under two different solvent conditions.[17,18] DS can allow one to determine the effect of temperature, pH, denaturants, or other solvents on the environment of aromatic residues in proteins. Because of the small changes observed in absorption spectroscopy, DS provides a sensitive method of quantifying absorption changes. In normal absorption spectroscopy measured

[15] G. H. Beaven and E. R. Holiday, *Adv. Prot. Chem.* **7,** 319 (1952).
[16] D. B. Wetlaufer, *Adv. Prot. Chem.* **17,** 303 (1962).
[17] T. T. Herskovits, *Methods Enzymol.* **11,** 748 (1967).
[18] J. W. Donovan, *Methods Enzymol.* **27,** 497 (1973).

in a double beam instrument, the absorption of the protein and buffer reference samples are measured simultaneously and subtracted one from the other. In DS, the absorption of two protein samples, which vary only in pH or solvent composition, can be measured simultaneously to give the difference spectrum. However, it is essential that the solvent without the protein has identical absorbance for both cuvettes in the wavelength range of interest.

The difference spectrum between the native and the unfolded state of proteins shows a unique pattern that depends on the relative number of Trp, Tyr, and Phe residues present in the protein, as well as their molecular environments in the folded state. The difference spectrum of Trp and Tyr exhibits two peaks each, at 291–294 and 281–284 nm for the former and at 285–288 and 278–281 nm for the latter.[13,17] Phenylalanine sometimes causes small ripples in the 250- to 260-nm region. Because the difference in the absorption spectrum of the native and unfolded protein is small, it is essential to have exactly the same concentration of protein in both sample and reference cell and to subtract the buffer contribution from the difference spectrum.

Solvent Perturbation

A classical method to probe the solvent accessibility of aromatic side chains in native proteins is to change the composition of the solvent by adding different concentrations of a nonpolar organic cosolute.[17,18] This method, known as solvent perturbation, is rarely used these days, but may be useful for certain proteins whose 3D structure has not been solved. Solvent perturbation takes advantage of the high sensitivity of aromatic residues to solvent polarity. Those residues that are exposed to the solvent will experience a change in absorbance that can be monitored by difference spectroscopy. By comparing the absorbance change of the perturbed protein sample with that of the unfolded protein it is possible to calculate the number of exposed aromatic residues. Also, a protein-free sample of the appropriate aromatic amino acids, at the same concentration as in the protein of interest, can serve as a reference for fully solvent-exposed residues.

Table III lists several perturbing solvents of reduced polarity. A solvent composition that is 80% water and 20% solvent of lower polarity is recommended, as a higher concentration of the solvent may induce conformational changes.[18] Using other methods (e.g., CD, fluorescence), it is important to verify that no conformational changes have occurred. Another consideration when choosing a solvent for perturbation experiments is the size of the perturbing molecule. This is especially important in the case of aromatic

TABLE III
COMMON ADDITIVES USED IN SOLVENT PERTURBATION STUDIES[a]

Perturbant	Mean diameter (Å)
Methanol	2.8
Dimethyl sulfoxide	4.0
Ethylene glycol	4.4
Glycercol	5.2
Glucose	7.2
Polyethylene glycol	9.2
Sucrose	9.4

[a] From T. T. Herskovits, *Methods Enzmol.* **11,** 748 (1967).

residues that are located in an active site or deep crevice of the protein, and consequently may be partially exposed to the solvent. In such cases, the perturbing molecules must be below a critical size to be able to reach the semiburied residue. A significant difference between the extent of perturbation by two molecules of different size indicates clearly the presence of aromatic residues in crevices. Furthermore, the diameter of the crevice for a particular residue can be determined by the size of the largest molecule that can perturb the chromophore inside the crevice.[17] Table III shows the mean diameter for some of the most commonly used perturbing solvents.

Experimental Considerations for Measuring Absorbance

To obtain accurate and consistent absorbance measurements, careful detail must be paid to such experimental aspects as cuvettes, baseline, protein concentration, and slit width.

Cuvettes need to be made of quartz for work in the UV region. In the visible region, quartz, glass, or disposable plastic cuvettes may be used. Cuvettes come in different sizes to allow working with a variety of sample volumes. While the light pathlength is generally 1 cm, the width may vary from 2 to 10 mm. Shorter path lengths can be used, especially when dealing with highly scattering or strongly fluorescent samples. Concerning cuvette volume, it is very useful to know the minimum volume of sample required to run a spectrum without interference of the sample meniscus. Therefore, it is important to determine the location where the light beam strikes the cuvette for a particular instrument. This is done by recording a spectrum for a sample of low volume and of successively larger sample volumes until no further absorbance change is observed.

A baseline should always be run with both cuvettes empty or filled with the same buffer to zero the instrument after the lamps have warmed up for at least 30 min (or until the signal remains constant). This is very important, as the cuvettes may not be from a matched set or one of them may contain residual impurities. A spectrum in the region of interest should be measured with the empty cuvettes after recording the baseline; the absorbance should read zero throughout the scan.

The protein concentration must be such as to preserve its linear relationship with absorbance (Beer–Lambert law). Usually, this means measuring samples with an OD < 1, but preferably between 0.1 and 0.3, where at least 50% of the incident light is transmitted and the relative contribution of stray light (unanticipated light that reaches the detector, beyond that which is isolated by the monochromator) is minimized. One potential source of errors in protein concentration determination is the formation of oligomers or higher order soluble aggregates. This can be checked by measuring the absorbance of a protein sample after successive twofold dilution. For example, a protein sample that shows an absorbance of 1 should have an absorbance of 0.5 and 0.25 after successive twofold dilution. If deviations from linearity occur, the sample can be diluted further until linearity is observed.

The slit width, which controls the spectral band width, is the main determinant of the signal-to-noise ratio. The spectral bandwidth is defined as the central half of the wavelengths passed by the exit slit of the monochromator and is a property of the instrument. An optimum absorption peak can be obtained when the slit width is adjusted so that the spectral bandwidth is $\leq$20% of the natural bandwidth (the width in nanometers, at half the height of the absorbance band under investigation).[19] A gradual increase of the slit width will result in a corresponding increase in the signal-to-noise ratio until the slit width becomes too large, at which point distortion and decreased intensity of absorbance bands can occur.[20] Therefore, a convenient way to find an optimal slit width is to record the spectrum of the sample several times using a larger width each time (starting from 0.2 nm). The optimum will be the largest slit width that does not distort or decrease the absorbance band.

Protein Aggregation

The presence of large protein particles or aggregates in solution will result in light scattering. Because light that is scattered is deviated from its

[19] R. K. Poole and C. L. Bashford, *in* "Spectrophotometry and Spectrofluorimetry" (D. A. Harris and C. L. Bashford, eds.), p. 23. IRL Press, Oxford, 1987.

[20] F. Z. Schmid, *in* "Protein Structure: A Practical Approach" (T. E. Creighton, ed.), p. 251. Oxford Univ. Press, Oxford, 1990.

normal trajectory, it will not reach the detector and will result in erroneously larger absorbance measurements. Therefore, much care should be exercised to minimize the presence of light-scattering particles in the cuvette. The best approach is to filter or centrifuge all buffers and stock solutions. In protein solutions, absorbance above 310 nm is usually an indication of light scattering. If the size of the scattering particles is smaller than the incident wavelength, the light scattered is proportional to λ^{-4}. Hence, the apparent absorbance caused by light scattering can be corrected by extrapolating the linear part of an A vs λ^{-4} plot to the wavelength of interest and subtracting this value from the total absorbance.[21] However, for larger particles the intensity of scattered light can depend on λ^{-4} to λ^{-2}.[22] Therefore, in highly turbid solutions it is more appropriate to correct for light scattering by plotting log A vs log λ at wavelengths above 310 nm and extrapolating to the region of interest.[23,24] This approach can also be used to correct the difference spectra of highly turbid samples.[18] If a λ^{-4} dependence is assumed, the simplest method to correct for light scattering at 278 and 280 nm is to multiply by two the absorbance at 333 and 331 nm, respectively.[24] Although protein aggregation is sometimes unavoidable, it is preferable to minimize light scattering by selecting the proper medium (ionic strength, pH, temperature) and by filtering or centrifuging the sample rather than correcting for light scattering when analyzing the spectrum.

Fluorescence

For certain molecules the absorption of light is followed by the fluorescence (emission) of light of lower energy (i.e., longer wavelength). Most fluorescent molecules are aromatic molecules with a rigid and asymmetric structure. In proteins, Trp, Tyr, and Phe serve as intrinsic probes, but the fluorescence of Phe is small compared to that of Tyr and Trp because of its low quantum yield (Table II). The "efficiency" of fluorescence is determined by the quantum yield (Φ), which is defined as the ratio of the number of quanta emitted to the number of quanta absorbed. The quantum yield of a molecule can range from 0 (nonfluorescent) to 1 (100% "efficiency"). For Trp and Tyr, Φ is 0.20 and 0.14, respectively (Table II); however, Trp fluorescence is more than five times more sensitive than Tyr, due mainly to the much higher absorption coefficient of Trp. Also, when Tyr and Trp are both present in proteins, there is usually efficient resonance

[21] A. P. Demchenko, "Ultraviolet Spectroscopy of Proteins." Springer-Verlag, Berlin, 1986.
[22] R. D. Camerini-Otero and L. A. Day, *Biopolymers* **17,** 2241 (1978).
[23] E. Schauenstein and H. Bayzer, *J. Polym. Sci.* **16,** 45 (1955).
[24] A. F. Winder and W. L. G. Gent, *Biopolymers* **10,** 1243 (1971).

energy transfer (defined later) from Tyr to Trp, and the fluorescence of the latter will dominate.[25,26] To excite Trp residues only, the excitation wavelength should be set at 295 nm.

Along with the intrinsic fluorescence of proteins, extrinsic fluorescent molecules can be used to report on ligand binding, conformational changes, or association.[27] Extrinsic fluorophores may bind noncovalently to the protein or may be chemically attached at specific sites in a protein, mainly by reaction with free thiol or amino groups.[27] When attaching a fluorescent probe to a protein, it is important to verify that it does not cause a change in protein conformation, that it is bound to a specific site, and that it is sensitive to protein conformational changes.[28] Common fluorescent probes used in protein studies include 1-anilino-8-naphthalene sulfonate (ANS), dansyl chloride, rhodamine, and fluorescein.[27] In particular, ANS, which binds nonconvalently to proteins, has been widely used to probe exposed hydrophobic regions in proteins, especially in molten globules. The term molten globule describes a compact denatured state of certain proteins that retain native-like secondary structure, but have fluctuating tertiary structure.[29–31] A hallmark property of the molten globules is their ability to bind ANS.[32–35] In aqueous solution, ANS fluoresces weakly, but in a nonpolar environment, such as the interior of a protein, its quantum yield increases markedly and the spectrum shifts toward a shorter wavelength.

Fluorescent labels attached to proteins can serve as molecular rulers, providing information on molecular distances between fluorophores through a phenomenon known as fluorescence resonance energy transfer (FRET).[36–39] FRET involves the nonradiative energy transfer of the excitation energy from a donor to an acceptor. Three important criteria for

[25] L. Brand and B. Witholt, *Methods Enzymol.* **11,** 776 (1967).
[26] L. Stryer, *Science* **162,** 526 (1968).
[27] J. Slavik, "Fluorescent Probes in Cellular and Molecular Biology." CRC Press, Boca Raton, FL, 1994.
[28] D. Freifelder, "Physical Biochemistry." Freeman, New York, 1982.
[29] D. A. Dolgikh, R. I. Gilmanshin, E. V. Brazhnikov, V. E. Bychkova, G. V. Semisotnov, S. Y. Venyaminov, and O. B. Ptitsyn, *FEBS Lett.* **136,** 311 (1981).
[30] M. Ohgushi and A. Wada, *FEBS Lett.* **164,** 21 (1983).
[31] O. B. Ptitsyn, *Adv. Prot. Chem.* **47,** 83 (1995).
[32] S. Tandon and P. M. Horowitz, *J. Biol. Chem.* **264,** 9859 (1989).
[33] Y. Goto and A. L. Fink, *Biochemistry* **28,** 945 (1989).
[34] Y. Goto, N. Takahishi, and A. L. Fink, *Biochemistry* **29,** 3480 (1990).
[35] O. B. Ptitsyn, R. H. Pain, G. V. Semisotnov, E. Zerovnik, and O. I. Razgulyaev, *FEBS Lett.* **262,** 20 (1990).
[36] I. Z. Steinberg, *Annu. Rev. Biochem.* **40,** 83 (1971).
[37] R. H. Fairclough and C. R. Cantor, *Methods Enzymol.* **48,** 347 (1978).
[38] L. Stryer, *Annu. Rev. Biochem.* **47,** 819 (1978).
[39] P. Wu and L. Brand, *Anal. Biochem.* **218,** 1 (1994).

efficient energy transfer are that the acceptor and donor dipoles be close to alignment (although a significant degree of misalignment is tolerated), that the emission spectrum of the donor overlaps with the absorption spectrum of the acceptor, and that they be close to each other in space.[40] The most common example of energy transfer in proteins is that which occurs from Tyr to Trp and results in the absence of a Tyr emission spectrum. Another simple example involves the transfer of energy from Trp and/or Tyr to an extrinsic fluorophore when the latter binds a protein, thereby providing evidence for ligand binding.[27,41,42]

Interpretation of Protein Fluorescence

Fluorescence spectroscopy has the advantage of being a very sensitive technique compared to absorbance, not only requiring much less sample (Table I), but exhibiting a much greater signal distinction between the folded and the unfolded protein. The fluorescence emission maxima, quantum yield, and excited state lifetime of a fluorophore are affected by its molecular neighborhood. In particular, fluorophores are sensitive to the polarity of their surrounding environment. Hence, fluorescence is useful in monitoring protein conformational changes. For Tyr or Trp fluorescence to be sensitive to conformational changes in a protein molecule, they must experience a change in their molecular environment. Typically, the fluorophore must be buried within the hydrophobic protein interior, protected from the solvent. The maximum fluorescence wavelength for a fully buried Trp is $\leq$320 nm, compared to $\approx$350 nm when fully exposed. The red shift of Trp fluorescence when exposed to a polar environment is due to a decrease in the excited state energy caused by interaction with the solvent.[25,43] In contrast, the fluorescense maxima of Tyr is $\approx$303 nm and does not change with solvent conditions. Due to this difference, the λ_{max} for Trp fluorescence can provide qualitative information on how buried the residue is in the native protein. Additional information is provided by the fluorescence intensity (or Φ), which can be either higher or lower in the native state compared to the unfolded state, and is strictly protein dependent. A buried Trp will usually have a higher fluorescence intensity unless it is quenched by nearby polar residues or by energy transfer. In principle, an exposed Trp residue may also serve as a conformational probe if its

[40] T. Föster, *Disc. Faraday Soc.* **27,** 7 (1959).

[41] J. R. Brocklehurst, R. B. Freedman, D. J. Hancok, and G. K. Radda, *Biochem, J.* **116,** 721 (1970).

[42] G. K. Radda, *Biochem. J.* **122,** 385 (1971).

[43] E. Lippert, W. Lueder, F. Moll, W. Naegele, H. Boos, H. Prigge, and I. Seibold-Blankestein, *Angew. Chem.* **73,** 695 (1961).

fluorescence in the native state is quenched by nearby amino or carboxylic groups.[21]

The effect on a protein's Φ or λ_{max}, of changes in solvent polarity or pH, or the binding of a small molecule can provide information on the microenvironment of Trp or Tyr residues. As in absorbance, solvent additives can be used to probe the solvent accessibility of aromatic residues. If the addition of a fluorescence quencher (a molecule that decreases the fluorescence of a fluorophore) such as cesium ions (positively charged), iodide ions (negatively charged), or acrylamide (neutral) quenches the fluorescence of a Trp or Tyr in a protein, that residue must be near the surface of the protein.[44] If there is no quenching, it is likely that the residue is in the interior of the protein or in a crevice inaccessible to the quencher. Occasionally, cesium and iodide ions will quench the fluorescence of a protein to a different extent, indicating that at least one aromatic residue is in a highly positively or negatively charged region of the protein. Another way to probe the environment of Trp residues is to vary the pH. The fluorescence of Trp is quenched by nearby protonated acidic groups.[25,45] Thus, if the pK_a of a pH-induced transition monitored by fluorescence is the same as the pK_a of ionizable side chains, (e.g., histidine, cysteine, or aspartic/glutamic acid), then the group must be near Trp. Of course, other methods must be used to verify that the change in pH did not induce a conformational change in the protein. Finally, the binding of small molecules to a protein site can also serve as a probe for the environment of Trp residues. This will only be the case if there is no major conformational change that exposes a buried Trp. If the fluorescence of a Trp residue is quenched by the binding of a small molecule, then the Trp is in or very near the binding site. If there is also a decrease in λ_{max}, this indicates that the Trp has been shielded from the solvent in the complex.

Fluorescence Polarization and Anisotropy

A special application that can provide information about molecular motion and therefore about complex formation and molecular size is fluorescence polarization or anisotropy.[46,47] In fluorescence polarization experiments, a polarizer is placed in the path of the exciting light, and a second polarizer is placed between the sample and the detector, with its axis either perpendicular or parallel to the former. When a fluorophore is excited with

[44] M. R. Eftink and C. A. Ghiron, *Anal. Biochem.* **114,** 199 (1981).
[45] A. White, *Biochem. J.* **71,** 217 (1960).
[46] G. Weber, *Biochem. J.* **51,** 145 (1952).
[47] G. Weber, *Adv. Prot. Chem.* **8,** 415 (1953).

polarized light, its fluorescence will be unpolarized or partially polarized, depending on several factors, including[48] (1) the presence of energy transfer, (2) the polarization of the excitation light, (3) the lifetime of the excited state, and (4) the mobility of the fluorophore. Because energy transfer can occur, although with less probability, when the dipoles of the acceptor and donor molecules are not parallel, energy transfer between two chromophores that differ in orientation will result in fluorescence depolarization. When the polarization plane of the excitation light and the electric dipole moment of the fluorophore are parallel to each other the fluorophore is preferentially excited; however, they need not be perfectly aligned for excitation to occur as the probability of absorption is proportional to $\cos^2 \theta$, where θ is the angle between the plane of polarization and the electric dipole moment. The lifetime of the excited state of the fluorophore must be long enough to be able to experience changes in rotational motion of the fluorophore. In proteins, the lifetimes of the excited state of intrinsic fluorophores are often so short that they can only be used with small proteins (MW $< 20{,}000$), which tumble faster. For this reason, polarization experiments usually involve extrinsic fluorophores. Finally, fluorophore mobility is the main factor that determines whether the emitted light is polarized or not. The fluorescence emission will be polarized if the fluorophore remains relatively stationary during the lifetime of the excited state. Because large molecular complexes tumble slowly relative to the lifetime of the fluorophore, they depolarize the light only slightly and therefore exhibit a relatively high polarization. In contrast, if the emitter is rotating very rapidly, such that its orientation changes substantially within the lifetime of the excited state, the "memory" of the initial light direction is lost, resulting in a decreased polarization of the emitted light. Thus, fluorescence polarization is a sensitive way of monitoring the binding of small fluorophores to macromolecules, as well as protein–protein or protein–DNA interactions, all of which decrease the polarization of the emitted light.[49]

The intensity of polarized light transmitted by a polarizer depends on the relative orientation of the incident light and the axis of the polarizer, with values of ≈ 1 or 0 when they are oriented parallel or perpendicular to each other, respectively. For partial polarization, as in the case of protein fluorescence measurements, the polarization of emission is defined as

$$p = (I_V - GI_H)/(I_V + GI_H) \quad (4)$$

[48] C. R. Cantor and S. N. Timasheff, *in* "The Proteins" (H. Neurath and R. L. Hill, eds.), p. 145. Academic Press, New York, 1982.

[49] J. J. Hill and C. A. Royer, *Methods Enzymol.* **278,** 390 (1997).

where V and H represent the vertical or horizontal orientation of the emission polarizer with respect to the vertically oriented excitation polarizer and I refers to the fluorescence emission. A correction factor, G, for the difference in the transmission of the excitation and emission polarizers is defined as I_V/I_H when the excitation polarizer is oriented in the horizontal orientation.[27,50] It is very common to use fluorescence anisotropy (a) instead of polarization because the former is normalized to the total intensity ($I_V + 2GI_H$) of the emitting systems.[51] Anisotropy is defined as

$$a = (I_V - GI_H)/(I_V + 2GI_H) \quad (5)$$

where the terms have the same meaning as in polarization. Four measurements are needed to calculate the anisotropy (or polarization) of a sample: the vertical and horizontal components of the emission with the excitation polarizer in both vertical and horizontal (to calculate G) positions.

The basic rule when interpreting fluorescence polarization data is that polarization decreases with increased mobility. There are several ways in which a fluorophore will experience a change in its mobility. Any time an extrinsic fluorophore binds to a macromolecule it will experience a decrease in mobility, dependent not only on the mobility of the entire macromolecule, but also on the mobility of the binding site. The mobility of a fluorophore bound to a protein will decrease on protein oligomerization or aggregation, making p or a measurements particularly useful in monitoring interactions between subunits in proteins. Even in the case of protein conformational change, a fluorophore may experience a change in polarization as the protein unfolds if the fluorophore is in a rigid environment within the native structure.

Experimental Considerations for Measuring Fluorescence

Sample Preparation

Sample purity is of utmost importance, as fluorescent contaminants will interfere with the results. Avoid plastic containers, which may be a source of fluorescent impurities. Plastic containers can be tested for impurities by adding water to them and recording a fluorescence scan in the region of interest. The absorbance of the protein sample should be ≤ 0.1, as a high concentration of fluorophore may lead to an inner filter effect (see later). Finally, samples should be filtered or centrifuged before recording the fluorescence to remove particles that may scatter light.

[50] R. F. Chen and R. L. Bowman, *Science* **147,** 729 (1965).
[51] A. Jablonski, *Acta Physiol. Pol.* **16,** 471 (1957).

Cuvettes

Because fluorescence occurs in all directions, the detector is usually positioned at a right angle to the exciting light so that only light emitted from the sample reaches the detector. This requires that the cuvettes be transparent on all sides. Normally, the inner dimensions of fluorescence cuvettes are 10 × 10 mm, but semimicrocuvettes (4 × 10 mm) can also be used to save the sample. The cuvettes should be cleaned thoroughly and handled with gloves, as oil released by the fingers (i.e., fingerprints) will fluoresce. A solution of 50% (v/v) nitric acid is recommended for thorough cleaning.[20] For routine cleaning between samples, rinsing with water followed by ethanol should suffice.

Instrument Setup

A few instrumental parameters require adjusting to set up the instrument: excitation and emission wavelengths, excitation and emission band width, scan speed, and photomultiplier voltage. The desired excitation wavelength is usually the wavelength of maximum absorbance. Occasionally, one may prefer to excite at a different wavelength to selectively excite one fluorophore over another. For example, the fluorescence of Trp residues can be selectively monitored by exciting at 295 nm, where the absorbance of Tyr is virtually zero. The choice of excitation and emission band width will depend on the signal desired *vs* the sample sensitivity to light. Most instruments allow band widths from 2 to 20 nm. The signal will increase with larger excitation or emission band width as more light will reach the sample and the detector, respectively. The optimum excitation and emission band width should be determined by trial and error. Small excitation band widths (2–5 nm) are recommended to protect the sample from photobleaching. The emission band width can be larger (5–10 nm) to maximize the emission signal; however, a large band width may result in less spectral resolution. The photomultiplier voltage should be set to 80–85% of the scale with the most fluorescent sample to be used. For reproducibility, it is a good idea to measure the fluorescence of a standard (e.g., *N*-acetyltryptophanamide) at various times during a long experiment to check the stability of the excitation source.

Raman Peak of Water

Unlike Rayleigh and other types of scattering, Raman-scattered light has a longer wavelength (lower energy) than excitation light. Raman bands are distinguished easily from fluorescence bands because their λ_{max} changes with the excitation wavelength, whereas fluorescence does not.[20] All solvents containing C–H or O–H bonds exhibit Raman bands. For water, the Raman peak caused by O–H bond vibration is weak and occurs at a specific wavelength, depending on the exciting wavelength.[20] Often, the Raman peak of water will overlap with the protein fluorescence spectrum and should be subtracted from data. An advantage of the Raman peak of water is that it can be used to monitor instrument performance by keeping a log of the intensity of the peak recorded at a specific instrumental setting. A decrease in signal-to-noise ratio over time is a warning that the lamp may be deteriorating.

Photobleaching

The bright light sources used in fluorescence experiments may lead to the phenomenon known as photobleaching, where the number of photons absorbed (and, therefore, the number emitted) fall in time as the number of ground state molecules also falls. Photobleaching may be caused by the excitation rate exceeding the rate of decay of the excited state, or worse, by chemical decomposition of the fluorophore caused by photochemical reaction in the excited state.[52] The propensity for photobleaching depends on the fluorophore being analyzed. To minimize the possibility of photobleaching it is necessary to use the smallest excitation slit width that will provide adequate signal and close the excitation shutter in between measurements. Selection of a longer wavelength (lower energy light) within the absorption band may also help.

Inner Filter Effect

If the excitation light is absorbed strongly by the sample, and therefore does not reach the center of the cuvette, the fluorophore molecules in the center of the cuvette will not be fully excited. Because of this, the emitted light will be less than it should be and the linear dependence of fluorescence intensity on concentration will be lost. This phenomenon is known as the inner filter effect.[25] The emitted light is also attenuated by reabsorption before it leaves the cuvette, but it is a small effect compared to the absorption of excitation light. To avoid inner filter effect, the total absorbance of the sample, not just the protein, should be ≤ 0.1 at the excitation wavelength.

Temperature Effects

Fluorescence is very sensitive to temperature changes and its intensity decreases as temperature increases. For Tyr, the fluorescence emission

[52] C. L. Bashford, *in* "Spectrophotometry and Spectrofluorimetry" (D. A. Harris and C. L. Bashford, eds.), p. 1. IRL Press, Oxford, 1987.

decreases about 1% per degree increase in temperature.[20] For Trp, the temperature effect is even more pronounced. Therefore, it is essential to have a constant temperature throughout the experiment and allow sufficient time for temperature equilibration.

Because of the steep temperature dependence of fluorescence, the temperature-induced denaturation of proteins is rarely monitored by fluorescence. However, if corrections are made for the temperature dependence of Trp and Tyr fluorescence, it is possible to analyze temperature denaturation curves monitored by fluorescence.[27] A problem may arise from the fact that the temperature dependence is different for native and unfolded protein; however, a simulation by Etfink[53] showed that a linear dependence of fluorescence on temperature can be assumed with little effect on thermodynamic parameters.

Aggregation Problems

Protein aggregation is detrimental to fluorescence measurements because the protein particles will scatter light into the detector and give rise to an altered signal that will erroneously suggest a conformational change. This problem is particularly relevant when working with proteins that are prone to aggregate, as is the case with amyloidogenic peptides and proteins. The contribution of light scattering to fluorescence data can be determined by the existence of a peak at the excitation wavelength of an emission spectrum or by the existence of a peak at the emission wavelength of an excitation spectrum.[21] No generic solution exists to this problem. Hence, sample conditions, including pH, temperature, salt concentration, and protein concentration, need to be optimized to avoid aggregation. Soluble aggregates may also interfere, as well as much smaller aggregates if the fluorescent residues experience a change in environment. Thus, it is recommended to employ more than one technique to verify that there are no unwanted intermolecular interactions and to perform experiments at various protein concentrations to distinguish between intramolecular and intermolecular events. Fluorescence can be a useful technique in monitoring the formation of soluble aggregates, as long as there is a difference in signal between the aggregate and that of the native and unfolded states. Such is the case of the core domain of p53, where soluble aggregates give rise to a strong emission signal with a maximum of 340 nm, which is different from that of the unfolded (350 nm) and the native (very weak signal) protein.[54]

[53] M. R. Eftink, *Biophys. J.* **66,** 482 (1994).

[54] A. N. Bullock, J. Henckel, B. S. DeDecker, C. M. Johnson, P. V. Nikolova, M. R. Proctor, D. P. Lane, and A. R. Fersht, *Proc. Natl. Acad. Sci. U.S.A.* **94,** 14338 (1997).

Circular Dichroism

The differential absorbance of left and right circularly polarized light by a chromophore gives rise to the phenomenon known as circular dichroism (CD). The asymmetry of the peptide bond and aromatic residues within a folded protein makes CD a very useful technique for monitoring the secondary and tertiary structure of proteins in solution, respectively.[20,55–60] A similar technique, optical rotatory dispersion (ORD), measures the ability of an optically active compound to rotate plane-polarized light.[56–58] Because ORD is rarely used these days, this section deals only with CD.

When right (R) and left (L) circularly polarized light interacts with a chromophore, the extent to which they are absorbed will depend on the optical activity of the chromophore. For an optically inactive chromophore the difference in absorbance between the R and the L circularly polarized light will be zero, whereas for an optically active chromophore it will be non-zero. The degree of optical activity of a molecule is determined by its asymmetry. Asymmetric molecules are not superimposable on their mirror image, just as the right and left hands are mirror images of each other, yet cannot be superimposed. For an optically active molecule, the differential absorption (ΔA) of R and L circularly polarized light can be expressed in terms of the absorption coefficients, ε_R and ε_L:

$$\varepsilon_L - \varepsilon_R = \Delta\varepsilon \tag{6}$$

where $\Delta\varepsilon$ is the CD of the chromophore. However, ellipticity is usually plotted when measuring the CD of a sample. Ellipticity, θ is related to ΔA by the following equation:

$$\theta = 33\Delta A \text{ degrees} \tag{7}$$

where A is the absorbance as defined by Eq. (2). A plot of θ vs wavelength is called a CD spectrum.

[55] W. C. Johnson, Jr., *in* "Methods of Biochemical Analysis" (D. Glick, ed.), p. 61. Wiley, New York, 1985.

[56] A. J. Adler, N. J. Greenfield, and G. D. Fasman, *Methods Enzymol.* **27,** 675 (1973).

[57] P. M. Bayley, *in* "An Introduction to Spectroscopy for Biochemists" (S. B. Broen, ed.), p. 148. Academic Press, London, 1980.

[58] C. R. Cantor and P. R. Schimmel, "Biophysical Chemistry, Part II." Freeman, San Francisco, 1980.

[59] R. W. Woody, *Methods Enzymol.* **246,** 34 (1995).

[60] A. Rodger and B. Nordén, "Circular Dichroism and Linear Dichroism." Oxford Univ. Press, Oxford, 1997.

CD data obtained as θ are usually converted and presented as molar ellipticity $[\theta]$ or mean residue ellipticity $[\theta]_{MRE}$:

$$[\theta] = 100\theta M/lc \tag{8}$$

$$[\theta]_{MRE} = 100\theta M/N_A lc \tag{9}$$

where θ is the observed ellipticity in degrees, M is the molecular weight, N_A is the number of amino acids in the protein or peptide, l is the path length in centimeters, and c is the concentration in milligrams per milliliter. The units of $[\theta]$ and $[\theta]_{MRE}$ are degrees cm^2/dmole. By combining Eq. (2), (7) and (8), $[\theta]$ can be expressed in terms of $\Delta\varepsilon$:

$$[\theta] = 3300\,\Delta\varepsilon \tag{10}$$

Analysis of Protein Secondary Structure by CD

The same chromophores that are responsible for the absorption bands in proteins also give rise to the CD signal. The environment of the peptide bond, which absorbs at ≤230 nm, is determined by the secondary structure of the protein. Therefore, the far-UV CD spectrum of a protein can provide qualitative as well as quantitative information about the identify and extent of a secondary structure in a protein. Aromatic residues can also contribute to the far-UV CD spectrum, but usually their contribution is insignificant unless the protein has many aromatic residues and very little α-helix structure.[61] CD spectroscopy has been widely used to estimate the 2° structure composition of proteins, and the different methods have been reviewed extensively.[59,62–64] Even though CD cannot compete with NMR and X-ray in terms of atomic resolution, it can provide valuable structural information when NMR and X-ray data are not available. CD can also be used to provide 2° structure information in the initial stages of studying a protein or when little sample is available. From a practical viewpoint, CD is less labor intensive, more cost effective, and does not require the level of expertise necessary for NMR or X-ray structure determination.

Distinct CD spectra have been determined for "pure" 2° structures (Fig. 1). The α-helix CD is characterized by two negative bands of similar magnitude with maxima at 222 and 208 nm. There is also a stronger positive band at about 192 nm. The β-sheet CD is much weaker than the α-helix CD and has a negative band at 217 ± 5 nm and a positive band near 195

[61] R. W. Woody, *Biopolymers* **17,** 1451 (1978).
[62] J. T. Yang, C.-S. C. Wu, and H. M. Martinez, *Methods Enzymol.* **130,** 208 (1986).
[63] W. C. Johnson, Jr., *Proteins* **7,** 205 (1990).
[64] M. C. Manning, *J. Pharm. Biomed. Anal.* **7,** 1103 (1989).

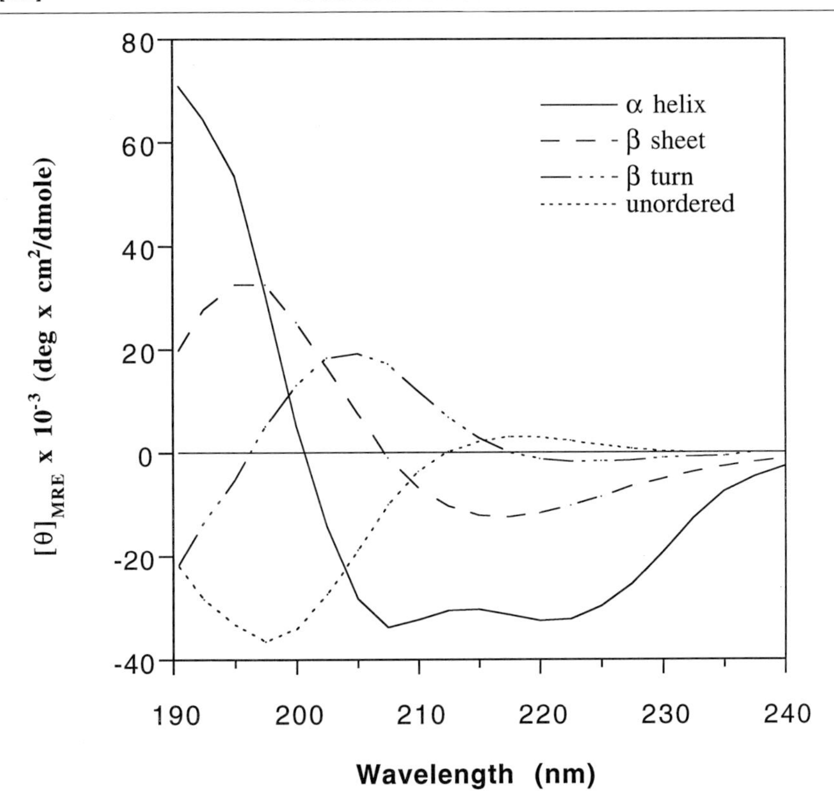

FIG. 1. Representative peptide circular dichroism spectra for α helix, β sheet, general β turn, and unordered structure. Curves were generated by plotting data from J. Reed and T. A. Reed, *Anal. Biochem.* **257,** 36 (1997).

nm. The β-turn CD is very variable because of the different types of turns found in proteins. A general β-turn CD has been described by Brahms and Brahms[65] (Fig. 1), which has a positive band at about 205 nm and a negative band near 190 nm. The unordered conformations (often called random coil) describe all those conformations that do not fit the well-characterized 2° structures. The net CD of unordered conformations has a strong negative band at about 200 nm and a weak positive band near 218 nm. Sometimes there is a weak positive band at 220–230 nm.

In principle, the CD spectrum of a native protein can be viewed as the sum of the appropriate fractions of each 2° structure component. Therefore, protein CD spectra are usually analyzed by expressing them as a linear

[65] S. Brahms and J. Brahms, *J. Mol. Biol.* **138,** 149 (1980).

combination of standard spectra of a specific 2° structure. In nearly all the methods used to analyze the 2° structure of proteins from their CD spectra, the CD of a group of proteins of known 2° structure are used as a reference set.[59,62–64] Mathematical methods are used to extract basis spectra of the various 2° structures and are then linearly combined to reconstruct the CD spectrum of the protein of interest. The proportion of the different basis spectra that gives the best fit to the measured CD spectrum indicates the percentage of the 2° structure present in the protein. Among the 2° structures, the α helix is the best determined by all methods, with an average accuracy of 95%. β sheets, β turns, and unordered conformations are not as well determined; the average accuracy is 85, 75, and 70%, respectively.[59] To obtain accurate values, it is very important to collect data to at least 185 nm and preferably to 178 nm.[66] The 2° structure of proteins can also be determined by IR absorption and Raman frequency. In particular, CD and IR appear to complement each other because CD is better at determining α-helix content, whereas the IR amide I band of β sheets is well resolved and can provide more accurate data for β sheet.[67]

Several sources of error may interfere with the analysis of the protein 2° structure by CD spectroscopy.[59,64] First, it is assumed that the crystal structure and solution conformation of proteins are identical. While this is probably true for most proteins, there are exceptions, including the proteins insulin, α-bungarotoxin, the complement component C3a, and the avian pancreatic polypeptide.[64] Second, the 2° structure of the protein being analyzed may differ substantially from the average structures in the reference set. Third, the assumption that tertiary contacts do not contribute to the far-UV CD may not always apply, as interaction between different elements of the 2° structure within the native conformation may result in nonadditivity. Using theoretical calculations on tropomyosin fragments, Cooper and Woody[68] have demonstrated that the influence of one helix on another can change the intensity of the CD spectrum. Fourth, the contribution of aromatic residues and disulfide bonds to the far-UV CD signal may be significant in some cases.[61,69] Fifth, CD spectra of α helices and β sheets depend on their lengths and, in the case of the latter, on the width and twist. Deviation from the average length or width compared to the reference protein set may result in inaccurate 2° structure determination. This is particularly relevant when trying to predict the structure of peptides

[66] P. Manavalan and W. C. Johnson, Jr., *Anal. Biochem.* **167,** 76 (1987).

[67] W. K. Surewicz and H. H. Mantsch, *in* "Spectroscopic Methods for Determining Protein Structure in Solution" (H. A. Havel, ed.). VCH Publishers, New York, 1996.

[68] T. M. Cooper and R. W. Woody, *Biopolymers* **30,** 657 (1990).

[69] E. H. Strickland, *CRC Crit. Rev. Biochem.* 113 (1974).

in solution, especially those that adopt the β sheet structure. In addition to the effect of length, width, and twist, other factors are likely to contribute significantly to the CD spectra of "β peptides" and interfere with the analysis of 2° structure, such as their tendency to aggregate and their weak CD signal, especially when coupled to the CD contribution of aromatic residues, which are much more common in β sheets than in α helices.

The available methods for analyzing far-UV CD spectra of proteins to determine their 2° structure content have not been as successful with peptides. This is probably due to a greater contribution in peptides of all the potential sources of errors mentioned earlier, in addition to their much greater flexibility. Model peptide spectra have also been used for estimating the 2° composition of peptides, but there are also problems with this approach, as most of them are homopolymers or copolymers of two or three different residues.[65] The side chain composition of model peptides is not as important for α helical peptides whose CD spectra are nearly independent of the nature of the side chain, but is very important for peptides adopting β sheet structure, as their CD spectra are sensitive to the specific side chain.[70] Reed and Reed[71] addressed the lack of side chain variability among model peptides by constructing model spectra from an average of CD spectra of different model peptides and then weight averaging the contribution of each according to the frequency with which their side chain(s) is found in proteins (Fig. 1). They report an improvement in the analysis of the peptide 2° structure over previous methods.

Monitoring Changes in Protein Tertiary Structure by CD

The aromatic side chains of Trp, Tyr, and Phe give rise to CD bands in the near-UV region, with maxima at around 280–290, 275–280, and 250–260, respectively. Disulfide bonds also generate a CD signal in the near-UV region (250–270 nm), but their contribution is usually much weaker than that of aromatic residues. For a protein to exhibit CD in the near-UV region, its aromatic residues must be in an asymmetric environment, usually as a result of being in a fixed geometry relative to the peptide backbone. For a change to occur in the near-UV CD signal of a protein, there must be a change in the freedom of rotation about the C_α, C_β, or C_γ bonds of the aromatic residues; such a change requires a conformational change in the protein. Hence, CD spectra in the near-UV region provide a "fingerprint" of the overall fold of a protein. Unlike far-UV CD, the near-UV–CD spectrum cannot be interpreted to obtain detailed structural

[70] R. W. Woody, *J. Polym. Sci. Macromol. Rev.* **12,** 181 (1977).

[71] J. Reed and T. A. Reed, *Anal. Biochem.* **254,** 36 (1997).

information and therefore it is most useful as a probe for global conformational changes. Total disappearance of the near-UV CD signal suggests that the protein has unfolded; however, this assumption has to be confirmed by other methods (e.g., far-UV CD), as local unfolding or an increase in protein dynamics can cause such a change. A classical example is that of the molten globules, which are known for lacking their native near-UV CD signal, but are not totally unfolded, as indicated by their native-like far-UV CD spectra.[31]

Using CD to Monitor Protein Folding and Stability

In contrast to absorbance and fluorescence spectroscopy, which monitor the environment of aromatic residues, CD is able to unambiguously distinguish between changes in 2° and 3° structure. CD measurements in the far-UV region (240–180 nm) are the method of choice to monitor qualitatively and quantitatively the secondary structure of peptides and proteins, whereas measurements in the near-UV region (300–250 nm) can provide information about the protein tertiary or quaternary structure. By comparing far-UV and near-UV CD data from equilibrium denaturation or folding kinetic studies, conformational events involving the secondary structure can be distinguished from those involving the tertiary structure.

The reversible temperature-induced denaturation of proteins is very often used to determine the stability of a protein and the effect of mutation on protein stability. Because fluorescence is temperature dependent, far-UV CD is the spectroscopic method of choice to monitor the stability of a protein against temperature denaturation. An additional advantage of CD over fluorescence and absorbance is that the former can monitor the changes in secondary structure directly, whereas the latter two can only monitor the environment of the aromatic residues. Thus, a change in protein dynamics or local conformation without significant protein denaturation may result in a UV or fluorescence signal change that may suggest erroneously that the protein is being unfolded by temperature.

Experimental Considerations for Measuring CD

Instrument Calibration

Most CD instruments need to be calibrated on a regular basis using a standard sample. The most widely used standard is (+)−10-camphorsulfonic acid (CSA). The CD of CSA measured at 290.5 and 192.5 nm should give a $\Delta\varepsilon$ of approximately 2.36 M^{-1} cm^{-1} and −4.9 to −4.7 M^{-1} cm^{-1}, respectively.[72,73] The ratio of the CD signals at the two wavelengths, $\Delta\varepsilon 192.5/\Delta\varepsilon 290.5$, provides a good estimate of instrument performance at low wavelengths; a value of 2 or greater is expected for a well-performing instrument.[72,73] It is important to determine the concentration of CSA by measuring its absorbance at 285 nm ($\varepsilon = 34.5\ M^{-1}\ \text{cm}^{-1}$). A 1-mg/ml solution of CSA in a 5-cm cell has an absorbance of 0.743 and an ellipticity of 33.5 in a 1-mm cell.[63]

Sample

The buffer of choice should have an absorbance band that does not interfere with CD measurements. This is especially important when measuring the CD in the far-UV region (<220 nm) where many buffers and salts have strong absorbance.[20] Phosphate and borate buffers are excellent to use in the far-UV region, whereas samples containing urea or GuHCl absorb strongly in the far-UV, and their CD spectra can only be measured to about 210 nm. The total absorbance of the sample, including the protein, buffer, and cell, should be ≈1 for the best signal-to-noise ratio and should always stay below 2.0. The goal when preparing a sample for CD is to minimize the absorbance of the solvent and maximize the absorbance of the protein.

Baseline

Because the baseline is usually not straight in a CD experiment, it is important to do a baseline correction. This is done by collecting the CD spectrum of the solvent and subtracting it from the sample spectrum. It is important that the instrument be warmed up for at least 30 min and that both solvent and sample spectra are collected under the same conditions.

Cells

Cells used for CD work are made of quartz and come in cylindrical and rectangular shapes. Cylindrical cells are often preferred because they usually have less birefringence (baseline CD). CD cells must be tested for birefringence by measuring the CD in the region of interest with the empty cell. The CD signal should coincide with the instrument baseline without the cell.

CD cells of various pathlengths are available to decrease or increase sample absorbance. For far-UV CD, cells with a width of 0.05–2 mm are

[72] J. Y. Cassim and J. T. Yang, *Anal. Lett.* **10,** 1195 (1977).
[73] J. P. Hennessey, *Anal. Biochem.* **125,** 177 (1982).

used, whereas for near-UV CD measurements, where the CD signal is one to three orders of magnitude weaker, cell width ranges from 1 to 10 cm. For far- and near-UV CD measurements the choice of cell width will usually be determined by the need to decrease and increase the total sample absorbance, respectively, while still being able to use the desired protein concentration. For the determination of secondary structure, measurements below 190 nm are required, and a 0.05- to 0.1-mm-wide cell is necessary.[63] Similarly, if a low protein concentration must be used to avoid aggregation, CD measurements in the near-UV CD region can be done in larger cells (5 or 10 cm).

Instrumental Parameters

The slit width and time constant (time over which the instrument averages) play an important role in the signal-to-noise ratio. Because the signal-to-noise ratio increases with the square root of the time constant, the latter should be as large as possible. A good rule of thumb is that the scan speed (nm/sec) times the time constant (sec) should be less than 0.33 nm.[73] The optimal combination of scan speed and time constant can be obtained by increasing the time constant (starting from 0.5 sec) at a constant scan speed until the spectrum begins to deteriorate. Because it can take more than 1 hr to obtain a high-quality CD spectrum, to avoid spectral drift it is better to run faster spectra (with baseline runs in between) and average them using the instrument software. For qualitative information, CD spectra can be obtained in much less time, depending on the protein concentration.

The slit width should be as large as possible for optimal signal-to-noise ratio. Because too large a slit width can distort the spectrum, it is necessary to determine the optimum value. The optimal slit width can be determined as described in the UV/Vis Spectroscopy section of this article.

Instrument Purging

The intense light from the lamp will interact with oxygen to create ozone, which will damage the CD optics. Therefore, CD instruments must be protected against oxygen when the lamps are on by flushing with nitrogen for 15–20 min before turning on the source and continual flushing throughout the experiment. The boil off from a liquid nitrogen tank is recommended at a flow rate of 15–20 liters/min.[63] It is a good practice not to use the last 10–15% of the tank as any impurities will be found at higher concentration. An instrument that "dies" below 200 nm is likely to have had its optics ruined by oxygen due to lack of or inadequate flushing with nitrogen.

Aggregation Problems

As in the case of UV/Vis and fluorescence measurements, the presence of aggregates can be deleterious to obtaining accurate CD data. Aggregates will not only scatter light, but will most likely change the environment and conformation of the peptide backbone and aromatic residues. In fact, protein aggregation can affect the CD spectrum before there is any turbidity in the sample. The histone octamer provides a good example of the interference of protein aggregation with data interpretation. Different CD spectra obtained for the histone octamer in 2.0 *M* sodium chloride and 2.3 *M* ammonium sulfate led to the erroneous conclusion that structural changes in the protein were caused by solvent conditions.[74] Reexamination of data revealed that the difference in CD spectra was due to aggregation, as no appreciable difference in their CD spectra was observed when the octamer in each solution was centrifuged at 30,000 rpm for 30 min.[75]

It is not possible to predict the effect of the oligomerization state on CD spectra of proteins. In certain cases, oligomer dissociation results in no change in the CD, whereas in other cases there is a significant change. For instance, the dissociation of the human relaxin dimer does not alter the far-UV CD, whereas the near-UV CD is altered significantly in the region assigned to Tyr residues.[76] In contrast, the dissociation of insulin hexamers into monomers is accompanied by a small but significant change in the far-UV CD, consistent with a small decrease in the α-helix structure and a 15% increase in β sheet.[77]

Occasionally, far- and near-UV CD can be used to monitor the formation of soluble protein aggregates if the aggregates have a unique signal. Also, it may be possible to monitor the CD signal of soluble aggregates as a function of time, pH, temperature, or any other variable and to determine conformational changes associated with aggregation. This approach has been used to monitor the aggregation of bovine growth hormone (bHG) and the amyloid β (Aβ) protein. Aggregation of partially denatured bGH exhibits a characteristic spectral band at 300 nm in the near-UV CD,[78] and Lehrman *et al.*[79] used this signal to determine the association constant of bGH aggregation. Similarly, Barrow *et al.*[80] used far-UV CD not only to

[74] K. Park and G. D. Fasman, *Biochemistry* **26,** 8042 (1987).

[75] A. D. Baxevanis, J. E. Godfrey, E. N. Moudrianakis, K. Park, and G. D. Fasman, *Biochemistry* **29,** 973 (1990).

[76] S. J. Shire, L. A. Holladay, and E. Rinderknecht, *Biochemistry* **30,** 7703 (1991).

[77] S. G. Melberg and W. C. Johnson, Jr., *Proteins* **8,** 280 (1990).

[78] H. A. Havel, E. W. Kauffman, S. M. Plaisted, and D. N. Brems, *Biochemistry* **25,** 6533 (1986).

[79] S. R. Lehrman, J. L. Tuls, H. A. Havel, R. J. Haskell, S. D. Putman, and C. C. Tomich, *Biochemistry* **30,** 5777 (1991).

[80] C. J. Barrow, A. Yasuda, P. T. M. Kenny, and M. G. Zagorski, *J. Mol. Biol.* 1075 (1992).

monitor the α helix to β sheet conformational change that precedes the aggregation of Aβ into amyloid, but also to determine the rate at which the aggregates formed before they precipitated from solution.

If the sample has been filtered or centrifuged, there is no simple method to eliminate the contribution of light scattering caused by protein aggregation from the CD spectrum of a protein. However, light-scattering effects can be minimized by increasing the detector acceptance angle; this can be achieved by placing the sample very close to the detector to minimize the amount of scattered light that misses the detector.[81]

Conclusion

The spectroscopic techniques discussed in this article can provide complementary information on the structure of proteins. The technique of choice will depend on many factors, including instrument availability, protein concentration needed, solvent conditions, protein availability, and the structural information desired. By combining UV/Vis, CD, and fluorescence spectroscopy, it is possible to obtain a wealth of complementary information about protein 2°, 3°, and 4° structure under various solvent conditions. These can further be coupled with FTIR, Raman, and NMR spectroscopy to provide a comprehensive description of the structure of the protein. Even when using multiple techniques to study the structure of proteins, it is important to be aware of, and test for, the possibility of spectrum distortion caused by soluble aggregates. Protein aggregation does not always cause turbidity and, therefore, its presence may not be obvious by visual inspection of the sample or the collected spectrum. Aggregation can be ruled out easily by performing experiments at different protein concentrations or by using size exclusion chromatography under the same experimental conditions.

Acknowledgements

I thank Dr. Joyce H. Diwan, Dr. Jane F. Koretz, Dr. Joseph Warden, Dr. Daniel Moriarty, and Charmi Miller for comments on the manuscript.

[81] M. Cascio, P. A. Glazer, and B. A. Wallace, *Biochem. Biophys. Res. Commun.* **162,** 1162 (1989).

[40] Probing Conformations of Amyloidogenic Proteins by Hydrogen Exchange and Mass Spectrometry

By EWAN J. NETTLETON and CAROL V. ROBINSON

Introduction

Evidence has suggested that the formation of amyloid fibrils may involve the structural rearrangement of native monomeric protein through partially folded intermediates.[1] For example, the kinetics of amyloid fibril formation *in vitro* are most favorable for wild-type transthyretin under solution conditions that destabilize the native state tetramer of the protein.[2] Furthermore, naturally occurring variants of human lysozyme,[3] transthyretin,[4] and immunoglobulin light chain V_L domain[5] have been shown to destabilize the native states of these proteins. Destabilization allows these proteins to unfold more readily, generating partially folded states that are proposed as amyloidogenic intermediates. These aggregation-prone intermediates are difficult to study by classical structural biology techniques as they are too dynamic for X-ray crystallography and are not amenable to nuclear magnetic resonance (NMR) measurements directly because the concentration necessary for their study promotes aggregation. In this article, hydrogen exchange measured by electrospray (ES) mass spectrometry (MS) is used to probe the solution dynamics of amyloidogenic proteins.

The technique of hydrogen exchange labeling is an immensely powerful tool in the elucidation of both protein structure and folding patterns in solution.[6–9] This process typically involves the exchange of labile amide, side chain, and terminal protons on a protein with solvent deuterons. The technique exploits the fact that hydrogen atoms buried in the core of the protein or involved in hydrogen bonding interactions do not exchange

[1] R. Wetzel, *Cell* **86,** 699 (1996).
[2] Z. Lai, W. Colon, and J. W. Kelly, *Biochemistry* **35,** 6470 (1996).
[3] D. Booth, M. Sunde, V. Bellotti, C. V. Robinson, W. L. Hutchinson, P. E. Fraser, P. N. Hawkins, C. M. Dobson, S. E. Radford, C. C. F. Blake, and M. B. Pepys, *Nature* **385,** 787 (1997).
[4] S. L. McCutchen, W. Colon, and J. W. Kelly, *Biochemistry* **32,** 12119 (1993).
[5] L. Helms and R. Wetzel, *J. Mol. Biol.* **257,** 77 (1996).
[6] A. Hvidt and S. O. Nielsen, *Adv. Prot. Chem.* **21,** 287 (1966).
[7] C. Woodward, I. Simon, and E. Tuchsen, *Mol. Cell. Biochem.* **48,** 135 (1982).
[8] R. L. Baldwin, *Curr. Opin. Struct. Biol.* **3,** 84 (1993).
[9] Y. W. Bai, T. R. Sosnick, L. Mayne, and S. W. Englander, *Science* **269,** 192 (1995).

0076-6879/99 $30.00

readily with solvent deuterons. Labile protons on the surface of the protein will exchange more readily than those buried deep in the solvent-excluded core of the protein.

Protection from exchange or slowly exchanging sites generally arises from backbone amides involved in hydrogen-bonding interactions in α helices or β sheets. Hydrogen exchange is facilitated by structural fluctuations of the protein that transiently expose the hydrophobic core to solvent. Consequently hydrogen exchange can be used as a sensitive probe to study the conformational dynamics within proteins. An in-depth discussion of the factors affecting hydrogen exchange rates is beyond the scope of this article, but is presented elsewhere.[10,11]

Hydrogen exchange kinetics may be followed by a variety of techniques. Traditionally, NMR has been the method of choice because it can locate the sites where exchange has taken place within the primary sequence of the protein yielding residue-specific information.

NMR does have some limitations: high, nonphysiological protein concentrations are generally required, only certain pH ranges may be used for direct measurements, and a molecular weight limit of the protein that can be assigned readily (typically up to 30 kDa).

MS is not subject to these constraints and can measure the increase in mass, which results from the exchange of protons for heavier deuterons, as a function of time.[12–15] Some of the advantages of using MS are lower concentrations of protein are required, often reflecting physiological protein concentrations; wider ranges in pH may be used for direct measurement, with positive- and negative-ion electrospray for acidic and basic protein solutions, respectively; and hydrogen exchange of very high mass protein systems can be followed (e.g., the conformation of a protein bound within the cavity of an 800-kDa molecular chaperone complex has been shown to resemble a partially folded molten globule state).[14]

Not only is it possible to measure the extent of incorporation of deuterons for a protein with time by ES-MS but also the distribution of deuterium in the ensemble of protein molecules. This dispersal of deuterium is evident

[10] J. Clarke, L. S. Itzhaki, and A. R. Fersht, *Trends Biochem. Sci.* **22,** 284 (1997).

[11] E. W. Chung, E. J. Nettleton, C. J. Morgan, M. Gross, A. Miranker, S. E. Radford, C. M. Dobson, and C. V. Robinson, *Prot. Sci.* **6,** 1316 (1997).

[12] V. Katta and B. T. Chait, *Rapid Commun. Mass Spectrom.* **5,** 214 (1991).

[13] A. Miranker, C. V. Robinson, S. E. Radford, R. T. Aplin, and C. M. Dobson, *Science* **262,** 896 (1993).

[14] C. V. Robinson, M. Gross, S. J. Eyles, J. J. Ewbank, M. Mayhew, F. U. Hartl, C. M. Dobson, and S. E. Radford, *Nature* **372,** 646 (1994).

[15] Q. Yi and D. Baker, *Prot. Sci.* **5,** 1060 (1996).

from the distribution of masses recorded.[11,13] For example, where more than one protein conformation is present in solution, some conformations may exchange more rapidly, leading to a broader distribution of masses than for a protein containing only natural abundance isotopes. Consequently, the width of the peak in the mass spectrum of protein undergoing hydrogen exchange can be used to gain information about the number of different protein conformers in solution. Moreover, peptic digestion at low pH enables the location of hydrogen exchange protection to specific regions of the protein.[16]

The important characteristic of the ES-MS method that renders it particularly amenable to the study of amyloid-forming proteins is its large dynamic range; the technique allows measurements to be made over a range of different solution conditions from millimolar to nanomolar protein concentration. Moreover, measurements can be made in the presence or absence of volatile salts and at a range of different pH values and temperatures (see later). An important criterion for measuring partially folded states is the requirement that in solution the monomeric protein is present and is not part of a larger aggregate. Not only does ES-MS allow protein to be analyzed at a low concentration to prevent extensive aggregation, it also offers the possibility of detecting higher oligomers acting as a probe for the aggregation of monomeric protein. Under solution conditions where proteins are known to aggregate, mass spectra are obtained in which protein oligomers are clearly visible. This provides the opportunity to probe hydrogen-exchange protection in these higher oligomers and to compare this with that of the monomeric protein under the same solution conditions.

Examples of three very different amyloid-forming proteins—variants of human lysozyme, wild type, and variants of human transthyretin and bovine insulin—are used to illustrate the type of information available with this methodology. These proteins have widely different structures: lysozyme is a two domain protein containing both an α-helical and a β-sheet structure whereas transthyretin is essentially β sheet and the monomeric form of insulin is composed of three α helices. In addition the proteins form fibrils both *in vitro* and *in vivo* under different solution conditions. For example, *in vitro* fibrils are formed from insulin at 70° and $pH_2 \cdot O$[17] and from Ile56Thr lysozyme at 4° and pH 8.0.[3]

[16] Z. Zhang and D. L. Smith, *Prot. Sci.* **2,** 522 (1993).

[17] I. Langmuir and D. Waugh, *J. Am. Chem. Soc.* **62,** 2771 (1940).

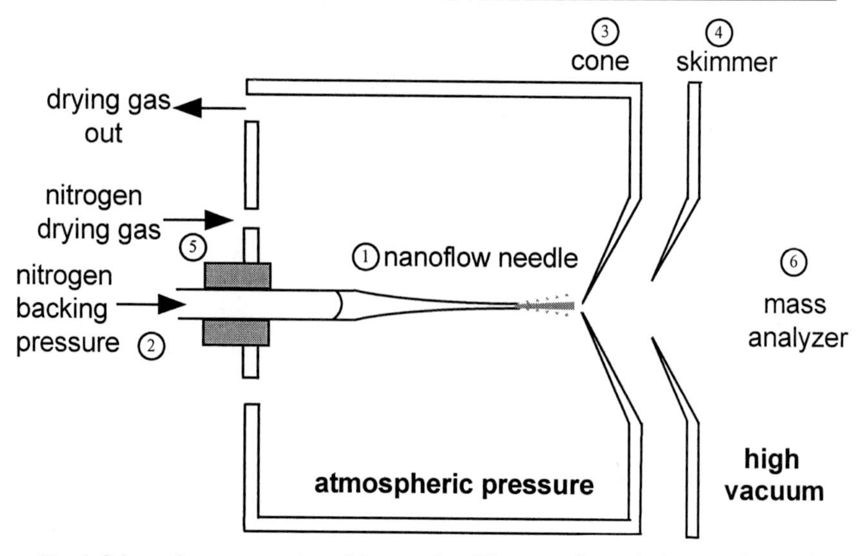

FIG. 1. Schematic representation of the nanoflow ES process for analyzing protein samples. Protein solution (1–2 μl, of 10–20 μM concentration) is placed in the nanoflow ES needle (1) prepared in house from a borosilicate capillary of external diameter of 1.0 mm and internal diameter 0.5 mm using a micropipette puller (Sutter Instrument Co.). The capillary is drawn out on a Model P-97 Flaming/Brown micropippette puller (Sutter Instrument Co.) and is gold coated with a SEM-coating system (Polaron). The tips are broken manually under a microscope before analysis. This process results in an internal diameter of between 1 and 10 μm at the tapered tip.[18] Nitrogen gas is used to apply a backing pressure (10–30 psi) to the sample during analysis to ensure a stable flow of viscous D_2O protein solutions (2). Mass spectrometer conditions are set typically with a needle voltage of 1.6 kV (1), cone voltages typically in the region of 60–120 V (3) are employed, and the difference between the capillary and the skimmer voltage is maintained at 5 V (4). The source is operated without heating and with a low-flow, rate-drying gas (20 liter/hr) (5). Data are acquired using a quadrupole mass analyzer (Micromass Platform II) (6). A Micromass Platform II mass spectrometer operating under MassLynx2.3 was used to acquire mass spectral data.

Sample Introduction

The essential features of the nanoflow ES process used in the analysis of transthyretin and insulin samples are shown in Fig. 1.

[18] M. S. Wilm and M. Mann, *Int. J. Mass Spectrom. Ion Proc.* **136,** 167 (1994).

Sample Preparation

Human Lysozyme and Variants

Human lysozyme and variants are expressed and purified as described previously.[3] Prior to MS, proteins are washed extensively in water at pH 3.8 using Centricon 10 concentrators (Amicon, Danvers, MA) and then equilibrated in water at pH 5.0 at a concentration of 20 μM. Hydrogen exchange is initiated by a 10-fold dilution from protein in water at pH 5.0 and 37° into D_2O (Aldrich, Milwaukee, WI) at pH 5.0 and 37°. No attempt is made to correct for pH differences in H_2O and D_2O. All pH readings in D_2O are represented by pH and are glass electrode readings. In all analyses, ultrapure water (ELGA maxima system) was used.

Human Transthyretin and Variants

Human transthyretin and Val30Met, Leu55Pro variants are purified as described previously[4] and stored in 10 m*M* potassium phosphate, 100 m*M* KCl, 1 m*M* EDTA, pH 7.4. Samples are prepared for MS by dialyzing extensively first against aqueous 10 m*M* ammonium acetate at pH 4.0 and then against 10 m*M* ammonium acetate at pH 7.0. Desalted samples are then buffer exchanged extensively using Centricon 10 concentrators (Amicon) with water pH 4.5 with HCl (Aldrich). The protein concentration is determined by bicinchoninic acid assay reagents (Pierce, Rockford, IL).

Hydrogen exchange is initiated by 100-fold dilution into D_2O, pH 4.5, to give a final concentration of 0.005 mg/ml (0.09 μM). Aliquots of the exchange solution are removed periodically and analyzed by nanoflow electrospray. A different needle is employed for each sample.

Bovine Insulin

Bovine insulin is purchased from Sigma Chemical Company. Hydrogen exchange is initiated by the dissolution of insulin (dry powder) into D_2O at pH 2.0 with DCl (Florochem) to give a final concentration of 2 m*M*.

Hydrogen Exchange

Hydrogen exchange is measured by either dissolving lyophilized protein or by buffer exchanging the protein into water at the appropriate pH. The hydrogen exchange reaction is initiated by dilution (100-fold if concentration permits) into D_2O at the same pH. The intrinsic rate for hydrogen exchange is at a minimum between pH 2.0 and 3.0.[19] Optimal conditions

[19] Y. W. Bai, J. S. Milne, L. Mayne, and S. W. Englander, *Prot. Struct. Funct. Genet.* **17,** 75 (1993).

for MS from aqueous solution are generally below neutral pH as this gives rise to a higher proton concentration and hence increases the efficiency in forming positively charged protein ions. For monitoring the hydrogen exchange of amyloidogenic proteins, the choice of pH is largely dictated by the conditions under which fibrils form. For example, the formation of fibrils from transthyretin occurs more readily below pH 5.0, whereas below pH 4.0 a partially folded form of transthyretin has been proposed.[2] The hydrogen exchange kinetics of transthyretin are measured at pH 4.5. Insulin fibrils form readily at pH 2.0, and this pH is therefore used for the measurement of hydrogen exchange kinetics, whereas pH 5.0 is used for human lysozyme. Hydrogen exchange may be measured at elevated temperatures, as in the case of the lysozyme and transthyretin discussed later where enhanced conformational dynamics are more apparent in the variants compared to the wild-type proteins at 37° than at 22° (data not shown). The appropriate concentration of protein is an important consideration as an association to higher oligomers will lead to increased protection from hydrogen exchange and, in the case of transthyretin, will populate the tetrameric form.

Mass Spectrometry

All mass spectra are recorded on a Micromass Platform II without source heating and in the absence of organic cosolvents. Each sample is filtered using Anopore filters (0.2 μm) (Whatman, Clifton, NJ) immediately prior to MS. For human lysozyme and variants, conventional ES is used. Ten microliters of protein solution (20 μM) is introduced via a Rheodyne injector into a continuous flow of 90% D_2O, pH 3.8 (with formic acid), at a flow rate of 10 μl/min. For the analysis of transthyretin and insulin, nanoflow ESMS is employed. Two microliters of protein solution (0.09 μM for transthyretin and 2 mM for insulin) is introduced via a nanoflow ES interface from a borosilicate gold-plated needle at 20°. All mass spectra are calibrated using hen lysozyme (Sigma) in H_2O at pH 5.0. The estimated error in hydrogen exchange experiments, calculated from the standard deviation of the charge states, is ±2 Da. The number of protected sites for hydrogen exchange is calculated by subtraction of the measured mass from the sum of the molecular mass of protein, and the number of labile sites is deduced from the amino acid sequence. Data are acquired in triplicate to confirm the reproducibility of the results.

Data Analysis

The average masses of protein samples are calculated from at least three charge states. The masses measured are centroid values, which take into account the asymmetry of the peaks arising from the natural isotope distribution. All mass spectra presented are the raw data, either on a m/z scale or transformed onto a mass scale, and with minimal smoothing, without any resolution enhancement. Hydrogen-exchange kinetics are analyzed using the scientific software package Sigma Plot (Jandell Scientific Ltd.). A random coil exchange profile, for an unstructured protein of the same sequence under identical experimental conditions, is calculated from near-neighbor inductive effects.[20] Simulated spectra are produced using Sigma Plot and are calculated from a Gaussian distribution of the natural abundance isotopes.

Human Lysozyme and Variants

The hydrogen-exchange properties of wild-type and variant human lysozymes are investigated at pH 5.0 and 37° (Fig. 2). Of the 262 exchange-labile hydrogens in wild-type human lysozyme, 130 are located on backbone amides and at the termini and 132 are in side chains. A stable core of 55 ± 2 remain protected from exchange after 90 min under these conditions. In contrast, however, in the Asp67His and Ile56Thr variants, 250 ± 2 protons have undergone hydrogen exchange after 90 min, leaving only 12 ± 2 sites remaining protected. Comparison of hydrogen-exchange data over a 4-hr time period (Fig. 2), confirms that the two variants are much less protective against hydrogen exchange than the wild type, with Ile56Thr being the least protective. Results show that the hydrogen-exchange protection in the wild-type protein is dramatically different from that of the two variants. Given that X-ray analysis reveals that both variant proteins have a native-like fold,[3] the lack of protection in the variant proteins is explained by an increased amplitude and frequency of native state fluctuations. This increased conformational dynamics allows solvent molecules to penetrate the interior of the protein, leading to an increased hydrogen-exchange rate in the variants. This result, together with data from other biophysical techniques, suggests that a transient population of amyloidogenic proteins in a partially folded state is crucial for their structural conversion to the fibrillar form.[3]

Wild-Type and Variant Transthyretins

In its physiologically active form, transthyretin is a tetramer containing four identical monomers consisting of 127 amino acid residues. The monomers are composed essentially of β sheet and associate to form a tetramer

[20] Y. W. Bai, J. S. Milne, L. Mayne, and S. W. Englander, *Methods Enzymol.* **259,** 344 (1995).

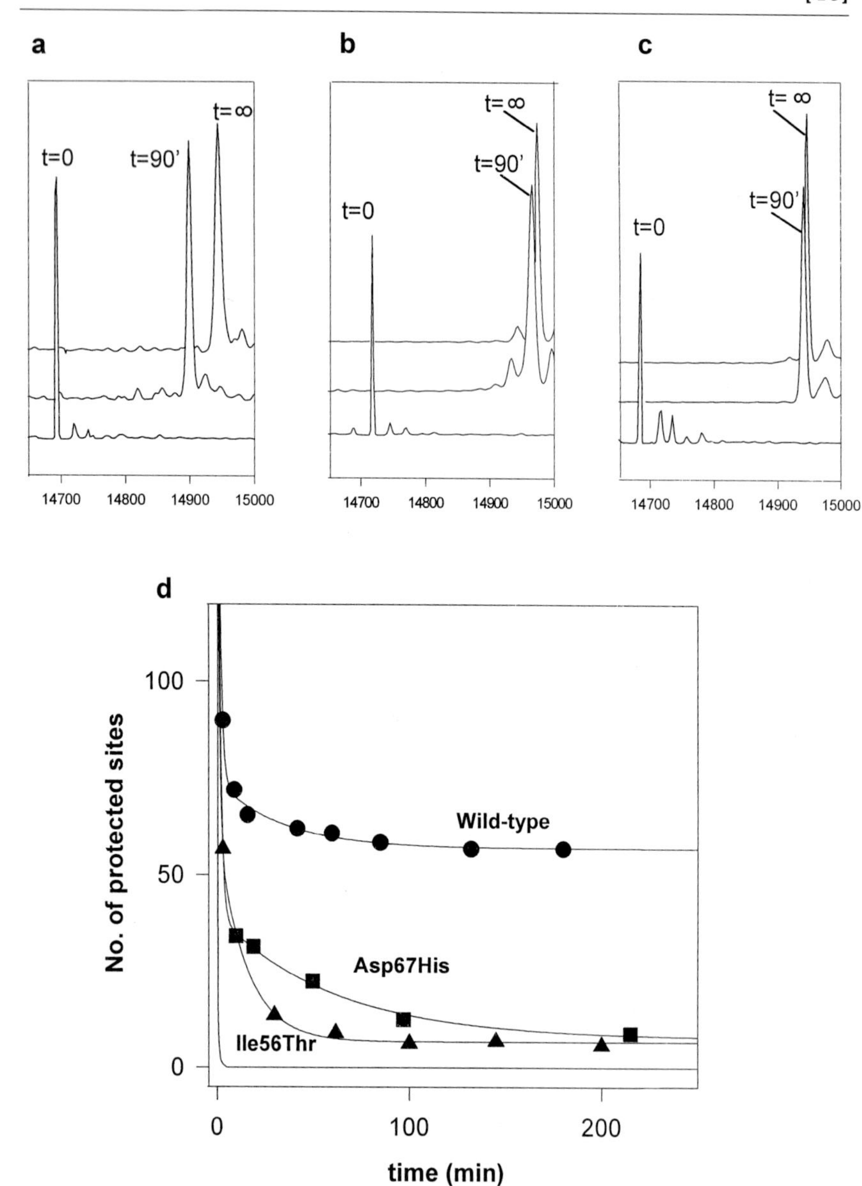

FIG. 2. Mass spectra and hydrogen exchange profiles of human lysozyme and variants at 37° and pH 5.0. Mass-transformed ES mass spectra of wild-type (a), Asp67His (b), and Ile55Thr (c) showing the fully protonated protein (t = 0), fully deuterated protein (t = ∞), and partially exchanged protein after 90 min of hydrogen exchange. The kinetic profile of hydrogen exhange (d) of (●) wild-type, (■) Asp67His, (▲) Ile56Thr human lysozymes, and (—) random coil simulation of hydrogen exchange from a fully unstructured state under the same solution conditions.[19]

with a large central channel, the binding sites for thyroid hormones. Previous studies using analytical ultracentrifugation to examine wild-type transthyretin under different solution conditions have shown that the protein dissociates to a monomer under acidic conditions, below pH 5.0.[4,21] Cross-linking followed by SDS–PAGE has shown that after incubation under fibril-forming conditions, the proportion of tetrameric transthyretin in solution is reduced for both Leu55Pro and Val30Met transthyretins. We have shown that it is possible to preserve intersubunit protein interactions in tetrameric TTR in the gas phase and to compare the strength of interactions in the wild-type protein with those in the Leu55Pro and Val30Met variants.[22] Using the same nanoflow conditions, ESI mass spectra show the presence of only monomeric and solvent-bound tetrameric species (Fig. 3A). The proportion of wild-type tetramer is, however, greater than that observed for the Val30Met variant. These results are in line with data from solution studies[4] and are consistent with the proposal that variant transthyretins form amyloid more readily via unfolding of the monomeric form to yield the amyloidogenic intermediate.

The structure and dynamics of the monomeric protein were investigated at nanomolar concentration by hydrogen exchange. At this protein concentration, both the wild type and Val30Met variant are in their monomeric form so that contributions from subunit interactions can be ruled out. The hydrogen-exchange kinetic profiles of the two proteins are shown in Fig. 3B. Significant differences in the hydrogen exchange properties are observed. Of the 207 labile sites in transthyretin, 89 are located in side chains and 118 arise from backbone amides. In monomeric wild-type TTR, an average of 55 ± 1 sites remain protected from exchange with solvent after 20 hr at 37°. The extent of hydrogen-exchange protection in the Val30Met variant, however, is remarkably different, with only 20 ± 1 sites remaining after 20 hr. In this study the loss of hydrogen-exchange protection was not assigned to specific β stands. The extent of protection observed, however, is consistent with the loss of some structural elements in the Val30Met transthyretin. Computer-modeling studies have suggested that partial disruption of the β-sandwich structure, through disorder in the "edge strands," may occur in variant transthyretins.[23,24] These results are consistent with this model, as a stable hydrogen-bonded structure in the remaining β strands would give a total of 20 hydrogen-bonded amides. This

[21] H. A. Lashuel, Z. Lai, and J. W. Kelly, *Biochemistry,* **37,** 17851 (1998).

[22] E. Nettleton, M. Sunde, V. Lai, J. Kelly, C. Dobson, and C. Robinson, *J. Mol. Biol.* **281,** 553 (1998).

[23] J. W. Kelly and P. T. Lansbury, *Amyloid Int. J. Exp. Clin. Invest.* **1,** 186 (1994).

[24] L. Serpell, G. Goldstein, I. Dacklin, E. Lundgren, and C. C. Blake, *Amyloid Int. J. Exp. Clin. Invest.* **3,** 75 (1996).

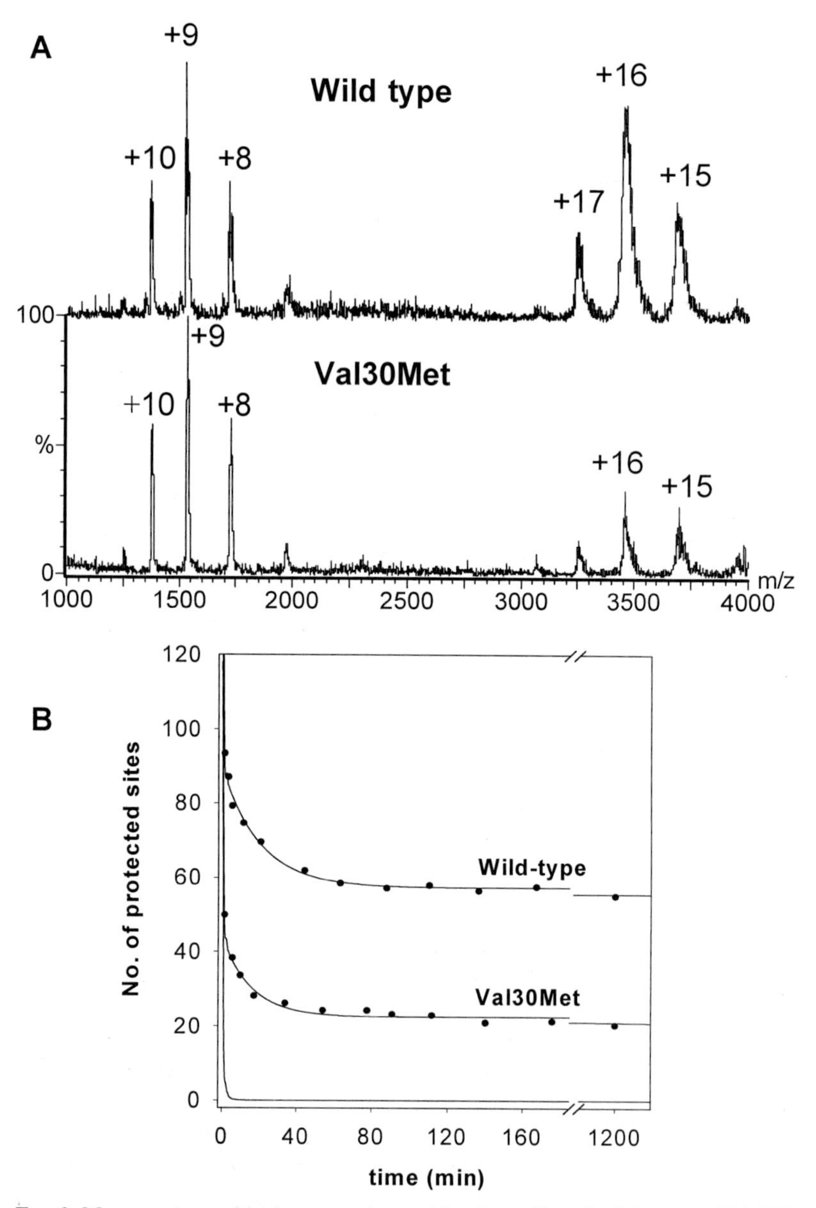

FIG. 3. Mass spectra and hydrogen-exchange kinetic profiles of wild-type and Val30Met transthyretins. (A) Nanoflow ES mass spectra of wild-type and Val30Met transthyretins obtained under identical instrument settings. Charge states labeled +15, +16, and +17 correspond in mass to the solvent-bound tetramer, whereas charge states labeled +8, +9, and +10 are assigned to monomeric transthyretins. There are no other species present. (B) Hydrogen-exchange kinetic profiles of wild-type and Val30Met transthyretin and a random coil simulation (—) of hydrogen exchange from a fully unstructured state under the same solution conditions at 37° and pH 4.5.

number is in excellent agreement with the 20 ± 1 hydrogens found to remain protected from exchange in the Val30Met transthyretin after 20 hr at pH 4.5 and 37°.

Insulin

In addition to the dissociation of the transthyretin tetramer, it is also possible to study protein association by nanoflow ES-MS. Under solution conditions where bovine insulin forms fibrils *in vitro,* higher oligomers are observed that extend beyond the mass to charge range capabilities of the quadrupole mass spectrometer used in these experiments (Fig. 4A). While many overlapping charge states are observed, it is possible to identify species containing up to eight insulin monomers associated under these conditions. These species are observed reproducibly under fibril-forming conditions (pH 2.0 and 2 m*M*) and are not present at a significant level when solutions at higher pH or lower protein concentrations are analyzed. Although the peak widths are much broader than for the monomeric species, hydrogen-exchange experiments suggest that these species are in equilibrium with the monomeric protein, as increased protection in higher oligomers is not observed (data not shown).

Hydrogen-exchange measurements of the monomer under fibril-forming conditions, pH 2.0 at 2 m*M,* are shown in Fig. 4. The increase in mass as deuterium is incorporated as a function of time is shown for the molecular ion of bovine insulin and is analyzed in terms of the peak width of the exchanging protein. The peak width of a molecular ion arises from the presence of natural abundance isotopes in the protein, principally the contribution from 1% of ^{13}C, and the resolution capabilities of the mass spectrometer. The distribution of masses of a protein undergoing hydrogen exchange is further complicated by the dispersal of deuterium among protein molecules.[11] This distribution of deuterium in proteins undergoing hydrogen exchange can, however, be used to gain valuable insights into different exchange mechanisms in proteins. In previous work, using model proteins, random exchange and protected core models were proposed to explain hydrogen exchange in different protein conformers.[11] Two such models are compared with data obtained for the hydrogen exchange of bovine insulin under fibril-forming conditions in Fig. 4B.

The hydrogen-exchange profile of insulin shows that a stable core of 23 of the backbone amide sites remain protected after 3 hr at pH 2.0. This could arise from different scenarios: either 23 of the 83 exchange labile sites remain protected in all of the molecules (core protection model) or the 60 deuterons incorporated could be distributed randomly among all 83 sites (random-exchange model). Comparison of the two models with data

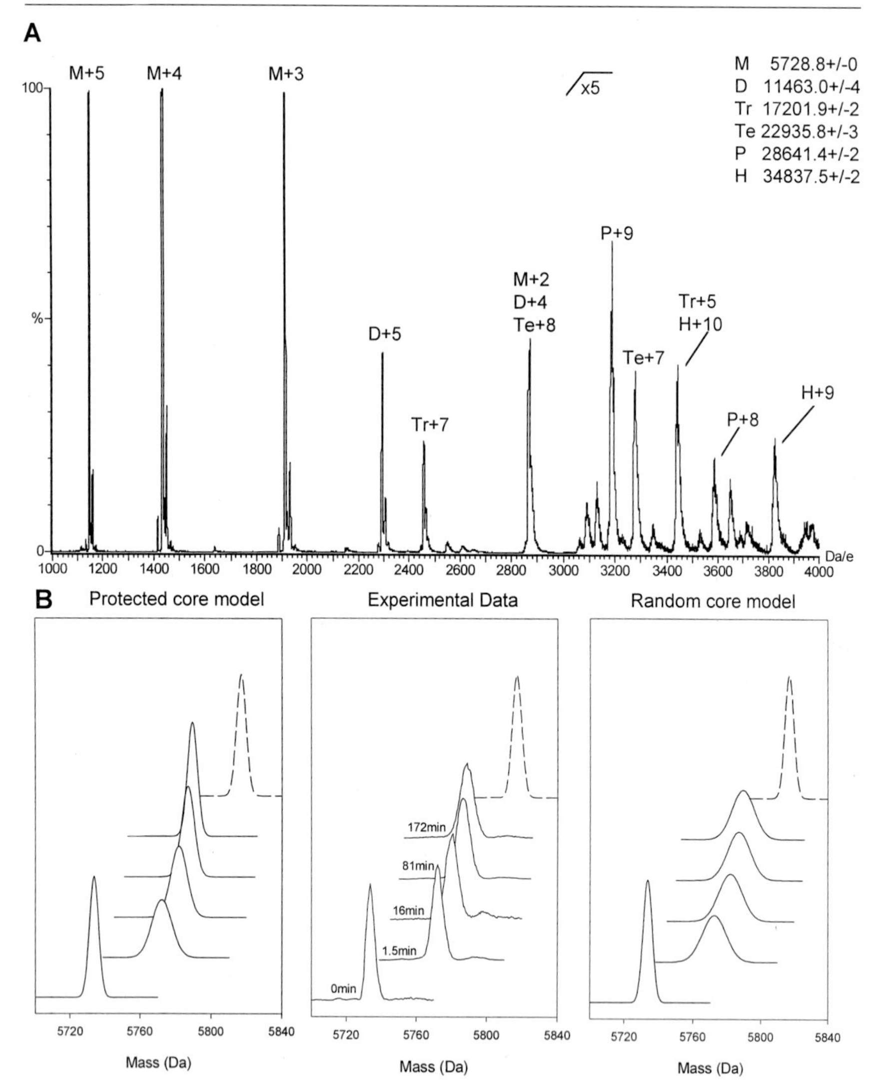

FIG. 4. (A) ES mass spectrum of bovine insulin at 2 m*M* concentration and at pH 2.0 showing the presence of higher oligomers that can be assigned on the basis of their mass charger ratios as monomer (M), dimer (D), trimer (Tr), tetramer (Te), pentamer (P), and hexamer (H). (B) Mass-transformed data for the hydrogen exchange of bovine insulin at pH 2.0 and 2 m*M* concentration. Experimental data are compared with a random exchange model in which all 83 labile sites exchange at the same rate and a protected core model in which 23 sites remain protected from solvent throughout the time course of exchange.

shows that during the initial stages of the hydrogen-exchange process, the mass distribution of data is narrower than predicted from either model. This narrow distribution is observed until 81 min, at which time the mass distribution of experimental data becomes broader. This broadening persists until the end point of the exchange.

These observations suggest that during the early part of the exchange process, before 81 min, a subset of hydrogens, presumably labile side chains, are exchanging more rapidly than would be predicted by either model. At later times, after 81 min, data are broader than the protected core model and yet narrower than the random-exchange model. These results strongly suggest that more than one conformation exists in solution after 81 min, giving rise to a broad distribution of masses. It is likely, however, that some elements of structure are retained, as the mass distribution is narrower than that predicted for the random distribution model. An alternative explanation could be that protein aggregation leads to a broader range of exchanging species. This possibility has been eliminated on the basis of hydrogen-exchange profiles of higher oligomers not showing increased protection against exchange (data not shown), suggesting that they do not form persistent hydrogen-bonded structure in solution, presumably as they are in dynamic equilibrium with the monomer protein. Data presented here are consistent with the rapid exchange of side chains followed by the slower formation of at least two distinct conformations with different hydrogen-exchange kinetics.

Future Perspectives

The technique of hydrogen exchange monitored by ES-MS was first described in 1991,[25] was later coupled to pulse-labeling techniques to elucidate protein-folding pathways,[13] and was extended to probe the effect of molecular chaperones on protein folding.[14] The technique has been applied to the study of amyloid-forming proteins, including monomeric[3] and tetrameric species,[22] and in both cases has shown that the monomeric protein exhibits increased conformational dynamics under fibril-forming conditions. The future for this technique, in probing amyloidogenic intermediates, lies in part in its ability to monitor small quantities of protein at low concentration and in its potential to localize hydrogen exchange to specific regions of the protein structure.[16] The potential of this method to probe interactions with molecular chaperones, proposed to mediate interactions between these aggregation prone intermediates, may provide further insight into the putative role of accessory proteins in stabilizing amyloidogenic intermediates.

[25] V. Katta and B. T. Chait, *Rapid Commun. Mass Spectrom.* **5,** 214 (1991).

Moreover, the introduction of commercial ES time-of-flight mass spectrometers will undoubtedly extend the mass-to-charge range available, allowing the study of higher mass range protein oligomers and enabling hydrogen exchange measurements that probe their persistence in solution.

Acknowledgments

We thank Margaret Sunde and Christopher Dobson for helpful discussions and acknowledge with thanks collaborations with Mark Pepys and Jeffery Kelly. This is a contribution from the Oxford Centre for Molecular Sciences, which is funded by the BBSRC, EPSRC, and MRC. EJN is grateful for financial support from Glaxo Wellcome and CVR thanks the Royal Society for support.

Section IV

Cellular and Organismic Consequences of Protein Deposition

[41] Yeast Prion [Ψ^+] and Its Determinant, Sup35p

By TRICIA R. SERIO, ANIL G. CASHIKAR, JAHAN J. MOSLEHI, ANTHONY S. KOWAL, and SUSAN L. LINDQUIST

Introduction

[*PSI*$^+$] and [*URE3*] are two non-Mendelian genetic elements of the yeast *Saccharomyces cerevisiae* that appear to be inherited through an unusual mechanism: the continued propagation of an alternate protein conformation. The protein determinants of these elements, Sup35p for [*PSI*$^+$][1,2] and Ure2p for [*URE3*],[3,4] have the unique ability to exist in at least two different, stable conformations *in vivo.*[4–8] Although the spontaneous generation of one conformer is rare, this alternate form, once acquired, becomes predominant, influencing the other conformer to change states.[5] This self-perpetuation of protein conformation is the key to the non-Mendelian inheritance of both [*PSI*$^+$] and [*URE3*]. In addition, the [*Het-S*] phenotype of *Podospora anserina,* another fungus, may be inherited by a similar mechanism.[9] This article focuses on both *in vivo* and *in vitro* methods used to analyze [*PSI*$^+$], the most extensively studied member of this group.

Genetics of [*PSI*$^+$] Inheritance

[*PSI*$^+$] was originally described in 1965 by Cox as a translation infidelity factor.[10] Translation terminated efficiently at nonsense codons in strains classified as [*psi*$^-$], whereas [*PSI*$^+$] strains were capable of omnipotently

[1] S. M. Doel, S. J. McCready, C. R. Nierras, and B. S. Cox, *Genetics* **137,** 659 (1994).
[2] M. D. Ter-Avanesyan, A. R. Dagkesamanskaya, V. V. Kushnirov, and V. N. Smirnov, *Genetics* **137,** 671 (1994).
[3] R. B. Wickner, *Science* **264,** 566 (1994).
[4] D. C. Masison and R. B. Wickner, *Science* **270,** 93 (1995).
[5] M. M. Patino, J. J. Liu, J. R. Glover, and S. Lindquist, *Science* **273,** 622 (1996).
[6] S. V. Paushkin, V. V. Kushnirov, V. N. Smirnov, and M. D. Ter-Avanesyan, *EMBO J.* **15,** 3127 (1996).
[7] S. V. Paushkin, V. V. Kushnirov, V. N. Smirnov, and M. D. Ter-Avanesyan, *Science* **277,** 381 (1997).
[8] I. L. Derkatch, Y. O. Chernoff, V. V. Kushnirov, S. G. Inge-Vechtomov, and S. W. Liebman, *Genetics* **144,** 1375 (1996).
[9] V. Coustou, C. Deleu, S. Saupe, and J. Begueret, *Proc. Natl. Acad. Sci. U.S.A.* **94,** 9773 (1997).
[10] B. Cox, *Heredity* **20,** 505 (1965).

0076-6879/99 $30.00

suppressing nonsense mutations by increasing the rate at which ribosomes read through stop codons.[10] The [PSI^+] phenotype, translational read through, has been monitored most frequently in yeast strains harboring nonsense mutations in metabolic enzymes by growth on defined medium lacking the product of that pathway. In these cases, [psi^-] yeast strains are auxotrophic for that nutrient and will not grow in the absence of supplements, whereas [PSI^+] yeast strains are at least partially prototrophic for the nutrient.

This convenient method of screening for [PSI^+] formed the basis of early genetic characterizations, which revealed the unique properties of [PSI^+] inheritance. Crosses between haploid [psi^-] and [PSI^+] strains yield [PSI^+] diploids, indicating that the [PSI^+] state is dominant *in vivo* (Fig. 1A).[10] Surprisingly, the meiotic haploid progeny of these [PSI^+] diploids are all [PSI^+], demonstrating that [PSI^+] was propagated by a non-Mendelian mode of inheritance (Fig. 1A).[10] This idea is supported further

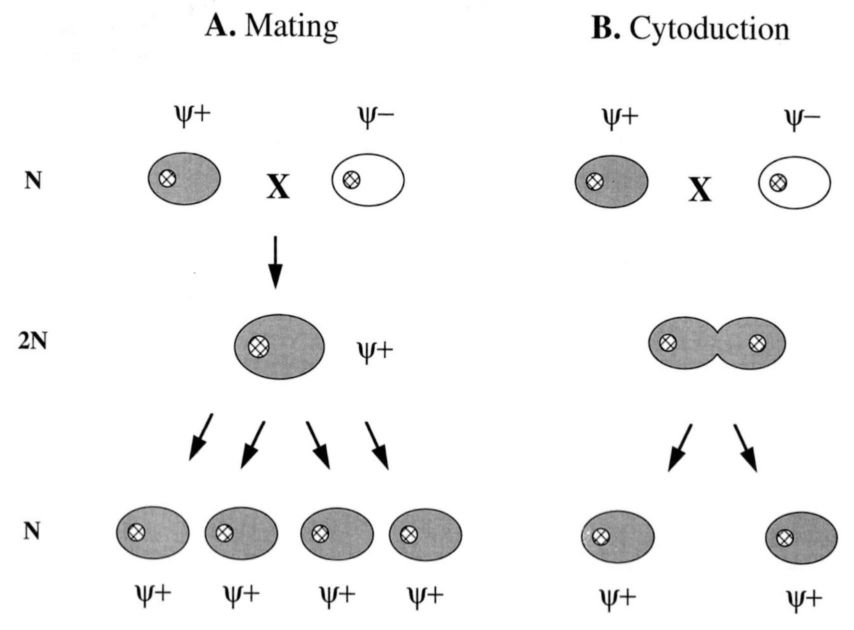

FIG. 1. Genetic analysis of [PSI^+] inheritance. (A) Shown are yeast cells that are [PSI^+] (Ψ^+, gray) or [psi^-] (Ψ^-, white) with hatched nuclei. Mating (X) produces a [PSI^+] diploid (Ψ^+, gray), and subsequent sporulation (four arrows) yields four haploid segregants that are all [PSI^+], (Ψ^+, gray). (B) [PSI^+] and [psi^-] yeast cells as in (A). Cytoduction allows cytoplasmic mixing in the absence of nuclear fusion (note separate hatched nuclei). Segregants are all [PSI^+] (Ψ^+, gray). Ploidy for both (A) and (B) is shown to the left of the figure (N = haploid, $2N$ = diploid).

by the observation that [*PSI*$^+$] could be transmitted to susceptible strains by cytoduction experiments in which cytoplasmic mixing occurs in the absence of nuclear fusion (Fig. 1B).[11]

Efforts to link [*PSI*$^+$] to extrachromosomal plasmids or cytoplasmically propagated nucleic acids proved fruitless[12,13] and were complicated further by another puzzling aspect of [*PSI*$^+$] inheritance. [*PSI*$^+$] is a metastable genetic element; [*PSI*$^+$] is lost at a strain-specific characteristic low rate (see later) through mitotic division, but it can also reappear spontaneously in these [*psi*-] strains.[14] This "reversible curing" is inconsistent with a nucleic acid-directed inheritance model. Furthermore, treatments that are nonmutagenic to nucleic acids, such as growth in the presence of 5 m*M* guanidine hydrochloride are efficient at curing [*PSI*$^+$].[14]

The nature of the [*PSI*$^+$] element remained a mystery for nearly 30 years until Wickner suggested that [*PSI*$^+$] and [*URE3*] were propagated by alternate protein conformations rather than nucleic acids.[3] Soon after this idea was proposed, a link between [*PSI*$^+$] propagation and the molecular chaperone heat shock protein 104 (Hsp104) was established. Either deletion or transient overexpression of Hsp104 is sufficient to convert yeast strains from [*PSI*$^+$] to [*psi*-].[5,15] That the transient overexpression of a molecular chaperone, whose only known function is to alter the physical state of substrate proteins, could induce a heritable change in phenotype in yeast provides one of the strongest arguments to date in support of a protein-only mode of inheritance for [*PSI*$^+$].

Sup35p, Protein Determinant of [*PSI*$^+$]

Identification of the yeast protein Sup35p as the determinant of [*PSI*$^+$] provided the first step in understanding the basis of the [*PSI*$^+$] phenotype and its propagation. Sup35p is the yeast homolog of the eukaryotic release factor eRF3.[16–18] Sup35p forms a functional translation termination complex

[11] B. S. Cox, M. F. Tuite, and C. S. McLaughlin, *Yeast* **4,** 159 (1988).

[12] M. F. Tuite, P. M. Lund, A. B. Futcher, M. J. Dobson, B. S. Cox, and C. S. McLaughlin, *Plasmid* **8,** 103 (1982).

[13] C. S. Young and B. S. Cox, *Heredity* **28,** 189 (1972).

[14] M. F. Tuite, C. R. Mundy, and B. S. Cox, *Genetics* **98,** 691 (1981).

[15] Y. O. Chernoff, S. L. Lindquist, B. Ono, S. G. Inge-Vechtomov, and S. W. Liebman, *Science* **268,** 880 (1995).

[16] S. A. Didichenko, M. D. Ter-Avanesyan, and V. N. Smirnov, *Eur. J. Biochem.* **198,** 705 (1991).

[17] I. Stansfield, K. M. Jones, V. V. Kushnirov, A. R. Dagkesamanskaya, A. I. Poznyakovski, S. V. Paushkin, C. R. Nierras, B. S. Cox, M. D. Ter-Avanesyan, and M. F. Tuite, *EMBO J.* **14,** 4365 (1995).

[18] G. Zhouravleva, L. Frolova, X. Le Goff, R. Le Guellec, S. Inge-Vechtomov, L. Kisselev, and M. Philippe, *EMBO J.* **14,** 4065 (1995).

with the yeast eRF1 homolog, Sup45p.[17,18] Together, Sup35p and Sup45p direct the faithful termination of translation at stop codons in [*psi*$^-$] cells. In [*PSI*$^+$] cells, however, the Sup35p function is compromised, leading to the nonsense suppression phenotype. This epigenetic loss of Sup35p activity may result from a unique type of Sup35p aggregation.[5,6] These aggregates have only been isolated from [*PSI*$^+$] strains and are lost by treatments that cure [*PSI*$^+$], such as the deletion or overexpression of Hsp104.[5]

Work by several groups has been instrumental in elucidating the regions of Sup35p important for translation termination and [*PSI*$^+$] propagation. Sup35p is composed of three regions—N, M, and C (Fig. 2A)—based on amino acid composition and homology to other translation factors.[19] The amino-terminal region, N [amino acids (aa) 1–123], has an unusual amino acid composition and distribution, with 78% of all residues being glycine (G), tyrosine (Y), asparagine (N), or glutamine (Q). This region contains six imperfect repeats of the sequence QGGYQ(Q)QYNP.[19] N is the prion-determining domain of Sup35p: it is necessary for the propagation of [*PSI*$^+$],[1,2] and transient overexpression of this region alone is sufficient to induce new [*PSI*$^+$] elements in all [*psi*$^-$] strains expressing full-length Sup35p.[8,20,21] N has a high propensity for self-association: when expressed as an isolated domain in yeast, it is always aggregated.[5] In addition, N has been shown to be highly amyloidogenic *in vitro*. It is insoluble in physiological buffers and forms amyloid even in the presence of denaturant.[22,23] The glutamine-rich repeats present in N play a central role in the self-assembly of N both *in vivo* and *in vitro*.[22,24,25]

The M region of Sup35p (aa 124–253) is highly charged. Notably, the charged residues are strongly biased to two amino acids: glutamic acid (18%) and lysine (19%), with no arginines present and aspartic acid comprising only a minor fraction (5%). Although not essential for the induction of [*PSI*$^+$], M appears to enhance the solubility of the prion-determining N region, profoundly altering its behavior both *in vivo* and *in vitro*

[19] V. V. Kushnirov, M. D. Ter-Avanesyan, M. V. Telckov, A. P. Surguchov, V. N. Smirnov, and S. G. Inge-Vechtomov, *Gene* **66,** 45 (1988).

[20] Y. O. Chernoff, M. V. Ptyushkina, M. G. Samsonova, G. I. Sizonencko, Y. I. Pavlov, M. D. Ter-Avanesyan, and S. G. Inge-Vechtomov, *Biochimie* **74,** 455 (1992).

[21] Y. O. Chernoff, I. L. Derkach, and S. G. Inge-Vechtomov, *Curr. Genet.* **24,** 268 (1993).

[22] J. R. Glover, A. S. Kowal, E. C. Schirmer, M. M. Patino, J. J. Liu, and S. Lindquist, *Cell* **89,** 811 (1997).

[23] C. Y. King, P. Tittmann, H. Gross, R. Gebert, M. Aebi, and K. Wuthrich, *Proc. Natl. Acad. Sci. U.S.A.* **94,** 6618 (1997).

[24] A. H. DePace, A. Santoso, P. Hillner, and J. S. Weissman, *Cell* **93,** 1241 (1998).

[25] J. Liu and S. Lindquist, *Nature,* in press (1999).

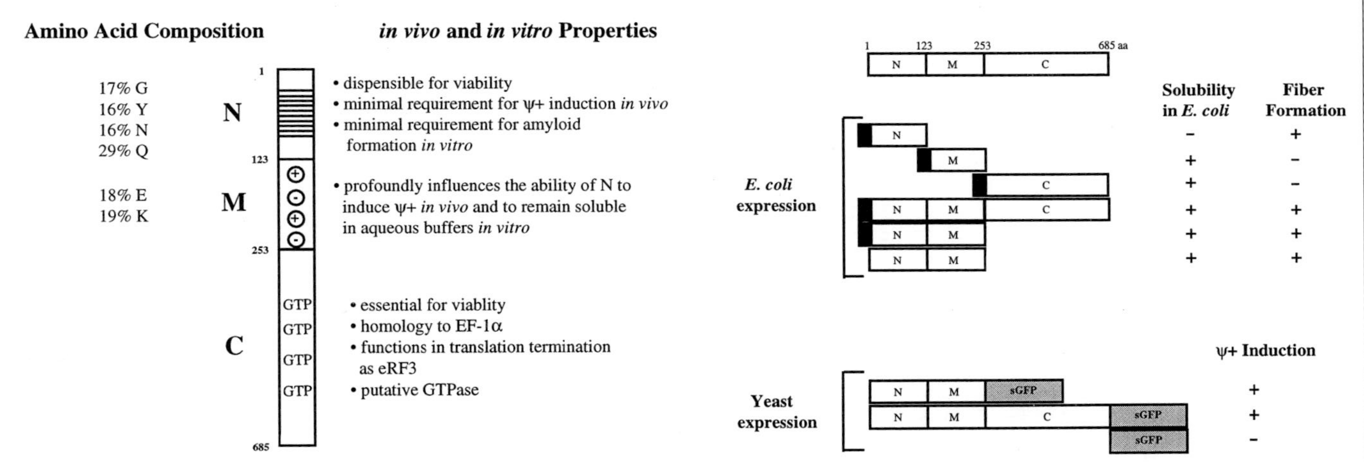

FIG. 2. (A) Biological and physical characteristics of Sup35p. The schematic diagram of Sup35p depicts amino-terminal (N), middle (M), and carboxy-terminal regions. (C). The amino acid boundaries of these regions, as well as their unusual amino acid compositions, are indicted to the left of the diagram (G, glycine; Y, tyrosine; N, asparagine; Q, glutamine; E, glutamic acid; and K, lysine). Horizontal lines indicate the nonapeptide repeats of N, the plus and minus symbols reflect the charged character of M, and GTP indicates consensus GTP-binding sites in C. A brief summary of the roles of these regions in [*PSI*$^+$] metabolism and translational termination is indicated to the right of the diagram. (B) Constructs for the expression of Sup35p and subfragments in *E. coli* and yeast. Schematic diagrams of expression constructs for N, M, C, or NM in *E. coli* are shown in the top half along with their solubilities in *E. coli* lysates and their ability to form amyloid fibers *in vitro*. Black rectangles indicate the position of a 10-residue His tag. Schematic diagrams of constructs used for the expression of NMsGFP, NMCsGFP, or sGFP from an extrachromosomal plasmid in yeast are indicated in the bottom half. The abilities of these constructs to induce Ψ^+ in the presence of full-length, wild-type Sup35p expressed from the genome are indicated to the right. A schematic of full-length Sup35p, including amino acid residue numbers, is shown at the top for reference.

(Fig. 2B).[22,26] In contrast to N, NM can exist in multiple states *in vivo,* modeling the differences in Sup35p solubility that are characteristic of [PSI^+] and [psi^-] strains.[5,6,27] In addition, purified NM forms amyloid slowly in physiologic buffers.[22] NM assembly is accelerated by the addition of preformed NM amyloid or lysates from [PSI^+] but not [psi^-] cells,[22] linking the properties of amyloid formation *in vitro* to the propagation of the [PSI^+] state *in vivo.* This ability of NM in one conformation to influence the physical state of the same protein in another conformation is the basis of protein-conformation self-perpetuation and the protein-only mode of inheritance for [PSI^+].

The carboxy-terminal region of Sup35p, C (aa 254–686), has sequence homology to the yeast translation elongation factor, EF-1α.[19] This region complexes with Sup45p,[17,18] contains several putative GTP-binding sites,[19] and functions in translational termination.[17,18,28] Unlike N and M, this region is essential for viability.[29] The epigenetic modulation of Sup35p carboxy terminus activity in translation termination is the [PSI^+] phenotype. Consequently, the carboxy terminus of Sup35p, linked to N and M, must be expressed in all [PSI^+] strains.

Analysis of [PSI^+] *in Vivo*

A guide to general yeast genetic techniques may be found elsewhere in this series.[30] This section discusses variations on those techniques that are particular to the study of [PSI^+] *in vivo.*

Reversibly curable nonsense suppression, exhibiting non-Mendelian, cytoplasmic inheritance, is the most commonly used test for [PSI^+]. Suppression of nonsense mutations in auxotrophic markers, such as *ade2-1* or *adel-14,* resulting in growth on defined medium lacking adenine (SD-Ade), has been the most convenient and well-accepted assay for [PSI^+]. Although the suppression of nonsense mutations in any metabolic enzyme and selection in the same manner are equally useful, analysis of strains harboring nonsense mutations in the adenine biosynthetic pathway provides an additional color selection assay. If grown on complete medium (YPD), [psi^-] strains carrying a nonsense mutation in the adenine pathway form red

[26] I. L. Derkatch, M. E. Bradley, P. Zhou, Y. O. Chernoff, and S. W. Liebman, *Genetics* **147,** 507 (1997).

[27] T. R. Serio and S. L. Lindquist, unpublished observation.

[28] L. Frolova, X. Le Goff, H. H. Rasmussen, S. Cheperegin, G. Drugeon, M. Kress, I. Arman, A. L. Haenni, J. E. Celis, M. Philippe *et al., Nature* **372,** 701 (1994).

[29] M. D. Ter-Avanesyan, V. V. Kushnirov, A. R. Dagkesamanskaya, S. A. Didichenko, Y. O. Chernoff, S. G. Inge-Vechtomov, and V. N. Smirnov, *Mol. Microbiol.* **7,** 683 (1993).

[30] C. Guthrie and G. Fink, eds., *Methods Enzymol.* **194** (1991).

colonies, whereas isogenic [PSI^+] strains form white or pink colonies (see later). This color readout of [PSI^+] is diminished if the complete medium is supplemented with extra adenine or the antibiotic tetracycline[27]; therefore, growth on minimal media (SD-Ade) should always be assessed in parallel.

The growth assay, as well as the color assay, for [PSI^+] described earlier is influenced by the "strength" of suppression, the efficiency with which nonsense codons are read through, and the mitotic stability of the phenotype. In most cases, growth of a [PSI^+] strain on SD-Ade, for example, will not be observed for 7–10 days at 25°, whereas a wild-type yeast strain, prototrophic for adenine, will grow within 2 days. For some markers, the efficiency of read through may not produce enough product to support growth in the absence of supplements. The strength of suppression can be increased, in some cases, by growth at lower temperatures (25° versus 30°) or on alternate carbon sources (i.e., ethanol). These characteristics are yeast strain specific but should be considered when monitoring [PSI^+].

In addition, others have described a variation in [PSI^+] suppression within a single genetic background.[26] For example, when a single [PSI^+] yeast strain containing a [PSI^+] -suppressible nonsense mutation in the *ADE1* gene is plated on YPD, most colonies are white; however, a few colonies that are different shades of pink are also observed. These isolates are characterized by different strengths of suppression and different mitotic stabilities, with white being the strongest in both cases. If one colony of any color is picked and replated, most colonies will maintain that same color, but colonies of different colors are also observed as before. The continued ability of different isolates to produce colonies with a spectrum of suppression strengths suggests that the variation between isolates is non-Mendelian in nature. For this reason, [PSI^+] variants isolated from a single genetic background are referred to as "strains" of [PSI^+].[26] [PSI^+] strains, which are isogenic, should not be confused with yeast strains that are genetically distinct.

Monitoring read through of nonsense mutations may also be complicated by differences in yeast strain backgrounds. For example, not all nonsense codons will be suppressed in all [PSI^+] yeast strains. Although the context of the nonsense mutation certainly plays a role in the efficiency of suppression,[31,32] additional *trans*-acting factors, known as allosuppressors, have also been implicated.[33–35] For example, [PSI^+] was originally described

[31] S. Mottagui-Tabar, M. F. Tuite, and L. A. Isaksson, *Eur. J. Biochem.* **257,** 249 (1998).

[32] B. Bonetti, L. Fu, J. Moon, and D. M. Bedwell, *J. Mol. Biol.* **251,** 334 (1995).

[33] S. W. Liebman, J. W. Stewart, and F. Sherman, *J. Mol. Biol.* **94,** 595 (1975).

[34] B. I. Ono, J. W. Stewart, and F. Sherman, *J. Mol. Biol.* **132,** 507 (1979).

[35] B. I. Ono, J. W. Stewart, and F. Sherman, *J. Mol. Biol.* **128,** 81 (1979).

as a factor capable of suppressing the *ade2-1* (UAA) allele in the presence of the allosuppressor *SUQ5,* a serine-inserting, UAA-specific tRNA.[33] Allosuppressors have been isolated in some but not all [PSI^+] strains, but interactions with *trans* regulators should be keep in mind when initially characterizing new strains for [PSI^+] status. For this reason, transmission to a [PSI^+]-susceptible strain by cytoduction is perhaps the most reliable test.

Although commonly used, the nonsense suppression assay for [PSI^+] is complicated by the need to support growth. To circumvent this problem, a quantitative nonsense suppression assay has also been developed that can be employed in any strain regardless of the auxotrophic markers available.[36] This assay is dependent on expression of a translational fusion between phosphoglycerate kinase and β-galactosidase. The two open reading frames are expressed in a single transcriptional unit but are separated by one of the three nonsense codons. Suppression of the nonsense mutation leads to the production of β-galactosidase, and the level of suppression may be quantitated by activity of this enzyme.

When working with [PSI^+], it is of the utmost importance to continually reconfirm the [PSI^+] status of yeast strains by a combination of experimental tests in addition to nonsense suppression. Suppression should be curable by treatment with guanidine hydrochloride or by the deletion and overexpression of Hsp104.[5,14,15] In addition, suppression should be dominant in diploids[10] and transmissible by cytoduction to susceptible yeast strains.[11] These additional analyses will avoid isolation of revertants of the nonsense mutation being monitored in a growth or color assay as well as mutations in other factors important for translational fidelity.

Characterization of mutations within the Sup35 coding sequence has been used to increase our understanding of the *cis*-acting requirements for [PSI^+] induction and propagation in the presence of a wild-type copy of Sup35.[1,24,25,37] Although additional work of this type will continue to increase our knowledge, several potential pitfalls are important to avoid. For example, work has indicated that while some mutations in Sup35p are capable of forming aggregates *in vivo* and/or amyloid *in vitro,* they do not induce new [PSI^+] elements or support [PSI^+] propagation in the presence of wild-type Sup35p *in vivo.*[24,25,38] These experiments indicate that compatibility between endogenous and exogenous Sup35p, as well as the efficiency of protein incoporation into aggregates, is critical for suppression and heritability. Consequently, both of these factors have the potential to complicate the analysis of Sup35p mutants. Replacement of endogenous Sup35p with

[36] M. Firoozan, C. M. Grant, J. A. Duarte, and M. F. Tuite, *Yeast* **7,** 173 (1991).
[37] S. J. McCready, B. S. Cox, and C. S. McLaughlin, *Mol. Gen. Genet.* **150,** 265 (1977).
[38] N. V. Kochneva-Pervakhova, *EMBO J.* **17,** 5805 (1998).

generated mutants may aid in their analysis. Because Sup35 is an essential gene,[29] replacement of mutant alleles with wild-type sequences provides additional insight into the functional state of the molecule.

Finally, analysis of [*PSI*$^+$] induction is complicated by a non-Mendelian factor, [*PIN*$^+$].[26] [*PIN*$^+$] stands for [*PSI*$^+$] inducible. [*PSI*$^+$] can be induced in [*PIN*$^+$] [*psi*$^-$] yeast strains by the overexpression of any fragment of Sup35p containing N, whereas yeast strains that are [*pin*$^-$][*psi*$^-$] can become [*PSI*$^+$] only by overexpression of the N region of Sup35p alone. Curing of [*PSI*$^+$] by growth on guanidine hydrochloride produces both [*pin*$^-$] and [*PIN*$^+$] isolates, whereas curing by the overexpression of Hsp104 seems to only produce [*PIN*$^+$] strains.[26] In addition, the [*PIN*$^+$] status may spontaneously change to [*pin*$^-$], especially if strains are kept at 4° for extended periods of time.[39] Care should therefore be taken to ensure that strains are susceptible to [*PSI*$^+$] induction with wild-type copies of Sup35p in parallel with any unknowns.

Analysis of Sup35p in Yeast Cells

The first biochemical evidence that the [*PSI*$^+$] phenotype was propagated by a protein-only mode of inheritance was provided by analysis of the physical state of Sup35p in [*psi*$^-$] and [*PSI*$^+$] cells.[5,6] Sup35p is largely soluble in [*psi*$^-$] cells but is mostly insoluble in [*PSI*$^+$] cells. Conversion between these physical states, invoked by transient changes in either the concentration of Sup35p or the molecular chaperone Hsp104,[5,15] results in a heritable change in phenotype between [*PSI*$^+$] and [*psi*$^-$] or vice versa. Two types of analyses, one in living yeast cells and one in yeast extracts, have been seminal in providing support for [*PSI*$^+$] as a yeast prion.[5]

Analysis of Sup35p Tagged with Green Fluorescent Protein in Living Yeast Cells

Translational fusions of full-length Sup35p or NM to the green fluorescent protein (GFP)[40,41] or GFP containing amino acid substitutions (S65T, V164A) that increase fluorescence and decrease self-association (superglow

[39] I. Derkatch and S. Liebman, unpublished observation.

[40] S. R. Kain, M. Adams, A. Kondepudi, T. T. Yang, W. W. Ward, and P. Kitts, *Biotechniques* **19,** 650 (1995).

[41] A. B. Cubitt, R. Heim, S. R. Adams, A. E. Boyd, L. A. Gross, and R. Y. Tsien, *Trends Biochem. Sci.* **20,** 448 (1995).

GFP, sGFP)[42,43] have provided an accurate model system for monitoring the aggregation state of full-length Sup35p in living cells.[5] Following short induction times for expression of these fusion proteins, fluorescence is diffuse in [*psi*$^-$] strains and coalesces rapidly in [*PSI*$^+$] strains, whereas fluorescence from GFP or sGFP alone remains diffuse in both [*psi*$^-$] and [*PSI*$^+$] cells (Fig. 3A). We typically observed a single focus with GFP fusions, but multiple foci are visible with sGFP, most likely due to the enhanced fluorescence and decreased propensity for self-aggregation of sGFP. This system accurately mimics all aspects of [*PSI*$^+$] metabolism.[5] Prolonged overexpression of NMGFP (or NMsGFP) or Sup35GFP (or Sup35sGFP) in [*psi*$^-$] strains will ultimately lead to the coalescence of Sup35p and the induction of new [*PSI*$^+$] elements,[5] but in the time courses described next, an unambiguous difference in fluorescence between [*psi*$^-$] and [*PSI*$^+$] strains is observed.

Construction of Expression Plasmids. We have constructed low- and high-copy number plasmids for either the constitutive or the inducible expression of Sup35p and NM fused to GFP or sGFP,[5,27] but we will limit the discussion here to the copper-inducible low-copy number vectors expressing translational fusions to sGFP. The parent plasmids for these constructs are the ampicillin-resistant, URA3+, CEN plasmid: pRS316.[44] The copper-inducible CUP1 promoter[45] was cloned between the *Eco*RI and the *Bam*HI sites of this plasmid by polymerase chain reaction (PCR) to generate 316CUP1. A cassette for expression of sGFP from the CUP1 promoter was also generated by PCR and cloned between the *Sac*I and the *Sac*II sites of 316CUP1 to yield 316CG. Finally, full-length Sup35p, lacking its natural stop codon, or NM was amplified by PCR and cloned between the *Bam*HI and the *Sac*I sites of 316CG to yield the expression vectors Sup35GFP (Sup35sGFP) or CNMG (NMsGFP), respectively.

Induction and Analysis of Sup35p Fusions to sGFP. Plasmids are transformed into isogenic [*PSI*$^+$] and [*psi*$^-$] yeast strains by the lithium acetate method, for example, and selected for on SD medium lacking uracil. Under certain circumstances, low-level expression from the CUP1 promoter in the absence of added copper has been observed. Others have described a modified synthetic medium containing a yeast nitrogen base lacking copper or the use of a specific copper chelator, bathocuproine disulfonate (BSC).[46] In our experience, however, uninduced expression levels in standard SD

[42] R. Heim, A. B. Cubitt, and R. Y. Tsien, *Nature* **373,** 663 (1995).

[43] J. A. Kahana and P. A. Silver, *in* "Current Protocols in Molecular Biology" (F. M. Ausubel *et al.,* eds.) p. 9.6.13. Wiley, New York, 1996.

[44] R. S. Sikorski and P. Hieter, *Genetics* **122,** 19 (1989).

[45] D. J. Thiele, *Mol. Cell. Biol.* **8,** 2745 (1988).

[46] S. Labbe and D. Thiele, *Methods Enzymol.* **306,** 145 (1999).

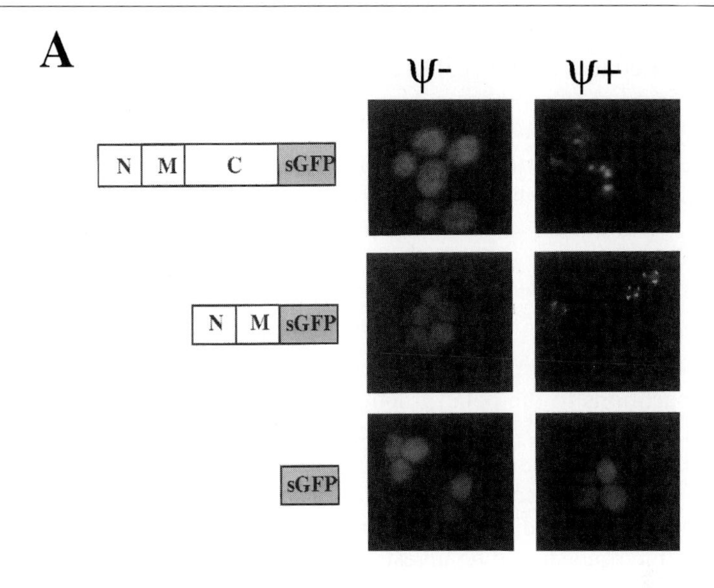

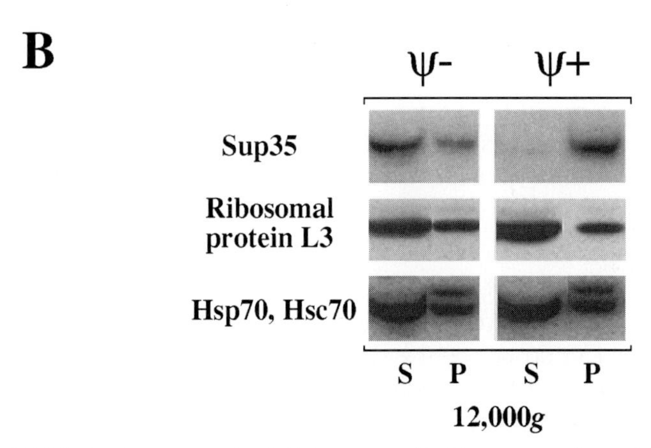

FIG. 3. (A) Monitoring aggregation in living yeast cells. Fluorescence from sGFP or NMC (NMCsGFP) and NM (NMsGFP) tagged with sGFP was monitored in 74-D694 Ψ^- and Ψ^+ strains 4 hr after induction with copper sulfate. At this point, coalesced fluorescence is observed only in the Ψ^+ strain from NMCsGFP or NMsGFP. (B) Solubility of Sup35p in yeast lysates. Immunoblots of yeast lysates from 74-D694 [*psi*$^-$] or [*PSI*$^+$] strains following differential centrifugation at 12,000*g* are shown, Polyclonal rabbit antisera to Sup35p or ribosomal protein L3 or a rat monoclonal antibody to Hsp70, Hsc70 (MAb 7.10) were used as probes. S, supernatant fraction; P, pellet.

are below the level of detection for fluorescence microscopy. Single colonies are inoculated into a liquid culture in SD–Ura at 30° with constant agitation on a roller drum (60 rpm) to a density of ~1 × 10^6 cells/ml. Copper sulfate ($CuSO_4$) is added to a final concentration of 50 μM to induce expression from the CUP1 promoter under the same growth conditions. Cells are viewed under blue light at 100× magnification 4 hr after induction.

Differential Centrifugation Analysis of Sup35p in Yeast Extracts

Another method to analyze the physical state of Sup35p in yeast is differential centrifugation. In the most simple case, Sup35p from [PSI^+] lysates partitions to the pellet fraction whereas Sup35p from [psi^-] lysates is found predominantly in the soluble fraction (Fig. 3B).[5] Unlike the behavior of Sup35p, the location of other proteins, such as the ribosomal protein L3 or Hsp70 and Hsc70, is not altered in a [PSI^+]-dependent manner. Because of the great difficulty in working with aggregation-prone proteins, considerable time should be invested to ensure reproducibility with known samples before analyzing the behavior of unknowns. The method described here involves centrifugation of the lysate alone,[5] whereas methods described by others have utilized sucrose cushions and gradients with similar results.[6] As with all fractionations, introduction of either cushions or gradients diminishes cross-contamination between fractions. However, the high-speed centrifugation step included in the sucrose cushion method to remove unbroken cells has the disadvantage of removing some fraction of Sup35p aggregates from [PSI^+] lysates. The degree of purity and yield required for subsequent analysis should, therefore, dictate the method employed. All of these methods are also suitable for the analysis of Sup35p fragments expressed in yeast.[5,6,27]

Culture Growth and Extract Preparation

1. Grow yeast cultures to midlog phase [~1–5 × 10^7 cells/ml in complete (YPD) or ~2–4 × 10^6 cells/ml in synthetic (SD) medium] at 30° with constant shaking (250 rpm). This procedure, however, has yielded the same results with stationary-phase cultures. Fifteen minutes prior to collection, add cyclohexamide to 200 μg/ml to stabilize polysomes and allow newly synthesized proteins to achieve their characteristic conformations.
2. Cool the cells on ice for 15 min and then harvest by low-speed centrifugation (2000*g*, 5 min, 4°). Discard the supernatant and wash the cell pellet once with an equal volume of cold water containing 200 μg/ml cyclohexamide.

3. Wash the pellet once with an equal volume of lysis buffer [50 m*M* Tris–HCl (pH 7.5), 5 m*M* $MgCl_2$, 10 m*M* KC1, 0.1 m*M* EDTA (pH 8.0), 1 m*M* dithiothreitol (DTT), 100 μg/ml cyclohexamide, 1 m*M* benzamidine, 2 m*M* phenylmethylsulfonyl fluoride (PMSF), 10 μg/ml leupeptin, 2 μg/ml pepstatin A, 100 μg/ml ribonuclease A].
4. Transfer cells to a 1.5-ml microcentrifuge tube and pellet at (2000*g*, 5 min, 4°). If desired, the cells may be flash frozen and stored at −80° at this point.
5. Resuspend pellet in lysis buffer at a concentration of ~3×10^6 cells/μl. Add an equal volume of 425- to 600-μm acid-washed glass beads (Sigma, St. Louis, MO).
6. Homogenize cells in a Mini Bead Beater 8 (Biospect Products) for approximately 4 min at 4°. Monitor cell breakage by light microscopy.
7. Puncture the bottom of the tube with a 18-gauge needle and place a smaller tube into a 12 × 75-mm round-bottom polypropylene tube (Falcon; Becton Dickinson, Franklin Lakes, NJ). Centrifuge at 100*g* for 1 min at 4° to separate lysate from glass beads.
8. Wash glass beads twice with 1/2 volume of lysis buffer originally used. Combine washes with lysate from step 7.
9. Preclear the lysate at 2,500*g* for 10 min at 4° to pellet unbroken cells. Remove the supernatant to a new tube without disturbing the pellet.
10. Determine the protein concentration of the lysate. Typical yields are 5–15 mg/ml using Bio-Rad (Richmond, CA) protein assay reagent with bovine serum albumin (BSA) as a standard. Lysates may be flash frozen and stored at −80° at this point.

Differential Centrifugation Analysis

1. Separate aggregates from soluble protein in a fraction of the lysate by centrifugation at 6000–12,000*g* for 10 min at 4°.
2. Remove supernatant to a tube containing an appropriate volume of 6× sample buffer [350 m*M* Tris–HCl (pH 6.8), 30% (v/v) glycerol, 10% (w/v) SDS, 600 m*M* DTT, 0.12% (w/v) bromphenol blue] to give a 1× concentration. This is the supernatant fraction.
3. Resuspend the pellet in the same volume of lysis buffer used in step 1 of this section. Transfer resuspended pellet to a new tube containing an appropriate volume of 6× sample buffer to give a 1× concentration. This is the pellet fraction.
4. Incubate supernatant and pellet samples as well as a total lysate sample at 100° in a water bath for 10 min.

5. Separate proteins on a 10% SDS–PAGE (25 mA/gel), electrotransfer to Immobilon-P (Millipore, Bedford, MA), and analyze by immunoblotting. A total of 36 μg of protein/lane yields a sufficient Sup35p signal for detection with our antiserum.[5] In addition, we typically analyze fractionation of ribosomes by immunoblotting with an antibody to the ribosomal protein L3.[5] Detection with either ^{125}I-labeled protein A (Amersham, Arlington Heights, IL) followed by autoradiography or protein A–peroxidase (Boehringer-Mannheim, Indianapolis, IN) followed by ECL (enhanced chemiluminescance, Amersham) yield, similar results.

Analysis of Sup35p Purified from *Escherichia coli*

Recent work has linked the process of amyloid formation *in vitro* to the propagation of [PSI^+] *in vivo*. Fragments of Sup35p capable of inducing [PSI^+] *in vivo* form amyloid *in vitro*.[22,23] Lysates from [PSI^+] but not [psi^-] strains accelerate the formation of amyloid *in vitro*, as do preformed fibers.[22,23] Deletions within the prion-determining N region,[22,24] as well as specific point mutations,[24] slow the process of assembly into amyloid[22,24] as well as block the induction of new [PSI^+] elements.[24] Similarly, the expansion of repeated sequences in the N region accelerates amyloid formation *in vitro* and increases the efficiency of [PSI^+] induction *in vivo*.[25]

We have assessed the abilities of N, M, NM and NMC, expressed and purified from *E. coli*, to form amyloid *in vitro* (Fig. 1B). Although fragments containing the N region are capable of this ordered assembly, we will only discuss the purification and analysis of the assembly process here for NM and NMC, as these fragments most accurately reflect [PSI^+] metabolism *in vivo*.[5,26] Because NM and NMC form amyloid under native conditions,[22] we routinely purify these fragments under denaturing conditions (8 *M* urea) to maintain the protein in a uniform state that is more amenable to studying the assembly process. The importance of obtaining a denatured, uniform solution of protein prior to initiating a kinetic analysis of amyloid formation cannot be stressed enough. Others have reported that protein purified in 6 *M* urea must be cleared of amyloid by filtration to obtain reproducible results.[24] Procedures for purification under denaturing conditions in 8 *M* urea will be discussed here.

Expression Constructs

The expression of all cloned fragments of Sup35p is driven by T7 polymerase. Fragments were cloned into either pJC45 encoding an amino-terminal 10 residue histidine tag (His_{10}) or pJC25, lacking a tag.[47] These

[47] J. Clos and S. Brandau, *Prot. Express. Purif.* **5,** 133 (1994).

plasmids are high copy number, containing the pUC origin of replication, and have a consensus T7 promoter and the *lacI* operator at the 5′ end of the multiple cloning site. Sup35p fragments were inserted between the *Nde*I and the *Bam*HI sites, allowing in-frame fusion to His_{10} in the case of pJC45. The plasmids impart ampicillin resistance.

Bacterial Growth and Induction

Each construct is expressed in BL21 [DE3] pAP *lacI*^{q}. This strain contains a [DE3] lysogen for the high-level expression of T7 polymerase following induction with isopropyl-β-D-thiogalactopyranoside (IPTG; Sigma). The strain also expresses a low level of the *lacI* product to repress leaky expression of the polymerase and, therefore, the target protein. The strain is kanamycin resistant.

Competent *E. coli* (BL21 [DE3] pAP *lacI*^{q}) are transformed with expression plasmids by electroporation and selected on LB plates containing 50 μg/ml kanamycin and 200 μg/ml ampicillin. Fresh transformants should always be used for the expression of full-length Sup35p, as prolonged passage of the expression plasmid in this strain leads to a high degree of proteolysis.[27] Proteolysis is not observed with NM, however, and glycerol stocks of the expression strain may be stored for months at −80°.[27]

A single colony of BL21 [DE3] pAP *lacI*^{q} containing the expression construct is inoculated into 1 liter of Circle Grow medium (Bio 101, Vista, CA) and incubated at 37° with constant shaking (300 rpm) until an OD_{600} of 0.8 is reached (approximately 7 hr). IPTG is added to 1 m*M*, and the culture is incubated further under the same conditions for 2 hr. Bacteria are then collected by centrifugation (3000*g*, 10 min, 4°). The pellet may be stored at −80° or processed immediately. Again, storage of the pellet should be minimal for the purification of full-length Sup35p.

Bacterial Lysis

Cell pellets are lysed in 50 ml lysis buffer H [20 m*M* Tris–HCl (pH 8.0), 8 *M* urea] for each liter of culture for all His_{10} proteins or in lysis buffer N [10 m*M* Tris–HCl (pH 7.2), 8 *M* urea] for nontagged NM. High-grade urea (Boehringer Mannheim) is prepared freshly for each purification to minimize covalent modifications to the protein due to the production of cyanate ions.[48] The resuspended pellet is incubated for 30 min at 25° with occasional agitation. The lysate is then precleared by centrifugation at 30,000*g* for 20 min at 10°.

[48] G. M. Means and R. E. Feeney, "Chemical Modifications of Proteins." Holden-Day, San Francisco, 1971.

Purification of His-Tagged NMC or NM

Buffers

Lysis buffer H: 20 m*M* Tris–HCl (pH 8.0), 8 *M* urea
Ni wash buffer: 20 m*M* Tris–HCl (pH 8.0), 8 *M* urea, 40 m*M* imidazole
Ni elution buffer: 20 m*M* Tris–HCl (pH 8.0), 8 *M* urea, 400 m*M* imidazole
Q wash buffer H: 20 m*M* Tris–HCl (pH 8.0), 8 *M* urea, 100 m*M* NaCl
Q elution buffer H: 20 m*M* Tris–HCl (pH 8.0), 8 *M* urea, 300 m*M* NaCl

Procedure. All steps are carried out at 25°.

Precleared supernatants from cell lysates are applied to a 50-ml Ni^{2+}-nitrilotriacetic acid agarose column (Ni^{2+}-Nta; Qiagen, Valencia, CA) preequilibrated with lysis buffer H at a flow rate of approximately 3 ml/min. The column is washed with 5 bed volumes of Ni wash buffer, and the protein is eluted in a single step with 100 ml of Ni elution buffer. The eluate is applied directly onto a 20-ml Q Sepharose Fast Flow column (Pharmacia, Piscataway, NJ) preequilibrated with Ni elution buffer. The column is washed with 5 bed volumes of Q wash buffer H, and the protein is eluted in a single step with 45 ml of Q elution buffer H. The purification and purity of the final product are analyzed by 10% SDS–PAGE followed by staining with Coomassie Brilliant Blue R-250. The predicted molecular weight of NM is 28,500; however, due to the presence of the highly charged M region, NM migrates aberrantly by SDS–PAGE at ~45,000.

This procedure is equally effective in purifying His_{10}NM if the columns are reversed.[27] In this case, the Ni^{2+}-Nta agarose column must be preequilibrated with Q elution buffer H. We prefer to use the protocol described here, however, because it is effective in removing trace metals leached from the Ni^{2+}-Nta agarose resin,[49] imidazole, and carboxy-terminal truncations of the expressed protein, which affect the kinetics of fiber assembly profoundly.

Purification of Nontagged NM

Buffers

Lysis buffer N: 10 m*M* Tris–HCl (pH 7.2), 8 *M* urea
Q wash buffer N: 10 m*M* Tris–HCl (pH 7.2), 8 *M* urea, 85 m*M* NaCl
Q elution buffer N: 10 m*M* Tris–HCl (pH 7.2), 8 *M* urea, 150 m*M* NaCl
HA preequilibration buffer: 10 m*M* Tris–HCl (pH 7.2), 8 *M* urea, 150 m*M* NaCl

[49] L. G. Hom and L. E. Vokman, *BioTechniques* **25,** 20 (1998).

HA wash buffer I: 1 m*M* potassium phosphate (pH 6.8), 8 *M* urea, 1 *M* NaCl

HA wash buffer II: 25 m*M* potassium phosphate (pH 6.8), 8 *M* urea

HA elution buffer: 8 *M* urea, 125 m*M* potassium phosphate

Procedure. All steps are carried out at 25°.

The precleared supernatant from cell lysis is applied to a 20-ml Q Sepharose Fast Flow column (Pharmacia) preequilibrated with lysis buffer N at a flow rate of 3 ml/min. The column is washed with 5 bed volumes of Q wash buffer N, and the protein is eluted in 3 volumes of Q elution buffer N.

The eluate from the Q Sepharose is loaded directly onto a 25-ml Macro Prep Ceramic Hydroxyapatite Type I 40-μm colum (Bio-Rad) preequilibrated with HA preequilibration buffer. The column is washed with 2 bed volumes of HA wash buffer I and then with two bed volumes of HA wash buffer II. The protein is eluted using a linear gradient of potassium phosphate (pH 6.8) from 25 to 125 m*M* (equal volumes of HA wash buffer II and HA elution buffer). Fractions (5 ml) are analyzed by 12.5% SDS–PAGE (loading 10 μl per lane) followed by staining with Coomassie Brilliant Blue R-250. Fractions containing purified NM are pooled and concentrated using one of the following methods.

Quantitation and Yields

Sup35p is stained poorly by Coomassie Brilliant Blue G-250, which binds primarily to arginine residues. Protein determination methods based on binding to this dye, such as Bradford, are, therefore, unreliable for the quantitation of protein yields. Sup35p staining by Coomassie Brilliant Blue R-250, however, is a reliable method for detecting the protein following gel electrophoresis. We routinely determine the concentration of His_{10}NM by the microbicinchoninic acid method (Micro-BCA; Pierce, Rockford, IL), using BSA as a standard. Alternately, we quantitate the protein concentration directly from the absorbance at 276 nm in 8 *M* urea using an extinction coefficient (ε) of 29,000 for NM. Typically, 50 mg of NM is obtained from a 1-liter culture.

Concentration and Storage of Purified Protein

Analysis of the amyloid assembly process by NM, discussed later, requires concentration of the protein to at least 30 mg/ml to allow ample dilution of denaturant while maintaining a sufficient protein concentration for fiber formation. This may be accomplished in multiple ways, and the method should be chosen based on the length of storage time required.

For short-term storage of NM, we routinely filter concentrate NM using Biomax Ultrafree-15 concentrators with a 10,000 molecular weight cutoff (Millipore). Column fractions containing NM are pooled and concentrated at 1500*g* for approximately 2.5 hr at 6°. The protein may be stored in this state for approximately 1 week at 4°.

For long-term storage, we methanol precipitate NM to remove urea and store the precipitate at −80°. Anhydrous methanol (100%) is added to elulates containing NM on ice at a ratio of 5:1. The mixture is incubated on ice for 30 min, and the precipitate is collected by centrifugation at 14,000*g* for 30 min at 4°. The pellet is then washed with 100% methanol (1/2 volume of supernatant) and collected by centrifugation again. The supernatant is removed, and the pellet is stored under 70% (v/v) methanol (1/2 volume of supernatant) at −80°. It is convenient to perform the precipitation in microcentrifuge tubes, as fractionating the protein after precipitation is less accurate.

Prior to use, the precipitated protein is collected by centrifugation at 14,000*g* for 30 min at 4°. The methanol is removed carefully, and the pellet is damp-dried under vacuum without heat for 5 min. The precipitated protein is resuspended in freshly made lysis buffer H to yield approximately a 30-mg/ml solution. The protein concentration should always be confirmed by one of the methods described previously.

Polymerization Reactions

The most detailed information regarding the assembly of Sup35p into amyloid has been gleaned from a study of the NM fragment.[22] Full-length Sup35p will form amyloid *in vitro,* but the process is more cumbersome, as quantitative recovery of the protein in the amyloid form requires slow dialysis from denaturant (2 *M* stepwise decreases in urea until no denaturant remains). In addition, full-length Sup35p in an unpolymerized form binds to the diagnostic amyloid dye, Congo red, eliminating this assay from the repertoire available for monitoring amyloid formation. Therefore, our discussion of amyloid assembly here will be restricted to a characterization of His_{10}NM.

General Considerations and Reaction Conditions. Multiple factors influence the efficiency with which His_{10}NM will form amyloid *in vitro.* Among these, protein concentration and sufficient dilution from denaturant are the most crucial. We have found that polymerization reactions in the micromolar range for His_{10}NM form amyloid within a reasonable time frame (30–90 hr).[22] In addition, we suggest at least a 100-fold dilution from denaturant into aqueous buffer, as excess denaturant slows or inhibits polymerization.[27]

We have observed polymerization of $His_{10}NM$ over a wide range of buffer, salt, temperature, and detergent conditions.[27] The effects of these conditions on $His_{10}NM$ assembly are minor, twofold at most.[27] In general, however, reactions proceed most efficiently within the pH range of 6.5–7.0 at 18°. Molar concentrations of monovalent salt and Triton-X 100 up to 10% (v/v) do not alter the process at all, but even 0.05% (w/v) SDS is sufficient to inhibit polymerization.[27] All of the following assays were conducted in Congo red binding buffer [CRBB: 5 m*M* potassium phosphate (pH 7.4), 150 m*M* NaCl]. For these analyses, $His_{10}NM$ is diluted directly into CRBB with gentle vortexing to a concentration of 5 μM and is then incubated at 25° without agitation.

Assembly of $His_{10}NM$ into amyloid may be accelerated by several conditions. For example, the addition of 1/50 volume of $His_{10}NM$ fibers preformed from a 5-μM solution of protein or yeast lysates from [*PSI*$^+$] strains will decrease the time of fiber formation to 10–12 hr.[22] Sonication of preformed fibers greatly increases their capacity to seed the assembly of freshly diluted $His_{10}NM$, further decreasing the polymerization to 2 hr.[27] Alternately, constant gentle agitation on a roller drum (60 rpm) accelerates the assembly of $His_{10}NM$ into amyloid to roughly 2 hr at micromolar concentrations.[27]

Analysis of NM Amyloid Assembly

We, and others,[22,23] have developed a number of tools to study the assembly of $His_{10}NM$ into amyloid *in vitro*. We have monitored the assembly of this protein into an ordered amyloid by spectroscopy, dye binding, sedimentation, and microscopy. These techniques are described next.

Binding to 8-Anilino-1-naphthalenesulfonic Acid. 8-Anilino-1-naphthalenesulfonic acid (ANS; Aldrich, Milwaukee, WI) is a spectroscopic probe that exhibits low fluorescence in aqueous solutions and high fluorescence in hydrophobic environments, with a concomitant blue shift in the wavelength of maximum emission (λ_{max}).[50] Folding intermediates, such as molten globules, exhibit increased ANS fluorescence relative to either denatured or fully folded proteins.[50] ANS binding to $His_{10}NM$ over a time course serves as a monitor of fiber assembly. Solutions of 5 μM $His_{10}NM$ and 10 μM ANS are excited at 370 nm, and fluorescence emission is monitored between 420 and 570 nm at a 5-nm bandwidth. Surprisingly, structured $His_{10}NM$ fibers exhibit a 10-fold increase in ANS fluorescence accompanied by a ~40-nm blue shift in the λ_{max} of emission to 484 nm compared to unpolymerized $His_{10}NM$ (Fig. 4A). This increased fluorescence may indicate the presence of an exposed hydrophobic pocket(s) or groove(s) in mature $His_{10}NM$ fibers.

[50] L. S. Stryer, *J. Mol. Biol.* **13,** 482 (1965).

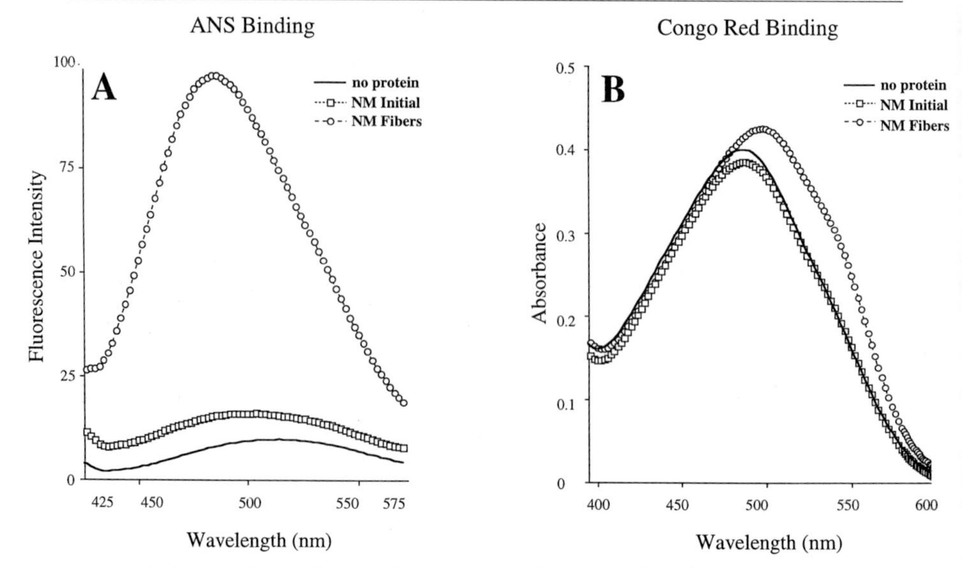

FIG. 4. Dye binding and spectral properties of His-tagged NM. (A) Binding of ANS to His-tagged NM. Fluorescence emission spectrum of ANS alone or bound to unpolymerized His-tagged NM or His-tagged NM assembled into amyloid fibers (B) Absorption spectrum of Congo red alone or bound to unpolymerized His-tagged NM or amyloid fibers of His-tagged NM.

Congo Red Binding. Similar to many other amyloidogenic proteins, His_{10}NM fibers bind to the diagnostic dye, Congo red (Sigma).[51] Monitoring Congo red binding over an assembly time course is a sensitive probe for fiber formation (Fig. 4C). The absorbance of a solution of 1 μM His_{10}NM and 10 μM Congo red in CRBB is monitored between 400 and 600 nm. His_{10}NM fibers exhibit a spectral shift in absorbance, with a new peak at 540 nm, in comparison with unpolymerized protein or Congo red alone. The amount of Congo red bound to His_{10}NM may be calculated using the following equation:

$$\text{mole Congo red bound/liter solution} = (A_{540}/25{,}295) - (A_{477}/46{,}306)$$

where A_{540} and A_{477} refer to the absorbance at 540 and 477 nm, respectively.[51] Under these conditions, His_{10}NM binds roughly 4.4 moles of Congo red per mole of protein with a K_d of 250 nM.[22] In addition, fibers of His_{10}NM stained with a solution of Congo red exhibit apple-green birefringence when viewed by polarized light.[23]

[51] W. E. Klunk, J. W. Pettegrew, and D. J. Abraham, *J. Histochem. Cytochem.* **37,** 1273 (1989).

Special consideration should be given to maintain identical buffer conditions when comparing different samples, as the quantity of Congo red binding to proteins is altered by pH, denaturant, and metals.[27]

SDS Solubility. Assembly of His_{10}NM into amyloid may also be monitored by the degree of solubilization in 2% (w/v) SDS. Unpolymerized protein remains soluble in 2% (w/v) SDS. In contrast, once amyloid has formed, these structures are largely insoluble in 2% SDS at room temperature. We have utilized this difference in SDS solubility, combined with SDS–PAGE, as an assay to monitor fiber formation. SDS sample buffer is added to a 1× concentration to two 20-μl aliquots of a 5 μM polymerization reaction. One sample is boiled in a water bath for 10 min, whereas the other sample is incubated at room temperature. The samples are then separated on a 10% SDS–PAGE gel that is subsequently stained with Coomassie Brilliant Blue R-250 (Fig. 5A). The same amount of unpolymerized His_{10}NM enters the gel whether or not the sample has been boiled. In contrast, His_{10}NM fibers only enter the gel in boiled samples.

Limited Proteolysis of NM. Limited proteolysis of His_{10}NM with chymotrypsin and V8 provides sensitive probes for domain-specific structural changes during amyloid assembly. The N region contains 20 tyrosine residues, which are high-affinity sites for cleavage with the protease chymotrypsin, whereas the M region contains none. Conversely, the M region contains 23 glutamic acids, which are high-affinity sites for cleavage with V8 protease, whereas the N region contains none. Alterations in the digestion pattern reflect either a change in conformation or accessibility for either the N (chymotrypsin) or the M (V8) region.

New batches of proteases should be titrated with known samples (both fibers and freshly diluted His_{10}NM). Samples (20 μl) of a 5 μM solution of NM in CRBB are incubated with either chymotrypsin (~1/250, w/w) or V8 (~1/25, w/w) at 37° for 15 min. Proteases are freshly resuspended at a concentration of 1 mg/ml in 1 mM HCl for chymotrypsin (Boehringer Mannheim) or 1 mg/ml in water for V8 (Endoproteinase Glu-C; Boehringer Mannheim). The reaction is terminated by adding SDS sample buffer to 1× and boiling for 10 min in a water bath to inactivate proteases. Digestion products are then separated on 10% SDS–PAGE gels that are subsequently stained with Coomassie Brilliant Blue R-250. V8 cleavage of His_{10}NM fibers produces a characteristic digestion pattern that is distinct from that of unpolymerized His_{10}NM (Fig. 5B). In contrast, His_{10}NM fibers exhibit resistance to chymotrypsin digestion (Fig. 5C), whereas unpolymerized His_{10}NM is cleaved rapidly.

Sedimentation Analysis. Another convenient assay for monitoring His_{10}NM assembly is differential sedimentation. In contrast to unpolymer-

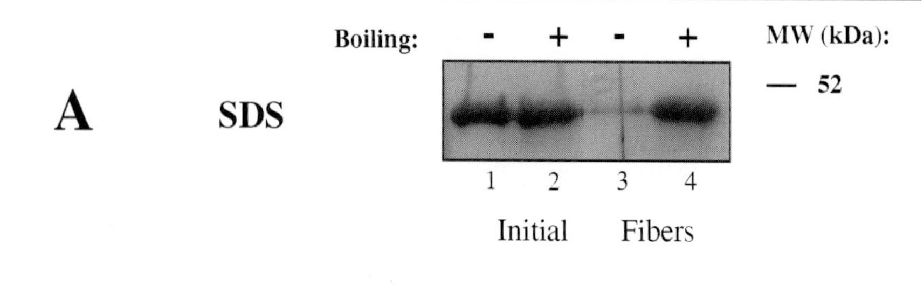

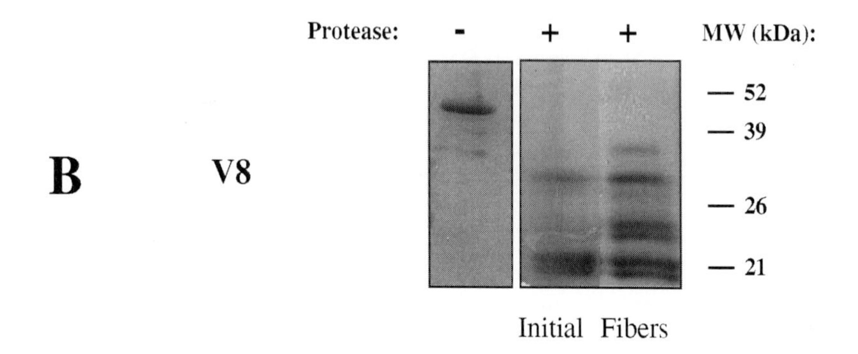

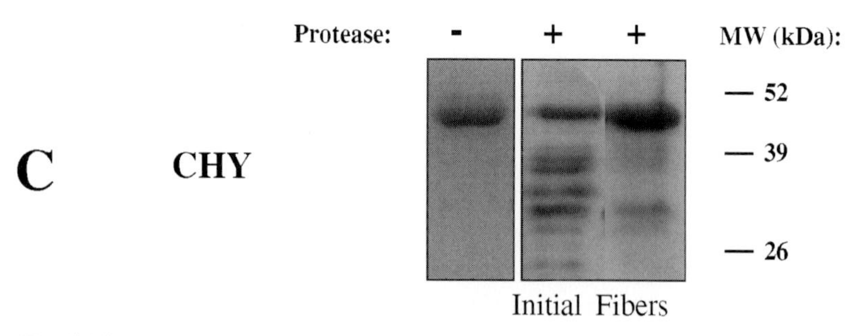

FIG. 5. Biochemical analysis of His-tagged NM amyloid fibers. (A) SDS solubility. Coomassie Brilliant Blue-stained 10% SDS–polyacrylamide gels of unpolymerized His-tagged NM (Initial; lanes 1 and 2) or amyloid (Fibers; lanes 3 and 4) incubated in 2% (w/v) SDS without (−) or with (+) boiling are shown. (B and C) Limited proteolysis with V8 and chymotrypsin, respectively. Unpolymerized His-tagged NM (Initial) or amyloid (Fibers) were digested with V8 protease (B) or chymotrypsin (C) and separated on a 10% SDS–polyacrylamide gel that was subsequently stained with Coomassie Brilliant Blue. His-tagged NM incubated in the absence of protease—is shown for comparison. The positions of molecular weight standards are indicated at the right.

ized $His_{10}NM$,[22] $His_{10}NM$ fibers will sediment at high speeds. Samples are centrifuged at 100,000*g* for 10 min at 4° and are then analyzed by 10% SDS–PAGE. The supernatant is removed following centrifugation, and SDS sample buffer is added to 1×. An equal volume of 1× SDS sample buffer is added to the pellet. Both samples are boiled for 10 min in a water bath and loaded onto a 10% SDS–PAGE, which is subsequently stained with Coomassie Brilliant Blue or transferred to Immobilon-P (Millipore) for quantitative Western blot analysis using ^{125}I-labeled protein A (Amersham). Partitioning between the supernatant and the pellet fractions is indicative of the assembly state, unpolymerized or polymerized, respectively.

Alternately, $His_{10}NM$ assembly may be monitored using radiolabeled protein. $His_{10}NM$ is radiolabeled with [^{35}S]methionine, purified, and added to a polymerization reaction (10,000 cpm/50 μl of reaction, supplemented with unlabeled NM to 5 μM). Samples (50 μl) are removed and separated into supernatant and pellet fractions as described earlier for unlabeled protein. Following centrifugation, the soluble counts remaining in the supernatant are measured in a scintillation counter as an indication of the extent of the reaction.

The labeling procedure follows, and the protein is purified by one of the methods described previously.

1. Grow a single colony of BL21 [DE3] pAP *lacI*^{q} harboring the expression plasmid to an OD_{600} of 0.2 at 37° at 300 rpm in 1 liter of Circle Grow medium (Bio 101) supplemented with 50 μg/ml kanamycin and 200 μg/ml ampicillin.
2. Collect the cells by centrifugation at 1500*g* for 10 min at 4°.
3. Resuspend the pellet in 1 liter of M9 medium (with $MgCl_2$ substituted for $MgSO_4$) supplemented with antibiotics as described earlier. Incubate at 37°, 300 rpm for 1 hr.
4. Collect the cells by centrifugation, as described previously.
5. Resuspend the pellet in 50 ml of M9 (with $MgCl_2$) supplemented with antibiotics as described earlier. Add 3.5 mCi of Tran ^{35}S label (NEN-Dupont, Wilmington, DE) and IPTG to 1 m*M*. Incubate at 37°, 300 rpm for 2 hr.
6. Collect cells by centrifugation as described earlier and store at −80° or proceed with purification. A typical specific activity is 2–3 × 10^4 cpm/μg protein, with a total yield of roughly 10 μg.

Electron Microscopy of NM

The most important method, to date, for identifying the presence of amyloid is by microscopy. We have utilized transmission electron micros-

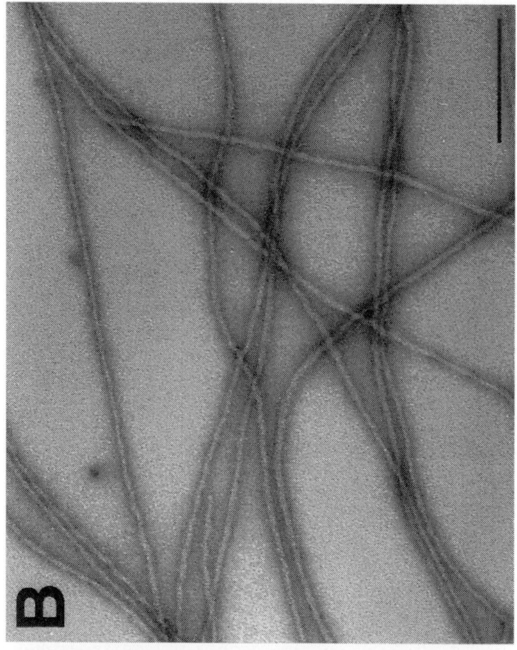

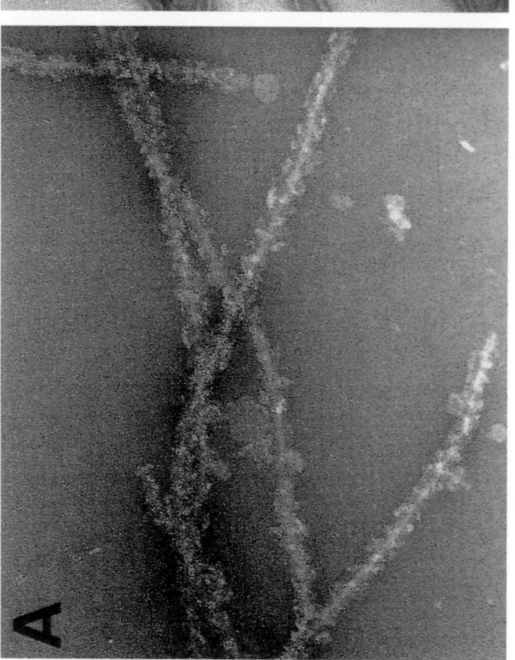

FIG. 6. Electron microscopy of fibers formed by His-tagged NMC and His-tagged NM. Protein samples were prepared for visualization by electron microscopy as described. (A) A field of His-tagged NMC incubated in a high salt buffer consisting of 30 m*M* Tris–HCl (pH 8.0), 1.2 *M* NaCl, 10 m*M* $MgCl_2$, 2 m*M* GTP, 280 m*M* imidazole, and 5 m*M* 2-mercaptoethanol. The base fiber structure has an approximate diameter of 10.6 ± 1.0 nm and displays an amorphous structure along its side, presumably the C domain. (B) A field of His-tagged NM dissolved in 20 m*M* Tris–HCl (pH 8.0), 150 m*M* NaCl, and 4 *M* urea and then diluted into 20 m*M* Tris–HCl (pH 8.0); 1.2 *M* NaCl. Fibers from His-tagged NM exhibit a smooth appearance, whose approximate average diameter is 11.5 ± 1.5 nm. Scale bar: 200 nm.

copy (TEM), scanning transmission electron microscopy (STEM), and atomic force microscopy (AFM) to monitor the assembly of NM into amyloid. Although each of these techniques provides distinct information about the structure and size of complexes formed by $His_{10}NM$, our discussion here will be limited to EM due to the general accessibility of this technique.

We routinely negatively stain $His_{10}NM$ fibers[52] for EM analysis. Protein (5 μl of a 5 μM solution) is applied to a glow-discharged 400 mesh carbon-coated copper grid (Ted Pella, Redding, CA). Protein is allowed to absorb to the grid for 30 sec and is then immediately stained with 200 μl of 2% (w/v) aqueous uranyl acetate. Excess liquid is removed from the grid with a filter paper wick, and they are then allowed to air dry. Samples are observed in a Philips (Eindhoven, The Netherlands) CM 120 transmission electron microscope at an accelerating voltage of 120 kV in low-dose mode. Samples are viewed at a magnification of 40,000×, and images are recorded on Kodak (Rochester, NY) SO 163 film.

Fibers formed by $His_{10}NM$ (Fig. 6A) have an apparent average diameter of 11.5 ± 1.5 nm.[22] The structure of fibers formed from $His_{10}Sup35p$ is sensitive to buffer conditions. Fibers formed in moderate ionic strength buffer (20 mM potassium phosphate (pH 7.5), 20 mM KCl, 5 mM $MgCl_2$, 2.5 mM 2-mercaptoethanol) are smooth and have an average diameter of 17 ± 2.0 nm.[22] In high ionic strength buffer (Fig. 6B), fibers of full-length $His_{10}Sup35p$ are more extended, revealing an interior rod (diameter 10.6 ± 1.0 nm) and an amorphous outer layer.

Conclusion

The study of amyloidogenic proteins is complex both *in vivo* and *in vitro*. Each assay presented in this article provides unique information about the physical state of Sup35p, but these techniques also have inherent pitfalls. We suggest that a robust characterization of transitions in Sup35p physical states both *in vivo* and *in vitro* requires analysis with a combination of techniques.

Acknowledgments

We are indebted to the many individuals whose work, referenced herein, has established the basic genetic and biochemical methodologies described. This work was supported by the Cancer Research Fund of the Damon Runyon-Walter Winchell Foundation Fellowship, DRG-1436 to TRS, the National Institutes of Health, GM025874 to SLL, and the Howard Hughes Medical Institute.

[52] E. Spiess, H. P. Zimmermann, and H. Lundsdorf, *in* "Electron Microscopy in Molecular Biology: A Practical Approach" (J. Sommerville and U. Scheer, eds.), p. 147. IRL Press Limited, Oxford, 1987.

[42] Senescence-Accelerated Mouse

By KEIICHI HIGUCHI, MASANORI HOSOKAWA, and TOSHIO TAKEDA

Introduction

We have succeeded in developing senescence-prone inbred strains (SAMP) with accelerated senescence and age-associated pathologies and also senescence-resistant inbred strains (SAMR) with normal aging except for manifestation of nonthymic lymphomas and histiocytic sarcoma. Advances in biomedical research depend to a considerable extent on the availability of relevant and appropriate animal models, particularly animals that have not been subjected to experimental manipulation. This is especially true for a model of aging in which aging progresses insidiously and irreversibly without apparent causes. This is the case with the SAM model in which selective breeding is the only manipulation. This article describes a brief history of senescence-accelerated mouse (SAM) development, aging characteristics, pathobiological phenotypes, genetic background, and amyloidosis.

History of Senescence-Accelerated Mouse Development

In 1968, several pairs of AKR/J strain mice were donated by the Jackson Laboratory (Bar Harbor, ME) to the Department of Pathology, Chest Disease Research Institute, Kyoto University. While continuing sister–brother mating to maintain this inbred strain under conventional conditions, we became aware that in certain litters most of the mice had an inherited form of senescence phenomena, such as loss of activity, hair loss and lack of glossiness, periophthalmic lesions, increased lordokyphoses, and early death without growth retardation, malformation, limb palsy, and other neurological signs such as tremors and convulsions. In 1975, five litters of mice with severe exhaustion were selected as the progenitors of the senescence-prone series (P series). Litters in which the aging process was normal were selected as progenitors of the senescence-resistant series (R series). Thereafter, based on data of the grading score of senescence, life span, and pathologic phenotypes, selective breeding was carried out. From each of the selected litters of mice with severe exhaustion, five different series, designated P-1, P-2, P-3, P-4, and P-5, were obtained. From each of the selected litters of mice with normal aging, three different series, designated R-1, R-2, and R-3, were obtained. Aging characteristics observed in the P series, compared with those in the R series, revealed that a characteris-

0076-6879/99 $30.00

tic feature of aging, common to all the P series, is "accelerated senescence": early onset and irreversible advance of senescence after a period of normal development, as described later. Thus, the model was named senescence-accelerated mouse in 1981.[1]

With the advancement of generations, breeding among P-4, P-5, and R-3 mice has been unsuccessful; however, it has been possible to establish several new strains that fulfill the criteria of inbred strains: successful sister–brother mating over 20 generations with a stable homozygosity and stable expression of pathobiologic phenotypes.[2,3] Furthermore, establishment of specific pathogen-free (SPF) SAM strains using some of these series and strains has been successful at Takeda Chemical Ind., Ltd., Osaka and Kagoshima University. At present, senescence-prone inbred strains (SAMP) consist of SAMP1, SAMP1//Ka, SAMP1TA, SAMP2, SAMP3, SAMP6, SAMP6/Ta, SAMP7, SAMP8, SAMP8/Ta, SAMP9, SAMP10, SAMP10//Ta, and SAMP11 and senescence-resistant inbred strains (SAMR) consist of SAMR1, SAMR1TA, SAMR4, and SAMR5. Breeding pairs of most of the SAM strains are now available for investigators if application is approved by the Council for SAM Research, which was established in 1984.

Characteristic Features of SAM

Aging Characteristics

Characteristic features of aging in the SAM model were analyzed based on data from the following parameters.

1. Growth Curve. The growth pattern of body weight was observed using mice of P and R series. In both series, the body weight increased steeply up to 14 weeks of age and there was no difference between P and R series in both sexes. Thereafter, the rate of increase was slight, particularly in the P series of both sexes; after 16 weeks of age in females and 28 weeks in males, the difference between R and P series became apparent.[1]
2. Grading Score of Senescence. A scoring system was designed to represent changes in the behavior and appearance of the mice to evaluate the degree of senescence in the mice of SAMP (SAMP1,

[1] T. Takeda, M. Hosokawa, S. Takeshita, M. Irino, K. Higuchi, T. Matsushita, Y. Tomita, K. Yasuhira, H. Hamamoto, K. Shimizu, M. Ishii, and T. Yamamuro, *Mech. Ageing Dev.* **17,** 183 (1981).

[2] T. Takeda, M. Hosokawa, and K. Higuchi, *J. Am. Geriatr. Soc.* **39,** 911 (1991).

[3] T. Takeda, M. Hosokawa, and K. Higuchi, *ILAR J.* **38,** 109, (1997).

SAMP2, and SAMP3) and SAMR (SAMR1, SAMR2, and SAMR3). The items examined included 11 categories of clinical signs and gross lesions considered associated with the aging process: reactivity, passivity, glossiness of hair, skin coarseness, hair loss, skin ulcer, periophthalmic lesions, corneal opacity, corneal ulcers, cataracts, and lordokyphosis. The degree of senescence in each category was graded 0 to 4 according to detailed criteria devised in our laboratory. Grades 0 and 4 correspond to the best and the worst part of the scale, respectively. For example, passivity was evaluated based on the escape reaction from pinching of the nuchal skin or from hanging by the forelimb and graded as follows: grade 0, natural escape reaction to pinching; grade 1, decreases in escape reaction to pinching; grade 2, loss of escape reaction to pinching, but preserved righting reaction to manual turnover; grade 3, neither escape reaction to pinching nor righting reaction, but escape reaction to hanging by the forelimb; grade 4, escape reaction nil. The score in each category was summed to obtain the grading score for each animal. The grading scores for each animal were averaged in each litter and the means of each litter were also averaged in each SAMP and SAMR.[1] The score increased gradually with age in both SAMP and SAMR. In young mice of SAMP and SAMR there was no difference. However, differences between them became obvious after the age of 6 months. The grading score in SAMP leveled out earlier than in SAMR because of a larger increasing rate in SAMP. However, there were no differences between the maximum grading scores of SAMP and SAMR.[4]

3. Survivorship Curve. The survival rate was calculated each month after birth using mice of the P and R series. The pattern of the curves, related to the survivors in the P and R series, differed somewhat; the rate of decline of survival in the P series became greater after 6 months of age compared with the rate in the R series (Fig. 1).[1]
4. Gompertz Function. Using the same litters of the P and R series used for the survivorship curve, Gompertz function[5] was obtained from the death rate each month after birth. When the age-specific death rate (%) is expressed as Y and age (months after birth) as X, the equation is $\log Y = -0.224 + 0.133X$ for the P series and $\log Y = -0.199 + 0.083X$ for the R series. The slope was significantly lower for the R series.[1]

[4] M. Hosokawa, R. Kasai, K. Higuchi, S. Takeshita, K. Shimizu, H. Hamamoto, A. Honma, M. Irino, K. Toda, A. Matsumura, M. Matsushita, and T. Takeda, *Mech. Ageing Dev.* **26,** 91 (1984).

[5] B. Gompertz, *Philos. Trans. R. Soc. London* **115,** 513 (1825).

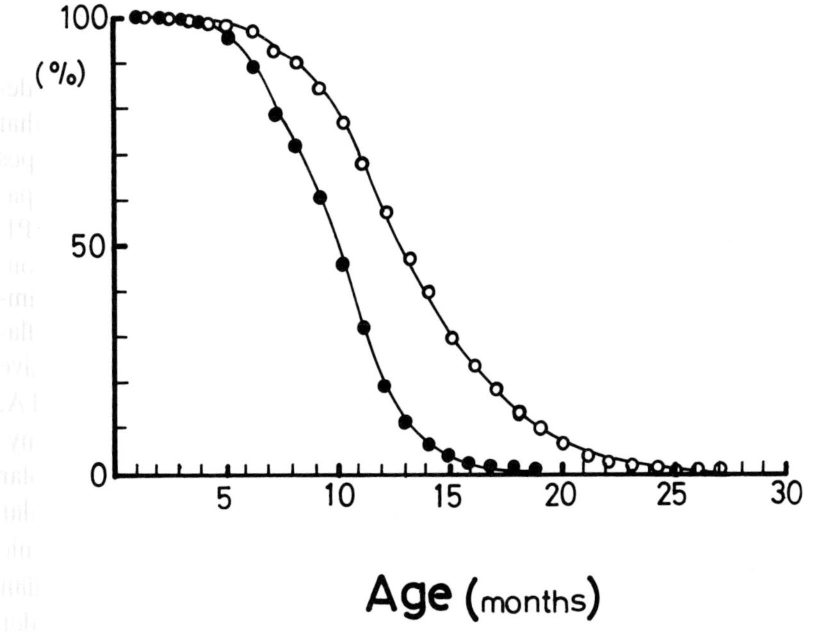

FIG. 1. The survival rate was calculated at each month after birth using 493 and 377 litters of the P and R series, respectively. The rate of decline of survival in the P series is greater with advancing age compared to the rate in the R series. (●) Curve of the P series and (○) that of the R series.

5. Life Span. The mean life span of the P and R series was also calculated. The mean life span in the P series was 9.7 months and that in the R series was 13.3 months.[1]

Based on these findings on aging characteristics, we inferred that the manifestation of senescence in SAMP did not occur in the developmental stage, but in an accelerated manner following normal development. Thus, the patterns of aging in SAMP and SAMR seem to relate to an accelerated senescence rather than to premature aging or senescence and to normal aging, respectively.[1–3] Studies show that the grading score of senescence at 8 months of age of the SAMP strains was 7.97, about twice that of SAMR (3.94), and that the median survival time of SAMP was 9.7 months, about 40% shorter than that of SAMR (16.3 months).[6]

[6] T. Takeda, M. Hosokawa, and K. Higuchi, *in* "The SAM Model of Senescence" (T. Takeda, ed.), p. 15. Excepta Medica, Amsterdam, 1994.

Pathobiological Phenotypes

Routine postmortem examinations and a series of systematically designed studies conducted on living as well as dead SAM mice revealed that SAMP and SAMR strains manifest various pathobiological phenotypes that are often characteristic enough to differentiate the strains.[3,7] The pathobiological phenotypes are senile (AApoAII) amyloidosis in SAMP1, SAMP1//Ka, SAMP2, SAMP7, SAMP9, SAMP10, and SAMP11, contracted kidney in SAMP1, SAMP2, SAMP10, and SAMP11, impaired immune response in SAMP1, SAMP2, SAMP8, and SAMP8/Ta, hyperinflation of the lungs in SAMP1, hearing impairment in SAMP1, hypertensive vascular disease in SAMP1, deficits in learning and memory in SAMP1TA, SAMP8, SAMP8/Ta, SAMP10, and SAMP10//Ta, secondary (AA) amyloidosis in SAMP2 and SAMP6, cataracts in SAMP2 and SAMP9, alveolar bone loss in SAMP2, degenerative joint disease of the temporomandibular joint in SAMP3, senile osteoporosis in SAMP6 and SAMP6/Ta, thymic lymphoblastic lymphoma in SAMP7 and SAMP9, abnormality of circadian rhythms in SAMP8, SAMP8/Ta, and SAMP10//Ta, emotional disorder (reduced anxiety-like behavior) in SAMP8/Ta, emotional disorder (depressive behavior) in SAMP10//Ta, brain atrophy in SAMP10, ovarian cysts in SAMP1, nonthymic lymphomas in SAMR1 and SAMR4, histiocytic sarcoma in SAMR1 and SAMR4, and colitis in SAMR5.

It is noteworthy that most of these pathobiological phenotypes are age-associated pathologies, with the incidence and severity increasing with advancing age. Furthermore, most "age-dependent" geriatric disorders, which were described as "a direct consequence of physiologic senescence" by Cotran *et al.*,[8] are included in the pathobiological phenotypes in SAM: osteoporosis, osteoarthtitis (degenerative joint disease), cataracts, hyperinflation of lungs, and hearing impairment. This evidence provides support for our proposal that the SAM model is a valid one and useful for aging research. Actually, various efforts are currently being made using the SAM model to clarify the pathogenic mechanisms of age-associated pathobiologies and, through these studies, to establish effective methods to modulate or ameliorate the advance of these pathologies.[9] A detailed description of each pathology is not attempted here.

[7] T. Takeda, T. Matsushita, M. Kurozumi, K. Takemura, K. Higuchi, and M. Hosokawa, *Exp. Gerontol.* **32,** 117 (1997).

[8] R. S. Cotran, V. Kumar, and S. L. Robbins, *in* "Robbins Pathologic Basis of Disease," p. 543. Saunders, Philadelphia, 1989.

[9] M. Hosokawa, M. Umezawa, K. Higuchi, and T. Takeda, *J. Anti-Aging Med.* **1,** 27 (1998).

Genetic Background

Judging from the characteristic pathologic findings in most of the SAM strains, i.e., a low incidence of thymic lymphoma and a high incidence of amyloidosis compared to AKR/J mice, it is assumed that the genetic background of SAM strains has deviated from the authentic AKR/J strain. Data of 17 biochemical and 10 immunogenetic markers revealed that there was at least one genotype in each SAM strain that differed from that in the AKR/J strain.[6] Data from a series of Southern hybridization experiments using oligonucleotide probes designed to recognize the endogenous mouse retrovirus sequences indicated that each SAM strain is genetically distinct and contains certain amounts of genetic information derived from strains other than AKR/J.[10]

Taking these results together, it is reasonable to assume that SAM strains are related inbred strains developed by accidental outbreeding between the AKR/J strain and one or more other strains that occurred before 1975 when the progenitors of the P and R series were selected and selective breeding began. At present, it is not clear what strain(s) was mated with AKR/J.

Senile Amyloidosis in Senescence-Accelerated Mouse

Severe senile amyloidosis is one of the most characteristic age-associated disorders in some of the SAMP strains, and these strains are a valuable model for investigation of the pathogenesis of amyloidosis and the development of effective therapeutic modalities.

Senile amyloidosis is a form of amyloidosis in which age is the only risk factor, i.e., the incidence and severity of amyloid deposition increase with age without any apparent predisposing conditions, particularly infectious diseases in aging animals and humans. Amyloid deposition has been noted to occur spontaneously in a variety of animal species.[11] Among them, mice appear be the most suitable model for the following reasons: (1) short life span (2 to 3 years), (2) uniform genetic background of inbred strains and sufficient information on genetics, (3) ease of making genetically modified animals, e.g., transgenic, knockout, and congenic mice, (4) ease of handling in the laboratory in which environmental conditions are strictly controlled, and (5) ease of obtaining a large number of animals. In 1986, the amyloid

[10] H. Kitado, K. Higuchi, and T. Takeda, *J. Gerontol.* **49,** B247 (1994).

[11] K. Higuchi and T. Takeda, *in* "Amyloidosis" (J. Marrink and M. H. Van Rijswijk, eds.), p. 283. Martinus Nijhoff, Dordrecht, The Netherlands, 1986.

fibril protein was isolated from SAMP1 mice, the complete amino acid sequence was determined, and the whole apolipoprotein A-II (apoA-II), an apoprotein in serum high-density lipoproteins (HDL), was revealed to be deposited as amyloid fibrils.[12,13] The molecular weight of apoA-II is 8700 and the protein with a blocked N-terminal. Based on these findings, the amyloid fibrils of mouse senile amyloidosis were termed AApoAII. AApoAII is present universally in aged mice of various strains.[14]

The morphological features of AApoAII—congophilia, green birefringence of Congo red-stained sections under polarized microscopy, and non-branching, twisted wire-like fibril structure under electron microscopy—are completely in accordance with criteria for amyloid deposition. Amyloidosis is systemic, and all organs, except bone and brain parenchyma, are involved.[15] The earliest AApoAII deposits are identified in the primary and secondary papillae of tongue, the lamina propria and submucosa of the small intestine, and the glandular portion and squamous-glandular junction of the stomach. With advancing age, AApoAII deposits extend into collecting tubules in the papillae of the kidneys, perimedullary zone of the adrenal cortex, alveolar septa of the lungs, interstitium of the heart muscles, the interstitium of the thyroid gland, papillary layer of the dermis, interstitium of the testis, corpora lutea, atretic follicle and interstitium of the ovaries, around portal veins and in the spaces of Disse of periportal sinusoid of the liver, the marginal zone around the lymphoid follicles of the spleen, and blood vessels of the whole body. At the final stage, the liver and spleen are enlarged and the kidneys are contracted with severe amyloid deposition. The percentage of AApoAII amyloid-positive areas reaches over 60% of the liver tissue and the weight of the liver increases to over 5 g.[16] Rather large amounts of AApoAII amyloid fibrils are able to be isolated and purified relatively easily from the livers and spleens enlarged with severe amyloid deposition in old mice.

[12] T. Yonezu, K. Higuchi, S. Tsunasawa, S. Takagi, F. Sakiyama, and T. Takeda, *FEBS Lett.* **203,** 149 (1986).

[13] K. Higuchi, T. Yonezu, K. Kogishi, A, Matsumura, S. Takeshita, K. Higuchi, A., Kohno, M. Matsushita, M, Hosokawa, and T. Takeda, *J. Biol. Chem.* **261,** 12834 (1986).

[14] K. Higuchi, H. Naiki, K. Kitagawa, M. Hosokawa, and T. Takeda, *Virch. Arch. B Cell Pathol.* **60,** 231 (1991).

[15] K. Higuchi, A. Matsumura, A. Honma, S. Takeshita, K. Hashimoto, M. Hosokawa, Y. Yasuhira, and T. Takeda, *Lab. Invest.* **48,** 231 (1983).

[16] H. Naiki, K. Higuchi, K. Matsushima, A. Shimada, W. H. Chen, M. Hosokawa, and T. Takeda, *Lab. Invest.* **62,** 768 (1990).

Isolation and Purification of ApoA-II Amyloid Fibrils, ApoA-II Monomer, and ProapoA-II Protein

Mouse Strain

Although senile amyloidosis is common in most strains of mice, severe senile amyloidosis has been reported in only a few strains, i.e., SJL/J, LLC, SAMP1, and PS strains.[17–19] For efficient isolation of AApoAII amyloid fibrils, old mice (>15 months old) of strains in which severe amyloid deposition is a characteristic phenotype should be used. Three variants of apoA-II protein (types A, B, and C) with different amino acid substitutions at four positions (5, 20, 26, and 38 amino acids from the N-terminal) are predicted from the nucleotide sequences of apoA-II cDNA among inbred strains of mice (Table I).[20–23] Each type of apoA-II gene is identifiable by the digestion of polymerase chain reaction (PCR)-amplified apoA-II DNA with the use of restriction fragment length polymorphism (RFLP) for restriction enzymes, *Cfr*13I and *Msp*I.[23] A part of the apoA-II gene, including the second and third exons, is amplified by PCR using two apoA-II-specific primers (5′-TGAAGCTTCTCGCAATGGTCGCACTGCTGGT-3′ and 5′-AGTCATGCTCTGAAAGTACTGTGTG-3′). The mixture in a final volume of 50 μl containing 0.5 μg mouse genomic DNA with conditions as specified in the Gene Taq protocol (Nippon Gene, Osaka Japan) can be amplified in a thermal cycler (PHC-3, Techne Ltd., Cambridge England) as follows: 1 cycle at 94° for 3 min and 35 cycles at 94° for 1 min, 60° for 2 min, and 72° for 2 min with a final extension step at 72° for 10 min. PCR products were digested with *Cfr*131 and *Msp*I, respectively, and were resolved by 4.0% agarose gel electrophoresis and stained by ethidium bromide. After digestion with *Cfr*13I, PCR products of the type A and type B apoA-II genes (*Apoa2^a* and *Apoa2^b*) show three bands (272, 106, and

[17] M. A. Scheinberg, E. S. Cathcart, J. W. Eastcott, M. Skinner, M. D. Benson, T. Shirahama, and M. Bennet, *Lab. Invest.* **35,** 47 (1976).

[18] C. K. Chai, *Am. J. Pathol.* **85,** 49 (1976).

[19] A. C. Koeger, A. Branellec, G. Hirbec, Y. Blouquit, C. L. Mouriquand, A. Sobel, and G. Lagrue, *Pathol. Biol.* **32,** 959 (1984).

[20] K. Higuchi, T. Yonezu, S. Tsumasawa, F. Sakiyama, and T. Takeda, *FEBS Lett.* **207,** 23 (1986).

[21] T. Kunisada, K. Higuchi, S. Aota, T. Takeda, and H. Yamagishi, *Nucleic Acids Res.* **14** 5729 (1986).

[22] T. Yonezu, M. Toda, H. Yamagishi, K. Higuchi, and T. Takeda, *Gene* **84,** 187 (1990).

[23] K. Higuchi, K. Kitagawa, H. Naiki, K. Hanada, M. Hosokawa, and T. Takeda, *Biochem. J.* **279,** 427 (1991).

TABLE I

GENETIC TYPES OF APOA-II AND AAPOAII AMYLOIDOSIS AMONG DIFFERENT INBRED MOUSE STRAINS

Type	Gene	Amino acid substitutions at				PCR[a]		Strain	Senile amyloidosis
		5	20	29	38	*Cfr*131	*Msp*I		
A	*Apoa2*[a]	Pro	Asp	Met	Ala	1	1	SAMP3, SAMP8, AKR/J, B10.BR, C57BL/6J, NSY	Mild
B	*Apoa2*[b]	Pro	Glu	Val	Val	2	1	SAMP6, SAMR1, SAMR4, BALB/*c*, CBA/N, CSH/He, DDD, NZB/N	Light
C	*Apoa2*[c]	Gln	Glu	Val	Ala	2	2	SAMP1, SAMP2, SAMP7, SAMP9, SAMP10, SAMP11, A/J, SJL/J, LLC, SM/J	Severe

[a] The pattern of restriction fragments observed for PCR products of the apoA-II gene after digestion with *Cfr*131 and *Msp*I is designated with a "1" or a "2". For *Cfr*131, the 1 pattern corresponds to 272-, 106-, and 46-bp fragments whereas the 2 pattern corresponds to 272- and 152-bp fragments. For *Msp*I, the 1 pattern corresponds to 296-, 83-, and 44-bp fragments whereas the 2 pattern corresponds to 296- and 128-bp fragments.

46 bp), whereas PCR products of type C (*Apoa2*^{c}) apoA-II gene show two bands (272 and 152 bp). After digestion with *Msp*I, the PCR product of the type A apoA-II gene shows three bands (296, 84, and 44 bp), whereas PCR products of type B and type C apoA-II genes show two bands (296 and 128 bp). The molecular type of apoA-II was determined in various strains of mice and was correlated with the susceptibility of each strain to senile amyloidosis. The SJL/J, A/J, LLC, SAMP1, SAMP2, SAMP7, SAMP9, SAMP10, and SAMP11 strains with a high incidence of and severe senile amyloidosis have type C apoA-II with glutamine at position 5, whereas the SAMR1, SAMR4, SAMP6, BALB/c, CBA/N, C3H/He, DDD, and NZB strains with a very low incidence of amyloidosis have type B apoA-II with proline at position 5. SAMP8, AKR/J, B10BR, C57BL/6J, and NSY strains with type A apoA-II, which has methionine at position 26, show a mild amyloid deposition in the intestines, kidneys, tongue, and lungs (Table I). Severe AApoAII amyloidosis in *Apoa2*^{c} strains is transmitted in an autosomal dominant manner with a strong gene dosage effect.[24] Congenic mice (R1.P1-*Apoa2*^{c}) have the amyloidogenic type C apoA-II gene of SAMP1 on the genetic background of SAMR1.[25] As compared with donor SAMP1 strains, severe amyloid deposition is present in the congenic R1.P1-*Apoa2*^{c} strain.[26] Because R1.P1-*Apoa2*^{c} mice have much better reproductivity and live longer than SAMP1 mice, R1.P1-*Apoa2*^{c} mice are more convenient for obtaining livers and spleens with severe AApoAII amyloid deposition for isolation of amyloid fibrils.

There is no evidence for sex difference in the incidence and degree of AApoAII amyloidosis. AApoAII amyloidosis is influenced by environmental conditions. In particular, AApoAII deposition is suppressed in mice raised in SPF conditions. AApoAII amyloid fibrils are mixed frequently with AA fibrils in several strains. Proper identification of the amyloid protein requires immunohistochemical or biochemical characterization of amyloid fibrils using specific antisera against apoA-II and AA protein.[15]

Purification of AApoAII Amyloid Fibrils

According to the methods of Pras,[27] the AApoAII amyloid fibril fraction is isolated as a water suspension from the livers and spleens of R1.P1-*Apoa2*^{c} mice of 15 months of age and up. Livers (~3.0 g) with heavy

[24] H. Naiki, K. Higuchi, A. Shimada, T. Takeda, and K. Nakakuki, *Lab. Invest.* **68,** 332 (1993).

[25] K. Higuchi, H. Kitado, K. Kitagawa, K. Kogishi, H. Naiki, and T. Takeda, *FEBS Lett.* **317,** 207 (1993).

[26] K. Higuchi, H. Naiki, K. Kitagawa, H. Kitado, K. Kogishi, T. Mastushita, and T. Takeda, *Lab. Invest.* **72,** 75 (1995).

[27] M. Pras, D. Zucker-Franklin, A. Rimon, and E. C. Franklin, *J. Exp. Med.* **130,** 777 (1969).

amyloid deposits are homogenized in ice-cold 30 ml distilled water for 30 sec with an ultradispenser (Ultra-Turrax T25, Janke & Kunkel Gmph Company, KG, Staufen, Germany), and homogenization is repeated three times at 90-sec intervals. After centrifugation of homogenate at 20,000 rpm at 4° for 20 min, the supernatant is discarded. This procedure must be repeated 10–30 times before the optical density at 280 nm of supernatant drops below 0.2. The pellet of the final centrifugation is rinsed with 30 ml of distilled water by 30 sec of homogenization and centrifugation at 20,000 rpm for 20 min. The amyloid fibril fraction can be extracted in 20 ml of ice-cold distilled water by homogenization three times × 30 sec and centrifugation at 20,000 rpm, 4° for 20 min. This extraction procedure must be repeated 3–6 times for the optical density of supernatant drops below 0.2. Isolated amyloid fibril fractions can be purified further as the pellet of ultracentrifugation at 100,000*g* and 4° for 1 hr. Sodium dodecyle sulfate–polyacrylamide gel electrophoresis (SDS–PAGE) revealed that this fibril fraction contains apoA-II protein as a major protein component but small amounts of amyloid associated proteins, such as proapoA-II and apoE. Sucrose gradient ultracentrifugation can be used as the final steps of purification of natural amyloid fibrils as described elsewhere[28] and in another article of this volume. Purified AApoAII fibrils show typical amyloid fibrils on electron microscopy but still have a small proportion of contaminants of fragments of cell membranes, organelles, and lipids. These fibrils can be used as seeds for *in vitro* polymerization (fibrillization) of the amyloid protein, which is described in another article. AApoAII amyloid deposition can be accelerated by the intravenous injection of AApoAII fibrils *in vivo.* One milliliter of a 1-mg/ml solution of AApoAII fibrils in distilled water is put into an Eppendorf tube (size, 1.5 ml) and sonicated on ice for 90 sec by a microtip-equipped Astrason ultrasonic processor W-380 (Heat System-Ultrasonics, Inc., Farmingdale, New York) at maximum power. This procedure is repeated 10 times at 90-sec intervals. Sonicated AApoAII and AA fibrils are used immediately for injection into mice. AApoAII fibrils (0.1 mg/0.1 ml/mouse) induce amyloid fibril formation without lag times and systemic amyloidosis should be observed after 3 to 4 months from the injection in mouse strains with type C apoA-II.

Isolation and Purification of ApoA-II

The ApoA-II protein monomer can be isolated from (1) AApoAII amyloid fibrils and (2) mouse serum HDL by nondenatured alkaline urea PAGE. Denatured 2.5 mg of crude AApoAII amyloid fibrils isolated by

[28] H. Naiki, K. Higuchi, M. Hosokawa, and T. Takeda, *Anal. Biochem.* **177,** 244 (1989).

ultracentrifugation at 100,000*g* and solubilize in 1.0 ml of 6 *M* guanidine hydrochloride, 0.1 *M* Tris–HCl, pH 10.0, 50 m*M* dithiothreitol (DTT, 1.0 mg/ml) for 24 hr with gentle stirring at room temperature. Denatured AApoAII fibrils should be dialyzed quickly against 10 m*M* NH_4HCO_3 solution and lyophilized. These guanidine-denatured amyloid fibrils (GDAM) are dissolved in a sample buffer of alkaline urea PAGE containing 6 *M* urea.[13,30] HDL with type C apoA-II can be prepared from serum of R1.P1-*Apoa2[c]* or other strains that have type C apoA-II by preparative ultracentrifugation according to a procedure described previously[13] with the some modification. The VLDL + LDL fraction ($d < 1.063$ g/ml) is removed from 1.8 ml mouse serum by ultracentrifugation at 100,000 rpm for 5 hr at 16° using a RP100ATA rotor (Hitachi, Ltd., Tokyo, Japan) after adjustment of the specific gravity to 1.063 by adding 0.9 ml of 0.195 *M* NaCl and 2.44 *M* NaBr solution. HDL is separated as lipoproteins in the density interval 1.063–1.219 g/ml following ultracentrifugation of the bottom solution (1.8 ml) whose gravity is increased by adding 0.9 ml of 0.195 *M* NaCl, 7.65 *M* NaBr solution at 100,000 rpm for 8 hr at 16° using a RP100ATA rotor. HDL is dialyzed against 0.15 *M* NaCl containing 0.05% EDTA adjusted to pH 8.0 at 4° and is concentrated to 10 mg protein/ml in a dialysis bag (Collodion filters, Sartorious). Concentrated HDL is delipidated with ethanol/ethyl ether at −10°[29] and is dissolved in a sample buffer of alkaline urea PAGE.

Alkaline urea PAGE is performed by the method of Davis[30] on a 15 × 10 × 0.2-cm slab gel with 7.5% acrylamide. GDAM (2.5 mg) or delipidated HDL (10 mg) is dissolved in 1.0 ml sample buffer and incubated for 6 hr at room temperature and is applied onto a gel and electrophoresed overnight at 6 mA and 4°. ApoA-II-containing bands located by stained reference gels are excised, diced into small pieces, sustained in 10 m*M* NH_4HCO_3, and extracted three times with stirring for 12 hr at 4°. The eluted apoA-II is dialyzed against the same elution buffer, lyophilized, and kept at −70° before use as an apoA-II monomer for *in vitro* fibrillization, making antibody, and so on.

Isolation of ProapoA-II

From the AApoAII amyloid fibril fraction isolated from old mouse liver, we have identified and purified a more basic amyloid protein than apoA-II. Amino-terminal sequencing of the protein revealed that the prosegment of five amino acid residues (alanine–leucine–valine–lysine–arginine) extends from the amino-terminal glutamine residue of mature

[29] A. Scanu, *J. Lipid Res.* **7,** 295 (1996).
[30] B. J. Davis, *Ann. N.Y. Acad. Sci.* **121,** 404 (1994).

apoA-II protein. The relative abundance of this proapoA-II protein to mature apoA-II in the amyloid fibril fractions in the livers and spleens with severe amyloidosis is around 15%. This figure suggests that propoA-II is concentrated almost 10-fold in amyloid fibril fractions compared with the contents in the serum. ProapoA-II prefers to aggregate and is insoluble compared with mature apoA-II. Thus proapoA-II may play a key role in the initialization of mouse senile amyloidosis.[31]

ProapoA-II can be purified from GDAM of the AApoAII fibril fraction using alkaline urea PAGE as described previously. ProapoA-II protein can be isolated as a minor and clearly separated band that migrates more slowly than major mature apoA-II because of its higher isoelectric point (5.76 vs 4.94), which is produced by two basic amino acid residues in the propeptide. Usually we can isolate about 100 μg proapoA-II protein from 2.5 mg of AApoAII amyloid fractions. Specific antiserum was prepared against purified proapoA-II, and this antiserum reacts with proapoA-II specifically and does not cross-react with mature apoA-II. However, polyclonal antiserum against mature apoA-II and GDAM of AApoAII reacts with proapoA-II.[31]

AApoAII amyloidosis has not been documented to date in humans and animals other than mice. Humans and primates have dimeric apoA-II bound by a disulfide bond between cysteines at position 6, which are replaced with aspartic acid in mice. The relationship between accelerated amyloidosis in the *Apoa2*^{c} strain and light senile amyloidosis in aged *Apoa2*^{a} and *Apoa2*^{b} mice resembles the relationship between systemic amyloidosis in familial amyloid polyneuropathy patients with mutant transthyretin and senile systemic amyloidosis in aged people. Thus AApoAII amyloidosis is an excellent animal model for both senile and hereditary amyloidosis. Usefulness in all sorts of experiments makes mice and AApoAII amyloidosis the appropriate model system for the investigation of the mechanism of amyloid fibril formation, an age-associated dramatic change in the conformation of proteins.

[31] K. Higuchi, K. Kogishi, J. Wang, C. Xia, T. Chiba, T. Matsushita, and M. Hosokawa, *Biochem. J.* **325,** 653 (1997).

[43] Detection of Polyglutamine Aggregation in Mouse Models

By STEPHEN W. DAVIES, KIRUPA SATHASIVAM, CARL HOBBS, PATRICK DOHERTY, LAURA MANGIARINI, EBERHARD SCHERZINGER, ERICH E. WANKER, and GILLIAN P. BATES

Introduction

Polyglutamine [poly(Q)] expansion has been found to cause a number of late onset, inherited, neurodegenerative disorders, which include Huntington's disease (HD), dentatorubral pallidoluysian atrophy (DRPLA), spinal and bulbar muscular atrophy (SBMA), and the spinocerebellar ataxias (SCA1, 2, 3, 6, and 7).[1] Comparison of the genetics and molecular biology of these diseases shows that a number of features predominate, namely a dominant pattern of inheritance, anticipation, and the broadly comparable sizes of the normal and expanded repeat ranges. Proteins harboring these poly(Q) expansions are unrelated, mostly of unknown function, and have extensively overlapping expression patterns.[1,2]

A mouse model of HD has been generated by the introduction into the mouse germ line of exon 1 of the human HD gene, containing highly expanded CAG repeats, under the control of the HD promoter.[3,4] Three lines develop a progressive neurological phenotype: R6/1, $(CAG)_{115}$, R6/2, $(CAG)_{145}$, and R6/5, $(CAG)_{128-156}$, the symptoms of which include a movement disorder and weight loss with similarities to HD. Immunocytochemistry with antibodies to the truncated exon 1 protein and to ubiquitin identified neuronal intranuclear inclusions (NII) in brains from all symptomatic lines.[5] At the ultrastructural level, inclusions can be observed in the absence of immunolabeling as a pale staining granular and fibrillar structure.[5] Line R6/2 has an age of onset of approximately 2 months and the disease progresses rapidly such that few mice are kept beyond 14 weeks. Only a single nuclear inclusion is detected per neuron. Inclusions are first

[1] G. P. Bates, L. Mangiarini, and S. W. Davies, *Brain Pathol.* **8,** 699 (1998).

[2] R. Wells and S. Warren, "Genetic Instabilities and Hereditary Neurological Diseases." Academic Press, San Diego, 1998.

[3] L. Mangiarini, K. Sathasivam, M. Seller, B. Cozens, A. Harper, C. Hetherington, M. Lawton, Y. Trottier, H. Lehrach, S. W. Davies, and G. P. Bates, *Cell* **87,** 493 (1996).

[4] L. Mangiarini, K. Sathasivam, A. Mahal, R. Mott, M. Seller, and G. P. Bates, *Nature Genet.* **15,** 197 (1997).

[5] S. W. Davies, M. Turmaine, B. A. Cozens, M. DiFiglia, A. H. Sharp, C. A. Ross, E. Scherzinger, E. E. Wanker, L. Mangiarini, and G. P. Bates, *Cell* **90,** 537 (1997).

METHODS IN ENZYMOLOGY, VOL. 309
0076-6879/99 $30.00

identified in the cortex at 3.5 weeks and in the striatum at 4.5 weeks prior to the onset of symptoms.[5] In contrast, selective cell death is not observed in these structures until 16–17 weeks of age, suggesting that the symptoms are due to a neuronal dysfunction rather than as a result of neurodegeneration (S. W. Davies *et al.*, unpublished data). In the two lines with the slower progression, R6/1 and R6/5, inclusions are also sometimes observed in neurites and occasionally in astrocytes (S. W. Davies *et al.*, unpublished data).

Insights into the structure of the poly(Q) inclusions have come from the *in vitro* analysis of glutathione *S*-transferase (GST)-exon 1 huntingtin fusion proteins.[6] It was found that GST fusion proteins with $(CAG)_{83}$ and $(CAG)_{121}$ spontaneously formed highly insoluble aggregates with a fibrillar morphology. Removal of the (GST) tag from proteins with $(CAG)_{51}$ also resulted in amyloid-like fibrils that showed green birefringence when stained with Congo red and observed under polarized light. However, this aggregation was not observed with proteins containing $(CAG)_{20}$ or $(CAG)_{30}$ repeats in the nonpathogenic range. This is consistent with the proposal by Perutz *et al.*[7] that poly(Q) stretches can form polar zippers via H bonding in a cross β-pleated amyloid-like structure.

Neuronal inclusions are now clearly established as the pathological signature of poly(Q) disease and have been identified in postmortem brains from HD,[8,9] SCA1,[10] SCA3,[11] SCA7,[12] and DRPLA[9,13] patients. In all cases they are composed of the specific poly(Q) containing protein and ubiquitin. Inclusions in affected neurons of SCA1 patients and an SCA1 transgenic mouse model stain positive for antibodies that detect the 20S proteasome

[6] E. Scherzinger, R. Lurz, M. Turmaine, L. Mangiarini, B. Hollenbach, R. Hasenbank, G. P. Bates, S. W. Davies, H. Lehrach, and E. E. Wanker, *Cell* **90,** 549 (1997).

[7] M. F. Perutz, T. Johnson, M. Suzuki, and J. T. Finch, *Proc. Natl. Acad. Sci. U.S.A.* **91,** 5355 (1994).

[8] M. DiFiglia, E. Sapp, K. O. Chase, S. W. Davies, G. P. Bates, J.-P. Vonsattel, and N. Aronin, *Science* **277,** 1990 (1997).

[9] M. W. Becher, J. A. Kotzuk, A. H. Sharp, S. W. Davies, G. P. Bates, D. L. Price, and C. A. Ross, *Neurobiol. Dis.* **4,** 387 (1998).

[10] P. J. Skinner, B. T. Koshy, C. J. Cummings, I. A. Klement, K. Helin, A. Servadio, H. Y. Zoghbi, and H. T. Orr, *Nature* **389,** 971 (1997).

[11] H. L. Paulson, M. K. Perez, Y. Trottier, J. Q. Trojanowski, S. H. Subramony, S. S. Das, P. Vig, J.-L. Mandel, K. H. Fischbeck, and R. N. Pittman, *Neuron* **19,** 1 (1997).

[12] M. Holmberg, C. Duyckaerts, A. Durr, G. Cancel, I. Gourfinkel-An, P. Damier, B. Faucheux, Y. Trottier, E. C. Hirsch, Y. Agid, and A. Brice, *Hum. Mol. Genet.* **7,** 913 (1998).

[13] S. Igarashi, R. Koide, T. Shimohata, M. Yamada, Y. Hayashi, H. Takano, H. Date, M. Oyake, T. Sato, A. Sato, S. Egawa, T. Ikeuchi, H. Tanaka, R. Nakano, K. Tanaka, I. Hozumi, T. Inuzuka, H. Takahashi, and S. Tsuji, *Nature Genet.* **18,** 111 (1998).

and the molecular chaperone HDJ-2/HSDJ.[14] Similarly, inclusions in HD patient brains and transgenic mice contain the 20S proteasome and several components of the 19S and 11S activation complexes (S. W. Davies *et al.*, unpublished data).

The first steps in the molecular pathogenesis of HD are beginning to be unraveled. The huntingtin protein is subjected to an initial cleavage step that releases an N-terminal fragment. The precise nature of this fragment and the cleavage mechanism are unknown, although caspase 3 has been proposed to fill this role.[15] The N-terminal fragment can move into the nucleus, but once again, the mechanism by which this occurs is unknown. Nuclear inclusions have been described in juvenile onset HD with a frequency of 38–50% in cortical neurons, whereas dystrophic neurites (axonal inclusions) were reported to predominate in the adult form of the disease.[8] Inclusions are observed in brain regions not traditionally associated with neurodegeneration.[9] Therefore, the presence of an inclusion does not necessarily lead to cell death and symptoms may be caused by dysfunction in brains regions not previously associated with HD. Dystrophic neurites were also identified in layer VI of the cortex of an individual who was presymptomatic for HD and had died of other causes.[8] This one case suggests that, as in mice, inclusions may form prior to the onset of symptoms. Although the evidence is currently circumstantial, the early detection of inclusions suggests that they may be causative of the disease symptoms.

A major application of the R6 transgenic lines will be to test drugs that slow down or prevent poly(Q) aggregation. In order to demonstrate causality, it will be necessary to show that by slowing down or preventing the formation of inclusions, the onset of symptoms is correspondingly prevented or delayed. This article provides protocols by which the appearance of aggregates can be monitored.

Mice and Antibodies

Transgenic mouse lines R6/1 and R6/2 [Jackson Codes: B6CBA-TgN(HDexon1)61 and B6CBA-TgN(HDexon1)62] can be obtained from The Induced Mutant Resource (The Jackson Laboratory, 600 Main Street, Bar Harbor, ME 04609 Fax: 207-288-6149 or 6079). The antiubiquitin anti-

[14] C. J. Cummings, M. A. Mancini, B. Antalffy, D. B. DeFranco, H. T. Orr, and H. Y. Zoghbi, *Nature Genet.* **19,** 148 (1998).

[15] Y. P. Goldberg, D. W. Nicholson, D. M. Rasper, M. A. Kalchman, H. B. Koide, R. K. Graham, M. Bromm, P. Kazemi-Esfarjani, N. A. Thornberry, J. P. Vaillancourt, and M. R. Hayden, *Nature Genet.* **13,** 442 (1996).

body is from Dako and is used at a dilution of 1:2000. Currently, only rabbit polyclonal antibodies raised either against a peptide corresponding to the first 17 amino acids of the huntingtin protein (Ab1 and AP78)[8,9] or against fusion proteins (HD1 and CAG53b)[6] have been used to detect the exon 1 huntingtin protein. These reagents are not available commercially and it is currently not possible to cite a nonfinite source. In all cases, the antibodies were not affinity purified and were used at the following dilutions: Ab1, 1 in 600; AP78, 1 in 1000; HD1, 1 in 2000; and CAG53b, 1 in 2000.[5]

Detection of Neuronal Inclusions by Immunocytochemistry

Neuronal intranuclear inclusions and the less frequent neurite inclusions can be detected in the R6 transgenic lines by both immunocytochemistry at the resolution of light microscopy and electron microscopy.[5] At the ultrastructural level they can also be seen readily in the absence of immunostaining.[5] Figure 1 shows cortical sections from symptomatic R6/2 mice (~150 CAG repeats in the pathogenic range) and HDex27 mice (18 CAG repeats in the nonpathogenic range) immunolabeled with antibodies to the exon 1 huntingtin protein and to ubiquitin.

Reagents. In all cases, water is deionized and distilled.

1. 0.4 *M* sodium phosphate buffer (pH 7.4). For 2 liters: 23.72 g $NaH_2PO_4 \cdot 2H_2O$ and 92 g Na_2HPO_4, make up to 2 liters with water, and filter sterilize.
2. PLP fixative. For 1 liter, depolymerize 20 g of paraformaldehyde (TAAB Laboratories Equipment Ltd., Aldermaston, Berks) in 500 ml of water with 1.0 g NaOH pellets at 65° and filter. Dissolve 13.65 g L-lysine (monohydrochloride, Sigma, St. Louis, MO) in 360 ml distilled water with 140 ml of 0.4 *M* sodium phosphate buffer. Before perfusion, mix the two solutions and add 2.13 g of sodium metaperiodate (Sigma).
3. Gelatinization of slides: Prepare gelatin solution by dissolving 5 g gelatin (BDH, Lutterworth, Leics) and 0.5 g chromium potassium sulfate (Sigma) in 500 ml water with heating and filter. Dip a rack of slides in the solution and shake off excess. Dry in a 37° incubator and store at room temperature. This treatment helps the specimen adhere to the slide.
4. Nissl substance with thionin: 1% thionin (Sigma) in 0.125 *M* sodium acetate buffer (pH 5.5–6). For 1 liter, dissolve 10 g thionin with 17 g sodium acetate trihydrate in water, adjust to pH 5.5–6 with dilute acetic acid, add water to 1 liter, and filter sterilize.
5. EM fixative: 4% paraformaldehyde and 0.1% glutaraldehyde in

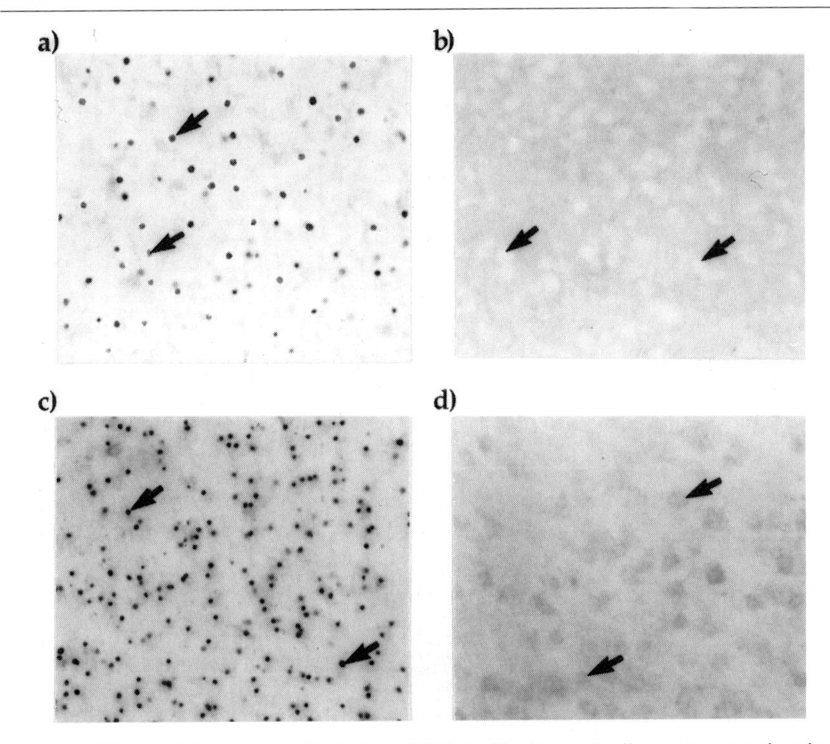

FIG. 1. Neuronal intranuclear inclusions (NII) in Huntington's disease transgenic mice. Immunocytochemical localization in the cerebral cortex: (a) huntingtin exon 1 protein in a symptomatic R6/2 mouse (~150 CAG repeats), (b) huntingtin exon 1 protein in an HDex27 mouse (18 CAG repeats), (c) ubiquitin in a symptomatic R6/2 mouse (~150 CAG repeats), and (d) ubiquitin in an HDex27 mouse (18 CAG repeats). The densely stained NII in symptomatic mice (a and c) contrasts with the huntingtin immunonegative nucleus (b) or the ubiquitin immunoreactive nucleus (d) in symptom-free mice carrying repeats in the nonpathogenic range.

0.1 *M* sodium phosphate buffer (pH 7.4). For 1 liter, dissolve 40 g paraformaldehyde (TAAB Laboratories Equipment Ltd.) in 500 ml water by heating to not more than 65° with five to six NaOH pellets. Add 250 ml 0.4 *M* sodium phosphate buffer, pH 7.4, 246 ml water, and 4 ml 25% glutaraldehyde solution (TAAB). Filter and check the final pH.

6. Osmium tetroxide solution: 1% osmium tetroxide (TAAB) in 0.1 *M* sodium phosphate buffer, pH 7.4. Store at 4°. Wear gloves and use a fume hood for preparation and use.
7. Uranyl acetate solution: 0.1% uranyl acetate (TAAB) in 0.125 *M* sodium acetate, pH 5.5–6.0. Wear gloves and use a fume hood for preparation and use.

8. Toluidine blue stain: 1% aqueous solution of toluidine blue (Sigma)
9. Buffer for primary antibody: 0.9% NaCl, 0.1 *M* Tris–HC1, pH 7.4, 0.02% sodium azide, 0.3% Triton X-100. For 100 ml: 10 μl 1 *M* Tris–HCl, pH 7.4, 0.9 g NaCl, 1 ml sodium azide (2% solution in sodium phosphate buffer 0.1 *M*), make up to 100 ml with water and add 300 μl Triton X-100 (BDH). Stir to mix.
10. ABC reagent: ABC kit, Vector Laboratories (rabbit IgG). 1 ml ABC reagent: 10 μl reagent A (contains avidin), 10 μl reagent B (contains biotinylated horseradish peroxidase), 980 μl Tris–HCl, pH 7.4.
11. DAB reagent: For 100 ml, dissolve 25 mg diaminobenzidine (DAB, 3,3′-diaminobenzidine, Sigma) in 100 ml 0.1 *M* Tris–HCl, pH 7.4, containing 33 μl hydrogen peroxide (Sigma). This solution should be colorless; if it is pale brown, reprepare the solution.

Tissue Processing for Light Microscopy

Procedure

1. Anesthetize mouse with an overdose of sodium pentobarbitone (100 mg/kg Sagatal, Rhône Mérieux Ltd.) by intraperitoneal injection and perfuse through the left cardiac ventricle with 35–50 ml of PLP fixative.
2. Remove brain carefully from the cranium and place into PLP fixative for 4–6 hr at 4°.
3. Transfer brain to 30% sucrose in 0.1 *M* Tris–HCl, pH 7.4, for 48 hr at 4°.
4. Mount the brain in Tissue-Tek OCT compound (Miles Laboratories, Elkhart, IN), freeze with powdered solid CO_2, and section in the coronal, saggital, or horizontal plane at 40 μm on a sledge microtome.
5. Immunostain sections free floating (as described later) before mounting onto gelatinized glass slides and drying overnight.
6. Stain alternate sections for Nissl substance with thionin. Dehydrate through an ascending series of alcohols (70% for 5 min, 90% for 5 min, 95% for 5 min, two changes of absolute ethanol for 10 min each). Transfer the slides to Histoclear (National Diagnostics Lab.) for 15 min. Clear in Histoclear again for a further 15 min and coverslip the slides in DPX mounting medium (Fisher Scientific).

Tissue Processing for Electron Microscopy

Procedure

1. Anesthetize mouse as described earlier and perfuse through the left cardiac ventricle with 35–50 ml EM fixative.

2. Remove brain from the skull and place in fresh EM fixative overnight at 4°.
3. Using an Oxford vibratome (Lancer), cut coronal sections through the cerebellum, striatum, and cerebral cortex (50–200 μm) and collect in serial order in 0.1 *M* sodium phosphate buffer, pH 7.4.
4. Osmicate the sections for 30 min in osmium tetroxide solution and then stain in uranyl acetate for 15 min at 4°. All osmium tetroxide and uranyl acetate work should be performed wearing gloves and carried out in a fume hood.
5. Dehydrate sections through an ascending series of alcohols (as described earlier), clear by three incubations in propylene oxide (BDH) each for 10 min, and embed in Araldite epoxy resin (Agar Scientific Ltd., Stanstead, Essex) between two sheets of Melanex (ICI).
6. Cut semithin (1 μm) sections with glass knives. Stain with toluidine blue for 1 min and examine under light microscopy.
7. Cut adjacent thin (70–80 nm) sections with a diamond knife on a Reichert Ultracut ultramicrotome.
8. Collect sections on mesh grids coated with a thin Formvar film (Sigma) counterstained with a saturated solution of Reynold's lead citrate for 4 min and view in a Jeol 1010 electron microscope.

Immunocytochemistry for Light Microscopy

Procedure

1. Incubate sections free floating in an Eppendorf tube containing primary antibody at 4° for 72 hr without shaking.
2. Transfer sections between solutions using a paint brush. Wash the sections in 0.1 *M* Tris–HCl, pH 7.4, for 45 min.
3. Incubate the sections in 10% normal goat serum (ABC kit, Vector Laboratories) for 45 min.
4. Wash the sections in 0.1 *M* Tris–HCl, pH 7.4, for 45 min.
5. Transfer the sections to biotinylated secondary antibody (1:200 in 0.1 *M* Tris–HCl, pH 7.4) for 2 hr at room temperature.
6. Wash the sections in 0.1 *M* Tris–HCl, pH 7.4, for 45 min.
7. Transfer sections to ABC reagent for 2 hr at room temperature.
8. Wash the sections in 0.1 *M* Tris–HCl, pH 7.4, for 45 min.
9. Visualize sites of peroxidase enzyme activity by incubating sections in DAB reagent for 10–30 min in small glass petri dish with shaking. An individual section is removed from the DAB solution and viewed under a 10× objective for positive staining. When an appropriate level of staining is obtained, sections are transferred to 0.1 *M* Tris–HCl, pH 7.4, for 30 min.

10. Mount sections onto gelatinized slides. Allow to dry overnight on the bench at room temperature.
11. Dehydrate through an ascending series of alcohols (70% for 5 min, 90% for 5 min, 95% for 5 min, two changes of absolute ethanol for 10 min each). Transfer the slides to Histoclear (National Diagnostic Lab.), incubate for 15 min, and replace with fresh Histoclear. The slides can remain in this solution until mounting.
12. Mount the slides with DPX mounting medium (Fisher Scientific). Leave to dry on the bench overnight, and peel away excess mounting medium the following day.

Immunocytochemistry for Electron Microscopy

Procedure. Procedures for processing sections for electron microscopy are the same as those for light microscopy with the exception that all stages use 0.1 *M* sodium phosphate buffer, pH 7.4, instead of 0.1 *M* Tris–HCl, pH 7.4.

Detection of Neuronal Inclusions by Filter Retardation Assay

The presence of inclusions can also be detected on Western blots and by the filter retardation assay.[6] On Western blots, inclusions remain in the wells as highly insoluble aggregates that barely enter the stacking gel. The filter retardation assay is documented elsewhere in this volume as an analytical approach for the study of *in vitro* aggregation.[16] It is described here as a method of determining the inclusion content of transgenic mouse brains. Figure 2 shows the purification of nuclear inclusions by the filter retardation assay from a longitudinal series of R6/2 transgenic mouse brains. Using this approach, it is possible to detect the presence of intranuclear inclusions at 4 weeks of age, consistent with the time at which it is first possible to detect them by immunocytochemistry.

Reagents. In all cases, water is deionized and distilled.

1. 2.5 *M* sucrose: Dissolve 855 g of sucrose in water by stirring overnight at 60°. Adjust the volume to 1 liter, filter sterilize, and store at 4°.
2. 1 *M* triethanolamine hydrochloride (pH 7.5): Dissolve 185.7 g of triethanolamine (Sigma) in 800 ml of water, adjust to pH 7.5 with NaOH, make up to 1 liter, filter sterilize, and store at 4°.

[16] E. E. Wanker, E. Scherzinger, V. Heiser, A. Sittler, H. Eickhof, and H. Lehrach, *Methods Enzymol.* **309** [24] 1999 (this volume).

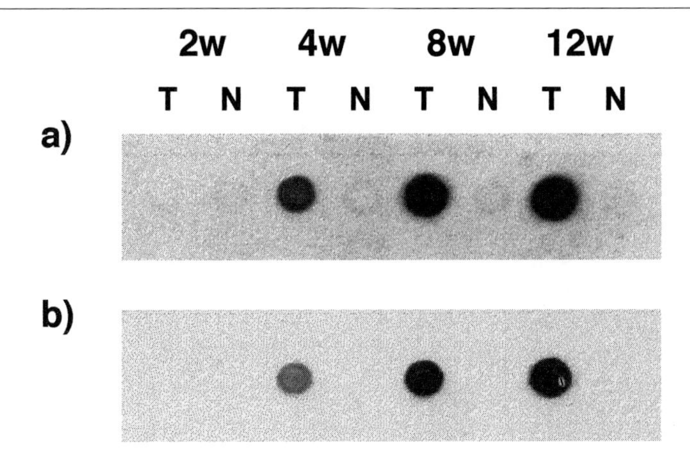

FIG. 2. Isolation of inclusions from R6/2 brains by the filter retardation assay. Aggregates were prepared from 2×10^6 nuclei per brain from R6/2 mice (T) and nontransgenic litter controls (N) at ages ranging from 2 to 12 weeks. Aggregates were trapped by the filter retardation assay and detected by immunoprobing with (a) the HD1 antibody[6] that detects the huntingtin exon 1 protein and (b) ubiquitin. Exposure times were (a) 50 sec and (b) 3 min.

3. 1 *M* dithiothreitol (DTT): Dissolve 0.077 g DTT (Promega, Madison, WI) in 0.5 ml water and keep on ice. Prepare fresh each time.
4. 0.1 *M* Phenylmethylsulfonyl fluoride (PMSF): Dissolve 17.4 mg PMSF (Sigma) in 1 ml ethanol, dissolve at 65°, and keep on ice. Because PMSF is toxic, wear gloves. Prepare fresh each time.
5. Buffer 1: 575 m*M* sucrose, 25 m*M* KCL, 50 m*M* triethanolamine hydrochloride (pH 7.5), 5 m*M* $MgCl_2$, 1 m*M* DTT, and 0.5 m*M* PMSF. For 100 ml: 23 ml 2.5 *M* sucrose, 1.25 ml 2 *M* KCl, 5 ml 1 *M* triethanolamine hydrochloride (pH 7.5), 0.5 ml 1 *M* $MgCl_2$, 0.1 ml 1 *M* DTT, 0.5 ml 0.1 *M* PMSF, and water to 100 ml. Prepare fresh and keep on ice until use.
6. Buffer 2: 2.3 *M* sucrose, 0.025 *M* KCl, 50 m*M* triethanolamine hydrochloride (pH 7.5), 5 m*M* $MgCl_2$, 1 m*M* DTT, and 0.5 m*M* PMSF. For 25 ml: 23 ml 2.5 *M* sucrose, 310 μl 2 *M* KCl, 1.25 ml 1 *M* triethanolamine hydrochloride (pH 7.5), 125 μl 1 *M* $MgCl_2$, 25 μl 1 *M* DTT, 125 μl 100 m*M* PMSF, and water to 25 ml. Prepare fresh and keep on ice until use.
7. Buffer A: 0.1 *M* NaCl, 0.01 *M* $MgCl_2$, 1% Triton X-100. For 100 ml: 2.5 ml 4 *M* NaCl, 1 ml 1 *M* $MgCl_2$, 1 ml Triton X-100 (BDH), and water to 100 ml.

8. Buffer B: 2 *M* NaCl, 0.2 m*M* $MgCl_2$, and 0.01 *M* Tris–HCl, pH 7.4. For 100 ml: 50 ml 4 *M* NaCl, 20 μl 1 *M* $MgCl_2$, 1 ml 1 *M* Tris–HCl pH 7.4, and water to 100 ml.
9. PBS (×10): 1.37 *M* NaCl, 26 m*M* KCl, 100 m*M* Na_2HPO_4, and 17.6 m*M*olar KH_2PO_4. For 1 liter: 80 g NaCl, 2 g KCl, 14.4 g Na_2HPO_4, and 2.4 g KH_2PO_4. Dissolve in 800 ml water, adjust to pH 7.4 with HCl, and make up to 1 liter with water.
10. Blocking solution: 5% Marvel in Phosphate-buffered saline (PBS). For 100 ml: 5 g fat-free dry milk powder (Marvel), 10 ml 10× (PBS), make up to 100 ml with water.
11. DNase 1: 10 mg/ml in 0.15 *M* NaCl. For 10 ml, dissolve 100 mg DNase I (Sigma) and 375 μl 4 *M* NaCl in 10 ml water, aliquot, and store at −20°.

Isolation of Nuclei from Total Brain Based on a Protocol by Hallberg[17]

1. Dissect brain from a transgenic or control mouse, weigh, and homogenize immediately or place in a bijou tube and freeze immediately in liquid N_2. Tissues can be stored for long term at −70°.
2. Weigh the mouse brain in an ice-cold weighing boat and transfer to a 5-ml homogenizer (kept on ice). An adult mouse brain weighs approaching 400 mg, whereas a brain from a symptomatic R6/2 transgenic mouse is approximately 20% smaller.[5]
3. Add 1 ml of ice-cold buffer 1 per 400 mg of brain tissue and homogenize using 5–10 strokes of the pestle. Adjust the final concentration of DTT in the homogenate to 5 m*M* (5 μl of 1 *M* DTT per 1 ml of homogenate).
4. Centrifuge at 800*g* for 15 min at 4°. Remove the supernatant carefully (cytoplasmic fraction) and transfer to an ice-cold tube. Aliquot and freeze immediately on dry ice if to be stored at −70°.
5. Add 1 ml buffer 1 to the loose pellet and homogenize with five strokes of the pestle. Add a further 1 ml of buffer 1 (total volume 2.1 ml).
6. Add 2 volumes (4.2 ml) of buffer 2 and mix gently by inversion until the solution becomes uniform (total volume 6.7 ml).
7. Label an SW41 tube (Beckman) and leave to cool on ice. Add 0.5 ml buffer 2 to the bottom of the tube to act as a sucrose cushion. Gently lay the homogenate (6.7 ml) over the cushion and balance the tube. Spin at 124,000*g* for 1 hr at 4°. Remove the supernatant, cut off the top of the tube, and gently resuspend the pellet in 0.5 ml

[17] E. Hallberg, in "Cell Biology: A Laboratory Handbook" (J. E. Celis, ed.), p. 613. Academic Press, San Diego.

buffer 1 on ice. At this stage the pellet is a white, creamy color and difficult to resuspend.

8. Transfer to a cold microfuge tube. Spin at 800g for 15 min at 4°. Take off the supernatant and resuspend the pellet in 100 μl of buffer 1.
9. Check the quality of the nuclei preparation under the microscope and estimate the total number of nuclei recovered by counting with a hemocytometer. Approximately 15–20 million are frequently obtained from one brain. The nuclear and cytoplasmic fractions can be used for SDS–PAGE analysis and Western blotting or for use in the filter retardation assay.

Preparation of Aggregates

1. Use approximately 2×10^6 nuclei per assay. Make the volume up to 100 μl with buffer A containing 0.5 mM PMSF and 0.5 mg/ml DNase I. Mix well and leave at room temperature for 20 min.
2. Centrifuge at 100,000g for 30 min at room temperature (41000 rpm, Beckman TLA-45 rotor for the bench-top ultracentrifuge). Remove the supernatant carefully and resuspend the pellet in 100 μl buffer B. Add 24 μl of 10% SDS (final concentration 2% SDS), mix well, and heat at 98° for 3 min.
3. Centrifuge at 100,000g for 30 min at room temperature and resuspend the pellet in 100 μl water.
4. Centrifuge at 100,000g for 30 min at room temperature, remove supernatant, resuspend the pellet in 25 μl water, and use in the assay immediately, do not store.
5. Prewet a 3MM Whatman (Clifton, NJ) paper and the cellulose acetate membrane (0.2 μm, Schleicher & Schuell, Keene, NH) in water. Place the Whatman paper over the base of the dot-blot apparatus and then the cellulose acetate membrane. Assemble the dot-blot apparatus. Be careful not to overtighten or handle the membrane more than necessary as this may damage the membrane and cause artifacts in the immunoblotting step.
6. Add 200 μl of 0.1% SDS to each test well to equilibrate the membrane. Draw the solution through the membrane under vacuum (approximately 10 mbar) for 2 min.
7. Heat the 25-μl solution containing the aggregates at 98° for 5 min. We do this by transferring to a 0.2-ml PCR tube and heating in the PCR machine with a hot bonnet to prevent evaporation.
8. Transfer each sample to the corresponding dot-blot well with a Gilson pipette. It is important to avoid both generating air bubbles and touching the membrane with the pipette tip. Draw the samples through the membrane under vacuum as before.

9. Wash the sample wells with 200 μl of 0.1% SDS under vacuum and repeat this wash five times.
10. Dismantle the apparatus, mark the wells and the samples, and wash the membrane in PBS. The membrane is now handled like a Western blot.

Immunoblotting of Retardation Assay Filters

11. Transfer the filter to the blocking solution and leave at room temperature for 1 hr or at 4° overnight.
12. Transfer the filter to a solution of 0.5% fat-free dried milk (Marvel) in PBS containing the primary antibody at the appropriate dilution and incubate with rocking for 1–2 hr at room temperature.
13. Wash the membrane in PBS containing 0.2% Tween 20 (Sigma) for 10 min with rocking and repeat this wash four more times.
14. Dilute the secondary antibody according to the ECL protocol (enhanced chemiluminescence Amersham Life Sciences ECL kit) and repeat the washes as discribed earlier.
15. Rinse the membrane twice in PBS.
16. Incubate the membrane in ECL reagents according to the manufacturer's recommendations for 1 min. Drain the membrane to remove the excess reagents, enclose in Saran wrap, and expose to X-ray film. Develop the autoradiogram by standard procedures.

Detection of Inclusions in Skeletal Muscle

In all cases, despite wide or ubiquitous expression patterns, the expansion of a CAG/poly(Q) tract into the pathogenic range causes a neurodegenerative disease. This was also found to be true when a poly(Q) expansion was introduced into the mouse *hprt* gene by gene targeting and expressed in an ectopic context.[18] The neurodegenerative nature of these diseases is likely to arise as the consequence of the terminally differentiated status of neurons. A seeding and aggregation model[19] would predict that the absence of divisions allows the critical concentration necessary for nucleation to be achieved. Consistent with this line of reasoning, nuclear inclusions have been detected in skeletal muscle, also terminally differentiated, from R6/2 transgenic mice. The presence of inclusions in muscle is important as it may be possible to use these to assess the ability of pharmaceutical compounds to

[18] J. M. Ordway, S. Tallaksen-Greene, C.-A. Gutekunst, E. M. Bernstein, J. A. Cearley, H. W. Wiener, L. S. Dure IV, R. Lindsey, S. M. Hersch, R. S. Jope, R. L. Albin, and P. J. Detloff, *Cell* **91,** 753 (1997).

[19] P. T. Lansbury, *Neuron* **19,** 1151 (1997).

inhibit aggregation without an initial requirement that they can cross the blood–brain barrier.

Reagents. In all cases, water is deionized and distilled.

1. 0.4 *M* sodium phosphate buffer (pH 7.4): 23.72 g $NaH_2PO_4 \cdot 2H_2O$ and 92 g Na_2HPO_4. Make up to 2 liters with water and filter sterilize.
2. 4% paraformaldehyde: Dissolve 40 g of paraformaldehyde (TAAB Laboratories Equipment Ltd.) in 500 ml water by heating to 65° with five to six NaOH pellets and filter. Add 250 ml 0.4 *M* sodium phosphate buffer and then water to 1 liter.
3. Blocking solution: 3% bovine serum albumin (BSA) 0.1 *M* Tris–HCl (pH 7.4), 0.1% sodium azide. For 100 ml: 3 g BSA (Sigma), 10 ml of 1 *M* Tris–HCl, pH 7.4, 1 ml 10% sodium azide (Sigma) in 0.1 *M* sodium phosphate buffer and make up to 100 ml with water.
4. Buffer for primary antibody: 0.9% NaCl, 0.1 *M* Tris–HCl, pH 7.4, 0.02% sodium azide, and 0.3% Triton X-100. For 100 ml: 10 ml 1 *M* Tris–HCl, pH 7.4, 0.9 g NaCl, 1 ml sodium azide (2% solution in sodium phosphate buffer 0.1 *M*), Make up to 100 ml with water and add 300 μl Triton X-100. Stir to mix.
5. ABC reagent: ABC kit, Vector Laboratories (rabbit IgG). 1 ml ABC reagent: 10 μl reagent A (contains avidin), 10 μl reagent B (contains biotinylated horseradish peroxidase), 980 μl 0.1 *M* Tris–HCl, pH 7.4.
6. DAB reagent: For 100 ml, dissolve 25 mg 3,3′-diaminobenzidine (DAB; Sigma) in 100 ml 0.1 *M* Tris–HCl, pH 7.4, containing 33 μl hydrogen peroxide (Sigma). This solution should be colorless; if it is pale brown, reprepare the solution.
7. Methyl green stain: dissolve 1 g methyl green (Sigma) in 100 ml 0.1 *M* sodium acetate, pH 4.2. Extract with chloroform to remove impurities. Add 100 ml chloroform to 100 ml 1% methyl green, mix well, and allow the phases to separate in a separation funnel. Repeat until the organic phase is clear (three extractions is generally sufficient).

Procedure

1. Dissect the quadriceps muscle and transfer to an aluminum foil cup containing a little of the frozen specimen embedding medium [Cryomatrix (Shandon), Life Science International Ltd.]. Position the muscle in the desired orientation (transverse or longitudinal) and cover with embedding medium, being careful to avoid air bubbles. Allow the muscle to rest for 5 min.
2. Cool isopentane in a Pyrex beaker until it has half-solidified by standing the beaker in a liquid N_2 bath (in a polystyrene container).

Freeze the muscle specimen by steadily lowering the embedded muscle into the isopentane (this minimizes the formation of ice crystal artifacts in the myofibers). Do not immerse too quickly as the block may crack. Transfer to a universal tube and maintain on dry ice. Specimens can be stored at −70° until they are required for sectioning.

3. Prepare silane-coated microscope slides as follows:

   a. 250 ml methanol for 5 min
   b. 250 ml 2% APES* [3-aminopropyltriethoxysilane (Sigma)] in methanol for 5 min
   c. 250 ml water, a good rinse
   d. 250 ml water, a good rinse (change the water after every four racks)
   e. 250 ml methanol, final rinse
   f. Allow slides to air dry under the fume hood

4. Cut 8-μm-thick sections in a cryostat and position on slides. Allow to air dry on the bench or under a fan. At this stage, slides can be stored for up to 6 months at −70°.
5. Fix the tissue in 4% paraformaldehyde for 30 min.
6. Rinse with water twice and transfer to 0.1 *M* Tris–HCl (pH 7.4) for 15 min.
7. Tap off excess solution and ring the sections with a water-repellent pen (Dako). Do not allow the sections to dry.
8. Place the slide in a humidity chamber (e.g., petri dish with wet 3MM Whatman paper) and cover the specimen with blocking solution for 15 min.
9. Dilute the primary antibody in the primary antibody buffer. Tip off the blocking solution and cover the specimen with the primary antibody. Place in the humidity chamber, cover, and incubate for 1–2 hr at room temperature or overnight at 4°.
10. Wash the slides in 0.1 *M* Tris–HCl (pH 7.4) for 30 min.
11. Dilute the biotinylated secondary antibody 1 : 200 in 0.1 *M* Tris–HCl, pH 7.4. Cover the specimen with the secondary antibody and incubate for 30–90 min in the moisture chamber at room temperature. Meanwhile, prepare the ABC reagent.
12. Wash the slides in 0.1 *M* Tris–HCl, pH 7.4, for 15–30 min.
13. Incubate the slides with ABC for 30–90 min.
14. Wash the slides in 0.1 *M* Tris–HCl (pH 7.4) for 15–30 min during which time prepare the DAB solution.

* APES is very toxic. Wear gloves and work in a fume hood.

15. Transfer the slides to the DAB solution and allow 10 min for the reaction.
16. Quench the reaction by immersing the slides in 0.1 *M* Tris–HCl, pH 7.4, for 15–30 min.
17. Counterstain nuclei using 1% methyl green in acetate buffer for 5 min.
18. Give three quick rinses in water (overextensive washing will easily remove the stain). Allow the slides to completely air dry for 30–60 min.
19. Transfer the slides to Histoclear (National Diagnostic Lab.), incubate for 15 min, and replace with fresh Histoclear. The slides can remain in this solution until mounting.
20. Mount the slides with DPX mounting medium (Fisher Scientific). Avoid trapping air bubbles between the slide and cover glass. Leave to dry on the bench overnight, and peel away excess mounting medium the following day.

Acknowledgements

This work was supported by grants from the Wellcome Trust, the Deutsche Forschungsgemeinschaft, the European Union (BMH4 CT96 0244), the Hereditary Disease Foundation (in the form of an award to GB from Harry Lieberman), the Huntington's Disease Society of America, and the Special Trustees of Guy's Hospital.

[44] A Mouse Model for Serum Amyloid A Amyloidosis

By MARK S. KINDY and FREDERICK C. DE BEER

Introduction

Amyloid A (AA) amyloidosis is associated with rheumatic diseases, certain chronic inflammatory diseases, chronic infections, and some malignant disorders. Crohn's disease and rheumatoid arthritis (particularly juvenile rheumatoid arthritis) are frequently complicated with amyloidosis[1]; whereas, in ulcerative colitis and systemic lupus erythematosis, amyloidosis is a rare confounding problem. Differences seen in these disease states are most likely due to altered regulation in the cytokine-mediated changes in amyloid precursor protein serum amyloid A (SAA) production. Renal

[1] M. A. Gertz and R. A. Kyle, *Medicine* (*Baltimore*) **70,** 246 (1991).

0076-6879/99 $30.00

carcinomas and Hodgkin's disease are known to be associated with fevers and systemic inflammatory features leading to serum amyloid A induction. AA amyloid incidence has decreased considerably since the early 1980s as a result of therapeutic interventions that reduce inflammation.[2] Nevertheless, our capability to maintain individuals suffering from various types of trauma has contributed to an increase in systemic AA amyloidosis, particularly due to chronic urinary tract infections. Idiopathic inflammatory disease (rheumatoid arthritis, juvenile rheumatoid arthritis, and chronic infections) is the leading cause of AA amyloidosis in the Western Hemisphere. Generally, 50% of patients with amyloidosis succumb to the disease within the first 5 years after diagnosis and an additional 25% between 5 and 15 years.

The deposition of AA amyloid involves multiple organ systems, usually without any initial symptomatic presentations. Approximately 25% of individuals with amyloidosis ultimately present with renal or glomerular involvement.[3] The typical clinical presentation is proteinurea due to the deposition of amyloid fibrils in the glomerulus. This results in a nephrotic syndrome leading to renal insufficiency or end-stage renal failure, the most common cause of death in these patients. Deposits can usually be detected in the spleen with some involvement of the liver, adrenals, and, to a lesser degree, in the heart, gut, and other tissues. Dialysis and renal transplantation can exacerbate the deposition of amyloid fibrils in other organs.[4] Therefore, there is a growing need to understand the process of systemic amyloid deposition and its complications.

The mouse amyloid model provides an ideal system for understanding amyloid fibrillogenesis.[5] Serum amyloid A proteins are a group of highly homologous apoproteins associated with high-density lipoproteins (HDLs) and are the precursors of AA amyloid.[6] Serum amyloid A proteins are the most dramatically increased acute-phase proteins in mice with levels increasing up to 1000-fold (from 1–5 to 1000 μg/ml) within 24 hr of an inflammatory response.[7] In the mouse, SAA_1 and SAA_2 are induced in the liver by circulating cytokines [interleukin-1 (IL-1), IL-6, and tumor necrosis factor (TNF)] in response to amyloid induction paradigms (Fig. 1). These two isotypes differ in only 9 of 103 amino acids; however, only SAA_2 is

[2] M. Pras, *Scand. J. Rheumatol.* **27,** 92 (1998).

[3] A. Montoli, E. Minola, F. Stabile, C. Grillo, L. Radaelli, D. Spanti, E. Luccarelli, C. Spata, and L. Minetti, *Am. J. Nephrol.* **15,** 142 (1995).

[4] S.-Y. Tan, A. Irish, C. G. Winearls, E. A. Brown, P. E. Gower, E. J. Clutterbuck, S. Madhoo, J. P. Lavender, and M. B. Pepys, *Kidney Int.* **50,** 282 (1996).

[5] R. L. Meek, J. S. Hoffman, and E. P. Benditt, *J. Exp. Med.* **163,** 499 (1986).

[6] A. Husebekk, B. Skogen, and G. Husby, *Scand. J. Immunol.* **25,** 375 (1987).

[7] K. P. W. J. McAdam, and J. D. Sipe, *J. Exp. Med.* **144,** 1121 (1976).

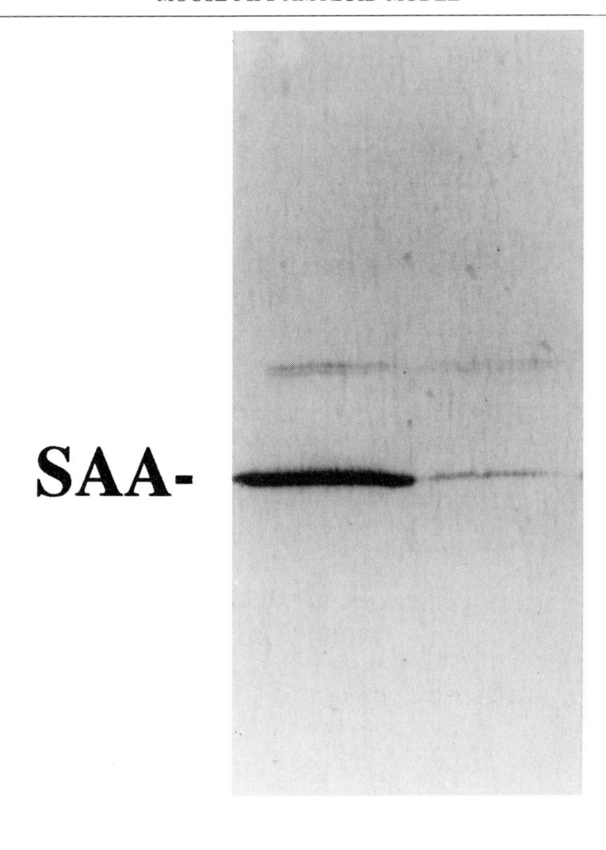

FIG. 1. Induction of SAA proteins following amyloid-enhancing factor (AEF) and silver nitrate (SN) injection. Immunocytochemical staining of apoSAA in C57BL6 mice: 10 μg of plasma electrophoresed onto a 5–20% acrylamide gel; lane 1, plasma from mice 24 hr after injected with AEF/SN; lane 2, plasma from normal mice.

selectively deposited into amyloid fibrils (Fig. 2). SAA expression in the CE/J mouse species is an exception in that gene duplication did not occur and the CE/J variant is a hybrid molecule sharing features of SAA_1 and SAA_2 (Fig. 2).[8] Even though the CE/J protein overall has features similar to the SAA_2 protein, it is not deposited into amyloid fibrils. Therefore, the

[8] M. C. de Beer, F. C. de Beer, W. D. McCubbin, C. M. Kay, and M. S. Kindy, *J. Biol. Chem.* **268,** 20606 (1993).

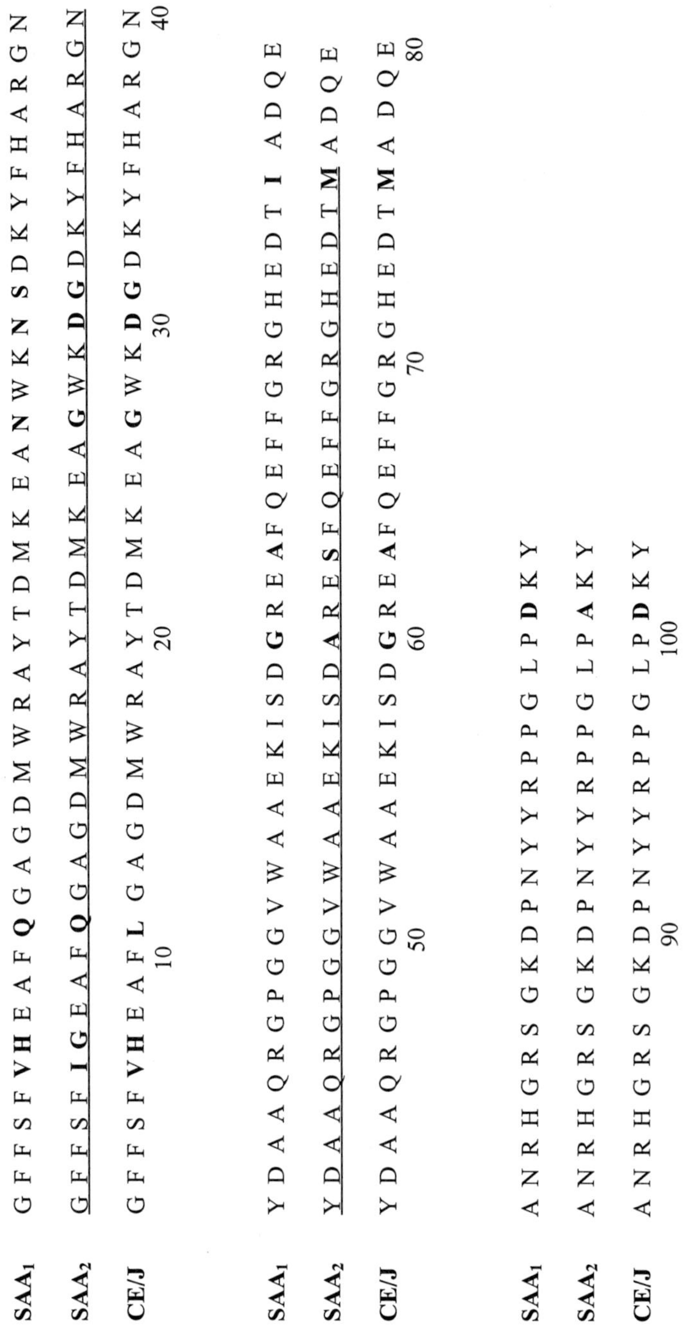

FIG. 2. Amino acid sequences of mouse serum amyloid A proteins. Mouse SAA_1, SAA_2, and CE/J protein sequences are indicated. The first amino acid is +1 from the mature protein. Boldface letters indicate differences between the individual mouse proteins. The underbar on SAA_2 indicates the predominant AA fragment found in mouse amyloid fibrils. As shown in the figure, the predominant peptide that is found in mouse AA amyloid is a 76 amino acid fragment of the SAA_2 protein.

CE/J mouse does not develop amyloid.[9] Differences in the CE/J protein that confer amyloid resistance appear to be at the amino acid positions 6 and 7, where it shows similarity with the nonamyloidogenic SAA_1 protein (Fig. 2).[8,10] Linkage analysis with amyloid-resistant mice (CE/J) and amyloid-susceptible mice (CBA/J) demonstrated that the CE/J isoform confers protection against amyloid deposition even in the presence of the SAA_2 isoform.[11] These data indicate valuable information that can be obtained from inbred strains of mice on the structural properties and specific interactions that may occur in the process of amyloidogenesis.

Our current understanding of the processes involved in AA amyloidosis is that it is the result of a two-stage mechanism (Fig. 3).[12] First, the generation of AA amyloid requires the synthesis of a precursor protein in sufficient quantities to allow for deposition. This is referred to as the preamyloid phase, which can persist for several days to months, depending on the stimulus and the levels of SAA expression. In response to a chronic or recurrent acute inflammatory condition, macrophage activation leads to the synthesis and secretion of cytokines (IL-1, IL-6, and TNF). These cytokines result in the hepatic synthesis of acute-phase proteins, specifically SAA proteins that are secreted and associated with HDL particles.[7] Second, in the amyloid phase, the continuous maintenance of high levels of the precursor is required to generate a nidus or fibrillar network (also referred to as amyloid-enhancing factor, AEF) onto which amyloid can be deposited.[13] The production of AEF results in the rapid deposition of AA fibrils, which can occur in days. Amyloid deposition first occurs in the spleen, liver, and kidney, and eventually amyloid can be detected in every organ, with the possible exception of the brain. However, studies have demonstrated the presence of SAA proteins in the brains of patients with Alzheimer's disease.[14,15]

Amyloid deposition in the mouse is extremely rapid, with extensive deposits detectable following amyloid induction paradigms. Injection of a modified casein solution results in amyloid deposition in various organs in

[9] J. D. Sipe, I. Carreras, W. A. Gonnerman, E. S. Cathcart, M. C. de Beer, and F. C. de Beer, *Am. J. Pathol.* **143,** 1480 (1993).

[10] H. Patel, J. Bramall, H. Waters, M. C. de Beer, and P. Woo, *Biochem. J.* **318,** 1041 (1996).

[11] W. A. Gonnerman, R. Elliot-Bryant, I. Carreras, J. D. Sipe, and E. S. Cathcart, *J. Exp. Med.* **181,** 2249 (1995).

[12] I. Kushner and D. L. Rzewnicki, *Clin. Rheumatol.* **8,** 513 (1994).

[13] M. A. Axelrad, R. Kisilevsky, J. Willmer, S. J. Chen, and M. Skinner, *Lab. Invest.* **47,** 139 (1982).

[14] J. S. Liang, J. A. Sloane, J. M. Wells, C. R. Abraham, R. E. Fine, and J. D. Sipe, *Neurosci. Lett.* **225,** 73 (1997)

[15] M. S. Kindy, J. Yu, J. T. Gao, and H. Zhu, *J. Alzheimer's Dis.,* in press (1999).

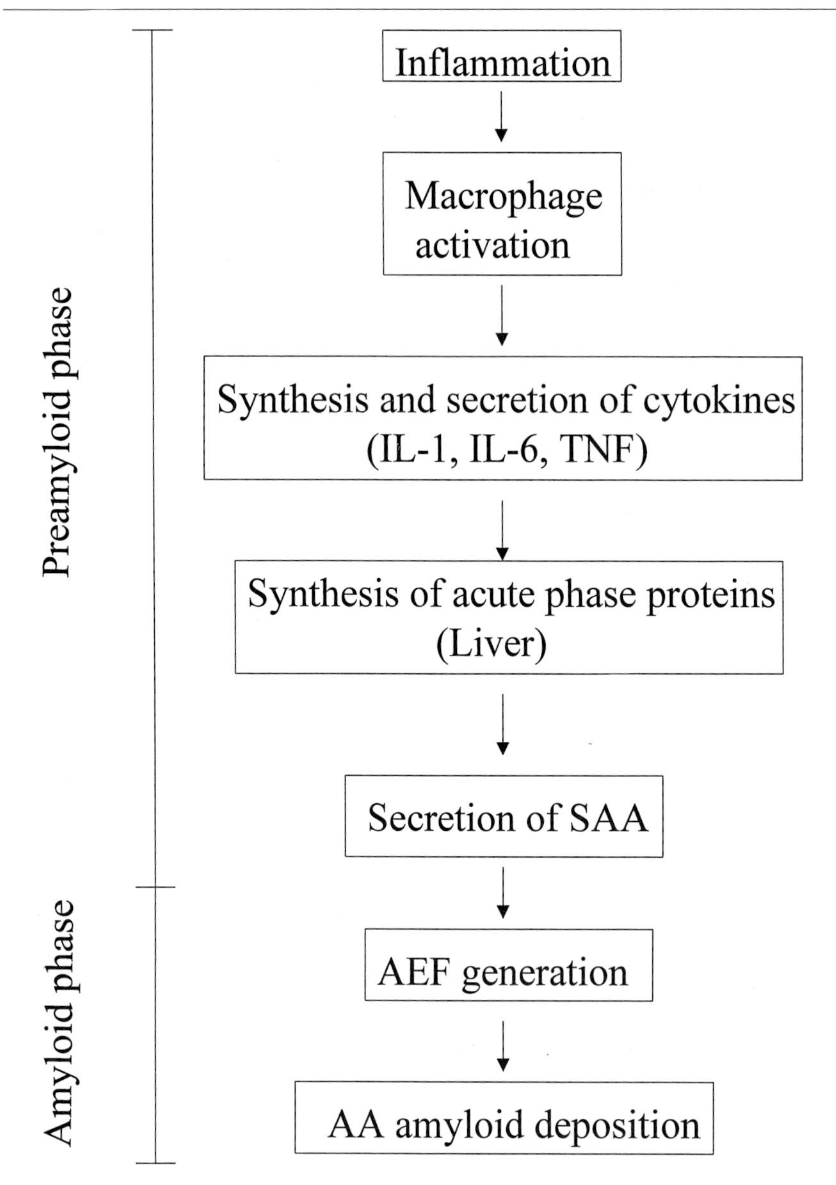

FIG. 3. Flow diagram of the events in amyloidogenesis. Sequence of events in amyloid formation in both mouse and human. Generation of an inflammatory response through the injection of casein in the mouse or infection in human results in the activation of a pathway leading to the deposition of amyloid fibrils. The preamyloid phase culminates in the synthesis and secretion of SAA. The amyloid phase originates from the generation of amyloid-enhancing factor (AEF; preformed amyloid fibrils, proteoglycans, etc.) to the deposition of amyloid fibrils in the tissue.

21 days after initiation. The use of AEF and silver nitrate has reduced this time to 3 days.[13] In addition, mouse AA amyloid has serum amyloid P component (SAP) and heparan sulfate proteoglycan (HSPG) associated with the amyloid fibrils.[16,17] Studies have shown that the mouse apolipoprotein E (apoE) is associated with mouse AA amyloid.[18] Kisilevsky *et al.*[19] showed that a series of small anionic sulfates and sulfonates interfered with the SAA–HSPG interaction and were effective in not only inhibiting amyloidogenesis but in reversing the process *in vivo.* These studies indicate the importance of HSPG in the events associated with amyloid formation. The generation of SAP and apoE knockout mice showed that inactivation of these genes resulted in delayed and reduced deposition of amyloid.[20,21] As suggested earlier, the involvement of accessory factors in the evolution of amyloid diseases needs to be examined as a possible mechanism in the development of therapeutic strategies for treatment of these diseases.

This article describes the methods used in the generation of AA amyloid in the mouse as a model of amyloid disease. Several different protocols have been used in the deposition of AA amyloid in mice. The first requires the continuous injection of a modified casein solution over an extended period of time to initiate amyloid deposition. The second utilizes AEF and silver nitrate to give extensive amyloid deposits within 3–5 days.

Induction of AA Amyloid in Mouse

Initial studies on amyloid induction in the mouse were performed using casein and azocasein with multiple injections over a several week period.[22] For induction of amyloid with casein or azocasein, 0.5 ml of a prepared 10% solution of sterile azocasein is injected subcutaneously daily for up to 21 days (or sometimes the injections are performed 5 days a week for up to 4 weeks). Injection of the azocasein results in a sterile abscess, which will cause an inflammatory reaction inducing SAA in the mouse. Using this paradigm, significant amounts of amyloid develop over the 21-day period. Initial deposition does not begin for several days after the injections

[16] M. B. Pepys, *in* "Amyloidosis" (J. Marrink and M. van Ryswijk, eds.), p. 43. Martinus Nijhoff, Boston, 1986.

[17] A. D. Snow, J. Willmer, and R. Kisilevsky, *Lab. Invest.* **56,** 170 (1987).

[18] M. S. Kindy, A. R. King, G. Perry, M. C. de Beer, and F. C. de Beer, *Lab. Invest.* **73,** 469 (1995).

[19] R. Kisilevsky, L. J. Lemieux, P. E. Faser, X. Kong, P. G. Hultin, and W. A. Szarek, *Nature Med.* **2,** 143 (1996).

[20] M. Botto, P. N. Hawkins, M. C. M. Bicerstaff, J. Herbert, A. E. Bygrave, A. McBride, W. L. Hutchinson, G. A. Tennent, M. J. Walport, and M. B. Pepys, *Nature Med.* **3,** 855 (1997).

[21] M. S. Kindy and D. J. Rader, *Am. J. Pathol.* **152,** 1387 (1998).

[22] D. T. Janigan, *Am. J. Pathol.* **55,** 379 (1969).

have commenced, delaying the process. This is due to the synthesis and initial deposition of the amyloid fibrils onto which amyloid can form.

Subsequently, amyloid is now induced in the mouse model by the injection of AEF (100 μg protein) intravenously via the tail vein followed by a subcutaneous injection of 0.5 ml of a 2% solution of silver nitrate (sterile preparation in water).[23] The AEF is injected into the tail vein as follows: the mouse is placed in a mouse holder and the tail is heated for 5 to 10 sec in a water bath or beaker of water heated to 50°. This will bring the tail vein to the surface and allow for easier injection of AEF. The tail is removed from the water bath and swabbed with alcohol to sterilize and then AEF (100 μg in 100 μl) is injected with a 29-gauge tuberculin syringe into the tail vein. Immediately following the tail vein injection, the animals are treated with a subcutaneous injection of sterile silver nitrate (0.5 ml of a 2% solution). This allows for extensive deposition of amyloid within 3–5 days. Studies have shown that in the AEF and silver nitrate injection paradigm, amyloid deposits can be detected as early as 12 hr posttreatment. This allows for a rapid model of amyloid formation to test the effects of gene mutations on amyloid formation and the effects of inhibitors of amyloid deposition.

AA amyloid deposits first develop in the spleen, possibly due to the metabolic effects of the spleen. Secondary deposition occurs in the liver, heart, kidneys, lungs, intestine, and every major organ except the brain. If the inflammatory stimulus is maintained for extended periods of time, the amyloid deposits will continue to expand, resulting in organ failure and death.

Amyloid-Inducing Agents

Preparation of Casein

Various experimental protocols have been utilized to induce amyloid deposition in animals over the years. It was originally assumed that amyloidosis was due to chronic infections or the presence of bacterial toxins. Subsequent to these findings, it was shown that the injection of sodium caseinate gave rise to experimental amyloidosis in mice. The hypothesis was that the presence of large quantities of protein material led to insolubility and deposition of the amyloid substance. Since that time, a modified form of casein, azocasein, has been used extensively for the generation of amyloid in the mouse model.[24] The preparation of azocasein is time-

[23] M. Axelrad, R. Kisilevsky, and S. Beswetherick, *Am. J. Pathol.* **78,** 277 (1975).
[24] D. T. Janigan and R. L. Druet, *Am. J. Pathol.* **48,** 1013 (1966).

consuming and there can be variability if the proper casein is not used in the preparation protocol. Hammerstein-grade casein from ICN (Costa Mesa, CA) has been used successfully in the production of azocasein that will consistently give amyloid deposits in mice.

A solution of sulfanilic acid (20 g, sodium salt, Sigma Chemical Co., St. Louis, MO) is prepared in 400 ml of cold distilled-deionized water containing 12 ml of 5 *N* NaOH. To this solution, 4.4 g of $NaNO_3$ is added and allowed to dissolve. Next, 32 ml of 5 *N* HCl, followed by 32 ml of 5 *N* NaOH, is added and the solution is stirred continuously in the cold. Casein (100 g) is dissolved in 2 liters of a 1% (w/v) $NaHCO_3$ solution prepared in distilled-deionized water and cooled on ice to 4°. The casein solution is added slowly to the sulfanilic acid to allow the casein to remain in solution. If the casein is added too rapidly, it will precipitate out of solution. The combination will turn a deep red color once the casein goes into solution and should be mixed for 2–3 hr in the cold. The azocasein is precipitated out of solution by bringing the pH to 2–3 with the addition of 0.2 *N* HCl, added slowly to allow for a flocculent precipitate to form. The precipitate (yellowish in color) can be filtered through cheesecloth in a large filter funnel. The azocasein is dissolved in a large volume of 0.2 *N* NaOH (1 liter) and is subjected to two additional rounds of precipitation and solubilization. The final precipitate is resuspended in 0.2 *M* $NaHCO_3$ and dialyzed in 10- to 12-kDa molecular mass dialysis tubing for 3 days against cold distilled-deionized water, changing the water twice a day. The azocasein solution is lyophilized completely (several days) and stored at −80° as a powder. For injection of the azocasein, a 10% solution in 0.01 *M* $NaHCO_3$ (5 g in 50 ml) is prepared in 50-ml polypropylene tubes and adjusted to a final pH of 6.5. The solution is sterilized by the addition of 1% diethyl pyrocarbonate.

Preparation of Amyloid-Enhancing Factor

Over time, amyloid deposition was considered to be a result of or related to an immunological response in human and animal models. The general principle has been that the subcutaneous injection of proteinaceous material would cause an immunologic reaction and result in amyloid formation. The involvement of the immune system in the amyloidogenic process suggested the presence of factors that may contribute to amyloid deposition. Transfer experiments using tissue and cells from amyloidogenic animals demonstrated the presence of an enhancing factor in the amyloid tissue.[25,26] Subse-

[25] F. Hardt and P. Ranlov, *Int. Rev. Exp. Pathol.* **49,** 273 (1976).

[26] I. Keizman, A. Rimon, E. Sohar, and J. Gafni, *Acta Pathol. Microbiol. Scand.* (*A*) **233,** 172 (1970).

quent studies by Kisilevsky and co-workers[27] have shown that the amyloid tissue contained amyloid-enhancing activity, which facilitated the deposition of amyloid in the mouse. Extraction of AEF from amyloidotic tissue (spleens and livers) and subsequent injection into mice with an inflammatory stimulus reduce the time of amyloid deposition from 4 weeks to 3 days. The composition of AEF is still not known; however, it is thought to be composed of preformed amyloid fibrils and proteoglycans. Studies have strengthened this hypothesis. Injection of amyloidogenic peptides from SAA and transthyretin proteins, as well as from Aβ peptides, resulted in the rapid accumulation and deposition of AA amyloid.[28,29] The utilization of SAA and other amyloidogenic peptides indicates that the amyloidogenic activity of AEF is due to amyloid fibrils. AEF appears to act as a seed or nucleus onto which the amyloid can form.

Amyloid-enhancing factor is prepared from the spleens and livers of animals that were injected with azocasein or AEF and silver nitrate to establish amyloid deposits, as described previously.[30] Animals injected with casein for 4 weeks or AEF and silver nitrate for 2 weeks are taken for AEF preparation. Using sterile techniques, the spleens and livers are removed from the amyloidotic mice (about 25 animals) and homogenized in 10 volumes of phosphate-buffered saline (PBS) containing protease inhibitors [1 m*M* phenylmethylsulfonyl fluoride (PMSF), 1 μM leupeptin; 5 μg aprotinin/ml] using a homogenizer (Polytron, Brinkmann Instruments, Westbury, NY). The tissue is homogenized on a low setting until a homogeneous suspension is obtained. The homogenate is centrifuged at 10,000 rpm for 20 min at 4° in a superspeed centrifuge. The pellet is resuspended in 10 volumes of PBS plus protease inhibitors (low-speed homogenization to disrupt the pellet) and centrifuged repeatedly until the optical density of the supernatant at 280 nm is less than 0.1. The final pellet is homogenized/extracted with sterile/distilled water (10 times the volume of the pellet). The sample is centrifuged at 16,000 rpm for 1 hr at 4°. The first supernatant can be discarded. The water extractions are repeated; the second and third supernatants are saved and quantified for protein content. Samples of AEF are stored frozen at −80° and tested for activity as described previously.

Detection of AA Amyloid

Several methods have been used in the detection of AA amyloid. The most frequently used method is Congo red staining in which the Congo

[27] R. Kisilevsky, M. D. Benson, M. A. Axelrad, and L. Boudreau, *Lab. Invest.* **41,** 206 (1979).

[28] K. Johan, G. Westermark, U. Engstrom, A. Gustausson, P. Jultman, and P. Westermark, *Proc. Natl. Acad. Sci. U.S.A.* **95,** 2558 (1998).

[29] M. S. Kindy, unpublished results (1999).

[30] R. Kisilevsky and L. Boudreau, *Lab. Invest.* **48,** 33 (1983).

red dye will stain amyloid structures specifically (the cross β-pleated sheet arrangement).[31] Congo red appears to bind to linear polymers that have a similar structure to cellulose.[24] Tissue sections should be fixed in 10% (w/v) buffered formalin and sections should be 6–12 μm in thickness.

Congo Red Staining

Paraffin-embedded sections are deparaffinized and hydrated to water (two changes for 5 min each in xylene, 100, 95, 80, 50, and 30% ethanol, then distilled water). The sections are stained in a 1% solution of Congo red for 1 hr at room temperature (1 g Congo red, color index number 22120 containing 98% dye [Sigma, St. Louis, MO], in 100 ml of distilled water). The Congo red solution should be stirred for several hours at room temperature to allow the dye to dissolve completely, then filtered through a Whatman (Clifton, NJ) No. 2 filter. This solution should be used within 1 month, as beyond this time the dye will not bind well to the amyloid. The sections are rinsed in running water and differentiated in alkaline alcohol (0.01% sodium hydroxide in 50% ethyl alcohol) for 3–5 sec and rinsed for 5 min in running water. It is easy to overdifferentiate the sections; if the sections are pale, repeat the Congo red staining. Counterstain with Mayer's hematoxylin for 5 min, wash for 15 min in water, dehydrate to xylene (two changes for 5 min in 50, 80, 95, and 100% ethanol followed by two changes in xylene), and mount with Permount (Fisher, Pittsburgh, PA). Results of the staining protocol are amyloid stains red to pink, nuclei stain blue, elastic fibers stain light red, and most other structures are not stained. When examined under light microscopy with polarizing filters (light-polarizing microscope filter set 31-52-62-75) the amyloid material shows an apple-green birefringence surrounding the splenic follicles (Fig. 4A). In the spleen, amyloid deposits are detected surrounding the splenic follicles, beginning as small amounts of congophilic material that accumulate over time in the presence of a continuous inflammatory response. In Fig. 4, the amyloid present in the spleen creates a dramatic effect with large deposits around the follicles. In other organs, the amyloid is more diffuse, with vascular amyloid appearing in the veins and arteries, as well as general deposits in the tissue.

Immunocytochemical Detection

A second method of staining amyloid is by immunocytochemical analysis using rabbit antimouse SAA or AA antibodies.[18] After generation of amyloid, the tissue (spleen, liver, heart, etc.) is drop fixed in formalin and embedded in paraffin. Alternatively, the animal can be perfused with cold

[31] H. Puchtler, F. Sweat, and M. Levine, *J. Histochem. Cytochem.* **10,** 355 (1962).

A

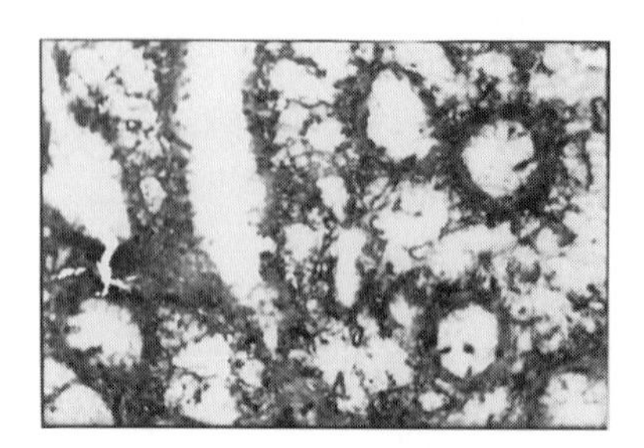

B

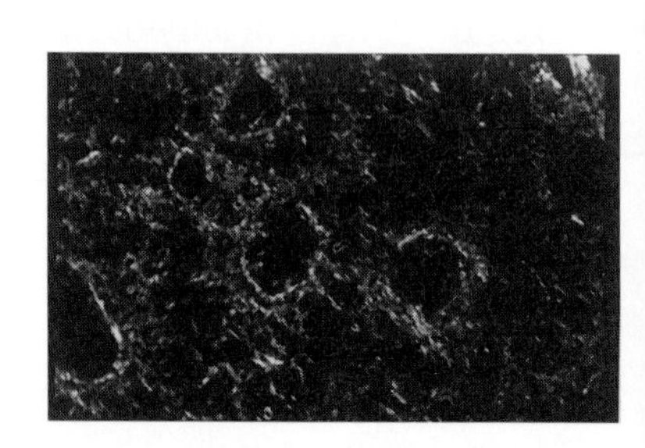

FIG. 4. Detection of AA amyloid in the spleen. (A) Congo red staining of an amyloidotic spleen and visualization using bipolar filters of apple-green birefringent material. (B) Immunocytochemical staining of amyloid. An antibody against mouse SAA proteins was used as the primary antibody. Detection was performed with a peroxidase secondary antibody and diaminobenzidine hydrochloride. Dark staining indicates AA fibrils and amyloid deposits.

saline (25 ml) followed by 4% paraformaldehyde (25 ml) and the tissues harvested. Paraformaldehyde (PF, 4 g) is suspended in saline and adjusted to a pH of 8.0 and heated to 60° to allow the PF to dissolve. The solution is cooled and adjusted to pH 7.4 for perfusion. Saline and paraformaldehyde are perfused by means of a 30-cm^3 syringe and 23-gauge needle inserted into the heart while the animal is anesthetized. After perfusion, the tissues are removed and submerged in fresh 4% (w/v) paraformaldehyde for 24 hr and then in 30% sucrose (30 g sucrose in 100 ml saline). After 24 to 48 hr, the tissue can be frozen in OCT medium (Tissue-Tek, Torrance, CA) and stored at −80°. Eight- to 10-μm sections of tissue are prepared and mounted on Superfrost Plus microscope slides (Fisher, Pittsburgh, PA) coated with 1% gelatin for immunocytochemistry. For slide preparation, 1 g of gelatin is dissolved in 100 ml water by heating and stirring. Slides are dipped in gelatin and allowed to air dry. Paraffin sections are deparaffinized and hydrated to water (as described earlier) and incubated for 3 min in 100% formic acid to help disrupt amyloid fibrils (this allows for better detection of the amyloid fibrils by exposing the epitopes on the AA for antibody binding). Sections are incubated in hydrogen peroxide (3% in methanol for 30 min) to remove any endogenous peroxidase activity in the tissue and hydrated (two changes of 5 min each) to Tris-buffered saline (TBS; 0.1 *M* Tris, 0.150 *M* NaCl, pH 7.4). Sections are incubated in 15% goat serum (in TBS) to block nonspecific sites and are incubated in the presence of the primary antibody at a dilution of 1 : 2000 (rabbit antimouse SAA antibody). After incubation at room temperature for 1 hr or overnight at 4°, the sections are washed (3 × 15 min in TBS plus 15% goat serum) and incubated with the secondary antibody (peroxidase-conjugated goat antirabbit; 1 : 5000; Sigma Chemical Co., St. Louis, MO) for 1 hr at room temperature. Sections are washed as before and are developed in enzyme buffer containing diaminobenzidine hydrochloride (50 μg/ml) until staining appears (2 to 10 min). Staining is stopped by rinsing in water and counterstaining for 10 sec in hematoxylin (Sigma, St. Louis, MO). Slides are dehydrated to xylene (reverse of process for hydration, as described earlier) and mounted with Permount. Figure 4B illustrates the presence of immunoreactive AA protein detected in the amyloid deposits. As with the Congo red staining, the splenic deposits are detected around the follicles. In animals with extensive amyloid, the deposits can extend from each follicle to overlap and appear to be a solid sheet of amyloid.

^{123}I-Labeled SAP Detection of Amyloid

Amyloid in mouse and human can be monitored specifically by using quantitative scintigraphy with ^{123}I-labeled human serum amyloid P (SAP) component or technetium-99m pyrophosphate.[32,33] SAP is used for the general detection of amyloid deposits, and pyrophosphate has been used in detecting amyloid in cardiac amyloidosis. We will focus on the SAP detection system because it is more relevant to these studies. Human SAP is isolated by affinity chromatography to 99% purity from sterile serum heated to 56° for 30 min.[34] Human sera (2 liters) is passed over a 5 × 85-cm column of Sepharose 4B equilibrated with Tris–Ca buffer (0.01 *M* Tris-buffered 0.14 *M* NaCl, pH 8.0, containing 0.002 *M* $CaCl_2$ and 0.1 g/liter NaN_3) at 100 ml/hr at 4°. The column is washed in Tris buffer until the effluent is zero at A_{280} and the bound protein elutes Tris buffer with EDTA (0.01 *M* EDTA). The eluted protein is passed over Sepharose-anti-NHS (normal human serum, 10 ml) followed by Blue-Sepharose (1 ml) to remove contaminating serum proteins. The samples are concentrated to 5–10 ml and gel filtered on a 2.6 × 100-cm column of Sephacryl S-300 (Amersham Pharmacia Biotech, Piscataway, NJ) eluted with Tris–EDTA at a flow rate of 16 ml/hr (5.8-ml fractions). Peak SAP as determined by Western blot analysis or electroimmunoassay is pooled, concentrated, and stored under liquid nitrogen. SAP is labeled with iodine-123 using *N*-bromosuccinimide and is purified by gel filtration on Sephadex G-25 (Sigma, St. Louis, MO). Approximately 100 μg of labeled SAP (550 μCi), administered by intravenous injection, will bind specifically but in a reversible fashion to amyloid fibrils *in vivo*. Mice injected with ^{123}I-labeled SAP are subjected to whole body scintigraphy. Animals are anesthetized [chloral hydrate (350 mg/kg) and xylazine (4 mg/kg)] and scanned with a Toshiba gamma camera and pinhole collimator for 1–4 min at $t = 0$ to $t = 24$ hr. The labeled SAP will

[32] P. N. Hawkins, J. P. Lavender, and M. B. Pepys, *N. Engl. J. Med.* **323,** 508 (1990).

[33] P. N. Hawkins, M. J. Myers, A. A. Epenetos, D. Caspi, and M. B. Pepys, *J. Exp. Med.* **167,** 903 (1988).

[34] F. C. de Beer and M. B. Pepys, *J. Immunol. Methods* **50,** 17 (1982).

localize to the amyloid deposits dependent on the amount of amyloid present in the tissue. Thus, one can determine the percentage of the injected dose that is retained in the body. In addition, the organs can be removed and counted in a gamma counter to determine the amount of labeled SAP that localizes to that organ (i.e., spleen). This number corresponds to the histological grade described later.

Quantification of Amyloid Load (Amyloid Deposition)

The quantification of amyloid varies dramatically from estimated semi-quantitative analysis to quantitative analysis by determining the proportion of the area of the sections occupied by amyloid material. Original analysis was performed by visual examination to determine the amount of amyloid based on a crude scale of 0 to 4.[35] The grade is as follows: grade 0, no amyloid; grade 1, trace amounts of amyloid in the perifollicular zone; grade 2, narrow ring of amyloid in the perifollicular zone; grade 3, broad bands of amyloid in the perifollicular zone; and grade 4, broad bands of amyloid with bridging between the follicles. To quantitate the amount of amyloid present in the tissue of mice, the following protocol is used.[19,21] A standard set of amyloid-containing tissues is generated (0, 10, 20, 30, 40, 50% of tissue containing amyloid). These are used as reference points to determine the amount of amyloid in a given tissue. Standard sections are examined under the microscope (Nikon, Garden City, NY; using polarizing filters to generate birefringence) and recorded on videotape (Sony, New York, NY). The images are digitized and transferred to a Macintosh computer for analysis. The digitized images are analyzed (Kontron, Prism, or MCID image analysis) for color (intensity and area, operator specified area of interest) under low power (20×). This generates a standard for comparison. Experimental tissue sections are analyzed and compared to the standards for the quantitation of amyloid. Approximately 20 sections from each spleen are examined by image analysis and averaged for amyloid quantification. Data generated from these studies are calculated as the percentage of amyloid in the tissue (i.e., the area of amyloid present in the spleen to the total area). Alternatively, sections are stained with fluorescence-labeled anti-SAA or SAP (1/1000 dilution) antibodies and images are analyzed on a confocal laser scanning microscope (Molecular Dynamics, Sunnyvale, CA) using ImageSpace ScanControl software and a SGI UNIX workstation (Summit, NJ). In most cases the spleens are analyzed for amyloid; however, we will take other tissues to verify the levels of amyloid in the mice. In addition, some animals are injected with ^{125}I-labeled SAP (labeled as described for ^{123}I-labeled SAP) for the quantitative assessment of amyloid.

[35] E. S. Cathcart, C. A. Leslie, S. N. Meydani, and K. C. Hayes, *J. Immunol.* **139,** 1850 (1987).

Animals are injected with 0.6 μCi of SAP (0.4 μCi/μg) at $t = 0$ and killed at $t = 24$ hr, the animals are bled out, and the organs weighed and subjected to gamma counting. SAP is cleared rapidly into the amyloid tissue, and quantitation by homogenization and scintillation counting is very reproducible.

Inhibition of Amyloid Fibrillogenesis

The mouse AA model is ideal for studying fibrillogenesis and developing inhibitors of fibril formation. As discussed earlier, many similarities exist between the mouse model of AA amyloid and other amyloid diseases, and this model can be used as an *in vivo* screening technique to identify compounds that may prevent or slow the process of amyloid fibril development. Several studies have shown the utility of this model. Kisilevsky *et al.*[19,36] predicted that the interactions between amyloid fibrils and HSPG were important and, by disruption of these contacts, may prevent amyloid deposition. Oral administration of low molecular weight anionic sulfates or sulfonates reduced amyloid development in the mouse model. In addition, they interfered with Aβ fibril formation *in vitro*. Many other potential compounds that may prove to be important in the inhibition of amyloid fibrillogenesis or in the regression of amyloid deposits can be tested in this model. These compounds include methyl 4,6-*O*[(*R*)-1-carboxyethylidene]β-D-galactopyranoside (M*O*βDG, interferes with SAP interaction with fibrils) and 4′-iodo-4′-deoxydoxorubicin (I-DOX, may interfere with fibril deposition), which have been shown to interfere with fibril formation and may prove to be effective *in vivo*.[37,38]

Concluding Remarks

Experimental approaches designed to characterize the mechanisms of amyloid deposition, components involved in the process, and inhibitors that block amyloid deposition should ideally include a systematic evaluation of the entire spectrum of amyloid diseases. The described methodology provides a strategy for evaluating amyloidogenesis in an *in vivo* paradigm that has most of the components present in all amyloid diseases.

Amyloidogenic diseases demonstrate a high degree of similarity and one can be used to help understand the basic principles of all the disorders. The AA model demonstrates the same effect seen in APP transgenic mice.

[36] R. Kisilevsky, *Drugs Aging* **2,** 75 (1996).

[37] G. A. Tennent, L. B. Lovat, and M. B. Pepys, *Proc. Natl. Acad. Sci. U.S.A.* **92,** 4299 (1995).

[38] L. Gianni, V. Bellotti, A. M. Gianni, and G. Merlini, *Blood* **86,** 855 (1995).

In both models, inactivation of the apoE gene resulted in a decrease in amyloid deposition.[21,39] Also, the generation of a SAP knockout mouse may show greater similarities in the two models.[20] The unique features of the AA model are the rapid development of the disease, and the ease in studying the role of specific factors and testing of drugs.

Acknowledgments

The authors thank Drs. Maria C. de Beer and Deneys van der Westhuyzen for critical reading of the manuscript, Drs. Jin Yu and Hong Wu, Mr. Darin Cecil, Ms. Amy King, Ms. Connie Gerardot, and Mr. John Cranfill for expert technical assistance during various phases of the investigations. This work was supported by National Institutes of Health Grants NS-32221 and AG-12981 (MSK), AG-10886 and the Veterans Affairs Medical Research Fund (FCD), and the Sanders-Brown Center on Aging.

[39] K. R. Bales, T. Verina, R. C. Dodel *et al., Nature Genet.* **17,** 263 (1997).

[45] Toxicity of Protein Aggregates in PC12 Cells: 3-(4,5-Dimethylthiazol-2-yl)-2,5-diphenyltetrazolium Bromide Assay

By MARK S. SHEARMAN

Introduction

Senile plaques, a neuropathological feature of Alzheimer's disease (AD), consist primarily of an insoluble aggregate of amyloid-β peptide (Aβ), a 40–43 amino acid peptide.[1] Dense core plaques of Aβ deposited in AD brain are typically surrounded by dystrophic neurites,[2] an observation that led to the proposal that Aβ itself may be neurotoxic.[3,4]

Although studied intensively in the early 1990s, the biochemical mechanisms that underlie the neurotoxicity of Aβ remain uncertain.[5] One observation that remains consistent, however, is that amyloidogenic peptides such

[1] C. L. Masters, G. Simms, N. A. Weinman, G. Multhaup, B. L. McDonald, and K. Beyreuther, *Proc. Natl. Acad. Sci. U.S.A.* **82,** 4245 (1985).

[2] D. M. A. Mann, P. O. Yates, and B. Marcyniuk, *Neurosci. Lett.* **56,** 51 (1985).

[3] R. E. Tanzi, P. H. St. George-Hyslop, and J. F. Gusella, *Trends Neurosci.* **12,** 152 (1989).

[4] D. J. Selkoe, *Neuron* **6,** 487 (1991).

[5] L. L. Iversen, R. M. Mortishire-Smith, S. J. Pollack, and M. S. Shearman, *Biochem. J.* **311,** 1 (1995).

0076-6879/99 $30.00

as Aβ, human amylin, and calcitonin—at least in the aggregated state that is associated with their neurotoxicity—potently inhibit cellular reduction of the tetrazolium redox dye 3-(4,5-dimethylthiazol-2-yl)-2,5-diphenyltetrazolium bromide (MTT).[6–10] Despite the fact that the biochemical basis for this inhibitory activity still remains unclear, it appears to be a reliable early indicator of the mechanism of amyloid peptide cytotoxicity.

Materials and Methods

Materials

MTT (M2128) and phenazine ethosulfate (PES; P4544) are from Sigma Chemical Company Ltd. (Dorset, UK) or from Lancaster Synthesis Ltd. (Lancashire, UK). 5-(3-Carboxymethoxyphenyl)-2-(4,5-dimethylthiazolyl)-3-(4-sulfophenyl)tetrazolium, inner salt (MTS) is from Promega (Southampton, UK). Synthetic Aβ and amylin peptides (TFA salt, >95% purity, approximately 80% peptide content) are available from a number of vendors. The peptides used in this work were purchased from California Peptide Research Inc. (California).

Cell Culture

Rat pheochromocytoma PC12 cells can be obtained from the American Type Culture Collection (Rockville, MD, CRL 1721). They are cultured routinely in Dulbecco's modified Eagle's medium (DMEM) containing 10% (v/v) fetal bovine serum, 5% (v/v) horse serum, 1% (v/v) glutamine, and 1% (v/v) penicillin/streptomycin (all obtained from Gibco-BRL, UK). In their undifferentiated form they grow as loosely adherent cells at low density, which are dislodged easily for subculturing or plating into 96-well plates. This is the preferred state. At higher densities, they tend to grow as large clumps in suspension and are more difficult to handle. For experimental purposes, exponentially growing cells are plated at approximately 5000 cells/well/100 μl fresh culture medium in 96-well tissue culture plates, although this number can be increased to as much as 20,000 cells/well with good results.

[6] C. J. Pike, A. J. Walencewicz, C. G. Glabe, and C. W. Cotman, *Eur. J. Pharmacol.* **207,** 367 (1991).

[7] A. Lorenzo, B. Razzaboni, G. C. Weir, and B. A. Yankner, *Nature* **368,** 756 (1994).

[8] M. S. Shearman, C. I. Ragan, and L. L. Iversen, *Proc. Natl. Acad. Sci. U.S.A.* **91,** 1470 (1994).

[9] D. Schubert, C. Behl, R. Lesley, A. Brack, R. Dargusch, Y. Sagara, and H. Kimura, *Proc. Natl. Acad. Sci. U.S.A.* **92,** 1989 (1995).

[10] M. S. Shearman, S. R. Hawtin, and V. T. Tailor, *J. Neurochem,* **65,** 218 (1995).

Peptides

Peptide stocks (HPLC purified, as described in the manufacturer's data sheets) are prepared by dissolution in sterile deionized water (Milli-Q, Millipore Bedford, MA, conductivity 18.2 MΩ) at a concentration of 1.2 m*M*. It is particularly important for the longer peptide species to assess the relative aggregation state of the peptides on initial dissolution, and various techniques are available to do this. A straightforward and inexpensive method is the Congo red dye binding assay of Klunk *et al.*[11] The Aβ(1–40) peptides obtained from California Peptide Research consistently gave low Congo red binding values (<0.1 mol/mol), although this is not the case for all suppliers. One possibility to correct for this potential inconsistency is to resuspend the peptides in the β-structure-breaking solvent hexafluoro-2-propanol and then relyophilize them.

It is well documented that the neurotoxicity of Aβ and other amyloidogenic peptides is associated with their ability to form stable aggregates in aqueous solution. Aggregation of the peptides can be achieved by prolonged (several days) incubation of the stock solutions at 37° with vortexing or sonication. Because the aggregation process is a nucleation-dependent phenomenon, this can be accelerated considerably (minutes to hours) by the addition of a preaggregated peptide "seed."

After manipulation of the aggregation state of the peptides has been achieved, serial dilutions are carried out in sterile Milli-Q water, and a final 10-fold dilution is made into the culture medium. Peptides are exposed to cells for a period of 2–48 hr.

MTT Reduction

Stock solutions of MTT are prepared in DMEM at 5 mg/ml (12.07 m*M*). Dissolution of the dye is facilitated by warming and stirring or brief sonication. The solution is filter sterilized (0.2 μm), aliquoted, and stored at −20°. Under these conditions, the solution is stable for more than 6 months if care is taken to protect it from light.

Measurement of cellular MTT reduction is carried out by a modification of the method described by Hansen and colleagues.[12] Following the appropriate incubation time with peptide, MTT (10 μl) is added to a final concentration of 0.42 mg/ml, and incubation is continued for a further 4–6 hr (depending on the initial cell plating density). Cell lysis buffer [100 μl/well; 20% (v/v) sodium dodecyl sulfate (SDS), 50% (v/v) *N,N*-dimethylformamide, pH 4.7] is then added. Dissolution of the purple MTT formazan

[11] W. E. Klunk, J. W. Pettegrew, and D. J. Abraham, *J. Histochem. Cytochem.* **37,** 1273 (1989).
[12] M. B. Hansen, S. E. Nielson, and K. Berg, *J. Immunol. Methods* **119,** 203 (1989).

crystals can be achieved by repeated pipetting or by allowing the plate to remain at 37° overnight. Following mixing, colorimetric determination of MTT formazan product formation is made at 570 nm. Absorbance values obtained following the addition of peptide vehicle instead of peptide at the start of the experiment are taken as 100%. Complete inhibition of cell function (0%) is defined as the absorbance value obtained following the addition at the start of the experiment of 0.1% (v/v) Triton X-100 to lyse the cells or from an equivalent volume of culture medium alone.

MTS Reduction

MTS reduction is an easy and straightforward alternative to lactate dehydrogenase activity measurements for quantifying cell viability. Stock solutions of MTS are prepared in Dulbecco's phosphate-buffered saline (PBS) at 3.3 mg/ml, filter sterilized, aliquoted, and stored at −20°. The electron coupling agent phenanzine ethosulfate (PES) is also prepared in PBS at 0.86 mg/ml (2.5 m*M*) and is treated as described earlier. Both solutions are stable for more than 6 months if care is taken to protect them from light. Immediately prior to use, the two solutions are mixed at a ratio of 10:1.

The measurement of cellular MTS reduction is carried out essentially as described in Promega technical bulletin NO. 169 (Madison, WI), which is based on the methodology described by Cory.[13] Following the appropriate incubation time with peptide, the MTS and PES mixture (10 μl) is added to the cells to give final concentrations of 0.25 mg/ml and 6.5 μg/ml, respectively. Incubation is continued for another 2–3 hr (determined by the initial cell plating density), and the colorimetric determination of MTS formazan product, which is soluble in aqueous solutions, is made at 492 nm. Absorbance values obtained following the addition of peptide vehicle instead of peptide at the start of the experiment are taken as 100%. Complete inhibition of cell function (0%) is defined as the absorbance value obtained following the addition at the start of the experiment of 0.1% (v/v) Triton X-100 to lyse the cells or from an equivalent volume of culture medium alone.

Experimental Results

The time course for MTT reduction following vehicle or 10 μM Aβ(25-35) treatment is shown in Fig. 1. The optimum incubation time with MTT to quantify the inhibitory effect is between 4 and 6 hr. Control

[13] A. H. Cory, *Cancer Commun.* **3,** 207 (1991).

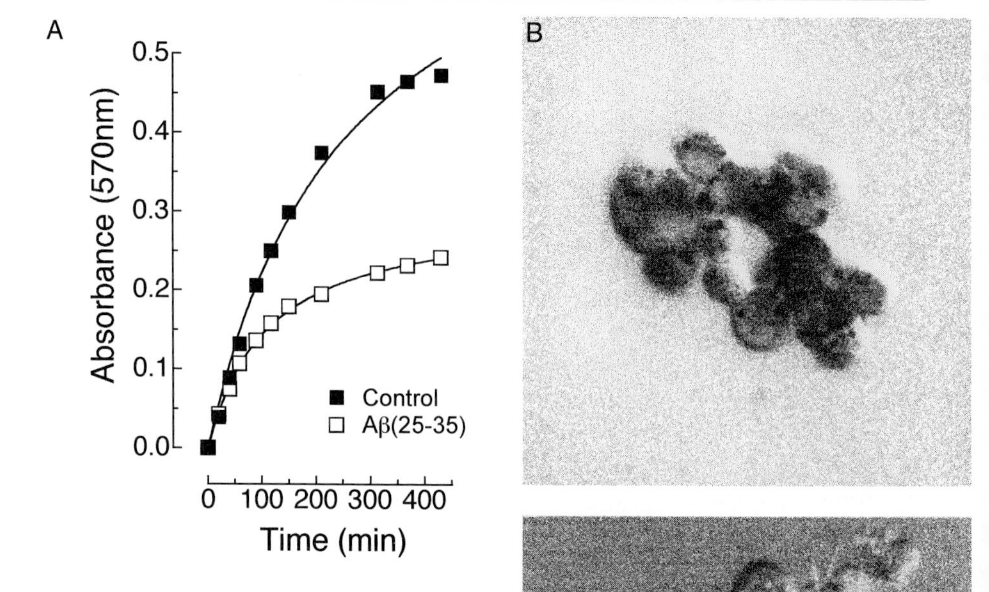

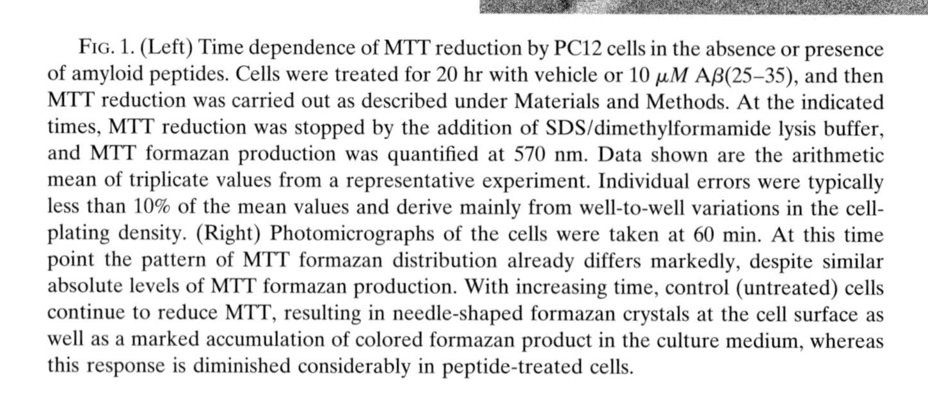

FIG. 1. (Left) Time dependence of MTT reduction by PC12 cells in the absence or presence of amyloid peptides. Cells were treated for 20 hr with vehicle or 10 μM Aβ(25–35), and then MTT reduction was carried out as described under Materials and Methods. At the indicated times, MTT reduction was stopped by the addition of SDS/dimethylformamide lysis buffer, and MTT formazan production was quantified at 570 nm. Data shown are the arithmetic mean of triplicate values from a representative experiment. Individual errors were typically less than 10% of the mean values and derive mainly from well-to-well variations in the cell-plating density. (Right) Photomicrographs of the cells were taken at 60 min. At this time point the pattern of MTT formazan distribution already differs markedly, despite similar absolute levels of MTT formazan production. With increasing time, control (untreated) cells continue to reduce MTT, resulting in needle-shaped formazan crystals at the cell surface as well as a marked accumulation of colored formazan product in the culture medium, whereas this response is diminished considerably in peptide-treated cells.

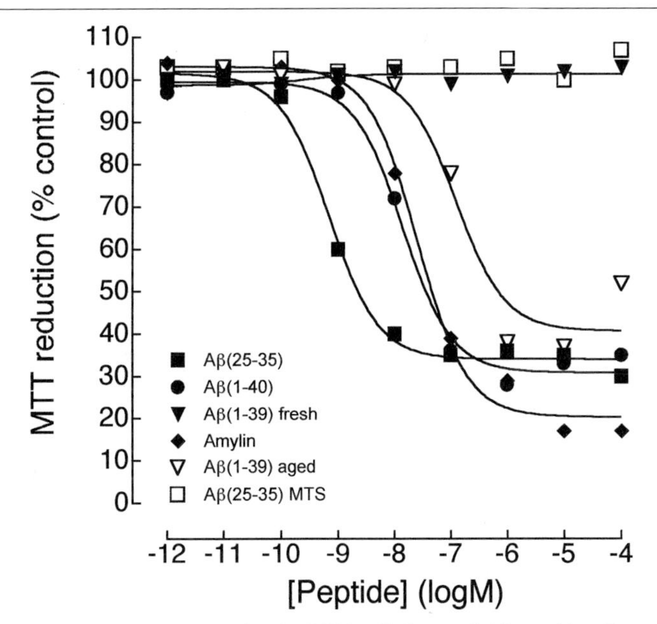

FIG. 2. Inhibition of MTT reduction in PC12 cells by amyloid peptides. Data shown are expressed as percentage control values (vehicle alone) and are the arithmetic mean of triplicate values from three to nine experiments. Individual errors are typically less than 10% of the mean values and derive mainly from well-to-well variations in the cell-plating density. Open squares show MTS reduction.

cells (Fig. 1, upper photomicrograph) are characterized by MTT formazan crystals (purple in color) deposited focally inside the cells at an early stage. Prolonged incubation of control cells results in needle-shaped MTT formazan crystals, also forming at the cell surface, as well as MTT formazan formation in the culture medium itself. In contrast to this, Aβ(25–35)-treated cells (Fig. 1, lower photomicrograph) do not show these punctate intracellular deposits of MTT formazan, but instead form a small number of needle-shaped crystals at the cell surface.

Data from a typical experiment to assess the inhibitory potency of amyloid peptides are illustrated in Fig. 2. For highly active peptides, inhibition of MTT reduction in PC12 cells occurs at the equivalent of nanomolar concentrations and is maximal between 1 and 10 μM.[14]

Although added routinely for 18–20 hr, the response of the cells to Aβ peptides is rapid and can be detected within as little as 2 hr after peptide

[14] These are not true concentrations, as the peptide essentially exists as insoluble β-fibrillar aggregates. It is more correct to consider the activity of the peptides as an x-fold dilution of the stock solution.

application. This inhibitory activity is characteristic of amyloidogenic peptides such as Aβ and human amylin that form stable oligomeric aggregates in solution and is not seen with "fresh" (i.e., nonaggregated, non-Congo red binding) peptide. It is also a characteristic of MTT reduction, but not MTS reduction (or other indicators of cell viability such as the release of lactate dehydrogenase activity), confirming that this "cytotoxic" response does not represent acute cell death.[8,10]

Comments

The sensitivity of different cell lines to Aβ has been found to vary considerably, with, in the author's experience, PC12 and the human epithelial carcinoma HeLa cells being the most responsive.[5] Based on the fact that a number of amyloidogenic peptides of different primary sequence similarly inhibit MTT reduction, it seems likely that the specificity of action of Aβ on certain cells is related to the cell surface expression of recognition sites for peptide β fibril structure rather than sequence. Further experiments utilizing radiolabeled Aβ peptides or cell clones deficient in particular cell surface molecules may help to resolve this question.

Despite the widespread use of the MTT assay as a measure of cell viability, it is clear that its reduction can be inhibited by Aβ peptides, at least acutely, without necessarily causing cell death.[8] The relationship between the mechanisms involved in the Aβ-mediated inhibition of MTT reduction, such as cellular redox activity, and those resulting in cell death is, however, supported by the observation that both processes depend on the Aβ peptide being in a β fibril-aggregated conformation and that peptides that do not inhibit MTT reduction are not cytotoxic.[7,8,15] Early studies showed that succinate-dependent MTT reduction by rat liver homogenates could be inhibited by respiratory chain inhibitors,[16] resulting in the assumption that the mitochondrion was the primary site of cellular MTT reduction. A more recent study came to the same conclusion,[17] but data arguing strongly against this have been presented.[10] It has also been suggested that Aβ exerts its effect by enhancing MTT formazan-dependent cell death.[18] Perhaps the most significant insight has come from two studies by Liu and colleagues,[15,19] which show that Aβ is capable of inhibiting MTT reduction

[15] Y. Liu and D. Schubert, *J. Neurochem.* **69,** 2285 (1997)

[16] T. F. Slater, B. Sawyer, and U. Sträuli, *Biochim. Biophys. Acta* **77,** 383 (1963).

[17] I. Kaneko, N. Yamada, Y. Sakaruba, M. Kamenosono, and S. Tutumi, *J. Neurochem.* **65,** 2585 (1995).

[18] C., Hertel, N., Hauser, R. Schubenel, B., Seilheimer, and J. A. Kemp, *J. Neurochem.* **67,** 272 (1996).

[19] Y. Liu, D. A. Peterson, H. Kimura, and D. Schubert, *J. Neurochem.* **69,** 581 (1997).

by dramatically enhancing MTT formazan vesicular exocytosis from the cell. It is not yet known what transduction pathway mediates this effect.

In summary, inhibition of PC12 MTT reduction is an inexpensive, simple, reliable, and reproducible indicator of the cytotoxic potential of Aβ peptides.

[46] Inflammatory Responses to Amyloid Fibrils

By STEPHEN L. YATES, JUNE KOCSIS-ANGLE, PAULA EMBURY, and KURT R. BRUNDEN

Introduction

A body of literature has established that a glial-mediated inflammatory response occurs in the immediate vicinity of senile plaques within the Alzheimer's disease (AD) brain. Both activated microglial cells and reactive astrocytes are found associated intimately with senile plaques,[1,2] and these cells have been shown to express increased amounts of interleukin-1 (IL-1)[3] and IL-6.[4] Microglial-derived IL-1 can lead to astrocyte activation and the subsequent production of additional inflammatory and acute-phase molecules, including IL-6, α-antichymotrypsin, complement C3, and nitric oxide.[5–7] A combination of these bioactive molecules could contribute to the neuronal pathology that leads to the progressive dementia of AD.

Cell culture studies suggest that microglial cytokine release in AD results from the interaction of these cells with the multimeric Aβ fibrils that form senile plaques. Using either rodent microglia or macrophage, several laboratories[8–10] have demonstrated that fibrillar Aβ peptides cause these cells to release increased amounts of tumor necrosis factor-α (TNF-α) and/or

[1] S. Haga, K. Adai, and T. Ishii, *Acta Neuropathol.* **77,** 569 (1989).
[2] P. L. McGeer, S. Itagaki, H. Tago, and E. G. McGeer, *Neurosci. Lett.* **9,** 195 (1987).
[3] W. S. T. Griffin, L. C. Stanley, C. Ling, L. White, V. MacLeod, L. J. Perrot, C. L. White III, and C. Araoz, *Proc. Natl. Acad. Sci. U.S.A* **86,** 7611 (1989).
[4] S. Strauss, J. Bauer, U. Ganter, U. Jonas, M. Berger, and B. Volk, *Lab. Invest.* **66,** 223 (1992).
[5] S. R. Barnum, J. L. Jones, and E. N. Benveniste, *Glia* **7,** 225 (1993).
[6] S. Das and H. Potter, *Neuron* **14,** 447 (1995).
[7] S. C. Lee, D. W. Dickson, and C. F. Brosnan, *Brain Behav. Immun.* **9,** 345 (1995).
[8] D. M. Araujo and C. W. Cotman, *Brain Res.* **569,** 141 (1992).
[9] L. Meda, M. A. Cassatella, G. I. Szendrel, L. Otvos, P. Baron, M. Villalba, D. Ferrare, and F. Rossi, *Nature* **374,** 647 (1995).
[10] S. D. Yan, X. Chen, J. Fu, M. Chen, H. Zhu, A. Roher, T. Slattery, L. Zhao, M. Nagashima, J. Morser, A. Migheli, P. Nawroth, D. Stern, and A. M. Schmidt, *Nature* **382,** 685 (1996).

0076-6879/99 $30.00

IL-1β. Our laboratory has examined the effects of Aβ on cytokine expression by the THP-1 human monocyte cell line. THP-1 cells acquire a microglia-like morphology when treated with lipopolysaccharide (LPS), thus serving as a model of primary human microglia, which are difficult to obtain in large quantities. LPS-treated THP-1 cells respond to the fibrillar forms of Aβ(1–42) or Aβ(1–40) by increasing their release of both IL-1β[11] and TNF-α (unpublished results). The IL-1β found in THP-1 cell culture medium after the addition of Aβ fibrils results from increased synthesis of the IL-1β precursor protein by these cells.[12] Interestingly, nonfibrillar preparations of amyloid peptides do not induce an increase in cytokine synthesis and release by LPS-treated THP-1 cells.[11]

The mechanism by which Aβ fibrils cause enhanced expression of proinflammatory cytokines from macrophage and microglia has not been completely resolved. Evidence shows that Aβ may interact with cell surface receptors, including receptors for advanced glycosylated end products (RAGE)[10] and scavenger receptors.[13] However, it is still unclear whether either of these receptors is involved in the upregulation of cytokine expression caused by Aβ fibrils.[14] Of interest in this regard is the observation that human amylin fibrils elicit an elevation in IL-1β synthesis and release from THP-1 cells that is comparable to that caused by fibrillar Aβ,[12] whereas the nonfibril-forming rat amylin has no effect on cytokine release (Fig. 1). Amylin, or islet amyloid polypeptide (IAPP), is a 37 amino acid peptide that forms amyloid deposits in the pancreas of individuals with type II diabetes.[15] Amylin has little sequence homology to Aβ,[16] although human amylin does form the crossed β fibril structure that is common to all amyloidogenic peptides and proteins.[17] Rat amylin, however, which is >90% homologous to human amylin, does not form these characteristic β fibrils.[16] The finding that both human amylin and Aβ fibrils cause LPS-differentiated THP-1 cells to increase the expression of IL-1β suggests that peptide interaction with a cell surface receptor is unlikely to result from recognition of a

[11] D. Lorton, J.-M. Kocsis, L. King, K. Madden, and K. R. Brunden, *J. Neuroimmunol.* **67,** 21 (1996).

[12] K. R. Brunden, J.-M. Kocsis, P. B. Embury, M. D. Dority, L. H. Burgess, and S. L. Yates, *Soc. Neurosci. Abstr.* **23,** 1634 (1997).

[13] J. El Khoury, S. E. Hickman, C. A. Thomas, L. Cao, S. C. Silverstein, and J. D. Loike, *Nature* **382,** 716 (1996).

[14] D. R. McDonald, K. R. Brunden, and G. E. Landreth, *J. Neurosci.* **17,** 2284 (1997).

[15] G. J. S. Cooper, A. C. Willis, A. Clark, R. C. Turner, R. B. Sim, and K. B. M. Reid, *Proc. Natl. Acad. Sci. U.S.A.* **84,** 8628 (1987).

[16] P. C. May, L. N. Boggs, and K. S. Fuson, *J. Neurochem.* **61,** 2330 (1993).

[17] P. Westermark, U. Engstrom, K. H. Johnson, G. T. Westermark, and C. Betsholtz, *Proc. Natl. Acad. Sci. U.S.A.* **87,** 5036 (1990).

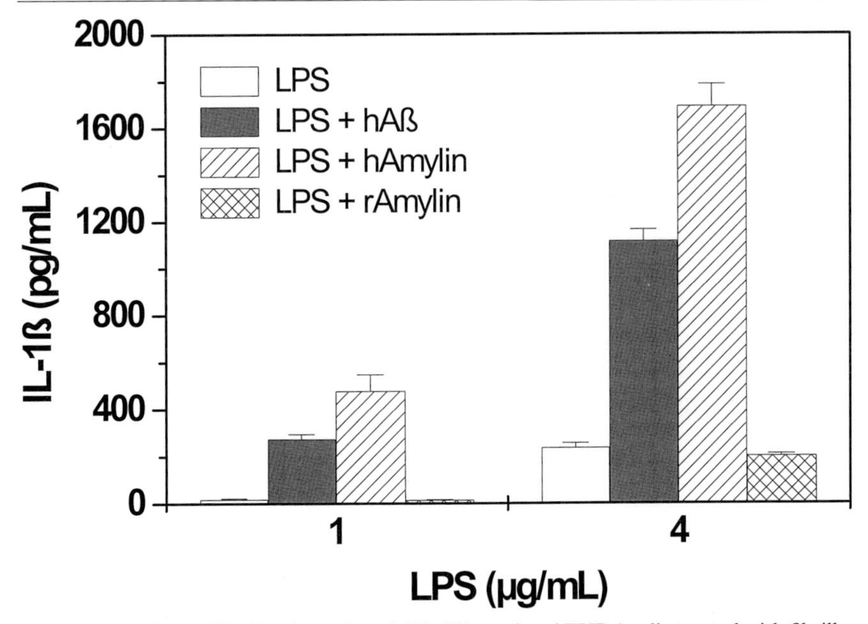

FIG. 1. Analysis of IL-1β release from LPS-differentiated THP-1 cells treated with fibrillar human Aβ(1–40) and amylin, as well as nonfibrillar rat amylin. THP-1 cells were treated with LPS alone (1 or 4 μg/ml) or LPS plus 43.5 ng/μl Aβ(1–40) or 11.7 ng/μl amylin for 48 hr. Mature IL-1β released into the culture media was assayed by ELISA as described in the text. We have previously documented that nonfibrillar Aβ has no effect on IL-1β synthesis and release.[11,12]

common sequence motif and may instead depend on shared conformational parameters between these amyloid peptides.

The following sections describe the methodology utilized in our laboratory to characterize amyloid peptide preparations and the assays used to quantify Aβ- and amylin-induced cytokine synthesis and release by THP-1 monocytes.

Characterization of Aβ Peptides

Principle

Amyloidogenic peptides, including human Aβ and amylin, assemble into multimeric fibrils of defined diameter (see article 18 of this volume). This process is thermodynamically driven and can be affected by peptide concentration, pH, and salts. Commercial preparations of lyophilized Aβ show considerable variation in the degree of fibril content immediately on solubilization. Therefore, it is necessary to establish this parameter prior to proceeding with the cellular studies described in this article. The use of thioflavin T to quantitate peptide fibril content has been described in an earlier article of this volume and elsewhere.[18] The following is an abbreviated method for the use of thioflavin T in assessing the fibrillar nature of Aβ and amylin prior to their use in cellular cytokine assays.

Materials

1. Human recombinant Aβ(1–40) and/or Aβ(1–42) [(Bachem Inc., Torrance, CA) or equivalent]
2. Human recombinant amylin (Bachem Inc. or equivalent)
3. Thioflavin T (Sigma, St. Louis, MO)

Procedures

Preparation of Fibrillar Aβ and Amylin Stocks

1. Suspend lyophilized Aβ(1–40), Aβ(1–42), or amylin in deionized water to a final concentration of 8.7, 9.1, and 7.8 μg/μl, respectively.
2. Once fully dissolved, split the solution into aliquots (typically 25 μl), lyophilize, and store at 4° for future use.
3. At the time of use, dissolve a single aliquot of the stored peptide in deionized water as described in step 1 and store at 4°.
4. Determine the fibril content of this sample as a function of storage time, as described later. Aβ(1–40) that has been dissolved in deionized water usually reaches maximal fluorescence after 1–7 days of storage, with the rate varying with different peptide lots. Aβ(1–42) and amylin preparations typically reach maximal fluorescence more rapidly than Aβ(1–40).

Determination of Aβ and Amylin Fibril Content

1. At regular intervals (e.g., daily), mix 2 μl of Aβ or amylin stock solution with 2 μl of 30 μ*M* thioflavin T and 2.0 ml of deionized water.
2. Analyze the samples using a fluorimeter with an excitation wavelength of 450 nm and an emission wavelength of 482 nm. The thioflavin T fluorescence signal that results from the dye binding to amyloid fibrils can be plotted against the length of time that the peptide stocks have been stored at 4°. We typically define fibrillar Aβ and amylin as a preparation that has reached >80% of its maximal thioflavin T fluorescence.

[18] H. LeVine, *Prot. Sci.* **2,** 404 (1993).

THP-1 Cell Culture and Plating for Cytokine Assays

Principle

THP-1 cells are a human monocyte cell line derived from the peripheral blood of a 1-year-old male with acute monocytic leukemia.[19] THP-1 cells can be differentiated into macrophage-like cells using bacterial LPS. In this differentiated state, THP-1 cells are morphologically and functionally reminiscent of microglia and thus can be used to study the possible effects of Aβ on the induction of cytokine synthesis and release in AD.

Materials

1. THP-1 human monocyte cell line (American Type Culture Collection, Rockville, MD)
2. RPMI-1640 media [(GIBCO-BRL, Gaithersburg, MD) or equivalent]
3. Lipopolysaccharide (Sigma or equivalent; *E. coli* serotype O26:B6)
4. Fetal bovine serum, heat inactivated [HyClone (Logan, UT) or equivalent]
5. 2-Mercaptoethanol (Sigma or equivalent)
6. Penicillin/streptomycin solution [(GIBCO-BRL) or equivalent; 5000 U/ml penicillin G and 5 mg/ml streptomycin]
7. Trypan blue (Sigma or equivalent; 0.4% solution)
8. Phosphate-buffered saline (PBS)
9. 96-well tissue culture-treated microtiter plates, flat bottom

Solutions

1. 2-Mercaptoethanol (100× stock for growth medium): 17.8 ml RPMI-1640 medium, 0.2 ml penicillin/streptomycin, 2.0 ml fetal bovine serum, and 7.0 μL 2-mercaptoethanol. Sterilize this solution using a 0.2-μm filter. The stock can be stored for several months at 2–6°.
2. 2-Mercaptoethanol (100× stock for cytokine assay medium): 19.4 ml RPMI-1640 medium, 0.2 ml penicillin/streptomycin, 0.4 ml fetal bovine serum, and 7.0 μl 2-mercaptoethanol. Sterilize this solution using a 0.2-μm filter. The stock can be stored for several months at 2–6°.
3. THP-1 growth medium (for stock cultures): 881 ml RPMI-1640 medium, 9 ml penicillin/streptomycin, 100 ml fetal bovine serum, and 10 ml 2-mercaptoethanol solution. Store at 2–6°.
4. THP-1 assay medium (for cytokine release assay): 192 ml RPMI-

[19] S. Tsuchiya, M. Yamabe, Y. Yamaguchi, Y. Kobayashi, T. Konno, and K. Tada, *Int. J. Cancer* **26,** 171 (1980).

1640 medium, 2 ml penicillin/streptomycin, 4 ml fetal bovine serum, and 2 ml 2-mercaptoethanol solution. Store at 2–6°.

5. Lipopolysaccharide stock: Lyophilized lipopolysaccharide is solubilized in sterile water to a concentration of 1 mg/ml. Aliquots can be stored indefinitely at −20°, whereas solutions are stable for approximately 1 month at 2–6°.

Procedures

Cell Culture

1. Frozen THP-1 cell stocks from the vendor should be thawed and diluted as specified by the vendor's instructions.
2. Maintain THP-1 culture stocks at a cell density of $0.15–1.0 \times 10^6$ cells/ml in tissue culture flasks containing THP-1 growth medium within a humidified CO_2 incubator at 37°. The cells are typically split every 3–4 days according to standard cell culture protocols.

Plating for Cytokine Assays

1. For each 96-well plate, remove a volume of culture stock containing 2.0×10^6 cells and centrifuge at 450*g* for 5.5 min at room temperature in a sterile centrifuge tube.
2. Remove the supernatant (being careful not to dislodge the cell pellet), rinse the cells with an equal volume of RPMI-1640 medium (without additives), and centrifuge as in step 1.
3. Repeat step 2.
4. Following the second rinse, carefully remove the supernatant and resuspend the cell pellet in a volume of assay medium that results in a density of at least 1.5×10^5 cells/ml.
5. Perform a viable cell count using a hemocytometer and trypan blue as per standard cell culture protocols.[20]
6. Dilute the cells with assay medium to a final concentration of 1.5×10^5 cells/ml. A total of 1.5×10^6 cells (10 ml volume) is needed for each 96-well plate.
7. Add a volume of LPS stock solution to the cell suspension that yields a final LPS concentration of 0.1–10.0 μg/ml. It is necessary to characterize each lot of LPS to determine the optimal concentration to use for the cytokine assays. To determine the optimal LPS concentration, we recommend that a pilot study be performed in which THP-1 cells are treated with several LPS concentrations in

[20] R. I. Freshney, "Culture of Animal Cells: A Manual of Basic Techniques," p. 245. Wiley-Liss, New York, 1987.

the absence and presence of Aβ, following the steps outlined later. The ideal LPS concentration is that which results in (1) the release of >200 pg/ml of IL-1β into the culture medium after treating cells with LPS and Aβ or amylin and (2) a more than fivefold increase in IL-1β released into the culture media following Aβ or amylin addition relative to that seen with LPS alone.

8. Multiply the number of assay wells that are to receive amyloid peptide by 0.11 ml to determine the volume of LPS-treated cell suspension to transfer to a separate sterile tube. Typically, add 0.5 μl of stock Aβ or amylin solution per 0.1 ml of cell suspension (a 200-fold dilution of the original peptide stocks). Additional peptide concentrations can be examined by varying the amount of Aβ or amylin added to the LPS-treated cell suspension.
9. Add 0.1 ml of either the LPS-treated or LPS + amyloid peptide-treated cell suspension (1.5×10^4 cells) to each well of a 96-well tissue culture plate, ensuring that the cells are fully mixed during the dispensing step. It is recommended that each experimental condition be performed in at least triplicate samples.
10. Incubate the cells (undisturbed) for 24 to 48 hr at 37° in a 5% CO_2 incubator.
11. At the end of the incubation period, remove 50 μl of medium from each well for the subsequent measurement of cytokine. The culture medium can be frozen once prior to ELISA analysis. Subsequent freeze/thaw cycles may result in diminished levels of detectable cytokines.

Preparation of THP-1 Cell Lysates for Intracellular Cytokine Determination

1. Plate and treat cells as described in the previous section.
2. At the end of the incubation period, freeze the media and cells at −80°. Alternatively, remove an aliquot of media for the assessment of released cytokine levels before freezing.
3. Thaw the lysed cells at room temperature.
4. Remove the lysed cells and media from each well and centrifuge at 16,000*g* for 6 min (4°).
5. Determine lysate cytokine levels as described in the following section.

Analysis of Cytokine Levels

Principle

Enzyme-linked immunosorbent assays (ELISA) are a sensitive and effective methodology for the measurement of intracellular and released cytokines from cell cultures. In general, these assays utilize an immobilized capture antibody to bind the cytokine of interest to the well of a 96-well plate and a separate detection antibody to quantitate the bound cytokine. Several commercial ELISA kits for a large variety of human, rat, and mouse cytokines are available from a number of sources. These kits are generally in a 96-well format and come with all required reagents (e.g., cytokine standards, antibodies, colorimetric dyes). It is recommended that the culture medium samples be diluted over a range of 1:4–1:20 during the initial testing to determine a dilution that will fall within the linear detection range of the assay (typically 5–500 pg/ml).

If cytokine assays are to be utilized routinely, it may be more cost effective to develop an in-house ELISA. We typically quantify IL-1β levels in cell culture medium with the following assay. For the quantitation of intracellular precursor IL-1β in cell lysates, we recommend using an ELISA kit from R&D Systems (Minneapolis, MN).

Materials

1. 96-well medium-binding plates
2. Recombinant human IL-1β (R&D Systems or equivalent)
3. Mouse antihuman IL-1β monoclonal antibody (R&D Systems or equivalent)
4. Goat antihuman IL-1β polyclonal antibody (R&D Systems or equivalent)
5. Peroxidase-labeled antigoat IgG [(Vector Laboratories, Burlingame, CA) or equivalent; typically 1 mg/ml]
6. Tween 20 (Bio-Rad, Richmond, CA or equivalent)
7. TMB peroxidase substrate and solution [(Kierkegaard & Perry (Gaithersburg, MD) or equivalent]
8. Dry nonfat milk
9. PBS
10. Concentrated phosphoric acid

Solutions

1. Wash buffer: 1000 ml PBS and 0.5 ml Tween 20. Buffer can be stored at 4° for up to 1 week.
2. Blocking solution: 995 ml PBS and 50 g dry nonfat milk. This solution should be prepared on the day of use.
3. IL-1β standards: Prepare a stock solution containing 1 ng/ml of IL-1β in THP-1 assay medium. Divide into 1-ml aliquots and store frozen at −80°.

4. Monoclonal antibody solution: Add 1.0 ml of PBS to a 0.5-mg vial of mouse antihuman IL-1β monoclonal antibody. Divide into 0.2-ml aliquots and store at $-80°$ for up to 6 months or at 4° for up to 1 month. Prior to use in an ELISA, dilute the stock solution to 1.5 μg/ml in PBS.
5. Polyclonal antibody solution: Add 1.0 ml of PBS to a 1.0-mg vial of goat antihuman IL-1β polyclonal antibody. Divide into 0.2-ml aliquots and store at $-80°$ for up to 6 months or at 4° for up to 1 month. Prior to use in an ELISA, dilute the stock solution to 1.5 μg/ml in a 1:1 mixture of PBS and PBS + 5% nonfat milk.
6. Peroxidase-labeled antibody solution: Mix 5 μl of peroxidase-conjugated antigoat IgG (1 mg/ml) with 5 ml of PBS for each 96-well plate. Add 5 ml of blocking solution to yield a final 1:2000 dilution of antibody in PBS + 2.5% nonfat milk.
7. TMB substrate: Mix 5 ml of TMB substrate solution with 5 ml of hydrogen peroxide solution (included with package) as per the supplier's instructions for each 96-well plate. This solution should be prepared immediately before use.
8. 1 *M* phosphoric acid stopping solution: Mix 67.8 ml of concentrated phosphoric acid with deionized water to 1000 ml.

Procedures

Human IL-1β ELISA

1. The day before culture medium samples are to be assayed for cytokine levels, coat the needed number of 96-well medium-binding plates with 0.1 ml per well of monoclonal antibody solution (1.5 μg/ml). Cover the plates with parafilm and incubate overnight at 4°.
2. The following day, aspirate each well and subsequently wash each well with 0.4 ml of wash buffer.
3. Repeat step 2 five times.
4. After the last wash, invert the plate and blot briskly onto paper towels. It is important to remove as much liquid as possible from the ELISA wells after the washings before proceeding to the next step.
5. Add 0.2 ml of blocking solution and incubate for 1 hr at room temperature.
6. Prepare IL-1β standards by making a twofold dilution series of the 1-ng/ml IL-1β stock, in THP-1 assay medium, to cover a range of 500 to 1.95 pg/ml. Culture medium samples should be diluted at 1:4 to 1:20 during the initial testing to find an appropriate dilution that provides absorbance values in the detection range of the ELISA.

7. Remove the blocking solution and add 0.2 ml of either standards or diluted media samples to triplicate wells. Incubate at room temperature for 2 hr.
8. Wash the plates as described in step 2 a minimum of three times and blot the plate after the last wash as described in step 4.
9. Add 0.1 ml of the polyclonal antibody solution (1.5 μg/ml) to each well and incubate for 1 hr at room temperature.
10. Wash the plates as described in step 2 a minimum of three times and blot the plate after the last wash as described in step 4.
11. Add 0.1 ml of peroxidase-conjugated antigoat IgG solution (1:2000) to each well and incubate for 1 hr.
12. Wash the plates as described in step 2 a minimum of five times and blot the plate after the last wash as described in step 4.
13. Add the TMB color substrate (0.1 ml) to each well and incubate for 20 min at room temperature.
14. Stop the color reaction by adding 0.1 ml of 1 *M* phosphoric acid per well.
15. Determine the absorbance of each well at 450 nm in a spectrophotometric plate reader within 30 min after the addition of the phosphoric acid stopping solution, using a reference (background) wavelength of 570 nm.

Data Analysis

1. Determine the mean absorbance values of triplicate wells for each concentration of IL-1β standard.

2. Plot the log of the mean absorbance values against the log of the IL-1β concentration to generate a linear standard curve. The correlation coefficient of this curve is typically >0.95.

3. Determine the mean absorbance values of triplicate wells for each experimental condition.

4. Determine the concentration of IL-1β in each sample by extrapolation from the standard curve.

5. Calculate the actual IL-1β concentration in the sample wells by multiplying the value obtained in the ELISA by the sample dilution used in the assay.

Summary

Our laboratory has routinely used the methodologies described here to characterize the effects of fibrillar Aβ and amylin on cytokine synthesis and secretion by LPS-differentiated THP-1 cells. Because LPS-treated THP-1 cells resemble macrophage and microglia, this assay system represents an *in vitro* model of the potential interactions between Aβ-containing senile plaques and microglia in the AD brain. As such, these methodologies should prove useful in the identification of compounds that inhibit this Aβ-induced inflammatory response.

[47] Impairment of Membrane Transport and Signal Transduction Systems by Amyloidogenic Proteins

By MARK P. MATTSON

Introduction

Proteins located within and associated with the plasma membrane and organellar membranes serve vital functions in essentially all cells. Two important types of membrane-associated proteins in neural cells are those involved in transporting various molecules (e.g., ions, nutrients, and neurotransmitters) across the membrane and those involved in transducing signals from different ligands (e.g., receptors for neurotransmitters, cytokines, and growth factors). Transporters in the plasma membrane include ion-motive ATPases such as Na^+, K^+-ATPase and Ca^{2+}-ATPase, glucose transporters, and amino acid transporters. Membrane-associated proteins involved in signal transduction are remarkably complex, but can be placed into different categories such as receptors, guanosine triphosphate (GTP)-binding proteins, and effector proteins. Some membrane receptors are linked to GTP-binding proteins (e.g., muscarinic cholinergic receptors), others have intrinsic tyrosine kinase activity (e.g., neurotrophic factor receptors), and others are ion channels (e.g., ionotropic excitatory amino acid receptors). As can be appreciated from other articles in this volume, amyloidogenic peptides can have a variety of adverse effects on cells ranging from modest dysfunction to overt cell death. This article describes methods for assessing the effects of amyloidogenic peptides on membrane transport and signal transduction systems and presents data illustrating the kinds of data that have arisen from such approaches. The focus will be on the action of amyloid β-peptide (Aβ), a 40–42 amino acid peptide that accumulates as diffuse and fibrillar deposits in the brains of individuals with Alzheimer's disease.[1] However, available data suggest that the mechanisms described in this article also apply to other amyloidogenic proteins and their associated disorders, including amylin in diabetes, β_2-microglobulin in systemic amy-

[1] M. P. Mattson, *Physiol. Rev.* **77,** 1081 (1997).

0076-6879/99 $30.00

loidosis, and prion proteins in Creutzfeld-Jacob disease and related disorders.[2]

Analyses of Effects of Amyloidogenic Proteins on Membrane Lipid Peroxidation

Background

Because oxidative stress and membrane lipid peroxidation appear to be key mediators of the adverse effects of amyloidogenic peptides on membrane transporter systems, methods for quantifying markers of membrane lipid peroxidation are briefly introduced.[2a] Studies of rat hippocampal cell cultures showed that Aβ can induce the accumulation of superoxide, hydrogen peroxide, and peroxynitrite in neurons and that antioxidants (e.g., vitamin E and nordihydroguaiaretic acid) can protect neurons against Aβ toxicity.[3–6] The mechanism whereby Aβ induces membrane lipid peroxidation is tightly correlated with the propencity of the peptide to form fibrils[7] and likely involves the Fe^{2+}-dependent formation of peptide radicals associated with methionine residue 35 of the peptide.[2a] Electron paramagnetic resonance (EPR) spectroscopy studies showed that Aβ induces lipid peroxidation in neocortical synaptosomes and isolated plasma membrane and mitochondrial membrane preparations.[8] Nonamyloidogenic peptides, including Aβ-related peptides with reversed or scrambled sequences and rat amylin, do not induce oxidative stress in cultured cells and are not cytotoxic.[2,3]

Exposure of cultured hippocampal neurons[9] and cortical synaptosomes[10] to Aβ results in production of the toxic aldehyde 4-hydroxynonenal as detected by high-performance liquid chromatography analyses and by Western blot and immunocytochemical analyses using antibodies against

[2] M. P. Mattson and Y. Goodman, *Brain Res.* **676,** 219 (1995).

[2a] D. A. Butterfield, S. M. Yater, S. Varadarajar, and T. Koppal, *Methods Enzymol.* **309** [48] 1999 (this volume).

[3] Y. Goodman and M. P. Mattson, *Exp. Neurol.* **128,** 1 (1994).

[4] Y. Goodman, M. R. Steiner, S. M. Steiner, and M. P. Mattson, *Brain Res.* **654,** 171 (1994).

[5] M. P. Mattson, Y. Goodman, H. Luo, W. Fu, and K. Furukawa, *J. Neurosci. Res.* **49,** 681 (1997).

[6] J. N. Keller, M. S. Kindy, F. W. Holtsberg, D. K. St Clair, H.-C. Yen, A. Germeyer, S. M. Steiner, A. J. Bruce-Keller, J. B. Hutchins, and M. P. Mattson, *J. Neurosci.* **18,** 687 (1998).

[7] M. P. Mattson, *Nature Struct. Biol.* **2,** 926 (1995).

[8] A. J. Bruce-Keller, J. G. Begley, W. Fu, D. A. Butterfield, D. E. Bredesen, J. B. Hutchins, K. Hensley, and M. P. Mattson, *J. Neurochem.* **70,** 31 (1998).

[9] R. J. Mark, M. A. Lovell, W. R. Markesbery, K. Uchida, and M. P. Mattson *J. Neurochem.* **68,** 255 (1997).

[10] J. N. Keller, Z. Pang, J. W. Geddes, J. G. Begley, A. Germeyer, G. Waeg, and M. P. Mattson, *J. Neurochem.* **69,** 273 (1997).

4-hydroxynonenal–protein adducts (Figs. 1 and 2). 4-Hydroxynonenal covalently modifies proteins on cysteine, lysine, and histidine residues via a process called Michael addition and also undergoes Schiff base chemistry with proteins. Exposure of cultured rat hippocampal neurons and cortical synaptosomes to Aβ or Fe^{2+} (a well-established inducer of membrane lipid peroxidation) results in generation of a variety of 4-hydroxynonenal–protein adducts of various molecular weights. 4-Hydroxynonenal appears to be a key mediator of neuronal injury and death induced by amyloidogenic peptides. As evidence, exposure of cultured hippocampal neurons to Aβ or Fe^{2+} results in the production of 4-hydroxynonenal at levels of 1–10 μM, and exposure of the same neurons to the same concenrations of pure 4-hydroxynonenal results in neuronal degeneration and death.[9]

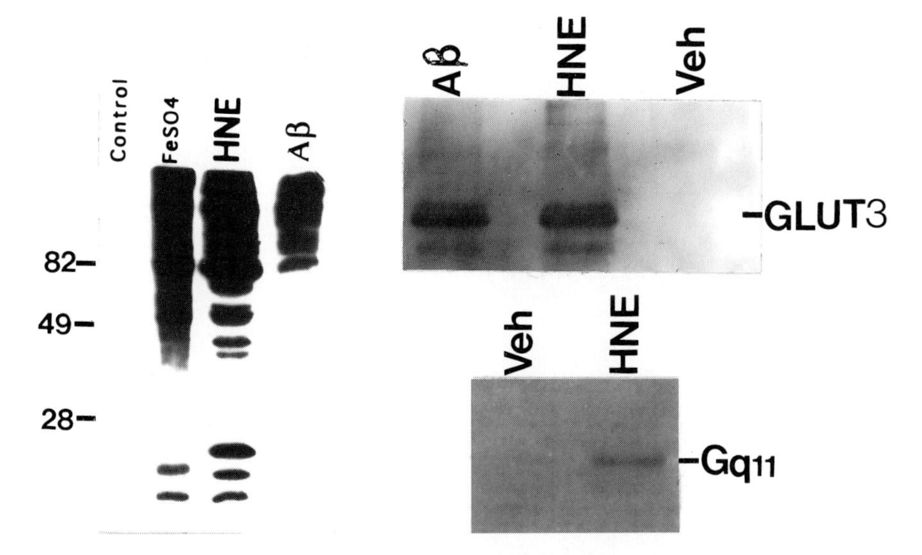

FIG. 1. Covalent modification of neuronal proteins, including a glucose transport protein and a GTP-binding protein, by the lipid peroxidation product 4-hydroxynonenal. (Left) Rat cortical synaptosomes were exposed for 2 hr to vehicle, 10 μM $FeSO_4$, 10 μM 4-hydroxynonenal, or 50 μM Aβ25–35. Proteins in synaptosomal homogenates were subjected to Western blot analysis using an antibody against 4-hydroxynonenal–protein conjugates. Note that $FeSO_4$, 4-hydroxynonenal, and Aβ25–35 each induce the presence of many different proteins that are modified covalently by 4-hydroxynonenal. Modified from Keller *et al.*[10] (Upper right) Cortical cultures were exposed for 4 hr to 10 μM Aβ25–35 (Aβ), 10 μM 4-hydroxynonenal (HNE), or vehicle (Veh). Cell proteins were immunoprecipitated with an antibody against the glucose transporter GLUT3, and the antibody-bound proteins were separated by SDS–PAGE, transferred to nitrocellulose, and immunoreacted with an HNE antibody. Modified from Mark *et al.*[13] (Lower right) Cultured cortical neurons were exposed to vehicle (0.2% ethanol) or 20 μM 4-hydroxynonenal for 1 hr. Proteins in cell homogenates were then immunoprecipitated with an antibody against the GTP-binding protein Gq11 and subjected to Western blot analysis using an antibody against 4-hydroxynonenal–protein conjugates.

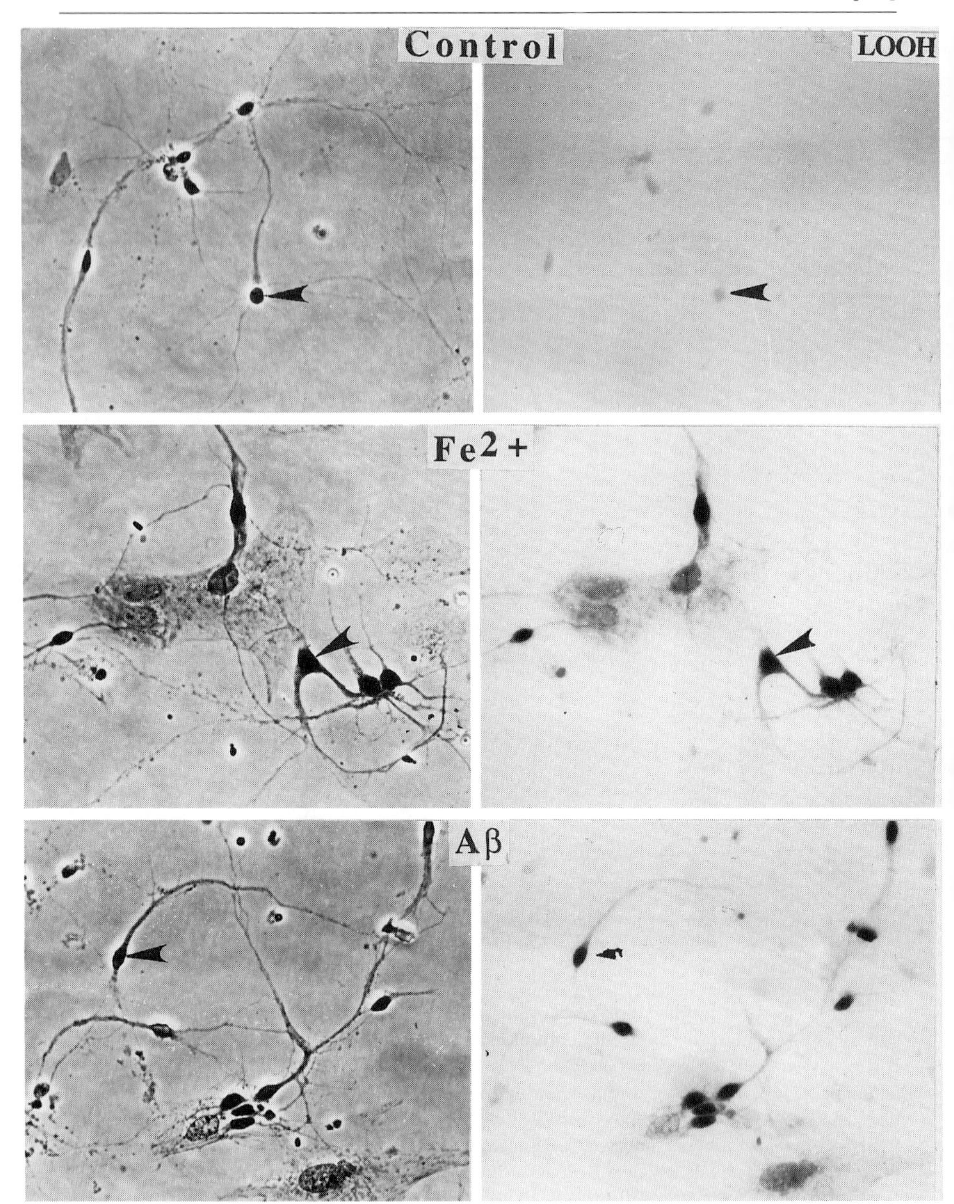

FIG. 2. Localization of lipid peroxidation products by immunocytochemistry in cultured embryonic rat hippocampal neurons. Phase-contrast micrographs (left) and bright-field micrographs showing lipid peroxide immunoreactivity (right) in an untreated control culture, a culture exposed to 5 μM $FeSO_4$ (Fe^{2+}) for 4 hr, and a culture exposed to 10 μM Aβ25–35 for 8 hr. Cells were immunostained with antibody against lipid peroxides. Note the presence of lipid peroxide immunoreactivity in neurons exposed to Fe^{2+} and Aβ25–35 (arrowheads).

Moreover, glutathione (a tripeptide with a cysteine residue that is known to bind and detoxify 4-hydroxynonenal) protects neurons against the toxicities of Aβ, Fe^{2+}, and HNE.[9,11] Amyloidogenic peptides have been shown to kill different types of cells by apoptosis, a type of cell death in which the cells shrink, and nuclear chromatin condensation and DNA fragmentation occurs. Lipid peroxiation plays a role in such apoptotic death as antioxidants that suppress membrane lipid peroxidation including vitamin E, propyl gallate, and 17β-estradiol prevent Aβ-induced apoptosis.

Procedures

Relative levels of malondialdehyde, a lipid released from oxidized membranes, are determined using the TBARS assay as described previously.[12] Following experimental treatment, trichloroacetic acid is added to suspensions of intact cells or isolated membranes to a final concentration of 5%. Precipitates are then incubated at 95° for 30 min in the presence of 2-thiobarbituric acid (0.335% in 50% glacial acetic acid). The solutions are then cooled, and fluorescence in the butanol-extractable lipid fraction is determined using a Cytofluor 2350 (Millipore Bedford, MA) fluorescent plate reader (518 nm excitation and 588 nm emission). Values are determined in a per microgram protein basis.

Levels and cellular localization of 4-hydroxynonenal–protein conjugates can be assessed using antibodies raised against 4-hydroxynonenal-modified proteins. The antibodies can be used for immunoprecipitation, Western blot analysis, and immunohistochemistry.[9,13,14] For immunoprecipitation, cells or membranes are treated with test agents and then homogenized in RIPA buffer [50 m*M* Tris–HCl, 10% glycerol, 1% Triton X-100, 150 m*M* NaCl, 100 m*M* NaF, 5 m*M* EDTA, 2 m*M* phenylmethylsulfonyl fluoride (PMSF), 1 m*M* sodium orthovanadate, and 1 μg/ml leupeptin, pH 7.5]. Protein extracts (typically 400 μg) are denatured in 0.1% sodium dodecyl sulfate (SDS) by heating at 60° for 5 min and then brought to a final volume of 1 ml with ice-cold RIPA buffer (final SDS concentration $<0.015\%$). A protein of interest is immunoprecipitated using a specific antibody (typically 0.02–0.1 μg antibody/ml extract). The antibody/lysate solution is shaken overnight at 4°. The affinity complex is then precipitated using 300 μg/ml protein A conjugated to acrylic beads, the mixture is incubated 6 hr at 4°

[11] I. Kruman, A. J. Bruce-Keller, D. E. Bredesen, and M. P. Mattson, *J. Neurosci.* **17,** 5089 (1997).
[12] Y. Goodman, A. J. Bruce, B. Cheng, and M. P. Mattson, *J. Neurochem.* **66,** 1836 (1996).
[13] R. J. Mark, Z. Pang, J. W. Geddes, K. Uchida, and M. P. Mattson, *J. Neurosci.* **17,** 1046 (1997).
[14] 14. A. J. Bruce-Keller, Y.-J. Li, M. A. Lovell, P. J. Kraemer, D. S. Gary, R. R. Brown, W. R. Markesbery, and M. P. Mattson, *J. Neuropathol. Exp. Neurol.* **57,** 257 (1988).

under constant rotation, and the beads are pelleted by centrifugation. The pellet is washed three times with ice-cold RIPA buffer and resuspended in 2× Laemmli sample buffer. Samples are boiled for 5 min and subjected to Western blot analysis. For Western blot analysis, proteins are separated by SDS–PAGE gel, transferred to a nitrocellulose sheet, and incubated in blocking solution [e.g., 5% milk or 2% bovine serum albumin (BSA)]. The blot is then incubated with 4-hydroxynonenal antibody, followed by alkaline phosphatase-conjugated antimouse secondary antibody. Blots are then processed with an alkaline phosphatase detection kit (Bio-Rad, Richmond, CA). For immunocytochemistry of cultured cells, the cells are fixed for 30 min in PBS containing 4% paraformaldehyde (4°). Membranes are permeabilized by exposing the fixed cells to PBS containing 0.2% (v/v) Triton X-100. Cells are then incubated sequentially in PBS solutions containing blocking serum (1% normal goat serum; 30 min), 4-hydroxynonenal antibody (3 hr), biotinylated antimouse secondary antibody (Vector Labs, Burlingame, CA; 1 hr), avidin–peroxidase complex (Vector Labs; 30 min), and diaminobenzidine tetrahydrochloride (Sigma, St. Louis, MO; 5 min). For densitometric analysis, images of Western blots are scanned into a Macintosh computer, and the intensities of the immunoreactive bands are quantified using Image 1.47 NIH software. For staining of tissue sections, 30-μm sections (free floating) are incubated overnight at 4° in the presence of blocking solution (5% horse serum in PBS containing 0.2% Triton X-100). Sections are then incubated for 24 hr in the presence of primary antibody (4°), followed by a 4-hr incubation in the presence of biotinylated secondary antibody at room temperature with continued incubation overnight at 4°. The sections are then treated with ABC reagent (Vector Laboratories) for 1 hr, followed by a 20-min incubation in a solution containing 0.1% (w/v) hydrogen peroxide, 0.05% (w/v) diaminobenzidine tetrahydrochloride, 0.025% (w/v) cobalt chloride, and 0.02% (w/v) nickel ammonium sulfate. Sections are then placed on slides, dehydrated, and mounted for examination.

Spin-labeling techniques can also be used to assess lipid peroxidation in isolated membranes or intact cells.[8] We have used the nitroxyl stearate spin labels (5-NS and 12-NS; Sigma) for this purpose. Detailed methods for these procedures can be found in the article by Butterfield in this volume.[2a]

Evaluation of Effects of Amyloidogenic Proteins on Function of Membrane Ion-Motive ATPases and Glucose and Glutamate Transporters

Background

It was appreciated some time ago that amyloidogenic peptides, including Aβ, human amylin, and β_2-microglobulin, can disrupt cellular calcium homeostasis, resulting in progressive increases in cytoplasmic-free calcium levels.[2,15] In neurons, a striking result was that subtoxic levels of Aβ greatly increased cellular vulnerability to glutamate toxicity. Glutamate is the major excitatory neurotransmitter in the brain, and overactivation of glutamate receptors can damage and kill neurons by causing massive calcium influx through *N*-methyl-D-aspartate receptor channels and voltage-dependent calcium channels. The latter mechanism of cell death, called excitotoxicity, is believed to contribute greatly to the neurodegenerative process in a variety of disorders, including Alzheimer's disease and prion disorders.[1] Treatment of cultured neurons (as well as nonneuronal cells such as vascular endothelial cells) with intracellular calcium chelators (e.g., BAPTA-AM) or agents that block voltage-dependent calcium channels (e.g., nifedipine) can protect the cells from being killed by amyloidogenic peptides, demonstrating a central role for calcium overload in the cell death process.[1]

In studies aimed at establishing the mechanism whereby amyloidogenic proteins disrupt cellular ion homeostasis, it was found that Aβ impairs the function of Na^+,K^+-ATPase and Ca^{2+}-ATPase in cultured rat hippocampal neurons and human hippocampal synaptosomes.[16] Vitamin E, propyl gallate, and 17β-estradiol (antioxidants that suppress membrane lipid peroxidation) prevented impairment of ion-motive ATPases by Aβ, indicating an important role for lipid peroxidation in the mechanism of action of Aβ (Fig. 4). Exposure of cultured neurons and synaptosomes to Fe^{2+} mimicked the inhibitory effects of Aβ on Na^+,K^+-ATPase and Ca^{2+}-ATPase activities, showing that the induction of lipid peroxidation was sufficient to impair the function of these ion-motive ATPases. Exposure of cultured hippocampal neurons to 4-hydroxynonenal also impaired Na^+,K^+-ATPase activity and increased neuronal vulnerability to excitotoxicity, suggesting a role for this aldehyde in Aβ-induced impairment of ion-motive ATPase activities.[9] Other membrane-associated Mg^{2+}-ATPase were not affected adversely by Aβ, suggesting that ion-motive ATPases are particularly vulnerable to oxidative stress. Collectively, these findings suggest a scenario in which amyloidogenic peptides induce membrane lipid peroxidation and 4-hydroxynonenal production, resulting in impairment of ion-motive ATPases (Fig. 2). Impairment of the Na^+,K^+-ATPase leads to membrane depolarization and thereby sensitizes neurons to excitotoxicity. Impairment of the plasma membrane Ca^{2+}-ATPase compromises the ability of the cells to restore intracellular calcium levels to basal levels and thereby contributes to calcium overload.

[15] M. P. Mattson, B. Cheng, D. Davis, K. Bryant, I. Lieberburg, and R. E. Rydel, *J. Neurosci.* **12,** 376 (1992).

[16] R. J. Mark, K. Hensley, D. A. Butterfield, and M. P. Mattson, *J. Neurosci.* **15,** 6239 (1995).

Glucose tranport into cells is critical for maintenance of ATP levels and is particularly important for neurons, a cell type that relies on a constant supply of glucose.[17] Glucose uptake in the brain is mediated by the specific membrane transport proteins GLUT-1 (located mainly in endothelial cells) and GLUT-3 (located in neurons). Glucose uptake into brain cells is known to be reduced greatly in Alzheimer's disease patients, and data suggest that activities of glucose transporters are decreased in Alzheimer's disease brain. Aβ may play a role in impairing glucose transport in Alzheimer's disease because exposure of cultured hippocampal neurons and cortical synaptosomes to Aβ results in time- and dose-dependent decreases in glucose transport.[10,16] Antioxidants that suppress membrane lipid peroxidation, including vitamin E and 17β-estradiol, prevented impairment of glucose transport by Aβ, indicating an important role for membrane lipid peroxidation in the action of Aβ. Exposure of cultured neurons and synaptosomes to Fe^{2+} mimicked the adverse effect of Aβ on glucose uptake, showing that induction of lipid peroxidation is sufficient to impair glucose transport. Exposure of cultured hippocampal neurons to 4-hydroxynonenal also impaired glucose transport.[16] In an additional experiment in the latter study, cultured cortical neurons were exposed to Aβ or Fe^{2+}, cells were homogenized, and proteins in the homogenate were immunoprecipitated with an antibody to GLUT-3. The precipitated proteins were separated by electrophoresis, transferred to a nitrocellulose sheet, and then reacted with an antibody against 4-hydroxynonenal (Fig. 1). Results showed that 4-hydroxynonenal was covalently bound to GLUT-3, suggesting that a direct interaction of 4-hydroxynonenal with the glucose transport protein may be responsible for the observed impairment of glucose transport by Aβ. In addition to impairing glucose transport in neurons, Aβ impaired glucose transport in cultured vascular endothelial cells,[18] suggesting that any circulating amyloidogenic peptide has the potential to impair endothelial cell membrane transporters.

Another membrane transporter that is affected adversely by amyloidogenic proteins is the glutamate transporter. Glutamate transporters are expressed mainly in glial cells, with levels of expression being particularly high in astrocytes.[19] Exposure of rat cortical synaptosomes to Aβ results in concentration- and time-dependent decreases in radiolabeled glutamate uptake into the nerve endings.[10] Fe^{2+} also inhibits glutamate transport, and

[17] I. A. Simpson, S. J. Vannucci, and F. Maher, *Biochem. Soc. Trans.* **22,** 671 (1994).

[18] E. M. Blanc, M. Toborek, R. J. Mark, B. Hennig, and M. P. Mattson, *J. Neurochem.* **68,** 1870 (1997).

[19] J. D. Rothstein, M. Dykes-Hoberg, C. A. Pardo, L. A. Bristol, L. Jin, R. W. Kuncl, Y. Kanai, M. A. Hediger, Y. Wang, J. P. Schielke, and D. F. Welty, *Neuron* **16,** 675 (1996).

vitamin E and 17β-estradiol prevent impairment of glutamate transport by Aβ, indicating a central role for lipid peroxidation (Fig. 3). Immunoprecipitation–Western blot analyses showed that 4-hydroxynonenal covalently modifies the astrocyte glutamate transporter GLT-1.[20] The latter study also showed that 4-hydroxynonenal causes cross-linking of GLT-1 monomers to form multimers, strongly suggesting that 4-hydroxynonenal mediates the impairment of glutamate transport by amyloidogenic peptides (Fig. 1). By removing glutamate from the extracellular fluid, astrocytes play an important role in protecting neurons against excitotoxicity. Impairment of astrocyte glutamate transport by Aβ may therefore contribute to excitotoxic neuronal injury in Alzheimer's disease.

Procedures

Ion-Motive ATPase Activity Assays. The method allows quantification of three different Mg^{2+}-dependent ATPase activities (Na^+,K^+-ATPase activity, Ca^{2+}-ATPase activity, and ouabain/Ca^{2+}-insensitive ATPase activity) in the same sample. Activities are measured in quadruplicate in covered 96-well microtiter plates at 37° on a shaker. Ninety microliters of assay buffer (18 m*M* histidine, 18 m*M* imidazole, 80 m*M* NaCl, 15 m*M* KCl, 3 m*M* $MgCl_2$, and 0.1 m*M* EGTA; pH 7.1.) containing 3 μg of membrane protein is added to each well. The assay is started with the addition of 10 μl ATP (final concentration 3 m*M*), and after 60 min, the reaction is terminated by the addition of 25 μl of 5% SDS. Inorganic phosphate levels, quantified using Fiske–Subbarow reagent, are used as a measure of ATPase activity. Na^+,K^+-ATPase activity is determined by subtracting the ouabain (0.2 m*M*)-sensitive activity from the overall Mg^{2+}-ATPase activity level. The Ca^{2+}-ATPase activity is determined by subtracting activity measured in the presence of Ca^{2+} and ouabain from that determined in the absence of Ca^{2+} (no added Ca^{2+} plus 0.1 m*M* EGTA) and the presence of ouabain. The plates are read on the Bio-Tek EL-340 plate reader (Winooski, VT) at 630 nm. Absorbance values are converted to activity values by linear regression using a standard curve obtained with increasing sodium monobasic phosphate concentrations.

Glucose and Glutamate Transport Assays Prior to the addition of experimental treatments, cultures and synaptosomes are switched to Locke's buffer (m*M*: NaCl, 154; KCl, 5.6; $CaCl_2$, 2.3; $MgCl_2$, 1.0; $NaHCO_3$, 3.6; glucose, 10; HEPES 5; pH 7.2). For glucose transport measurements, the assay is started by the addition of 1.5 μCi of 2-[^{3}H]deoxyglucose (New England Nuclear, Boston, MA), and for glutamate transport measurements

[20] E. M. Blanc, J. N. Keller, S. Fernandez, and M. P. Mattson, *Glia* **22,** 149 (1998).

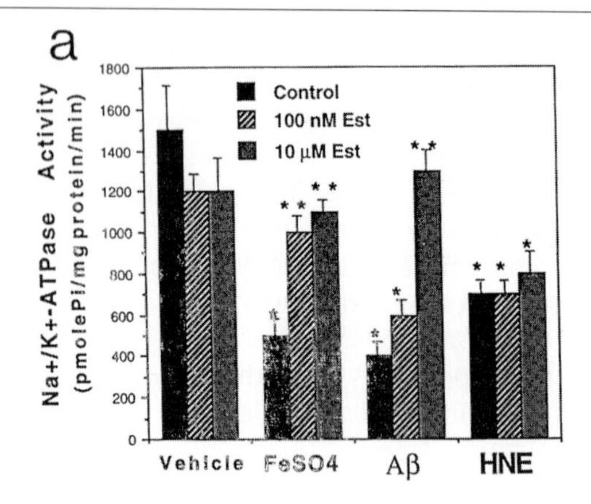
a
Control
100 nM Est
10 μM Est
Na+/K+-ATPase Activity
(pmolePI/mgprotein/min)
Vehicle
FeSO4
Aβ
HNE

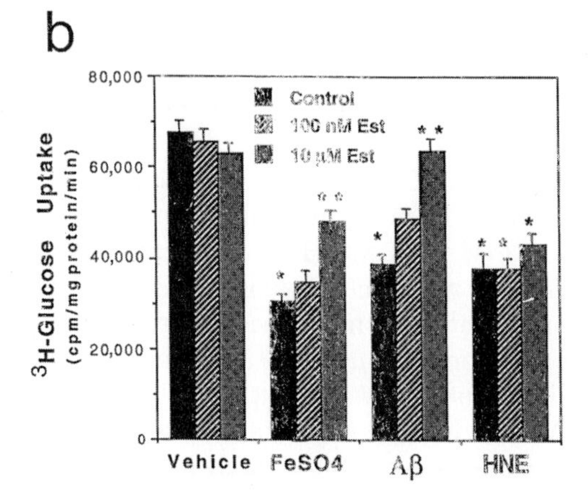
b
Control
100 nM Est
10 μM Est
3H-Glucose Uptake
(cpm/mgprotein/min)
Vehicle
FeSO4
Aβ
HNE

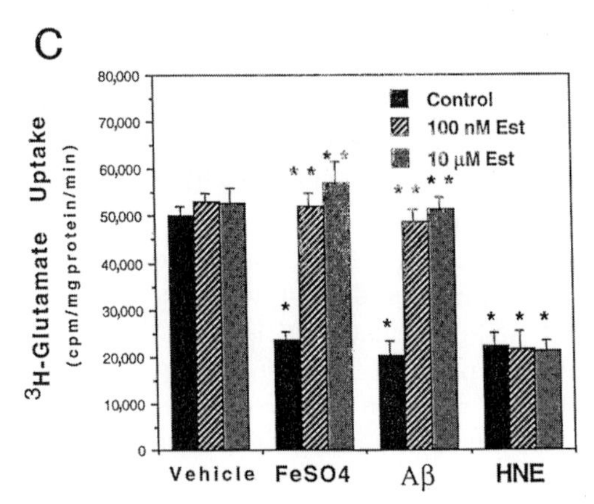
c
Control
100 nM Est
10 μM Est
3H-Glutamate Uptake
(cpm/mgprotein/min)
Vehicle
FeSO4
Aβ
HNE

the assay is started by the addition of [^{3} H]glutamate (0.1 μCi/ml; specific activity 0.2 μCi/ml) (Amersham). Cultures or synaptosomes are incubated at 37°. For cultured cells the assay is stopped 5 min later by aspiration of the supernatant and rapid washing with PBS (three rinses, 5–7 sec/rinse). Cells are lysed in 200 μl of a 0.5 *N* NaOH/0.05% SDS solution; 10 μl is used for protein determination (Pierce BCA kit) and the remainder is counted in a Packard 2500TR liquid scintillation counter. For synaptosomes the assay is stopped by rapidly washing the preparations three times in Locke's buffer using Whatman filters in a vacumn filtration apparatus. The filters are then placed in scintillation vials containing Scintiverse and radioactivity determined by scintillation spectroscopy. Data are expressed as cpm radiolabeled glucose or glutamate/mg protein/min. Specificity of glucose transport is established using the specific glucose transport inhibitor phloretin (100 μM), whereas specificity of glutamate transport is assessed using excess cold glutamate.

Assessments of Effects of Amyloidogenic Proteins on Signaling via GTP-Binding Protein-Coupled Receptors

Background

Intercellular signaling is critical for the development, maintenance, and proper functioning of essentially all organ systems. The plasma membrane is the site at which the vast majority of intercellular signals are received and transduced into an intracellular response. Cells possess membrane-associated receptors for a diverse array of signals, including growth factors, cytokines, hormones, neurotransmitters, and cell adhesion molecules. Amyloidogenic proteins can have profound adverse effects on such membrane signal transduction systems, even at subtoxic levels of exposure. For example, exposure of cultured rat cortical neurons to Aβ resulted in an impair-

FIG. 3. Estrogen attenuates Aβ- and Fe^{2+}-induced impairment of membrane transporters for sodium, glucose, and glutamate in cortical synaptosomes. Synaptosomes were pretreated for 1 hr with 0.2% ethanol (control) or the indicated concentrations of 17β-estradiol (Est) and were then exposed for 3 hr to vehicle, 50 μM $FeSO_4$, 50 μM Aβ25–35, or 10 μM 4-hydroxynonenal (HNE). Na^+,K^+-ATPase activity (a), glucose transport (b), and glutamate transport (c) were then quantified. Values are the mean and SEM of determinations made in four separate preparations. $*p < 0.01$ compared to value in vehicle-treated control synaptosomes; $**p < 0.01$ compared to corresponding control value.

ment of muscarinic cholinergic signaling.[21] Muscarinic receptors are functionally linked to a GTP-binding protein called Gq11, which transduces a signal that activates the enzyme phospholipase C, resulting in inositol phospholipid hydrolysis and liberation of the second messengers diacylglycerol and inositol triphosphate (IP_3). Analyses of the state of activation of each of these steps in the signaling pathway following the exposure of cells to Aβ revealed that Aβ impairs coupling of the muscarinic receptor to Gq11. The adverse effect of Aβ on muscarinic signaling was abolished in neurons treated with vitamin E and was mimicked by treatment of neurons with Fe^{2+}, indicating that lipid peroxidation was responsible.[21] Exposure of cultured cortical neurons to 4-hydroxynonenal also impaired muscarinic signaling, and immunoprecipitation–Western blot analysis showed that 4-hydroxynonenal covalently modified Gq11[22] (Figs. 1 and 4). Other signaling pathways that employ membrane receptors linked to GTP-binding proteins may also be sensitive to impairment by amyloidogenic peptides, as indicated by the abilities of Fe^{2+} and 4-hydroxynonenal to disrupt coupling of metabotropic glutamate receptors to their cognate GTP-binding protein.[22] In addition, Aβ impaired thrombin signaling in cultured neural cells,[23] suggesting a role for amyloidogenic proteins in disrupting cellular responses to this important signaling molecule in a variety of cell types. Interestingly, exposure of cultured fibroblasts to Aβ resulted in a decrease in levels of a GTP-binding protein called Cp20, which appears to regulate K^+ channel activity, and levels of Cp20 are decreased in fibroblasts taken from Alzheimer's patients.[24] In addition to impairing signal transduction, amyloidogenic peptides may activate certain downstream signaling processes. For example, exposure of culture neural cells to Aβ25–35 induced activation of phospholipase C as indicated by increased inositol phosphate production.[25] Thus, amyloidogenic peptides can have adverse effects on a variety of membrane transporters and signaling pathways via a mechanism involving increased oxidative stress and membrane lipid peroxidation.

Procedures

GTPase Activity Assay. Neocortical cells or synaptosomes are homogenized in 750 μl of 50 mM Na^+–K^+–phosphate buffer (pH 7.4), 10 mM

[21] J. Kelly, K. Furukawa, S. W. Barger, R. J. Mark, M. R. Rengen, E. M. Blanc, G. S. Roth, and M. P. Mattson, *Proc. Natl. Acad. Sci. U.S.A.* **93,** 6753 (1996).

[22] E. M. Blanc, J. F. Kelly, R. J. Mark, and M. P. Mattson, *J. Neurochem.* **69,** 570 (1997).

[23] M. P. Mattson and J. G. Begley, *Amyloid* **3,** 28 (1996).

[24] C. S. Kim, Y. F. Han, R. Etcheberrigaray, T. J. Nelson, J. L. Olds, T. Yoshioka, and D. L. Alkon, *Proc. Natl. Acad. Sci. U.S.A.* **28,** 3060 (1995).

[25] J. N. Singh, G. Sorrentino, and J. N. Kanfer, *J. Neurochem.* **69,** 252 (1997).

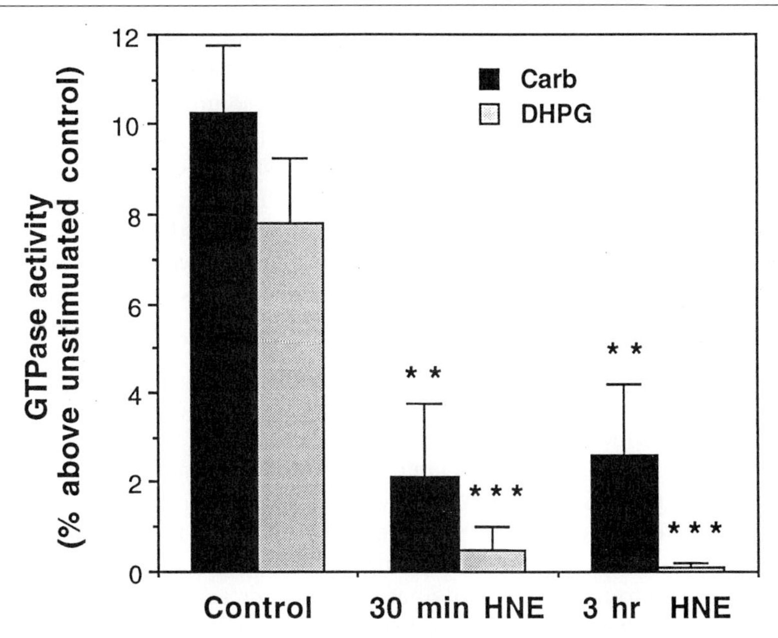

FIG. 4. Impairment of functional coupling of muscarinic cholinergic and metabotropic glutamatergic receptors to GTP-binding protein activation following exposure of cortical neurons to 4-hydroxynonenal. Cultures were exposed for 3 hr to vehicle (control) or for 30 min or 3 hr to 20 μM HNE. Low K_m GTPase activity stimulated by 1 mM carbachol or 100 μM DHPG was then quantified. Values are expressed as the percentage increase above unstimulated control cultures and represent the mean and SEM (n = 2–8 separate experiments). $^{**}p < 0.01$, $^{***}p < 0.001$ compared to control value. Modified from Blanc *et al.*[22]

EDTA, and either 0.1 mM PMSF or 4-(2-aminoethyl)benzenesulfonyl fluoride hydrochloride (Pefabloc SC), with 8–10 strokes in a Potter–Elvejhem # 20 glass homogenizer with a Teflon plunger. Low K_m GTPase activity is determined using a modification of the methods of Cassel and Selinger.[26] Twenty-five microliters of crude cell membrane preparation (20 μg protein) is added to 75 μl of a reaction mixture consisting of 20 mM Tris–HCl (pH 7.4), 0.3 μM [γ-^{32}P]GTP, 5 mM $MgCl_2$, 1 mM ATP, 2 mM AppNHp, 10 mM phosphocreatine, 60 units of creatine phosphokinase, 0.1 mM EDTA, 0.1 mM EGTA, 2 mM dithiothreitol, and 100 mM NaCl; cold GTP (100 μM) and/or agonist (e.g., 1 mM of the muscarinic receptor agonist carbachol or 100 μM of the metabotropic glutamate receptor agonist DHPG) is added to some reactions (the concentrations of agonist that are E_{max} for the

[26] D. Cassell and Z. Selinger, *Proc. Natl. Acad. Sci. U.S.A.* **75,** 4155 (1978).

assay should be determined empirically). Upon addition of protein, samples are vortexed immediately and incubated at 37° for 10 min. The reaction is stopped by the addition of 900 μl of an ice-cold solution of 5% activated charcoal and phosphoric acid (pH 2.5). Samples are centrifuged at 11,950g for 5 min. A 100-μl aliquot of the supernatant is removed and mixed with 4 ml of scintillation fluid and counted. Low K_m GTPase activity is calculated by subtracting total activity measured in the presence of 100 μM unlabeled GTP from basal and carbachol-stimulated GTPase activity. The change in low K_m GTPase activity stimulated by carbachol is expressed as the percentage increase over basal activity.

Conclusions

Amyloidogenic peptides can exert both subtle and striking effects on the function of membrane-associated proteins that subserve transport and signal transduction functions. Actions of the amyloidogenic peptides appear to be mediated, in the main, by increased membrane lipid peroxidation. Quantitative assays for measurements of ion-motive ATPase activities, glucose transporter activity, glutamate transporter activity, and activation of GTP-binding proteins can provide useful end points for determining the impact of amyloidogenic peptides on cell function.

[48] Amyloid β-Peptide-Associated Free Radical Oxidative Stress, Neurotoxicity, and Alzheimer's Disease

By D. ALLAN BUTTERFIELD, SERVET M. YATIN, SRIDHAR VARADARAJAN, and TANUJA KOPPAL

Introduction

Free radicals, including reactive oxygen species (ROS) and reactive nitrogen species (RNS), react with various biological moieties in the cell to induce altered structure and function.[1] Often included in the term ROS is H_2O_2, although this nonpolar molecule is not a free radical; rather, in the presence of Fe^{2+} or Cu^{+}, hydroxyl radical ($\cdot OH$) is formed.

Oxidative stress, a natural sequelae of living and increasing in age in an aerobic environment, can be defined as a toxic condition characterized by chemical and/or functional alterations of biomolecules caused by an

[1] B. Halliwell and M. M. C. Gutteridge, "Free Radicals in Biology and Medicine." Clarendon Press, Oxford 1989.

0076-6879/99 $30.00

overproduction of free radicals, ROS and/or RNS, or as the result of a diminution of free radial defense mechanisms. For example, protein oxidation, a key marker of oxidative stress (see later), increases exponentially with age.[2] Further, oxidative stress handling enzymes, such as superoxide dismutase (SOD), catalase, and glutathione peroxidase, decline with aging.

Aging and age-related neurodegenerative disorders, such as Alzheimer's disease (AD), Parkinson's disease (PD), amyotrophic lateral sclerosis (ALS), and stroke, are associated with free radical oxidative stress.[3,4] Oxidative stress is thought to occur in other neurodegenerative disorders, including diffuse Lewy body disease and Wilson's disease, and traumatic brain injury as well.[1]

Sources of ROS and RNS

Excluding ionizing radiation-derived and photochemically derived free radicals, ROS and RNS can be formed *in vivo* chemically or biologically. Some of the processes by which free radicals arise *in vivo* have been reviewed.[1,2,5] Examples of ROS and RNS free radicals include ·OH, $O_2\cdot^-$, $HO_2\cdot$, NO·, ROO·, and RO·. Several ROS and RNS, although not free radicals, can form free radicals or cause oxidative stress by subsequent reaction, e.g., peroxynitrite ($ONOO^-$), H_2O_2, HOCl, O_3, singlet oxygen ($^1O_2$), and ROOH. Hydroxyl radicals can be formed by Fenton chemistry or by decomposition of peroxynitrite. SOD action on superoxide radical anion is a source of H_2O_2 needed for the Fenton reaction, and $O_2\cdot^-$ is partially reduced oxygen formed in mitochondrial electron transport pathways, by mixed function oxidases, or by the respiratory bursts associated with neutrophils, polymorpholeukocytes, and macrophages peripherally and microglia in the central nervous system.[5] Overstimulation of *N*-methyl-D-aspartate (NMDA) receptors can lead to increased intracellular $O_2\cdot^-$.[6] There are other free radicals, often sulfur based, that will not be discussed here; rather, the emphasis of this article is on ROS and RNS associated with amyloid β-peptide (Aβ).

Evidence of Oxidative Stress in Brain Cells: Markers for Oxidative Stress

Although there may be others, the principal markers of oxidative stress caused by ROS and RNS in brain cells are (a) protein oxidation, evidenced

[2] E. R. Stadtman, *Science* **257,** 1220 (1992).
[3] D. A. Butterfield, *Chem. Res. Toxicol.* **10,** 495 (1997).
[4] W. R. Markesbery, *Free Radic. Biol. Med.* **23,** 134 (1997).
[5] D. A. Butterfield and E. R. Stadtman, *Adv. Cell Aging Gerontol.* **2,** 116 (1997).
[6] M. Lafon-Cazal, S. Pietri, M. Culcasi, and J. Bockaert, *Nature* **364,** 537 (1992).

by increased protein carbonyl content[2,5]; (b) lipid peroxidation, evidenced by a number of methods, including, but not limited to, formation of reactive aldehydic compounds (e.g., malondialdehyde) or alkenals (4-hydroxy-2-*trans*-nonenal, HNE),[7] release of free fatty acids, presumably by oxidative stress activation of phospholipase A_2,[8] and by reduction of paramagnetism of lipid bilayer resident stearic acid spin probes in electron paramagnetic resonance (EPR) studies[8–10]; (c) formation of excess intracellular ROS, often evidenced by ROS-sensitive fluorescence dyes[11,12]; (d) the presence of 3-nitrotyrosine, a marker for $ONOO^-$ reaction with proteins[13]; (e) inhibition of the activity of oxidatively prone enzymes such as glutamine synthetase (GS) and creatine kinase (CK)[2,5,14–17]; (f) upregulation of message for and/or alterations in the activity of oxidative stress handling enzymes, evidenced by message and/or activity analyses[18,19]; and (g) prevention and/or significant modulation of these markers for oxidative stress and neuronal death by pretreatment of brain cells with appropriate free radical antioxidants.[1] Although important, oxidative stress modification of DNA bases, oxidative stress-induced modification of behavior, or other more psychological or biological manifestations of oxidative stress will not be described here. Each of these markers will be described briefly and the respective methods for their analysis will be given.

Aβ and Oxidative Stress in Alzheimer's Disease Brain

Alzheimer's disease is the major dementing disorder of the elderly, affecting 4–5 million Americans currently and projected to involve 14 mil-

[7] H. Esterbauer, R. J. Schaur, and H. Zollner, *Free Radic. Biol. Med.* **11,** 41 (1991).

[8] T. Koppal, R. Subramaniam, J. Drake, M. R. Prasad, and D. A. Butterfield, *Brain Res.* **786,** 270 (1998).

[9] D. A. Butterfield, K. Hensley, M. Harris, M. Mattson, and J. Carney, *Biochem. Biophys. Res. Commun.* **200,** 710 (1994).

[10] A. J. Bruce-Keller, J. G. Begley, W. Fu, D. A. Butterfield, D. E. Bredesen, J. B. Hutchins, K. Hensley, and M. P. Mattson, *J. Neurochem.* **70,** 31 (1998).

[11] C. Behl, J. B. Davis, R. Lesley, and D. Schubert, *Cell* **77,** 817 (1994).

[12] S. M. Yatin, M. Aksenov, and D. A. Butterfield, *Neurochem. Res.* **24,** 427 (1999).

[13] J. S. Beckman, J. Chen, H. Ischiropoulous, and J. P. Crow, *Methods Enzymol.* **233,** 229 (1994).

[14] M. Y. Aksenov, M. V. Aksenova, J. M. Carney, and D. A. Butterfield, *Free Radic. Res.* **27,** 267 (1997).

[15] D. A. Butterfield, K. Hensley, P. Cole, R. Subramaniam, M. Aksenov, M. Aksenova, P. M. Bummer, B. E. Haley, and J. M. Carney, *J. Neurochem.* **68,** 2451 (1997).

[16] K. Hensley, J. Carney, M. Mattson, M. Aksenova, M. Harris, J. F. Wu, R. Floyd, and D. A. Butterfield, *Proc. Natl. Acad. Sci. U.S.A.* **91,** 3270 (1994).

[17] C. D. Smith, J. M. Carney, P. E. Starke-Reed, C. N. Oliver, E. R. Stadtman, and R. A. Floyd, *Proc. Natl. Acad. Sci. U.S.A.* **88,** 10540 (1991).

[18] M. Y. Aksenov, H. M. Tucker, P. Nair, M. V. Aksenova, D. A. Butterfield, S. Estus, and W. R. Markesbery, *J. Mol. Neurosci.,* in press (1999).

[19] M. A. Lovell, W. D. Ehmann, S. M. Butler, and W. R. Markesbery, *Neurology* **45,** 1594 (1995).

lion Americans early in the new millennium.[20] Clinically, AD patients present a progressive dementia characterized by cognitive and memory loss, and AD brain exhibits several characteristic pathological findings, including loss of cortical neurons with a corresponding loss of synapses, neurofibrillary tangles, composed mostly of hyperphosphorylated tau (a cytoskeletal protein), and senile (neuritic) plaques (SP), composed of a central core of aggregated Aβ surrounded by dystrophic neurites and many other moieties.[20]

The molecular basis for the etiology and pathogenesis of AD remains uncertain, but strong evidence has led to greater, although not total, acceptance of the amyloid hypothesis for AD.[21] Evidence from many approaches supports the notion of a central role for Aβ in AD.[22] Briefly, mutations in the amyloid precursor protein (APP), from which Aβ is derived, lead to excess deposition of Aβ and development of AD; persons with Down's syndrome invariably develop AD after sufficient time, and APP is coded for on chromosome 21, the locus of trisomy in Down's syndrome; the presenilin proteins, presenilin-1 and presenilin-2, coded for on chromosomes 14 and 1, respectively, may be involved in APP processing, and mutations in these proteins are associated with early onset familial AD; mice overexpressing human APP develop a subset of pathology associated with AD. Although different progenitors of excess Aβ deposition are operative, a common final pathway may be the excess presence of Aβ, with a consequent increased opportunity for oxidative stress and, therefore, death of neurons.

Oxidative stress is increased in AD brain (for a review, see Markesbery[4]). For example, TBARS, a measure of aldehydes formed from lipid peroxidation, are increased in AD brain.[19] The lipid peroxidation product, HNE, is also increased in ventricular fluid and in frontal cortex in AD brain,[4] as is free fatty acid release,[4] consistent with increased lipid peroxidation in AD brain. Protein oxidation is also elevated in AD brain, in those areas in which amyloid-rich SP are found but not in SP-poor cerebellum.[17,23] Consonant with a proposed role of Aβ in oxidative stress found in AD brain,[3,23,24] APP overexpressing mice, in which excess Aβ is deposited in brain, show evidence of brain-resident oxidative stress,[25] and high-dose

[20] R. Katzman and R. Saitoh, *FASEB J.* **4,** 278 (1991).

[21] D. J. Selkoe, *J. Biol. Chem.* **271,** 18295 (1996).

[22] D. L. Price and S. S. Sisoda, *Annu. Rev. Neurosci.* **21,** 479 (1998).

[23] K. Hensley, N. Hall, R. Subramaniam, P. Cole, M. Harris, M. Aksenov, M. Aksenova, S. P. Gabbita, J. F. Wu, J. M. Carney, M. Lovell, W. R. Markesbery, and D. A. Butterfield, *J. Neurochem.* **65,** 2146 (1995).

[24] C. Behl and F. Holsboer, *Fortschr. Neurol. Psychiatr.* **66,** 113 (1998).

[25] M. A. Smith, K. Hirai, K. Hsiao, M. A. Pappolla, P. L. Harris, S. L. Siedlak, M. Tabaton, and G. Perry, *J. Neurochem.* **70,** 2212 (1998).

vitamin E treatment of AD patients is reported to significantly delay institutionalization.[26] Our laboratory postulated a relationship between a central role of Aβ and increased oxidative stress in AD brain. We developed an Aβ-associated free radical oxidative stress hypothesis for neurotoxicity in AD brain that is consistent with most pathological, epidemiological, and therapeutic aspects of this disorder (for a review, see Butterfield[3]).

As evidenced by many laboratories employing multiple markers of oxidative stress outlined in this article and prevention or modulation of the same by appropriate antioxidants, Aβ clearly appears to be associated with free radical oxidative stress.[3] In the absence of effective therapy to slow or prevent the development of AD, the number of AD cases early in the next millennium will be enormous, with consequent implications to patients, families, health care facilities, and the national health care budget.[20] The strength of the Aβ-associated free radical oxidative stress model for neurotoxicity in AD brain, most of the predictions of which have been observed and confirmed,[3,4] is that it unifies much of the literature on AD. (a) The age of onset of AD[20] is consistent with age-related oxidative stress. (b) The many different proteins, enzymes, transport systems, lipids, and so on, that are altered in AD brain can be considered consequences of free radical attack on, conformational change in, and dysfunction of these moieties.[3,4] (c) The presence of markers of oxidative stress in AD brain regions that are rich in Aβ-containing SP but not in SP-poor cerebellum[3,4,17,23] reflects Aβ-associated free radical oxidative stress. (d) Aβ causes increased intracellular Ca^{2+}, and the consequent deleterious effects on the neuron lead to neurotoxicity.[27] (e) Other factors, such as presenilin mutations, lead to more Aβ deposition, increased oxidative stress, and disrupted Ca^{2+} homeostasis.[22,28] (f) Apolipoprotein E_4, which is a risk factor for AD and which contains no SH groups, would be predicted to participate less in protection against oxidative stress.[29] Other inducers of oxidative stress, e.g., redox metals, advanced glycation end products, and activated microglia, could cause aggregation of and increase toxicity of Aβ.[4] There are doubtless other factors to be considered. However, the extensive and comprehensive evidence from our laboratory and those of others strongly supports a central role for Aβ-associated free radical oxidative stress in AD neurotoxicity

[26] M. Sano, C. Ernesto, R. G. Thomas, M. R. Klauber, K. Schafer, M. Grundman, P. Woodbury, J. Growdon, C. W. Cotman, E. Pfeiffer, L. S. Schneider, and L. J. Thal, *N. Engl. J. Med.* **336,** 1216 (1997).

[27] M. P. Mattson, S. W. Barger, B. Cheng, I. Lieberburg, V. L. Smith-Swintosky, and R. E. Rydel, *Trends Neurosci.* **16,** 409 (1993).

[28] D. Blacher and R. E. Tanzi, *Arch. Neurol.* **55,** 294 (1998).

[29] T. J. Montine, V. Amaranth, M. E. Martin, W. J. Srittmatter, and D. G. Graham, *Am. J. Pathol.* **148,** 89 (1996).

and offers a unifying hypothesis for brain cell death in this disorder for which rational therapeutic strategies can be designed.

Methods for Assessing Aβ-Associated Free Radical Oxidative Stress

Assessing Whether a Particular Lot of Synthetic Aβ Is Useful for Study

Lot-to-lot variability of the neurotoxicity and other properties of synthetic samples of Aβ peptides has been reported,[30–32] and this variability is not due to sample impurities[31]; rather, structural differences in lyophilized samples may be important.[30,31] Therefore, the potency of all samples of synthetic Aβ peptides should first be assessed before meaningful interpretations can be obtained. Unfortunately, this is not done routinely, possibly because some of the expensive peptides necessarily must be consumed for such analysis.

Nevertheless, unless one knows the potency of a particular lot of peptide, uncertain results will be obtained. We showed a correlation between EPR spin-trapping experiments with *N-tert*-butyl-α-phenylnitrone (PBN) and Aβ and the degree of inhibition of GS activity by Aβ,[32] and we reported that Aβ(1–40) hippocampal neuronal toxicity is related to spin trapping, protein carbonyl, and other indices of oxidative stress.[33] Thus, we recommend EPR spin trapping, GS assay, and neuronal toxicity procedures (see later) to quantify the oxidative potency of synthetic Aβ peptides prior to beginning any experiment with a new lot of synthetic peptide. Although we describe these methods in the context of Aβ-associated oxidative stress, we contend that some type of potency analysis should be undertaken in any experiments involving synthetically obtained Aβ. Given that EPR is not routinely available in all laboratories, then at least the GS assay and cell toxicity studies should be performed to determine if a particular lot of Aβ is capable of oxidative stress. In view of the now widely accepted notion that Aβ is associated with oxidative stress, in the absence of assays to determine the toxic and oxidative potency of synthetic Aβ, claims that Aβ is not toxic or does not induce oxidative stress must be viewed with caution.

[30] P. C. May, B. D. Gitter, D. C. Waters, L. K. Simmons, G. W. Becker, J. S. Small, and P. M. Robinson, *Neurobiol. Aging* **13,** 605 (1992).

[31] L. K. Simmons, P. C. May, K. J. Tomaselli, R. E. Ryel, K. S. Fuson, E. R. Brigham, S. W. Wright, I. Lieberburg, G. W. Becker, D. N. Brems, and W. Y. Li, *Mol. Pharmacol.* **45,** 373 (1994).

[32] K. Hensley, M. Aksenova, J. M. Carney, M. Harris, and D. A. Butterfield, *NeuroReport* **6,** 489 (1995).

[33] M. E. Harris, K. Hensley, D. A. Butterfield, R. A. Leedle, and J. M. Carney, *Exp. Neurol.* **131,** 193 (1995).

Related to commercial lots of Aβ sometimes showing no or low reactivity, in a patented procedure from this laboratory,[34] we show in favorable cases how to regain some oxidative stress activity of apparently unreactive Aβ. Briefly, the solvent activation restructuring process (SARP) is a chemical process for modifying (specifically increasing) the chemical reactivity, biological functioning, an aggregational characteristic of amyloid peptides, which possesses a structural component of chemical reactivity.[34] As noted earlier, others showed that the lot-to-lot variability of synthetic Aβ is due to structural factors and not to impurities.[30,31] SARP offers a means to improve the reactivity and usefulness of benign synthetic Aβs, thereby enhancing their value as research materials. Consistent with reported structural differences in variable lots of synthetic Aβ,[30,31] we hypothesize that important reactive sites on Aβ are sterically blocked or otherwise inaccessible in nontoxic commercial lots of the peptide. SARP involves dissolving the purified peptide solid in an oxygen-free, appropriate organic structure-modulating solvent [e.g., trifluoroethanol (TFE), dimethyl sulfoxide (DMSO), and others] under conditions of gentle heating followed by rapid and controlled removal of the solvent by lyophilization or by N_2 gas.[34] The procedure is as follows: highly purified TFE (Aldrich) is thoroughly degassed by N_2 sparge prior to use. Samples of Aβ are dissolved in deoxygenated TFE to a concentration of 0.15 mg/ml and incubated in glass vials for 2 hr at 40–45°. The TFE/Aβ solution is initially cloudy but becomes clear after about 1 hr. The solution is allowed to equilibrate to room temperature, and the TFE is then removed by rapid evaporation under a stream of N_2. (We sometimes experienced difficulty with lyophilization as frozen TFE melts rather than sublimes under refrigerated vacuum.) The treated peptide had severalfold greater EPR signal intensity following spin trapping with PBN, a significantly increased ability to inactivate GS, and formed fibrils much more readily than the synthetic peptide prior to SARP,[34] suggesting that this structure-enhancing solvent has made inaccessible reactive sites on Aβ accessible for reaction.

In Vitro Models

Cortical Synaptosomes. The use of synaptosomes is physiologically relevant to study Aβ-induced oxidative stress because Aβ deposits have been reported in the synaptic region in AD brain, and the loss of synapses is thought to be an early event in AD brain damage. Synaptosomal membranes contain a large number of mitochondria, transporters, and ion-motive ATPases, making them vulnerable to oxidative stress.

[34] K. Hensley, D. A. Butterfield, J. M. Carney, and M. Aksenov, U. S. Patent, 5,840,838 (1998).

Cortical synaptosomal membranes used in our studies are obtained from male Mongolian gerbils housed in the University of Kentucky Central Animal Facility under 12-hr light/12 hr dark conditions and fed standard rodent laboratory chow (Purina) *ad libitum* in the home cage. Animals are decapitated during the light phase, and the brain is removed rapidly and placed on ice. The cortical mantle is dissected free, taking care to exclude the hippocampus and striatum, and the telencephalon is separated from underlying white matter. Cortices from two to six animals are then pooled together and suspended in approximately 20 ml of ice-cold isolation buffer [0.32 *M* sucrose containing protease inhibitors (4 μg/ml leupeptin, 4 μg/ml pepstatin, 5 μg/ml aprotinin, 20 μg/ml type II-S soybean trypsin inhibitor), 0.2 m*M* phenylmethylsulfonyl fluoride (PMSF), 2 m*M* ethylenediaminetetraacetic acid (EDTA), 2 m*M* ethylene glycol bistetraacetic acid (EGTA), and 20 m*M* 4-(2-hydroxyethyl)-l-piperazineethanesulfonic acid (HEPES)]. The samples are homogenized by 12 passes in a Wheaton 30-ml motor-driven Potter-type homogenizer with a Teflon pestle.

Synaptosomes are purified from homogenized cortices via ultracentrifugation across discontinuous sucrose gradients.[35] Crude homogenate (in isolation buffer as described earlier) is centrifuged at 4°, 1500*g* for 10 min, the pellet is discarded, and the supernatant is respun at 20,000*g* for 10 min. The resulting pellet is carefully dispersed, and this suspension is layered on top of sucrose density gradients containing 12 ml each of 1.18 *M* sucrose, 1.0 *M* sucrose, and 0.85 *M* sucrose, plus 2 m*M* EDTA, 2 m*M* EGTA, and 10 m*M* HEPES (pH 8.0 for 0.85 and 1.0 *M* sucrose solutions and pH 8.5 for 1.18 *M* solution). Samples are then spun at 4° and 82,500*g* for 120 min in an SW28 rotor in a Beckman L7-55 refrigerated ultracentrifuge. Synaptosomes are removed from the 1.18 *M*/1.0 *M* interface and washed three times in lysing buffer (10 m*M* HEPES, 2 m*M* EDTA, and 2 m*M* EGTA, pH 7.4). The washings are performed by suspending the synaptosomes in ~30 ml lysing buffer and centrifuging them in a Sorvall refrigerated centrifuge at 16,500 rpm for 10 min at 4°. After each spin the supernatant is discarded and the pellet is resuspended in 30 ml lysing buffer. This procedure results in relatively pure synaptosomes with few mitochondria.[35] After washings the synaptosomal pellet is resuspended in about 1 ml lysing buffer, and the protein concentration of the synaptosomal suspension is determined.[36] The protein concentration of each sample is then adjusted to the required concentration. For the oxidation protocol executed, treat-

[35] S. A. Umhauer, D. T. Isbell, and D. A. Butterfield, *Anal. Lett.* **25,** 1201 (1992).

[36] O. H. Lowry, N. J. Rosebrough. A. L. Farr, and R. J. Randall, *J. Biol. Chem.* **193,** 265 (1951).

ments are performed in duplicate on samples taken from the same synaptosome suspension.

Hippocampal Neuronal Cultures. Several regions of the brain are found to be affected in AD, but damage is especially severe in the hippocampus, where a high density of amyloid containing SP is located and many aspects of memory are processed. Hence, the use of hippocampal neuronal cells is appropriate for assessing Aβ-related oxidative damage.

Hippocampal neuronal cultures used in our studies are prepared from 18-day-old Sprague–Dawley rat fetuses. Rat fetal hippocampi are dissected and incubated for 15 min in a solution of 2 mg/ml trypsin in Ca^{2+}- and Mg^{2+}-free Hanks' balanced salt solution (HBSS) buffered with 10 m*M* HEPES (GIBCO, Grand Island, NY). The tissue is then exposed for 2 min to soybean trypsin inhibitor (1 mg/ml in HBSS) and rinsed three times in HBSS. Cells are dissociated by trituration and distributed into 60-mm^2 polyethyleneimine-coated plastic culture dishes (Fisher, Pittsburgh, PA). Initial plating densities are 75–100 cells/mm^2. At the time of plating, the culture dishes contain 2 ml of Eagle's minimum essential medium (MEM; GIBCO) supplemented with 100 ml/liter fetal bovine serum (Sigma, St. Louis, MO), 1 m*M* L-glutamine, 20 m*M* KCl, 1 m*M* pyruvate, and 40 m*M* glucose. After a 5-hr period to allow cell attachment, the original medium is removed and replaced with 1.6 ml of fresh medium of the same composition. After a 24-hr period, 1.2 ml MEM is replaced with 1.0 ml of neurobasal medium (GIBCO) supplemented with 2% (v/v) B-27 or N-2 (GIBCO) depending on the experiment, 2 m*M* L-glutamine (GIBCO), and 0.5% (w/v) D-(+)-glucose. On the fifth day, two-thirds of the neurobasal medium is replaced with freshly prepared neurobasal medium of the same composition. Cultures are maintained at 37° in a 5% CO_2/95% (v/v) room air-humidified incubator at all times. The neuronal cultures are used for the experiments between 9 and 11 days. In the authors' opinion, the use of 9- to 11-day-old cultures ensures that all major systems are expressed [i.e., if one were to use 2-day cultures (as some have reported), NMDA and other receptors may not be expressed] and, given that AD is an age-related neurodegenerative disorder, the authors' view is that sufficiently aged samples should be employed. An advantage of the use of B27/neurobasal medium is that glial growth is reported to be less than 0.5% of the nearly pure neuronal population[37]; hence, potentially confounding effects related to mixed populations of cells are limited.

EPR Methods Employed

EPR is a highly sensitive technique for studying free radical systems. It is particularly suitable for studying biological systems because it operates

[37] G. J. Brewer, J. R. Torricelli, E. K. Evega, and P. J. Price, *J. Neurosci. Res.* **35,** 567 (1993).

in the nanosecond time scale at which many biological motions occur and it can be used to study opaque samples such as biological membranes, which cannot be studied easily by conventional spectrophotometric methods. Further, this method does not suffer from the light-scattering effects of optical spectroscopy.

Spin Labeling. Membranes are generally nonparamagnetic species and, hence, paramagnetic protein- and lipid-specific spin labels have been used to study changes in the physical state and conformation of proteins and lipids by EPR.[38] Spin labels are compounds with a stable unpaired electron, such as nitroxides. Changes in the motion of the spin label in membranes can be used to monitor changes in the microenvironment of proteins and lipids

Spin Labeling with the Protein-Specific Spin Label MAL-6. The thiol-specific spin label, 2,2,6,6-tetramethyl-4-maleimidopiperidin-1-oxyl (MAL-6), binds covalently to proteins. Depending on the site of protein attachment, the motion of the spin label is either relatively free or highly restricted, and the resulting EPR spectra reflect these two extremes of spin label location on proteins.[38] For example, MAL-6 bound to surface-localized sulfhydryl residues has relatively unhindered motion, and the spin label is referred to being in a weakly immobilized site. The resonance line associated with this set of SH-bound spin labels is relatively narrow (Fig. 1); in contrast, MAL-6 bound to SH sites localized to narrow pockets or clefts within the protein three-dimensional structure has a relatively hindered motion, and the spin label is referred to as being in a strongly immobilized site.[38] The resonance line associated with this later set of SH-bound spin labels is relatively broad with consequently low amplitude (Fig. 1). The ratio of the low-field EPR signal amplitudes of the weakly (W) to strongly (S) immobilized protein-bound spin labels (the W/S ratio) is a highly sensitive means of studying changes in the protein microenvironment.[38] A decrease in the W/S ratio is an indirect measure of increased protein oxidation, resulting from increased protein cross-linking, increased protein–protein interactions, changes in conformation, or in accessibility to spin label binding sites[38] and has been well characterized in several models of oxidative stress such as hyperoxia, ischemia–reperfusion injury, accelerated senescence, and Fe^{2+}/H_2O_2-induced hydroxyl free radical-mediated oxidation (reviewed in Butterfield[3]). Aβ caused a significant reduction in the W/S ratio of MAL-6-labeled cortical synaptosomal membranes, an effect that was inhibited by vitamin E.[3,39]

For EPR spin-labeling studies with MAL-6, the synaptosomal membrane pellet is treated with the agent mediating oxidative stress, with the

[38] D. A. Butterfield, *Biol. Magn. Reson.* **4,** 1 (1982).
[39] R. Subramaniam, T. Koppal, M. Green, S. Yatin, B. Jordan, and D. A. Butterfield, *Neurochem. Res.* **23,** 1403 (1998).

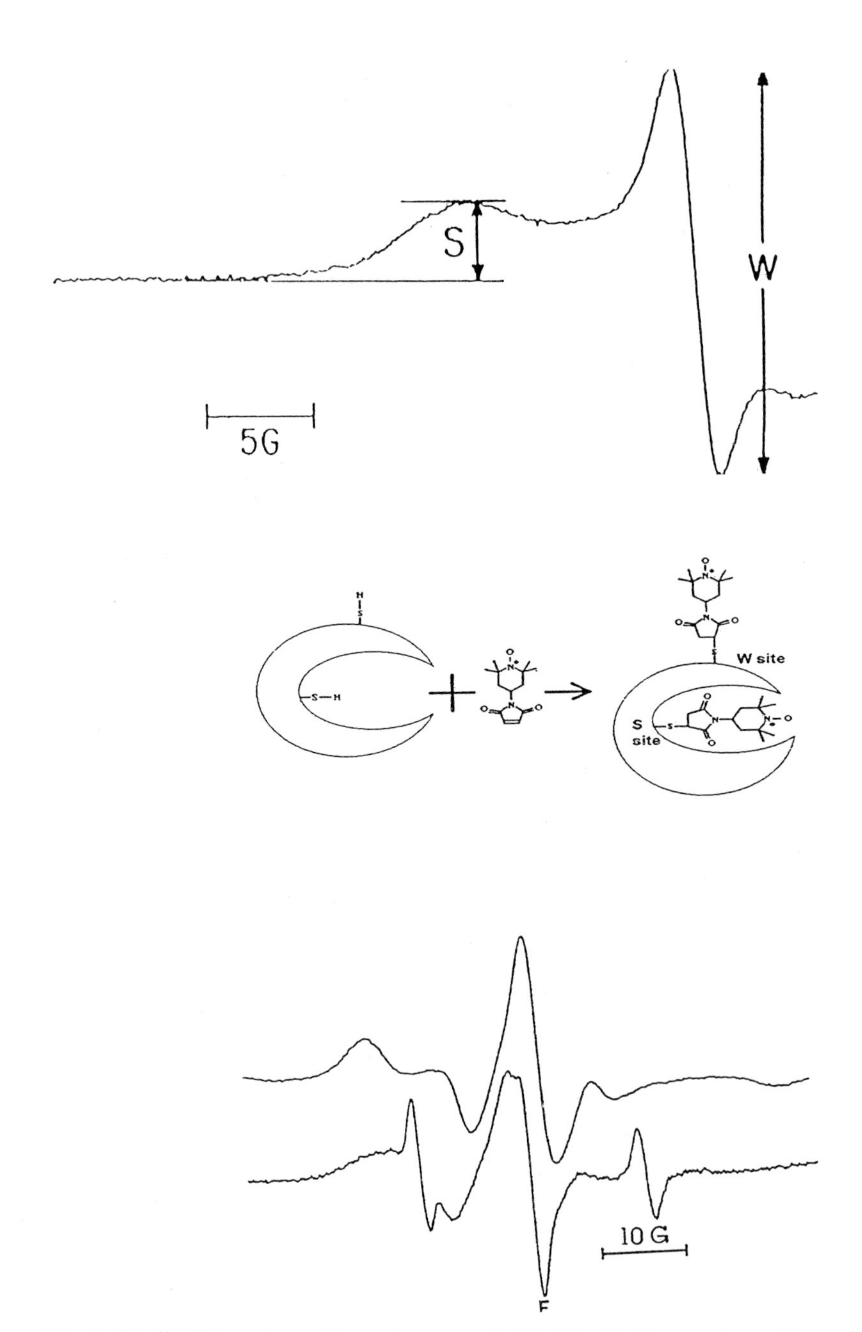

FIG. 1. (Top) EPR spectrum of synaptosomes spin labeled with the protein-specific spin label MAL-6. Only the low-field resonance lines, indicating the W and S resonances, are shown. (Middle) A schematic diagram showing the origin of weakly (W) and strongly (S) immobilized spin label-binding sites is also presented. (Bottom) EPR spectrum of synaptosomes spin labeled with the lipid-specific spin labels 5-NS and 12-NS. Note the appearance of free and bound spin label in the latter.

time and dose of treatment being determined previously. The treatment usually involves incubating the pellet with the toxicological agent of study in 1 ml lysing buffer. Following this incubation, the homogenate is centrifuged in a table-top refrigerated Eppendorf microcentrifuge for 4 min at 14,000 rpm. One milliliter of a freshly prepared MAL-6 solution is then added to the treated pellet. The MAL-6 solution is prepared by dissolving 2.5 mg of MAL-6 in 50 ml of lysing buffer (the dissolution requires at least 45 min). An alternate method involves dissolving 2.5 mg of MAL-6 in 100 μl of acetonitrile, which is then added to 50 ml of lysing buffer and stirred for a few minutes before addition to the synaptosomal pellet. This concentration of MAL-6 (about 40 μM) is far lower than the 1 mM concentration of *N*-ethylmaleimide that will disrupt cytoskeletal proteins, and this low spin label concentration will not lead to spin–spin interactions.[38] After incubation of the synaptosomal proteins with MAL-6 for 14–18 hr at 4°, the pellets are washed in the table-top microcentrifuge by spinning for 4 min at 14,000 rpm to remove the excess, unreacted spin label. The washings are repeated six times by discarding the supernatant and resuspending the pellet each time in lysing buffer. The number of washes is determined by obtaining the EPR spectrum of the final wash to show the absence of any free spin label. Samples are always kept on ice except prior to spectral acquisition when they are at room temperature for 10–15 min. Both cytoskeletal and transmembrane proteins are labeled with MAL-6.[35]

Spectral Acquisition. EPR spectra are obtained on a Bruker 300 or Bruker EMX EPR spectrometer equipped with computerized acquisition and analysis capabilities, with the instrument parameters as follows: microwave frequency, 9.78 GHz; microwave power, 20 mW; modulation frequency, 100 kHz; modulation amplitude, 0.30 G; and time constant, 1.28 msec.

Spin Labeling with Lipid-Specific Spin Probes 5-NS and 12-NS. 5-Nitroxyl stearate (5-NS) and 12-nitroxyl stearate (12-NS) are hydrophobic spin probes that insert into the lipid bilayer.[38] Measurement of the half-width at half-height (HWHH) of the low-field resonance line in the EPR spectrum of these spin probes is used as a measure of the order and motion (fluidity) of the lipid bilayer. The theory of using HWHH for an index of membrane fluidity is beyond the scope of this article, but, in brief, as the lipid-resident spin probe reorients in the bilayer such that the *p* orbital of the nitroxide (in which there is the greatest probability of finding the unpaired electron of the spin probe) changes from parallel to perpendicular orientation with respect to the external magnetic field, the line width changes.[38] Analogous to chemical exchange, as the frequency of orientation increases (as would occur in a more fluid bilayer), the HWHH increases, and vice versa.[38]

In order to examine changes in lipid fluidity, lysed and washed synaptosomes, with a protein concentration of 3 mg/ml, are added to the tube containing either 5-NS or 12-NS to give a 1:40 5- or 12-NS/lipid molar ratio. The spin probes, 5-NS or 12-NS (10 mg), are dissolved in 1 ml of chloroform. Of this stock, 160 μl is then again diluted to 1 ml with chloroform. Fifteen microliters of this latter spin label solution in chloroform is carefully added to the bottom of a borosilicate culture tube, and the chloroform is slowly evaporated under a stream of nitrogen. The tube is then covered with Parafilm and stored in the freezer until needed. Before use the tubes are brought to room temperature. Samples are incubated at room temperature with occasional gentle shaking. After exactly 30 min of incubation with the spin label, the EPR spectrum is obtained as discussed earlier. Lipid-specific spin labels also are used to detect free radical processes that occur in the lipid phase of brain cells[8–10] (see later).

EPR Spin Trapping. Free radicals are unstable transient species and hence EPR spin trapping methods generally are required to detect them. In spin-trapping studies, the transient free radicals react with a nonparamagnetic species (the trap), such as PBN, to form a stable, paramagnetic (EPR detectable) moiety, usually a nitroxide.[38]

In order to detect Aβ-associated free radicals, the Aβ peptides are solubilized at 1 mg/ml concentrations in PBS buffer (150 m*M* NaCl, 5 m*M* sodium phosphate, pH 7.4) containing PBN (50 m*M*). The PBN is synthesized as described in the literature[40] and purified several times by repeated sublimations. Only pure PBN that is EPR silent even on incubation at 37° in buffer solutions for up to 3 days is used in the experiments. In studies used to inhibit possible metal-catalyzed reactions, the chelator deferroxamine mesylate is dissolved to 2 m*M* in the buffer prior to peptide addition. In a further attempt to minimize any role of redox metal ions, the buffer is stirred overnight with Chelex 100 beads, a metal-chelating agent, and decanted prior to use. The peptide solution is incubated in a water bath at 37° for periods between 0 and 24 hr. Three hundred microliters of this solution is placed into a EPR flat cell at appropriate time points for EPR analysis, which is performed on a Bruker 300 or Bruker EMX EPR spectrometer as described earlier. Typical instrumental parameters are microwave power, 20 mW; modulation amplitude, 0.30 G; gain, 1×10^5; and conversion time, 10.28 msec. As noted, Aβ leads to PBN spin adducts that are detected by EPR.[3,41]

[40] R. Huie and W. R. Cherry, *J. Org. Chem.* **50,** 1531 (1985).

[41] T. Tomiyama, A. Shoji, K. Kataoka, Y. Suwa, S. Asano, H. Kaneko, and N. Endo, *J. Biol. Chem.* **271,** 6839 (1996).

Glutamine Synthetase and Creatine Kinase Assays

GS. Glutamine synthetase (GS) is one of the most oxidatively sensitive cytosolic enzymes found in the astrocytes of mammalian brain.[2] The function of GS is to convert glutamate to glutamine, thereby preventing glutamate-induced NMDA receptor-mediated excitoxicity.

$$\text{Glutamate} + \text{ammonia} + \text{ATP} + \text{Mg}^{2+} \rightarrow \text{glutamine} + \text{ADP} + \text{P}_i \quad (1)$$

Oxidation of a single histidine residue found in the metal-binding region of the enzyme can lead to complete loss of enzyme activity.[2] Hence, activity of GS is a useful marker of protein oxidation. Further, diminished GS activity occurs in AD brain,[17,23] and Aβ inactivates this enzyme.[15,16]

The assay of GS activity is based on the principle that when ammonia is replaced by hydroxylamine in the just-described reaction, the product is γ-glutamylhydroxamate, which gives a characteristic brown color on addition of iron(III) chloride. The enzyme catalyzes a γ-glutamyl transfer reaction, which requires catalytic quantities of nucleotide (ATP or ADP) and also inorganic phosphate (or arsenate, As_i), and this transfer reaction may be followed by the use of the iron(III) chloride reagent.

$$\text{L-Glutamine} + \text{NH}_2\text{OH} + \text{ADP (ATP)} \rightarrow \text{L-}\gamma\text{-glutamylhydroxamate} + \text{NH}_3 \quad (2)$$

$$\text{L-}\gamma\text{-Glutamylhydroxamate} + \text{FeCl}_3 \rightarrow \text{brown color}$$

Gerbil cortical homogenate is prepared as described earlier, and the clear supernatant (cytosolic fraction) obtained after centrifugation at 10,000*g* for 10 min at 4° is recovered and used for the enzymatic assay. Protein concentration in the supernatant is measured by the Lowry *et al.*[36] method and is diluted to a final concentration of 0.3–0.4 mg/ml. GS activity is determined by the method of Rowe *et al.*[42] as modified by Miller *et al.*[43] Assay mixtures incubated at 37° for 30 min to 24 hr contain (in 1 ml): 50 m*M* imidazole hydrochloride (pH 6.8), 50 m*M* NH_2OH (pH 6.8), 100 m*M* L-glutamine, 25 m*M* potassium dihydrogen arsenate (pH 6.8), 0.2 m*M* ADP, 0.5 m*M* $MnCl_2$, enzyme solution containing 0.04 mg protein, and, for studies with Aβ, 1 mg peptide sample. Reactions are initiated by the addition of ADP and arsenate and are terminated by the addition of 1 ml of iron(III) chloride solution (0.37 *M* $FeCl_3$, 0.67 *N* HCl, and 0.2 *M* trichloroacetic acid). The γ-glutamylhydroxamate formed is determined by comparing

[42] W. B. Rowe, R. A. Remzio, V. P. Wellner, and A. Meister, *Methods Enzymol.* **17,** 900 (1970).
[43] R. E. Miller, R. Haddenburg, and H. Gersham, *Proc. Natl. Acad. Sci. U.S.A.* **57,** 1418 (1978).

the absorbance at 505 nm to that observed when authentic γ-glutamylhydroxamate is substituted for the enzyme. Controls, lacking ADP and arsenate, are used to correct for nonspecific activity. All assays are performed in duplicate and averaged for each sample. A unit of glutamine synthetase activity is defined as that amount of activity that catalyzes the synthesis of 1 μmol of γ-glutamylhydroxamate per minute under the specified assay conditions. Specific activity is expressed as units per milligram of protein.

CK. Creatine kinases (CK, EC 2.7.3.2) are a family of isoenzymes catalyzing the reversible transfer of a phosphoryl group between ATP and creatine.

$$\text{Creatine} + \text{ATP} = \text{creatine phosphate} + \text{ADP} \qquad (3)$$

Being a sulfhydryl-containing enzyme, CK is susceptible to oxidative inactivation.[2] In vertebrates, CK often exists in four isoforms, produced by separate genes: brain CK (BB CK), muscle CK, ubiquitous mitochondrial CK, and sarcomeric mitochondrial CK. Our laboratory assays BB CK to study the effects of Aβ.[12] This isoform is detected by enzyme activity staining following nondenaturing gel electrophoresis using a commercial kit (CIBA Corning Diagnostic Corp., Palo Alto, CA). One unit of CK activity is calculated as the amount of enzyme that will convert 1.0 μmol of creatine to creatine phosphate per minute at 37°, pH 9.0.

One microliter of crude cell extract (1 mg/ml total protein) is applied to each lane of 1.0% agarose Corning Multitrac CK isoenzyme gel/8 and electrophoresed in CK isoenzyme buffer (Corning) at 90 V for 20 min. The enzyme pattern is developed by layering the plates with CK reagent (Corning). The gel is incubated at 37° for 20 min, and the enzyme bands are visualized with UV irradiation.

Neurotoxicity Determinations

Aβ causes oxidative stress as manifested by increased lipid peroxidation and protein oxidation, which in turn causes synaptic damage, alterations in cytoskeletal proteins, and induces apoptosis, all of which lead to neuronal death.[3] Lipid peroxidation and protein oxidation lead to altered Ca^{2+} homeostasis, resulting in a significantly, and irreversibly, increased intracellular accumulation of Ca^{2+}, which in turn causes additional oxidative stress. Further, an increased intracellular accumulation of Ca^{2+} leads to synaptic damage, alterations in cytoskeletal proteins, and apoptosis, indicating a large and complex feedback loop. Mitochondrial impairment, manifested, for example, by decreased activity of cytochrome oxidase, and itself caused by oxidative stress, leads to free radical leakiness[1,3,5] and to an increased intracellular accumulation of Ca^{2+}, which, as noted earlier, can cause neuronal death.[3] Other pathways for neurotoxicity involving Aβ-associated oxidative stress are possible, including, among others, sequelae of oxidative stress-induced transcription factors, e.g., NF-κB, inflammatory response, advanced glycation end products, and increased levels of redox-active trace metals.[4] The following methods have been employed for determinations of Aβ-induced neurotoxicity.[12,33,39]

MTT Reduction Assay for Mitochondrial Function. The viability of cells can be detected using a tetrazolium dye, [3-(4,5-dimethylthiazol-2-yl)-2,5-diphenyl] tetrazolium bromide (MTT), that is reduced to blue formazan by mitochondrial dehydrogenases in living cells but not by dead cells.[44] The MTT dye (Sigma) is freshly prepared at 5 mg/ml in phosphate-buffered saline (PBS) in a darkened tube. A 30-μl aliquot of MTT is added to plates that contain 300 μl medium (1/10; MTT/cell medium) and plates are incubated at 37° for 4 hr. When the incubation is complete, the unreacted dye and the medium are removed by inverting the plate. Add 1 ml of 0.04 *M* HCl in 2-propanol into each well to solubilize MTT formazan. Complete solubilization is achieved by pipetting the solution several times. Absorbance is measured by UV–VIS spectrophotometry at a wavelength of 570 nm and a reference wavelength of 630 nm.

Trypan Blue Exclusion Assay. Trypan blue exclusion is performed by counting the neurons that internalize the dye. Exclusion of the dye corresponds to cell survival. Cells are rinsed three times with 1 ml PBS (pH 7.4). Trypan blue (Sigma) (0.4%) is added to cells along with 300 μl PBS and incubated for 10 min. Sixteen different areas are counted under the microscope for blue staining, which, if present, index dead cells.

Morphometric Cell Counting Assay. Morphological criteria for cell survival correlate with Trypan blue dye staining methods.[45] Viable neurons have neurites that are uniform in diameter and smooth in appearance and somata that are smooth and round to oval shape. However, nonviable degenerating neurons have fragmented and beaded neurites and soma that are rough, swollen, vacuolated, and irregular in shape. In addition to these morphological changes, we observed that degenerated neurons detach from the culture substrate.[33] Viable neurons (eight regions of 1 mm^2/plate) typically are counted prior to and at the desired time points following the experimental treatment. Cultures are visualized and photographed with a phase-contrast Nikon Diaphot inverted microscope.

[44] T. Mossman, *J. Immunol. Methods* **65,** 55 (1983).
[45] M. P. Mattson and S. B. Kater, *Int. J. Dev. Neurosci.* **6,** 439 (1988).

ROS Detection

Anionic 2′,7′-dichlorofluorescin (DCF) is formed from nonpolar 2′,7′-dichlorofluorescin diacetate (DCF-DA) intracellularly after the latter crosses the cell membrane and is acted on by endogenous esterases, and, because of the resulting negative charge, DCF cannot diffuse out of the cell. Oxidation of DCF by ROS (mostly peroxide and peroxyl radicals, but also possibly superoxide) gives fluorescence, whose intensity is related to the amount of ROS present. Intracellular reactive oxygen species in neurons are measured by a DCF assay with some modifications to adapt it to fluorescence microscopy.[12,33] Cells are loaded with DCFH-DA (Molecular Probes, Eugene, OR) by incubating in the non-CO_2 incubator for 50 min in the presence of 100 μM of the dye in HBSS containing 10 m*M* HEPES and 10 m*M* glucose. Cells are washed three times with warm MSF buffer, and fluorescence visualization is accomplished using a confocal laser scanning microscope (Molecular Dynamics, Sarastro 2000) coupled to an inverted microscope (Nikon). Fluorescence is excited at 488 nm and emission filtered using a 510-nm barrier filter. Cells scanned are chosen randomly. Measurement of the fluorescence intensity is performed using Imagespace software, which averages fluorescence values of individual pixels composing the cell image.

Protein Oxidation (Protein Carbonyl) Determinations

ROS and RNS can cause oxidation of amino acid residues of protein side chains, increase protein–protein cross-linking, and oxidize the peptide backbone resulting in protein fragmentation.[2,5] Carbonyl groups are the most widely used marker of protein oxidation and can result from the direct oxidation of many amino acids, such as histidine, lysine, arginine, proline, and threonine, among others.[2,5] Also, reactions of proteins with the lipid peroxidation product HNE, or reaction of reducing sugars or their oxidation products with lysine by glycation or glycoxidation, can result in carbonyl formation.[2,5] The carbonyl assays were delveloped as general assays of oxidative protein damage. The basis of the assays is the reaction of 2,4-dinitrophenylhydrazine (DNPH) with protein carbonyls to give hydrazone derivatives in acidic media,[46] which then can be determined in several ways.

$$\text{Protein-C=O} + \text{H}_2\text{N-NH-2,4-DNP} \rightarrow \text{Protein-C=N-NH-2,4-DNP} + \text{H}_2\text{O} \quad (4)$$

[46] R. L. Levine, J. A. Williams, E. R. Stadtman, and E. Shacter, *Methods Enzymol.* **233,** 346 (1994).

Spectrophotometric Method. This method requires large amounts of DNPH reagent, and the excess reagent must be removed for the spectrophotometric determination of the protein-bound hydrazone. This method also requires relatively large amounts of proteins compared to other methods. Protein concentration can be measured according to the Lowry *et al.*[36] method, and 0.1–0.5 mg protein is optimal for this procedure.

Protein carbonyl content is measured according to Levine *et al.*[46] Samples are treated with 0.2% DNPH in 2 *M* HCl (2 *M* HCl is used for the blank) and are allowed to stand at room temperature for 1 hr, with vortexing at 5-min intervals. After incubation, hydrazone derivatives are extracted with 10% (v/v) trichloroacetic acid (TCA) and centrifuged by using a tabletop centrifuge (11,000*g*) for 3 min at room temperature. The supernatant is discarded. Protein pellets are washed three times with ethanol/ethyl acetate (1:1, v/v) to remove free reagent. Samples are then vortexed, incubated at room temperature for 10 min and centrifuged (11,000*g*) for 10 min at room temperature for each wash. After discarding the supernatant of the last wash, the precipitated protein is dissolved in 0.6 ml of 6 *M* guanidine hydrochloride in 20 m*M* sodium phosphate buffer (pH 6.5) by incubation at 37° for 15 min. Insoluble materials are removed by centrifugation at 11,000*g* for 3 min at room temperature, and the absorbance is measured at optimal wavelength (360–390 nm) against the complementary blank using a millimolar absorptivity of 21.0 mM^{-1} cm^{-1}. This widely used method had been criticized by some as underestimating the number of carbonyls present, possibly due to the heterogeneous phase reactions that occur in the assay.[47] However, even if this is the case, if the assay shows increased protein carbonyls (increased protein oxidation), then one can state with confidence that oxidation has occurred.

Immunochemical Method. To determine the level of protein oxidation using immunochemistry and Western blot analysis, an oxidized protein detection kit (Oxyblot, ONCOR, Gaithersburg, MD) is employed. The method is based on immunochemical detection of protein carbonyl groups derivatized with DNPH[46] (see reaction described earlier) and requires low amounts of protein, in the range of several micrograms, compared to spectrophotometric detection, which requires 100 times more. A further advantage of this method is that the pattern of oxidized proteins is obtained, and the protein band(s) corresponding to the desired molecular weight can be scanned individually. The samples are treated with 20 m*M* DNPH in 10% trifluoroacetic acid and derivatization-control solution and are incubated for 15–30 min. After derivatization and neutralization with 2 *M* Tris/30% glycerol (neutralization solution, Oxyblot kit) and 19% 2-mercaptoethanol, cell proteins are separated by SDS–PAGE.

[47] G. Cao and R. G. Cuttler, *Arch. Biochem. Biophys.* **320,** 106 (1995).

Oxidized bovine serum albumin (BSA), with a known concentration of carbonyls (20 nmol of carbonyl/mg of protein determined spectrophotometrically as described earlier), is treated with DNPH and loaded as a standard (1 pmol of protein carbonyl per lane) with each set of the samples.[14] For the standard preparation, BSA (standard for gel filtration chromatography, Sigma) is dissolved in double-deionized water at 2 mg/ml and oxidized by Fe^{2+}/H_2O_2 (100 μM/1 mM) for 2 hr at 37°. The reaction is stopped with deferroxamine mesylate, and the small molecular weight substances are removed from the protein passage through a Sephadex G-25 desalting column.[14] The concentration of carbonyl groups per milligram of pure BSA protein is determined by the colorimetric carbonyl assay described earlier.[46] Polyacrylamide gel electrophoresis (SDS–PAGE) is performed in minislabs (0.75 × 60 × 70 mm, 12% total acrylamide) according to method of Laemmli.[48] Following electrophoresis, proteins are transferred to nitrocellulose for further immunoblotting analysis according to the procedure adapted from Glenney.[49] The transfer of proteins to nitrocellulose after SDS–PAGE, using a transfer buffer of Tris–glycine, pH 8.5, with 20% methanol, is complete in 2 hr. After the transfer, membranes are blocked in 3% BSA (in PBS with sodium azide, 0.01% and Tween 20, 0.2%) for 1 hr at room temperature. Rabbit anti-DNPH antibody from the ONCOR Oxyblot kit (1 : 150 working dilution) is used as a primary antibody. Secondary antibodies (antirabbit IgG conjugated with alkaline phosphatase, Sigma) are diluted in the blocking solution 1 : 15,000 and are incubated with the membrane for 1 hr at 37°. Membranes are washed after every step in washing buffer (PBS with 0.01% sodium azide and 0.2% Tween 20) for 10 min at room temperature. Washed membranes are developed using BCIP-NBT solution (SigmaFast tablets, Sigma). Western blots are analyzed using computer-assisted imaging (MCID/M4 software from Imaging Research). Data are reported as percentages of corresponding vehicle-treated values. Increased protein carbonyls in hippocampal neurons treated with Aβ and their inhibition by vitamin E were detected using this method.[12,39]

Histofluorescence Method. The method uses biotin-4-amidobenzoic hydrazide to bind carbonyls followed by confocal fluorescence visualization.[33] With this method the regions of oxidative damage (cellular compartmentalization of the carbonyl groups) in individual neurons can be localized. Cells are fixed in 1,4-phenylene diisothiocyanate [5 mg/ml in polyethylene glycol (PEG) 200] for 30–45 min, followed by one rinse in PEG 200 and then one rinse in methanol. Cells are covered in 50% PEG 200/50% PBS for 1 hr to hydrolyze the remaining isothiocyanate groups. Cells are then covered

[48] U. Laemmli, *Nature* **227,** 680 (1970).

[49] J. R. Glenney, *Anal. Biochem.* **156,** 315 (1986).

in a 50% acetate/50% PEG 200 buffer (200 mM sodium acetate, 0.5% phenol, 5 mM EDTA, PEG 200, pH 5.5) for 5 min before the addition of biotin-4-amidobenzoic hydrazide (1.9 mg/ml in acetate buffer). After 1 hr, residual hydrazide is removed by rinsing the cells three times with PBS. This is followed by incubation with [5-[(4,6-dichlorotriazin-2-yl)amino]fluorescein]-conjugated streptavidin. Fluorescence is amplified using a 1:2000 dilution of biotinylated antistreptavidin antibody. Cells are incubated with a PBS solution containing 1% BSA for 5 min followed by a 30-min incubation with antibody at 40°. The cells are then covered in Vectashield (Vector, Marion, IA) mounting medium to retard photobleaching and coverslipped. Cells are imaged by confocal laser scanning microscopy (Molecular Dynamics, Sarastro 2000, Sunnyvale, CA) coupled to an inverted microscope (Nikon). The excitation wavelength used is 488 nm and emission is filtered using a 510-nm barrier filter. Measurement of the fluorescence intensity is accomplished using Imagespace software, which averages fluorescence values of individual pixels composing the cell image. Only the biotin–streptavidin complex leads to fluorescence. Thus, if Aβ induces protein oxidation or lipid peroxidation, with subsequent reaction of the HNE formed with proteins, both of which oxidative processes lead to protein carbonyls, then the biotinylated hydrazine forms the Schiff base, and interaction of fluorescently labeled streptavidin leads to fluorescence. In the absence of Aβ or in the presence of the antioxidant propyl gallate, no fluorescence was found.[33]

Lipid Peroxidation Determinations

Lipid peroxidation is found to occur in the AD brain.[4] Polyunsaturated fatty acids (PUFAs) in the phospholipids of the membrane lipid bilayer are most susceptible to oxidation, and the brain, which utilizes a high percentage of inspired oxygen, contains a significant amount of PUFAs. Free radicals attack unsaturated sites in the PUFAs, and in the presence of paramagnetic oxygen (which is highly soluble in the lipid bilayer), PUFAs are converted to lipid peroxides or hydroperoxides.

Reduction of Paramagnetism of Stearic Acid Spin Probes. The lipid-specific spin labels 5-NS and 12-NS are used to detect free radical processes that occur in membrane lipids. The principle of the method is that a peptide-associated free radical, when interacting with the unpaired electron on lipid-specific spin probes, causes a loss of paramagnetism of the latter. Consequently, a loss of signal intensity occurs that can be quantified easily.

In order to assess Aβ-associated free radicals in synaptosomal membranes, the membranes are first labeled with the stearate spin probes as described earlier. Then 0.5 ml of the peptide solution (1 mg/ml) in PBS that has been preincubated at 37° for 6 hr is added to 0.5 ml of the spin-labeled synaptosomal homogenate (8 mg/ml protein concentration). The EPR spectrum is obtained at appropriate time points, as described earlier, and the loss of signal intensity over time is used as a measure for detecting the presence of free radicals in the membrane lipid bilayer. Vitamin E (5 m*M*) is added to the synaptosomal homogenate 0.5 hr prior to peptide addition in experiments designed to assess the free radical-scavenging ability of vitamin E (see later). In contrast to the EPR spectrum obtained with 5-NS incorporated into synaptosomal or PC-12 culture systems,[8–10] 12-NS, with its paramagnetic center located deeper in the bilayer, consistently partitions between the lipid and the aqueous phases,[38] and one can resolve both a bound and a free population of spin probe in the 12-NS EPR spectrum. The intensity of bound population greatly predominates in untreated samples, with the three sharp lines of the free population providing only a minor contribution to the total integrated spectral intensity. Thus, loss of signal from the bound 12-NS is consistent with a free radical attack on the paramagnetic center of this lipid bilayer resident spin probe.

HNE Analysis Using Western Blotting. HNE is formed from free fatty acids, such as arachidonic acid, that are released following lipid peroxidation.[7] Hence, HNE can serve as another marker of lipid peroxidation. For HNE determinations, synaptosomal membranes are oxidized by Aβ or other oxidative stress agents, and an aliquot of the sample is added to a sample buffer containing SDS and mercaptoethanol and is boiled at 100° for 5 min. The volume of the sample buffer is such that the final concentration of protein in the mixture is 1 μg/ml.

Sixty micrograms of sample protein and 10 μl of prestained molecular weight standards obtained from Bio-Rad are loaded onto 10% polyacrylamide gels and separated using 100 V for 2 hr. The proteins are then transferred to a nitrocellulose paper by Western blotting techniques. A potential of 100 V is applied and the transfer is carried out for an hour. After the transfer is completed the paper is blocked with milk solution for an hour, and later the primary antibody solution (polyclonal HNE antibody raised in rabbit, specific for hemiacetal groups of lysine-bound HNE) is added to the nitrocellulose paper. The paper is incubated in diluted primary antibody solution (1:4000) for 18–20 hr at 4°. The primary antibody is then removed by washing, and the secondary antibody (polyclonal goat antirabbit IgG, coupled to alkaline phosphatase) is added. A semiquantitative determination of the proteins present is performed using densitometry to calculate the optical densities of the protein bands. Alternatively, and likely a more sensitive method, chemiluminescence can be used with the secondary antibody. The protein bands are developed using BCIP/NBT coloring agents from Bio-Rad).

Free Fatty Acid Analysis. Lipid peroxidation triggers phospholipases that cleave the oxidized fatty acids from the phospholipid backbone, resulting in the formation of "free" fatty acids (FFA); hence, FFA measures act as a marker of lipid peroxidation.[8]

Free fatty acids are measured in synaptosomal membrane lipids treated with Aβ in the presence and absence of vitamin E as described earlier. Lipids, being hydrophobic, are separated from proteins and other hydrophilic components by extraction into an organic phase. FFA from the lipid extract are then separated from a mixture of phospholipids and triglycerides using thin-layer chromatography (TLC). For analysis by GC, the samples must be volatile, and, hence, the FFA are converted to their respective methyl esters using methanol and concentrated sulfuric acid. FFA are eluted from the GC column depending on their molecular weight and degree of unsaturation, and the concentration of the individual FFA is determined by measuring the area under the peak and comparing it to the area of the internal standard of known concentration.

Extraction of the lipids is performed by the method of Bligh and Dyer.[50] All solvents and extracts are bubbled with nitrogen periodically during the study to prevent autoxidation of the lipids. One milligram (protein concentration) of the synaptosomal homogenate is added to 3 ml of the solvent mixture (chloroform : methanol, 1 : 2, v/v) containing 0.005% butylated hydroxytoluene and 10 nmol final concentration of heptadecanoic acid (17 : 0, an internal standard for FFA estimation). The lipids are extracted with chloroform and 0.9% saline (final composition, 1 : 1 : 0.9, v/v/v) and are reextracted with 2 ml of chloroform. The lower chloroform layer containing lipids is saved after each extraction, and the pooled layers are dried under nitrogen. The dried extract is redissolved in 200 μl of chloroform, and half of this is spotted on silica gel G TLC plates for FFA analysis. The respective regions are scraped into 1.5 ml methanol with 0.005% BHT. Fatty acids in all the fractions are converted to their methyl esters by adding 45 μl of concentrated sulfuric acid to each tube containing 1.5 ml methanol, and the sample is heated to 65° for 30 min. The methyl esters are then dried under nitrogen and reconstituted with hexane to obtain the right concentration for quantitation on a Hewlett-Packard gas chromatograph.

The concentration of fatty acid (nmol/mg protein) equals [(area of fatty acid peak/area of peak of internal standard) × dilution factor × concentration of internal standard]/mg protein.

Summary

Given the increasing evidence of oxidative stress in AD brain[3,4] and studies from different perspectives that appear to show a converging, central role for Aβ in the pathogenesis and etiology of AD,[21,22,28] insight into

[50] E. G. Bligh and W. J. Dyer, *J. Biochem. Physiol.* **37,** 911 (1959).

Aβ-associated free radical oxidative stress will likely lead to a greater understanding of AD and, potentially, to better therapeutic strategies in this disorder. This article outlined methods to investigate markers of oxidative stress induced by Aβ in brain membrane systems.[3] Especially important are markers for protein oxidation, lipid peroxidation, and ROS generation by Aβ. Oxidative stress and its sequelae are likely related to both necrotic and apoptotic mechanisms of neurotoxicity,[4] and Aβ-associated free radical oxidative stress may be of fundamental importance in Alzheimer's disease etiology and pathogenesis.[3,4] The methods described here provide some means for investigating this possibility.

Acknowledgments

This work was supported in part by grants from NIH (AG-05119; AG-10836). Useful discussions with Professors William Markesbery and Mark Mattson and Dr. John Carney are gratefully acknowledged.

Note Added in Proof

Methionine, residue 35 of Aβ(1–42), appears to be important in both *in vitro* and *in vivo* oxidative stress associated with Aβ.[51,52]

[51] S. M. Yatin, S. Varadarajan, C. D. Link, and D. A. Butterfield, *Neurobiol. Aging,* in press (1999).

[52] S. Varadarajan, S. Yatin, J. Kanski, F. Jahanshahi, and D. A. Butterfield, *Brain Res. Bull.,* in press (1999).

Author Index

Numbers in parentheses are footnote reference numbers and indicate that an author's work is referred to although the name is not cited in the text.

A

B

C

D

E

G

H

I

J

K

L

M

N

O

P

Q

R

S

T

U

V

W

X

Y

Z

Subject Index

A

B

D

E

F

G

H

I

N

Q

U

W

ISBN 0-12-182210-9